T0190288

Lecture Notes in Computer Science

Lecture Notes in Artificial Intelligence 13713

Founding Editor

Jörg Siekmann

The series Lecture Notes in Artificial Intelligence (LNAI) was established in 1988 as a topical subseries of LNCS devoted to artificial intelligence.

The series publishes state-of-the-art research results at a high level. As with the LNCS mother series, the mission of the series is to serve the international R & D community by providing an invaluable service, mainly focused on the publication of conference and workshop proceedings and postproceedings.

Massih-Reza Amini · Stéphane Canu ·
Asja Fischer · Tias Guns · Petra Kralj Novak ·
Grigorios Tsoumakas
Editors

Machine Learning and Knowledge Discovery in Databases

European Conference, ECML PKDD 2022 Grenoble, France, September 19–23, 2022 Proceedings, Part I

Editors
Massih-Reza Amini
Grenoble Alpes University
Saint Martin d'Hères, France

Stéphane Canu
INSA Rouen Normandy
Saint Etienne du Rouvray, France

Asja Fischer
Ruhr-Universität Bochum
Bochum, Germany

Tias Guns
KU Leuven
Leuven, Belgium

Petra Kralj Novak
Central European University
Vienna, Austria

Grigorios Tsoumakas
Aristotle University of Thessaloniki
Thessaloniki, Greece

ISSN 0302-9743 ISSN 1611-3349 (electronic)
Lecture Notes in Artificial Intelligence
ISBN 978-3-031-26386-6 ISBN 978-3-031-26387-3 (eBook)
https://doi.org/10.1007/978-3-031-26387-3

LNCS Sublibrary: SL7 – Artificial Intelligence

This Springer imprint is published by the registered company Springer Nature Switzerland AG
The registered company address is: Gewerbestrasse 11, 6330 Cham, Switzerland

Preface

The European Conference on Machine Learning and Principles and Practice of Knowledge Discovery in Databases (ECML–PKDD 2022) in Grenoble, France, was once again a place for in-person gathering and the exchange of ideas after two years of completely virtual conferences due to the SARS-CoV-2 pandemic. This year the conference was hosted for the first time in hybrid format, and we are honored and delighted to offer you these proceedings as a result.

The annual ECML–PKDD conference serves as a global venue for the most recent research in all fields of machine learning and knowledge discovery in databases, including cutting-edge applications. It builds on a highly successful run of ECML–PKDD conferences which has made it the premier European machine learning and data mining conference.

This year, the conference drew over 1080 participants (762 in-person and 318 online) from 37 countries, including 23 European nations. This wealth of interest considerably exceeded our expectations, and we were both excited and under pressure to plan a special event. Overall, the conference attracted a lot of interest from industry thanks to sponsorship, participation, and the conference's industrial day.

The main conference program consisted of presentations of 242 accepted papers and four keynote talks (in order of appearance):

- Francis Bach (Inria), Information Theory with Kernel Methods
- Danai Koutra (University of Michigan), Mining & Learning [Compact] Representations for Structured Data
- Fosca Gianotti (Scuola Normale Superiore di Pisa), Explainable Machine Learning for Trustworthy AI
- Yann Le Cun (Facebook AI Research), From Machine Learning to Autonomous Intelligence

In addition, there were respectively twenty three in-person and three online workshops; five in-person and three online tutorials; two combined in-person and one combined online workshop-tutorials, together with a PhD Forum, a discovery challenge and demonstrations.

Papers presented during the three main conference days were organized in 4 tracks, within 54 sessions:

- Research Track: articles on research or methodology from all branches of machine learning, data mining, and knowledge discovery;
- Applied Data Science Track: articles on cutting-edge uses of machine learning, data mining, and knowledge discovery to resolve practical use cases and close the gap between current theory and practice;
- Journal Track: articles that were published in special issues of the journals *Machine Learning* and *Data Mining and Knowledge Discovery*;

– Demo Track: short articles that propose a novel system that advances the state of the art and include a demonstration video.

We received a record number of 1238 abstract submissions, and for the Research and Applied Data Science Tracks, 932 papers made it through the review process (the remaining papers were withdrawn, with the bulk being desk rejected). We accepted 189 (27.3%) Research papers and 53 (22.2%) Applied Data science articles. 47 papers from the Journal Track and 17 demo papers were also included in the program. We were able to put together an extraordinarily rich and engaging program because of the high quality submissions.

Research articles that were judged to be of exceptional quality and deserving of special distinction were chosen by the awards committee:

– Machine Learning Best Paper Award: "*Bounding the Family-Wise Error Rate in Local Causal Discovery Using Rademacher Averages*", by Dario Simionato (University of Padova) and Fabio Vandin (University of Padova)
– Data-Mining Best Paper Award: "*Transforming PageRank into an Infinite-Depth Graph Neural Network*", by Andreas Roth (TU Dortmund), and Thomas Liebig (TU Dortmund)
– Test of Time Award for highest impact paper from ECML–PKDD 2012: "*Fairness-Aware Classifier with Prejudice Remover Regularizer*", by Toshihiro Kamishima (National Institute of Advanced Industrial Science and Technology AIST), Shotaro Akashi (National Institute of Advanced Industrial Science and Technology AIST), Hideki Asoh (National Institute of Advanced Industrial Science and Technology AIST), and Jun Sakuma (University of Tsukuba)

We sincerely thank the contributions of all participants, authors, PC members, area chairs, session chairs, volunteers, and co-organizers who made ECML–PKDD 2022 a huge success. We would especially like to thank Julie from the Grenoble World Trade Center for all her help and Titouan from Insight-outside, who worked so hard to make the online event possible. We also like to express our gratitude to Thierry for the design of the conference logo representing the three mountain chains surrounding the Grenoble city, as well as the sponsors and the ECML–PKDD Steering Committee.

October 2022

Massih-Reza Amini
Stéphane Canu
Asja Fischer
Petra Kralj Novak
Tias Guns
Grigorios Tsoumakas
Georgios Balikas
Fragkiskos Malliaros

Organization

General Chairs

Massih-Reza Amini	University Grenoble Alpes, France
Stéphane Canu	INSA Rouen, France

Program Chairs

Asja Fischer	Ruhr University Bochum, Germany
Tias Guns	KU Leuven, Belgium
Petra Kralj Novak	Central European University, Austria
Grigorios Tsoumakas	Aristotle University of Thessaloniki, Greece

Journal Track Chairs

Peggy Cellier	INSA Rennes, IRISA, France
Krzysztof Dembczyński	Yahoo Research, USA
Emilie Devijver	CNRS, France
Albrecht Zimmermann	University of Caen Normandie, France

Workshop and Tutorial Chairs

Bruno Crémilleux	University of Caen Normandie, France
Charlotte Laclau	Telecom Paris, France

Local Chairs

Latifa Boudiba	University Grenoble Alpes, France
Franck Iutzeler	University Grenoble Alpes, France

Proceedings Chairs

Wouter Duivesteijn	Technische Universiteit Eindhoven, the Netherlands
Sibylle Hess	Technische Universiteit Eindhoven, the Netherlands

Industry Track Chairs

Rohit Babbar	Aalto University, Finland
Françoise Fogelmann	Hub France IA, France

Discovery Challenge Chairs

Ioannis Katakis	University of Nicosia, Cyprus
Ioannis Partalas	Expedia, Switzerland

Demonstration Chairs

Georgios Balikas	Salesforce, France
Fragkiskos Malliaros	CentraleSupélec, France

PhD Forum Chairs

Esther Galbrun	University of Eastern Finland, Finland
Justine Reynaud	University of Caen Normandie, France

Awards Chairs

Francesca Lisi	Università degli Studi di Bari, Italy
Michalis Vlachos	University of Lausanne, Switzerland

Sponsorship Chairs

Patrice Aknin	IRT SystemX, France
Gilles Gasso	INSA Rouen, France

Web Chairs

Martine Harshé	Laboratoire d'Informatique de Grenoble, France
Marta Soare	University Grenoble Alpes, France

Publicity Chair

Emilie Morvant	Université Jean Monnet, France

ECML PKDD Steering Committee

Annalisa Appice	University of Bari Aldo Moro, Italy
Ira Assent	Aarhus University, Denmark
Albert Bifet	Télécom ParisTech, France
Francesco Bonchi	ISI Foundation, Italy
Tania Cerquitelli	Politecnico di Torino, Italy
Sašo Džeroski	Jožef Stefan Institute, Slovenia
Elisa Fromont	Université de Rennes, France
Andreas Hotho	Julius-Maximilians-Universität Würzburg, Germany
Alípio Jorge	University of Porto, Portugal
Kristian Kersting	TU Darmstadt, Germany
Jefrey Lijffijt	Ghent University, Belgium
Luís Moreira-Matias	University of Porto, Portugal
Katharina Morik	TU Dortmund, Germany
Siegfried Nijssen	Université catholique de Louvain, Belgium
Andrea Passerini	University of Trento, Italy
Fernando Perez-Cruz	ETH Zurich, Switzerland
Alessandra Sala	Shutterstock Ireland Limited, Ireland
Arno Siebes	Utrecht University, the Netherlands
Isabel Valera	Universität des Saarlandes, Germany

Program Committees

Guest Editorial Board, Journal Track

Richard Allmendinger	University of Manchester, UK
Marie Anastacio	Universiteit Leiden, the Netherlands
Ira Assent	Aarhus University, Denmark
Martin Atzmueller	Universität Osnabrück, Germany
Rohit Babbar	Aalto University, Finland

Jaume Bacardit	Newcastle University, UK
Anthony Bagnall	University of East Anglia, UK
Mitra Baratchi	Universiteit Leiden, the Netherlands
Francesco Bariatti	IRISA, France
German Barquero	Universität de Barcelona, Spain
Alessio Benavoli	Trinity College Dublin, Ireland
Viktor Bengs	Ludwig-Maximilians-Universität München, Germany
Massimo Bilancia	Università degli Studi di Bari Aldo Moro, Italy
Ilaria Bordino	Unicredit R&D, Italy
Jakob Bossek	University of Münster, Germany
Ulf Brefeld	Leuphana University of Lüneburg, Germany
Ricardo Campello	University of Newcastle, UK
Michelangelo Ceci	University of Bari, Italy
Loic Cerf	Universidade Federal de Minas Gerais, Brazil
Vitor Cerqueira	Universidade do Porto, Portugal
Laetitia Chapel	IRISA, France
Jinghui Chen	Pennsylvania State University, USA
Silvia Chiusano	Politecnico di Torino, Italy
Roberto Corizzo	Università degli Studi di Bari Aldo Moro, Italy
Bruno Cremilleux	Université de Caen Normandie, France
Marco de Gemmis	University of Bari Aldo Moro, Italy
Sebastien Destercke	Centre National de la Recherche Scientifique, France
Shridhar Devamane	Global Academy of Technology, India
Benjamin Doerr	Ecole Polytechnique, France
Wouter Duivesteijn	Technische Universiteit Eindhoven, the Netherlands
Thomas Dyhre Nielsen	Aalborg University, Denmark
Tapio Elomaa	Tampere University, Finland
Remi Emonet	Université Jean Monnet Saint-Etienne, France
Nicola Fanizzi	Università degli Studi di Bari Aldo Moro, Italy
Pedro Ferreira	University of Lisbon, Portugal
Cesar Ferri	Universität Politecnica de Valencia, Spain
Julia Flores	University of Castilla-La Mancha, Spain
Ionut Florescu	Stevens Institute of Technology, USA
Germain Forestier	Université de Haute-Alsace, France
Joel Frank	Ruhr-Universität Bochum, Germany
Marco Frasca	Università degli Studi di Milano, Italy
Jose A. Gomez	Universidad de Castilla-La Mancha, Spain
Stephan Günnemann	Institute for Advanced Study, Germany
Luis Galarraga	Inria, France

Corrado Loglisci	Università degli Studi di Bari Aldo Moro, Italy
Nuno Lourenço	University of Coimbra, Portugal
Claudio Lucchese	Ca'Foscari University of Venice, Italy
Brian MacNamee	University College Dublin, Ireland
Davide Maiorca	University of Cagliari, Italy
Giuseppe Manco	National Research Council, Italy
Elio Masciari	University of Naples Federico II, Italy
Andres Masegosa	University of Aalborg, Denmark
Ernestina Menasalvas	Universidad Politecnica de Madrid, Spain
Lien Michiels	Universiteit Antwerpen, Belgium
Jan Mielniczuk	Polish Academy of Sciences, Poland
Paolo Mignone	Università degli Studi di Bari Aldo Moro, Italy
Anna Monreale	University of Pisa, Italy
Giovanni Montana	University of Warwick, UK
Gregoire Montavon	Technische Universität Berlin, Germany
Amedeo Napoli	LORIA, France
Frank Neumann	University of Adelaide, Australia
Thomas Nielsen	Aalborg Universitet, Denmark
Bruno Ordozgoiti	Aalto-yliopisto, Finland
Panagiotis Papapetrou	Stockholms Universitet, Sweden
Andrea Passerini	University of Trento, Italy
Mykola Pechenizkiy	Technische Universiteit Eindhoven, the Netherlands
Charlotte Pelletier	IRISA, France
Ruggero Pensa	University of Turin, Italy
Nico Piatkowski	Technische Universität Dortmund, Germany
Gianvito Pio	Università degli Studi di Bari Aldo Moro, Italy
Marc Plantevit	Université Claude Bernard Lyon 1, France
Jose M. Puerta	Universidad de Castilla-La Mancha, Spain
Kai Puolamaki	Helsingin Yliopisto, Finland
Michael Rabbat	Meta Platforms Inc, USA
Jan Ramon	Inria Lille Nord Europe, France
Rita Ribeiro	Universidade do Porto, Portugal
Kaspar Riesen	University of Bern, Switzerland
Matteo Riondato	Amherst College, USA
Celine Robardet	INSA Lyon, France
Pieter Robberechts	KU Leuven, Belgium
Antonio Salmeron	University of Almería, Spain
Jorg Sander	University of Alberta, Canada
Roberto Santana	University of the Basque Country, Spain
Michael Schaub	Rheinisch-Westfälische Technische Hochschule, Germany

Erik Schultheis	Aalto-yliopisto, Finland
Thomas Seidl	Ludwig-Maximilians-Universität München, Germany
Moritz Seiler	University of Münster, Germany
Kijung Shin	KAIST, South Korea
Shinichi Shirakawa	Yokohama National University, Japan
Marek Smieja	Jagiellonian University, Poland
James Edward Smith	University of the West of England, UK
Carlos Soares	Universidade do Porto, Portugal
Arnaud Soulet	Université de Tours, France
Gerasimos Spanakis	Maastricht University, the Netherlands
Giancarlo Sperli	University of Campania Luigi Vanvitelli, Italy
Myra Spiliopoulou	Otto von Guericke Universität Magdeburg, Germany
Jerzy Stefanowski	Poznan University of Technology, Poland
Giovanni Stilo	Università degli Studi dell'Aquila, Italy
Catalin Stoean	University of Craiova, Romania
Mahito Sugiyama	National Institute of Informatics, Japan
Nikolaj Tatti	Helsingin Yliopisto, Finland
Alexandre Termier	Université de Rennes 1, France
Luis Torgo	Dalhousie University, Canada
Leonardo Trujillo	Tecnologico Nacional de Mexico, Mexico
Wei-Wei Tu	4Paradigm Inc., China
Steffen Udluft	Siemens AG Corporate Technology, Germany
Arnaud Vandaele	Université de Mons, Belgium
Celine Vens	KU Leuven, Belgium
Herna Viktor	University of Ottawa, Canada
Marco Virgolin	Centrum Wiskunde en Informatica, the Netherlands
Jordi Vitria	Universität de Barcelona, Spain
Jilles Vreeken	CISPA Helmholtz Center for Information Security, Germany
Willem Waegeman	Universiteit Gent, Belgium
Markus Wagner	University of Adelaide, Australia
Elizabeth Wanner	Centro Federal de Educacao Tecnologica de Minas, Brazil
Marcel Wever	Universität Paderborn, Germany
Ngai Wong	University of Hong Kong, Hong Kong, China
Man Leung Wong	Lingnan University, Hong Kong, China
Marek Wydmuch	Poznan University of Technology, Poland
Guoxian Yu	Shandong University, China
Xiang Zhang	University of Hong Kong, Hong Kong, China

Ye Zhu	Deakin University, USA
Arthur Zimek	Syddansk Universitet, Denmark
Albrecht Zimmermann	Université de Caen Normandie, France

Area Chairs

Fabrizio Angiulli	DIMES, University of Calabria, Italy
Annalisa Appice	University of Bari, Italy
Ira Assent	Aarhus University, Denmark
Martin Atzmueller	Osnabrück University, Germany
Michael Berthold	Universität Konstanz, Germany
Albert Bifet	Université Paris-Saclay, France
Hendrik Blockeel	KU Leuven, Belgium
Christian Böhm	LMU Munich, Germany
Francesco Bonchi	ISI Foundation, Turin, Italy
Ulf Brefeld	Leuphana, Germany
Francesco Calabrese	Richemont, USA
Toon Calders	Universiteit Antwerpen, Belgium
Michelangelo Ceci	University of Bari, Italy
Peggy Cellier	IRISA, France
Duen Horng Chau	Georgia Institute of Technology, USA
Nicolas Courty	IRISA, Université Bretagne-Sud, France
Bruno Cremilleux	Université de Caen Normandie, France
Jesse Davis	KU Leuven, Belgium
Gianmarco De Francisci Morales	CentAI, Italy
Tom Diethe	Amazon, UK
Carlotta Domeniconi	George Mason University, USA
Yuxiao Dong	Tsinghua University, China
Kurt Driessens	Maastricht University, the Netherlands
Tapio Elomaa	Tampere University, Finland
Sergio Escalera	CVC and University of Barcelona, Spain
Faisal Farooq	Qatar Computing Research Institute, Qatar
Asja Fischer	Ruhr University Bochum, Germany
Peter Flach	University of Bristol, UK
Eibe Frank	University of Waikato, New Zealand
Paolo Frasconi	Università degli Studi di Firenze, Italy
Elisa Fromont	Université Rennes 1, IRISA/Inria, France
Johannes Fürnkranz	JKU Linz, Austria
Patrick Gallinari	Sorbonne Université, Criteo AI Lab, France
Joao Gama	INESC TEC - LIAAD, Portugal
Jose Gamez	Universidad de Castilla-La Mancha, Spain
Roman Garnett	Washington University in St. Louis, USA
Thomas Gärtner	TU Wien, Austria

Andrea Passerini	University of Trento, Italy
Mykola Pechenizkiy	TU Eindhoven, the Netherlands
Dino Pedreschi	University of Pisa, Italy
Robert Peharz	Graz University of Technology, Austria
Julien Perez	Naver Labs Europe, France
Franz Pernkopf	Graz University of Technology, Austria
Bernhard Pfahringer	University of Waikato, New Zealand
Fabio Pinelli	IMT Lucca, Italy
Visvanathan Ramesh	Goethe University Frankfurt, Germany
Jesse Read	Ecole Polytechnique, France
Zhaochun Ren	Shandong University, China
Marian-Andrei Rizoiu	University of Technology Sydney, Australia
Celine Robardet	INSA Lyon, France
Sriparna Saha	IIT Patna, India
Ute Schmid	University of Bamberg, Germany
Lars Schmidt-Thieme	University of Hildesheim, Germany
Michele Sebag	LISN CNRS, France
Thomas Seidl	LMU Munich, Germany
Arno Siebes	Universiteit Utrecht, the Netherlands
Fabrizio Silvestri	Sapienza, University of Rome, Italy
Myrá Spiliopoulou	Otto-von-Guericke-University Magdeburg, Germany
Yizhou Sun	UCLA, USA
Jie Tang	Tsinghua University, China
Nikolaj Tatti	Helsinki University, Finland
Evimaria Terzi	Boston University, USA
Marc Tommasi	Lille University, France
Antti Ukkonen	University of Helsinki, Finland
Herke van Hoof	University of Amsterdam, the Netherlands
Matthijs van Leeuwen	Leiden University, the Netherlands
Celine Vens	KU Leuven, Belgium
Christel Vrain	University of Orleans, France
Jilles Vreeken	CISPA Helmholtz Center for Information Security, Germany
Willem Waegeman	Universiteit Gent, Belgium
Stefan Wrobel	Fraunhofer IAIS, Germany
Xing Xie	Microsoft Research Asia, China
Min-Ling Zhang	Southeast University, China
Albrecht Zimmermann	Université de Caen Normandie, France
Indre Zliobaite	University of Helsinki, Finland

Program Committee Members

Amos Abbott	Virginia Tech, USA
Pedro Abreu	CISUC, Portugal
Maribel Acosta	Ruhr University Bochum, Germany
Timilehin Aderinola	Insight Centre, University College Dublin, Ireland
Linara Adilova	Ruhr University Bochum, Fraunhofer IAIS, Germany
Florian Adriaens	KTH, Sweden
Azim Ahmadzadeh	Georgia State University, USA
Nourhan Ahmed	University of Hildesheim, Germany
Deepak Ajwani	University College Dublin, Ireland
Amir Hossein Akhavan Rahnama	KTH Royal Institute of Technology, Sweden
Aymen Al Marjani	ENS Lyon, France
Mehwish Alam	Leibniz Institute for Information Infrastructure, Germany
Francesco Alesiani	NEC Laboratories Europe, Germany
Omar Alfarisi	ADNOC, Canada
Pegah Alizadeh	Ericsson Research, France
Reem Alotaibi	King Abdulaziz University, Saudi Arabia
Jumanah Alshehri	Temple University, USA
Bakhtiar Amen	University of Huddersfield, UK
Evelin Amorim	Inesc tec, Portugal
Shin Ando	Tokyo University of Science, Japan
Thiago Andrade	INESC TEC - LIAAD, Portugal
Jean-Marc Andreoli	Naverlabs Europe, France
Giuseppina Andresini	University of Bari Aldo Moro, Italy
Alessandro Antonucci	IDSIA, Switzerland
Xiang Ao	Institute of Computing Technology, CAS, China
Siddharth Aravindan	National University of Singapore, Singapore
Héber H. Arcolezi	Inria and École Polytechnique, France
Adrián Arnaiz-Rodríguez	ELLIS Unit Alicante, Spain
Yusuf Arslan	University of Luxembourg, Luxembourg
André Artelt	Bielefeld University, Germany
Sunil Aryal	Deakin University, Australia
Charles Assaad	Easyvista, France
Matthias Aßenmacher	Ludwig-Maxmilians-Universität München, Germany
Zeyar Aung	Masdar Institute, UAE
Serge Autexier	DFKI Bremen, Germany
Rohit Babbar	Aalto University, Finland
Housam Babiker	University of Alberta, Canada

Antonio Bahamonde	University of Oviedo, Spain
Maroua Bahri	Inria Paris, France
Georgios Balikas	Salesforce, France
Maria Bampa	Stockholm University, Sweden
Hubert Baniecki	Warsaw University of Technology, Poland
Elena Baralis	Politecnico di Torino, Italy
Mitra Baratchi	LIACS - University of Leiden, the Netherlands
Kalliopi Basioti	Rutgers University, USA
Martin Becker	Stanford University, USA
Diana Benavides Prado	University of Auckland, New Zealand
Anes Bendimerad	LIRIS, France
Idir Benouaret	Université Grenoble Alpes, France
Isacco Beretta	Università di Pisa, Italy
Victor Berger	CEA, France
Christoph Bergmeir	Monash University, Australia
Cuissart Bertrand	University of Caen, France
Antonio Bevilacqua	University College Dublin, Ireland
Yaxin Bi	Ulster University, UK
Ranran Bian	University of Auckland, New Zealand
Adrien Bibal	University of Louvain, Belgium
Subhodip Biswas	Virginia Tech, USA
Patrick Blöbaum	Amazon AWS, USA
Carlos Bobed	University of Zaragoza, Spain
Paul Bogdan	USC, USA
Chiara Boldrini	CNR, Italy
Clément Bonet	Université Bretagne Sud, France
Andrea Bontempelli	University of Trento, Italy
Ludovico Boratto	University of Cagliari, Italy
Stefano Bortoli	Huawei Research Center, Germany
Diana-Laura Borza	Babes Bolyai University, Romania
Ahcene Boubekki	UiT, Norway
Sabri Boughorbel	QCRI, Qatar
Paula Branco	University of Ottawa, Canada
Jure Brence	Jožef Stefan Institute, Slovenia
Martin Breskvar	Jožef Stefan Institute, Slovenia
Marco Bressan	University of Milan, Italy
Dariusz Brzezinski	Poznan University of Technology, Poland
Florian Buettner	German Cancer Research Center, Germany
Julian Busch	Siemens Technology, Germany
Sebastian Buschjäger	TU Dortmund Artificial Intelligence Unit, Germany
Ali Butt	Virginia Tech, USA

Narayanan C. Krishnan	IIT Palakkad, India
Xiangrui Cai	Nankai University, China
Xiongcai Cai	UNSW Sydney, Australia
Zekun Cai	University of Tokyo, Japan
Andrea Campagner	Università degli Studi di Milano-Bicocca, Italy
Seyit Camtepe	CSIRO Data61, Australia
Jiangxia Cao	Chinese Academy of Sciences, China
Pengfei Cao	Chinese Academy of Sciences, China
Yongcan Cao	University of Texas at San Antonio, USA
Cécile Capponi	Aix-Marseille University, France
Axel Carlier	Institut National Polytechnique de Toulouse, France
Paula Carroll	University College Dublin, Ireland
John Cartlidge	University of Bristol, UK
Simon Caton	University College Dublin, Ireland
Bogdan Cautis	University of Paris-Saclay, France
Mustafa Cavus	Warsaw University of Technology, Poland
Remy Cazabet	Université Lyon 1, France
Josu Ceberio	University of the Basque Country, Spain
David Cechák	CEITEC Masaryk University, Czechia
Abdulkadir Celikkanat	Technical University of Denmark, Denmark
Dumitru-Clementin Cercel	University Politehnica of Bucharest, Romania
Christophe Cerisara	CNRS, France
Vítor Cerqueira	Dalhousie University, Canada
Mattia Cerrato	JGU Mainz, Germany
Ricardo Cerri	Federal University of São Carlos, Brazil
Hubert Chan	University of Hong Kong, Hong Kong, China
Vaggos Chatziafratis	Stanford University, USA
Siu Lun Chau	University of Oxford, UK
Chaochao Chen	Zhejiang University, China
Chuan Chen	Sun Yat-sen University, China
Hechang Chen	Jilin University, China
Jia Chen	Beihang University, China
Jiaoyan Chen	University of Oxford, UK
Jiawei Chen	Zhejiang University, China
Jin Chen	University of Electronic Science and Technology, China
Kuan-Hsun Chen	University of Twente, the Netherlands
Lingwei Chen	Wright State University, USA
Tianyi Chen	Boston University, USA
Wang Chen	Google, USA
Xinyuan Chen	Universiti Kuala Lumpur, Malaysia

Yuqiao Chen	UT Dallas, USA
Yuzhou Chen	Princeton University, USA
Zhennan Chen	Xiamen University, China
Zhiyu Chen	UCSB, USA
Zhqian Chen	Mississippi State University, USA
Ziheng Chen	Stony Brook University, USA
Zhiyong Cheng	Shandong Academy of Sciences, China
Noëlie Cherrier	CITiO, France
Anshuman Chhabra	UC Davis, USA
Zhixuan Chu	Ant Group, China
Guillaume Cleuziou	LIFO, France
Ciaran Cooney	AflacNI, UK
Robson Cordeiro	University of São Paulo, Brazil
Roberto Corizzo	American University, USA
Antoine Cornuéjols	AgroParisTech, France
Fabrizio Costa	Exeter University, UK
Gustavo Costa	Instituto Federal de Goiás - Campus Jataí, Brazil
Luís Cruz	Delft University of Technology, the Netherlands
Tianyu Cui	Institute of Information Engineering, China
Wang-Zhou Dai	Imperial College London, UK
Tanmoy Dam	University of New South Wales Canberra, Australia
Thi-Bich-Hanh Dao	University of Orleans, France
Adrian Sergiu Darabant	Babes Bolyai University, Romania
Mrinal Das	IIT Palakaad, India
Sina Däubener	Ruhr University, Bochum, Germany
Padraig Davidson	University of Würzburg, Germany
Paul Davidsson	Malmö University, Sweden
Andre de Carvalho	USP, Brazil
Antoine de Mathelin	ENS Paris-Saclay, France
Tom De Schepper	University of Antwerp, Belgium
Marcilio de Souto	LIFO/Univ. Orleans, France
Gaetan De Waele	Ghent University, Belgium
Pieter Delobelle	KU Leuven, Belgium
Alper Demir	Izmir University of Economics, Turkey
Ambra Demontis	University of Cagliari, Italy
Difan Deng	Leibniz Universität Hannover, Germany
Guillaume Derval	UCLouvain - ICTEAM, Belgium
Maunendra Sankar Desarkar	IIT Hyderabad, India
Chris Develder	University of Ghent - iMec, Belgium
Arnout Devos	Swiss Federal Institute of Technology Lausanne, Switzerland

Stefano Ferilli	University of Bari, Italy
Daniel Fernández-Sánchez	Universidad Autónoma de Madrid, Spain
Pedro Ferreira	Faculty of Sciences University of Porto, Portugal
Cèsar Ferri	Universität Politècnica València, Spain
Flavio Figueiredo	UFMG, Brazil
Soukaina Filali Boubrahimi	Utah State University, USA
Raphael Fischer	TU Dortmund, Germany
Germain Forestier	University of Haute Alsace, France
Edouard Fouché	Karlsruhe Institute of Technology, Germany
Philippe Fournier-Viger	Shenzhen University, China
Kary Framling	Umeå University, Sweden
Jérôme François	Inria Nancy Grand-Est, France
Fabio Fumarola	Prometeia, Italy
Pratik Gajane	Eindhoven University of Technology, the Netherlands
Esther Galbrun	University of Eastern Finland, Finland
Laura Galindez Olascoaga	KU Leuven, Belgium
Sunanda Gamage	University of Western Ontario, Canada
Chen Gao	Tsinghua University, China
Wei Gao	Nanjing University, China
Xiaofeng Gao	Shanghai Jiaotong University, China
Yuan Gao	University of Science and Technology of China, China
Jochen Garcke	University of Bonn, Germany
Clement Gautrais	Brightclue, France
Benoit Gauzere	INSA Rouen, France
Dominique Gay	Université de La Réunion, France
Xiou Ge	University of Southern California, USA
Bernhard Geiger	Know-Center GmbH, Germany
Jiahui Geng	University of Stavanger, Norway
Yangliao Geng	Tsinghua University, China
Konstantin Genin	University of Tübingen, Germany
Firas Gerges	New Jersey Institute of Technology, USA
Pierre Geurts	University of Liège, Belgium
Gizem Gezici	Sabanci University, Turkey
Amirata Ghorbani	Stanford, USA
Biraja Ghoshal	TCS, UK
Anna Giabelli	Università degli studi di Milano Bicocca, Italy
George Giannakopoulos	IIT Demokritos, Greece
Tobias Glasmachers	Ruhr-University Bochum, Germany
Heitor Murilo Gomes	University of Waikato, New Zealand
Anastasios Gounaris	Aristotle University of Thessaloniki, Greece

Antoine Gourru	University of Lyon, France
Michael Granitzer	University of Passau, Germany
Magda Gregorova	Hochschule Würzburg-Schweinfurt, Germany
Moritz Grosse-Wentrup	University of Vienna, Austria
Divya Grover	Chalmers University, Sweden
Bochen Guan	OPPO US Research Center, USA
Xinyu Guan	Xian Jiaotong University, China
Guillaume Guerard	ESILV, France
Daniel Guerreiro e Silva	University of Brasilia, Brazil
Riccardo Guidotti	University of Pisa, Italy
Ekta Gujral	University of California, Riverside, USA
Aditya Gulati	ELLIS Unit Alicante, Spain
Guibing Guo	Northeastern University, China
Jianxiong Guo	Beijing Normal University, China
Yuhui Guo	Renmin University of China, China
Karthik Gurumoorthy	Amazon, India
Thomas Guyet	Inria, Centre de Lyon, France
Guillaume Habault	KDDI Research, Inc., Japan
Amaury Habrard	University of St-Etienne, France
Shahrzad Haddadan	Brown University, USA
Shah Muhammad Hamdi	New Mexico State University, USA
Massinissa Hamidi	PRES Sorbonne Paris Cité, France
Peng Han	KAUST, Saudi Arabia
Tom Hanika	University of Kassel, Germany
Sébastien Harispe	IMT Mines Alès, France
Marwan Hassani	TU Eindhoven, the Netherlands
Kohei Hayashi	Preferred Networks, Inc., Japan
Conor Hayes	National University of Ireland Galway, Ireland
Lingna He	Zhejiang University of Technology, China
Ramya Hebbalaguppe	Indian Institute of Technology, Delhi, India
Jukka Heikkonen	University of Turku, Finland
Fredrik Heintz	Linköping University, Sweden
Patrick Hemmer	Karlsruhe Institute of Technology, Germany
Romain Hérault	INSA de Rouen, France
Jeronimo Hernandez-Gonzalez	University of Barcelona, Spain
Sibylle Hess	TU Eindhoven, the Netherlands
Fabian Hinder	Bielefeld University, Germany
Lars Holdijk	University of Amsterdam, the Netherlands
Martin Holena	Institute of Computer Science, Czechia
Mike Holenderski	Eindhoven University of Technology, the Netherlands
Shenda Hong	Peking University, China

Yupeng Hou	Renmin University of China, China
Binbin Hu	Ant Financial Services Group, China
Jian Hu	Queen Mary University of London, UK
Liang Hu	Tongji University, China
Wen Hu	Ant Group, China
Wenbin Hu	Wuhan University, China
Wenbo Hu	Tsinghua University, China
Yaowei Hu	University of Arkansas, USA
Chao Huang	University of Hong Kong, China
Gang Huang	Zhejiang Lab, China
Guanjie Huang	Penn State University, USA
Hong Huang	HUST, China
Jin Huang	University of Amsterdam, the Netherlands
Junjie Huang	Chinese Academy of Sciences, China
Qiang Huang	Jilin University, China
Shangrong Huang	Hunan University, China
Weitian Huang	South China University of Technology, China
Yan Huang	Huazhong University of Science and Technology, China
Yiran Huang	Karlsruhe Institute of Technology, Germany
Angelo Impedovo	University of Bari, Italy
Roberto Interdonato	CIRAD, France
Iñaki Inza	University of the Basque Country, Spain
Stratis Ioannidis	Northeastern University, USA
Rakib Islam	Facebook, USA
Tobias Jacobs	NEC Laboratories Europe GmbH, Germany
Priyank Jaini	Google, Canada
Johannes Jakubik	Karlsruhe Institute of Technology, Germany
Nathalie Japkowicz	American University, USA
Szymon Jaroszewicz	Polish Academy of Sciences, Poland
Shayan Jawed	University of Hildesheim, Germany
Rathinaraja Jeyaraj	Kyungpook National University, South Korea
Shaoxiong Ji	Aalto University, Finland
Taoran Ji	Virginia Tech, USA
Bin-Bin Jia	Southeast University, China
Yuheng Jia	Southeast University, China
Ziyu Jia	Beijing Jiaotong University, China
Nan Jiang	Purdue University, USA
Renhe Jiang	University of Tokyo, Japan
Siyang Jiang	National Taiwan University, Taiwan
Song Jiang	University of California, Los Angeles, USA
Wenyu Jiang	Nanjing University, China

Anna Krause	University of Würzburg, Germany
Alexander Kravberg	KTH Royal Institute of Technology, Sweden
Anastasia Krithara	NCSR Demokritos, Greece
Meelis Kull	University of Tartu, Estonia
Pawan Kumar	IIIT, Hyderabad, India
Suresh Kirthi Kumaraswamy	InterDigital, France
Gautam Kunapuli	Verisk Inc, USA
Marcin Kurdziel	AGH University of Science and Technology, Poland
Vladimir Kuzmanovski	Aalto University, Finland
Ariel Kwiatkowski	École Polytechnique, France
Firas Laakom	Tampere University, Finland
Harri Lähdesmäki	Aalto University, Finland
Stefanos Laskaridis	Samsung AI, UK
Alberto Lavelli	FBK-ict, Italy
Aonghus Lawlor	University College Dublin, Ireland
Thai Le	University of Mississippi, USA
Hoàng-Ân Lê	IRISA, University of South Brittany, France
Hoel Le Capitaine	University of Nantes, France
Thach Le Nguyen	Insight Centre, Ireland
Tai Le Quy	L3S Research Center - Leibniz University Hannover, Germany
Mustapha Lebbah	Sorbonne Paris Nord University, France
Dongman Lee	KAIST, South Korea
John Lee	Université catholique de Louvain, Belgium
Minwoo Lee	University of North Carolina at Charlotte, USA
Zed Lee	Stockholm University, Sweden
Yunwen Lei	University of Birmingham, UK
Douglas Leith	Trinity College Dublin, Ireland
Florian Lemmerich	RWTH Aachen, Germany
Carson Leung	University of Manitoba, Canada
Chaozhuo Li	Microsoft Research Asia, China
Jian Li	Institute of Information Engineering, China
Lei Li	Peking University, China
Li Li	Southwest University, China
Rui Li	Inspur Group, China
Shiyang Li	UCSB, USA
Shuokai Li	Chinese Academy of Sciences, China
Tianyu Li	Alibaba Group, China
Wenye Li	The Chinese University of Hong Kong, Shenzhen, China
Wenzhong Li	Nanjing University, China

Xiaoting Li	Pennsylvania State University, USA
Yang Li	University of North Carolina at Chapel Hill, USA
Zejian Li	Zhejiang University, China
Zhidong Li	UTS, Australia
Zhixin Li	Guangxi Normal University, China
Defu Lian	University of Science and Technology of China, China
Bin Liang	UTS, Australia
Yuchen Liang	RPI, USA
Yiwen Liao	University of Stuttgart, Germany
Pieter Libin	VUB, Belgium
Thomas Liebig	TU Dortmund, Germany
Seng Pei Liew	LINE Corporation, Japan
Beiyu Lin	University of Nevada - Las Vegas, USA
Chen Lin	Xiamen University, China
Tony Lindgren	Stockholm University, Sweden
Chen Ling	Emory University, USA
Jiajing Ling	Singapore Management University, Singapore
Marco Lippi	University of Modena and Reggio Emilia, Italy
Bin Liu	Chongqing University, China
Bowen Liu	Stanford University, USA
Chang Liu	Institute of Information Engineering, CAS, China
Chien-Liang Liu	National Chiao Tung University, Taiwan
Feng Liu	East China Normal University, China
Jiacheng Liu	Chinese University of Hong Kong, China
Li Liu	Chongqing University, China
Shengcai Liu	Southern University of Science and Technology, China
Shenghua Liu	Institute of Computing Technology, CAS, China
Tingwen Liu	Institute of Information Engineering, CAS, China
Xiangyu Liu	Tencent, China
Yong Liu	Renmin University of China, China
Yuansan Liu	University of Melbourne, Australia
Zhiwei Liu	Salesforce, USA
Tuwe Löfström	Jönköping University, Sweden
Corrado Loglisci	Università degli Studi di Bari Aldo Moro, Italy
Ting Long	Shanghai Jiao Tong University, China
Beatriz López	University of Girona, Spain
Yin Lou	Ant Group, USA
Samir Loudni	TASC (LS2N-CNRS), IMT Atlantique, France
Yang Lu	Xiamen University, China
Yuxun Lu	National Institute of Informatics, Japan

Massimiliano Luca	Bruno Kessler Foundation, Italy
Stefan Lüdtke	University of Mannheim, Germany
Jovita Lukasik	University of Mannheim, Germany
Denis Lukovnikov	University of Bonn, Germany
Pedro Henrique Luz de Araujo	University of Brasília, Brazil
Fenglong Ma	Pennsylvania State University, USA
Jing Ma	University of Virginia, USA
Meng Ma	Peking University, China
Muyang Ma	Shandong University, China
Ruizhe Ma	University of Massachusetts Lowell, USA
Xingkong Ma	National University of Defense Technology, China
Xueqi Ma	Tsinghua University, China
Zichen Ma	The Chinese University of Hong Kong, Shenzhen, China
Luis Macedo	University of Coimbra, Portugal
Harshitha Machiraju	EPFL, Switzerland
Manchit Madan	Delivery Hero, Germany
Seiji Maekawa	Osaka University, Japan
Sindri Magnusson	Stockholm University, Sweden
Pathum Chamikara Mahawaga	CSIRO Data61, Australia
Saket Maheshwary	Amazon, India
Ajay Mahimkar	AT&T, USA
Pierre Maillot	Inria, France
Lorenzo Malandri	Unimib, Italy
Rammohan Mallipeddi	Kyungpook National University, South Korea
Sahil Manchanda	IIT Delhi, India
Domenico Mandaglio	DIMES-UNICAL, Italy
Panagiotis Mandros	Harvard University, USA
Robin Manhaeve	KU Leuven, Belgium
Silviu Maniu	Université Paris-Saclay, France
Cinmayii Manliguez	National Sun Yat-Sen University, Taiwan
Naresh Manwani	International Institute of Information Technology, India
Jiali Mao	East China Normal University, China
Alexandru Mara	Ghent University, Belgium
Radu Marculescu	University of Texas at Austin, USA
Roger Mark	Massachusetts Institute of Technology, USA
Fernando Martínez-Plume	Joint Research Centre - European Commission, Belgium
Koji Maruhashi	Fujitsu Research, Fujitsu Limited, Japan
Simone Marullo	University of Siena, Italy

Elio Masciari	University of Naples, Italy
Florent Masseglia	Inria, France
Michael Mathioudakis	University of Helsinki, Finland
Takashi Matsubara	Osaka University, Japan
Tetsu Matsukawa	Kyushu University, Japan
Santiago Mazuelas	BCAM-Basque Center for Applied Mathematics, Spain
Ryan McConville	University of Bristol, UK
Hardik Meisheri	TCS Research, India
Panagiotis Meletis	Eindhoven University of Technology, the Netherlands
Gabor Melli	Medable, USA
Joao Mendes-Moreira	INESC TEC, Portugal
Chuan Meng	University of Amsterdam, the Netherlands
Cristina Menghini	Brown University, USA
Engelbert Mephu Nguifo	Université Clermont Auvergne, CNRS, LIMOS, France
Fabio Mercorio	University of Milan-Bicocca, Italy
Guillaume Metzler	Laboratoire ERIC, France
Hao Miao	Aalborg University, Denmark
Alessio Micheli	Università di Pisa, Italy
Paolo Mignone	University of Bari Aldo Moro, Italy
Matej Mihelcic	University of Zagreb, Croatia
Ioanna Miliou	Stockholm University, Sweden
Bamdev Mishra	Microsoft, India
Rishabh Misra	Twitter, Inc, USA
Dixant Mittal	National University of Singapore, Singapore
Zhaobin Mo	Columbia University, USA
Daichi Mochihashi	Institute of Statistical Mathematics, Japan
Armin Moharrer	Northeastern University, USA
Ioannis Mollas	Aristotle University of Thessaloniki, Greece
Carlos Monserrat-Aranda	Universität Politècnica de València, Spain
Konda Reddy Mopuri	Indian Institute of Technology Guwahati, India
Raha Moraffah	Arizona State University, USA
Pawel Morawiecki	Polish Academy of Sciences, Poland
Ahmadreza Mosallanezhad	Arizona State University, USA
Davide Mottin	Aarhus University, Denmark
Koyel Mukherjee	Adobe Research, India
Maximilian Münch	University of Applied Sciences Würzburg, Germany
Fabricio Murai	Universidade Federal de Minas Gerais, Brazil
Taichi Murayama	NAIST, Japan

Stéphane Mussard	CHROME, France
Mohamed Nadif	Centre Borelli - Université Paris Cité, France
Cian Naik	University of Oxford, UK
Felipe Kenji Nakano	KU Leuven, Belgium
Mirco Nanni	ISTI-CNR Pisa, Italy
Apurva Narayan	University of Waterloo, Canada
Usman Naseem	University of Sydney, Australia
Gergely Nemeth	ELLIS Unit Alicante, Spain
Stefan Neumann	KTH Royal Institute of Technology, Sweden
Anna Nguyen	Karlsruhe Institute of Technology, Germany
Quan Nguyen	Washington University in St. Louis, USA
Thi Phuong Quyen Nguyen	University of Da Nang, Vietnam
Thu Nguyen	SimulaMet, Norway
Thu Trang Nguyen	University College Dublin, Ireland
Prajakta Nimbhorkar	Chennai Mathematical Institute, Chennai, India
Xuefei Ning	Tsinghua University, China
Ikuko Nishikawa	Ritsumeikan University, Japan
Hao Niu	KDDI Research, Inc., Japan
Paraskevi Nousi	Aristotle University of Thessaloniki, Greece
Erik Novak	Jožef Stefan Institute, Slovenia
Slawomir Nowaczyk	Halmstad University, Sweden
Aleksandra Nowak	Jagiellonian University, Poland
Eirini Ntoutsi	Freie Universität Berlin, Germany
Andreas Nürnberger	Magdeburg University, Germany
James O'Neill	University of Liverpool, UK
Lutz Oettershagen	University of Bonn, Germany
Tsuyoshi Okita	Kyushu Institute of Technology, Japan
Makoto Onizuka	Osaka University, Japan
Subba Reddy Oota	IIIT Hyderabad, India
María Óskarsdóttir	University of Reykjavík, Iceland
Aomar Osmani	PRES Sorbonne Paris Cité, France
Aljaz Osojnik	JSI, Slovenia
Shuichi Otake	National Institute of Informatics, Japan
Greger Ottosson	IBM, France
Zijing Ou	Sun Yat-sen University, China
Abdelkader Ouali	University of Caen Normandy, France
Latifa Oukhellou	IFSTTAR, France
Kai Ouyang	Tsinghua University, France
Andrei Paleyes	University of Cambridge, UK
Pankaj Pandey	Indian Institute of Technology Gandhinagar, India
Guansong Pang	Singapore Management University, Singapore
Pance Panov	Jožef Stefan Institute, Slovenia

Apostolos Papadopoulos	Aristotle University of Thessaloniki, Greece
Evangelos Papalexakis	UC Riverside, USA
Anna Pappa	Université Paris 8, France
Chanyoung Park	UIUC, USA
Haekyu Park	Georgia Institute of Technology, USA
Sanghyun Park	Yonsei University, South Korea
Luca Pasa	University of Padova, Italy
Kevin Pasini	IRT SystemX, France
Vincenzo Pasquadibisceglie	University of Bari Aldo Moro, Italy
Nikolaos Passalis	Aristotle University of Thessaloniki, Greece
Javier Pastorino	University of Colorado, Denver, USA
Kitsuchart Pasupa	King Mongkut's Institute of Technology, Thailand
Andrea Paudice	University of Milan, Italy
Anand Paul	Kyungpook National University, South Korea
Yulong Pei	TU Eindhoven, the Netherlands
Charlotte Pelletier	Université de Bretagne du Sud, France
Jaakko Peltonen	Tampere University, Finland
Ruggero Pensa	University of Torino, Italy
Fabiola Pereira	Federal University of Uberlandia, Brazil
Lucas Pereira	ITI, LARSyS, Técnico Lisboa, Portugal
Aritz Pérez	Basque Center for Applied Mathematics, Spain
Lorenzo Perini	KU Leuven, Belgium
Alan Perotti	CENTAI Institute, Italy
Michaël Perrot	Inria Lille, France
Matej Petkovic	Institute Jožef Stefan, Slovenia
Lukas Pfahler	TU Dortmund University, Germany
Nico Piatkowski	Fraunhofer IAIS, Germany
Francesco Piccialli	University of Naples Federico II, Italy
Gianvito Pio	University of Bari, Italy
Giuseppe Pirrò	Sapienza University of Rome, Italy
Marc Plantevit	EPITA, France
Konstantinos Pliakos	KU Leuven, Belgium
Matthias Pohl	Otto von Guericke University, Germany
Nicolas Posocco	EURA NOVA, Belgium
Cedric Pradalier	GeorgiaTech Lorraine, France
Paul Prasse	University of Potsdam, Germany
Mahardhika Pratama	University of South Australia, Australia
Francesca Pratesi	ISTI - CNR, Italy
Steven Prestwich	University College Cork, Ireland
Giulia Preti	CentAI, Italy
Philippe Preux	Inria, France
Shalini Priya	Oak Ridge National Laboratory, USA

Ricardo Prudencio	Universidade Federal de Pernambuco, Brazil
Luca Putelli	Università degli Studi di Brescia, Italy
Peter van der Putten	Leiden University, the Netherlands
Chuan Qin	Baidu, China
Jixiang Qing	Ghent University, Belgium
Jolin Qu	Western Sydney University, Australia
Nicolas Quesada	Polytechnique Montreal, Canada
Teeradaj Racharak	Japan Advanced Institute of Science and Technology, Japan
Krystian Radlak	Warsaw University of Technology, Poland
Sandro Radovanovic	University of Belgrade, Serbia
Md Masudur Rahman	Purdue University, USA
Ankita Raj	Indian Institute of Technology Delhi, India
Herilalaina Rakotoarison	Inria, France
Alexander Rakowski	Hasso Plattner Institute, Germany
Jan Ramon	Inria, France
Sascha Ranftl	Graz University of Technology, Austria
Aleksandra Rashkovska Koceva	Jožef Stefan Institute, Slovenia
S. Ravi	Biocomplexity Institute, USA
Jesse Read	Ecole Polytechnique, France
David Reich	Universität Potsdam, Germany
Marina Reyboz	CEA, LIST, France
Pedro Ribeiro	University of Porto, Portugal
Rita P. Ribeiro	University of Porto, Portugal
Piera Riccio	ELLIS Unit Alicante Foundation, Spain
Christophe Rigotti	INSA Lyon, France
Matteo Riondato	Amherst College, USA
Mateus Riva	Telecom ParisTech, France
Kit Rodolfa	CMU, USA
Christophe Rodrigues	DVRC Pôle Universitaire Léonard de Vinci, France
Simon Rodríguez-Santana	ICMAT, Spain
Gaetano Rossiello	IBM Research, USA
Mohammad Rostami	University of Southern California, USA
Franz Rothlauf	Mainz Universität, Germany
Celine Rouveirol	Université Paris-Nord, France
Arjun Roy	Freie Universität Berlin, Germany
Joze Rozanec	Josef Stefan International Postgraduate School, Slovenia
Salvatore Ruggieri	University of Pisa, Italy
Marko Ruman	UTIA, AV CR, Czechia
Ellen Rushe	University College Dublin, Ireland

Dawid Rymarczyk	Jagiellonian University, Poland
Amal Saadallah	TU Dortmund, Germany
Khaled Mohammed Saifuddin	Georgia State University, USA
Hajer Salem	AUDENSIEL, France
Francesco Salvetti	Politecnico di Torino, Italy
Roberto Santana	University of the Basque Country (UPV/EHU), Spain
KC Santosh	University of South Dakota, USA
Somdeb Sarkhel	Adobe, USA
Yuya Sasaki	Osaka University, Japan
Yücel Saygın	Sabancı Universitesi, Turkey
Patrick Schäfer	Humboldt-Universität zu Berlin, Germany
Alexander Schiendorfer	Technische Hochschule Ingolstadt, Germany
Peter Schlicht	Volkswagen Group Research, Germany
Daniel Schmidt	Monash University, Australia
Johannes Schneider	University of Liechtenstein, Liechtenstein
Steven Schockaert	Cardiff University, UK
Jens Schreiber	University of Kassel, Germany
Matthias Schubert	Ludwig-Maximilians-Universität München, Germany
Alexander Schulz	CITEC, Bielefeld University, Germany
Jan-Philipp Schulze	Fraunhofer AISEC, Germany
Andreas Schwung	Fachhochschule Südwestfalen, Germany
Vasile-Marian Scuturici	LIRIS, France
Raquel Sebastião	IEETA/DETI-UA, Portugal
Stanislav Selitskiy	University of Bedfordshire, UK
Edoardo Serra	Boise State University, USA
Lorenzo Severini	UniCredit, R&D Dept., Italy
Tapan Shah	GE, USA
Ammar Shaker	NEC Laboratories Europe, Germany
Shiv Shankar	University of Massachusetts, USA
Junming Shao	University of Electronic Science and Technology, China
Kartik Sharma	Georgia Institute of Technology, USA
Manali Sharma	Samsung, USA
Ariona Shashaj	Network Contacts, Italy
Betty Shea	University of British Columbia, Canada
Chengchao Shen	Central South University, China
Hailan Shen	Central South University, China
Jiawei Sheng	Chinese Academy of Sciences, China
Yongpan Sheng	Southwest University, China
Chongyang Shi	Beijing Institute of Technology, China

Zhengxiang Shi	University College London, UK
Naman Shukla	Deepair LLC, USA
Pablo Silva	Dell Technologies, Brazil
Simeon Simoff	Western Sydney University, Australia
Maneesh Singh	Motive Technologies, USA
Nikhil Singh	MIT Media Lab, USA
Sarath Sivaprasad	IIIT Hyderabad, India
Elena Sizikova	NYU, USA
Andrzej Skowron	University of Warsaw, Poland
Blaz Skrlj	Institute Jožef Stefan, Slovenia
Oliver Snow	Simon Fraser University, Canada
Jonas Soenen	KU Leuven, Belgium
Nataliya Sokolovska	Sorbonne University, France
K. M. A. Solaiman	Purdue University, USA
Shuangyong Song	Jing Dong, China
Zixing Song	The Chinese University of Hong Kong, China
Tiberiu Sosea	University of Illinois at Chicago, USA
Arnaud Soulet	University of Tours, France
Lucas Souza	UFRJ, Brazil
Jens Sparsø	Technical University of Denmark, Denmark
Vivek Srivastava	TCS Research, USA
Marija Stanojevic	Temple University, USA
Jerzy Stefanowski	Poznan University of Technology, Poland
Simon Stieber	University of Augsburg, Germany
Jinyan Su	University of Electronic Science and Technology, China
Yongduo Sui	University of Science and Technology of China, China
Huiyan Sun	Jilin University, China
Yuwei Sun	University of Tokyo/RIKEN AIP, Japan
Gokul Swamy	Amazon, USA
Maryam Tabar	Pennsylvania State University, USA
Anika Tabassum	Virginia Tech, USA
Shazia Tabassum	INESCTEC, Portugal
Koji Tabata	Hokkaido University, Japan
Andrea Tagarelli	DIMES, University of Calabria, Italy
Etienne Tajeuna	Université de Laval, Canada
Acar Tamersoy	NortonLifeLock Research Group, USA
Chang Wei Tan	Monash University, Australia
Cheng Tan	Westlake University, China
Feilong Tang	Shanghai Jiao Tong University, China
Feng Tao	Volvo Cars, USA

Youming Tao	Shandong University, China
Martin Tappler	Graz University of Technology, Austria
Garth Tarr	University of Sydney, Australia
Mohammad Tayebi	Simon Fraser University, Canada
Anastasios Tefas	Aristotle University of Thessaloniki, Greece
Maguelonne Teisseire	INRAE - UMR Tetis, France
Stefano Teso	University of Trento, Italy
Olivier Teste	IRIT, University of Toulouse, France
Maximilian Thiessen	TU Wien, Austria
Eleftherios Tiakas	Aristotle University of Thessaloniki, Greece
Hongda Tian	University of Technology Sydney, Australia
Alessandro Tibo	Aalborg University, Denmark
Aditya Srinivas Timmaraju	Facebook, USA
Christos Tjortjis	International Hellenic University, Greece
Ljupco Todorovski	University of Ljubljana, Slovenia
Laszlo Toka	BME, Hungary
Ancy Tom	University of Minnesota, Twin Cities, USA
Panagiotis Traganitis	Michigan State University, USA
Cuong Tran	Syracuse University, USA
Minh-Tuan Tran	KAIST, South Korea
Giovanni Trappolini	Sapienza University of Rome, Italy
Volker Tresp	LMU, Germany
Yu-Chee Tseng	National Yang Ming Chiao Tung University, Taiwan
Maria Tzelepi	Aristotle University of Thessaloniki, Greece
Willy Ugarte	University of Applied Sciences (UPC), Peru
Antti Ukkonen	University of Helsinki, Finland
Abhishek Kumar Umrawal	Purdue University, USA
Athena Vakal	Aristotle University, Greece
Matias Valdenegro Toro	University of Groningen, the Netherlands
Maaike Van Roy	KU Leuven, Belgium
Dinh Van Tran	University of Freiburg, Germany
Fabio Vandin	University of Padova, Italy
Valerie Vaquet	CITEC, Bielefeld University, Germany
Iraklis Varlamis	Harokopio University of Athens, Greece
Santiago Velasco-Forero	MINES ParisTech, France
Bruno Veloso	Porto, Portugal
Dmytro Velychko	Carl von Ossietzky Universität Oldenburg, Germany
Sreekanth Vempati	Myntra, India
Sebastián Ventura Soto	University of Cordoba, Portugal
Rosana Veroneze	LBiC, Brazil

Jan Verwaeren	Ghent University, Belgium
Vassilios Verykios	Hellenic Open University, Greece
Herna Viktor	University of Ottawa, Canada
João Vinagre	LIAAD - INESC TEC, Portugal
Fabio Vitale	Centai Institute, Italy
Vasiliki Voukelatou	ISTI - CNR, Italy
Dong Quan Vu	Safran Tech, France
Maxime Wabartha	McGill University, Canada
Tomasz Walkowiak	Wroclaw University of Science and Technology, Poland
Vijay Walunj	University of Missouri-Kansas City, USA
Michael Wand	University of Mainz, Germany
Beilun Wang	Southeast University, China
Chang-Dong Wang	Sun Yat-sen University, China
Daheng Wang	Amazon, USA
Deng-Bao Wang	Southeast University, China
Di Wang	Nanyang Technological University, Singapore
Di Wang	KAUST, Saudi Arabia
Fu Wang	University of Exeter, UK
Hao Wang	Nanyang Technological University, Singapore
Hao Wang	Louisiana State University, USA
Hao Wang	University of Science and Technology of China, China
Hongwei Wang	University of Illinois Urbana-Champaign, USA
Hui Wang	SKLSDE, China
Hui (Wendy) Wang	Stevens Institute of Technology, USA
Jia Wang	Xi'an Jiaotong-Liverpool University, China
Jing Wang	Beijing Jiaotong University, China
Junxiang Wang	Emory University, USA
Qing Wang	IBM Research, USA
Rongguang Wang	University of Pennsylvania, USA
Ruoyu Wang	Shanghai Jiao Tong University, China
Ruxin Wang	Shenzhen Institutes of Advanced Technology, China
Senzhang Wang	Central South University, China
Shoujin Wang	Macquarie University, Australia
Xi Wang	Chinese Academy of Sciences, China
Yanchen Wang	Georgetown University, USA
Ye Wang	Chongqing University, China
Ye Wang	National University of Singapore, Singapore
Yifei Wang	Peking University, China
Yongqing Wang	Chinese Academy of Sciences, China

Yuandong Wang	Tsinghua University, China
Yue Wang	Microsoft Research, USA
Yun Cheng Wang	University of Southern California, USA
Zhaonan Wang	University of Tokyo, Japan
Zhaoxia Wang	SMU, Singapore
Zhiwei Wang	University of Chinese Academy of Sciences, China
Zihan Wang	Shandong University, China
Zijie J. Wang	Georgia Tech, USA
Dilusha Weeraddana	CSIRO, Australia
Pascal Welke	University of Bonn, Germany
Tobias Weller	University of Mannheim, Germany
Jörg Wicker	University of Auckland, New Zealand
Lena Wiese	Goethe University Frankfurt, Germany
Michael Wilbur	Vanderbilt University, USA
Moritz Wolter	Bonn University, Germany
Bin Wu	Beijing University of Posts and Telecommunications, China
Bo Wu	Renmin University of China, China
Jiancan Wu	University of Science and Technology of China, China
Jiantao Wu	University of Jinan, China
Ou Wu	Tianjin University, China
Yang Wu	Chinese Academy of Sciences, China
Yiqing Wu	University of Chinese Academic of Science, China
Yuejia Wu	Inner Mongolia University, China
Bin Xiao	University of Ottawa, Canada
Zhiwen Xiao	Southwest Jiaotong University, China
Ruobing Xie	WeChat, Tencent, China
Zikang Xiong	Purdue University, USA
Depeng Xu	University of North Carolina at Charlotte, USA
Jian Xu	Citadel, USA
Jiarong Xu	Fudan University, China
Kunpeng Xu	University of Sherbrooke, Canada
Ning Xu	Southeast University, China
Xianghong Xu	Tsinghua University, China
Sangeeta Yadav	Indian Institute of Science, India
Mehrdad Yaghoobi	University of Edinburgh, UK
Makoto Yamada	RIKEN AIP/Kyoto University, Japan
Akihiro Yamaguchi	Toshiba Corporation, Japan
Anil Yaman	Vrije Universiteit Amsterdam, the Netherlands

Hao Yan	Washington University in St Louis, USA
Qiao Yan	Shenzhen University, China
Chuang Yang	University of Tokyo, Japan
Deqing Yang	Fudan University, China
Haitian Yang	Chinese Academy of Sciences, China
Renchi Yang	National University of Singapore, Singapore
Shaofu Yang	Southeast University, China
Yang Yang	Nanjing University of Science and Technology, China
Yang Yang	Northwestern University, USA
Yiyang Yang	Guangdong University of Technology, China
Yu Yang	The Hong Kong Polytechnic University, China
Peng Yao	University of Science and Technology of China, China
Vithya Yogarajan	University of Auckland, New Zealand
Tetsuya Yoshida	Nara Women's University, Japan
Hong Yu	Chongqing Laboratory of Comput. Intelligence, China
Wenjian Yu	Tsinghua University, China
Yanwei Yu	Ocean University of China, China
Ziqiang Yu	Yantai University, China
Sha Yuan	Beijing Academy of Artificial Intelligence, China
Shuhan Yuan	Utah State University, USA
Mingxuan Yue	Google, USA
Aras Yurtman	KU Leuven, Belgium
Nayyar Zaidi	Deakin University, Australia
Zelin Zang	Zhejiang University & Westlake University, China
Masoumeh Zareapoor	Shanghai Jiao Tong University, China
Hanqing Zeng	USC, USA
Tieyong Zeng	The Chinese University of Hong Kong, China
Bin Zhang	South China University of Technology, China
Bob Zhang	University of Macau, Macao, China
Hang Zhang	National University of Defense Technology, China
Huaizheng Zhang	Nanyang Technological University, Singapore
Jiangwei Zhang	Tencent, China
Jinwei Zhang	Cornell University, USA
Jun Zhang	Tsinghua University, China
Lei Zhang	Virginia Tech, USA
Luxin Zhang	Worldline/Inria, France
Mimi Zhang	Trinity College Dublin, Ireland
Qi Zhang	University of Technology Sydney, Australia

Qiyiwen Zhang	University of Pennsylvania, USA
Teng Zhang	Huazhong University of Science and Technology, China
Tianle Zhang	University of Exeter, UK
Xuan Zhang	Renmin University of China, China
Yang Zhang	University of Science and Technology of China, China
Yaqian Zhang	University of Waikato, New Zealand
Yu Zhang	University of Illinois at Urbana-Champaign, USA
Zhengbo Zhang	Beihang University, China
Zhiyuan Zhang	Peking University, China
Heng Zhao	Shenzhen Technology University, China
Mia Zhao	Airbnb, USA
Tong Zhao	Snap Inc., USA
Qinkai Zheng	Tsinghua University, China
Xiangping Zheng	Renmin University of China, China
Bingxin Zhou	University of Sydney, Australia
Bo Zhou	Baidu, Inc., China
Min Zhou	Huawei Technologies, China
Zhipeng Zhou	University of Science and Technology of China, China
Hui Zhu	Chinese Academy of Sciences, China
Kenny Zhu	SJTU, China
Lingwei Zhu	Nara Institute of Science and Technology, Japan
Mengying Zhu	Zhejiang University, China
Renbo Zhu	Peking University, China
Yanmin Zhu	Shanghai Jiao Tong University, China
Yifan Zhu	Tsinghua University, China
Bartosz Zieliński	Jagiellonian University, Poland
Sebastian Ziesche	Bosch Center for Artificial Intelligence, Germany
Indre Zliobaite	University of Helsinki, Finland
Gianlucca Zuin	UFM, Brazil

Program Committee Members, Demo Track

Hesam Amoualian	WholeSoft Market, France
Georgios Balikas	Salesforce, France
Giannis Bekoulis	Vrije Universiteit Brussel, Belgium
Ludovico Boratto	University of Cagliari, Italy
Michelangelo Ceci	University of Bari, Italy
Abdulkadir Celikkanat	Technical University of Denmark, Denmark

Tania Cerquitelli	Informatica Politecnico di Torino, Italy
Mel Chekol	Utrecht University, the Netherlands
Charalampos Chelmis	University at Albany, USA
Yagmur Gizem Cinar	Amazon, France
Eustache Diemert	Criteo AI Lab, France
Sophie Fellenz	TU Kaiserslautern, Germany
James Foulds	University of Maryland, Baltimore County, USA
Jhony H. Giraldo	Télécom Paris, France
Parantapa Goswami	Rakuten Institute of Technology, Rakuten Group, Japan
Derek Greene	University College Dublin, Ireland
Lili Jiang	Umeå University, Sweden
Bikash Joshi	Elsevier, the Netherlands
Alexander Jung	Aalto University, Finland
Zekarias Kefato	KTH Royal Institute of Technology, Sweden
Ilkcan Keles	Aalborg University, Denmark
Sammy Khalife	Johns Hopkins University, USA
Tuan Le	New Mexico State University, USA
Ye Liu	Salesforce, USA
Fragkiskos Malliaros	CentraleSupelec, France
Hamid Mirisaee	AMLRightSource, France
Robert Moro	Kempelen Institute of Intelligent Technologies, Slovakia
Iosif Mporas	University of Hertfordshire, UK
Giannis Nikolentzos	Ecole Polytechnique, France
Eirini Ntoutsi	Freie Universität Berlin, Germany
Frans Oliehoek	Delft University of Technology, the Netherlands
Nora Ouzir	CentraleSupélec, France
Özlem Özgöbek	Norwegian University of Science and Technology, Norway
Manos Papagelis	York University, UK
Shichao Pei	University of Notre Dame, USA
Botao Peng	Chinese Academy of Sciences, China
Antonia Saravanou	National and Kapodistrian University of Athens, Greece
Rik Sarkar	University of Edinburgh, UK
Vera Shalaeva	Inria Lille-Nord, France
Kostas Stefanidis	Tampere University, Finland
Nikolaos Tziortziotis	Jellyfish, France
Davide Vega	Uppsala University, Sweden
Sagar Verma	CentraleSupelec, France
Yanhao Wang	East China Normal University, China

Zhirong Yang	Norwegian University of Science and Technology, Norway
Xiangyu Zhao	City University of Hong Kong, Hong Kong, China

Sponsors

ASML

Google

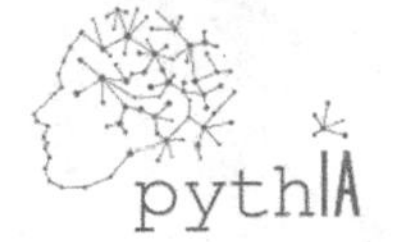

Contents – Part I

Anomaly Detection

Interpretability and Explainability

Ranking and Recommender Systems

Clustering and Dimensionality Reduction

Pass-Efficient Randomized SVD with Boosted Accuracy

Xu Feng, Wenjian Yu(✉), and Yuyang Xie

Department of Computer Science and Technology, BNRist, Tsinghua University, Beijing, China
{fx17,xyy18}@mails.tsinghua.edu.cn, yu-wj@tsinghua.edu.cn

Abstract. Singular value decomposition (SVD) is a widely used tool in data analysis and numerical linear algebra. Computing truncated SVD of a very large matrix encounters difficulty due to excessive time and memory cost. In this work, we aim to tackle this difficulty and enable accurate SVD computation for the large data which cannot be loaded into memory. We first propose a randomized SVD algorithm with fewer passes over the matrix. It reduces the passes in the basic randomized SVD by half, almost not sacrificing accuracy. Then, a shifted power iteration technique is proposed to improve the accuracy of result, where a dynamic scheme of updating the shift value in each power iteration is included. Finally, collaborating the proposed techniques with several accelerating skills, we develop a Pass-efficient randomized SVD (PerSVD) algorithm for efficient and accurate treatment of large data stored on hard disk. Experiments on synthetic and real-world data validate that the proposed techniques largely improve the accuracy of randomized SVD with same number of passes over the matrix. With 3 or 4 passes over the data, PerSVD is able to reduce the error of SVD result by three or four orders of magnitude compared with the basic randomized SVD and single-pass SVD algorithms, with similar or less runtime and less memory usage.

Keywords: Singular value decomposition · Shifted power iteration · Random embedding · Pass-efficient algorithm

1 Introduction

Truncated singular value decomposition (SVD) has broad applications in data analysis and machine learning, such as dimension reduction, matrix completion, and information retrieval. However, for the large and high-dimensional input data from social network analysis, natural language processing and recommender system, etc., computing truncated SVD often consumes tremendous computational resource.

This work is supported by NSFC under grant No. 61872206.

Supplementary Information The online version contains supplementary material available at https://doi.org/10.1007/978-3-031-26387-3_1.

M.-R. Amini et al. (Eds.): ECML PKDD 2022, LNAI 13713, pp. 3–20, 2023.
https://doi.org/10.1007/978-3-031-26387-3_1

A conventional method of computing truncated SVD, i.e. the first k largest singular values and corresponding singular vectors, is using `svds` in Matlab [3]. In `svds`, Lanczos bidiagonal process is used to compute the truncated SVD [3]. Although there are variant algorithms of `svds`, like `lansvd` in PROPACK [11], `svds` is still most robust and runs fastest in most scenarios. However, `svds` needs several times of k times passes over the matrix to produce result, which is not efficient to deal with large data matrices which cannot be stored in RAM. To tackle the difficulty of handling large matrix, approximate algorithms for truncated SVD have been proposed, which consume less time, less memory and fewer passes over input matrix while sacrificing little accuracy [2,6,9,13,15,20–22]. Among them, a class of randomized methods gains a lot of attention which is based on random embedding through multiplying a random matrix [14]. The randomized method obtains a near-optimal low-rank decomposition of the matrix, and has the performance advantages over classical methods, in terms of computational time, pass efficiency and parallelizability. The comprehensive presentation of relevant techniques and theories can be found in [9,14].

When data is too large to be stored in RAM, traditional truncated SVD algorithms are not efficient, if not impossible, to deal with data stored in slow memory (hard disk). The single-pass SVD algorithms [4,8,18,19,21] can tackle this difficulty. And, they are also suitable to handle streaming data. Among these algorithms, Tropp's single-pass SVD algorithm in [19] is the state-of-the-art. Although Tropp's single-pass SVD algorithm performs well on matrices with singular values decaying very fast, it results in large error while handling matrices with slow decay of singular values. Therefore, how to accurately compute the truncated SVD of a large matrix stored on hard disk on a computer with limited memory is a problem.

In this paper, we aim to tackle the difficulty of handling large matrix stored on hard disk or slow memory. We propose a pass-efficient randomized SVD (PerSVD) algorithm which accurately computes SVD of large matrix stored on hard disk with less memory and affordable time. Major contributions and results are summarized as follows.

- We propose a technique to reduce the number of passes over the matrix in the basic randomized SVD algorithm. It takes advantage of the row-major format of the matrix and reads it row by row to build $\mathbf{A\Phi}$ and $\mathbf{A}^{\mathrm{T}}\mathbf{A\Phi}$ with just one pass over matrix. With this algorithm, the passes over the matrix in the basic randomized SVD algorithm is reduced by half, with negligible loss of accuracy.
- Inspired by the shift technique in the power method [7], we propose to use the shift skill in the power iteration called shifted power iteration to improve the accuracy of results. A dynamic scheme of updating the shift value in each power iteration is proposed to optimize the performance of the shifted power iteration. This facilitates a pass-efficient randomized SVD algorithm, i.e. PerSVD, which accurately computes truncated SVD of large matrix on a limited-memory computer.
- Experiments on synthetic and real large data demonstrate that the proposed techniques are all beneficial to improve the accuracy of result with same number of passes over the matrix. With same 4 passes the over matrix, the

result computed with PerSVD is up to **20,318X** more accurate than that obtained with the basic randomized SVD. And, the proposed PerSVD with 3 passes over the matrix consumes just 16%–26% memory of Tropp's algorithm [19] while producing more accurate results (with up to **7,497X** reduction of error), with less runtime. For the FERET data stored as a 150GB file, PerSVD just costs 12 min and 1.9 GB memory to compute the truncated SVD ($k=100$) with 3 passes over the data.

2 Preliminaries

Below we follow the Matlab conventions to specify indices of matrix and functions.

2.1 Basics of Truncated SVD

The economic SVD of a matrix $\mathbf{A}\in\mathbb{R}^{m\times n}$ $(m\geq n)$ can be stated as:

$$\mathbf{A}=\mathbf{U}\mathbf{\Sigma}\mathbf{V}^{\mathrm{T}}, \tag{1}$$

where $\mathbf{U}=[\mathbf{u}_1,\mathbf{u}_2,\cdots,\mathbf{u}_n]$ and $\mathbf{V}=[\mathbf{v}_1,\mathbf{v}_2,\cdots,\mathbf{v}_n]$ are orthogonal or orthonormal matrices, representing the left and right singular vectors of $\mathbf{A}$, respectively. The $n\times n$ diagonal matrix $\mathbf{\Sigma}$ contains the singular values $(\sigma_1,\sigma_2,\cdots,\sigma_n)$ of $\mathbf{A}$ in descending order. The truncated SVD $\mathbf{A}_k$ can be derived, which is an approximation of $\mathbf{A}$:

$$\mathbf{A}\approx\mathbf{A}_k=\mathbf{U}_k\mathbf{\Sigma}_k\mathbf{V}_k^{\mathrm{T}},\quad k<\min(m,n), \tag{2}$$

where $\mathbf{U}_k$ and $\mathbf{V}_k$ are matrices with the first k columns of $\mathbf{U}$ and $\mathbf{V}$ respectively, and the diagonal matrix $\mathbf{\Sigma}_k$ is the $k\times k$ upper-left submatrix of $\mathbf{\Sigma}$. Notice that, $\mathbf{A}_k$ is the best rank-k approximation of $\mathbf{A}$ in both spectral and Frobenius norm [5].

2.2 Randomized SVD Algorithm with Power Iteration

The basic randomized SVD algorithm [9] can be described as Algorithm 1, where the power iteration in Step 3 through 6 is for improving the accuracy of result. In Algorithm 1, $\mathbf{\Omega}$ is a Gaussian random matrix. Although other kinds of random matrix can replace $\mathbf{\Omega}$ to reduce the computational cost of $\mathbf{A\Omega}$, they bring some sacrifice on accuracy. The orthonormalization operation "orth()" can be implemented with a call to a packaged QR factorization.

If there is no power iteration, the $m\times l$ orthonormal matrix $\mathbf{Q}=\mathrm{orth}(\mathbf{A\Omega})$ approximates the basis of dominant subspace of *range*($\mathbf{A}$), i.e., *span*$\{\mathbf{u}_1,\mathbf{u}_2,\cdots,\mathbf{u}_l\}$. So, $\mathbf{A}\approx\mathbf{Q}\mathbf{Q}^{\mathrm{T}}\mathbf{A}$, i.e. $\mathbf{A}\approx\mathbf{QB}$ according to Step 8. By performing the economic SVD, i.e. (1), on the "short-and-fat" $l\times n$ matrix $\mathbf{B}$, one finally obtains the approximate truncated SVD of $\mathbf{A}$. Using the power iteration in Steps 3-6, one obtains $\mathbf{Q}=\mathrm{orth}((\mathbf{A}\mathbf{A}^{\mathrm{T}})^p\mathbf{A\Omega})$. This makes $\mathbf{Q}$ better approximate the basis of dominant subspace of *range*$((\mathbf{A}\mathbf{A}^{\mathrm{T}})^p\mathbf{A})$, same as that of *range*($\mathbf{A}$), because $(\mathbf{AA}^{\mathrm{T}})^p\mathbf{A}$'s singular values decay more quickly than those of $\mathbf{A}$ [9]. Therefore, the resulted singular values and vectors have better accuracy, and the larger p makes more accurate results and more computational cost as well.

The orthonormalization is practically used in the power iteration steps to alleviate the round-off error in the floating-point computation. It can be performed after every other matrix-matrix multiplication to save computational cost with little sacrifice on accuracy [6,20]. This turns Steps 2-7 of Algorithm 1 to

2: $\mathbf{Q} \leftarrow \mathrm{orth}(\mathbf{A\Omega})$
3: **for** $j \leftarrow 1, 2, \cdots p$ **do**
4: $\quad\mathbf{Q} \leftarrow \mathrm{orth}(\mathbf{AA}^{\mathrm{T}}\mathbf{Q})$
5: **end for**

Notice that the original Step 7 in Algorithm 1 is dropped. We use the floating-point operation (*flop*) to specify the time cost of algorithms. Suppose the *flop* count of multiplication of $\mathbf{M} \in \mathbb{R}^{m\times l}$ and $\mathbf{N} \in \mathbb{R}^{l\times l}$ is $C_{mul}ml^2$. Here, C_{mul} reflects one addition and one multiplication. Thus, the *flop* count of Algorithm 1 is

$$\mathrm{FC}_1 = (2p+2)C_{mul}mnl + (p+1)C_{qr}ml^2 + C_{svd}nl^2 + C_{mul}ml^2, \quad (3)$$

where $(2p+2)C_{mul}mnl$ reflects $2p+2$ times matrix-matrix multiplication on $\mathbf{A}$, $(p+1)C_{qr}ml^2$ reflects $p+1$ times QR factorization on $m \times l$ matrix and $C_{svd}nl^2 + C_{mul}ml^2$ reflects SVD and matrix-matrix multiplication in Step 9 and 10.

2.3 Tropp's Single-Pass SVD Algorithm

On a machine with limited memory, single-pass SVD algorithms can be used to handle very large data stored on hard disk [4,8,18,19,21]. Although the

Algorithm 1. Basic randomized SVD with power iteration

Input: $\mathbf{A} \in \mathbb{R}^{m\times n}$, rank parameter k, oversampling parameter l ($l \geq k$), power parameter p
Output: $\mathbf{U} \in \mathbb{R}^{m\times k}$, $\mathbf{S} \in \mathbb{R}^{k\times k}$, $\mathbf{V} \in \mathbb{R}^{n\times k}$
1: $\mathbf{\Omega} \leftarrow \mathrm{randn}(n, l)$
2: $\mathbf{Q} \leftarrow \mathbf{A\Omega}$
3: **for** $j \leftarrow 1, 2, \cdots, p$ **do**
4: $\quad\mathbf{G} \leftarrow \mathbf{A}^{\mathrm{T}}\mathbf{Q}$
5: $\quad\mathbf{Q} \leftarrow \mathbf{AG}$
6: **end for**
7: $\mathbf{Q} \leftarrow \mathrm{orth}(\mathbf{Q})$
8: $\mathbf{B} \leftarrow \mathbf{Q}^{\mathrm{T}}\mathbf{A}$
9: $[\mathbf{U}, \mathbf{S}, \mathbf{V}] \leftarrow \mathrm{svd}(\mathbf{B}, \text{'econ'})$
10: $\mathbf{U} \leftarrow \mathbf{QU}$
11: $\mathbf{U} \leftarrow \mathbf{U}(:, 1\!:\!k), \mathbf{S} \leftarrow \mathbf{S}(1\!:\!k, 1\!:\!k), \mathbf{V} \leftarrow \mathbf{V}(:, 1\!:\!k)$

single-pass algorithm in [18] achieves lower computational complexity, it is more suitable for the matrix with fast decay of singular values and $m \ll n$. Tropp's single-pass SVD algorithm [19] is the state-of-the-art single-pass SVD algorithm, with lower approximation error compared with it's predecessors given the same sketch sizes. Tropp's single-pass SVD algorithm (Algorithm 2) first constructs several sketches of the input matrix $\mathbf{A}$ in Step 4-7. Then, QR factorization is performed in Step 8 and 9 to compute the orthonormal basis of row and column space of $\mathbf{A}$, respectively. Then, matrix $\mathbf{C}$ is computed to approximate the core of $\mathbf{A}$ in Step 10. Notice that'/' and'\' represent the left and right division in Matlab, respectively. This is followed by SVD in Step 11 for computing the singular values and vectors of $\mathbf{C}$. Finally, the singular vectors are computed by projecting the orthonormal matrices to the row and column spaces of original matrix in Step 12.

Algorithm 2. Tropp's single-pass SVD algorithm

Input: $\mathbf{A} \in \mathbb{R}^{m\times n}$, rank parameter k
Output: $\mathbf{U} \in \mathbb{R}^{m\times k}$, $\mathbf{S} \in \mathbb{R}^{k\times k}$, $\mathbf{V} \in \mathbb{R}^{n\times k}$
1: $r \leftarrow 4k+1$, $s \leftarrow 2r+1$
2: $\mathbf{\Upsilon} \leftarrow \text{randn}(r,m)$, $\mathbf{\Omega} \leftarrow \text{randn}(r,n)$, $\mathbf{\Phi} \leftarrow \text{randn}(s,m)$, $\mathbf{\Psi} \leftarrow \text{randn}(s,n)$
3: $\mathbf{X} \leftarrow \text{zeros}(r,n)$, $\mathbf{Y} \leftarrow \text{zeros}(m,r)$, $\mathbf{Z} \leftarrow \text{zeros}(s,s)$
4: **for** $j \leftarrow 1,2,\cdots,m$ **do**
5: $\quad\mathbf{a}_i$ is the i-th row of $\mathbf{A}$
6: $\quad\mathbf{X} \leftarrow \mathbf{X} + \mathbf{\Upsilon}(i,:)\mathbf{a}_i$, $\mathbf{Y}(i,:) \leftarrow \mathbf{a}_i\mathbf{\Omega}$, $\mathbf{Z} \leftarrow \mathbf{Z} + \mathbf{\Phi}(i,:)\mathbf{a}_i\mathbf{\Psi}^{\mathrm{T}}$
7: **end for**
8: $[\mathbf{Q},\sim] \leftarrow \text{qr}(\mathbf{Y},0)$
9: $[\mathbf{P},\sim] \leftarrow \text{qr}(\mathbf{X}^{\mathrm{T}},0)$
10: $\mathbf{C} \leftarrow ((\mathbf{\Phi Q})/\mathbf{Z})\backslash(\mathbf{\Psi P}^{\mathrm{T}})$
11: $[\mathbf{U},\mathbf{S},\mathbf{V}] \leftarrow \text{svd}(\mathbf{C},\text{'econ'})$
12: $\mathbf{U} \leftarrow \mathbf{QU}(:,1{:}k)$, $\mathbf{S} \leftarrow \mathbf{S}(1{:}k,1{:}k)$, $\mathbf{V} \leftarrow \mathbf{PV}(:,1{:}k)$

The peak memory usage of Algorithm 2 is $(m+n)(2r+s)\times 8 \approx 16(m+n)k\times 8$ bytes, which is caused by all matrices computed in Algorithm 2. And, the *flop* count of Algorithm 2 is

$$\begin{aligned}\mathrm{FC}_2 &= C_{mul}m(2nr+ns+s^2) + C_{qr}(m+n)r^2 + C_{mul}(m+n)rs + 2C_{solve}s^2k \\ &\quad + C_{svd}r^3 + C_{mul}(m+n)rk \\ &\approx C_{mul}m(2nr+ns+s^2) + C_{qr}(m+n)r^2 + C_{mul}(m+n)r(s+k) \\ &\approx 16C_{mul}mnk + 100C_{mul}mk^2 + 36C_{mul}nk^2 + 16C_{qr}(m+n)k^2,\end{aligned} \tag{4}$$

where $C_{mul}m(2nr+ns+s^2)$ reflects the matrix-matrix multiplication in Step 4-7, $C_{qr}(m+n)r^2$ reflects the QR factorization in Step 8 and 9, $C_{mul}(m+n)rs + 2C_{solve}s^2k$ reflects the matrix-matrix multiplication and the solve operation in Step 10 and $C_{svd}r^3 + C_{mul}(m+n)rk$ reflects the SVD and matrix-matrix multiplication in Step 11 and 12. When $k \ll \min(m,n)$, we can drop the $2C_{solve}s^2k$ and $C_{svd}r^3$ in FC_2.

It should be pointed out that the Tropp's algorithm does not perform well on matrices with slow decay of singular values, exhibiting large error on computed singular values/vectors. However, the matrices with slow decay of singular values are common in real applications.

3 Pass-Efficient SVD with Shifted Power Iteration

In this section, we develop a pass-efficient randomized SVD algorithm named PerSVD. Firstly, we develop a pass-efficient scheme to reduce the passes over $\mathbf{A}$ within basic randomized SVD algorithm. Secondly, inspired by the shift technique in the power method [7], we propose a technique of shifted power iteration to improve the accuracy of result. Finally, combining with the proposed shift updating scheme in each power iteration, we describe the pass-efficient PerSVD algorithm which is able to accurately compute SVD of large matrix on hard disk with less memory and affordable time.

3.1 Randomized SVD with Fewer Passes

Suppose $\mathbf{a}_i$ is the i-th ($i \leq m$) row of matrix $\mathbf{A} \in \mathbb{R}^{m\times n}$, and $\mathbf{\Phi} \in \mathbb{R}^{n\times l}$. To compute $\mathbf{Y} = \mathbf{A\Phi}$ and $\mathbf{W} = \mathbf{A}^{\mathrm{T}}\mathbf{A\Phi}$ with $\mathbf{a}_i$, we have

$$\mathbf{Y}(i,:) = \mathbf{a}_i\mathbf{\Phi}, \mathbf{W} = \mathbf{A}^{\mathrm{T}}\mathbf{Y} = \sum_{i=1}^{m} \mathbf{a}_i^{\mathrm{T}}\mathbf{Y}(i,:), \tag{5}$$

which reflects the data stored in the row-major format can be read once to compute $\mathbf{Y} = \mathbf{A\Phi}$ and $\mathbf{W} = \mathbf{A}^{\mathrm{T}}\mathbf{A\Phi}$. This idea is similar to that employed in [21,22]. Below, we derive that it can be repeatedly used to compute the $(\mathbf{A}^{\mathrm{T}}\mathbf{A})^p\mathbf{\Omega}$ in the basic randomized SVD algorithm with power iteration.

With $\mathbf{Y}$ and $\mathbf{W}$, we can develop the formulation of $\mathbf{Q}$ and $\mathbf{B}$ in Step 7 and 8 of Algorithm 1. Suppose $\mathbf{\Phi} = (\mathbf{A}^{\mathrm{T}}\mathbf{A})^p\mathbf{\Omega}$, $\mathbf{Y} = \mathbf{A\Phi}$, $\mathbf{W} = \mathbf{A}^{\mathrm{T}}\mathbf{A\Phi}$ and $[\mathbf{Q}, \tilde{\mathbf{S}}, \tilde{\mathbf{V}}] \leftarrow \text{svd}(\mathbf{Y}, \text{'econ'})$. Combining the fact $\mathbf{W} = \mathbf{A}^{\mathrm{T}}\mathbf{Y}$, we can derive

$$\mathbf{W} = \mathbf{A}^{\mathrm{T}}\mathbf{Y} = \mathbf{A}^{\mathrm{T}}\mathbf{Q}\tilde{\mathbf{S}}\tilde{\mathbf{V}}^{\mathrm{T}} \Rightarrow \mathbf{W}\tilde{\mathbf{V}}\tilde{\mathbf{S}}^{-1} = \mathbf{A}^{\mathrm{T}}\mathbf{Q} \tag{6}$$

which implies $\mathbf{Q}(\mathbf{W}\tilde{\mathbf{V}}\tilde{\mathbf{S}}^{-1})^{\mathrm{T}} = \mathbf{Q}\mathbf{Q}^{\mathrm{T}}\mathbf{A} \approx \mathbf{A}$. Because $\mathbf{Q}$ is the orthonormalization of $\mathbf{A}(\mathbf{A}^{\mathrm{T}}\mathbf{A})^p\mathbf{\Omega}$, this approximation performs with same accuracy as $\mathbf{A} \approx \mathbf{QB}$ in Algorithm 1. With (5) and (6) combined, the randomized SVD with fewer passes is proposed and described as Algorithm 3. Now, the randomized SVD with p power iteration just needs $p+1$ passes over $\mathbf{A}$, which reduces half of the passes in Algorithm 1.

The above deduction reveals Algorithm 3 is mathematically equivalent to the basic randomized SVD (the modified Algorithm 1) in exact arithmetic. In the practice considering numerical error, the computational results of the both algorithms are very close which means Algorithm 3 largely reduces the passes with just negligible loss of accuracy. Besides, it is easy to see that we can read multiple rows once to compute $\mathbf{Y}$ and $\mathbf{W}$ by (5).

Algorithm 3. Randomized SVD with fewer passes the over matrix

Input: $\mathbf{A} \in \mathbb{R}^{m \times n}$, k, l $(l \geq k)$, p
Output: $\mathbf{U} \in \mathbb{R}^{m \times k}$, $\mathbf{S} \in \mathbb{R}^{k \times k}$, $\mathbf{V} \in \mathbb{R}^{n \times k}$

1: $\mathbf{\Omega} \leftarrow \text{randn}(n, l)$
2: $\mathbf{Q} \leftarrow \text{orth}(\mathbf{\Omega})$
3: **for** $j = 1, 2, \cdots, p+1$ **do**
4: $\quad \mathbf{W} \leftarrow \text{zeros}(n, l)$
5: $\quad$ **for** $i = 1, 2, \cdots, m$ **do**
6: $\quad\quad \mathbf{a}_i$ is the i-th row of matrix $\mathbf{A}$
7: $\quad\quad \mathbf{Y}(i,:) \leftarrow \mathbf{a}_i\mathbf{Q}$, $\mathbf{W} \leftarrow \mathbf{W} + \mathbf{a}_i^{\mathrm{T}}\mathbf{Y}(i,:)$
8: $\quad$ **end for**
9: $\quad$ **if** $j == p+1$ **break**
10: $\quad \mathbf{Q} \leftarrow \text{orth}(\mathbf{W})$
11: **end for**
12: $[\mathbf{Q}, \tilde{\mathbf{S}}, \tilde{\mathbf{V}}] \leftarrow \text{svd}(\mathbf{Y}, \text{'econ'})$
13: $\mathbf{B} \leftarrow \tilde{\mathbf{S}}^{-1}\tilde{\mathbf{V}}^{\mathrm{T}}\mathbf{W}^{\mathrm{T}}$
14: $[\mathbf{U}, \mathbf{S}, \mathbf{V}] \leftarrow \text{svd}(\mathbf{B}, \text{'econ'})$
15: $\mathbf{U} \leftarrow \mathbf{QU}(:, 1:k), \mathbf{S} \leftarrow \mathbf{S}(1:k, 1:k), \mathbf{V} \leftarrow \mathbf{V}(:, 1:k)$

3.2 The Idea of Shifted Power Iteration

In Algorithm 3, the computation $\mathbf{Q} \leftarrow \mathbf{A}^{\mathrm{T}}\mathbf{AQ}$ in power iteration is the same as that in the power method for computing the largest eigenvalue and corresponding eigenvector of $\mathbf{A}^{\mathrm{T}}\mathbf{A}$. For the power method, the shift skill can be used to accelerate the convergence of iteration by reducing the ratio between the second largest eigenvalue and the largest one [7]. This inspires our idea of shifted power iteration. To derive our method, we first give two Lemmas [7].

Lemma 1. *For symmetric matrix* $\mathbf{A}$*, its singular values are the absolute values of its eigenvalues. And, for any eigenvalue* λ *of* $\mathbf{A}$*, the left singular vector corresponding to its singular value* $|\lambda|$ *is the normalized eigenvector for* λ*.*

Lemma 2. *Suppose matrix* $\mathbf{A} \in \mathbb{R}^{n \times n}$*, and a shift* $\alpha \in \mathbb{R}$*. For any eigenvalue* λ *of* $\mathbf{A}$*,* $\lambda - \alpha$ *is an eigenvalue of* $\mathbf{A} - \alpha\mathbf{I}$*, where* $\mathbf{I}$ *is the identity matrix. And, the eigenspace of* $\mathbf{A}$ *for* λ *is the same as the eigenspace of* $\mathbf{A} - \alpha\mathbf{I}$ *for* $\lambda - \alpha$*.*

Because $\mathbf{A}^{\mathrm{T}}\mathbf{A}$ is a symmetric positive semi-definite matrix, its singular value is its eigenvalue according to Lemma 1. We use $\sigma_i(\cdot)$ to denote the i-th largest singular value. Along with Lemma 2, we see that $\sigma_i(\mathbf{A}^{\mathrm{T}}\mathbf{A}) - \alpha$ is the eigenvalue of $\mathbf{A}^{\mathrm{T}}\mathbf{A} - \alpha\mathbf{I}$. Then, $|\sigma_i(\mathbf{A}^{\mathrm{T}}\mathbf{A}) - \alpha|$ is the singular value of $\mathbf{A}^{\mathrm{T}}\mathbf{A} - \alpha\mathbf{I}$ according to Lemma 1. This can be illustrated by Fig. 1. Notice that $|\sigma_i(\mathbf{A}^{\mathrm{T}}\mathbf{A}) - \alpha|$ is not necessarily the i-th largest singular value.

For Algorithm 1, the decay trend of the first l singular values of handled matrix affects the accuracy of result. If $\sigma_i(\mathbf{A}^{\mathrm{T}}\mathbf{A}) - \alpha > 0$, $(i \leq l)$, and they are the l largest singular values of $\mathbf{A}^{\mathrm{T}}\mathbf{A} - \alpha\mathbf{I}$, these shifted singular values

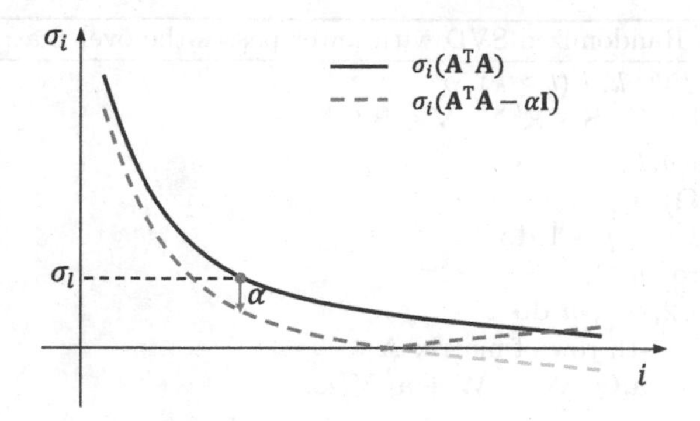

Fig. 1. The illustration of the singular value curves of $\mathbf{A}^{\mathrm{T}}\mathbf{A}$ and $\mathbf{A}^{\mathrm{T}}\mathbf{A}-\alpha\mathbf{I}$.

obviously exhibit faster decay (see Fig. 1). The following Theorem states when these conditions are satisfied.

Theorem 1. *Suppose positive number* $\alpha \leq \sigma_l(\mathbf{A}^{\mathrm{T}}\mathbf{A})/2$ *and* $i \leq l$. *Then,* $\sigma_i(\mathbf{A}^{\mathrm{T}}\mathbf{A} - \alpha\mathbf{I}) = \sigma_i(\mathbf{A}^{\mathrm{T}}\mathbf{A}) - \alpha$, *where* $\sigma_i(\cdot)$ *denotes the i-th largest singular value. And, the left singular vector corresponding to the i-th singular value of* $\mathbf{A}^{\mathrm{T}}\mathbf{A}-\alpha\mathbf{I}$ *is the same as the left singular vector corresponding to the i-th singular value of* $\mathbf{A}^{\mathrm{T}}\mathbf{A}$.

The complete proof is in Appendix A.1. The first statement of Theorem 1 is straightforward from Fig. 1. The second statement can be derived from the statements on relationships between eigenvectors and singular vectors in Lemma 1 and 2.

According to Theorem 1, if we choose a shift $\alpha \leq \sigma_l(\mathbf{A}^{\mathrm{T}}\mathbf{A})/2$ we can change the computation $\mathbf{Q}=\mathbf{A}^{\mathrm{T}}\mathbf{A}\mathbf{Q}$ to $\mathbf{Q}=(\mathbf{A}^{\mathrm{T}}\mathbf{A}-\alpha\mathbf{I})\mathbf{Q}$ in the power iteration, with the approximated dominant subspace unchanged. We called this *shifted power iteration*. For each step of shifted power iteration, this makes $\mathbf{AQ}$ approximate the basis of dominant subspace of $range(\mathbf{A})$ to a larger extent than executing an original power iteration step, because the singular values of $\mathbf{A}^{\mathrm{T}}\mathbf{A}-\alpha\mathbf{I}$ decay faster than those of $\mathbf{A}^{\mathrm{T}}\mathbf{A}$. Therefore, the shifted power iteration would improve the accuracy of the randomized SVD algorithm with same power parameter p. The remaining problem is how to set the shift α.

Consider the change of ratio of singular values from $\frac{\sigma_i(\mathbf{A}^{\mathrm{T}}\mathbf{A})}{\sigma_1(\mathbf{A}^{\mathrm{T}}\mathbf{A})}$ to $\frac{\sigma_i(\mathbf{A}^{\mathrm{T}}\mathbf{A}-\alpha\mathbf{I})}{\sigma_1(\mathbf{A}^{\mathrm{T}}\mathbf{A}-\alpha\mathbf{I})}$, for $i \leq l$. It is easy to see

$$\frac{\sigma_i(\mathbf{A}^{\mathrm{T}}\mathbf{A}-\alpha\mathbf{I})}{\sigma_1(\mathbf{A}^{\mathrm{T}}\mathbf{A}-\alpha\mathbf{I})}=\frac{\sigma_i(\mathbf{A}^{\mathrm{T}}\mathbf{A})-\alpha}{\sigma_1(\mathbf{A}^{\mathrm{T}}\mathbf{A})-\alpha}<\frac{\sigma_i(\mathbf{A}^{\mathrm{T}}\mathbf{A})}{\sigma_1(\mathbf{A}^{\mathrm{T}}\mathbf{A})}, \tag{7}$$

if the assumption of α in Theorem 1 holds. And, the larger value of α, the smaller the ratio $\frac{\sigma_i(\mathbf{A}^{\mathrm{T}}\mathbf{A}-\alpha\mathbf{I})}{\sigma_1(\mathbf{A}^{\mathrm{T}}\mathbf{A}-\alpha\mathbf{I})}$, meaning faster decay of singular values. Therefore, to maximize the effect of shifted power iteration on improving the accuracy, we should choose the shift α as large as possible while satisfying $\alpha \leq \sigma_l(\mathbf{A}^{\mathrm{T}}\mathbf{A})/2$.

Notice that calculating $\sigma_l(\mathbf{A}^{\mathrm{T}}\mathbf{A})$ is very difficult. Our idea is using the singular value of $\mathbf{A}^{\mathrm{T}}\mathbf{A}\mathbf{Q}$ at the first step of the power iteration to approximate $\sigma_l(\mathbf{A}^{\mathrm{T}}\mathbf{A})$ and set the shift α.

Lemma 3. *[10] Let $\mathbf{A}, \mathbf{C} \in \mathbb{R}^{m\times n}$ be given. The following inequality holds for the decreasingly ordered singular values of $\mathbf{A}$, $\mathbf{C}$ and $\mathbf{A}\mathbf{C}^{\mathrm{T}}$ $(1 \leq i, j \leq \min(m, n)$ and $i + j - 1 \leq \min(m, n))$*

$$\sigma_{i+j-1}(\mathbf{A}\mathbf{C}^{\mathrm{T}}) \leq \sigma_i(\mathbf{A})\sigma_j(\mathbf{C})\ , \tag{8}$$

and

$$\sigma_{i+j-1}(\mathbf{A}+\mathbf{C}) \leq \sigma_i(\mathbf{A}) + \sigma_j(\mathbf{C})\,. \tag{9}$$

Based on Lemma 3, i.e. (3.3.17) and (3.3.18) in [10], we prove Theorem 2

Theorem 2. *Suppose $\mathbf{Q} \in \mathbb{R}^{m\times l}(l \leq m)$ is an orthonormal matrix and $\mathbf{A} \in \mathbb{R}^{m\times m}$. Then, for any $i \leq l$*

$$\sigma_i(\mathbf{A}\mathbf{Q}) \leq \sigma_i(\mathbf{A}). \tag{10}$$

Proof. We append zero columns to $\mathbf{Q}$ to get an $m \times m$ matrix $\mathbf{C}^{\mathrm{T}} = [\mathbf{Q}, \mathbf{0}] \in \mathbb{R}^{m\times m}$. Since $\mathbf{Q}$ is an orthonormal matrix, $\sigma_1(\mathbf{C}) = 1$. According to (8) in Lemma 3,

$$\sigma_i(\mathbf{A}\mathbf{C}^{\mathrm{T}}) \leq \sigma_i(\mathbf{A})\sigma_1(\mathbf{C}) = \sigma_i(\mathbf{A}). \tag{11}$$

Because $\mathbf{A}\mathbf{C}^{\mathrm{T}} = [\mathbf{A}\mathbf{Q}, \mathbf{0}]$, for any $i \leq l$, $\sigma_i(\mathbf{A}\mathbf{Q}) = \sigma_i(\mathbf{A}\mathbf{C}^{\mathrm{T}})$. Then, combining (11) we can prove (10).

Suppose $\mathbf{Q} \in \mathbb{R}^{m\times l}$ is the orthonormal matrix in power iteration of Algorithm 3. According to Theorem 2,

$$\sigma_i(\mathbf{A}^{\mathrm{T}}\mathbf{A}\mathbf{Q}) \leq \sigma_i(\mathbf{A}^{\mathrm{T}}\mathbf{A}),\ i \leq l, \tag{12}$$

which means we can set $\alpha = \sigma_l(\mathbf{A}^{\mathrm{T}}\mathbf{A}\mathbf{Q})/2$ to guarantee the requirement of α in Theorem 1 for performing the shifted power iteration. In order to do the orthonormalization for alleviating round-off error and calculate $\sigma_l(\mathbf{A}^{\mathrm{T}}\mathbf{A}\mathbf{Q})$, we implement "orth(·)" with the economic SVD. This has similar computational cost as using QR factorization, and the resulting matrix of left singular vectors includes the orthonormal basis of same subspace.

So far, we can obtain the value of α at the first step of power iteration, and then we perform $\mathbf{Q} = (\mathbf{A}^{\mathrm{T}}\mathbf{A} - \alpha\mathbf{I})\mathbf{Q}$ in the following iteration steps. Combining the shifted power iteration with fixed shift value, we derive a randomized SVD algorithm with shifted power iteration as Algorithm 4.

3.3 Update Shift in Each Power Iteration

In order to improve the accuracy of Algorithm 4, we try to make the shift α as large as possible in each power iteration. Therefore, we further propose a dynamic scheme to set α, which updates α with larger values and thus increases the decay of singular values.

Algorithm 4. Randomized SVD with shifted power iteration

Input: $\mathbf{A} \in \mathbb{R}^{m\times n}$, k, l $(l \geq k)$, p
Output: $\mathbf{U} \in \mathbb{R}^{m\times k}$, $\mathbf{S} \in \mathbb{R}^{k\times k}$, $\mathbf{V} \in \mathbb{R}^{n\times k}$

1: $\mathbf{\Omega} \leftarrow \text{randn}(n, l)$
2: $\mathbf{Q} \leftarrow \text{orth}(\mathbf{\Omega})$
3: $\alpha \leftarrow 0$
4: **for** $j = 1, 2, \cdots, p+1$ **do**
5: $\quad\mathbf{W} \leftarrow \text{zeros}(n, l)$
6: $\quad$**for** $i = 1, 2, \cdots, m$ **do**
7: $\quad\quad\mathbf{a}_i$ is the i-th row of matrix $\mathbf{A}$
8: $\quad\quad\mathbf{Y}(i,:) \leftarrow \mathbf{a}_i\mathbf{Q}$, $\mathbf{W} \leftarrow \mathbf{W} + \mathbf{a}_i^{\mathrm{T}}\mathbf{Y}(i,:)$
9: $\quad$**end for**
10: $\quad$**if** $j == p+1$ **break**
11: $\quad[\mathbf{Q}, \hat{\mathbf{S}}, \sim] \leftarrow \text{svd}(\mathbf{W} - \alpha\mathbf{Q}, '\text{econ}')$
12: $\quad$**if** $\alpha == 0$ **then** $\alpha \leftarrow (\hat{\mathbf{S}}(l,l) + \alpha)/2$
13: **end for**
14: $[\mathbf{Q}, \tilde{\mathbf{S}}, \tilde{\mathbf{V}}] \leftarrow \text{svd}(\mathbf{Y}, '\text{econ}')$
15: $\mathbf{B} \leftarrow \tilde{\mathbf{S}}^{-1}\tilde{\mathbf{V}}^{\mathrm{T}}\mathbf{W}^{\mathrm{T}}$
16: $[\mathbf{U}, \mathbf{S}, \mathbf{V}] \leftarrow \text{svd}(\mathbf{B}, '\text{econ}')$
17: $\mathbf{U} \leftarrow \mathbf{Q}\mathbf{U}(:, 1:k), \mathbf{S} \leftarrow \mathbf{S}(1:k, 1:k), \mathbf{V} \leftarrow \mathbf{V}(:, 1:k)$

In the iteration steps of shifted power iteration, it is convenient to calculate the singular values of $(\mathbf{A}^{\mathrm{T}}\mathbf{A}-\alpha\mathbf{I})\mathbf{Q}$. The following Theorems state how to use it to approximate $\sigma_i(\mathbf{A}^{\mathrm{T}}\mathbf{A})$. So, in each iteration step we obtain a valid value of shift and update α with it if we have a larger α.

Theorem 3. *Suppose* $\mathbf{A} \in \mathbb{R}^{m\times n}, \mathbf{Q} \in \mathbb{R}^{n\times l}(l < n)$ *is an orthonormal matrix and* $0 < \alpha < \sigma_l(\mathbf{A}^{\mathrm{T}}\mathbf{A})/2$. *Then,*

$$\sigma_i((\mathbf{A}^{\mathrm{T}}\mathbf{A} - \alpha\mathbf{I})\mathbf{Q}) + \alpha \leq \sigma_i(\mathbf{A}^{\mathrm{T}}\mathbf{A}),\ i \leq l. \tag{13}$$

Proof. For any $i \leq l$,

$$\sigma_i((\mathbf{A}^{\mathrm{T}}\mathbf{A} - \alpha\mathbf{I})\mathbf{Q}) + \alpha \leq \sigma_i(\mathbf{A}^{\mathrm{T}}\mathbf{A} - \alpha\mathbf{I}) + \alpha = \sigma_i(\mathbf{A}^{\mathrm{T}}\mathbf{A}), \tag{14}$$

due to Theorem 2 and Theorem 1.

Theorem 4. *Suppose* $\mathbf{A} \in \mathbb{R}^{m\times n}$, $\mathbf{Q} \in \mathbb{R}^{n\times l}(l < n)$ *is an orthonormal matrix,* $\alpha^{(0)} = 0$ *and* $\alpha^{(u)} = (\sigma_l(\mathbf{A}^{\mathrm{T}}\mathbf{A}\mathbf{Q} - \alpha^{(u-1)}\mathbf{Q}) + \alpha^{(u-1)})/2$ *for any* $u > 0$. *Then,* $\alpha^{(0)}, \alpha^{(1)}, \alpha^{(2)}, \cdots$ *are in ascending order.*

Proof. We prove this Theorem using induction.

When $u = 1$, $\alpha^{(1)} = \sigma_l(\mathbf{A}^{\mathrm{T}}\mathbf{A}\mathbf{Q}) \geq \alpha^{(0)}$.

When $u > 1$, suppose $\alpha^{(u-1)} \geq \alpha^{(u-2)}$. Then, according to (9) in Lemma 3

$$\begin{aligned}\sigma_l(\mathbf{A}^{\mathrm{T}}\mathbf{A}\mathbf{Q} - \alpha^{(u-2)}\mathbf{Q}) &= \sigma_l(\mathbf{A}^{\mathrm{T}}\mathbf{A}\mathbf{Q} - \alpha^{(u-1)}\mathbf{Q} + (\alpha^{(u-1)} - \alpha^{(u-2)})\mathbf{Q}) \\ &\leq \sigma_l(\mathbf{A}^{\mathrm{T}}\mathbf{A}\mathbf{Q} - \alpha^{(u-1)}\mathbf{Q}) + \alpha^{(u-1)} - \alpha^{(u-2)}.\end{aligned} \tag{15}$$

Therefore,

$$\alpha^{(u-1)} = \frac{\sigma_l(\mathbf{A}^{\mathrm{T}}\mathbf{A}\mathbf{Q} - \alpha^{(u-2)}\mathbf{Q}) + \alpha^{(u-2)}}{2} \leq \frac{\sigma_l(\mathbf{A}^{\mathrm{T}}\mathbf{A}\mathbf{Q} - \alpha^{(u-1)}\mathbf{Q}) + \alpha^{(u-1)}}{2} = \alpha^{(u)}. \quad (16)$$

According to these equations, this Theorem is proven.

Remark 2. According to the proof of Theorem 4, it can be simply proven that $\alpha^{(0)}, \alpha^{(1)}, \alpha^{(2)}, \cdots$ are in ascending order when $\alpha^{(0)} \leq \alpha^{(1)}$, where $\alpha^{(0)} \geq 0$ and $\alpha^{(u)} = (\sigma_l(\mathbf{A}^{\mathrm{T}}\mathbf{A}\mathbf{Q} - \alpha^{(u-1)}\mathbf{Q}) + \alpha^{(u-1)})/2$ for any $u > 0$.

So, we can increase α in each shifted power iteration with the following steps.

11: **while** α dose not converge **do**
12: $[\sim, \hat{\mathbf{S}}, \sim] \leftarrow \mathrm{svd}(\mathbf{W} - \alpha\mathbf{Q}, \text{'econ'})$
13: **if** $\alpha > \hat{\mathbf{S}}(l,l)$ **then break**
14: $\alpha \leftarrow (\hat{\mathbf{S}}(l,l) + \alpha)/2$
15: **end while**

These steps are appended in the front of Step 11 in Algorithm 4. Among them, performing SVD consumes much time and can be optimized with the following trick. Suppose $\mathbf{C} = \mathbf{A}^{\mathrm{T}}\mathbf{A}\mathbf{Q} - \alpha\mathbf{Q}$. Then, the singular values of $\mathbf{C}$ are the square root of the eigenvalues of $\mathbf{C}^{\mathrm{T}}\mathbf{C}$. We can derive

$$\mathbf{C}^{\mathrm{T}}\mathbf{C} = \mathbf{Q}^{\mathrm{T}}\mathbf{A}^{\mathrm{T}}\mathbf{A}\mathbf{A}^{\mathrm{T}}\mathbf{A}\mathbf{Q} - 2\alpha\mathbf{Q}^{\mathrm{T}}\mathbf{A}^{\mathrm{T}}\mathbf{A}\mathbf{Q} + \alpha^2\mathbf{I} = \mathbf{W}^{\mathrm{T}}\mathbf{W} - 2\alpha\mathbf{Y}^{\mathrm{T}}\mathbf{Y} + \alpha^2\mathbf{I}. \quad (17)$$

Therefore, we can firstly compute two matrices $\mathbf{D}_1 = \mathbf{W}^{\mathrm{T}}\mathbf{W}$ and $\mathbf{D}_2 = \mathbf{Y}^{\mathrm{T}}\mathbf{Y}$ to replace SVD with $[\sim, \hat{\mathbf{S}}^2] \leftarrow \mathrm{eig}(\mathbf{D}_1 - 2\alpha\mathbf{D}_2 + \alpha^2\mathbf{I})$, which just applies eigenvalue decomposition on $l \times l$ matrix to update α.

Combining the techniques proposed in last subsections with the dynamic scheme to update α in each power iteration, we derive the PerSVD algorithm which is described as Algorithm 5. In Step 14 and 18, it checks if $\frac{\sigma_l((\mathbf{A}^{\mathrm{T}}\mathbf{A} - \alpha\mathbf{I})\mathbf{Q}) + \alpha}{2}$ is larger than α. For same setting of power parameter p, Algorithm 5 has comparable computational cost as Algorithm 3, but achieves results with better accuracy due to the usage of shifted power iteration. For matrix $\mathbf{Q}$ computed with Algorithm 5, we have derived a bound of $||\mathbf{Q}\mathbf{Q}^{\mathrm{T}}\mathbf{A} - \mathbf{A}||$ which reflects how close the computed truncated SVD is to optimal. And, we prove that the bound is smaller than that derived in [17] for $\mathbf{Q}$ computed with the basic randomized SVD algorithm. The complete proof is given in Appendix A.2. To accelerate Algorithm 5, we can use eigenvalue decomposition to compute the economic SVD or the orthonormal basis of a "tall-and-skinny" matrix in Step 2, 20 and 22. This is accomplished by using the eigSVD algorithm from [6], described in Appendix A.3.

3.4 Analysis of Computational Cost

Firstly, we analyze the peak memory usage of Algorithm 5. Because the SVD in Step 17 costs extra $2nl \times 8$ bytes memory except for the space of $\mathbf{W}$ and $\mathbf{Q}$,

Algorithm 5 . Pass-efficient randomized SVD with shifted power iteration (PerSVD)

Input: $\mathbf{A} \in \mathbb{R}^{m\times n}$, k, l $(l \geq k)$, p
Output: $\mathbf{U} \in \mathbb{R}^{m\times k}$, $\mathbf{S} \in \mathbb{R}^{k\times k}$, $\mathbf{V} \in \mathbb{R}^{n\times k}$

1: $\boldsymbol{\Omega} \leftarrow \text{randn}(n, l)$
2: $\mathbf{Q} \leftarrow \text{orth}(\boldsymbol{\Omega})$
3: $\alpha \leftarrow 0$
4: **for** $j = 1, 2, \cdots, p+1$ **do**
5: $\quad \mathbf{W} \leftarrow \text{zeros}(n, l)$
6: $\quad$ **for** $i = 1, 2, \cdots, m$ **do**
7: $\quad\quad \mathbf{a}_i$ is the i-th row of matrix $\mathbf{A}$
8: $\quad\quad \mathbf{Y}(i, :) \leftarrow \mathbf{a}_i\mathbf{Q}$, $\mathbf{W} \leftarrow \mathbf{W} + \mathbf{a}_i^{\mathrm{T}}\mathbf{Y}(i, :)$
9: $\quad$ **end for**
10: $\quad$ **if** $j == p+1$ **break**
11: $\quad \mathbf{D}_1 \leftarrow \mathbf{W}^{\mathrm{T}}\mathbf{W}$, $\mathbf{D}_2 \leftarrow \mathbf{Y}^{\mathrm{T}}\mathbf{Y}$
12: $\quad$ **while** α dose not converge **do**
13: $\quad\quad [\sim, \hat{\mathbf{S}}^2] \leftarrow \text{eig}(\mathbf{D}_1 - 2\alpha\mathbf{D}_2 + \alpha^2\mathbf{I})$
14: $\quad\quad$ **if** $\alpha > \hat{\mathbf{S}}(l, l)$ **then break**
15: $\quad\quad \alpha \leftarrow (\hat{\mathbf{S}}(l, l) + \alpha)/2$
16: $\quad$ **end while**
17: $\quad [\mathbf{Q}, \hat{\mathbf{S}}, \sim] \leftarrow \text{svd}(\mathbf{W} - \alpha\mathbf{Q}, \text{'econ'})$
18: $\quad$ **if** $\alpha < \hat{\mathbf{S}}(l, l)$ **then** $\alpha \leftarrow (\hat{\mathbf{S}}(l, l) + \alpha)/2$
19: **end for**
20: $[\mathbf{Q}, \tilde{\mathbf{S}}, \tilde{\mathbf{V}}] \leftarrow \text{svd}(\mathbf{Y}, \text{'econ'})$
21: $\mathbf{B} \leftarrow \tilde{\mathbf{S}}^{-1}\tilde{\mathbf{V}}^{\mathrm{T}}\mathbf{W}^{\mathrm{T}}$
22: $[\mathbf{U}, \mathbf{S}, \mathbf{V}] \leftarrow \text{svd}(\mathbf{B}, \text{'econ'})$
23: $\mathbf{U} \leftarrow \mathbf{Q}\mathbf{U}(:, 1:k), \mathbf{S} \leftarrow \mathbf{S}(1:k, 1:k), \mathbf{V} \leftarrow \mathbf{V}(:, 1:k)$

the memory usage in Step 17 is $(m + 4n)l \times 8$ bytes, which reflects the space of one $m \times l$ and two $n \times l$ matrices and the space caused by SVD operation. Because the SVD in Step 20 is replaced by eigSVD, the memory usage at Step 20 is $(2m + n)l \times 8$ bytes. Therefore, the peak memory usage of Algorithm 5 is $\max((m + 4n)l, (2m + n)l) \times 8$ bytes. Usually we set $l = 1.5k$, so the memory usage of Algorithm 2 ($16(m + n)k \times 8$ bytes) is several times larger than that of Algorithm 5 ($\max(1.5(m + 4n)k, 1.5(2m + n)k) \times 8$ bytes).

Secondly, we analyze the *flop* count of Algorithm 5. Because eigSVD is used in Step 2, 20 and 22, and the *flop* count of eigSVD on a $m \times l$ matrix is $2C_{mul}ml^2 + C_{eig}l^3$, the *flop* count of those computations is $C_{mul}(4n+2m)l^2 + 3C_{eig}l^3$. Because the computations in Step 12-16 are all about $l \times l$ matrices and $l \ll \min(m, n)$, we drop the *flop* count in Step 12-16. Therefore, the *flop* count of Algorithm 5 is

$$\begin{aligned}\mathrm{FC}_5 =& (2p+2)C_{mul}mnl + pC_{mul}(m+n)l^2 + pC_{svd}nl^2 + C_{mul}nl^2 + C_{mul}mlk \\ &+ C_{mul}(4n+2m)l^2 + 3C_{eig}l^3 \\ \approx& (2p+2)C_{mul}mnl + pC_{mul}(m+n)l^2 + pC_{svd}nl^2 + C_{mul}(mlk + 2ml^2 + 5nl^2)\end{aligned} \tag{18}$$

where $(2p+2)C_{mul}mn$ reflects $2p+2$ times matrix-matrix multiplication in Step 6-9, $pC_{mul}(m+n)l^2$ reflects the matrix-matrix multiplications in Step 11, $pC_{svd}nl^2$ reflects the SVD in Step 17, and $C_{mul}nl^2 + C_{mul}mlk$ reflects the matrix-matrix multiplications in Step 21 and 23.

Because $\min(m,n) \gg l$ and FC_1 and FC_5 both contain $(2p+2)C_{mul}mnl$ which reflects the main computation, the *flop* count of Algorithm 5 is similar with *flop* count of Algorithm 1 in (3) with the same p. Because the *flop* count of main computation in Algorithm 5 is $(2p+2)C_{mul}mnl$ and that of Algorithm 2 is $16C_{mul}mnk$, according to the fact $l = 1.5k$, $(2p+2)C_{mul}mnl = (3p+3)C_{mul}mnk$ is less than $16C_{mul}mnk$ when $p \leq 4$. If $p = 2$ or 3, we see that the total runtime of the two algorithms may be comparable, considering Algorithm 5 reads data multiple times from the hard disk and Algorithm 2 just reads data once.

4 Experimental Results

In this section, numerical experiments are carried out to validate the proposed techniques. We first compare Algorithms 1, 3, 4 and 5 (PerSVD), to validate whether the proposed schemes in Sect. 3 can remarkably reduce the passes with the same accuracy of results. Then, we compare PerSVD and Tropp's algorithm to show the advantage of PerSVD on accurately computing SVD of large matrix stored on hard disk[1].

We consider several matrices stored in the row-major format on hard disk as the test data, which are listed in Table 1. Firstly, two 40,000 × 40,000 matrices (denoted by Dense1 and Dense2) are synthetically generated. Dense1 is randomly generated with the i-th singular value following $\sigma_i = 1/i$. Then, Dense2 is randomly generated with the i-th singular value following $\sigma_i = 1/\sqrt{i}$, which reflects the singular values of Dense2 decay slower than those of Dense1. Then we construct two matrices from real-world data. We use the training set of MNIST [12] which has 60k images of size 28×28, and we reshape each image into a vector in size 784 to obtain the first matrix in size $60,000 \times 784$ for experiment. The second matrix is obtained from images of faces from FERET database [16]. As in [8], we add two duplicates for each image into the data. For each duplicate, the value of a random choice of 10% of the pixels is set to random numbers uniformly chosen from $0,1,\cdots,255$. This forms a 102,042 × 393,216 matrix, whose rows consist of images. We normalize the matrix by subtracting from each row its mean, and then dividing it by its Euclidean norm.

[1] The code is avaliable at https://github.com/XuFengthucs/PerSVD.

Table 1. Test matrices.

Matrix	# of rows	# of columns	Space usage on hard disk
Dense1	40,000	40,000	6.0 GB
Dense2	40,000	40,000	6.0 GB
MNIST	60,000	782	180 MB
FERET	102,042	393,216	150 GB

All experiments are carried out on a Ubuntu server with two 8-core Intel Xeon CPU (at 2.10 GHz) and 32 GB RAM. The proposed techniques, basic randomize SVD algorithm and Tropp's algorithm are all implemented in C with MKL [1] and OpenMP directives for multi-thread parallel computing. `svds` in Matlab 2020b is used for computing the accurate results and for error metrics. We set $l = 1.5k$, and each time we read k rows of matrix to avoid extra memory cost for all algorithms. All the programs are evaluated with wall-clock runtime and peak memory usage. To simulate the machine with limited computational resources, we just use 8 threads on one CPU to test the algorithms. Because the matrix FERET is too large to be loaded in Matlab, the experiments on FERET are without accurate results to compute the error metrics.

4.1 Error Metrics

Theoretical research has revealed that the randomized SVD with power iteration produces the rank-k approximation close to optimal. Under spectral (or Frebenius) norm the computational result ($\hat{\mathbf{U}}$, $\hat{\mathbf{\Sigma}}$ and $\hat{\mathbf{V}}$) has the following multiplicative guarantee:

$$\|\mathbf{A} - \hat{\mathbf{U}}\hat{\mathbf{\Sigma}}\hat{\mathbf{V}}^{\mathrm{T}}\| \leq (1+\epsilon)\|\mathbf{A} - \mathbf{A}_k\|, \tag{19}$$

with high probability. Based on (19), we use

$$\epsilon_{\mathrm{F}} = (\|\mathbf{A} - \hat{\mathbf{U}}\hat{\mathbf{\Sigma}}\hat{\mathbf{V}}^{\mathrm{T}}\|_F - \|\mathbf{A} - \mathbf{A}_k\|_F)/\|\mathbf{A} - \mathbf{A}_k\|_F, \quad \text{and} \tag{20}$$

$$\epsilon_{\mathrm{s}} = (\|\mathbf{A} - \hat{\mathbf{U}}\hat{\mathbf{\Sigma}}\hat{\mathbf{V}}^{\mathrm{T}}\|_2 - \|\mathbf{A} - \mathbf{A}_k\|_2)/\|\mathbf{A} - \mathbf{A}_k\|_2, \tag{21}$$

as first two error metrics to evaluate the accuracy of randomized SVD algorithms for Frobenius norm and spectral norm.

Another guarantee proposed in [15], which is stronger and more meaningful in practice, is:

$$\forall i \leq k, \quad |\mathbf{u}_i^{\mathrm{T}}\mathbf{A}\mathbf{A}^{\mathrm{T}}\mathbf{u}_i - \hat{\mathbf{u}}_i^{\mathrm{T}}\mathbf{A}\mathbf{A}^{\mathrm{T}}\hat{\mathbf{u}}_i| \leq \epsilon\sigma_{k+1}(\mathbf{A})^2, \tag{22}$$

where $\mathbf{u}_i$ is the i-th left singular vector of $\mathbf{A}$, and $\hat{\mathbf{u}}_i$ is the computed i-th left singular vector. This is called *per vector error* bound for singular vectors. In [15], it is demonstrated that the $(1+\epsilon)$ error bound in (19) may not guarantee any accuracy in the computed singular vectors. In contrary, the per vector guarantee (22) requires each computed singular vector to capture nearly as much variance as the corresponding accurate singular vector. Based on the per vector error bound (22), we use

$$\epsilon_{\text{PVE}} = \max_{i \le k} \frac{|\mathbf{u}_i^{\text{T}} \mathbf{A} \mathbf{A}^{\text{T}} \mathbf{u}_i - \hat{\mathbf{u}}_i^{\text{T}} \mathbf{A} \mathbf{A}^{\text{T}} \hat{\mathbf{u}}_i|}{\sigma_{k+1}(\mathbf{A})^2} \tag{23}$$

to evaluate the accuracy of randomized SVD algorithms. Notice that the metric (20), (21) and (23) were also used in [2] with name "Fnorm", "spectral" and "rayleigh(last)".

4.2 Comparison with Basic Randomized SVD Algorithm

In order to validate proposed techniques, we set different power parameter p and perform the basic randomized SVD (Algorithm 1), Algorithm 3 with pass-efficient scheme, Algorithm 4 with fixed shift value and the proposed PerSVD

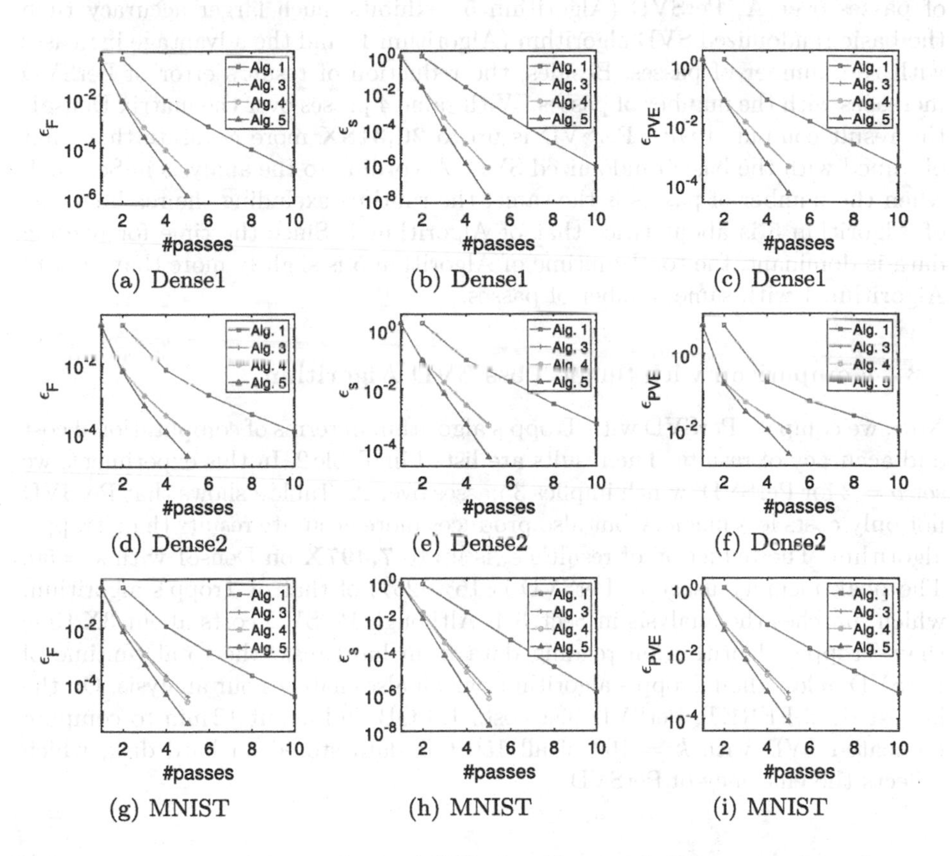

Fig. 2. Error curves of the randomized SVD algorithms with different number of passes ($k = 100$).

with shifted power iteration (Algorithm 5) on test matrices. The orthonormalization is used in the power iteration of Algorithm 1. With the accurate results obtained from svds, the corresponding error metrics (20), (21) and (23) are calculated to evaluate the accuracy. For the largest matrix FERET in Table 1, the accurate SVD cannot be computed with svds due to out-of-memory error. So, the results for it are not available. The curves of error metrics vs. the number of passes over the matrix **A** are drawn in Fig. 2. We set $k = 100$ in this experiment.

From Fig. 2 we see that, the dynamic scheme for setting the shift consistently (i.e. Algorithm 5) produces results with remarkably better accuracy than the scheme with and without the fixed shift. On Dense1 with 4 passes, the reduction of ϵ_s of Algorithm 5 is 14X and 13X compared with Algorithms 3 and 4, respectively. And, the results of Algorithm 4 are more accurate than the results of Algorithm 3 with same number of passes. Notice the error metrics of Algorithm 4 on Dense1 and Dense2 are less than those of Algorithm 3 although the curves are indistinguishable in Fig. 2. For example, on Dense1 with 5 passes, the ϵ_s of Algorithm 4 is 5.8×10^{-7} less than 6.2×10^{-7} of Algorithm 3, which reflects fixed shift value can improve limited accuracy of results. With the same number of passes over **A**, PerSVD (Algorithm 5) exhibits much larger accuracy than the basic randomized SVD algorithm (Algorithm 1) and the advantage increases with the number of passes. Besides, the reduction of result's error of PerSVD increases with the number of passes. With same 4 passes over the matrix Dense1, the result computed with PerSVD is up to **20,318X** more accurate than that obtained with the basic randomized SVD. According to the analysis in Sect. 3.4, when the number of passes is the same, the runtime excluding the reading time of Algorithm 5 is about twice that of Algorithm 1. Since the time for reading data is dominant, the total runtime of Algorithm 5 is slightly more than that of Algorithm 1 with same number of passes.

4.3 Comparison with Single-Pass SVD Algorithm

Now, we compare PerSVD with Tropp's algorithm in terms of computational cost and accuracy of results. The results are listed in Table 2. In this experiment, we set $p = 2$ for PerSVD, which implies 3 passes over **A**. Table 2 shows that PerSVD not only costs less memory but also produces more accurate results than Tropp's algorithm. The reduction of result's ϵ_s is up to **7,497X** on Dense1 with $k = 50$. The peak memory usage of PerSVD is 16%–26% of that of Tropp's algorithm, which matches the analysis in Sect. 3.4. Although PerSVD costs about 3X time than Tropp's algorithm on reading data from hard disk, the total runtime of PerSVD is less than Tropp's algorithm, which also matches our analysis. On the largest data FERET, PerSVD just costs 1.9 GB and about 12 min to compute truncated SVD with $k = 100$ of all 150 GB data stored on hard disk, which reflects the efficiency of PerSVD.

Table 2. The runtime, memory cost and result errors of the Tropp's algorithm and our PerSVD algorithm ($p = 2$). The unit of runtime is second.

Matrix	k	Tropp's algorithm						PerSVD ($p = 2$)					
		t_r	t	Memory	ϵ_F	ϵ_s	ϵ_{PVE}	t_r	t	Memory	ϵ_F	ϵ_s	ϵ_{PVE}
Dense1	50	4.7	27	592 MB	0.38	0.46	0.87	15	26	144 MB	4E–4	6E–5	0.009
	100	4.4	41	1133 MB	0.39	0.51	0.66	14	30	260 MB	4E–4	0.001	0.01
Dense2	50	4.4	27	591 MB	0.49	3.57	9.2	16	27	144 MB	7E–4	0.006	0.04
	100	4.4	50	1133 MB	0.51	3.36	9.2	14	31	260 MB	8E–4	0.02	0.04
MNIST	50	0.29	3.9	498 MB	0.32	0.42	0.67	0.90	1.4	81 MB	4E–4	0.001	0.008
	100	0.31	8.9	966 MB	0.14	0.10	0.24	0.79	1.5	156 MB	4E–4	3E–4	0.006
FERET	50	140	648	3.71 GB	–	–	–	296	597	0.96 GB	–	–	–
	100	178	1366	7.25 GB	–	–	–	293	703	1.90 GB	–	–	–

t_r means the time for reading the data, while t means the total runtime (including t_r).
"–" means the error metrics are not available for FERET matrix.

5 Conclusion

We have developed a pass-efficient randomized SVD algorithm named PerSVD to accurately compute the truncated top-k SVD. PerSVD builds on a technique reducing the passes over the matrix and an innovative shifted power iteration technique. It aims to handle the real-world data with slowly-decayed singular values and accurately compute the top-k singular triplets with a couple of passes over the data. Experiments on various matrices have verified the effectiveness of PerSVD in terms of runtime, accuracy and memory cost, compared with existing randomized SVD and single-pass SVD algorithms. PerSVD is expected to become a powerful tool for computing SVD of really large data.

References

1. Intel oneAPI Math Kernel Library. https://software.intel.com/content/www/us/en/develop/tools/oneapi/components/onemkl.html (2021)
2. Allen-Zhu, Z., Li, Y.: LazySVD: even faster SVD decomposition yet without agonizing pain. In: Advances in Neural Information Processing Systems, pp. 974–982 (2016)
3. Baglama, J., Reichel, L.: Augmented implicitly restarted Lanczos bidiagonalization methods. SIAM J. Sci. Comput. **27**(1), 19–42 (2005)
4. Boutsidis, C., Woodruff, D.P., Zhong, P.: Optimal principal component analysis in distributed and streaming models. In: Proceedings of the the 48th Annual ACM Symposium on Theory of Computing, pp. 236–249 (2016)
5. Eckart, C., Young, G.: The approximation of one matrix by another of lower rank. Psychometrika **1**(3), 211–218 (1936)
6. Feng, X., Xie, Y., Song, M., Yu, W., Tang, J.: Fast randomized PCA for sparse data. In: Proceedings of the 10th Asian Conference on Machine Learning (ACML), 14–16 Nov 2018, pp. 710–725 (2018)

7. Golub, G.H., Van Loan, C.F.: Matrix computations. JHU Press (2012)
8. Halko, N., Martinsson, P.G., Shkolnisky, Y., Tygert, M.: An algorithm for the principal component analysis of large data sets. SIAM J. Sci. Comput. **33**(5), 2580–2594 (2011)
9. Halko, N., Martinsson, P.G., Tropp, J.A.: Finding structure with randomness: Probabilistic algorithms for constructing approximate matrix decompositions. SIAM Rev. **53**(2), 217–288 (2011)
10. Horn, R.A., Johnson, C.R.: Topics in matrix analysis. Cambridge University Press (1991). https://doi.org/10.1017/CBO9780511840371
11. Larsen, R.M.: Propack-software for large and sparse SVD calculations. https://sun.stanford.edu/~rmunk/PROPACK (2004)
12. LeCun, Y., Bottou, L., Bengio, Y., Haffner, P.: Gradient-based learning applied to document recognition. Proc. IEEE **86**(11), 2278–2324 (1998)
13. Li, H., Linderman, G.C., Szlam, A., Stanton, K.P., Kluger, Y., Tygert, M.: Algorithm 971: an implementation of a randomized algorithm for principal component analysis. ACM Trans. Math. Softw. **43**(3), 1–14 (2017)
14. Martinsson, P.G., Tropp, J.A.: Randomized numerical linear algebra: foundations and algorithms. Acta Numer **29**, 403–572 (2020)
15. Musco, C., Musco, C.: Randomized block Krylov methods for stronger and faster approximate singular value decomposition. In: Advances in Neural Information Processing Systems, pp. 1396–1404 (2015)
16. Phillips, P.J., Moon, H., Rizvi, S.A., Rauss, P.J.: The FERET evaluation methodology for face-recognition algorithms. IEEE Trans. Pattern Anal. Mach. Intell. **22**(10), 1090–1104 (2000)
17. Rokhlin, V., Szlam, A., Tygert, M.: A randomized algorithm for principal component analysis. SIAM J. Matrix Anal. Appl. **31**(3), 1100–1124 (2010)
18. Shishkin, S.L., Shalaginov, A., Bopardikar, S.D.: Fast approximate truncated SVD. Numer. Linear Algebra Appl. **26**(4), e2246 (2019)
19. Tropp, J.A., Yurtsever, A., Udell, M., Cevher, V.: Streaming low-rank matrix approximation with an application to scientific simulation. SIAM J. Sci. Comput. **41**(4), A2430–A2463 (2019)
20. Voronin, S., Martinsson, P.G.: RSVDPACK: an implementation of randomized algorithms for computing the singular value, interpolative, and CUR decompositions of matrices on multi-core and GPU architectures. arXiv preprint arXiv:1502.05366 (2015)
21. Yu, W., Gu, Y., Li, J., Liu, S., Li, Y.: Single-pass PCA of large high-dimensional data. In: Proceedings of the International Joint Conference on Artificial Intelligence (IJCAI), pp. 3350–3356 (2017)
22. Yu, W., Gu, Y., Li, Y.: Efficient randomized algorithms for the fixed-precision low-rank matrix approximation. SIAM J. Matrix Anal. Appl. **39**(3), 1339–1359 (2018)

CDPS: Constrained DTW-Preserving Shapelets

Hussein El Amouri[1(✉)], Thomas Lampert[1(✉)], Pierre Gançarski[1(✉)], and Clément Mallet[2(✉)]

[1] ICube, University of Strasbourg, Strasbourg, France
{helamouri,lampert,gancarski}@unistra.fr
[2] Univ Gustave Eiffel, IGN, ENSG, Saint-Mande, France
clement.mallet@ign.fr

Abstract. The analysis of time series for clustering and classification is becoming ever more popular because of the increasingly ubiquitous nature of IoT, satellite constellations, and handheld and smart-wearable devices, etc. The presence of phase shift, differences in sample duration, and/or compression and dilation of a signal means that Euclidean distance is unsuitable in many cases. As such, several similarity measures specific to time-series have been proposed, Dynamic Time Warping (DTW) being the most popular. Nevertheless, DTW does not respect the axioms of a metric and therefore Learning DTW-Preserving Shapelets (LDPS) have been developed to regain these properties by using the concept of shapelet transform. LDPS learns an unsupervised representation that models DTW distances using Euclidean distance in shapelet space. This article proposes constrained DTW-preserving shapelets (CDPS), in which a limited amount of user knowledge is available in the form of must link and cannot link constraints, to guide the representation such that it better captures the user's interpretation of the data rather than the algorithm's bias. Subsequently, any unconstrained algorithm can be applied, e.g. K-means clustering, k NN classification, etc, to obtain a result that fulfils the constraints (without explicit knowledge of them). Furthermore, this representation is generalisable to out-of-sample data, overcoming the limitations of standard transductive constrained-clustering algorithms. CLDPS is shown to outperform the state-of-the-art constrained-clustering algorithms on multiple time-series datasets. An open-source implementation based on PyTorch is available From: https://git.unistra.fr/helamouri/constrained-dtw-preserving-shapelets, which takes full advantage of GPU acceleration.

This work was supported by the HIATUS (ANR-18-CE23-0025) and HERELLES (ANR-20-CE23-0022) ANR projects. We thank Nvidia Corporation for donating GPUs and the Centre de Calcul de l'Université de Strasbourg for access to the GPUs used for this research.

Supplementary Information The online version contains supplementary material available at https://doi.org/10.1007/978-3-031-26387-3_2.

M.-R. Amini et al. (Eds.): ECML PKDD 2022, LNAI 13713, pp. 21–37, 2023.
https://doi.org/10.1007/978-3-031-26387-3_2

Keywords: Shapelets · Semi-supervised learning · Clustering · Constrained-clustering · Time series · Learning representation

1 Introduction

The availability of time series data is increasing rapidly with the development of sensing technology and the increasing number of fields that uses such technology. This increase in data volume means that providing ground truth labels becomes difficult due to the time and cost needed. Labelling difficulty is exacerbated when making exploratory analyses and when working in nascent domains for which classes are not well defined. For that reason, supervised approaches such as classification become unfeasible and unsupervised clustering is often preferred. However, unsupervised approaches may lead to irrelevant or unreliable results since they have no knowledge about the user's requirements and are instead lead by the algorithm's bias. On the other hand, semi-supervised algorithms try to remove the rigid requirements of supervised approaches but retain the ability of a user to guide the algorithm to produce a meaningful output. This can be achieved by providing a set of constraints to the algorithm that encode some expert knowledge. These can take many forms but this work is concerned with must-link and cannot-link constraints since they are the easiest to interpret and provide.

Must-link and cannot-link constraints do not define what a sample represents (a class), instead they label pairs of samples as being the same (must-link), thus belong to the same cluster, or not (cannot-link). In this way the algorithm is guided to converge on a result that is meaningful to the user without explicitly, nor exhaustively labelling samples. Generally, time series are characterised by trend, shapes, and distortions either to time or shape [22] and therefore exhibit phase shifts and warping. As such, the Euclidean distance is unsuitable and several similarity measures specific to time-series have been proposed [17], for example compression-based measures [7], Levenshtein Distance [10], Longest Common Subsequnce [25] and Dynamic Time Warping (DTW) [19,20]. DTW is one of the most popular since it overcomes these problems by aligning two series through the computation of a cost function based on Euclidean distance [8], it is therefore known as an elastic measure [17]. Moreover, Paparrizos et al. show that DTW is a good basis for calculating embeddings, an approach that employs a similarity to construct a new representation. Time series also exhibit complex structure which are often highly correlated [22]. This makes their analysis difficult to achieve and time consuming, indeed several attempts to accelerate DTW's computation have been proposed [1,22]. Shapelets [30] offer a simpler approach to increase the accuracy of time-series analysis. Shapelets are phase-independent discriminative sub-sequences extracted or learnt to form features that map a time-series into a more discriminative representational space, therefore increasing the reliability and interpretability of downstream tasks. Since DTW does not respect the axioms of a metric, LDPS [13] extends shapelets to preserve DTW distances in a Euclidean embedding.

The contribution of this article is to introduce constrained DTW-preserving shapelets (CDPS), in which a time series representation is learnt to overcome time series distortions by approximating DTW and is influenced by a limited amount of user knowledge by providing constraints. Thus CDPS can model a user's interpretation, rather than being influenced by the algorithm's bias. Subsequently, any unconstrained algorithm can be applied to the embedding, e.g. K-means clustering, k-NN classification, etc, to obtain a result that fulfils the constraints (without explicit knowledge of them). The proposed embedding process is studied in a constrained clustering setting, on multiple datasets, and is compared to COP-KMeans [27], FeatTS [23], and unsupervised DTW-preserving shapelets [13].

The representational embedding that is learnt by CDPS is generalisable to out-of-sample data, overcoming the limitations of standard constrained-clustering algorithms such as COP-KMeans. It is interpretable, since the learnt shapelets can themselves be visualised as time-series. Finally, since CDPS results in a vectorial representation of the data, they and the constraints can be analysed using norm-base measures, something that is not possible when using DTW as a similarity measure [8]. This opens up the possibility of measuring constraint informativeness [3] and constraint consistency [26] in time-series clustering, and explaining and interpreting the constraints, which is a concern for future work. Such measures, and notions of density, are needed to develop novel interactive and active constrained clustering processes for time-series.

The rest of this article is organised as follows: in Sect. 2 related work is reviewed, in Sect. 3 the Constrained DTW-Preserving Shapelets (CDPS) algorithm is presented, in Sect. 4 CDPS is compared to constrained/semi-supervised and unconstrained approaches from the literature, and finally Sect. 6 presents the conclusions and future work.

2 Related Work

This section will present works related to shapelets and constrained clustering.

2.1 Shapelets

Shapelets are sub-sequences of time-series that were originally developed to discriminate between time-series using a tree based classifier [30,31]. As such, the shapelets themselves were chosen from a set of all possible sub-sequences of the set of time series being analysed, which is time consuming and exhaustive. Different approaches are proposed to increase the speed of finding shapelets. Rakthanmanon and Keogh [18] propose to first project the time-series into a symbolic representation to increase the speed of discovering the shapelets. Subsequently, Mueen et al. [14] introduce logical shapelets, which combines shapelets with complex rules of discrimination to increase the reliability of the shapelets and their ability to discriminate between the time-series. Sperandio [22] presents a detailed review of early shapelet approaches.

Lines et al. [12] proposed a new way of handling shapelets that separated classification from transformation. This was later extended by Hills et al. [6] to the shapelet transform, which transforms the raw data into a vectorial representation in which the shapelets define the representation space's bases. It was proved that this separation leads to more accurate classification results even with non-tree based approaches.

2.2 Learning Shapelets

In order to overcome the exhaustive search for optimal shapelets, Grabocka et al. [4] introduce the concept of learning shapelets in a supervised setting. In this approach the optimal shapelets are learnt by minimising a classification objective function. The authors consider shapelets to be features to be learnt instead of searching for a set of possible candidates, they report that this method provides a significant improvement in accuracy compared to previous search based approaches. Other supervised approaches have been proposed, Shah et al. [21] increase accuracy by learning more relevant and representative shapelets. This is achieved by using DTW similarity instead of Euclidean distance, since it is better adapted to measure the similarity between the shapelets and the time-series. Another approach for learning shapelets is to optimise the partial AUC [29], in which shapelets are learnt in conjunction with a classifier.

2.3 Unsupervised Shapelets

Zakaria et al. [32] introduced the first approach for clustering time-series with shapelets, called unsupervised-shapelets or u-shapelets. U-shapelets are those that best partition a subset of the time series from the rest of the data set. The shapelets are chosen from a set of all possible sub-sequences by partitioning the dataset and removing the time series that are similar to the shapelet, this process is repeated until no further improvements (i.e. partitions) can be made. It is therefore an exhaustive search, as were the early supervised approaches. U-shapelets have been used in several works since their initial introduction [24,33]. Since these unsupervised methods take a similar approach to the original supervised shapelets, they have the same drawbacks. To overcome these, Zhang et al. [34] propose to combine learning shapelets with unsupervised feature selection to learn the optimal shapelets. Learning DTW-preserving shapelets (LDPS) expands the learning paradigm for shapelets by integrating additional constraints on the learnt representation. In LDPS these constrain the representation space to model the DTW distances between the time-series, instead of focusing on learning shapelets that best discriminate between them.

A multitude of other unsupervised approaches to build an embedding space for time series exist (other than shapelets) and Paparrizos et al. [17] provide an extensive study of them. Generic Representation Learning (GRAIL) [15], Shift Invariant Dictionary Learning (SIDL) [35], Preserving Representation Learning method (SPIRAL) [9], and Random Warping Series (RWS) [28] are different approaches to building such representations. Since these are unsupervised they are not of concern in this article.

2.4 Constrained Clustering

Constrained clustering algorithms are those that add expert knowledge to the process such as COP-Kmeans [27] and Constraint Clustering via Spectral Regularization (CCSR) [11]. Constraints can be given in different forms such as cluster level constraints and instance level constraints. Must-link and cannot-link constraints between samples fall under the latter.

Many constrained clustering algorithms have been proposed, some of which have been adapted to time-series. For a full review, the reader is referred to [8]. Here, those relevant to this study are mentioned. COP-KMeans is an extension to k-Means that often offers state-of-the-art performance without the need to choose parameters [8]. Cluster allocations are validated using the constraint set at each iteration to verify that no constraints are violated. For use with time-series the DTW distance measure is often used along with an appropriate averaging method such as DTW barycenter averaging (DBA) [8] to calculate the cluster centres. Another semi-supervised approach developed specifically for clustering time series is FeatTS [23]. FeatTS uses a percentage of labeled samples to extract relevant features used to calculate a co-occurrence matrix from a graph created by the features. The co-occurrence matrix is then used to cluster the dataset.

Other approaches to time-series clustering exist, such as k-shape [16], however being unsupervised, these fall outside the scope of this article.

3 Constrained DTW-Preserving Shapelets

This section proposes Constrained DTW-Preserving Shapelets (CDPS), which learns shapelets in a semi-supervised manner using ML and CL constraints. Therefore allowing expert knowledge to influence the transformation learning process, while also preserving DTW similarity and interpretability of the resulting shapelets. Definitions and notations are presented in Sub-Sect. 3.1, and the algorithm in Sub-Sect. 3.3.

3.1 Definitions and Notations

Time series: *is an ordered set of real-valued observations. Let* $\mathcal{T} = \{T_1, T_2, \ldots, T_N\}$ *be a set of* N *uni-dimensional time series (for simplicity of notation, nevertheless CDPS is also applicable to multi-dimensional time series).* L_i *is the length of a time series such that* T_i *is composed of* L_i *elements (each time-series may have different lengths), such that*

$$T_i = T_{i,1}, \ldots, T_{i,L_i}. \quad (1)$$

A segment of a time series T_i *at the* m^{th} *element with length* L *is denoted as* $T_{i,m:L} = \{T_{i,m}, \ldots, T_{i,L}\}$.

Shapelet *is an ordered set of real-valued variables, with a length smaller than, or equal to, that of the shortest time series in the dataset. Let a Shapelet be denoted as* S *having length* L_k. *Let* $\mathcal{S} = \{S_1, \ldots, S_K\}$ *be a set of* K *shapelets, where* $S_k = S_{j,1:L_k}$. *In our work, the set* $\mathcal{S}$ *can have shapelets with different lengths, but for the simplicity we will use shapelets with same length in the formulation.*

Squared Euclidean Score *is the similarity score between a shapelet* S_k *and a time series sub-sequence* $T_{i,m:L_S}$, *such that*

$$D_{i,k,m} = \frac{1}{l} \sum_{x=1}^{l} (T_{i,m+x-l} - S_{k,x})^2. \tag{2}$$

Euclidean Shapelet Match *represents the matching score between shapelet* S_k *and a time series* T_i, *such that*

$$\overline{T}_{i,k} = \min_{m \in \{1:L_i - L_k + 1\}} D_{i,k,m}. \tag{3}$$

Shapelet transform *is the mapping of time series* T_i *using Euclidean shapelet matching with respect to the set of shapelets* $\mathcal{S}$. *Where the new vectorial representation is*

$$\overline{T}_i = \{\overline{T}_{i,1}, \ldots, \overline{T}_{i,K}\}. \tag{4}$$

Constraint Sets *Let* C_k *be the* k^{th} *cluster, ML be the set containing time series connected by a must link and CL the set such that they are connected by a cannot link. Thus,* $\forall\, T_i, T_j$ *such that* $i, j \in \{1, \ldots, N\}$ *and* $i \neq j$ *we have*

$$\mathrm{ML} = \{(i,j) | \forall\, k \in \{1, \ldots, K\}, T_i \in C_k \Leftrightarrow T_j \in C_k\}, \tag{5}$$

$$\mathrm{CL} = \{(i,j) | \forall\, k \in \{1, \ldots, K\}, \neg(T_i \in C_k \wedge T_j \in C_k)\}. \tag{6}$$

3.2 Objective Function

In order to achieve a guided constrained learning approach, a new objective function is introduced based on contrastive learning [5] that extends the loss function used in LDPS [13] to a semi-supervised setting. The loss between two time-series takes the form

$$\mathcal{L}(T_i, T_j) = \frac{1}{2} \left(DTW(T_i, T_j) - \beta Dist_{i,j}\right)^2 + \phi_{i,j}, \tag{7}$$

where $DTW(T_i, T_j)$ is the dynamic time warping similarity between time-series T_i and T_j, $Dist_{i,j} = ||\overline{T}_i - \overline{T}_j||_2$ is the similarity measure between T_i and T_j in the embedded space such that $|| \cdot ||_2$ is the L_2 norm, and β scales the time-series similarity (distance) in the embedded space to the corresponding DTW similarity. The term $\phi_{i,j}$ is inspired by the contrastive loss and is defined, such that

$$\phi_{i,j} = \begin{cases} \alpha Dist_{i,j}^2, & \text{if } (i,j) \in \mathrm{ML}, \\ \gamma \max(0, w - Dist_{i,j})^2, & \text{if } (i,j) \in \mathrm{CL}, \\ 0, & \text{otherwise}, \end{cases} \tag{8}$$

where α, γ are weights that regularise the must-link and cannot-link similarity distances respectively, and w is the minimum distance between samples for them to be considered well separated in the embedded space (after which, there is no influence on the loss) and is calculated using the following function $w = \max_{\forall i, \forall j}(DTW(T_i, T_j)) - \log(\frac{DTW(T_i,T_j)}{\max_{\forall i,\forall j}(DTW(T_i,T_j))})$, such that $i \neq j$.

The overall loss function is therefore defined, such that

$$\mathcal{L}(\mathcal{T}) = \frac{2}{K(K-1)} \sum_{i=1}^{K} \sum_{j=i+1}^{K-1} \mathcal{L}(T_i, T_j). \quad (9)$$

3.3 CDPS Algorithm

Algorithm 1. CDPS algorithm

Input: $\mathcal{T}$ a set of Time-series,
ML and CL constraint sets,
$L_{\min}$ minimum length of shapelets,
$S_{\max}$ maximum number of shapelet blocks,
n_{epochs}, s_{batch}, c_{batch}

Output: Set S of shapelets,
Embeddings of $\mathcal{T}$.

1: ShapeletBlocks ← GET_SHAPELET_BLOCKS($L_{\min}$, $S_{\max}$, L_i)
2: Shapelets ← INITIALIZE_SHAPELETS(ShapeletBlocks)
3: **for** $i \leftarrow 0$ to n_{epochs} **do**
4: **for** 1 to $|\mathcal{T}|/s_{\text{batch}}$ **do**
5: minibatch ← GET_BATCH($\mathcal{T}$, ML, CL, S_{batch}, C_{batch})
6: Compute the DTW between the $T_{i's}$ and $T_{j's}$ in minibatch
7: Update the Shapelets and β by descending the gradient $\nabla\mathcal{L}(T_i, T_j)$
8: Embeddings ← SHAPELET_TRANSFROM($\mathcal{T}$)

Algorithm 1 describes CDPS's approach to learning the representational embedding. In which ShapeletBlocks is a dictionary containing $S_{\max}$ pairs, {shapelet length; shapelet number}, where shapelet length is $L_{\min} \cdot b_{\text{ind}}$, $L_{\min}$ is the minimum shapelet length and $b_{\text{ind}} \in \{1, \ldots, S_{\max}\}$ is the index of the shapelet block. The number of shapelets for each block is calculated using the same approach as LDPS [13]: $10 \log(L_i - L_{\min} \cdot b_{\text{ind}}) \times 10$. The parameter C_{batch} defines the number of constraints in each training batch, the aim of this parameter is to increase the importance of the constrained time-series in face of the large number of the unconstrained time-series. INITIALIZE_SHAPELETS initialises the shapelets either randomly or rule-based. Here the following rule-based approach is taken: (1) Shapelets are initialised by drawing a number of time series samples then reshaping them into sub-sequences with length equal to that of the shapelets;

(2) k-means clustering is then performed on the sub-sequences and the cluster centers are extracted to form the initial shapelets. GET_BATCH generates batches containing both constrained and unconstrained samples. If there are insufficient constraints to fulfil C_{batch} then they are repeated. For speed and to take advantage of GPU acceleration, the above algorithm can be implemented as a 1D convolutional neural network in which each layer represents a shapelet block composed of all the shapelets having the same length followed by max-pooling in order to obtain the embeddings. The derivation of the gradient of $\mathcal{L}(\mathcal{T})$, $\nabla\mathcal{L}(T_i, T_j)$ (Algorithm 1, Line 7), is given in the supplementary material.

4 Evaluation

In this section CDPS is evaluated with respect to different constraint sets under two cases: the classical constrained clustering setting in which clusters are extracted from a dataset, called transductive learning; and the second, which is normally not possible using classical constrained clustering algorithms, in which the constraints used to learn a representation are generalised to an unseen test set, called inductive learning.

4.1 Experimental Setup

Algorithm 1 is executed using mini-batch gradient descent with a batch size $s_{\text{batch}} = 64$, $c_{\text{batch}} = 16$ constraints in each batch for the transductive setting, while $s_{\text{batch}} = 32$, $c_{\text{batch}} = 8$ for the inductive setting (since there are fewer samples). The influence of α and γ on accuracy were evaluated and the algorithm was found to be stable to variations in most of the cases and for that reason the value for both is fixed to 2.5. The minimum shapelet length $L_{min} = 0.15 \cdot L_i$, and the maximum number of shapelets $S_{max} = 3$ are taken to be the same as used in LDPS [13]. All models are trained for 500 epochs using the Adam optimiser.

K-means and COP-KMeans [27] are used as comparison methods (unconstrained and constrained respectively) since k-means based algorithms are the most widely applied in real-world applications, offering state-of-the-art (or close to state-of-the-art) performance. CDPS is also compared to FeatTS [23], which is a semi-supervised algorithm that extracts features and uses k-Mediods clustering.

Thirty-five datasets[1] chosen randomly from the UCR repository [2] are used for evaluation. The number of clusters is set to the number of classes in each dataset. The Normalised Mutual Information (NMI), which measures the coherence between the true and predicted labels, is measured to evaluate the resulting clusters with 0 indicating no mutual information and 1 a perfect correlation.

For the first use case, termed Transductive, the training and test sets of the UCR datasets are combined, this reflects the real-world transductive case in which a dataset is to be explored and knowledge extracted. In the second, termed

[1] Details on the datasets used are provided in the supplementary material.

Inductive, the embedding is learnt on the training set and its generalised performance on the test set is evaluated. This inductive use-case is something that is not normally possible when evaluating constrained clustering algorithms since clustering is a transductive operation and this highlights one of the key contributions of CDPS - the ability generalise constraints to unseen data. The third use case, highlights the importance of CDPS shapelets as features and their general ability to be integrated into any downstream algorithm. As such, FeatTS's semi-supervised statistical features are replaced with the dataset's CDPS embedding.

Each algorithm's performance is evaluated on each dataset with increasing numbers of constraints, expressed in percentages of samples that are subject to a constraint in the 5%, 15%, 25%. These represent a very small fraction of the total number of possible constraints, which is $\frac{1}{2}N(N-1)$. Each clustering experiment is repeated 10 times, each with a different random constraint set, and each clustering algorithm is repeated 10 times for each constraint set (i.e. there are 100 repetitions for each percentage of constraints[2]). The constraints are generated by taking the ground truth data, randomly selecting two samples, and adding an ML or CL constraint depending on their class labels until the correct number of constraints are created.

In the FeatTS comparison, both the train and test sets were used (i.e. transductive). FeatTS and CDPS were evaluated using 25% of the ground truth information: FeatTS takes this information in the form of labels; while CDPS in the form of ML/CL constraints, CDPS embeddings are generated and replace FeatTS's features (CDPS+FeatTS). The number of features used for FeatTS was 20, as indicated in the author's paper. With both feature sets, k-Mediods was applied on the co-occurrence matrix to obtain the final clustering [23].

4.2 Results

In this section the results of each use case (described in Sect. 4.1) are presented.

Transductive: Figure 1 shows the NMI scores for CDPS (Euclidean k-means performed on the CDPS embeddings) compared to k-means (on the raw time-series), COP-Kmeans (also on the raw time-series), and LDPS (Euclidean k-means on the LDPS embeddings). Unconstrained k-means and LDPS are presented as a reference for the constrained algorithms (COP-kmeans and CDPS respectively) to give insight into the benefit of constraints for each. It can be seen that overall LDPS and k-means offer similar performance.

It can also be seen that CDPS uses the information gained by constraints more efficiently, outperforming COP-Kmeans in almost all the different constraint fractions for most datasets.

[2] Note that it is not always possible for COP-KMeans to converge on a result due to constraint violations, although many initialisation were tried to obtain as many results as possible some of the COP-KMeans results represent fewer repetitions.

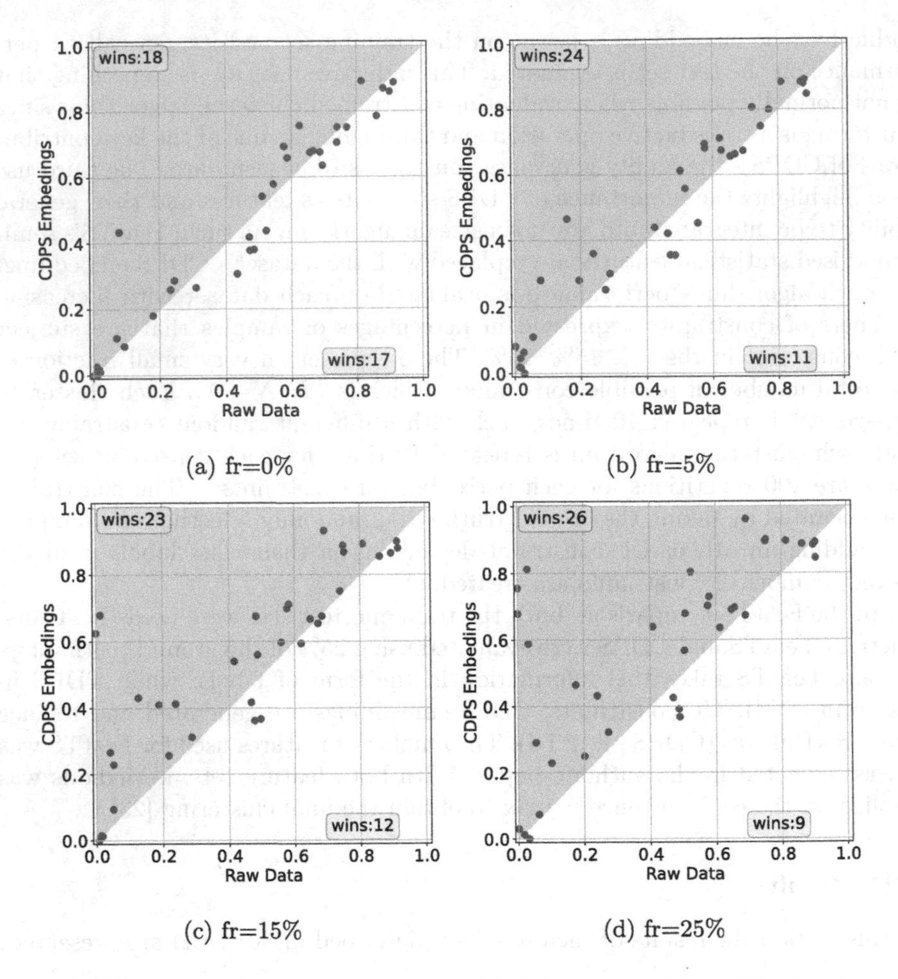

Fig. 1. A Transductive comparison between CDPS+kmeans and Raw-TS+CopKmeans with different constraint fractions.

It appears, nevertheless, that some datasets lend themselves to (unconstrained) k-means based algorithms since it outperforms LDPS. Nevertheless, CPDS exhibits an increase in performance as the number of constraints increase, whereas COP-Kmeans tends to stagnate. This can be seen as the cloud of points move upwards (CDPS score increases) as more constraints are given. We can also observe that for some datasets the constrained algorithms behave similarly with 5% constraints, i.e. the cloud of points in the lower left corner, but again CDPS benefits most from increasing the number of constraints and significantly outperforms COP-KMeans with larger constraint percentages.

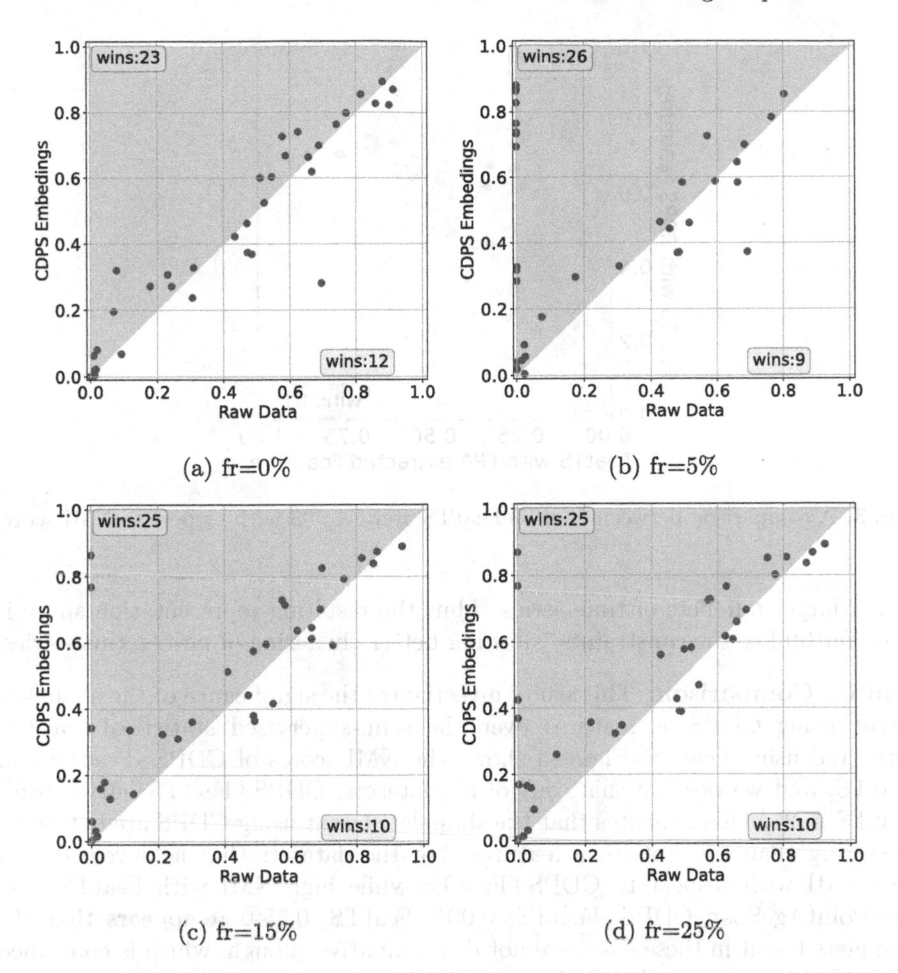

Fig. 2. An inductive comparison between CDPS+kmeans and Raw-TS+CopKmeans with different constraint fractions.

Inductive: Figure 2 presents the Inductive results, in which the embedding space is learnt on the training set and the generalisation performance evaluated on the unseen test set.

It should be noted that when training on the train set, there are significantly fewer constraints then when using the merged datasets for the same constraint percentage. It can therefore be concluded that even in the face of few data and constraints, CDPS is still able to learn a generalisable representation and attain (within a certain margin) the same clustering performance then when trained on the merged dataset. This is probably explained by the fact that having a smaller number of samples with few constraints means that they are repeated in the mini-batches (see Sect. 3.3), and this allows CDPS to focus on learning shapelets that are discriminative and preserve DTW rather than shapelets that

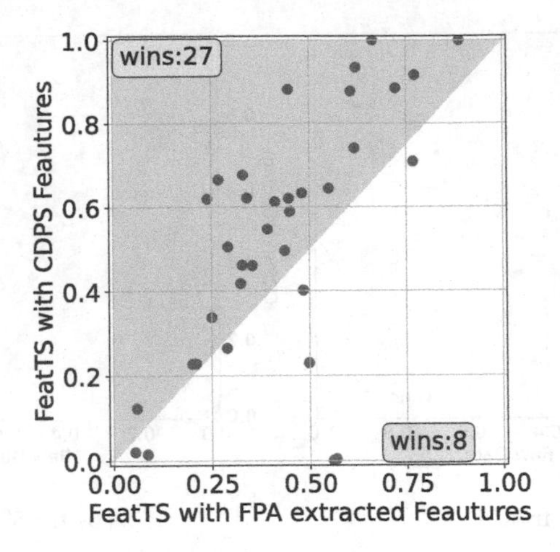

Fig. 3. A comparison between CDPS+FeatTS and FeatTS with respect to NMI score.

model larger numbers of time series. Thus the resulting representation space is more faithful to the constraints, allowing better clustering of unseen time-series.

FeatTS Comparison: This study investigates the significance of the shapelets learnt using CDPS as features over the semi-supervised statistical features extracted using FeatTS. Figure 3 shows the NMI scores of CDPS+FeatTS and FeatTS, and we observe that, out of 35 datasets, CDPS+FeatTS outperforms FeatTS in 27. This indicates that the shapelets learnt using CDPS are better for clustering than the statistical features. For the datasets that achieved around zero NMI with respect to CDPS+FeatTS while high NMI with FeatTS (e.g. GunPointAgeSpan CDPS+FeatTS: 0.001, FeatTS: 0.559) it appears that the shapelets learnt in these cases are not discriminative enough, which is confirmed by CDPS' low scores (CDPS+Kmeans: 0.004).

5 Discussion

Since LDPS only models DTW distance, the comparisons between it and k-means (Figs. 2a & 1a) give approximately equal performance. Nevertheless, CDPS is better able to exploit the information contained in the constraints when they are introduced, giving more accurate clustering results overall. Both LDPS and CDPS result in a metric space, which is beneficial for further analysis and processing.

Being a hard constraint algorithm, COP-KMeans offers no guarantee of convergence, which was evident in the presented study where several of the results were missing after multiple tries. This is due to the difficulty of clustering with an elastic distance measure such as DTW. In these experiments, all constraints can be considered as coherent since they are generated from the ground truth data, however, in real-world situations this problem would be exacerbated by inconsistent constraints, particularly considering time-series since these are very hard to label. CDPS, does not suffer from such limitations.

Although it was included in this study in order to have a comparison method, using COP-KMeans in an inductive use-case is not usual practice for a classical clustering algorithm. It was simulated by providing COP-KMeans with the combined 'training' dataset, its constraints, and the test data to be clustered. CDPS, on the other hand, offers a truly inductive approach in which new data can be projected into the resulting space, which inherently models user constraints. In this setting the difference between COP-KMeans is reduced, however, it should be noted that CDPS does not 'see' the training data during the inductive setup, whereas COP-KMeans used all the data to derive the clusters. This also exposes the infeasibility of using COP-KMeans in this way, the data needs to be stored, and accessed each time new data should be clustered, which will become computationally expensive as its size grows. Finally, CDPS's embedding can be used for tasks other than clustering (classification, generation, etc.).

CDPS's inductive complexity (once the space has been determined) is $\mathcal{O}(NL_kK)$, plus kmeans' complexity $\mathcal{O}(NkKi)$, where k is the number of clusters, i the number of iterations until convergence, and the complexity of COP-KMeans is $\mathcal{O}(NkKi|\mathrm{ML} \cup \mathrm{CL}|)$.

Overall, the CDPS algorithm leads to better clustering results since it is able to exploit the information brought to the learning process by the constraints. Relatively, it can be seen that the number of datasets in which CDPS outperforms COP-KMeans increases in line with the number of constraints. In absolute terms, COP-KMeans' performance tends to decrease as more constraints are introduced, and the opposite can be said for CDPS. These constraints bias CDPS to find shapelets that define a representation that respects them while retaining the properties of DTW. Although the focus of this article is not to evaluate whether clustering on these datasets benefits from constraints, it can be observed that generally better performance is found when constraints are introduced.

The studies in the previous section show that the transformed space not only preserves the desirable properties of DTW but also implicitly models the constraints given during training. Although it was not evaluated, it is also possible to use COP-Kmeans (constrained) clustering in the Inductive CDPS embedding, thus allowing another mechanism to integrate constraints after the embedding has been learnt. Although CDPS has several parameters, it has been shown that these do not need to be fine-tuned for each dataset to achieve state-of-the-art performance (although better performance may be achieved if this is done).

5.1 Model Selection

When performing clustering there is no validation data with which to determine a stopping criteria. It is therefore important to analyse the behaviour of CDPS during training to give some general recommendations.

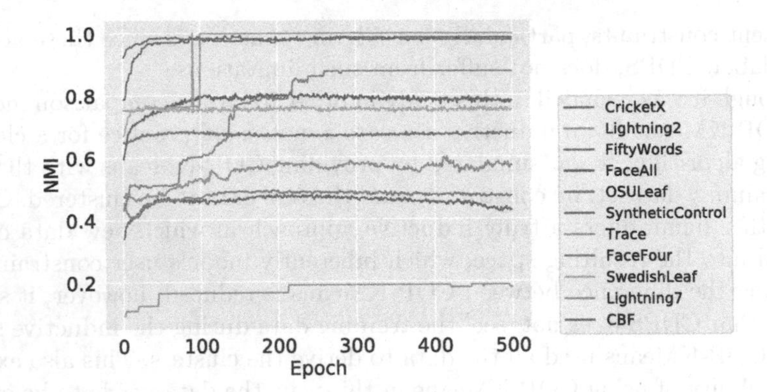

Fig. 4. Clustering quality (NMI) as a function of the number of epochs for each dataset, using a constraint fraction of 30%.

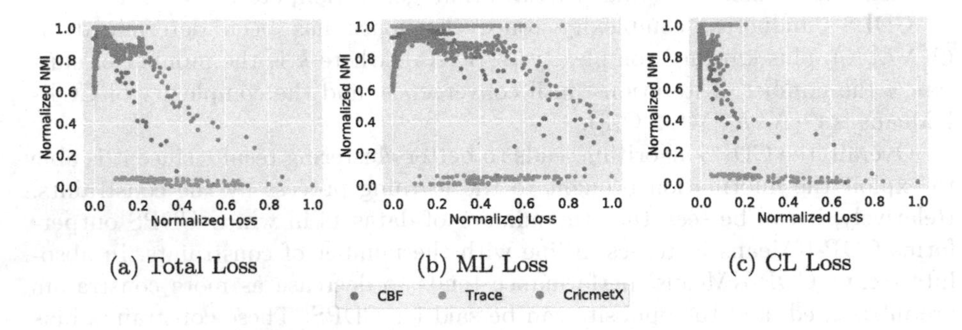

Fig. 5. Relationship between NMI and CDPS Loss for each dataset. To highlight the relationship between datasets, both loss and NMI have been scaled to between 0 and 1.

Figure 4 presents the CDPS clustering quality (NMI) as a function of the number of epochs for each dataset (using 30% constraints). It demonstrates that generally most of the models converge within a small number of epochs, with FaceFour taking the most epochs to converge. Moreover, the quality of the learnt representation does not deteriorate as the number of epochs increases, i.e. neither the DTW preserving aspect nor the constraint influence dominate the loss and diminish the other as epochs increase.

Figure 5 presents scatter-plots of the NMI and CDPS loss (both normalised to between 0 and 1) for several datasets. In addition to the total loss, both the ML and CL losses have been included. The general trend observed in the overall loss is that a lower loss equates to a higher NMI.

These show that the loss can be used as a model selection criterion without any additional knowledge of the dataset. For practical application, the embedding can be trained for a fixed large enough number of epochs (as done in this study) or until stability is achieved. This is in line with the typical manner in which clustering algorithms are applied.

6 Conclusions

This article has presented CDPS, an approach for learning shapelet based time-series representations that respect user constraints while also respecting the DTW similarity of the raw time-series. The constraints take the form of must-link and cannot-link pairs of samples provided by the user. The influence of the constraints on the learning process is ensured through the use of mini-batch gradient descent in which a fraction of each batch contains samples under constraint. The resulting space removes many limitations inherent with using the DTW similarity measure for time-series, particularly interpretability, constraint analysis, and the analysis of sample density. CDPS therefore paves the way for new developments in constraint proposition and incremental (active) learning for time-series clustering. The representations learnt by CDPS are general purpose and can be used with any machine learning task. The presented study focused on its use in constrained clustering. By evaluating the proposed method on thirty-five public datasets, it was found that using unconstrained k-means on CPDS representations outperforms COP-Kmeans, unconstrained k-means (on the original time-series), and LDPS with k-means. Also, CDPS is shown to outperform FeatTS that uses statistical features. It was also shown that the representation learnt by CDPS is generalisable, something that is not possible with classic constrained clustering algorithms and when applied to unseen data, CDPS outperforms COP-KMeans.

References

1. Cai, B., Huang, G., Xiang, Y., Angelova, M., Guo, L., Chi, C.H.: Multi-scale shapelets discovery for time-series classification. Int. J. Inf. Technol. Decis. Mak **19**(03), 721–739 (2020)
2. Dau, H.A., et al.: The UCR time series classification archive (October 2018). https://www.cs.ucr.edu/~eamonn/time_series_data_2018/
3. Davidson, I., Ravi, S.: Identifying and generating easy sets of constraints for clustering. In: AAAI Conference on Artificial Intelligence (AAAI), pp. 336–341 (2006)
4. Grabocka, J., Schilling, N., Wistuba, M., Schmidt-Thieme, L.: Learning time-series shapelets. In: International Conference on Knowledge Discovery & Data Mining (SIGKDD), pp. 392–401 (2014)
5. Hadsell, R., Chopra, S., LeCun, Y.: Dimensionality reduction by learning an invariant mapping. In: 2006 IEEE Computer Society Conference on Computer Vision and Pattern Recognition (CVPR). vol. 2, pp. 1735–1742 (2006)
6. Hills, J., Lines, J., Baranauskas, E., Mapp, J., Bagnall, A.: Classification of time series by shapelet transformation. Data Min. Knowl. Discov. **28**(4), 851–881 (2014)
7. Keogh, E., Lonardi, S., Ratanamahatana, C.A.: Towards parameter-free data mining. In: International Conference on Knowledge Discovery & Data Mining (SIGKDD), pp. 206–215 (2004)

8. Lampert, T., et al.: Constrained distance based clustering for time-series: a comparative and experimental study. Data Min. Knowl. Discov. **32**(6), 1663–1707 (2018). https://doi.org/10.1007/s10618-018-0573-y
9. Lei, Q., Yi, J., Vaculin, R., Wu, L., Dhillon, I.S.: Similarity preserving representation learning for time series clustering. In: Proceedings of the Twenty-Eighth International Joint Conference on Artificial Intelligence (IJCAI), pp. 2845–2851 (2017)
10. Levenshtein, V.: Binary codes capable of correcting deletions, insertions, and reversals. Sov. Phys. Dokl. **10**(8), 707–710 (1966)
11. Li, Z., Liu, J., Tang, X.: Constrained clustering via spectral regularization. In: 2009 IEEE Conference on Computer Vision and Pattern Recognition (CVPR), pp. 421–428. IEEE (2009)
12. Lines, J., Davis, L.M., Hills, J., Bagnall, A.: A shapelet transform for time series classification. In: Proceedings of the 18th ACM SIGKDD International Conference on Knowledge Discovery and Data Mining (SIGKDD), pp. 289–297 (2012)
13. Lods, A., Malinowski, S., Tavenard, R., Amsaleg, L.: Learning DTW-preserving shapelets. In: International Symposium on Intelligent Data Analysis (IDA) (2017)
14. Mueen, A., Keogh, E., Young, N.: Logical-shapelets: an expressive primitive for time series classification. In: Proceedings of ACM SIGKDD: International Conference on Knowledge Discovery and Data Mining (SIGKDD), pp. 1154–1162 (2011)
15. Paparrizos, J., Franklin, M.J.: GRAIL: efficient time-series representation learning. VLDB Endowment **12**(11), 1762–1777 (2019)
16. Paparrizos, J., Gravano, L.: k-shape: Efficient and accurate clustering of time series. In: Proceedings of the 2015 ACM SIGMOD International Conference on Management of Data (SIGMOD), pp. 1855–1870 (2015)
17. Paparrizos, J., Liu, C., Elmore, A.J., Franklin, M.J.: Debunking four long-standing misconceptions of time-series distance measures. In: Proceedings of the 2020 ACM SIGMOD International Conference on Management of Data (ACM SIGMOD), pp. 1887–1905 (2020)
18. Rakthanmanon, T., Keogh, E.: Fast shapelets: A scalable algorithm for discovering time series shapelets. In: Proceedings of the 2013 SIAM International Conference on Data Mining (SDM), pp. 668–676 (2013)
19. Sakoe, H., Chiba, S.: Dynamic-programming approach to continuous speech recognition. In: Proceedings of the International Cartographic Association ICA, pp. 65–69 (1971)
20. Sakoe, H., Chiba, S.: Dynamic programming algorithm optimization for spoken word recognition. IEEE Tans. Acoust. Speech Signal Process. **26**(1), 43–49 (1978)
21. Shah, M., Grabocka, J., Schilling, N., Wistuba, M., Schmidt-Thieme, L.: Learning DTW-shapelets for time-series classification. In: Proceedings of the 3rd IKDD Conference on Data Science (ACM IKDD CODS), pp. 1–8 (2016)
22. Sperandio, R.C.: Recherche de séries temporelles à l'aide de DTW-preserving shapelets. Ph.D. thesis, Université Rennes 1 (2019)
23. Tiano, D., Bonifati, A., Ng, R.: Feature-driven time series clustering. In: 24th International Conference on Extending Database Technology (EDBT), pp. 349–354 (2021)
24. Ulanova, L., Begum, N., Keogh, E.: Scalable clustering of time series with u-shapelets. In: Proceedings of the 2015 SIAM International Conference on Data Mining (SDM), pp. 900–908 (2015)
25. Vlachos, M., Hadjieleftheriou, M., Gunopulos, D., Keogh, E.: Indexing multidimensional time-series. VLDB J. **15**(1), 1–20 (2006)

26. Wagstaff, K., Basu, S., Davidson, I.: When is constrained clustering beneficial, and why? In: AAAI Conference on Artificial Intelligence (IAAI) (2006)
27. Wagstaff, K., Cardie, C., Rogers, S., Schrödl, S.: Constrained k-means clustering with background knowledge. In: Proceedings of the Eighteenth International Conference on Machine Learning (ICML). vol. 1, pp. 577–584 (2001)
28. Wu, L., Yen, I.E.H., Yi, J., Xu, F., Lei, Q., Witbrock, M.: Random warping series: a random features method for time-series embedding. In: 21st International Conference on Artificial Intelligence and Statistics (AISTATS), pp. 793–802 (2018)
29. Yamaguchi, A., Maya, S., Maruchi, K., Ueno, K.: LTSpAUC: learning time-series shapelets for optimizing partial AUC. In: Proceedings of the 2020 SIAM International Conference on Data Mining (SDM), pp. 1–9 (2020)
30. Ye, L., Keogh, E.: Time series shapelets: a new primitive for data mining. In: Proceedings of the 15th ACM SIGKDD International Conference on Knowledge Discovery and Data Mining (SIGKDD), pp. 947–956 (2009)
31. Ye, L., Keogh, E.: Time series shapelets: a novel technique that allows accurate, interpretable and fast classification. Data Min. Knowl. Discov. **22**(1), 149–182 (2011)
32. Zakaria, J., Mueen, A., Keogh, E.: Clustering time series using unsupervised-shapelets. In: 2012 IEEE 12th International Conference on Data Mining (ICDM), pp. 785–794 (2012)
33. Zakaria, J., Mueen, A., Keogh, E., Young, N.: Accelerating the discovery of unsupervised-shapelets. Data Min. Knowl. Discov. **30**(1), 243–281 (2016)
34. Zhang, Q., Wu, J., Yang, H., Tian, Y., Zhang, C.: Unsupervised feature learning from time series. In: International Joint Conferences on Artificial Intelligence (IJCAI), pp. 2322–2328 (2016)
35. Zheng, G., Yang, Y., Carbonell, J.: Efficient shift-invariant dictionary learning. In: International Conference on Knowledge Discovery & Data Mining ACM SIGKDD, pp. 2095–2104 (2016)

Structured Nonlinear Discriminant Analysis

Christopher Bonenberger[1,2(✉)], Wolfgang Ertel[1], Markus Schneider[1], and Friedhelm Schwenker[2]

[1] Ravensburg-Weingarten University of Applied Sciences (Institute for Artificial Intelligence), Weingarten, Germany
bonenbch@rwu.de

[2] University of Ulm (Institute of Neural Information Processing), James-Franck-Ring 89081 Ulm, Germany

Abstract. Many traditional machine learning and pattern recognition algorithms—as for example linear discriminant analysis (LDA) or principal component analysis (PCA)—optimize data representation with respect to an information theoretic criterion. For time series analysis these traditional techniques are typically insufficient. In this work we propose an extension to linear discriminant analysis that allows to learn a data representation based on an algebraic structure that is tailored for time series. Specifically we propose a generalization of LDA towards shift-invariance that is based on cyclic structures. We expand this framework towards more general structures, that allow to incorporate previous knowledge about the data at hand within the representation learning step. The effectiveness of this proposed approach is demonstrated on synthetic and real-world data sets. Finally, we show the interrelation of our approach to common machine learning and signal processing techniques.

Keywords: Linear discriminant analysis · Time series analysis · Circulant matrices · Representation learning · Algebraic structure

1 Introduction

Often, when being confronted with temporal data, machine learning practitioners use feature transformations. Yet, mostly these feature transformations are not adaptive but rely on decomposition with respect to a fixed basis (Fourier, Wavelet, etc.). This is because simple data-adaptive methods as principal component analysis (PCA, [12]) or linear discriminant analysis (LDA, [22]) often lead to undesirable results for stationary time series [20]. Both methods, PCA and LDA, are based on successive projections onto optimal one-dimensional subspaces. For time series analysis via LDA this leads to problems, especially when the data at hand is not locally coherent in the corresponding vector space [10]—which is likely to be the case for high-dimensional (long) time series. In this paper we propose an adaption of LDA that relies on learning with algebraic structures. More precisely, we propose to learn a projection onto a structured

M.-R. Amini et al. (Eds.): ECML PKDD 2022, LNAI 13713, pp. 38–54, 2023.
https://doi.org/10.1007/978-3-031-26387-3_3

multi-dimensional subspace instead of a single vector. The scope of this work is mainly to introduce the theoretical basics of *structured discriminant analysis* for time series, i.e., we focus on cyclic structures that incorporate shift-invariance and thus regularize supervised representation learning.

From an algebraic point of view the problem at hand is mainly an issue of data representation in terms of bases and frames [23]. In this setting we seek a basis (or a frame [5,15]) for the input space, which is optimal with respect to some information-theoretic criterion. When it comes to time series the vital point is, that these algorithms can be tailored to meet the conditions of temporal data. *Basis pursuit* methods like dictionary learning (DL, optimize over-complete representations with respect to sparsity and reconstruction error) are often altered in order to yield shift-invariance and to model temporal dependencies [8,17,21]. Also convolutional neural networks are implicitly equipped with a mechanism to involve algebraic structure in the learning process, because this way "the architecture itself realizes a form of regularization" [3]. In fact both, shift-invariant DL [8,21] and CNN, use cyclic structures, i.e., convolutions.

However, so far the idea of learning with cyclic structures has hardly been transferred to basic machine learning methods. The motive of this work is to transfer the idea of implicit shift-invariance to LDA by learning with algebraic structure. We strive for interpretable algorithms that go along with low computational complexity. This way we seek to bridge complex methods like convolutional neural networks and simple, well-understood techniques like LDA.

Recently [1] proposed a generalization of PCA that allows unsupervised representation learning with algebraic structure, which is tightly linked to methods like dynamic PCA [13], singular spectrum analysis [9] and spectral density estimation [1]. However, similarly to PCA this method does not allow to incorporate labeling information. Yet, representation learning can benefit from class-information. In this respect our main contribution is a formulation of linear discriminant analysis that involves cyclic structures, thus being optimized for stationary temporal data. We provide a generalization of this framework towards non-stationary time series and even arbitrary correlation structures. Moreover, the proposed technique is linked to classical signal processing methods.

2 Prerequisites

In the following we will briefly discuss the underlying theory of linear discriminant analysis, circulant matrices and linear filtering. In Sect. 2.2 we revisit principal component analysis and its generalization towards shift-invariance.

2.1 Circulant Matrices

We define a circulant matrix as a matrix of the form

$$\mathbf{G} = \begin{bmatrix} g_1 & g_D & g_{D-1} & g_{D-2} & \cdots & g_2 \\ g_2 & g_1 & g_D & g_{D-1} & \cdots & g_3 \\ g_3 & g_2 & g_1 & g_D & \cdots & g_4 \\ \vdots & \vdots & \vdots & \vdots & \ddots & \vdots \\ g_D & g_{D-1} & g_{D-2} & g_{D-3} & \cdots & g_1 \end{bmatrix} \in \mathbb{R}^{D \times D}, \tag{1}$$

i.e., the circulant $\mathbf{G}$ is fully defined by its first column vector $\mathbf{g}$. In short, the i-th row of a circulant matrix contains the first row right-shifted by $i-1$. In the following, we write circulant matrices as a matrix polynomial of the form

$$\mathbf{G} = \sum_{l=0}^{L} g_l \mathbf{P}^{l-1} \tag{2}$$

with

$$\mathbf{P} = \begin{bmatrix} 0 & 0 & 0 & \cdots & 1 \\ 1 & 0 & 0 & & 0 \\ 0 & 1 & 0 & & 0 \\ \vdots & & & \ddots & \vdots \\ 0 & 0 & \cdots & 1 & 0 \end{bmatrix} \in \mathbb{R}^{D \times D}. \tag{3}$$

Note that $\mathbf{P}$ itself is also a circulant matrix.

Left-multiplying a signal $\mathbf{x}$ with a circulant matrix is equivalent to

$$\mathbf{G}\mathbf{x} = \mathbf{F}^{-1}\mathbf{\Lambda}\mathbf{F}\mathbf{x} \tag{4}$$

where $\mathbf{F} \in \mathbb{R}^{D \times D}$ is the Fourier matrix with coefficients

$$[\mathbf{F}]_{j,k} = \frac{1}{\sqrt{D}} \exp\left(-2\pi \mathrm{i} \frac{(j-1)(k-1)}{D}\right) \tag{5}$$

where $\mathbf{\Lambda}$ is a diagonal matrix with the Fourier transform $\hat{\mathbf{g}} = \mathbf{F}\mathbf{g}$ of $\mathbf{g} \in \mathbb{R}^D$ on its diagonal and i is the complex number, i.e., $\mathrm{i}^2 = -1$. Hence, Eq. (4) describes a circular convolution

$$\mathbf{G}\mathbf{x} = \mathbf{F}^{-1}\mathbf{\Lambda}\hat{\mathbf{x}} = \mathbf{F}^{-1}(\hat{\mathbf{g}} \odot \hat{\mathbf{x}}) = \mathbf{g} \circledast \mathbf{x}.$$

Here, $\odot$ is the Hadamard product (pointwise multiplication) and $\circledast$ denotes the discrete circular convolution. Moreover, Eq. (4) describes the diagonalization of circulant matrices by means of the Fourier matrix.

2.2 (Circulant) Principal Component Analysis

Heading towards linear discriminant analysis, it is interesting to start with the Rayleigh quotient and its role in PCA (cf. [19]). The Rayleigh quotient of some vector $\mathbf{g} \in \mathbb{R}^D$ with respect to a symmetric matrix $\mathbf{S} \in \mathbb{R}^{D \times D}$ is defined as

$$\mathcal{R}(\mathbf{g}, \mathbf{S}) = \frac{\mathbf{g}^\mathsf{T}\mathbf{S}\mathbf{g}}{\mathbf{g}^\mathsf{T}\mathbf{g}}.$$

Maximizing $\mathcal{R}(\mathbf{g}, \mathbf{S})$ with respect to $\mathbf{g}$ leads to the eigenvalue problem $\mathbf{S}\mathbf{g} = \lambda\mathbf{g}$. The optimal vector $\mathbf{g}$ is the eigenvector of $\mathbf{S}$ with the largest corresponding eigenvalue.

Having a labeled data set $\{(\mathbf{x}_\nu, y_\nu)\}_{\nu=1,\dots,N}$ with observations $\mathbf{x} \in \mathbb{R}^D$ and corresponding labels $y \in \{1, \dots, C\}$, we define the overall data matrix as

$$\mathbf{X} = \begin{bmatrix} | & & | \\ \mathbf{x}_1 & \cdots & \mathbf{x}_N \\ | & & | \end{bmatrix} \in \mathbb{R}^{D \times N}.$$

Moreover we define class-specific data matrices $\mathbf{X}_c \in \mathbb{R}^{D \times N_c}$, where N_c is the number of observations $\mathbf{x}_\nu$ with corresponding label $y_\nu = c$.

The relation to principal component analysis becomes obvious when $\mathbf{S}$ is the empirical covariance matrix estimated from $\mathbf{X}$. Assuming zero-mean data, i.e., the expected value $\mathbb{E}\{\mathbf{x}\} = 0$, the matrix $\mathbf{S} = \mathbf{X}\mathbf{X}^\mathsf{T}$ is the empirical covariance matrix of $\mathbf{X}$. Thus

$$\max_{\mathbf{g} \in \mathbb{R}^D} \{\mathcal{R}(\mathbf{g}, \mathbf{S})\}$$

is equivalent to the linear constrained optimization problem

$$\max_{\mathbf{g} \in \mathbb{R}^D} \left\{ \left\| \mathbf{g}^\mathsf{T} \mathbf{X} \right\|_2^2 \right\} \text{ s.t. } \|\mathbf{g}\|_2^2 = 1, \tag{6}$$

which in turn formulates principal component analysis, where maximizing $\left\| \mathbf{g}^\mathsf{T} \mathbf{X} \right\|_2^2$ means to maximize variance (respectively power). As known the optimal *principal component vector(s)* are found from the eigenvalue problem (cf. [12])

$$\mathbf{S}\mathbf{g} = \lambda \mathbf{g}.$$

while classical PCA is based on a projection onto an optimal one-dimensional subspace [1] proposed a generalization of PCA which projects on a multi-dimensional subspace that is formed from cyclic permutations. This results in optimizing

$$\max_{\mathbf{g} \in \mathbb{R}^D} \left\{ \|\mathbf{G}\mathbf{X}\|_F^2 \right\} \text{ s.t. } \|\mathbf{g}\|_2^2 = 1, \tag{7}$$

with $\mathbf{G}$ being a κ-circulant matrix defined by the elements of $\mathbf{g}$ (see Section 2.1). The Frobenius norm $\|\mathbf{A}\|_F^2 = \operatorname{tr}\{\mathbf{A}^\mathsf{T}\mathbf{A}\}$. Solving Eq. (7) amounts to set the partial derivatives of the Lagrangian function

$$\mathbf{L}(\mathbf{g}, \lambda) = \operatorname{tr}\{\mathbf{X}^\mathsf{T}\mathbf{G}^\mathsf{T}\mathbf{G}\mathbf{X}\} + \lambda \left(\mathbf{g}^\mathsf{T}\mathbf{g} - 1\right)$$

to zero. Analogously to PCA this finally leads to the eigenvalue problem (see [1])

$$\mathbf{Z}\mathbf{g} = \lambda \mathbf{g}$$

with $[\mathbf{Z}]_{k,l} = \sum_\nu \mathbf{x}_\nu^\mathsf{T} \mathbf{P}^{l-k} \mathbf{x}_\nu$ using $\mathbf{P}$ as defined in Eq. (3), i.e., we write $\mathbf{G}$ as in Eq. (2).

2.3 Linear Discriminant Analysis

Relying on a decomposition of the overall empirical covariance matrix without using the class labels can result in a disadvantageous data representation. Yet, linear discriminant analysis exploits labeling information by maximizing the generalized Rayleigh quotient (cf. [22])

$$\mathcal{R}(\mathbf{g}, \mathbf{B}, \mathbf{W}) = \frac{\mathbf{g}^\mathsf{T}\mathbf{B}\mathbf{g}}{\mathbf{g}^\mathsf{T}\mathbf{W}\mathbf{g}}. \tag{8}$$

with the beneath-class scatter matrix

$$\mathbf{B} = \sum_{c=1}^{C} P_c \left(\overline{\mathbf{x}}_c - \overline{\mathbf{x}}_0\right) \left(\overline{\mathbf{x}}_c - \overline{\mathbf{x}}_0\right)^\mathsf{T} \tag{9}$$

and the within-class scatter matrix

$$\mathbf{W} = \sum_{c=1}^{C} P_c \left(\sum_{\nu \in \mathcal{I}_c} \left(\mathbf{x}_\nu - \overline{\mathbf{x}}_c\right) \left(\mathbf{x}_\nu - \overline{\mathbf{x}}_c\right)^\mathsf{T} \right), \tag{10}$$

where $\mathcal{I}_c$ is the index set for class c, i.e., $\mathcal{I}_c = \{\nu \in [1, \ldots, N] \mid y_\nu = c\}$. Moreover $\overline{\mathbf{x}}_c$ is the sample mean of observations from the class c and $\overline{\mathbf{x}}_0$ is the overall empirical mean value. The a priori class probabilities P_c have to be estimated as $P_c \approx N_c/N$. Note that the beneath-class scatter matrix is the empirical covariance matrix of class-specific sample mean values, while the within-class scatter matrix is a sum of the class-specific covariances. While typically $\operatorname{rank}\{\mathbf{W}\} = D$ the rank of the beneath-class scatter matrix is $\operatorname{rank}\{\mathbf{B}\} \leq C - 1$.

The expression in Eq. (8), also known as Fisher's criterion, measures the separability of classes. Maximizing the Rayleigh quotient in Eq. (8) with respect to $\mathbf{g}$ defines LDA. Hence the optimal one-dimensional subspace of $\mathbb{R}^D$, where the optimality criterion is class separability due to the Rayleigh quotient, is found from

$$\max_{\mathbf{g} \in \mathbb{R}^D} \left\{\mathbf{g}^\mathsf{T}\mathbf{B}\mathbf{g}\right\} \text{ s.t. } \mathbf{g}^\mathsf{T}\mathbf{W}\mathbf{g} = 1. \tag{11}$$

Again this constrained linear optimization problem is solved by setting the partial derivatives of the corresponding Lagrangian $\mathbf{L}(\mathbf{g}, \lambda)$ to zero. Analogously to PCA we find a (generalized) eigenvalue problem

$$\frac{\partial \mathbf{L}(\mathbf{g}, \lambda)}{\partial \mathbf{g}} = 0 \iff \mathbf{B}\mathbf{g} = \lambda \mathbf{W}\mathbf{g}. \tag{12}$$

Assuming that $\mathbf{W}^{-1}$ exists, then

$$\mathbf{W}^{-1}\mathbf{B}\mathbf{g} = \lambda \mathbf{g}. \tag{13}$$

The projection $\mathbf{X}^\perp \in \mathbb{R}^{C-1 \times N}$ of $\mathbf{X} \in \mathbb{R}^{D \times N}$ onto the optimal subspace defined by the eigenvectors $\mathbf{g}_1, \ldots, \mathbf{g}_{C-1}$ of $\mathbf{W}^{-1}\mathbf{B}$ belonging to the $C-1$ largest eigenvalues is

$$\mathbf{X}^{\perp} = \begin{bmatrix} - & \mathbf{g}_1^{\mathsf{T}} & - \\ & \vdots & \\ - & \mathbf{g}_{C-1}^{\mathsf{T}} & - \end{bmatrix} \begin{bmatrix} | & & | \\ \mathbf{x}_1 & \cdots & \mathbf{x}_N \\ | & & | \end{bmatrix},$$

i.e., $\mathbf{X}^{\perp}$ is the mapping of $\mathbf{X}$ into a feature space that is optimal with respect to class discrimination.

3 Structured Discriminant Analysis

This section presents the main contribution of the paper, namely we introduce a generalization of linear discriminant analysis that allows for learning with algebraic structure. Instead of the projection onto a one-dimensional subspace we propose to learn the coefficients of a multi-dimensional structured subspace that is optimal with respect to class discrimination.

3.1 Circulant Discriminant Analysis

As introduced in Sect. 2.2 [1] proposes to modify Eq. (6) using κ-circulant matrices, which generalizes PCA towards shift invariance. In the following, we will adopt this approach and modify linear discriminant analysis as defined by Eq. (11) using circulant structures, i.e., we seek the coefficients of a circulant matrix $\mathbf{G} \in \mathbb{R}^{D \times D}$ of the form

$$\mathbf{G} = \sum_{l=1}^{L} g_l \mathbf{P}^{l-1} \tag{14}$$

instead of $\mathbf{g}$. Again $\mathbf{P}$ performs a cyclic permutation, as defined in Eq. (3). An example is depicted in Fig. 1, panel (1).

In this regard we use $\tilde{\mathbf{x}}_\nu = \mathbf{x}_\nu - \overline{\mathbf{x}}_c$ (class affiliation of $\mathbf{x}_\nu$ is unambiguous) and $\tilde{\overline{\mathbf{x}}}_c = \overline{\mathbf{x}}_c - \overline{\mathbf{x}}_0$ as an abbreviation, i.e.,

$$\mathbf{B} = \sum_{c=1}^{C} P_c \tilde{\overline{\mathbf{x}}}_c \tilde{\overline{\mathbf{x}}}_c^{\mathsf{T}}.$$

and

$$\mathbf{W} = \sum_{c=1}^{C} P_c \sum_{\nu \in \mathcal{I}_c} \tilde{\mathbf{x}}_\nu \tilde{\mathbf{x}}_\nu^{\mathsf{T}}.$$

The coefficients $\mathbf{g} \in \mathbb{R}^L$ of $\mathbf{G}$ that go along with optimal class separation are found from the linear constrained optimization problem

$$\max_{\mathbf{g} \in \mathbb{R}^D} \left\{ \sum_{c=1}^{C} P_c \left\| \mathbf{G} \tilde{\overline{\mathbf{x}}}_c \right\|_2^2 \right\} \text{ s.t. } \sum_{c=1}^{C} P_c \sum_{\nu \in \mathcal{I}_c} \| \mathbf{G} \tilde{\mathbf{x}}_\nu \|_2^2 = 1, \tag{15}$$

which basically means to perform LDA on the data set $\mathbf{GX}$.

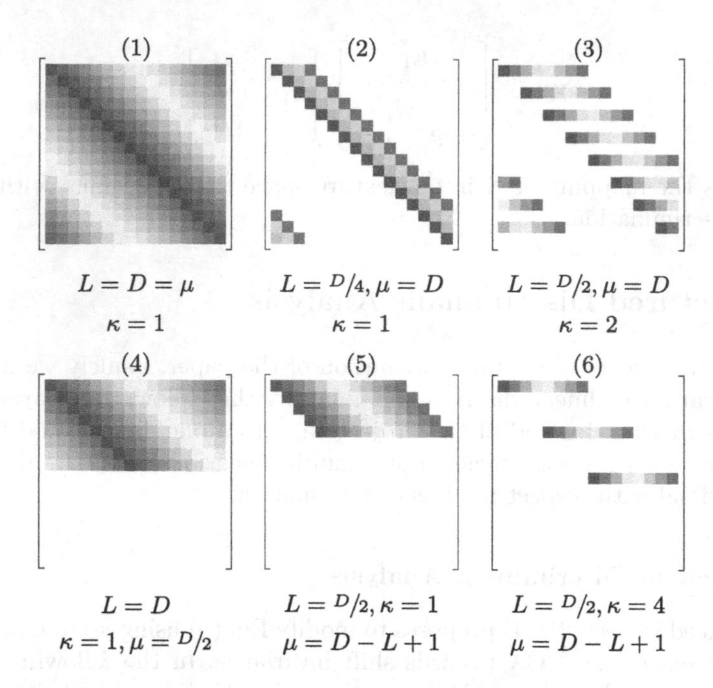

Fig. 1. Different examples on possible structures of $\mathbf{G}$ with $\mathbf{g}$ being a straight line from -1 to 1. These structures illustrate the dependencies associated to different parameter settings, e.g. in panel (1) each coordinate is related to each other while in (4) dependencies are restricted to $\pm D/4$.

In a geometrical understanding we seek a multi-dimensional cyclically structured subspace of $\mathbb{R}^D$ that is optimal with respect to class separability, while in classical LDA the sought subspace is one-dimensional. This implies that instead of $\|\mathbf{g}^\mathsf{T}\mathbf{x}\|_2^2$ (variance) we measure the length $\|\mathbf{G}\mathbf{x}\|_2^2$ of the projection onto the subspace defined by $\mathbf{G}$ (*total variation*[1] with respect to the variable under consideration, $\tilde{\bar{\mathbf{x}}}_c$ or $\tilde{\mathbf{x}}_\nu$). However, according to Eq. (4) $\|\mathbf{G}\mathbf{x}\|_2^2$ can also be understood as the power of the filtered signal $\mathbf{G}^\mathsf{T}\mathbf{x}$, while $\mathbf{G}$ is an optimally matched filter. In a two-class setting $\mathbf{G}$ can even be understood as a Wiener filter [24].

The Lagrangian for Eq. (15) is

$$\mathbf{L}(\mathbf{g}, \lambda) = \sum_{c=1}^{C} P_c \tilde{\bar{\mathbf{x}}}_c^\mathsf{T} \mathbf{G}^\mathsf{T} \mathbf{G} \tilde{\bar{\mathbf{x}}}_c - \lambda \sum_{c=1}^{C} P_c \sum_{\nu \in \mathcal{I}_c} \tilde{\mathbf{x}}_\nu^\mathsf{T} \mathbf{G}^\mathsf{T} \mathbf{G} \tilde{\mathbf{x}}_\nu - \lambda. \tag{16}$$

Due to $(\mathbf{P}^i)^\mathsf{T}\mathbf{P}^j = \mathbf{P}^{j-i}\ \forall i, j \in \mathbb{N}$ with $\mathbf{P}$ according to Eq. (14) we find

$$\mathbf{x}^\mathsf{T}\mathbf{G}^\mathsf{T}\mathbf{G}\mathbf{x} = \mathbf{x}^\mathsf{T}(g_1^2\mathbf{P}^0 + \cdots + g_1 g_L \mathbf{P}^{L-1} + \cdots + g_L g_1 \mathbf{P}^{-L+1} + \cdots + g_L^2 \mathbf{P}^0)\mathbf{x}$$

[1] The total variation is the trace of the covariance matrix (cf. [18]).

for any vector $\mathbf{x} \in \mathbb{R}^D$. The derivative with respect to g_k is

$$\frac{\mathrm{d}\mathbf{x}^\mathsf{T}\mathbf{G}^\mathsf{T}\mathbf{G}\mathbf{x}}{\mathrm{d}g_k} = \mathbf{x}^\mathsf{T}\sum_{l=1}^{L} g_l \left(\mathbf{P}^{l-k} + \mathbf{P}^{k-l}\right)\mathbf{x} = 2\mathbf{x}^\mathsf{T}\sum_{l=1}^{L} g_l \mathbf{P}^{l-k}\mathbf{x}. \tag{17}$$

The second equality in Eq. (17) is using the symmetry of real inner products and $(\mathbf{P}^i)^\mathsf{T} = \mathbf{P}^{-i}$, which leads to $\mathbf{x}^\mathsf{T}\mathbf{P}^{-i}\mathbf{x} = (\mathbf{P}^i\mathbf{x})^\mathsf{T}\mathbf{x} = \mathbf{x}^\mathsf{T}\mathbf{P}^i\mathbf{x}$. Using Eq. (17) the partial derivative of Eq. (16) w.r.t. g_k can be written as

$$\frac{\partial \mathbf{L}(\mathbf{g},\lambda)}{\partial g_k} = 2\sum_{c=1}^{C} P_c \overline{\mathbf{x}}_c^\mathsf{T} \textstyle\sum_l g_l \mathbf{P}^{l-k}\overline{\mathbf{x}}_c - 2\lambda \displaystyle\sum_{c=1}^{C} P_c \sum_{\nu\in\mathcal{I}_c} \tilde{\mathbf{x}}_\nu^\mathsf{T} \textstyle\sum_l g_l \mathbf{P}^{l-k}\tilde{\mathbf{x}}_\nu. \tag{18}$$

Setting Eq. (18) to zero leads to the generalized eigenvalue problem

$$\mathbf{Z}_B\mathbf{g} = \lambda \mathbf{Z}_W \mathbf{g}, \tag{19}$$

where

$$[\mathbf{Z}_B]_{k,l} = \sum_{c=1}^{C} P_c \tilde{\overline{\mathbf{x}}}_c^\mathsf{T} \mathbf{P}^{l-k} \tilde{\overline{\mathbf{x}}}_c \tag{20}$$

and

$$[\mathbf{Z}_W]_{k,l} = \sum_{c=1}^{C} P_c \sum_{\nu\in\mathcal{I}_c} \tilde{\mathbf{x}}_\nu^\mathsf{T} \mathbf{P}^{l-k} \tilde{\mathbf{x}}_\nu. \tag{21}$$

Analogously to classical LDA we find the eigenvalue problem

$$\mathbf{Z}_W^{-1}\mathbf{Z}_B\mathbf{g} = \lambda\mathbf{g}. \tag{22}$$

In contrast to LDA, for non-trivial $\tilde{\overline{\mathbf{x}}}_c$ rank$\{\mathbf{Z}_B\}$ = rank$\{\mathbf{Z}_W\}$ = L, which is due to the permutations in Eqs. (20) and (21). Every eigenvector $\mathbf{g}_q$ defines a circulant matrix $\mathbf{G}_q$ and a corresponding subspace. In accordance with Eq.(15) the length of the projection onto this subspace is

$$x^\perp = \|\mathbf{G}_q\mathbf{x}\|_2^2. \tag{23}$$

Note that using the nonlinear projection in Eq. (23) yields a nonlinear algorithm. Of course this is not a necessity, since it would be viable to proceed with the linear representation $\mathbf{G}_q\mathbf{x}$. However, Eq. (23) fits in with linear discriminant analysis and is easily interpretable in terms of linear filtering.

3.2 Computational Aspects for Circulant Structures

Since both matrices $\mathbf{Z}_B, \mathbf{Z}_W \in \mathbb{R}^{L\times L}$ have a symmetric Toeplitz structure[2] the computational complexity of Eqs. 20 and (21) can be reduced to $\mathcal{O}(L)$ as both matrices are fully determined by their first row vectors $\mathbf{z}_B, \mathbf{z}_W \in \mathbb{R}^L$ respectively.

[2] $[\mathbf{Z}]_{i,j}$ is constant for constant $i - j$ and $\mathbf{x}^\mathsf{T}\mathbf{P}^{j-i}\mathbf{x} = \mathbf{x}^\mathsf{T}\mathbf{P}^{i-j}\mathbf{x}$ (cf. [1]).

Beyond that, the term $\mathbf{x}^\top \mathbf{P}^{l-k}\mathbf{y}$ realizes a circular convolution $\mathbf{x} \circledast \mathbf{y}$. A circular convolution in turn can be expressed by means of the (fast) Fourier transform (FFT, cf. [24]), i.e., $\mathbf{x} \circledast \mathbf{y} = \mathbf{F}^{-1}(\mathbf{F}(\mathbf{x}) \odot \mathbf{F}(\mathbf{y}))$, where $\mathbf{F}$ denotes the Fourier transform and $\mathbf{F}^{-1}$ its inverse. This allows to compute $\mathbf{z}_B$ and $\mathbf{z}_W$ in $\mathcal{O}(D \log D)$ using the fast Fourier transform, i.e.,

$$[\mathbf{z}_B]_l = \left[\sum_{c=1}^{C} P_c \mathbf{F}^{-1}\left(\mathbf{F}\tilde{\bar{\mathbf{x}}}_c \odot \mathbf{F}\tilde{\bar{\mathbf{x}}}_c\right)\right]_l, \quad l = 1\ldots, L \tag{24}$$

and

$$[\mathbf{z}_W]_l = \left[\sum_{c=1}^{C} P_c \mathbf{F}^{-1} \sum_{\nu \in \mathcal{I}_c} \mathbf{F}\tilde{\mathbf{x}}_\nu \odot \mathbf{F}\tilde{\mathbf{x}}_\nu\right]_l, \quad l = 1\ldots, L. \tag{25}$$

Using these insights the projection according to Eq. (23) can be accelerated via

$$\mathbf{x}^\perp = \left\|\mathbf{F}^{-1}\left(\mathbf{F}\mathbf{g}_q \odot \mathbf{F}\mathbf{x}\right)\right\|_2^2, \tag{26}$$

where $\mathbf{g}_q$ has to be zero-padded such that $\mathbf{g}_q \in \mathbb{R}^D$.

Beneath the low complexity of estimating $\mathbf{Z}_B$ and $\mathbf{Z}_W$ via the FFT, there is a considerable reduction of computational complexity in solving Eq. (22) because L can be chosen much smaller than D. In fact, $L \ll D$ is typically a reasonable choice, because for large L the localization in frequency domain is inappropriately precise (cf. Figs. 4 and 5 and Sect. 3.3).

3.3 Harmonic Solutions

As can be seen from Figs. 4 and 5 for circulant structures with $L = D$ the optimal solution to Eq. (15) is Fourier mode. Investigating Eqs.(21) and (20) for $L = D$ we can see that both, $\mathbf{Z}_W$ and $\mathbf{Z}_B$ are circulant matrices for $L = D$. Generally both matrices have coefficients of the form

$$[\mathbf{Z}]_{k,l} = \mathbf{x}\mathbf{P}^{l-k}\mathbf{x},$$

with some $\mathbf{x} \in \mathbb{R}^D$. Hence, when $L = D$ the first row of $\mathbf{Z}$ is palindromic, i.e., $\mathbf{Z}_{k,l} = \mathbf{Z}_{k,D-l}$ because $\mathbf{P}^{-l} = \mathbf{P}^{D-l}$. Thus $\mathbf{Z}$, respectively $\mathbf{Z}_W$ and $\mathbf{Z}_B$ are symmetric circulant Toeplitz matrices and both admit an eigendecomposition according to Eq. (4), i.e.,

$$\mathbf{Z} = \begin{bmatrix} z_{1,1} & z_{1,2} & z_{1,3} & \cdots & z_{1,3} & z_{1,2} \\ z_{1,2} & z_{1,1} & z_{1,2} & \cdots & z_{1,4} & z_{1,3} \\ z_{1,3} & z_{1,2} & z_{1,1} & \cdots & z_{1,5} & z_{1,4} \\ \vdots & & & & & \vdots \\ z_{1,2} & z_{1,3} & z_{1,4} & \cdots & z_{1,2} & z_{1,1} \end{bmatrix} = \mathbf{F}^{-1}\mathbf{\Lambda}\mathbf{F} \in \mathbb{R}^{D \times D}.$$

Notably, the inverse of a circulant (Toeplitz) matrix is again a circulant matrix [14]. Thus we can conclude that for $L = D$ we have

$$\mathbf{Z}_W^{-1}\mathbf{Z}_B\mathbf{F} = \mathbf{F}\mathbf{\Lambda},$$

with $\mathbf{F}$ being the Fourier matrix (cf. Eq. (5)). We observe that independently of the data at hand the optimal solutions to Eq. (22) respectively Eq. (15) are Fourier modes, i.e., for stationary data Fourier modes maximize the Rayleigh quotient.

3.4 Truncated κ-Circulants

Although circulant structures are beneficial in terms of computational complexity their use is tied to the assumption of *stationarity*[3]. In the following we slightly change the definition of $\mathbf{G}$ to a more general "cyclic" matrix $\mathbf{\Gamma}$ in order to gain more flexibility when incorporating dependencies into the structure of $\mathbf{\Gamma}$. We refer to a cyclic matrix, when Eq. (4) is not full-filled, i.e., the matrix is based on cyclic permutations, but is not strictly circulant. [17] describes κ-circulants as down-sampled versions of simple circulant. This approach can be generalized to truncated κ-circulant matrices

$$\mathbf{\Gamma} = \mathbf{M} \sum_{l=1}^{L} g_l \mathbf{P}^{l-1} \tag{27}$$

with $\mathbf{M}$ performing the down-sampling (with a factor κ) and truncation of all rows following the μ-th row, i.e.,

$$[\mathbf{M}]_{i,j} = \begin{cases} 1 & \text{if } \mu \geq i = j \in [1, \kappa+1, 2\kappa+1, \cdots, \lfloor D/\kappa + 1 \rfloor \kappa] \\ 0 & \text{else.} \end{cases}$$

This especially allows to model dependencies for non-stationary data (see Sect. (4.2)). The idea of truncation is important, as it allows a simple handling of the boundaries by setting $\mu = D - L + 1$ (as known from singular spectrum analysis [2,9]). On the other hand using some $\mu > 1$ along with $L = D$ is the LDA-equivalent to dynamic PCA (cf. [1,2,13]) Setting $\kappa > 1$ implements down-sampling and is equivalent to *stride* in CNNs. Some examples are given in Fig. 1.[4] Using $\mathbf{\Gamma}$ instead of $\mathbf{G}$ leads to

$$[\mathbf{Z}_B]_{k,l} = \sum_{c=1}^{C} P_c \tilde{\bar{\mathbf{x}}}_c^\mathsf{T} \mathbf{P}^{1-l} \mathbf{M} \mathbf{P}^{k-1} \tilde{\bar{\mathbf{x}}}_c \tag{28}$$

and

$$[\mathbf{Z}_W]_{k,l} = \sum_{c=1}^{C} P_c \sum_{\nu \in \mathcal{I}_c} \tilde{\mathbf{x}}_\nu^\mathsf{T} \mathbf{P}^{1-l} \mathbf{M} \mathbf{P}^{k-1} \tilde{\mathbf{x}}_\nu . \tag{29}$$

All derivations are analogous to Sect. 3.1, except for the binary diagonal matrix $\mathbf{M}$ with $\mathbf{M}^\mathsf{T}\mathbf{M} = \mathbf{M}$.[5] Note that with $\mu = D$ and $\kappa = 1$ Eqs. (28) and (29)

---

[3] The distribution of stationary signals is invariant with respect to time ($\mathbb{E}\{x_t\}$ is constant for all t and the covariance $C(x_t, x_s)$ solely depends on the index/time difference $|t - s|$) [16].

[4] κ-circulant structures can also be used to model Wavelet-like structures [24].

[5] Hence we can fully simplify analogously to the step from Eq. (17) to Eq. (18).

are equal to Eqs. (20) and (21) (circulant discriminant analysis). Moreover, for $L = D$ and $\mu = 1$ (or $\kappa = D$) Eqs. (28) and (29) coincide with $\mathbf{B}$ and $\mathbf{W}$ (Eqs. (9) and (10)). In that sense, circulant discriminant analysis and classical LDA are a special case of truncated κ-circulant structures.

3.5 Non-cyclic Structures

Using circulant structures as proposed in Sect. 2.1 is an adequate approach for (weakly) stationary data sets. The generalization to truncated κ-circulant matrices allows to embed more complex dependencies into the structure of $\mathbf{G}$ (respectively $\mathbf{\Gamma}$) and hence allows to work with non-stationary data. Yet, when choosing the correlation structure, of course, one is not limited to cyclic structures. As the truncated κ-circulant structure in Eq. (27) can be given as $\mathbf{\Gamma} = \sum_l g_l \mathbf{M}\mathbf{P}^{l-1}$, clearly we can formulate Eq. (15) using an arbitrary structure $\mathbf{\Gamma}_A \in \mathbb{R}^{D \times D}$ which is modeled as

$$\mathbf{\Gamma}_A = \sum_{l=1}^{L} g_l \mathbf{\Pi}_l.$$

Here, the coefficients of $\mathbf{\Pi}_l$ model the dependencies of the i-th variable.

The corresponding solution is equivalent to the above derivations, i.e., we find the generalized eigenvalue problem of Eq. (19). However, the matrices $\mathbf{Z}_B$ and $\mathbf{Z}_W$ are defined as

$$[\mathbf{Z}_B]_{k,l} = \sum_{c=1}^{C} P_c \tilde{\bar{\mathbf{x}}}_c^{\mathsf{T}} \mathbf{\Pi}_l^{\mathsf{T}} \mathbf{\Pi}_k \tilde{\bar{\mathbf{x}}}_c$$

and

$$[\mathbf{Z}_W]_{k,l} = \sum_{c=1}^{C} P_c \sum_{\nu \in \mathcal{I}_c} \tilde{\mathbf{x}}_\nu^{\mathsf{T}} \mathbf{\Pi}_l^{\mathsf{T}} \mathbf{\Pi}_k \tilde{\mathbf{x}}_\nu.$$

This very general formulation also allows for more complex structures that can explicitly model statistical dependencies for non-temporal data. In the field of time series analysis this general approach can be used to build over-complete multi-scale models.

4 Examples and Interpretation

In this section we illustrate the proposed method at the example of different real-world and synthetic data sets.

4.1 (Quasi-)Stationary Data

As a start we use synthetic data generated from different auto-regressive moving average models (ARMA model, cf. [16]) corresponding to the different classes. In the left panel of Fig. 2 realizations from these four different models are depicted.

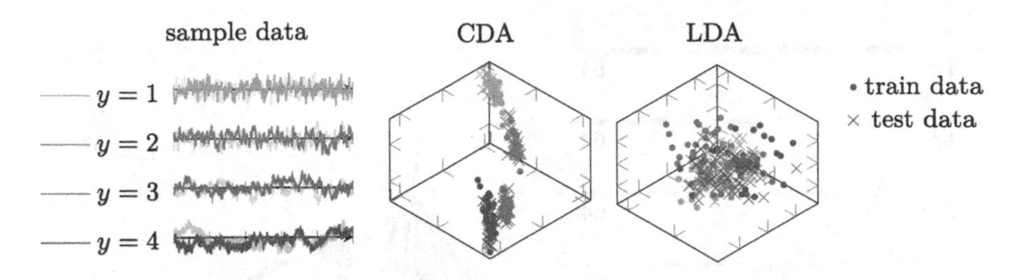

Fig. 2. A simple example demonstrating the performance of circulant discriminant analysis compared to classical linear discriminant analysis according to Sect. 4.1.

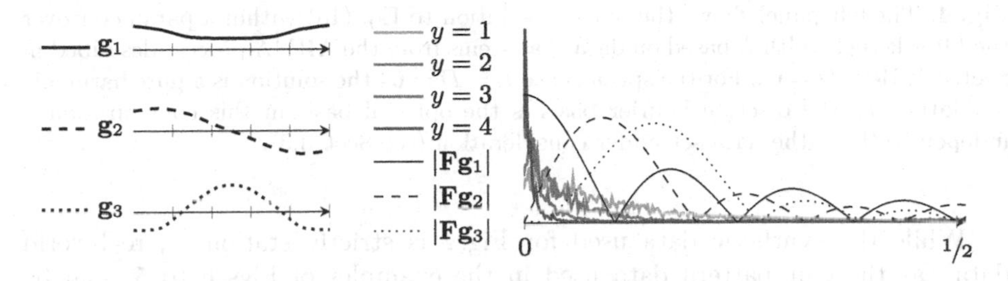

Fig. 3. This figure shows the first three eigenvectors of $\mathbf{Z}_W^{-1}\mathbf{Z}_B$ for the ARMA-process data set (cf. Section 4.1 and Fig. 2) and their spectrum. The right panel shows the spectral density of the four largest classes (colored) along with Fourier transformed (filter) coefficients. The x axis in the right panel is the frequency axis in half cycles per sample. (Color figure online)

For the sake of simplicity all model parameters are chosen depending on the class index, i.e., observations belonging to class c stem from a ARMA(p, q) model with $p = q = c$ and coefficients $\theta_i = \phi_i = 1/(c+1)$ for all $i = 1, \ldots, c$, where θ_i and ϕ_i are AR and MA coefficients respectively. Each class comprises $N_c = 50$ samples, with a 50/50 train-test split. The data dimension is $D = 256$. In the middle and right panel of Fig. 2 a comparison of circulant discriminant analysis (CDA, according to Sect. 3.1) and linear discriminant analysis based on this data is shown. For CDA we used $L = 8$. For both methods the projection onto the first three subspaces is used. More precisely, for CDA the projection is according to Eq. (26). Note that CDA is considerably faster than LDA, due to the computational simplifications proposed in Sect. 3.2.

In Fig. 6 the "user identification from walking activity" data set (cf. [4]) from the UCR machine learning repository ([7]) is used. The data set contains accelerometer data from 22 different individuals, each walking the same predefined path. For each class, there are x, y and z measurements of the accelerometer forming three time series. For further use, we use sub-series of equal length D from a single variable (acceleration in x-direction).

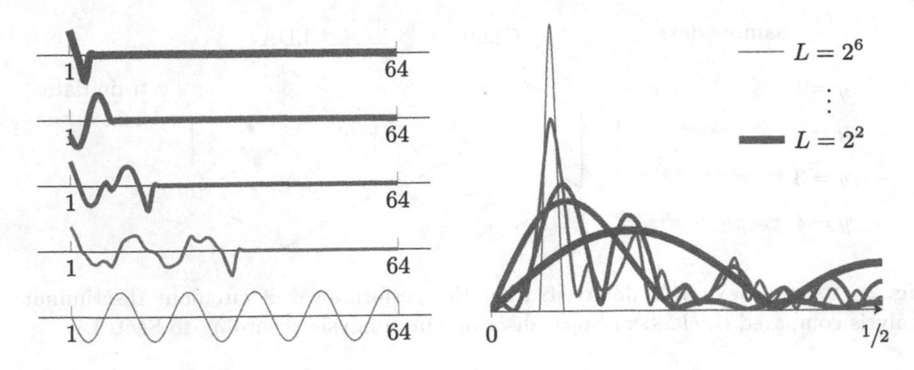

Fig. 4. The left panel shows the optimal solution to Eq. (15) within a parameter over the filter kernel width L based on data that stems from the ARMA process described in Sect. 4.1. Here $D = 64$. For the special case $L = D = 64$ the solution is a pure harmonic oscillation, as the discrete Fourier basis is the optimal basis in this configuration—independently of the data set under consideration (see Sect. 3.3).

While the synthetic data used for Fig. 2 is strictly stationary, real-world data—as the gait pattern data used in the examples of Figs. 6 to 5—can be assumed to be stationary on (small) intervals [11]. For the visualization in Fig. 6 we used a 50/50 train-test-split based on observations of length $D = 64$ which corresponds to approximately 2 seconds window width. For the sake of simplicity we used only one variable (the x-coordinate). The accuracies in Fig. 6 are based on a feature vector with 3 elements, i.e., classification is performed on the depicted data. The overall 1-nearest neighbor accuracy on the complete data set with 22 classes on a single variable (x-coordinate) is 46% (CDA) and 20% (LDA) respectively.

4.2 Non-stationary Data

In Sect. 3.4 we introduced κ-circulant structures, that account for non-stationary data. Here we demonstrate the use of such structures using the "Plane" data set (cf. [6]).

Often for time series the assumption of stationarity does not hold. In one example, the data at hand is triggered, i.e., all observations start at a certain point in time (space, ...). This results in non-stationarity, because distinct patterns are likely to be found at a certain index. The "Plane" data set from the UCR Time Series Archive (see [6]) is such a triggered data set. It contains seven different classes that encode the outline of different planes as a function of angle. The triggering stems from the fact, that the outline is captured using the identical starting angle. Hence, the "Plane" data set is an example for non-stationary data, that nevertheless shows temporal (spatial) correlations.

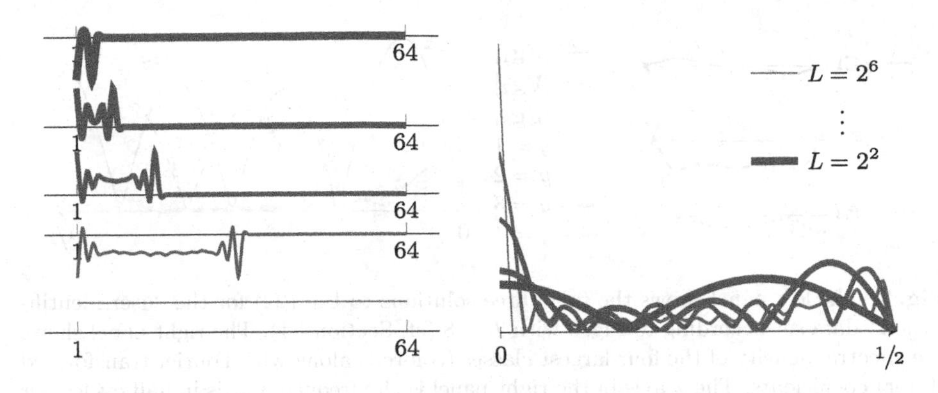

Fig. 5. Illustration of a parameter sweep over L according to Fig. 4. However, here the underlying dataset is the "user identification from walking activity" data set (cf. Figs. 6 and 7). Again for $L = D = 64$ we find a Fourier mode as optimal solution (cf. Section 3.3).

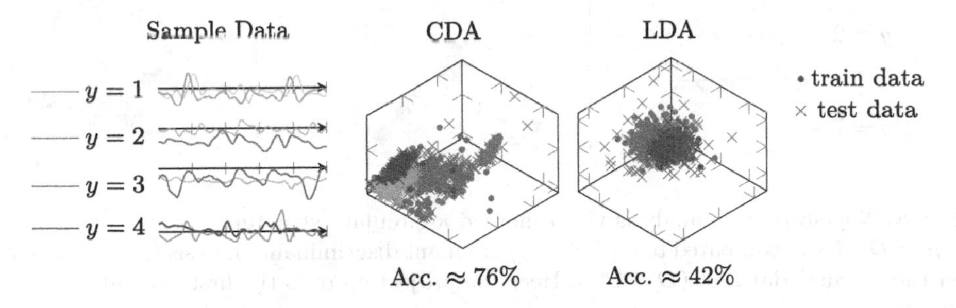

Fig. 6. Comparison of LDA and CDA at the example of the according to Sect. 4.1. For this figure the four largest classes of the data set are depicted.

Figure 8 shows a comparison of stationary and non-stationary parameter settings. The difference between these settings is shown in Fig. 1. A structure for stationary data is visualized in panel (2), while the non-stationary setting is shown in panel (5) of Fig. 1. The former is equivalent to a FIR-filter with data-adaptive coefficients, while the latter is similar to singular spectrum analysis. A detailed analysis of interrelations between theses techniques is provided in [2].

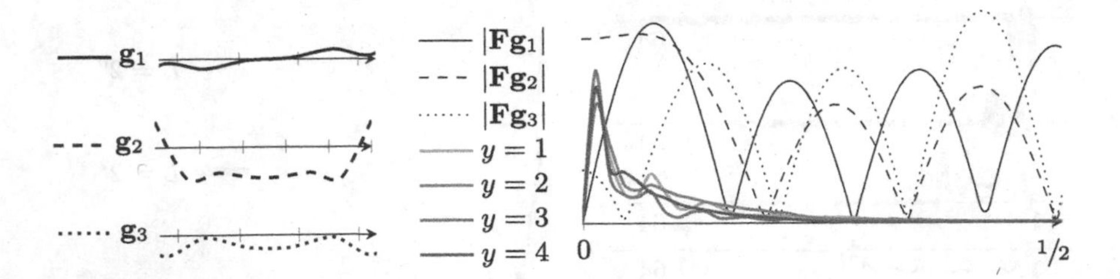

Fig. 7. The left panel shows the first three solutions to Eq. (22) for the "user identification" data set according to Fig. 6 with $L = 8$ (cf. Section 4.1). The right panel shows the spectral density of the four largest classes (colored) along with Fourier transformed (filter) coefficients. The x axis in the right panel is the frequency axis in half cycles per sample. (Color figure online)

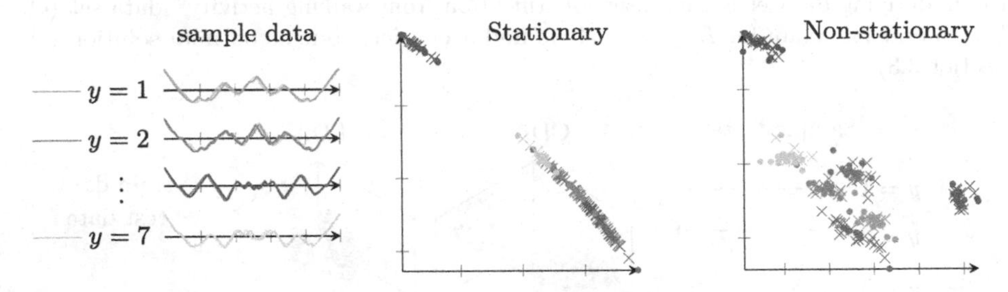

Fig. 8. Non-stationary analysis via truncated κ-circulant structures (using $L = 5, \kappa = 1, \mu = D-L+1$) compared to (stationary) circulant discriminant analysis ($L = 5$) based on the "Plane" data set ($D = 144$). Here the projection onto the first two subspace is shown.

5 Conclusion

Linear discriminant analysis is a core technique in machine learning and statistics. In this work we introduced an adaption of linear discriminant analysis that is optimized for stationary time series. This approach is based on the idea of projecting data onto cyclically structured subspaces, which is related to adaptive linear filtering. We generalize this approach towards non-stationary data and show how arbitrary correlation structures can be modeled. This reconnects to classical LDA, which is a special case of circulant discriminant analysis with truncated κ-circulants. The effectiveness of this approach is demonstrated on synthetic stationary data and temporal data from benchmark data sets. Finally, we discussed the connection between circulant discriminant analysis and linear filtering as well as Fourier analysis.

Acknowledgements. We are grateful for the careful review of the manuscript. We thank the reviewers for corrections and helpful comments. Additionally we would like to thank the maintainers of the UCI Machine Learning Repository and the UCR Time Series Archive for providing benchmark data sets.

References

1. Bonenberger, C., Ertel, W., Schneider, M.: κ-circulant maximum variance bases. In: Edelkamp, S., Möller, R., Rueckert, E. (eds.) KI 2021. LNCS (LNAI), vol. 12873, pp. 17–29. Springer, Cham (2021). https://doi.org/10.1007/978-3-030-87626-5_2
2. Bonenberger, C., Ertel, W., Schwenker, F., Schneider, M.: Singular spectrum analysis and circulant maximum variance frames. In: Advances in Data Science and Adaptive Analysis (2022)
3. Bouvrie, J.: Notes on convolutional neural networks (2006)
4. Casale, P., Pujol, O., Radeva, P.: Personalization and user verification in wearable systems using biometric walking patterns. Pers. Ubiquit. Comput. **16**(5), 563–580 (2012)
5. Christensen, O.: An introduction to frames and Riesz bases. ANHA, Springer, Cham (2016). https://doi.org/10.1007/978-3-319-25613-9
6. Dau, H.A., et al.: The UCR time series archive. IEEE/CAA J. Autom. Sinica **6**(6), 1293–1305 (2019)
7. Dua, D., Graff, C.: UCI machine learning repository (2017). http://archive.ics.uci.edu/ml
8. Garcia-Cardona, C., Wohlberg, B.: Convolutional dictionary learning: a comparative review and new algorithms. IEEE Trans. Comput. Imaging **4**(3), 366–381 (2018)
9. Golyandina, N., Zhigljavsky, A.: Singular spectrum analysis for time series. Springer Science & Business Media (2013)
10. Hastie, T., Tibshirani, R., Friedman, J.: The Elements of Statistical Learning. SSS, Springer, New York (2009). https://doi.org/10.1007/978-0-387-84858-7
11. Hoffmann, R., Wolff, M.: Intelligente Signalverarbeitung 1. Springer, Heidelberg (2014). https://doi.org/10.1007/978-3-662-45323-0
12. Jolliffe, I.T.: Principal component analysis, vol. 2. Springer (2002). https://doi.org/10.1007/b98835
13. Ku, W., Storer, R.H., Georgakis, C.: Disturbance detection and isolation by dynamic principal component analysis. Chemom. Intell. Lab. Syst. **30**(1), 179–196 (1995)
14. Lv, X.G., Huang, T.Z.: A note on inversion of toeplitz matrices. Appl. Math. Lett. **20**(12), 1189–1193 (2007)
15. Morgenshtern, V.I., Bölcskei, H.: A short course on frame theory. arXiv preprint arXiv:1104.4300 (2011)
16. Pollock, D.S.G., Green, R.C., Nguyen, T.: Handbook of time series analysis, signal processing, and dynamics. Elsevier (1999)
17. Rusu, C., Dumitrescu, B., Tsaftaris, S.A.: Explicit shift-invariant dictionary learning. IEEE Signal Process. Lett. **21**(1), 6–9 (2013)
18. Seber, G.A.: Multivariate observations. John Wiley & Sons (2009)
19. Serpedin, E., Chen, T., Rajan, D.: Mathematical foundations for signal processing, communications, and networking. CRC Press (2011)
20. Shumway, R.: Discriminant analysis for time series. Handbook Statist. **2**, 1–46 (1982)

21. Sulam, J., Papyan, V., Romano, Y., Elad, M.: Multilayer convolutional sparse modeling: pursuit and dictionary learning. IEEE Trans. Signal Process. **66**(15), 4090–4104 (2018)
22. Theodoridis, S., Koutroumbas, K.: Pattern recognition. Elsevier (2006)
23. Tosic, I., Frossard, P.: Dictionary learning: what is the right representation for my signal? IEEE Sig. Process. Mag. **28**, 27–38 (2011)
24. Vetterli, M., Kovačević, J., Goyal, V.K.: Foundations of signal processing. Cambridge University Press (2014)

LSCALE: Latent Space Clustering-Based Active Learning for Node Classification

Juncheng Liu(✉), Yiwei Wang, Bryan Hooi, Renchi Yang, and Xiaokui Xiao

School of Computing, National University of Singapore, Singapore, Singapore
{juncheng,y-wang,bhooi}@comp.nus.edu.sg, {renchi,xkxiao}@nus.edu.sg

Abstract. Node classification on graphs is an important task in many practical domains. It usually requires labels for training, which can be difficult or expensive to obtain in practice. Given a budget for labelling, active learning aims to improve performance by carefully choosing which nodes to label. Previous graph active learning methods learn representations using labelled nodes and select some unlabelled nodes for label acquisition. However, they do not fully utilize the representation power present in unlabelled nodes. We argue that the representation power in unlabelled nodes can be useful for active learning and for further improving performance of active learning for node classification. In this paper, we propose a latent space clustering-based active learning framework for node classification (LSCALE), where we fully utilize the representation power in both labelled and unlabelled nodes. Specifically, to select nodes for labelling, our framework uses the K-Medoids clustering algorithm on a latent space based on a dynamic combination of both unsupervised features and supervised features. In addition, we design an incremental clustering module to avoid redundancy between nodes selected at different steps. Extensive experiments on five datasets show that our proposed framework LSCALE consistently and significantly outperforms the state-of-the-art approaches by a large margin.

1 Introduction

Node classification on graphs has attracted much attention in the graph representation learning area. Numerous graph learning methods [6,11,15,27] have been proposed for node classification with impressive performance, especially on the semi-supervised setting, where labels are required for the classification task.

In reality, labels are often difficult and expensive to collect. To mitigate this issue, active learning aims to select the most informative data points which can lead to better classification performance using the same amount of labelled data. Graph neural networks (GNNs) have been used for some applications such as disease prediction and drug discovery [10,21], in which labels often have to be obtained through costly means such as chemical assays. Thus, these applications motivate research into active learning with GNNs.

Supplementary Information The online version contains supplementary material available at https://doi.org/10.1007/978-3-031-26387-3_4.

M.-R. Amini et al. (Eds.): ECML PKDD 2022, LNAI 13713, pp. 55–70, 2023.
https://doi.org/10.1007/978-3-031-26387-3_4

In this work, we focus on active learning for node classification on attributed graphs. Recently, a few GNN-based active learning methods [5,9,13,22,28] have been proposed for attributed graphs. However, their performance is still less than satisfactory in terms of node classification. These approaches do not fully utilize the useful representation power in unlabelled nodes and only use unlabelled nodes for label acquisition. For example, AGE [5] and ANRMAB [9] select corresponding informative nodes to label based on the hidden representations of graph convolutional networks (GCNs) and graph structures. These hidden representations can be updated only based on the labelled data. On the other hand, FeatProp [28] is a clustering-based algorithm which uses propagated node attributes to select nodes to label. However, these propagated node attributes are generated in a fixed manner based on the graph structure and node attributes, and are not learnable. In summary, existing approaches do not fully utilize the information present in unlabelled nodes. To utilize the information in unlabelled nodes, a straightforward method is to use features extracted from a trained unsupervised model for choosing which nodes to select. For example, FeatProp can conduct clustering based on unsupervised features for selecting nodes. However, as shown in our experimental results, it still cannot effectively utilize the information in unlabelled nodes for active learning.

Motivated by the limitations above, we propose an effective **L**atent **S**pace **C**lustering-based **A**ctive **LE**arning framework (hereafter **LSCALE**). In this framework, we conduct clustering-based active learning on a designed latent space for node classification. Our desired *active learning latent space* should have two key properties: 1) *low label requirements:* it should utilize the representation power from all nodes, not just labelled nodes, thereby obtaining accurate representations even when very few labelled nodes are available; 2) *informative distances:* in this latent space, intra-class nodes should be closer together, while inter-class nodes should be further apart. This can facilitate clustering-based active selection approaches, which rely on these distances to output a diverse set of query points.

To achieve these, our approach incorporates an unsupervised model (e.g., DGI [26]) on all nodes to generate unsupervised features, which utilizes the information in unlabelled nodes, satisfying the first desired property. In addition, we design a distance-based classifier to classify nodes using the representations from our latent space. This ensures that distances in our latent space are informative for active learning selection, satisfying our second desired property. To select nodes for querying labels, we leverage the K-Medoids clustering algorithm in our latent space to obtain cluster centers, which are the queried nodes. As more labelled data are received, the distances between different nodes in the latent space change based on a dynamic combination of unsupervised learning features and learnable supervised representations.

Furthermore, we propose an effective incremental clustering strategy for clustering-based active learning to prevent redundancy during node selection. Existing clustering-based active learning methods like [24,28] only select nodes in multiple rounds with a myopic approach. More specifically, in each round, they apply clustering over all unlabelled nodes and select center nodes for labelling.

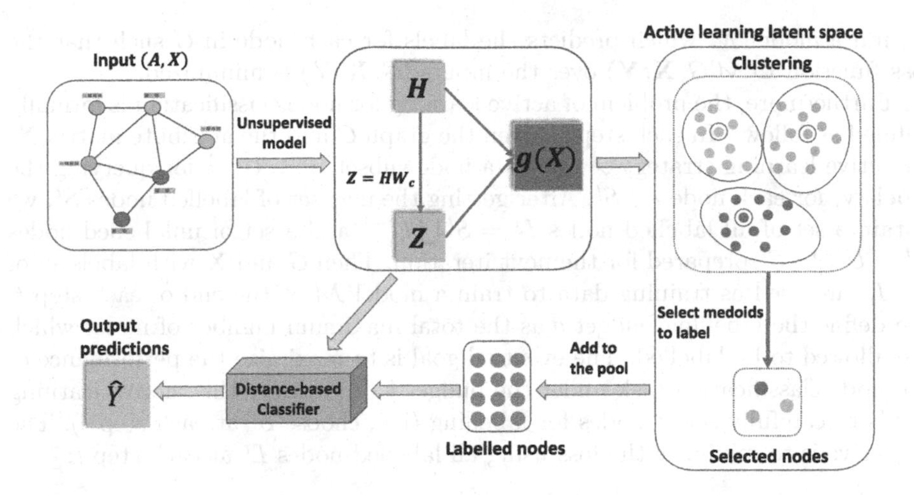

Fig. 1. Illustration of the proposed framework LSCALE.

However, the cluster centers tend to be near the ones obtained in the previous rounds. Therefore, the clustering can select redundant nodes and does not provide much new information in the later rounds. In contrast, our incremental clustering is designed to be aware of the selected nodes in the previous rounds and ensure newly selected nodes are more informative. Our contributions are summarized as follows:

- We propose a latent space clustering-based active learning framework (LSCALE) for node classification on attributed graphs. LSCALE contains a latent space with two key properties designed for active learning purposes: 1) *low label requirements*, 2) *informative distances*.
- We design an incremental clustering strategy to ensure that newly selected nodes are not redundant with previous nodes, which further improves the performance.
- We conduct comprehensive experiments on three public citation datasets and two co-authorship datasets. The results show that our method provides a consistent and significant performance improvement compared to the state-of-the-art active learning methods for node classification on attributed graphs.

2 Problem Definition

In this section, we present the formal problem definition of active learning for node classification. Let $G = (V, E)$ be a graph with node set V and edge set E, where $|V| = n$ and $|E| = m$. $\mathbf{X} \in \mathbb{R}^{n \times d}$ and $\mathbf{Y}$ represent the input node attribute matrix and label matrix of graph G, respectively. In particular, each node $v \in V$ is associated with a length-d attribute vector $\mathbf{x}_v$ and a one-hot label vector $\mathbf{y}_v$. Given a graph G and its associated attribute matrix $\mathbf{X}$, node classification aims to find a model $\mathcal{M}$ which predicts the labels for each node in G such that the loss function $\mathcal{L}(\mathcal{M}|G, \mathbf{X}, \mathbf{Y})$ over the inputs $(G, \mathbf{X}, \mathbf{Y})$ is minimized.

Furthermore, the problem of active learning for node classification is formally defined as follows. In each step t, given the graph G and the attribute matrix $\mathbf{X}$, an active learning strategy $\mathcal{A}$ selects a node subset $S^t \subseteq U^{t-1}$ for querying the labels $\mathbf{y}_i$ for each node $i \in S^t$. After getting the new set of labelled nodes S^t, we obtain a set of all labelled nodes $L^t = S^t \bigcup L^{t-1}$ and a set of unlabelled nodes $U^t = U^{t-1} \setminus S^t$ prepared for the next iteration. Then G and $\mathbf{X}$ with labels $\mathbf{y}_i$ of $i \in L^t$ are used as training data to train a model $\mathcal{M}$ at the end of each step t. We define the labelling budget b as the total maximum number of nodes which are allowed to be labelled. The eventual goal is to maximize the performance of the node classification task under the budget b. To achieve this, active learning needs to carefully select nodes for labelling (i.e., choose S^t at each step t). The objective is to minimize the loss using all labelled nodes L^t at each step t:

$$\min_{L^t} \mathcal{L}(\mathcal{M}, \mathcal{A}|G, \mathbf{X}, \mathbf{Y}) \tag{1}$$

3 Methodology

In this section, we introduce our active learning framework LSCALE for node classification in a top-down fashion. First, we describe the overview and the key idea of LSCALE. Then we provide the details of each module used in LSCALE.

The overview of our latent space clustering-based active learning framework is shown in Fig. 1. The most important aspect of our framework is to design a suitable *active learning latent space*, specifically designed for clustering-based active learning. Motivated by the limitations of previous methods, we design a latent space with two important properties: 1) *low label requirements:* the latent space representations can be learned effectively even with very few labels, by utilizing the representation power from all nodes (including unlabelled nodes) rather than only labelled nodes; 2) *informative distances:* in the latent space, distances between intra-class nodes should be smaller than distances between inter-class nodes. With the first property, LSCALE can learn effective node representations throughout the active learning process, even when very few labelled nodes have been acquired. The second property makes distances in our latent space informative with respect to active learning selection, ensuring that clustering-based active selection processes choose a diverse set of query points.

To satisfy the first property, we use an unsupervised learning method to learn unsupervised node representations $\mathbf{H} \in \mathbb{R}^{n \times d'}$ based on graphs and node attributes, where d' is the dimension of representations. After obtaining $\mathbf{H}$, we design a linear distance-based classifier to generate output predictions. In the classifier, we apply a learnable linear transformation on $\mathbf{H}$ to obtain hidden representations $\mathbf{Z}$. The distances between nodes are calculated by a dynamic combination of both $\mathbf{Z}$ and $\mathbf{H}$, which satisfies the second desired property. Clustering is performed on the latent space using the distances to select informative nodes. We additionally propose an incremental clustering method to ensure that the newly selected nodes are not redundant with the previously selected nodes. In summary, the framework contains a few main components:

- An unsupervised graph learning method to generate unsupervised node representations.
- An active learning latent space with two aforementioned properties: 1) *low label requirements*, 2) *informative distances.*
- An incremental clustering method to select data points as centroids, which we use as the nodes to be labelled, and prevent redundancy during node selection.

3.1 Active Learning Latent Space

To facilitate clustering-based active learning, we need a latent space with two desired properties. Therefore, we propose a distance-based classifier for generating representations from supervised signals and a distance function to dynamically consider supervised and unsupervised representations simultaneously.

Distance-Based Classifier. In our framework, we design a novel distance-based classifier to ensure that distances between nodes in our latent space can facilitate active learning further. Intuitively, a desired property of the latent space is that nodes from different classes should be more separated and nodes with the same class should be more concentrated in the latent space. Thus, it can help clustering-based active learning methods select representative nodes from different classes. To achieve this, we first map unsupervised features $\mathbf{H}$ to another set of features $\mathbf{Z}$ by a linear transformation:

$$\mathbf{Z} = \mathbf{H}\mathbf{W}_c, \tag{2}$$

where $\mathbf{W}_c \in \mathbb{R}^{d' \times l'}$ is the trainable linear transformation matrix, l' is the dimension of latent representations. Then we define a set of learnable class representations $\mathbf{c}_1, \mathbf{c}_2, ..., \mathbf{c}_K$, where K is the number of classes. The distance vector of node i is defined as:

$$\mathbf{a}_i = ||\mathbf{z}_i - \mathbf{c}_1||_2 \oplus ||\mathbf{z}_i - \mathbf{c}_2||_2 ... \oplus ||\mathbf{z}_i - \mathbf{c}_K||_2 \in \mathbb{R}^K, \tag{3}$$

where $\oplus$ is the concatenation operation and $|| \cdot ||_2$ is the L_2 norm. The j-th element $\hat{y}_{ij}$ in the output prediction $\hat{\mathbf{y}}_i$ of node i is obtained by the softmax function:

$$\hat{y}_{ij} = \mathsf{softmax}(\mathbf{a}_i)_j = \frac{\exp(||\mathbf{z}_i - c_j||_2)}{\sum_{k=0}^{K} \exp(||\mathbf{z}_i - c_k||_2)} \tag{4}$$

For training the classifier, suppose the labelled node set at step t is L^t. The cross-entropy loss function for node classification over the labeled node set is defined as:

$$\mathcal{L} = -\frac{1}{|L^t|} \sum_{i \in L^t} \sum_{c=1}^{K} y_{ic} \ln \hat{y}_{ic}, \tag{5}$$

where y_{ic} denotes the c-th element in the label vector $\mathbf{y}_i$.

With the guidance of labelled nodes and their labels via backpropagation, we can update the transformation matrix $\mathbf{W}_c$ and new features $\mathbf{Z}$ can capture the supervised information from labelled data. In addition, new features $\mathbf{Z}$ allow intra-class nodes more close and inter-class nodes more separate in the feature space. Through this distance-based classifier, the generated feature space allows the clustering-based active selection effectively select a diverse set of query nodes.

Distance Function. In LSCALE, the distance function determines the distances between nodes in the latent space for further clustering. We define our distance function as:

$$d(v_i, v_j) = ||g(\mathbf{X})_i - g(\mathbf{X})_j||_2, \tag{6}$$

where $g(\mathbf{X})$ is a mapping from node attributes $\mathbf{X}$ to new distance features. As previous graph active learning methods do not effectively utilize the unlabelled nodes, we aim to take advantage of unsupervised learning features and supervised information from labelled data. To this end, we combine unsupervised learning features $\mathbf{H}$ and supervised hidden representations $\mathbf{Z}$ in the distance function. A straightforward way to combine them is using concatenation of $\mathbf{H}$ and $\mathbf{Z}$: $g(\mathbf{X}) = \mathbf{H} \oplus \mathbf{Z}$. Noted that $\mathbf{H}$ and $\mathbf{Z}$ are in different spaces and may have different magnitudes of row vectors. We define the distance features as follows:

$$g(\mathbf{X}) = \alpha \cdot \mathbf{H}' \oplus (1 - \alpha) \cdot \mathbf{Z}', \tag{7}$$

where $\mathbf{H}'$ and $\mathbf{Z}'$ are $l2$-normalized $\mathbf{H}$ and $\mathbf{Z}$ respectively to make sure they have same Euclidean norms of rows. α can be treated as a parameter for controlling the dynamic combination of unsupervised features and supervised features.

Intuitively, $\mathbf{Z}$ can be unstable in the early stages as there are relatively few labelled nodes in the training set. So, in the early stages, we would like to focus more on unsupervised features $\mathbf{H}$, which are much more stable than $\mathbf{Z}$. As the number of labelled nodes increases, the focus should be shifted to hidden representations $\mathbf{Z}$ in order to emphasize supervised information. Inspired by curriculum learning approaches [2], we set an exponentially decaying weight as follows:

$$\alpha = \lambda^{|L^t|}, \tag{8}$$

where $|L^t|$ is the number of labelled nodes at step t. λ can be set as a number close to 1.0, e.g., 0.99. By using this dynamic combination of unsupervised learning features $\mathbf{H}$ and supervised hidden representations $\mathbf{Z}$, we eventually construct the latent space $g(\mathbf{X})$ which has the two important properties: 1) *low label requirements:* it utilizes the representation power from all nodes including unlabelled nodes; 2) *informative distances:* distances between nodes are informative for node selection. Thus, the latent space can facilitate selecting diverse and representative nodes in the clustering module.

Note that FeatProp [28] uses propagated node attributes as representations for calculating distances. The propagated node attributes are fixed and not learnable throughout the whole active learning process, which makes the node selection less effective. In contrast, our latent space is learned based on signals from both labelled and unlabelled data. In addition, it gradually shifts its focus to emphasize supervised signals as we acquire more labelled data.

3.2 Clustering Module

At each step, we use the K-Medoids clustering on our latent space to obtain cluster representatives. In K-Medoids, medoids are chosen from among the data points themselves, ensuring that they are valid points to select during active learning. So, after clustering, we directly select these medoids for labelling. This ensures that the chosen centers are well spread out and provide good coverage of the remaining data, which matches the intuition of active learning, since we want the chosen centers to help us classify as much as possible of the rest of the data. At each step t, the objective of K-Medoids is:

$$\sum_{i=1}^{n} \min_{j \in S^t} d(v_i, v_j) = \sum_{i=1}^{n} \min_{j \in S^t} ||g(\mathbf{X})_i - g(\mathbf{X})_j||_2 \quad (9)$$

Besides K-Medoids, common clustering methods used in the previous work are K-Means [5,9] and K-Centers [24]. K-Means cannot be directly used for selecting nodes in active learning as it does not return real sample nodes as cluster representatives.

Incremental Clustering. Despite these advantages of K-Medoids for active learning on graphs, a crucial drawback is that it is possible to select similar nodes for querying during multiple iterations. That is, newly selected nodes may be close to previously selected ones, making them redundant and hence worsening the performance of active learning. The reason is that the clustering algorithm only generates the representative nodes in the whole representation space without the awareness of previously selected nodes. To overcome this problem, we design an effective incremental clustering algorithm for K-Medoids to avoid selecting redundant nodes.

In our incremental clustering method, the key idea is that fixing previous selected nodes as some medoids can force the K-Medoids algorithm to select additional medoids that are dissimilar with the previous ones. We illustrate our incremental clustering method in Algorithm 1.

After calculating the distances for every node pair (Line 4), incremental K-Medoids is conducted (Line 5 to Line 15). Compared to the original K-Medoids, the most important modification is that only clusters with a medoid, which is not in the previous labelled nodes set (i.e., $m \notin L^{t-1}$), can update the medoid (Line 10-13). When all the medoids are the same as those in the previous iteration, the K-Medoids algorithm stops and keeps the medoids. For the medoids which are not the previous selected nodes, we put them in selected node set S^t, meanwhile we set labelled node set L^t and unlabelled node set U^t using S^t accordingly.

4 Experiments

The main goal of our experiments is to verify the effectiveness of our proposed framework LSCALE[1] We design experiments to answer the following research questions:

[1] The code can be found https://github.com/liu-jc/LSCALE.

Algorithm 1: Incremental K-Medoids clustering

Input: the set of previous labelled nodes L^{t-1},
the set of unlabelled nodes U^{t-1} as the pool,
the budget b^t of the current step.

1 $k \leftarrow |L^{t-1}| + b^t$;
2 Randomly select b^t nodes from U^{t-1};
3 Set selected b^t nodes and nodes in L^{t-1} as k initial medoids;
4 Compute $d(v_i, v_j)$ for every node pair (v_i, v_j) by Eq. (6);
5 **repeat**
6   **foreach** *node* $u \in U^{t-1}$ **do**
7   Assign u to the cluster with the closest medoid;
8   **end**
9   **foreach** *cluster* C *with medoid* m **do**
10   **if** $m \notin L^{t-1}$ **then**
11   Find the node m' which minimize the sum of distances to all other nodes within C;
12   Update node m' as the medoid of C;
13   **end**
14   **end**
15 **until** *all the medoids are not changed*;
16 Construct selected node set S^t using the medoids $m \notin L^{t-1}$;
17 $L^t \leftarrow S^t \bigcup L^{t-1}$; $U^t \leftarrow U^{t-1} \setminus S^t$;
18 **return** L^t, U^t, S^t

- **RQ1. Overall performance and effectiveness of unsupervised features**: How does LSCALE perform as compared with state-of-the-art graph active learning methods? Is utilizing unsupervised features also helpful for other clustering-based graph active learning methods?
- **RQ2. Efficiency:** How efficient is LSCALE as compared with other methods?
- **RQ3. Ablation study:** Are the designed dynamic feature combination and incremental clustering useful to improve the performance? How does our distance-based classifier affect the performance?

Datasets. To evaluate the effectiveness of LSCALE, we conduct the experiments on Cora, Citeseer [23], Pubmed [20], Coauthor-CS (short as Co-CS) and Coauthor-Physics (short as Co-Phy) [25]. The first three are citation networks while Co-CS and Co-Phy are two co-authorship networks. We describe the datasets in detail and summarize the dataset statistics in Supplement B.1.

Baselines. In the experiments, to show the compatibility with different unsupervised learning methods, we use two variants LSCALE-DGI and LSCALE-MVGRL, which use DGI [26] and MVGRL [12] as the unsupervised learning method, respectively. To demonstrate the effectiveness of LSCALE, we compare

Table 1. The averaged accuracies (%) and standard deviations at different budgets on citation networks.

Dataset	Cora			Citeseer			Pubmed		
Budget	10	30	60	10	30	60	10	30	60
Random	47.65 ± 7.2	65.19 ± 4.6	73.33 ± 3.1	37.76 ± 9.7	57.73 ± 7.1	66.38 ± 4.4	63.60 ± 6.8	74.17 ± 3.9	77.93 ± 2.4
Uncertainty	45.78 ± 4.6	56.34 ± 8.4	70.22 ± 6.0	27.65 ± 8.8	45.04 ± 8.2	59.41 ± 9.2	60.72 ± 5.7	69.64 ± 4.2	74.95 ± 4.2
AGE	41.22 ± 9.3	65.09 ± 2.7	73.63 ± 1.6	31.76 ± 3.3	60.22 ± 9.3	64.77 ± 9.1	66.96 ± 6.7	75.82 ± 4.0	80.27 ± 1.0
ANRMAB	30.43 ± 8.2	61.11 ± 8.8	71.92 ± 2.3	25.66 ± 6.6	47.56 ± 9.4	58.28 ± 9.2	57.85 ± 8.7	65.33 ± 9.6	75.01 ± 8.4
FeatProp	51.78 ± 6.7	66.49 ± 4.7	74.70 ± 2.7	39.63 ± 9.2	57.92 ± 7.2	66.95 ± 4.2	67.33 ± 5.5	75.08 ± 3.2	77.60 ± 1.9
GEEM	45.73 ± 9.8	67.21 ± 8.7	76.51 ± 1.6	41.10 ± 7.2	62.96 ± 7.8	70.82 ± 1.2	64.38 ± 6.7	76.12 ± 1.9	79.10 ± 2.3
DGI-Rand	62.55 ± 5.8	73.04 ± 3.8	78.36 ± 2.6	54.46 ± 7.6	67.26 ± 4.0	70.24 ± 2.4	73.17 ± 3.8	78.10 ± 2.8	80.28 ± 1.6
FeatProp-D	68.94 ± 5.7	75.47 ± 2.9	77.64 ± 2.0	61.84 ± 5.9	66.99 ± 3.6	68.97 ± 2.0	73.50 ± 4.7	77.36 ± 3.4	78.54 ± 2.3
LSCALE-D	70.83 ± 4.8	77.41 ± 3.5	80.77 ± 1.7	$\mathbf{65.60 \pm 4.7}$	$\mathbf{69.06 \pm 2.6}$	$\mathbf{70.91 \pm 2.2}$	$\mathbf{74.28 \pm 4.4}$	78.54 ± 2.8	80.62 ± 1.7
LSCALE-M	$\mathbf{72.71 \pm 3.9}$	$\mathbf{78.67 \pm 2.7}$	$\mathbf{82.03 \pm 1.8}$	64.24 ± 4.8	68.68 ± 3.2	70.34 ± 1.9	73.51 ± 4.9	$\mathbf{79.09 \pm 2.3}$	$\mathbf{81.32 \pm 1.7}$

two variants with the following representative active learning methods on graphs. **Random**: select the nodes uniformly from the unlabelled node pool; **Uncertainty**: select the nodes with the max information entropy according to the current model. **AGE** [5] constructs three different criteria based on graph neural networks to choose a query node. Combining these different criteria with time-sensitive variables to decide which nodes to selected for labelling. **ANRMAB** [9] proposes a multi-arm-bandit mechanism to assign different weights to the different criteria when constructing the score to select a query node. **FeatProp** [28] performs the K-Medoids clustering on the propagated features obtained by simplified GCN [27] and selects the medoids to query their labels. **GEEM** [22]: inspired by error reduction, it uses simplified GCN [27] to select the nodes by minimizing the expected error.

As suggested in [5,9,28], AGE, ANRMAB, FeatProp, Random, and Uncertainty use GCNs as the prediction model, which is trained after receiving labelled nodes at each step. GEEM uses the simplified graph convolution (SGC) [27] as the prediction model as mentioned in [22].

4.1 Experimental Setting

We evaluate LSCALE-DGI, LSCALE-MVGRL, and other baselines on node classification task with a transductive learning setup, following the experimental setup as in [9,28] for a fair comparison.

Dataset Splits. For each citation dataset, we use the same testing set as in [15], which contains 1000 nodes. For coauthor datasets, we randomly sample 20% nodes as the testing sets. From the non-testing set in each dataset, we randomly sample 500 nodes as a validation set and fix it for all the methods to ensure a fair comparison.

Experiment Procedure. In the experiments, we set the budget sizes differently for different datasets and we focus on the "batched" multi-step setting as in [24,28]. Each active learning method is provided a small set of labelled nodes as an initial pool. As in [28], we randomly select 5 nodes regardless of the

Table 2. The averaged accuracies (%) and standard deviations at different budgets on co-authorship networks.

Dataset	Co-Phy			Co-CS		
Budget	10	30	60	10	30	60
Random	74.80	86.48	90.70 ± 2.6	49.72	69.98	78.15 ± 3.6
Uncertainty	71.42	86.64	91.29 ± 2.0	42.38	57.43	65.66 ± 9.5
AGE	63.96	84.47	91.30 ± 2.0	27.20	70.22	76.52 ± 3.6
ANRMAB	68.47	84.19	89.35 ± 4.2	43.48	69.98	75.51 ± 2.4
FeatProp	80.23	86.83	90.82 ± 2.6	52.45	70.83	76.60 ± 3.9
GEEM	79.24	88.58	91.56 ± 0.5	61.63	75.03	82.57 ± 1.9
DGI-Rand	82.81	90.35	92.44 ± 1.5	64.07	78.63	84.28 ± 2.7
FeatProp-D	87.90	91.23	91.51 ± 1.6	67.37	77.65	80.33 ± 2.5
LSCALE-D	**90.38**	92.75	**93.70 ± 0.6**	**73.07**	**82.96**	**86.70 ± 1.7**
LSCALE-M	90.28	**92.89**	93.05 ± 0.7	71.16	81.82	85.79 ± 1.6

class as an initial pool. The whole active learning process is as follows: (1) we first train the prediction model with initial labelled nodes. (2) we use the active learning strategy to select new nodes for labelling and add them to the labelled node pool; (3) we train the model based on the labelled nodes again. We repeat Step (2) and Step (3) until the budget is reached and train the model based on the final labelled node pool. For clustering-based methods (i.e., FeatProp and LSCALE), 10 nodes are selected for labelling in each iteration as these methods depend on selecting medoids to label.

Hyperparameter Settings. For hyperparameters of other baselines, we set them as suggested in their papers. We specify hyperparameters of our methods in Supplement B.2.

4.2 Performance Comparison (RQ1)

Overall Comparison. We evaluate the performance by using the averaged classification accuracy. We report the results over 20 runs with 10 different random data splits. In Tables 1 and 2, we show accuracy scores of different methods when the number of labelled nodes is less than 60. Analysing Table 1 and 2, we have the following observations:

- In general, our methods LSCALE-DGI (short as LSCALE-D) and LSCALE-MVGRL (short as LSCALE-M) significantly outperform the baselines on the varying datasets, while they provide relatively lower standard deviations on most datasets. In particular, when the total budget is only 10, LSCALE-M provides remarkable improvements compared with GEEM by absolute values 26.9%, 23.3%, 11.5%, on Cora, Citeseer, and Co-CS respectively.
- With the budget size less than 30, the Uncertainty baseline always performs worse than the Random baseline for all datasets. Meanwhile, AGE and

ANRMAB do not have much higher accuracies on most datasets compared with the Random baseline. Both of the above results indicate that GCN representations, which are used in AGE, ANRMAB and the Uncertainty baseline for selecting nodes, are inadequate when having only a few labelled nodes.
- GEEM generally outperforms other baselines on all datasets, which might be attributed to its expected error minimization scheme. However, the important drawback of expected error minimization is the inefficiency, which we show later in Sec 4.3.

Supplement B.3 shows more results about how accuracy scores of different methods change as the number of labelled nodes increases. Supplement B.4 demonstrates how different hyperparameters affect the performance.

Effectiveness of Utilizing Unsupervised Features. Existing works overlook the information in unlabelled nodes, whereas LSCALE utilizes unsupervised features by using unsupervised learning on all nodes (including unlabelled ones). As we argue before, the information in unlabelled nodes is useful for active learning on graphs. To verify the usefulness, we design two additional baselines as follows:

- **FeatProp-DGI**: It replaces the propagated features with unsupervised DGI features in FeatProp to select nodes for labelling.
- **DGI-Rand**: It uses unsupervised DGI features and randomly selects nodes from the unlabelled node pool to label. For simplicity, it trains a simple logistic regression model with DGI features as the prediction model.

Regarding the effectiveness of unsupervised features, from Table 1 and 2, we have the following observations:

- On all datasets, FeatProp-DGI (short as FeatProp-D) consistently outperforms FeatProp, which indicates unsupervised features are useful for other clustering-based graph active learning approaches besides our framework.
- DGI-Rand also achieves better performance compared with AGE, ANRMAB, and GEEM, especially when the labelling budget is small (e.g., 10). This verifies again that existing approaches do not fully utilize the representation power in unlabelled nodes.
- DGI-Rand outperforms FeatProp-D when the labelling budget increases to 60. This observation shows that FeatProp-D cannot effectively select informative nodes in the late stage, which can be caused by redundant nodes selected in the late stage.
- While DGI-Rand and FeatProp-D use the representation power in unlabelled nodes, they are still consistently outperformed by LSCALE-D and LSCALE-M, which verifies the superiority of our framework.

4.3 Efficiency Comparison (RQ2)

We empirically compare the efficiency of LSCALE-D with that of four state-of-the-art methods (i.e., AGE, ANRMAB, FeatProp, and GEEM). Table 3 shows

Table 3. The total running time of different models.

Method	Cora	Citeseer	Pubmed	Co-CS	Co-Phy
AGE	208.7 s	244.1 s	2672.8 s	6390.5 s	745.5 s
ANRMAB	201.8 s	231.5 s	2723.3 s	6423.5 s	767.1 s
FeatProp	16.5 s	16.7 s	58.7 s	169.2 s	336.4 s
GEEM	3.1 h	5.2 h	1.8 h	52.5 h	46.2 h
LSCALE-D	13.1 s	15.6 s	53.4 s	59.8 s	131.3 s

the total running time of these models on different datasets. From Table 3, GEEM has worst efficiency as it trains the simplied GCN model $n \times K$ times (K is the number of classes) for selecting a single node. FeatProp and LSCALE-D are much faster than the other methods. The reason is that FeatProp and LSCALE-D both select several nodes in a step and train the classifier once for this step, whereas AGE, ANRMAB and GEEM all select a single node once in a step. Comparing LSCALE-D and FeatProp, LSCALE-D requires less time as the clustering in LSCALE-D is performed in the latent space where the dimension is less than that in the original attribute space used in FeatProp.

4.4 Ablation Study (RQ3)

Effectiveness of Dynamic Feature Combination and Incremental Clustering. We conduct an ablation study to evaluate the contributions of two different components in our framework: dynamic feature combination and incremental clustering. The results are shown in Fig. 3. DGI features is the variant without either dynamic combination or incremental clustering, and it only uses features obtained by DGI as distance features for the K-Medoids clustering algorithm. Dynamic_Comb uses the dynamic combination to obtain distance features for clustering. LSCALE is the full version of our variant with dynamic feature combination and incremental clustering. It is worth noting that DGI features can be considered as a simple method utilizing unsupervised features. Analysing Fig. 3, we have the following observations:

- Dynamic_Comb generally provides better performance than DGI features, which shows the effectiveness of our dynamic feature combination for distance features.
- LSCALE and Dynamic_Comb provide no much different performance when the number of labelled nodes is relatively low. However, LSCALE gradually outperforms Dynamic_Comb as the number of labelled nodes increases. This confirms that incremental clustering can select more informative nodes by avoiding redundancy between nodes selected at different steps.

In summary, the results verify the effectiveness and necessity of dynamic combination and incremental clustering.

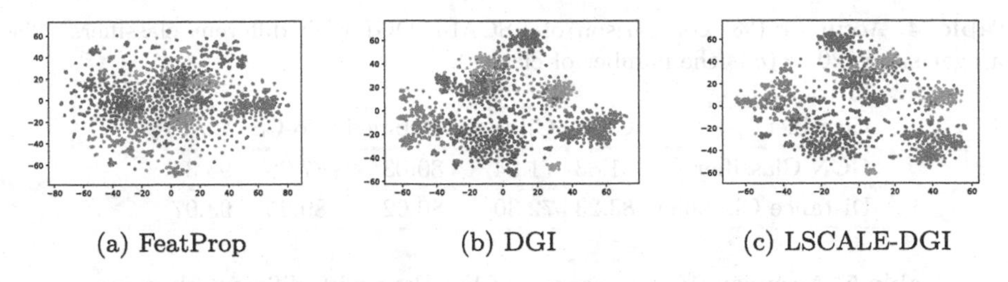

(a) FeatProp (b) DGI (c) LSCALE-DGI

Fig. 2. t-SNE visualization of different distance features on Cora dataset. Colors denote the ground-truth class labels. Our distance features have clearer separations between different classes. (Color figure online)

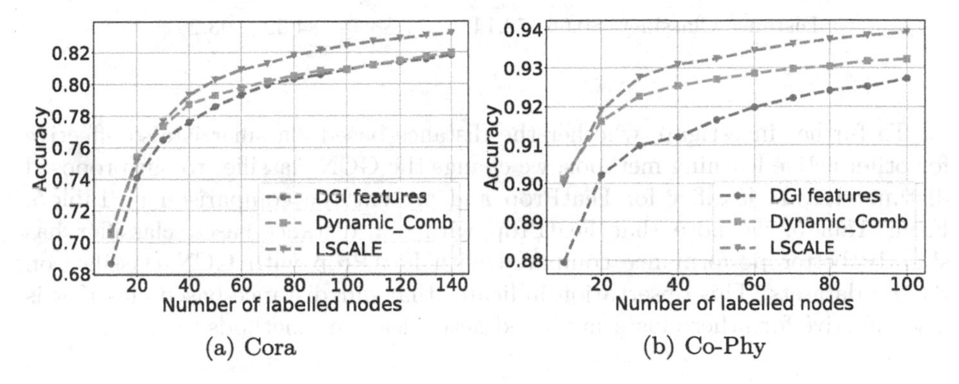

(a) Cora (b) Co-Phy

Fig. 3. Ablation study: dynamic feature combination and incremental clustering.

Effectiveness of Our Distance Features $g(\mathbf{X})$. Furthermore, we also qualitatively demonstrate the effectiveness of our distance feature $g(\mathbf{X})$. Figure 2 shows t-SNE visualizations [17] of FeatProp features, DGI features, and the distance features of LSCALE-DGI. The distance features are obtained by dynamically combining DGI features and supervised hidden representations on 20 labelled nodes. Recall that FeatProp uses propagated node attributes as distance features and DGI features are learned using an unsupervised method with unlabelled data. Compared with others, the distance features used in LSCALE-DGI have clearer boundaries between different classes, which satisfies our second desired property (*informative distances*) and further facilitates selecting informative nodes in the clustering algorithm.

Effectiveness of Distance-Based Classifier. We design a distance-based classifier in LSCALE to ensure that distances are informative in the active learning latent space. To demonstrate the effectiveness of the distance-based classifier, we replace it with a GCN classifier and show the comparison in Table 4. With the distance-based classifier, LSCALE can achieve better performance than that with a GCN classifier. This comparison shows the effectiveness of the designed distance-based classifier.

Table 4. Accuracy (%) comparison of LSCALE-DGI with different classifiers. The budget size is $20 \cdot c$ (c is the number of classes).

Classifier	Cora	Citeseer	Pubmed	Co-CS	Co-Phy
GCN Classifier	81.83	71.24	80.03	87.28	93.34
Distance Classifier	83.23	72.30	80.62	89.25	93.97

Table 5. Accuracy (%) comparison of FeatProp with different classifiers.

Classifier	Cora	Citeseer	Pubmed	Co-CS	Co-Phy
GCN Classifier	80.50	72.04	77.65	83.49	93.06
Distance Classifier	80.66	72.14	77.88	84.32	93.28

To further investigate whether the distance-based classifier is also effective for other active learning methods, we change the GCN classifier to our proposed distance-based classifier for FeatProp and present the comparison in Table 5. From Table 5, we note that FeatProp with our distance-based classifier has slightly better performance compared with FeatProp with GCN classifier on all the datasets. This observation indicates that our distance-based classifier is also effective for other clustering-based active learning methods.

5 Related Work

Active Learning on Graphs. For active learning on graphs, early works without using graph representations are proposed in [3,4,19], where the graph structure is used to train the classifier and calculate the query scores for selecting nodes. More recent works [1,8] study non-parametric classification models with graph regularization for active learning with graph data.

Recent works [5,9,22,28] utilize graph convolutional neural networks (GCNs) [15], which consider the graph structure and the learned embeddings simultaneously. AGE [5] design an active selecting strategy based on a weighted sum of three metrics considering the uncertainty, the the graph centrality and the information density. Improving upon the weight assignment mechanism, ANRMAB [9] designs a multi-armed bandit method with a reward scheme to adaptively assign weights for the different metrics.

Besides the metric-based active selection on graphs, FeatProp [28] uses a clustering-based active learning method, which calculates the distances between nodes based on representations of a simplified GCN model [27] and conducts a clustering algorithm (i.e., K-Medoids) for selecting representative nodes. A recent method GEEM [22] uses a simplified GCN [27] for prediction and maximizes the expected error reduction to select informative nodes to label. Rather than actively selecting nodes and training/testing on a single graph, [13] learns a selection policy on several labelled graphs via reinforcement learning and actively selects nodes using that policy on unlabelled graphs. [16,18] use adversarial learning and meta learning approaches for active learning on graphs. However, even with relatively complicated learning methods, their performance are similar with AGE [5] and ANRMAB [9]. [7] investigates active learning on heterogeneous graphs. [29] considers noisy oracle setting where labels obtained by an oracle can be incorrect. In this work, we focus on the homogeneous single-graph setting like in [5,9,22,28]. To tackle limitations of previous work on this setting, we have presented an effective and efficient framework that can utilize the representation power in unlabelled nodes and achieve better performance under the same labelling budget.

6 Conclusion

In this paper, we focus on active learning for node classification on graphs and argue that existing methods are still less than satisfactory as they do not fully utilize the information in unlabelled nodes. Motivated by this, we propose LSCALE, a latent space clustering-based active learning framework, which uses a latent space with two desired properties for clustering-based active selection. We also design an incremental clustering module to minimize redundancy between nodes selected at different steps. Extensive experiments demonstrate that our method provides superior performance over the state-of-the-art models. Our work points out a new possibility for active learning on graphs, which is to better utilize the information in unlabelled nodes by designing a feature space more suitable for active learning. Future work could propose new unsupervised methods which are more integrated with active learning process and enhance our framework further.

Acknowledgements. This paper is supported by the Ministry of Education, Singapore (Grant Number MOE2018-T2-2-091) and A*STAR, Singapore (Number A19E3b0099).

References

1. Aodha, O.M., Campbell, N.D.F., Kautz, J., Brostow, G.J.: Hierarchical subquery evaluation for active learning on a graph. In: CVPR (2014)
2. Bengio, Y., Louradour, J., Collobert, R., Weston, J.: Curriculum learning. In: ICML (2009)
3. Berberidis, D., Giannakis, G.B.: Data-adaptive active sampling for efficient graph-cognizant classification. IEEE Trans. Sig. Process. **66**, 5167–5179 (2018)
4. Bilgic, M., Mihalkova, L., Getoor, L.: Active learning for networked data. In: ICML (2010)
5. Cai, H., Zheng, V.W., Chang, K.C.: Active learning for graph embedding. arXiv preprint arXiv:1705.05085 (2017)
6. Chen, J., Ma, T., Xiao, C.: FastGCN: fast learning with graph convolutional networks via importance sampling. In: ICLR (2018)
7. Chen, X., Yu, G., Wang, J., Domeniconi, C., Li, Z., Zhang, X.: ActiveHNE: active heterogeneous network embedding. In: Proceedings of the Twenty-Eighth International Joint Conference on Artificial Intelligence, IJCAI-19 (2019)

8. Dasarathy, G., Nowak, R.D., Zhu, X.: S2: an efficient graph based active learning algorithm with application to nonparametric classification. In: COLT (2015)
9. Gao, L., Yang, H., Zhou, C., Wu, J., Pan, S., Hu, Y.: Active discriminative network representation learning. In: IJCAI (2018)
10. Gilmer, J., Schoenholz, S.S., Riley, P.F., Vinyals, O., Dahl, G.E.: Neural message passing for quantum chemistry. In: ICML, pp. 1263–1272 (2017)
11. Hamilton, W., Ying, Z., Leskovec, J.: Inductive representation learning on large graphs. In: NIPS, pp. 1024–1034 (2017)
12. Hassani, K., Khasahmadi, A.H.: Contrastive multi-view representation learning on graphs. In: ICML, pp. 4116–4126 (2020)
13. Hu, S., et al.: In: Advances in Neural Information Processing Systems (2020)
14. Kingma, D.P., Ba, J.: Adam: a method for stochastic optimization. In: ICLR (2015)
15. Kipf, T.N., Welling, M.: Semi-supervised classification with graph convolutional networks. In: ICLR (2016)
16. Li, Y., Yin, J., Chen, L.: Seal: semisupervised adversarial active learning on attributed graphs. IEEE Trans. Neural Netw. Learn. Syst. **32**, 3136–3147 (2020)
17. van der Maaten, L., Hinton, G.: Visualizing data using t-SNE. J. Mach. Learn. Res. **9**(86), 2579–2605 (2008)
18. Madhawa, K., Murata, T.: Metal: active semi-supervised learning on graphs via meta-learning. In: Asian Conference on Machine Learning, pp. 561–576. PMLR (2020)
19. Moore, C., Yan, X., Zhu, Y., Rouquier, J., Lane, T.: Active learning for node classification in assortative and disassortative networks. In: SIGKDD (2011)
20. Namata, G., London, B., Getoor, L., Huang, B.: Query-driven active surveying for collective classification. In: 10th International Workshop on Mining and Learning with Graphs (2012)
21. Parisot, S., et al.: Disease prediction using graph convolutional networks: application to autism spectrum disorder and Alzheimer's disease. Med. Image Anal. **48**, 117–130 (2018)
22. Regol, F., Pal, S., Zhang, Y., Coates, M.: Active learning on attributed graphs via graph cognizant logistic regression and preemptive query generation. In: ICML, pp. 8041–8050 (2020)
23. Sen, P., Namata, G., Bilgic, M., Getoor, L., Galligher, B., Eliassi-Rad, T.: Collective classification in network data. AI Mag. **29**, 93 (2008)
24. Sener, O., Savarese, S.: Active learning for convolutional neural networks: a core-set approach. In: ICLR (2018)
25. Shchur, O., Mumme, M., Bojchevski, A., Günnemann, S.: Pitfalls of graph neural network evaluation. arXiv preprint arXiv:1811.05868 (2018)
26. Veličković, P., Fedus, W., Hamilton, W.L., Liò, P., Bengio, Y., Hjelm, R.D.: Deep graph infomax. In: ICLR (2018)
27. Wu, F., Souza, A., Zhang, T., Fifty, C., Yu, T., Weinberger, K.: Simplifying graph convolutional networks. In: ICML, pp. 6861–6871 (2019)
28. Wu, Y., Xu, Y., Singh, A., Yang, Y., Dubrawski, A.: Active learning for graph neural networks via node feature propagation. In: Proceedings of NeurIPS 2019 Graph Representation Learning Workshop (GRL) (2019)
29. Zhang, W., et al.: Rim: reliable influence-based active learning on graphs. In: Advances in Neural Information Processing Systems, vol. 34 (2021)

Powershap: A Power-Full Shapley Feature Selection Method

Jarne Verhaeghe(✉), Jeroen Van Der Donckt, Femke Ongenae, and Sofie Van Hoecke

IDLab, Ghent University - imec, 9000 Ghent, Belgium
jarne.verhaeghe@ugent.be
http://predict.idlab.ugent.be/

Abstract. Feature selection is a crucial step in developing robust and powerful machine learning models. Feature selection techniques can be divided into two categories: filter and wrapper methods. While wrapper methods commonly result in strong predictive performances, they suffer from a large computational complexity and therefore take a significant amount of time to complete, especially when dealing with high-dimensional feature sets. Alternatively, filter methods are considerably faster, but suffer from several other disadvantages, such as (i) requiring a threshold value, (ii) many filter methods not taking into account intercorrelation between features, and (iii) ignoring feature interactions with the model. To this end, we present *powershap*, a novel wrapper feature selection method, which leverages statistical hypothesis testing and power calculations in combination with Shapley values for quick and intuitive feature selection. *Powershap* is built on the core assumption that an informative feature will have a larger impact on the prediction compared to a known random feature. Benchmarks and simulations show that *powershap* outperforms other filter methods with predictive performances on par with wrapper methods while being significantly faster, often even reaching half or a third of the execution time. As such, *powershap* provides a competitive and quick algorithm that can be used by various models in different domains. Furthermore, *powershap* is implemented as a plug-and-play and open-source *sklearn* component, enabling easy integration in conventional data science pipelines. User experience is even further enhanced by also providing an automatic mode that automatically tunes the hyper-parameters of the *powershap* algorithm, allowing to use the algorithm without any configuration needed.

Keywords: Feature selection · Shap · Benchmark · Simulation · Toolkit · Python · Open source

1 Introduction

In many data mining and machine learning problems, the goal is to extract and discover knowledge from data. One of the challenges frequently faced in these

M.-R. Amini et al. (Eds.): ECML PKDD 2022, LNAI 13713, pp. 71–87, 2023.
https://doi.org/10.1007/978-3-031-26387-3_5

problems is the high dimensionality and the unknown relevance of features [10]. Ignoring these challenges will more than often result in modeling obstacles, such as sparse data, overfitting, and the curse of dimensionality. Therefore, feature selection is frequently applied, among other techniques, to effectively reduce the feature dimensionality. The smaller subset of features has the potential to explain the problem better, reduce overfitting, alleviate the curse of dimensionality, and even facilitate interpretation. Furthermore, feature selection is known to increase model performance, increase computational efficiency, and increase the robustness of many models due to the dimensionality reduction [10].

In this work, we present a novel feature selection method, called *powershap*, that is a faster and easy-to-use wrapper method. The feature selection is realized by using Shapley values, statistical tests, and power calculations.

First, in Sect. 2, a short overview of the related work is given to show how *powershap* improves upon all these methods. Subsequently, in Sect. 3, the method and the design choices are explained as well as the resulting algorithm. Finally, the performance of *powershap* is compared to other state-of-the-art methods in Sects. 4 and 5 using both simulation and open-source benchmark datasets and the results are discussed in Sect. 6. Finally, the conclusions are summarized in Sect. 7.

2 Related Work

Feature selection approaches can be categorized into filter and wrapper methods. Filter methods select features by measuring the relevance of the feature using model-agnostic measures, such as statistical tests, information gain, distance, similarity, and consistency to the dependent variable (if available). These methods are model-independent as this category of feature selection does not rely on training machine learning models [6], resulting in a fast evaluation. However, the disadvantages of filter methods are that they frequently impose assumptions on the data, are limited to a single type of prediction, such as classification or regression, not all methods take inter-correlation between features into account, and often require a cut-off value or hyperparameter tuning [8]. Examples of these filter methods are rank, chi^2 test, f-test, correlation-based feature selection, Markov blanket filter, and Pearson correlation [6].

Wrapper methods measure the relevance of features using a specific evaluation procedure through training supervised models. Depending on the wrapper technique, models are trained on either subsets of the features or on the complete feature set. The trained models are then utilized to select the resulting feature subset, using the aforementioned performance metrics, or by ranking the inferred feature importances. In general, wrapper methods tend to provide smaller and more qualitative feature subsets than filter methods, as they take the interaction between the features, and between the model and the features, into account [4]. A major drawback of wrapper methods is the considerable time complexity associated with the underlying search algorithm, or in the case of feature importance ranking the hyperparameter tuning. Examples of wrapper methods are forward, backward, genetic, or rank-based feature importance feature selection.

In the interpretable machine learning field, one of the emerging and proven techniques to explain model predictions is SHAP [13]. This technique aims at quantifying the impact of features on the output. To do so, SHAP uses a game-theory inspired additive feature-attribution method based on Shapley Regression Values [13]. This method is model-agnostic and implemented for various models, e.g., linear, kernel-based, deep learning, and tree-based models. Although SHAP suffers from shortcomings, such as its TreeExplainer providing non-zero Shapley values to noise features, it is technically strong and very popular [11].

The strength of the SHAP algorithm facilitates the development of new feature selection methods using Shapley values. A simple implementation would be a rank-based feature selection, which ranks the different features based on their Shapley values on which a rank cut-off value determines the final feature set. However, there are more advanced methods available. One of these more advanced techniques is borutashap [7]. Borutashap is based on the Boruta algorithm that makes use of shadow features, i.e. features with randomly shuffled values. Boruta is built on the idea that a feature is only useful if it is doing better than the best-performing shuffled feature. To do so, Boruta compares the feature importance of the best shadow feature to all other features, selecting only features with larger feature importance than the highest shadow feature importance. Statistical interpretation is realized by repeating this algorithm for several iterations, resulting in a binomial distribution which can be used for p-value cut-off selection [9]. Borutashap improves on the underlying Boruta algorithm by using Shapley values and an optimized version of the shap TreeExplainer [7]. As such, implementations of Borutashap are limited to tree-based models only.

Another shap-based feature selection method using statistics is shapicant [3]. This feature selection method is inspired by the permutation-importance method, which first trains a model on the true dataset, and afterward, it shuffles the labels and retrains the model on the shuffled dataset. This process is repeated for a set amount of iterations, from which the average feature importances of both models are compared. If for a specific feature, the feature importance of the true dataset model is consistently larger than the importance of the shuffled dataset model, that feature is considered informative. Using a non-parametric estimation it is possible to assign a p-value to determine a wanted cut-off value [1]. Shapicant improves on this underlying algorithm by using Shapley values. Specifically, it uses both the mean of the negative and positive Shapley values instead of Gini importances, which are only positive and frequently used for tree-based model importances. Furthermore, shapicant uses out-of-sample feature importances for more accurate estimations and an improved non-parametric estimation formula [3].

Powershap draws inspiration from the non-parametric estimation of shapicant and the random feature usage in borutashap and improves upon all these state-of-the-art filter and wrapper algorithms resulting in at least comparable performances while being significantly faster.

3 Powershap

Powershap builds upon the idea that a known random feature should have, on average, a lower impact on the predictions than an informative feature. To realize feature selection, the *powershap* algorithm consists of two components: the *Explain* component and the core *powershap* component. First, in the *Explain* part, multiple models are trained using different random seeds, on different subsets of the data. Each of these subsets is comprised of all the original features together with one random feature. Once the models are trained, the average impact of the features (including the random feature) is explained using Shapley values on an out-of-sample dataset. Then, in the core *powershap* component, the impacts of the original features are statistically compared to the random feature, enabling the selection of all informative features.

3.1 Powershap Algorithm

In the *Explain* component, a single known random uniform (RandomUniform) feature is added to the feature set for training a machine learning model. Unlike the Boruta algorithm, where all features are duplicated and shuffled, only a single random feature is added. In some models, such as neural networks, duplicating the complete feature set increases the scale and thereby increases the time complexity drastically. Using the Shapley values on an out-of-sample subset of the data allows for quantifying the impact on the output for each feature. The Shapley values are evaluated on unseen data to assess the true unbiased impact [2]. As a final step, the absolute value of all the Shapley values is taken and then averaged (μ) to get the total average impact of each feature. Compared to shapicant, only a single mean value is used here, resulting in easier statistical comparisons. Furthermore, by utilizing the absolute Shapley values, the positive values and the negative values are added to the total impact, which could result in a different distribution compared to the Gini importance. This procedure is then repeated for I iterations, where every iteration retrains the model with a different random feature and uses a different subset of the data to quantify the Shapley values, resulting in an empirical distribution of average impacts that will further be used for the statistical comparison. In the codebase, the procedure explained above is referred to as the *Explain* function. The pseudocode of the *Explain* function is shown in Algorithm 1.

Given the average impact of each feature for each iteration, it is then possible to compare it to the impact of the random feature in the core powershap component. This comparison is quantified using the percentile formula shown in Eq. 1 where $\mathbf{s}$ depicts an array of average Shapley values for a single feature with the same length as the number of iterations, while x represents a single value, and $\mathbb{I}$ represents the indicator function. This formula calculates the fraction of iterations where x was higher than the average shap-value of that iteration and can therefore be interpreted as the p-value.

$$Percentile(\mathbf{s}, x) = \sum_{i}^{n} \frac{\mathbb{I}(x > s_i)}{n} \tag{1}$$

Algorithm 1: Powershap Explain algorithm

Function *Explain(I* $\leftarrow$ *Iterations, M* $\leftarrow$ *Model,* $\mathbf{D}^{n \times m} \leftarrow$ *Data, rs* $\leftarrow$ *Random seed)*

    $\mathbf{powershap}_{values} \leftarrow$size $[I, m+1]$
    for $i \leftarrow 1, 2, \ldots, I$ **do**
        $RS \leftarrow i + rs$
        $D^n_{random} \leftarrow$ RandomUniform(RS) $\in [-1, 1]$ size n
        $\mathbf{D}^{n \times m+1} \leftarrow \mathbf{D}^{n \times m} \cup D^n_{random}$
        $\mathbf{D}^{0.8n \times m+1}_{train}, \mathbf{D}^{0.2n \times m+1}_{val} \leftarrow$ split $\mathbf{D}$
        $M \leftarrow$ Fit $M(\mathbf{D}_{train})$
        $\mathbf{S}_{values} \leftarrow$ SHAP(M, $\mathbf{D}_{val}$)
        $\mathbf{S}_{values} \leftarrow |\mathbf{S}_{values}|$
        for $j \leftarrow 1, 2, \ldots, m+1$ **do**
            $\mathbf{powershap}_{values}[i][j] \leftarrow \mu(\mathbf{S}_{values}[\ldots][j])$
        end
    end
    return $\mathbf{powershap}_{values}$

Note that this formula provides smaller p-values than what should be observed, the correct empirical formula is $(1+\sum_i^n \mathbb{I}(x > s_i))/(n+1)$ as explained by North et al. [14]. This issue of smaller p-values mainly persists for lower number of iterations. However, *powershap* implements Eq. 1 as this anticonservative estimation of the p-value is desired behavior for the automatic mode (see Sect. 3.2). This formula enables setting a static cut-off value for the p-value instead of a varying cut-off value and results in fewer required iterations, while still providing correct results. This will be further explained at the end of Sect. 3.2.

As the hypothesis states that the impact of the random feature should be on average lower than any informative feature, all impacts of the random feature are again averaged, resulting in a single value that can be used in the percentile function. This results in a p-value for every original feature. This p-value represents the fraction of cases where the feature is less important, on average than a random feature. Given the hypothesis and these p-value calculations, a heuristic implementation of a one-sample one-tailed student-t smaller statistic test can be done, where the null hypothesis states that the random feature (H_1-distribution) is not more important than the tested feature (H_0-distribution) [12]. Therefore, the positive class in this statistical test represents a true null hypothesis. This heuristic implementation does not assume a distribution on the tested feature impact scores, in contrast to a standard student-t statistic test where a standard Gaussian distribution is assumed. Then, given a threshold p-value α, it is possible to find and output the set of informative features. The pseudocode of Algorithm 2 details how the core *powershap* feature selection method is realized.

Algorithm 2: Powershap core algorithm

Function *Powershap* ($I \leftarrow$ *Iterations*, $M \leftarrow$ *Model*, $\mathbf{F}_{set} \leftarrow F_1, \ldots, F_m$, $\mathbf{D} \leftarrow$ *Data size* $[n, m]$, $\alpha \leftarrow$ *required p-value*)
  $\mathbf{powershap}_{values} \leftarrow \mathbf{Explain}(I, M, \mathbf{D})$
  $S_{random} \leftarrow \mu(\mathbf{powershap}_{values}[...][m+1])$
  $\mathbf{P}^m \leftarrow$ initialize
  for $j \leftarrow 1, 2, \ldots, m$ **do**
    $\mathbf{P}[j] \leftarrow Percentile(\mathbf{powershap}_{values}[...][j], S_{random})$
  end
  return $\{F_i \mid \forall\, i : \mathbf{P}[i] < \alpha\}$

3.2 Automatic Mode

Running the *powershap* algorithm consisting of the *explain* and the *core* components, requires setting two hyperparameters: α the p-value threshold and I the number of iterations. When hyperparameter tuning, one should make a trade-off between runtime and quality. On the one hand, there should be enough iterations to avoid false negatives for a given α, especially with the anticonservative p-values. On the other hand, adding iterations increases the time complexity. To avoid the need for users to manually optimize these two hyperparameters, *powershap* also has an automatic mode. This automatic mode, automatically determines and optimizes the iteration hyperparameter I using statistical power calculation for α, hence the name *powershap*.

The statistical power of a test is $1 - \beta$, where β is the probability of false negatives. In this case, a false negative is a non-informative feature flagged as an informative one. If a statistical test of a tested sample outputs a p-value α, this represents the chance that the tested sample could be flagged as *significant* by chance given the current data. This is calculated using Eq. 2. If the data in the statistical test is small, it is possible to have a very low α but a large β, resulting in an output that cannot be trusted. Therefore, for a given α, the associated power should be as close to 1 as possible to avoid any false negatives. The power of a statistical test can be calculated using the cumulative distribution function F of the underlying tested distribution H_1 using Eq. 3. Figure 1 explains this visually. In the current context, H_0 could represent the random feature impact distribution and H_1 the tested feature impact distribution.

$$\alpha(x) = F_{H_0}(x) \tag{2}$$

$$Power(\alpha) = F_{H_1}\left(F_{H_0}^{-1}(\alpha)\right) \tag{3}$$

The power calculations require the cumulative distribution function F. However, the underlying distributions of the calculated feature impacts are unknown. In addition, calculating F heuristically does not enable calculating the required iteration hyperparameter, which is the goal of the automatic mode. *Powershap* circumvents this by mapping the underlying distributions to two standard

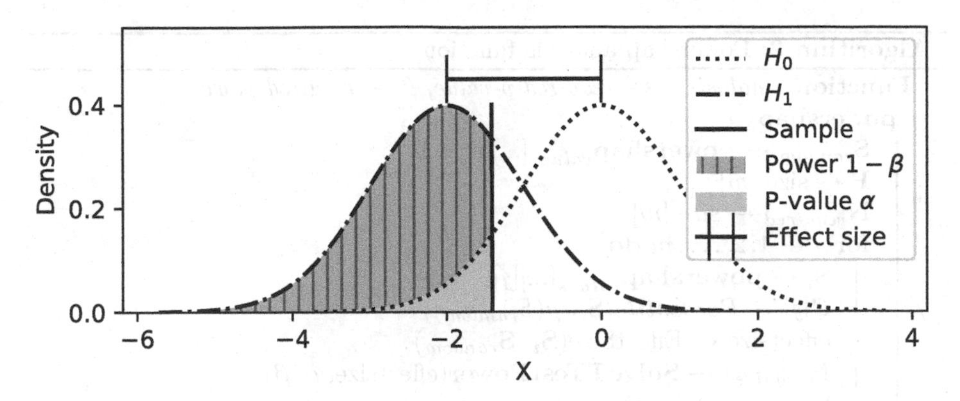

Fig. 1. Visualization of p-value, effect size, and power for a standard t-test.

student-t distributions as visualized in Fig. 1. It first calculates the pooled standard deviation, using Eq. 4, by averaging the standard deviations σ of both distributions. It then calculates the distance d between these two distributions, also called the effect size, in terms of this pooled standard deviation using the Cohen's d effect size as detailed in Eq. 5 [12]. Now, it is possible to define two standard student-t distributions with distance $\sqrt{I} \cdot d$ apart and $I - 1$ degrees of freedom, where I is the amount of *powershap* iterations. The standard central student-t F_{CT} and non-central student-t F_{NCT} cumulative distribution functions are then used to calculate the power of the statistical test according to Eq. 6. This equation can in turn be used in a heuristic algorithm to solve for I. *Powershap* uses the *solve_power* implementation of *statsmodels* to determine the required I from the TTestPower equation using brentq expansion for a provided required power [17]. The *powershap* pseudocode for the calculation of the effect size, power, and required iterations is shown in Algorithm 3.

$$PooledStd(\mathbf{s}_1, \mathbf{s}_2) = \frac{\sqrt{(\sigma^2(\mathbf{s}_1) + \sigma^2(\mathbf{s}_2))}}{2} \tag{4}$$

$$EffectSize(\mathbf{s}_1, \mathbf{s}_2) = \frac{\mu(\mathbf{s}_1) - \mu(\mathbf{s}_2)}{PooledStd(\mathbf{s}_1, \mathbf{s}_2)} \tag{5}$$

$$TTestPower(\alpha, I, d_{calc}) = F_{NCT}\left(F_{CT}^{-1}(\alpha, k = I-1), k = n-1, d = \sqrt{I} \cdot d_{calc}\right) \tag{6}$$

With the calculated required amount of iterations n, the automatic *powershap* algorithm can be executed. The pseudocode to enable the automatic mode is shown in Algorithm 4. As can be seen, this is an expansion of the core algorithm (see Algorithm 2) and starts with an initial ten iterations to calculate the initial p-value, effect sizes, power, and required iterations for all features. Then, it searches for the largest required number of iterations I_{max} of all tested features having a p-value below the threshold α. If I_{max} exceeds the already performed number of iterations I_{old}, automatic mode continues *powershap* for

Algorithm 3: Powershap analysis function

Function *Analysis*($\alpha \leftarrow$ *required p-value*, $\beta \leftarrow$ *required power*, $\mathbf{powershap}_{values}$)
- $\mathbf{S}_{random} \leftarrow \mathbf{powershap}_{values}[...][m+1]$
- $\mathbf{P} \leftarrow$ size $[m]$
- $\mathbf{N}_{required} \leftarrow$ size $[m]$
- **for** $j \leftarrow 1, 2, \ldots, m$ **do**
  - $\mathbf{S}_i \leftarrow \mathbf{powershap}_{values}[...][j]$
  - $\mathbf{P}[j] \leftarrow Percentile(\mathbf{S}_i, \mu(S_{random}))$
  - effectsize $\leftarrow$ EffectSize($\mathbf{S}_i$, $\mathbf{S}_{random}$)
  - $\mathbf{N}_{required} \leftarrow$ **SolveTTestPower**(effectsize, α, β)
- **end**
- **return** $\mathbf{P}$, $\mathbf{N}_{required}$

the extra required iterations. This process is repeated until the performed iterations exceed the required iterations. For optimization, when the extra required iterations ($I_{max} - I_{old}$) exceed ten iterations, the automatic mode first adds ten iterations and then re-evaluates the required iterations because the required iterations are influenced by the already performed iterations. Furthermore, it is also possible to provide a stopping criterion on the re-execution of *powershap* to avoid an infinite calculation. As a result the time complexity of the algorithm is linear in terms of the underlying model and shap explainer and can be formulated as $O(p[M_{n+1} + S(M_{n+1}])$, with n the amount of features, p the number of powershap iterations, S the shap explainer time, and M_x the model fit time for x features. For the automatic mode, by default, α is set to 0.01 while the required power is set to 0.99. This results in only selecting features that are more important than the random feature for all iterations. Furthermore, this also compensates for the anticonservative p-value and avoids as many false negatives as possible. Realizing the same desired behavior with the more accurate p-value estimation would require a varying α of $1/n$, complicating the power calculations and increasing the likelihood of false negatives. The resulting powershap algorithm is implemented in Python as an open-source plug-and-play *sklearn* compatible component to enables direct usage in conventional machine learning pipelines [15]. The codebase [1] already supports a wide variety of models, such as linear, tree-based, and even deep learning models. To assure the quality and correctness of the implementation, we tested the functionality using unit testing.

[1] The code, documentation, and more benchmarks can be found using the following link: https://github.com/predict-idlab/PowerSHAP.

Algorithm 4: Automatic Powershap algorithm version

Function *Powershap* ($M \leftarrow Model$, $\mathbf{F}_{set} \leftarrow F_1, \ldots, F_m$, $\mathbf{D}^{n \times m} \leftarrow Data$, $\alpha \leftarrow$ *required p-value*, $\beta \leftarrow$ *required power*)
    $\mathbf{powershap}_{values} \leftarrow \mathbf{Explain}(I \leftarrow 10, M, \mathbf{D}, rs \leftarrow 0)$
    $\mathbf{P}, \mathbf{N}_{required} \leftarrow \mathbf{Analysis}(\alpha, \beta, \mathbf{powershap}_{values})$
    $I_{max} \leftarrow \text{ceil}(\mathbf{N}_{required}[\text{MaxArg}(\mathbf{P} < \alpha)])$
    $I_{old} \leftarrow 10$
    while $I_{max} > I_{old}$ **do**
        if $I_{max} - I_{old} > 10$ **then**
            $\mathbf{auto}_{values} \leftarrow \mathbf{Explain}(I \leftarrow 10, M, \mathbf{D}, rs \leftarrow 0)$
            $I_{old} \leftarrow I_{old} + 10$
        else
            $\mathbf{auto}_{values} \leftarrow \mathbf{Explain}(I \leftarrow I_{max} - I_{old}, M, \mathbf{D}, rs \leftarrow 0)$
            $I_{old} \leftarrow I_{max}$
        end
        $\mathbf{powershap}_{values} \leftarrow \mathbf{powershap}_{values} \cup \mathbf{auto}_{values}$
        $\mathbf{P}, \mathbf{N}_{required} \leftarrow \mathbf{Analysis}(\alpha, \beta, \mathbf{powershap}_{values})$
        $I_{max} \leftarrow \text{ceil}(Max(\mathbf{N}_{\mathbf{required}}[i, \forall i : \mathbf{P}[i] < \alpha]))$
    end
    return $[F_i, \forall\, i : \mathbf{P}[i] < \alpha]$

4 Experiments

4.1 Feature Selection Methods

To facilitate a comparison with other feature selection techniques, we benchmark *powershap* together with other frequently used techniques on both synthetic and real-world datasets. In particular, *powershap* is compared with both filter and wrapper methods, and state-of-the-art shap-based wrapper methods. To provide a fair comparison, all methods, including *powershap*, were used in their default out-of-the-box mode without tuning. For *powershap*, this default mode is the automatic mode. Concerning filter methods, two methods were chosen: the chi-squared and f-test feature selection from the *sklearn*-library [15]. The chi-squared test measures the dependence between a feature and the classification outcome and assigns a low p-value to features that are not independent of the outcome. As the chi-squared test only works with positive values, the values are shifted in all chi-squared experiments such that all values are positive. This has no effect on tree-estimators as they are invariant to data scaling [12]. The F-test in *sklearn* is a univariate test that calculates the F-score and p-values on the predictions of a univariate fitted linear regressor with the target [15]. Both filter methods provide p-values that are set to the same threshold as *powershap*. As wrapper feature selection method, forward feature selection was chosen. This method is a greedy algorithm that starts with an empty set of features and trains a model with each feature separately. In every iteration, forward feature selection then adds the best feature according to a specified metric, often evaluated in cross-validation, until the metric stops improving. This is generally considered a strong method but has a very large time complexity [6]. *Powershap* is also compared to shapicant [3] and borutashap [7], two SHAP-based feature selection methods. The default machine learning model used for all datasets and all feature selection methods, including *powershap*, is a CatBoost gradient boosting tree-based estimator using 250 estimators with the overfitting detector enabled. For classification, the CatBoost model uses adjusted class weights to compensate for any potential class imbalance. The Catboost estimator often results in strong predictive performances out-of-the-box, without any hyper-parameter tuning, making it the perfect candidate for benchmarking and comparison [16]. All experiments are performed on a laptop with a Intel(R) Core(TM) i7-9850H CPU at 2.60 GHz processor and 16 GB RAM running at 2667 MHz, with background processes to a minimum.

4.2 Simulation Dataset

The methods are first tested on a simulated dataset to assess their ability to discern noise features from informative features. The used simulation dataset is created using the `make_classification` function of *sklearn*. This function creates a classification dataset, however, exactly the same can be done for obtaining a regression dataset (by using `make_regression`). The simulations are run using 20, 100, 250, and 500 total features to understand the performance on varying dimensions of feature sets. The ratio of informative features is varied as 10%, 33%, 50%, and 90% of the total feature set, allowing for assessing the quality of the selected features in terms of this ratio. The resulting simulation datasets each contain 5000 samples. Each simulation experiment was repeated five times with different random seeds. The number of redundant features, which are linear combinations of informative features, and the number of duplicate features were set to zero. Redundant features and duplicate features reduce the performance of models, but they cannot be discerned from true informative features as they are inherently informative. Therefore they are not included in the simulation dataset as the goal of *powershap* is to find informative features. The *powershap* method is compared to shapicant, chi^2, borutashap, and the f-test for feature selection on this simulation dataset. Due to time complexity constraints, forward feature selection was not included in the simulation benchmarking.

4.3 Benchmark Datasets

In addition to the simulation benchmark, the different methods are also evaluated on five publicly available datasets, i.e. three classification datasets: the Madelon [19], the Gina priori [20], and the Scene dataset [18], and two regression datasets: CT location [5] and Appliances [5]. The details of these datasets are shown in Table 1. The Scene dataset is a multi-label dataset, however,

Table 1. Properties of all datasets

Dataset	Type	Source	# features	train size	test size
Madelon	Classification	OpenML	500	1950	650
Gina priori	Classification	OpenML	784	2601	867
Scene	Classification	OpenML	294	1805	867
CT location	Regression	UCI	384	41347	12153
Appliances	Regression	UCI	30	14801	4934

a multi-label problem can always be reduced to a one-vs-all classification problem. Therefore only the label "Urban" was chosen here to assess binary classification performance.

Almost all of these datasets have a large feature set, ideal for benchmarking feature selection methods. The datasets are split into a training and test set using a 75/25 split. All methods are evaluated using both 10-fold cross-validation on the training set and 1000 bootstraps on the test set to assess the robustness of the performance. The test set is utilized to assess generalization beyond the validation set as wrapper methods tend to slightly overfit their validation set [6], while the training set is used for feature selection. The forward feature selection method was performed with 5-fold cross-validation and not 10-fold cross-validation due to the high time complexity. A validation set of 20% of the training set is used for shapicant, using the same validation size as *powershap* in Algorithm 1. The models are evaluated with the AUC metric for classification datasets and with the R^2 metric for regression datasets.

5 Results

5.1 Simulation Dataset

The results of the simulation benchmarking are shown in Fig. 2. Each row of subfigures shows the duration, the percentage of informative features found, and the number of selected noise features. These measures are shown for each feature selection method for varying feature set dimensions and varying amounts of informative features. As can be seen, the shapicant method is the slowest wrapper method while *powershap* is, without doubt, the fastest wrapper method. The filter methods are substantially faster than any of the wrapper methods, as they do not train models. Furthermore, *powershap* finds all informative features with a limited amount of outputted noise features up to the case with 250 total features with 50% (125) informative features, outperforming every other method. This can be explained by the model underfitting the data. Even with higher dimensional feature sets, *powershap* finds more informative features than the other methods. Interestingly, most methods do not output many noise features, except for shapicant in the experiment with 20 total and 10% informative features.

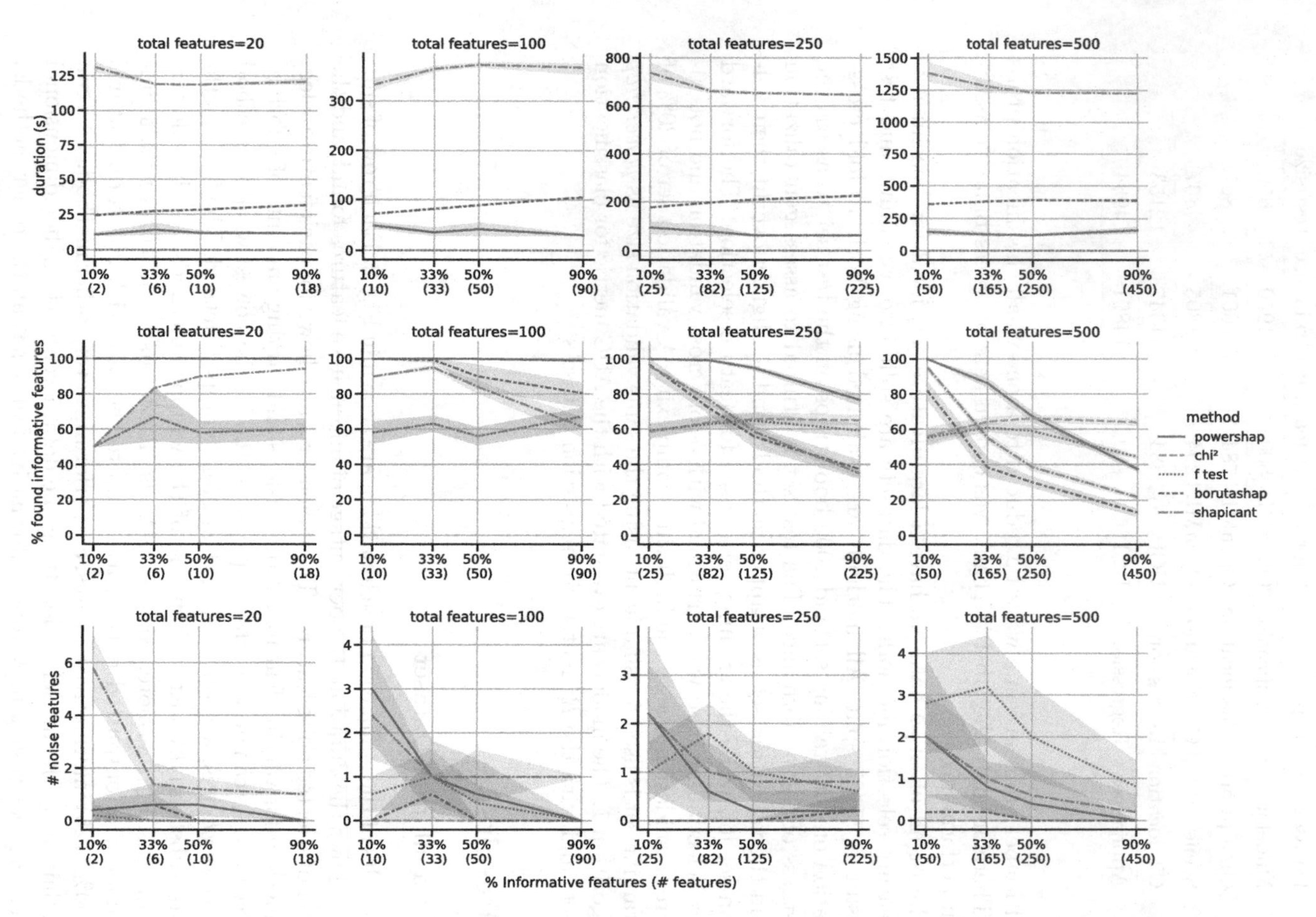

Fig. 2. Simulation benchmark results using the *make_classification sklearn* function for 5000 samples with five different *make_classification* random seeds.

5.2 Benchmark Datasets

Table 2 shows the duration of the feature selection methods and the size of the selected feature sets for each method on the different open-source datasets. Chi^2 does not apply to regression problems and is therefore not included in the results of the CT location and Appliances datasets. The table shows that *powershap* is again the fastest wrapper method, while the number of selected features is in line with the other methods. The filter methods tend to output more features, while forward feature selection outputs a more conservative set of features.

Table 2. Benchmarks results for duration and selected features. "default" indicates no feature selection or all features.

Duration (s)							
Dataset	powershap	borutashap	shapicant	forward	chi^2	f test	default
Madelon	132 s	186 s	632 s	10483 s	$< 1s$	$< 1s$	N/A
Gina priori	184 s	299 s	812 s	68845 s	$< 1s$	$< 1s$	N/A
Scene	115 s	220 s	749 s	12496 s	$< 1s$	$< 1s$	N/A
CT location	459 s	543 s	1553 s	56879 s	N/A	$< 1s$	N/A
Appliances	34 s	48 s	134 s	1913 s	N/A	$< 1s$	N/A
Selected features							
Madelon	22	10	30	8	43	18	500
Gina priori	105	37	106	26	328	405	784
Scene	36	14	56	15	93	220	294
CT location	123	162	74	75	N/A	350	384
Appliances	24	24	10	13	N/A	20	30

The performance of the selected feature sets for each classification benchmark dataset is shown in Fig. 3a and in Fig. 3b for the regression benchmarks. These figures show that *powershap* provides a steady performance on all datasets, consistently achieving the best or equal performance on both the cross-validation and test sets. However, even in cases with equal performance, *powershap* achieves these performances considerably quicker, especially compared to shapicant and forward feature selection. The CT location dataset performances show that forward feature selection tends to overfit on the cross-validation dataset while *powershap* is more robust.

6 Discussion

For the above test results, we used the default automatic *powershap* implementation. However, similar to many other feature selection methods, *powershap* can be further optimized or tuned. One of these optimizations is the use of a *convergence* mode to extract as many informative features as possible. In this mode, *powershap* continues recursively in automatic mode where in every recursive iteration, *powershap* re-executes but with any previously found and selected features excluded from the considered feature set. This process continues, until no more informative features can be found. The convergence mode is especially useful in use-cases with high dimensional feature sets or datasets with a large risk of underfitting as it reduces the feature set dimension each recursive iteration to facilitate finding new informative features. As a basic experiment on the simulation benchmark, using the convergence mode for 500 features and 90% (450) informative features, the percentage of found features increases from around 38% (170) to 73% (330) without adding noise features. However, the duration also increases to the same duration as shapicant.

Other possible optimizations are also applicable to other feature selection methods, such as applying backward feature selection after *powershap* to eliminate any noise features, redundant, or duplicate features. Another possibility is optimizing the used machine learning model to better match the dataset and rerun *powershap*, e.g. by using more CatBoost estimators for datasets with large sample sizes and high dimensional feature sets.

In the benchmarking results, there are datasets where including all features perform equally well or even better, such as in the case of the Gina prior test set. In these cases, the filter methods perform well but output large feature sets, while the forward feature selection performs the worst. Alternatively, *powershap* can be used here as a fast wrapper-based dimensionality reduction method to retain approximately the same performance with a much smaller feature set. As such, there will still be a trade-off for each use-case between filter and wrapper methods based on time and performance.

We are aware that the current design of the benchmarks has some limitations. For the simulation benchmark, the `make_classification` function uses by default a hypercube to create its classification problem, resulting in a linear classification problem, which is inherently easier to classify [15]. The compared filter methods were chosen by their most common usage and availability, however, these are fast and simple methods and are of a much lower complexity than *powershap*. The same argument could be made for our choice of the forward feature selection method (as wrapper method) compared to other methods such as genetic algorithm based solutions. Furthermore, wrapper methods, and thus also *powershap*, are highly dependent on the used model, as the feature selection quality suffers from modeling issues such as for example overfitting and underfitting. Therefore, the true potential achievable performances on the benchmark datasets may differ since every use-case and dataset requires its own tuned model to achieve optimal performance. Additionally, the cut-off values and hyperparameters of none of the methods were optimized and are either set to the same value as in *powershap* or used with their default values. This might impact the performance and could have skewed the benchmark results in both directions. However, choosing the same model and the same values for hyperparameters

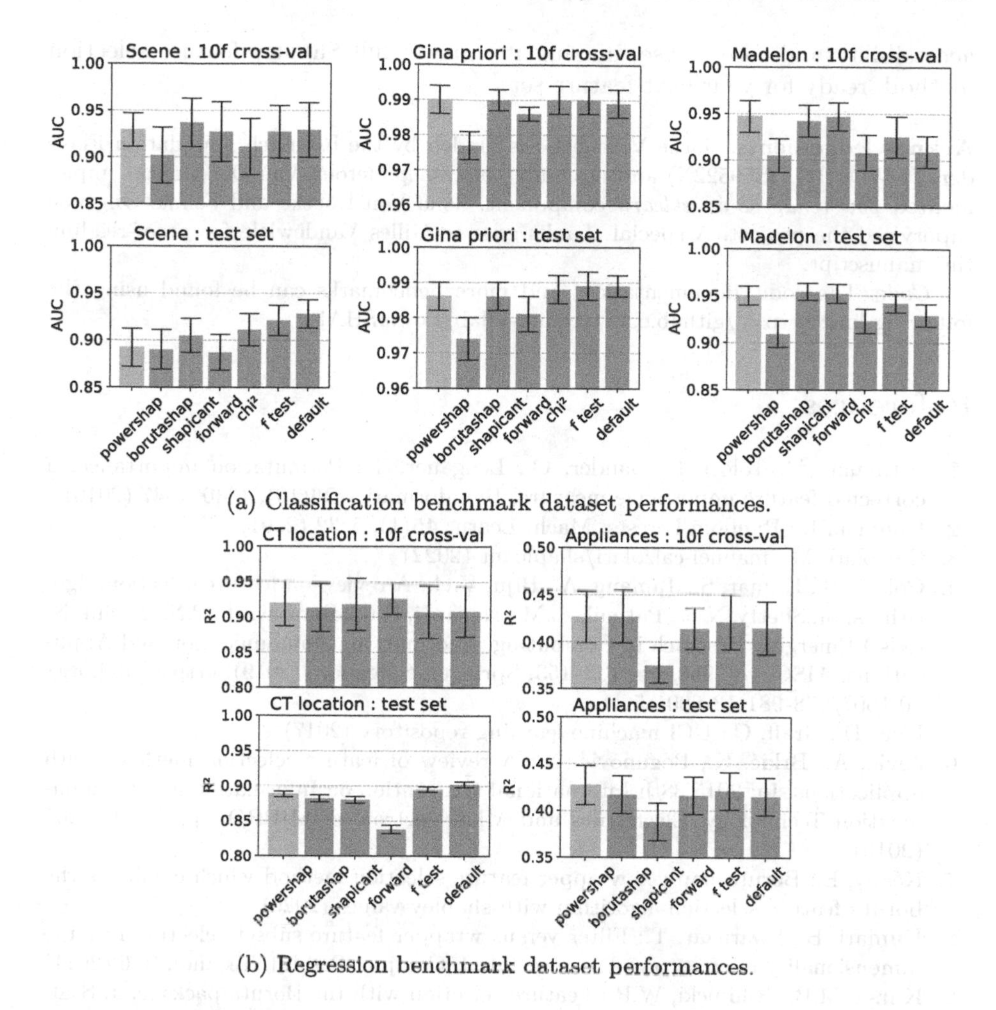

(a) Classification benchmark dataset performances.

(b) Regression benchmark dataset performances.

Fig. 3. Benchmark performances. The error bars represent the standard deviation.

(if possible) in all experiments, reduces potential performance differences and facilitates a fair enough comparison.

7 Conclusion

We proposed *powershap*, a wrapper feature selection method using Shapley values and statistical tests to determine the significance of features. *powershap* uses power calculations to optimize the number of required iterations in an automatic mode to realize fast, strong, and reliable feature selection. Benchmarks indicate that *powershap*'s performance is significantly faster and more reliable than comparable state-of-the-art shap-based wrapper methods. *Powershap* is implemented as an open-source plug-and-play *sklearn* component, increasing its accessibility and ease of use, making it a power-full Shapley feature selection method, ready for your next feature set.

Acknowledgements. Jarne Verhaeghe is funded by the Research Foundation Flanders (FWO, Ref. 1S59522N) and designed *powershap*. Jeroen Van Der Donckt implemented *powershap* as an *sklearn* component. Sofie Van Hoecke and Femke Ongenae supervised the project. A special thanks goes to Gilles Vandewiele for proof-reading the manuscript.

Code. The code, documentation, and more benchmarks can be found using the following link: https://github.com/predict-idlab/PowerSHAP.

References

1. Altmann, A., Toloşi, L., Sander, O., Lengauer, T.: Permutation importance: a corrected feature importance measure. Bioinformatics **26**(10), 1340–1347 (2010)
2. Breiman, L.: Random Forests. Mach. Learn. **45**(1), 5–32 (2001)
3. Calzolari, M.: manuel-calzolari/shapicant (2022)
4. Colaco, S., Kumar, S., Tamang, A., Biju, V.G.: A review on feature selection algorithms. In: Shetty, N.R., Patnaik, L.M., Nagaraj, H.C., Hamsavath, P.N., Nalini, N. (eds.) Emerging Research in Computing, Information, Communication and Applications. AISC, vol. 906, pp. 133–153. Springer, Singapore (2019). https://doi.org/10.1007/978-981-13-6001-5_11
5. Dua, D., Graff, C.: UCI machine learning repository (2017)
6. Jović, A., Brkić, K., Bogunović, N.: A review of feature selection methods with applications. In: 2015 38th International Convention on Information and Communication Technology, Electronics and Microelectronics (MIPRO), pp. 1200–1205 (2015)
7. Keany, E.: Borutashap : A wrapper feature selection method which combines the boruta feature selection algorithm with shapley values (2020)
8. Kumari, B., Swarnkar, T.: Filter versus wrapper feature subset selection in large dimensionality micro array: a review. Int. J. Comput. Sci. Inf. Technol. **2**, 6 (2011)
9. Kursa, M.B., Rudnicki, W.R.: Feature selection with the Boruta package. J. Stat. Softw. **36**(11), 1–13 (2010)
10. Li, J., Cheng, K., Wang, S., et al.: Feature selection: a data perspective. ACM Comput. Surv. **50**(6), 1–45 (2017)
11. Linardatos, P., Papastefanopoulos, V., Kotsiantis, S.: Explainable AI: a review of machine learning interpretability methods. Entropy **23**(1), 18 (2020)
12. Lomax, R.G.: An introduction to statistical concepts. In: Mahwah, N.J. (eds.): Lawrence Erlbaum Associates Publishers (2007)
13. Lundberg, S.M., Lee, S.I.: A unified approach to interpreting model predictions. In: Advances in Neural Information Processing Systems 30, pp. 4765–4774. Curran Associates, Inc. (2017)
14. North, B.V., Curtis, D., Sham, P.C.: A note on the calculation of empirical p values from monte Carlo procedures. Am. J. Hum. Genet. **71**(2), 439–441 (2002)
15. Pedregosa, F., Varoquaux, G., Gramfort, A., et al.: Scikit-learn: machine learning in Python. J. Mach. Learn. Res. **12**, 2825–2830 (2011)
16. Prokhorenkova, L., Gusev, G., et al.: CatBoost: unbiased boosting with categorical features. arXiv:1706.09516 (2019)

17. Seabold, S., Perktold, J.: statsmodels: Econometric and statistical modeling with python. In: 9th Python in Science Conference (2010)
18. Vanschoren, J.: OpenML: gina_priori. https://www.openml.org/d/1042
19. Vanschoren, J.: OpenML: madelon. https://www.openml.org/d/1485
20. Vanschoren, J.: OpenML: scene. https://www.openml.org/d/312

Automated Cancer Subtyping via Vector Quantization Mutual Information Maximization

Zheng Chen[1](✉), Lingwei Zhu[2], Ziwei Yang[3], and Takashi Matsubara[1]

[1] Osaka University, Osaka, Japan
chen.zheng.bn1@gmail.com
[2] University of Alberta, Edmonton, Canada
[3] Nara Institute of Science and Technology, Nara, Japan

Abstract. Cancer subtyping is crucial for understanding the nature of tumors and providing suitable therapy. However, existing labelling methods are medically controversial, and have driven the process of subtyping away from teaching signals. Moreover, cancer genetic expression profiles are high-dimensional, scarce, and have complicated dependence, thereby posing a serious challenge to existing subtyping models for outputting sensible clustering. In this study, we propose a novel clustering method for exploiting genetic expression profiles and distinguishing subtypes in an unsupervised manner. The proposed method adaptively learns categorical correspondence from latent representations of expression profiles to the subtypes output by the model. By maximizing the problem-agnostic mutual information between input expression profiles and output subtypes, our method can automatically decide a suitable number of subtypes. Through experiments, we demonstrate that our proposed method can refine existing controversial labels, and, by further medical analysis, this refinement is proven to have a high correlation with cancer survival rates.

Keywords: Cancer subtypes · Information maximization · Clustering

1 Introduction

Cancer is by far one of the deadliest epidemiological diseases known to humans: consider the breast cancer which is the most prevalent (incidence 47.8% worldwide) and the most well-studied cancer in the world [32], the 5-year mortality rate can still reach 13.6% [1]. Its heterogeneity is considered as the crux of limiting the efficacy of targeted therapies and compromising treatment outcomes since some tumors that differ radically at the molecular level might exhibit highly resemblant morphological appearance [22]. Increasing evidence from modern transcriptomic studies has supported the assumption that each specific cancer is composed of multiple categories (known as cancer subtypes) [4,33]. Reliably

Z. Chen and L. Zhu—Indicates joint first authors.

M.-R. Amini et al. (Eds.): ECML PKDD 2022, LNAI 13713, pp. 88–103, 2023.
https://doi.org/10.1007/978-3-031-26387-3_6

identifying cancer subtypes can significantly facilitate the prognosis and personalized treatment [21]. However, currently there is a fierce debate in the cancer community: given transcriptomic data of one cancer, authoritative resources put that there might be different number of subtypes from distinct viewpoints, that is, the fiducial definition of the subtypes is constantly undergoing calibration [12], suggesting for the majority of cancers the ground-truth labeling remains partially unavailable and awaits better definition.

In the data science community, the lack of ground truth for the cancer data can be addressed as a clustering problem [11], in which the clusters give a hint on the underlying subtypes. Such clustering methods rely crucially on the quality of the data and suitable representations. Modern subtyping methods typically leverage molecular transcriptomic expression profiles (expression profiles in short) which consist of genetic and microRNA (miRNA) expressions that characterize the cancer properties [21,26]. However, several dilemmas exist in the way of fully exploiting the power of expression profiles:

- *High-dimensionality*: the expression profiles are typically of $> 60,000$ dimensions; even after typical preprocessing the dimension can still be $> 10,000$.
- *Scarcity*: cancer data are scarce and costly. Even for the most well-studied breast cancer, the largest public available dataset consists of expression profiles from around only 1500 subjects [30];
- *Dependence*: expression profiles have complicated dependence: a specific expression might be under joint control of several genes, and sometimes such the joint regulation can be circular, forming the well-known gene regulation network [10].

To extract information from the inherently high-dimensional expression profiles for tractable grouping [9], traditional methods preprocess the data via variants of principal components analysis (PCA) or least absolute shrinkage and selection operator (LASSO) [3] for reducing the dimensionality of the data. However, expression profiles with such complicated dependence have already been shown to not perform well with PCA and LASSO [14], since many seemingly less salient features can play an important role in the gene regulation network. Motivated by the resurgence of deep learning techniques, recently the community has seen promising applications leveraging deep autoencoders (AEs) or variational AEs (VAEs) for compressing the data into a lower-dimensional latent space that models the underlying genetic regulation [33]. However, VAEs with powerful autoregressive decoders often ignore the latent spaces [8,25], which runs the risk of overfitting [28]. Furthermore, the latent representation is assumed to be continuous variables (usually Gaussian) [18,31], which is at odds with the inherently categorical cancer subtypes [5]. As a result, those subtyping models might have poor performance as well as generalization ability.

Aside from feature extraction, another issue concerns the grouping process itself. Given extracted features from the expression profiles, the above-mentioned methods usually apply similarity-based clustering algorithms such as K-means for subsequent grouping. However, such methods require strong assumptions on the data and are sensitive to representations [27]: one will have to define a similarity metric for the data (often Euclidean) and find appropriate transformations (such as logarithm transform) as informative features. Unsuitable choices of the metric and transformation can greatly degrade the model performance. Recently, mutual information has been gaining huge popularity in deep representation learning as a replacement for similarity metrics [6,13]: it is the unique measure of relatedness between a pair of variables invariant to invertible transformations of the data, hence one does not need to find a *right* representation [20]. Better yet, if two genes share more than one bit of information, then the underlying mechanism must be more subtle than just *on* and *off*. Such subtlety and more general dependence can be captured by the mutual information [27].

In this paper, we propose a novel, generally applicable clustering method that is capable of fully exploiting the expression profiles and outputting sensible cancer subtyping solutions. Besides tackling the above-mentioned problems in a unified and consistent manner, the proposed method has an intriguing property of automatically adjusting the number of groups thanks to its special architecture, which stands as a sheer contrast to prior methods that predetermine the number of groups by domain knowledge. Before introducing the proposed architecture in Sect. 3, we summarize our contributions as follows:

- (Algorithmic) Inspired by recent work, we propose a novel clustering method vector quantization regularized information maximization (VQ-RIM) for cancer subtyping. VQ-RIM maximizes mutual information in the categorical VQ-VAE model, which results in a combination of VAE reconstruction loss and mutual information loss. (Sect. 3.2)
- (Effective) We compare the clustering results of VQ-RIM against existing ground truth labels (together with controversial labels from the entirety of labels) on different cancer datasets and find that VQ-RIM concords well with the ground truth, which verifies the correctness of VQ-RIM. (Sects. 4.1 and 4.2)
- (Medical) Extensive experiments on distinct cancers verify that VQ-RIM produces subtyping that consistently outperform the controversial labels in terms of enlarged separation of between-group life expectancies. The clearer separation suggests VQ-RIM is capable of better capturing the underlying characteristics of subtypes than controversial labels. We believe such results are far-reaching in providing new insights into the unsettled debate on cancer subtyping.(Sect. 4.2)

2 Related Work

Feature Extraction for Subtyping. Building a model suitable for cancer subtyping is non-trivial as a result of the cancer data scarcity. High dimensionality and data scarcity pose a great challenge to automated models for generating reliable clustering results [31]. Conventionally, the problem is tackled by leveraging classic dimension reduction methods such as PCA [3]. However, since the progress of cancers is regulated by massive genes in a complicated manner (which themselves are under the control of miRNAs), brute-force dimension reduction might run the risk of removing informative features [15]. On the other hand, recently popular AE-based models [21,33], especially VAEs, construct the feature space by reconstructing the input through a multi-dimensional Gaussian posterior distribution in the latent space [31]. The latent posterior learns to model the underlying causalities, which in the cancer subtyping context corresponds to modeling the relationship among expression profiles such as regulation or co-expression [33]. Unfortunately, recent investigation has revealed that VAEs with powerful autoregressive decoders easily ignore the latent space. As a result, the posterior could be either too simple to capture the causalities; or too complicated so the posterior distribution becomes brittle and at the risk of posterior collapse [2,25]. Moreover, the Gaussian posterior is at odds with the inherently categorical cancer subtypes [5].

In this paper, we propose to leverage the categorical VQ-VAE to address the aforementioned issues: (i) VQ-VAE does not train its decoder, preventing the model from ignoring its latent feature space resulting from an over-powerful decoder; (ii) VQ-VAE learns categorical correspondence between input expression profiles, latent representations, and output subtypes, which theoretically suggests better capability of learning more useful features. (iii) the categorical latent allows the proposed model to automatically set a suitable number of groups by plugging in mutual information maximization classifier, which is not available for the VAEs.

Information Maximization for Subtyping. Cancer subtyping is risk-sensitive since misspecification might incur an unsuitable treatment modality. It is hence desired that the clustering should be *as certain as possible for individual prediction, while keeping subtypes as separated as possible* [7,11]. Further, to allow for subsequent analysis and further investigation of medical experts, it is desired that the method should output probabilistic prediction for each subject. In short, we might summarize the requirements for the subtyping decision boundaries as (i) should not be overly complicated; (ii) should not be located at where subjects are densely populated; (iii) should output probabilistic predictions. These requirements can be formalized via the information-theoretic objective as maximizing the mutual information between the input expression profiles and the output subtypes [19,29]. Such objective is problem-agnostic, transformation-invariant, and unique for measuring the relationship between pairs of variables. Superior performance over knowledge-based heuristics has been shown by exploiting such an objective [27].

3 Method

3.1 Problem Setting

Let $\mathcal{X}$ be a dataset $\mathcal{X} = \{\boldsymbol{x}_1, \dots, \boldsymbol{x}_N\}$, where $\boldsymbol{x}_i \in \mathbb{R}^d, 1 \leq i \leq N$ are d-dimensional vectors consisting of cancer expression profiles. For a given $\boldsymbol{x}$, our goal lies in determining a suitable cancer subtype $y \in \{1, 2, \dots, K\}$ given $\boldsymbol{x}$, where K is not fixed beforehand and needs to be automatically determined. Numeric values such as $y = 1, \dots, K$ do not bear any medical interpretation on their own and simply represent distinct representations due to the underlying data. It is worth noting while a label set $\mathcal{Y}$ is available, it comprises a small subset of ground-truth labels $\mathcal{Y}_{gt} := \{y_{gt}\}$ that have been medically validated and a larger portion of *controversial* labels $\mathcal{Y}_c := \{y_c\}$, with $\mathcal{Y}_{gt} = \mathcal{Y} n \mathcal{Y}_c$. Our approach is to compare the clustering result y of the proposed method against ground truth labels y_{gt} to see if they agree well, as a first step of validation. We then compare y against controversial labels y_c and conduct extensive experiments to verify that the proposed method achieves improvement upon the subtyping given by y_c. Our goal is to unsupervisedly learn a discriminative classifier D which outputs conditional probability $P(y|\boldsymbol{x}, D)$. Naturally it is expected that $\sum_{k=1}^{K} P(y = k|\boldsymbol{x}, D) = 1$ and we would like D to be probabilistic so the uncertainty associated with assigning data items can be quantitized. Following [28], we assume the marginal class distribution $P(y|D)$ is close to the prior $P(y)$ for all k. However, unlike prior work [19,28] we do not assume the amount of examples per class in $\mathcal{X}$ is uniformly distributed due to the imbalance of subtypes in the data.

3.2 Proposed Model

Information Maximization. Given expression profiles of subject $\boldsymbol{x}$, the discriminator outputs a K-dimensional probability logit vector $D(\boldsymbol{x}) \in \mathbb{R}^K$. The probability of $\boldsymbol{x}$ belonging to any of the K subtypes is given by the softmax parametrization:

$$P(y = k|\boldsymbol{x}, D) = \frac{e^{D_k(\boldsymbol{x})}}{\sum_{k=1}^{K} e^{D_k(\boldsymbol{x})}},$$

where $D_k(\boldsymbol{x})$ denotes the k-th entry of the vector $D(\boldsymbol{x})$. Let us drop the dependence on D for uncluttered notation. It is naturally desired that the individual prediction be as certain as possible, while the distance between the predicted subtypes as large as possible. This consideration can be effectively reflected by the mutual information between the input expression profiles and the output prediction label. Essentially, the mutual information can be decomposed into the following two terms:

$$\hat{I}(\boldsymbol{x}, y) := \underbrace{-\sum_{k=1}^{K} P(y = k) \log P(y = k)}_{\widehat{\mathcal{H}}(P(y))} + \underbrace{\alpha \sum_{i=1}^{N} \frac{1}{N} \sum_{k=1}^{K} P(y = k|\boldsymbol{x}_i) \log P(y = k|\boldsymbol{x}_i)}_{-\widehat{\mathcal{H}}(P(y|\mathcal{X}))}. \tag{1}$$

which are the marginal entropy of labels $\widehat{\mathcal{H}}(P(y = k))$ and the conditional entropy $\widehat{\mathcal{H}}(P(y|\mathcal{X}))$ approximated by N Monte Carlo samples $\boldsymbol{x}_i, i \in \{1, \dots, N\}$. α is an adjustable parameter for weighting the contribution, setting $\alpha = 1$ recovers the standard mutual information formulation [19]. This formulation constitutes the *regularized information maximization (RIM)* part of the proposed method. The regularization effect can be seen from the following:

- Conditional entropy $\widehat{\mathcal{H}}(P(y|\mathcal{X}))$ encourages confident prediction by minimizing uncertainty. It effectively captures the modeling principles that decision boundaries should not be located at dense population of data [11].
- Marginal entropy $\widehat{\mathcal{H}}(P(y))$ aims to separate the subtypes as far as possible. Intuitively, it attempts to keep the subtypes *uniform*. Maximizing only $\widehat{\mathcal{H}}(P(y|\mathcal{X}))$ tends to produce degenerate solutions by removing subtypes [6,19], hence $\widehat{\mathcal{H}}(P(y))$ serves as an effective regularization for ensuring non-trivial solutions.

Categorical Latents Generative Feature Extraction. Recent studies have revealed that performing RIM alone is often insufficient for obtaining stable and sensible clustering solutions [6,20,28]: Discriminative methods are prone to overfitting spurious correlations in the data, e.g., some entry A in the expression profiles might appear to have direct control over certain other entries B. The model might naïvely conclude that the appearance of B shows positive evidence of A. However, such relationship is in general not true due to existence of complicated biological functional passways: Such pathways have complex (sometimes circular) dependence between A and B [24]. Since discriminative methods model $P(y|\boldsymbol{x})$ but not the data generation mechanism $P(\boldsymbol{x})$ (and the joint distribution $P(\boldsymbol{x}, y)$) [11], such dependence between genes and miRNAs might not be effectively captured by solely exploiting the discriminator, especially given the issues of data scarcity and high dimensionality.

A generative model that explicitly captures the characteristics in $P(\boldsymbol{x})$ is often introduced as a rescue for leveraging RIM-based methods [13,23,28]. Such methods highlight the use of VAEs for modeling the latent feature spaces underlying input $\mathcal{X}$: given input $\boldsymbol{x}$, VAEs attempt to compress it to a lower-dimensional latent $\boldsymbol{z}$, and reconstruct $\tilde{\boldsymbol{x}}$ from $\boldsymbol{z}$. Recently there has been active research on leveraging VAEs for performing cancer subtyping [31,33]. However, existing literature leverage continuous latents (often Gaussian) for tractability, which is at odds with the inherently categorical cancer subtypes. Furthermore, VAEs often ignore the latents which implies the extracted feature space is essential dismissed and again runs the risk of overfitting [2].

We exploit the recent vector quantization variational auto-encoder (VQ-VAE) [25] as the generative part of the proposed architecture. The categorical latents of VQ-VAE are not only suitable for modeling inherently categorical cancer subtypes, but also avoids the above-mentioned latent ignoring problem [18]. In VQ-VAE, the latent embedding space is defined as $\{e_i\} \in \mathbb{R}^{M \times l}$, where M denotes the number of embedding vectors and hence a M-way categorical distribution. $l < d$ is the dimension of each latent embedding vector $e_i, i \in \{1, \dots, M\}$. VQ-VAE maps input $\boldsymbol{x}$ to a latent variable $\boldsymbol{z}$ via its encoder $z_e(\boldsymbol{x})$ by performing a nearest neighbor search among the embedding vectors e_i, and output a reconstructed vector $\tilde{\boldsymbol{x}}$ via its decoder $\boldsymbol{z}_q$. VQ-VAE outputs a deterministic posterior distribution q such that

$$q(\boldsymbol{z} = k|\boldsymbol{x}) = \begin{cases} 1, & \text{if } k = \arg\min_j ||z_e(\boldsymbol{x}) - e_j||_2^2 \\ 0, & \text{otherwise} \end{cases} \tag{2}$$

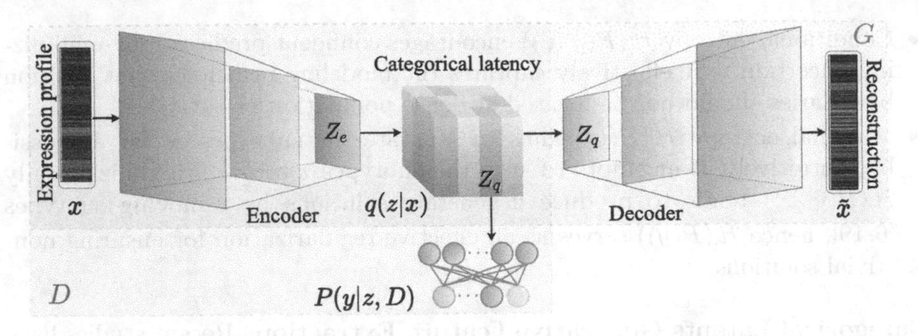

Fig. 1. Overview of the proposed system. D denotes the discriminator, G denotes the generator.

The decoder does not possess gradient and is trained by copying the gradients from the encoder. The final output of the decoder is the log-posterior probability $\log P(\boldsymbol{x}|\boldsymbol{z}_q)$ which is part of the reconstruction loss.

Architecture and Optimization. We propose a novel model for clustering expression profiles as shown in Fig. 1. The model consists of a discriminator denoted as D that maximizes the mutual information and a generator G that aims to reconstruct the input via modeling a categorical underlying latent feature space spanned by $\{e_i\}$. D and G are deeply coupled via the latent embeddings $\boldsymbol{z}$, which is made possible through the fact the decoder of VQ-VAE does not possess gradients and hence the embedding space can be controlled by only the encoder and the discriminator. In prior work, the generator is often architecturally independent from the discriminator and is only weakly related through loss functions [13,20,28]. Intuitively, one can consider the proposed model attempts to simultaneously minimize reconstruction loss as well as maximize the mutual information:

$$\mathcal{L} := \underbrace{\widehat{\mathcal{H}}(P(y)) - \widehat{\mathcal{H}}(P(y|\boldsymbol{z})) - R(\lambda)}_{\mathcal{L}_D} + \underbrace{\log P(\boldsymbol{x}|\boldsymbol{z}_q) + ||\mathrm{sg}[\boldsymbol{z}_e] - e||_2 + ||\boldsymbol{z}_e - \mathrm{sg}[e]||_2}_{\mathcal{L}_G} \tag{3}$$

where $\mathcal{L}_D, \mathcal{L}_G$ denote the discriminator loss and the generator loss, respectively. $R(\lambda)$ is a possible regularizer that controls the weight growth, e.g. $R(\lambda) := \frac{\lambda}{2}||w^T w||_2^2$, where w denotes the weight parameters of the model. $\mathrm{sg}[\cdot]$ denotes the stop gradient operator.

Automatically Setting Number of Subtypes. The proposed model can automatically determine suitable number of subtypes by exploiting hidden information contained in the expression profiles which is not available to conventional methods such as K-means relying on prior knowledge. The automatic subtyping is made possible via the deeply coupled latents and the discriminator: the multi-layer perceptron in the discriminator outputs the logarithm of posterior distribution $\log q(\boldsymbol{z}|\boldsymbol{x})$. However, by definition of Eq. (2) the posterior is deterministic, which suggests $\log q(\boldsymbol{z}|\boldsymbol{x})$ must either be 0 or tend to $-\infty$. The subsequent softmax layer hence outputs:

$$P(y = k|\boldsymbol{z}) = \begin{cases} \frac{q(z=k|\boldsymbol{x})}{\sum_{k=1}^{K} q(z=k|\boldsymbol{x})}, & \text{if } k = \arg\min_j ||z_e(\boldsymbol{x}) - e_j||_2^2 \\ 0, & \text{otherwise} \end{cases} \tag{4}$$

We can set K to a sufficient large integer $\tilde{K}$ initially that covers the maximum possible number of subtypes. Since the nearest neighbor lookup of VQ-VAE typically only updates a small number of embeddings e_j, by Eq. (4) we see for any unused $e_i, i \neq j$ the clustering probability is zero, which suggests the number of subtypes K will finally narrow down to a much smaller number $K \ll \tilde{K}$.

4 Experiments

The expression profile data used in this study were collected from the world's largest cancer gene information database Genomic Data Commons (GDC) portal. All of the used expression data were generated from cancer samples prior to treatment.

We utilized the expression profiles of three representative types of cancer for experiments:

- Breast invasive carcinoma (BRCA): BRCA is the most prevalent cancer in the world. Its expression profiles were collected from the Illumina Hi-Seq platform and the Illumina GA platform.
- Brain lower grade glioma (LGG): the expression profiles were collected from the Illumina Hi-Seq platform.
- Glioblastoma multiforme (GBM): the expression profiles were collected from the Agilent array platform. Results on this dataset are deferred to the appendix.

These datasets consist of continuous-valued expression profiles (feature length: 11327) of 639, 417 and 452 subjects, respectively. Additional experimental results and hyperparameters can be seen in Appendix Section A available at https://arxiv.org/abs/2206.10801.

The experimental section is organized as follows: we first compare the clustering results with the ground truth labels $\mathcal{Y}_{gt}$ in Sect. 4.1 to validate the proposed method. We show in Sect. 4.2 that VQ-RIM consistenly re-assigns subjects to different subtypes and produces one more potential subtype with enlarged separation in between-group life expectancies, which in turn suggests VQ-RIM is capable of better capturing the underlying characteristics of subtypes. Extensive ablation studies on both the categorical generator (VQ-VAE) and the information maximizing discriminator (RIM) are performed to validate the proposed architecture in Sect. 4.3. We believe the VQ-RIM subtyping result is far-reaching and can provide important new insights to the unsettled debate on cancer subtyping.

4.1 Ground Truth Comparison

For validating the correctness of VQ-RIM, we show an example in Fig. 2, i.e., the Basal-like cancer subtype of BRCA that has been well-studied and extensively validated by human experts and can be confidently subtyped, which can be exploited as the ground-truth labels $\mathcal{Y}_{gt}$.

However, other subtypes lack such well-verified labels and are regarded as the controversial labels $\mathcal{Y}_c$. The left subfigure of Fig. 2 shows the two principal axes of Basal-like expression profiles after PCA. The blue triangles in the right subfigure indicates the difference between $\mathcal{Y}_{gt}$ and the VQ-RIM result. It can be seen that VQ-RIM agrees well with the ground truth.

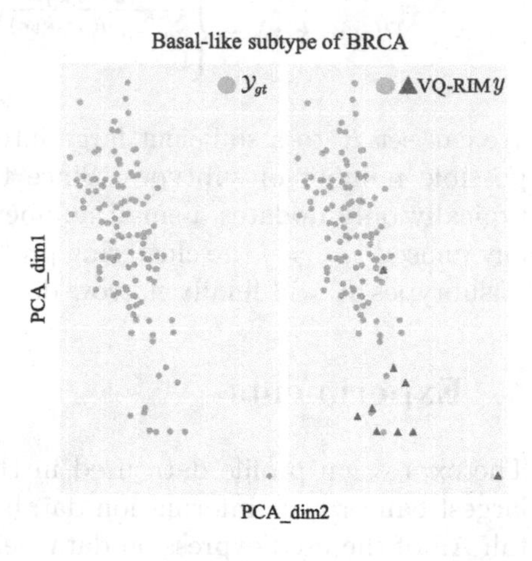

Fig. 2. Comparison between $\mathcal{Y}_{gt}$ and the VQ-RIM label y on the Basal-like subtype of BRCA.

4.2 Controversial Label Comparison

Subtype Comparison. We compare existing controversial labels $\mathcal{Y}_c$ with the clustering results of VQ-RIM in Fig. 3. VQ-RIM output sensible decision boundaries that separated the data well and consistently produced one more subtype than $\mathcal{Y}_c$. As confirmed in Sect. 4.1, the Basal-like subtype concorded well with the VQ-RIM Cluster A. On the other hand, other subtypes exhibited significant differences: controversial labels seem to compactly fit into a fan-like shape in the two-dimensional visualization. This is owing to the human experts' heuristics in subtyping: intuitively, the similarity of tumors in the clinical variables such as morphological appearance often renders them being classified into an identical subtype. However, cancer subtypes are the result of complicated causes on the molecular level. Two main observations can be made from the BRCA VQ-RIM label: (1) Luminal A was divided into three distinct clusters C,D,E. Cluster E now occupies the left and right wings of the fan which are separated by Cluster B and C; (2) A new subtype Cluster F emerged from Luminal B, which was indistinguishable from Cluster E if naïvely viewed from the visualization. This counter-intuitive clustering result confirmed the complexity of cancer subtypes in expression profiles seldom admits simple representations as was done in the controversial labels. A similar conclusion holds as well for other datasets such as LGG: IDH mut-codel was divided into two distinct subtypes (Cluster A, B), among which the new subtype Cluster A found by VQ-RIM occupied

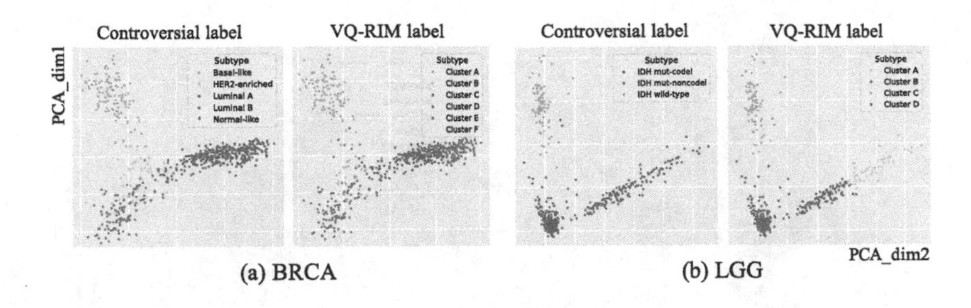

Fig. 3. PCA visualization of the first two principal axes for BRCA and LGG.

the right wing of IDH mut-codel. In later subsections, the one more cluster and re-assignment of VQ-RIM are justified by analyzing the subtype population and from a medical point of view. Due to page limit, we provide analysis focusing on BRCA only.

Label Flows. The controversial labels might run the risk of *over-simplifying assignment* which refers to that in the regions overlapped with several distinct subtypes, controversial labels put all subjects into one of them without further identifying their sources. Such assignment can be illustrated by Fig. 4. Here, Fig. (4a) plots the sample distribution with darker colors indicating denser population of samples. It is visible that the samples can be assigned to five clusters. However, by injecting subtyping label information it is clear from Fig. (4b) that in the lower left corner there existed strong overlaps of three different subtypes. Controversial labels assigned them to a single subtype Luminal A. VQ-RIM, on the other hand, was capable of separating those three subtypes. This separation can be seen from Fig. (4c) which compares the two labeling when setting the number of VQ-RIM subtypes to 5 in accordance with controversial labels, or to 6 by setting K to a sufficiently large value and automatically determines the suitable number of subtypes. In either case, VQ-RIM consistently separated the Luminal A into three distinct subtypes: (B,C,E) in 5 subtypes case and (C^*, D^*, E^*) in the 6 subtypes case. In the next subsection, we verify the effectiveness of such finer-grained subtyping by performing survival analysis.

Medical Evaluation. To demonstrate the clinical relevance of the identified subtypes, we perform subtype-specific survival analysis by the Kaplan-Meier (KM) estimate. KM estimate is one of the most frequently used statistical methods for survival analysis, which we use to complementarily validate the VQ-RIM labels from a clinical point of view [16]. KM compares survival probabilities in a given length of time between different sample groups. The KM estimator is given by: $\widehat{S} = \prod_{i:t_i<t} \frac{n_i - d_i}{n_i}$, where n_i is the number of samples under observation at time i and d_i is the number of individuals dying at time i. The survival analysis graph is plotted between estimated survival probabilities (on Y-axis)

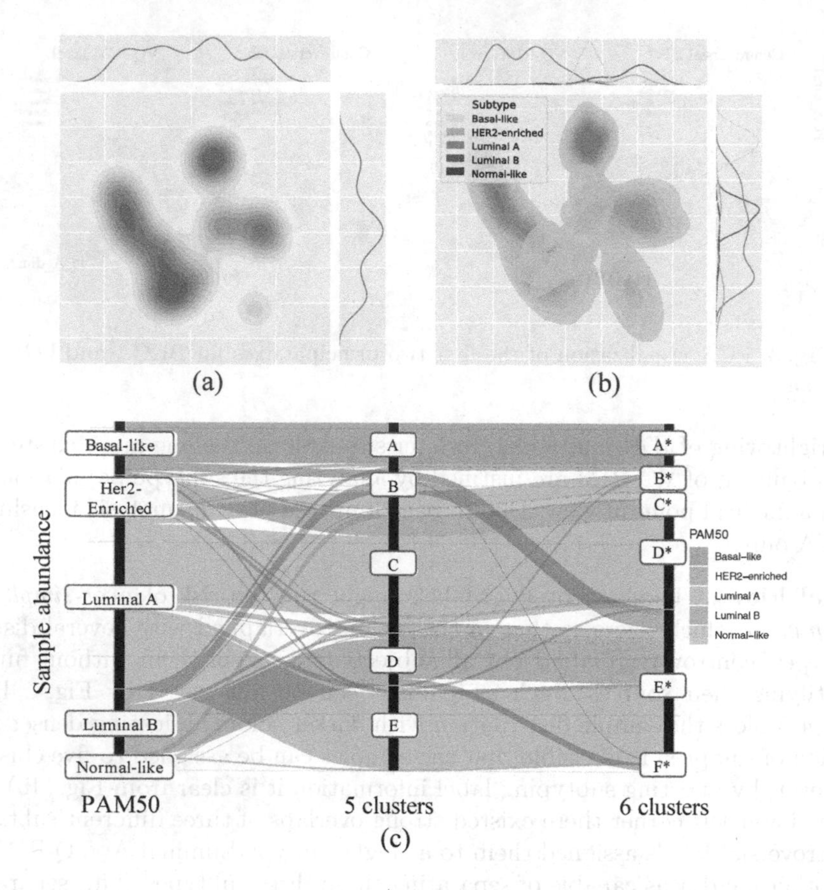

Fig. 4. (a) t-SNE visualization of the sample distribution on BRCA. (b) t-SNE of the samples with controversial labels. (c) label flows from the controversial labels (left) to VQ-RIM 5 subtypes (mid) and 6 subtypes (right).

and the time passed after samples entry into the study (on X-axis), where the survival curve is drawn as a step function and falls only when a subject dies.

We can compare curves for different subtypes by examining gaps between the curves in horizontal or vertical direction. A vertical gap means that at a specific time point, samples belonging to one subtype had a greater fraction of surviving, while a horizontal one means that it takes longer for these samples to experience a certain fraction of deaths. The survival curves can also be compared statistically by testing the null hypothesis, i.e. there is no difference regarding survival situation among different groups, which can be tested by classical methods like the log-rank test and the Cox proportional hazard test.

Figure 5 shows the KM survival analysis graph for BRCA samples, based on the PAM50 subtyping system and VQ-RIM subtypes. Compared with the PAM50, the survival curves of VQ-RIM subtypes are more significantly

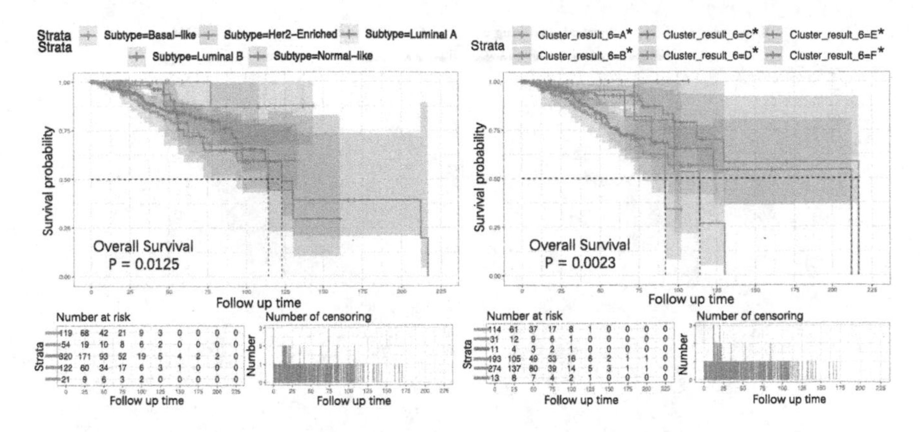

Fig. 5. Kaplan-Meier survival analysis within each identified subtype group (right) compared with original subtyping system (left) as a baseline. The line in different colors represent patients from different subtypes. P-value was calculated by Kaplan-Meier analysis with the log-rank test.

separated. Log-rank test also shows that there is significant difference in between-group survival with a smaller p-value of 0.0023 compared against the PAM50 ($\mathcal{Y}_c$). Smaller p-values indicate better subtyping results. We indicate the subtype-specific median survival time with dashed lines. It is visible that VQ-RIM performed better in identifying subtypes with large median survival time differences.

4.3 Ablation Study

In this section, we conduct comprehensive ablation experiments to further strengthen the effectiveness of VQ-RIM. Specifically, we validate the VQ part and RIM part respectively by comparing VQ-RIM against the following combinations:

- we replace the VQ part with AE and VAEs with continuous latent which have been exploited for subtyping in [31,33]. The expression profiles are compressed into continuous latent feature spaces for subsequent RIM clustering.
- we replace the RIM part with existing classic clustering algorithms such as K-Means, spectral clustering, and Gaussian mixture models [3]. Categorical latent variables from VQ-VAE are fed into them for subtyping.

Results of the ablation studies can be seen by inspecting Fig. 6 row-wise and column-wise, respectively. All methods applied $\mathcal{Y}_c$ (PAM50) for labeling. By inspecting the results row-wise, a first observation is that for all clustering methods AE and VAE tended to disperse data points. On the other hand, it can be seen from VQ-VAE clustering that subjects from distinct subtypes compactly located in lower-dimensional spaces. Column-wise inspection indicates that compared to other clustering methods, RIM tended to more cohesively aggregate the in-group points: by contrast, it is visible from the VAE row that only Basal-like

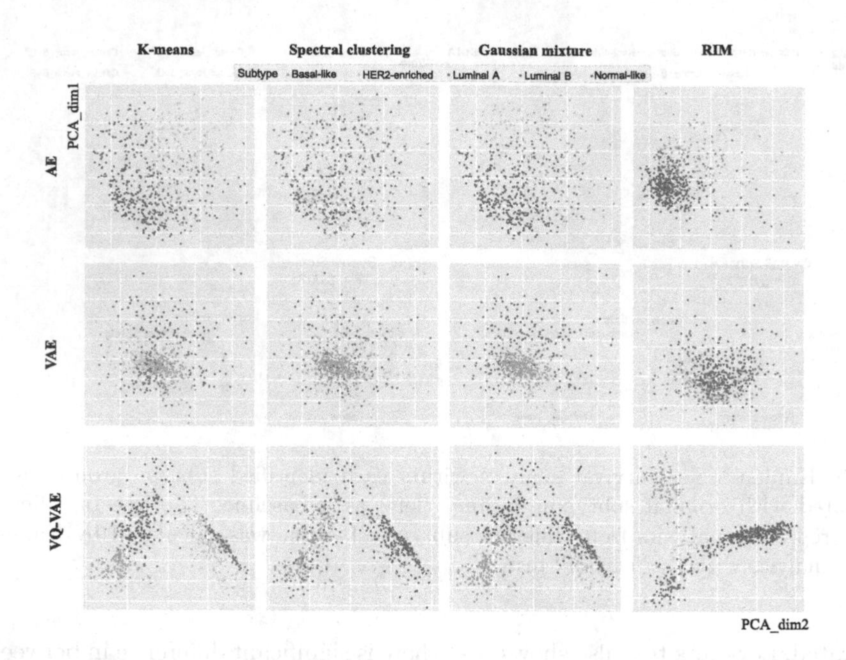

Fig. 6. PCA visualization of the first two principal axes for BRCA using different clustering methods. The number of cluster number is determined by $\mathcal{Y}_c$ (PAM50).

subjects were cohesively grouped. Among all compared methods, VQ-RIM stood out as the subjects were located in lower dimensional spaces from which clear decision boundaries emerged. The clear separation of VQ-RIM can be attributed to the underlying nearest neighbor search: such search essentially performed a preliminary grouping on the data, which greatly facilitated later separation. This observation is consistent with the recent finding of pseudo-labeling that explicitly performs K-means in creating preliminary clusters [20]. Besides the aforementioned qualitative evaluation, we also quantitatively measure in Table 1 the scores of all the clustering results in Fig. 6 by using the three well-accepted metrics: Normalized Mutual Information (NMI), Sihouette coefficient scores (Silhouette) and p-value of survival analysis [17].

However, in Sect. 4.2, the labeling $\mathcal{Y}_c$ might not be the best medically even if the clustering result accords well with human intuition. In Fig. 7 we focus on VQ-VAE, and set the number of clusters for RIM to a sufficiently large value and let RIM automatically determines a suitable number of subtypes. We term this strategy automatic VQ-VAE in the last row of Table 1.

For other clustering algorithms, the number of clusters is determined from the Silhouette coefficient scores and the elbow method [17]. It is visible that clustering algorithms other than RIM tended to reduce the number of subtypes for higher scores. By contrast, VQ-RIM produced one more subtype. This automatic VQ-RIM clustering was superior from a medical perspective since it achieved

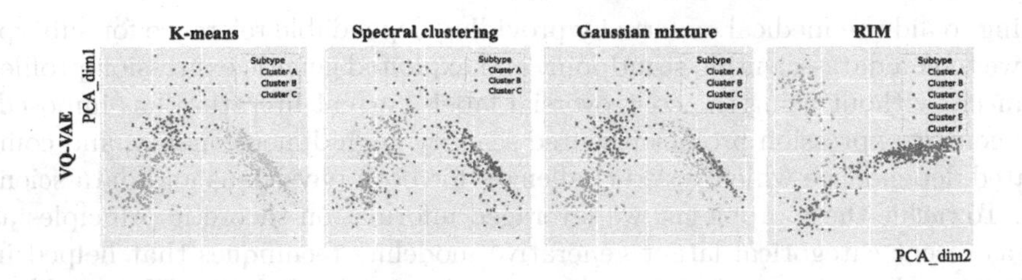

Fig. 7. The number of clustering is determined by combining the results from the elbow method, not used for RIM.

greatest subtyping result as demonstrated by the smallest p-value of 0.0023. Furthermore, algorithmically it is better than plain VQ-RIM as it achieved the highest scores of 0.63 and 0.54 of NMI and Silhouette among all ablation choices.

Table 1. Metrics used for measuring the clustering results of Figs. 6 and 7. The top three rows show the number of clustering determined by $\mathcal{Y}_c$ (PAM50), while the last row shows the number of clustering is automatically determined.

Generator	Discriminator	NMI ↑	Silhouette ↑	p-value ↓
AE +	K-Means	0.34	0.13	0.0861
	Spectral clustering	0.01	0.01	0.1523
	Gaussian mixtures	0.34	0.11	0.0734
	RIM	0.31	0.13	0.0834
VAE +	K-Mmeans	0.29	0.17	0.0812
	Spectral clustering	0.06	0.06	0.1382
	Gaussian mixtures	0.28	0.17	0.0899
	RIM	0.33	0.22	0.0732
VQ-VAE +	K-Means	0.33	0.29	0.0154
	Spectral clustering	0.05	0.04	0.0194
	Gaussian mixtures	0.44	0.24	0.0166
	RIM	0.55	0.29	0.0042
VQ-VAE +(Automatic)	K-Means	0.42	0.29	0.0145
	Spectral clustering	0.06	0.05	0.0188
	Gaussian mixtures	0.51	0.32	0.0132
	RIM	**0.63**	**0.54**	**0.0023**

5 Discussion and Conclusion

In this paper we were concerned with the cancer subtyping problem that aimed to disclose the difference between subtypes within a specific cancer. Existing literature is having an unsettled debate over the subtyping problem, with various definition and suitable number of subtypes put forward from distinct viewpoints.

Aiming to aid the medical experts by providing dependable reference for subtyping, we took a data-scientific standpoint and exploited genetic expression profiles of cancers without using the controversial labels current literature has imposed. Such genetic expression profiles featured scarcity, high dimensionality, and complicated dependence which posed challenges for both physicians and data scientists. To tackle these problems, we leveraged information-theoretic principles as well as recent categorical latent generative modeling techniques that helped in minimizing clustering confusion and maximizing interpretability. The resultant novel model: Vector Quantization Regularized Information Maximization (VQ-RIM) can better reveal the intrinsic difference between cancer genetic expression profiles and based on which automatically decide a suitable number of subtypes. The experiment on ground-truth BRCA cancer verified the correctness of VQ-RIM, while more extensive experiments on multiple authoritative datasets consisting of various cancers showed the difference between VQ-RIM results and the controversial labels. By comprehensive analysis from both data scientific and medical views, we illustrated that the different subtyping result yielded by VQ-RIM consistently outperformed existing ones in terms of survival analysis, and contributed important new insights into the unsettled debate.

The future work consists of two interesting directions: (1) to further validate the effectiveness of VQ-RIM, comprehensive experiments on all available cancer datasets and comparison with their existing labeling might be necessary. (2) the VQ-RIM architecture might not only work well with cancer data but also be generally applicable on radically different data such as images, voices that inherently exploit discrete nature of the data.

Acknowledgement. This work was supported by JST Mirai Program (JPMJMI20B8) and JST PRESTO (JPMJPR21C7), Japan.

References

1. Cancer today - IARC. https://gco.iarc.fr/today/home. Accessed: 2021
2. Alemi, et al.: Fixing a broken ELBO. In: ICML-18, pp. 159–168 (2018)
3. Alexe, G., et al.: Analysis of breast cancer progression using principal component analysis and clustering. J. Biosci. **32**, 1027–1039 (2007). https://doi.org/10.1007/s12038-007-0102-4
4. Bair, E., Tibshirani, R.: Machine learning methods applied to DNA microarray data can improve the diagnosis of cancer. ACM SIGKDD Explor. Newslett. **5**, 48–55 (2003)
5. Berger, et al.: A comprehensive pan-cancer molecular study of gynecologic and breast cancers. Cancer cell. **33**, 690–705 (2018)
6. Boudiaf, M., Ziko, I., Rony, J., Dolz, J., Piantanida, P., Ben Ayed, I.: Information maximization for few-shot learning. In: NIPS-20, pp. 2445–2457 (2020)
7. Bridle, J., Heading, A., MacKay, D.: Unsupervised classifiers, mutual information and phantom targets. In: NIPS, pp. 1096–1101 (1991)
8. Chen, X., et al.: Variational lossy autoencoder. In: ICLR-17 (2017)
9. Dagdia, Z.C., et al.: Rough set theory as a data mining technique: a case study in epidemiology and cancer incidence prediction. In: ECML-PKDD, pp. 440–455 (2018)

10. Dizaji, K.G., Wang, X., Huang, H.: Semi-supervised generative adversarial network for gene expression inference. In: KDD 2018, pp. 1435–1444 (2018)
11. Grandvalet, Y., Bengio, Y.: Semi-supervised learning by entropy minimization. In: NIPS-04, pp. 529–536 (2004)
12. Heiser, L.M., Sadanandam, A., et al.: Subtype and pathway specific responses to anticancer compounds in breast cancer. PNAS **8**, 2724–2729 (2012)
13. Hu, W., et al.: Learning discrete representations via information maximizing self-augmented training. In: ICLR-17, pp. 1558–1567 (2017)
14. Jiang, B., Ding, C., Bin, L.: Covariate-correlated lasso for feature selection. In: ECML-PKDD 2014, pp. 595–606 (2014)
15. Jolliffe, I.T., Cadima, J.: Principal component analysis: a review and recent developments. Philos. Trans. Royal Society A: Math. Phys. Eng. Sci. **374**(2065), 20150202 (2016)
16. Kaplan, E.L., Meier, P.: Nonparametric estimation from incomplete observations. J. Am. Stat. Assoc. **282**, 457–481 (1958)
17. Kaufman, L., Rousseeuw, P.J.: Finding groups in data: an introduction to cluster analysis. John Wiley & Sons (2009)
18. Kingma, D.P., Welling, M.: Auto-Encoding Variational Bayes. In: ICLR-14, pp. 1–9 (2014)
19. Krause, A., Perona, P., Gomes, R.: Discriminative clustering by regularized information maximization. In: NIPS-10, pp. 775–783 (2010)
20. Liang, et al.: Do we really need to access the source data? Source hypothesis transfer for unsupervised domain adaptation. In: ICML-20, pp. 6028–6039 (2020)
21. Liang, C., Shang, M., Luo, J.: Cancer subtype identification by consensus guided graph autoencoders. Bioinformatics **24**, 4779–4786 (2021)
22. Liu, C.C., et al.: LRP6 overexpression defines a class of breast cancer subtype and is a target for therapy. PNAS **11**, 5136–41 (2010)
23. Löwe, S., O' Connor, P., Veeling, B.: Putting an end to end-to-end: gradient-isolated learning of representations. In: NIPS-19, pp. 1–13 (2019)
24. Mangan, S., Alon, U.: Structure and function of the feed-forward loop network motif. PNAS **21**, 11980–11985 (2003)
25. van den Oord, A., Vinyals, O., Kavukcuoglu, K.: Neural discrete representation learning. In: NIPS-17, pp. 6309–6318 (2017)
26. de Ronde, J., Wessels, L., Wesseling, J.: Molecular subtyping of breast cancer: ready to use? Lancet Oncol. **4**, 306–307 (2010)
27. Slonim, N., Atwal, G.S., Tkačik, G., Bialek, W.: Information-based clustering. Proceed. Nat. Acad. Sci. PNAS **51**, 18297–18302 (2005)
28. Springenberg, J.T.: Unsupervised and semi-supervised learning with categorical generative adversarial networks. In: ICLR-16, pp. 1–12 (2016)
29. Tschannen, M., Djolonga, J., Rubenstein, P.K., Gelly, S., Lucic, M.: On mutual information maximization for representation learning. In: ICLR-20, pp. 1–12 (2020)
30. Weinstein, et al.: The cancer genome atlas pan-cancer analysis project. Nat. Genetics **45**, 1113–1120 (2013)
31. Withnell, et al.: XOmiVAE: an interpretable deep learning model for cancer classification using high-dimensional omics data. Brief. Bioinf. **22**(6), bbab315 (2021)
32. Xi, J., et al.: Tolerating data missing in breast cancer diagnosis from clinical ultrasound reports via knowledge graph inference. In: KDD 2021, pp. 3756–3764 (2021)
33. Yang, B., Xin, T.T., Pang, S.M., Wang, M., Wang, Y.J.: Deep Subspace Mutual Learning for cancer subtypes prediction. Bioinformatics **21**, 3715–3722 (2021)

Wasserstein t-SNE

Fynn Bachmann[1,2(✉)], Philipp Hennig[2], and Dmitry Kobak[2]

[1] University of Hamburg, Hamburg, Germany
fsvbach@gmail.com

[2] University of Tübingen, Tübingen, Germany
{philipp.hennig,dmitry.kobak}@uni-tuebingen.de

Abstract. Scientific datasets often have hierarchical structure: for example, in surveys, individual participants (samples) might be grouped at a higher level (units) such as their geographical region. In these settings, the interest is often in exploring the structure on the unit level rather than on the sample level. Units can be compared based on the distance between their means, however this ignores the within-unit distribution of samples. Here we develop an approach for exploratory analysis of hierarchical datasets using the Wasserstein distance metric that takes into account the shapes of within-unit distributions. We use t-SNE to construct 2D embeddings of the units, based on the matrix of pairwise Wasserstein distances between them. The distance matrix can be efficiently computed by approximating each unit with a Gaussian distribution, but we also provide a scalable method to compute exact Wasserstein distances. We use synthetic data to demonstrate the effectiveness of our *Wasserstein t-SNE*, and apply it to data from the 2017 German parliamentary election, considering polling stations as samples and voting districts as units. The resulting embedding uncovers meaningful structure in the data.

Keywords: Wasserstein metric · t-SNE · Dimensionality reduction · Election data · Hierarchical data · Optimal transport

1 Introduction

We consider dimensionality reduction for the purpose of data visualization, for the situation in which each 'data point' is a probability distribution, or a set of samples from it. This situation naturally arises when the data have *hierarchical structure*, i.e. the individual samples can be grouped at a higher level. Throughout this work we will use the word 'unit' to refer to this grouping level; for each 'unit' there is a number of 'samples' in the data (Fig. 1). For example, in a social science survey, participants can be seen as samples and their countries of origin can be seen as units. For exploratory analysis, the interest may often be in the relationships between units (countries), rather than samples (participants).

A common approach for data exploration is to visualize the dataset as a 2D embedding, using dimensionality reduction algorithms such as PCA, MDS, t-SNE [14] or UMAP [15]. These algorithms are designed to get vectors as input,

M.-R. Amini et al. (Eds.): ECML PKDD 2022, LNAI 13713, pp. 104–120, 2023.
https://doi.org/10.1007/978-3-031-26387-3_7

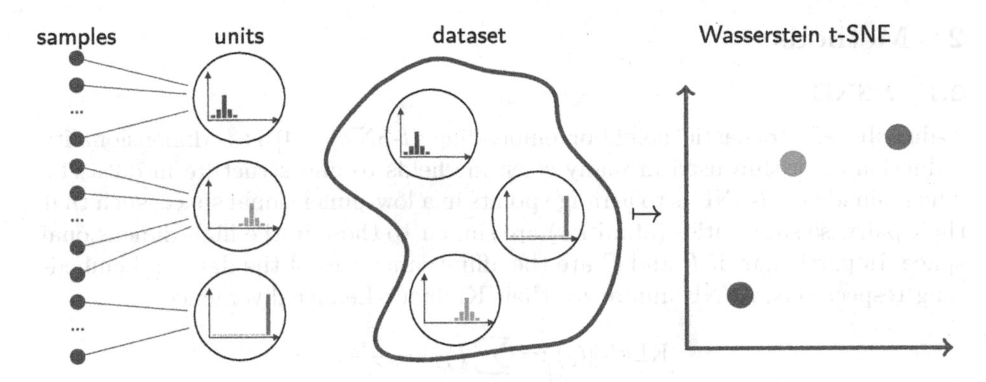

Fig. 1. Hierarchical data. Individual *samples* can be grouped into *units*. Each unit forms a probability distribution over its samples. In our *Wasserstein t-SNE* approach, units in the dataset are compared using the Wasserstein metric to construct a pairwise distance matrix, which is then embedded in two dimensions using the t-SNE algorithm. Units with similar probability distributions end up close together in the 2D embedding.

and compute pairwise distances (e.g. Euclidean) between the input vectors. However, when analyzing units in a hierarchical dataset, each single unit forms an entire probability distribution over its samples, and cannot be represented by one vector. A simple approach would be to collapse all within-unit distributions to their means, and then apply any standard dimensionality reduction algorithm. However, this procedure can loose important information, particularly when some of the units share the same mean but have different shape.

Here we propose to use the Wasserstein metric [9] to compute pairwise distances between units. The Wasserstein distance has got recent attention in applications to Generative Adversarial Networks [1] or discriminant analysis [4] where it was used to compare probability densities with different support. The Wasserstein metric is convenient because there exists a closed-form solution for Gaussian distributions [2]. Using the Gaussian approximation, it is possible to efficiently construct the pairwise distance matrix between units in a hierarchical dataset. This distance matrix can then be used for downstream analysis, such as clustering or dimensionality reduction. Our focus here will be on t-SNE embeddings.

In the first part of this work we use simulated data to demonstrate the effectiveness of our *Wasserstein t-SNE*. In the second part we apply the same method to real-world data, in particular the data from the 2017 German parliamentary election. Here, samples correspond to polling stations while the units correspond to voting districts. We use the Gaussian approximation but also compute the exact Wasserstein distances, using an efficient linear programming approach.

The Python implementation of Wasserstein t-SNE is available on GitHub at fsvbach/WassersteinTSNE and as a package on PyPi WasserteinTSNE. The analysis code reproducing all figures in this paper can be found on GitHub at fsvbach/wassersteinTSNE-paper together with the analyzed data.

2 Methods

2.1 *t*-SNE

T-distributed stochastic neighbor embedding (*t*-SNE) [14] is a dimensionality reduction algorithm used in many scientific fields to find structure in datasets. The main idea of *t*-SNE is to arrange points in a low-dimensional space, such that their pairwise similarities (affinities) are similar to those in the high-dimensional space. In particular, if P and Q are the affinity matrices of the data and embedding respectively, *t*-SNE minimizes their Kullback-Leibler divergence

$$\mathrm{KL}(P\|Q) := \sum_{ij} P_{ij} \log \frac{P_{ij}}{Q_{ij}}.$$

The affinity matrix P is constructed from the pairwise distances d_{ij} by Gaussian kernels with bandwidth σ_i

$$P_{j|i} = \frac{\exp(-d_{ij}^2/2\sigma_i^2)}{\sum_{k\neq i}\exp(-d_{ik}^2/2\sigma_i^2)}$$

such that all perplexities of the conditional distributions equal some predefined value. In most *t*-SNE implementations this parameter defaults to 30 which we leave untouched in our experiments. As a reminder, if $p(x)$ is a discrete probability density function, the perplexity of p is given by

$$\mathcal{P}(p) := 2^{H(p)} = \prod_x p(x)^{-p(x)}.$$

The affinity matrix P is then symmetrized by

$$P_{ij} = \frac{P_{j|i} + P_{i|j}}{2n}.$$

In the low-dimensional space, the affinity matrix Q is based on the pairwise distances between the embedding vectors $\mathbf{y}_i$, using the *t*-distribution kernel:

$$Q_{ij} = \frac{\left(1 + \|\mathbf{y}_i - \mathbf{y}_j\|^2\right)^{-1}}{\sum_{k\neq l}\left(1 + \|\mathbf{y}_k - \mathbf{y}_l\|^2\right)^{-1}}.$$

The *t*-SNE algorithm minimizes the Kullback-Leibler divergence $\mathrm{KL}(P\|Q)$ with respect to the coordinates $\mathbf{y}_i$. The embedding is initialized randomly, or using another algorithm such as PCA [11]. The optimization is done with gradient descent, i.e. the points move along the gradient until convergence. This results in a local minimum where no point can be moved without yielding a worse embedding.

When interpreting *t*-SNE embeddings it is important to keep in mind that the algorithm puts emphasis on close points, i.e., similar data points are embedded close to each other. The opposite does not hold: points that are embedded far from each other are not necessarily far from each other in the original space.

In this work we use the implementation of `openTSNE` [17], and keep all parameters at their default values.

2.2 Wasserstein Metric

The Wasserstein metric [9] is a natural choice to compare probability distributions. It can be used to compare densities which do not have the same support, as long as a distance measure of their support is given. The downside of the Wasserstein distance is its computational complexity, which is linked to optimal transport [16,20].

Definition 1. *Let (M,d) be a metric space. The p-Wasserstein distance of two distributions μ and ν is defined as*

$$W_p(\mu,\nu) := \left(\inf_{\gamma\in\Gamma(\mu,\nu)} \int_{M\times M} d(x,y)^p \mathrm{d}\gamma(x,y)\right)^{\frac{1}{p}}$$

where Γ is the set of all couplings of μ and ν.

In computer science the 1-Wasserstein metric is also known as *Earth Mover's Distance*, because if one imagines the probability distributions as piles of earth, then $W_p(\mu,\nu)$ represents the minimal amount of work necessary to transfer this mass from μ to ν. This intuition also explains why the probability distributions must be defined on a metric space M, because we have to measure how far two points are away from each other, i.e. how far the mass has to be transported.

In general, the p-Wasserstein distance for continuous distributions is hard to compute [20]. But there exists a closed-form solution for the 2-Wasserstein metric for multivariate Gaussian distributions [2] (also known as Fréchet Inception Distance [8]). If μ,ν are two Gaussian distributions $\mathcal{N}_i(m_i,C_i)$ with means m_i and covariance matrices C_i, the 2-Wasserstein distance between them is given by

$$\begin{aligned} W_2(\mu,\nu)^2 &= \|m_1-m_2\|_2^2 + \operatorname{tr}\left(C_1+C_2-2\left(C_2^{1/2}C_1C_2^{1/2}\right)^{1/2}\right) \\ &= \|m_1-m_2\|_2^2 + \operatorname{tr}\left(C_1+C_2-2\left(C_2C_1\right)^{1/2}\right). \end{aligned}$$

The first term here is the Euclidean distance between the means, while the second term defines a metric on the space of covariance matrices [2]. By introducing a hyperparameter $\lambda\in[0,1]$ we can put emphasis either on the means or on the covariances. We therefore propose a convex generalization of the 2-Wasserstein distance for Gaussians:

$$\tilde{W}(\mu,\nu)^2 = (1-\lambda)\cdot\|m_1-m_2\|_2^2 + \lambda\cdot\operatorname{tr}\left(C_1+C_2-2\left(C_2C_1\right)^{1/2}\right). \qquad (\star)$$

This reduces to the Euclidean distance between the means for $\lambda=0$ and to the distance between covariance matrices for $\lambda=1$. The 2-Wasserstein distance corresponds to $\lambda=0.5$ (up to a scaling factor).

In closed form, the 2-Wasserstein distance between two Gaussians can be computed in polynomial time. The matrix multiplication and the eigenvalue decomposition (for taking the square root) have $\mathcal{O}(d^3)$ complexity, where d is the number of features. If there are n units in the dataset, the $n\times n$ pairwise distance matrix can be computed in $\mathcal{O}(n^2d^3)$ time.

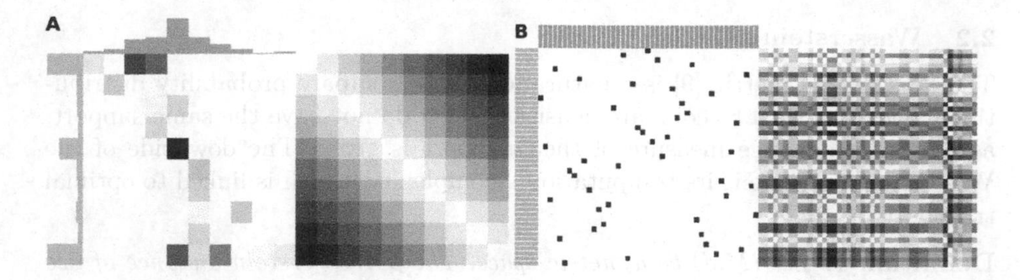

Fig. 2. Wasserstein distance as a linear program. (**A**) The optimal transport map γ of two probability distributions ν (orange) and μ (blue) is shown. The heatmap represents the cost matrix C. (**B**) The same distributions can be visualized as a collection of samples, which have different support. The distance between samples ν_i and μ_j is given in the cost matrix entry C_{ij}. The size of the optimization variable γ is then upper bounded by the product of the sample sizes (Color figure online).

2.3 Linear Programming

We are also interested in computing exact Wasserstein distances without relying on the Gaussian approximation. Since real-life datasets contain discrete samples, this is possible by discretizing Definition 1 to

$$W_p(\mu,\nu)^p := \min_{\gamma\in\Gamma(\mu,\nu)} \sum_{M\times M} d(m_i,m_j)^p \cdot \gamma(m_i,m_j),$$

which is equivalent to the following linear program [16]:

$$\begin{array}{rl|rl} \textbf{primal form:} & & \textbf{dual form:} & \\ \text{minimize} \quad z = & \mathbf{c}^T\mathbf{x}, & \text{maximize} \quad \tilde{z} = & \mathbf{b}^T\mathbf{y}, \\ \text{so that } \mathbf{A}\mathbf{x} = & \mathbf{b} & \text{so that } \mathbf{A}^T\mathbf{y} \leq & \mathbf{c}. \\ \text{and} \quad \mathbf{x} \geq & \mathbf{0} & & \end{array} \qquad (\star\star)$$

The vectorized matrix $\mathbf{c}$ defines the transport cost, i.e., $c_{ij} = d(m_i,m_j)^p$ represents the L_p-distance of the points m_i and m_j, where M is the discrete metric space on which the probability distributions are defined. The optimization variable $\mathbf{x}$ represents the vectorized transport plan γ as in Fig. 2A. Each entry of $\mathbf{x}$ must be non-negative. The constraint $\mathbf{A}\mathbf{x} = \mathbf{b}$ is set up such that it is satisfied if the marginals of γ equal the densities μ,ν. The primal form in $(\star\star)$ yields an explicit transport plan while the dual form has less variables and is faster. Due to the strong duality of a linear program the resulting solution z is the same. In practice we therefore use the dual form to compute exact Wasserstein distances.

Simplex algorithms and interior-point methods can solve real-world linear programs with a unique solution in polynomial time [7]. However, the exact complexity depends on the constraint matrix in the problem formulation. In general, the runtime of a linear program depends on the size of the optimization variable. In our case this is given by the product of the support sizes of the two

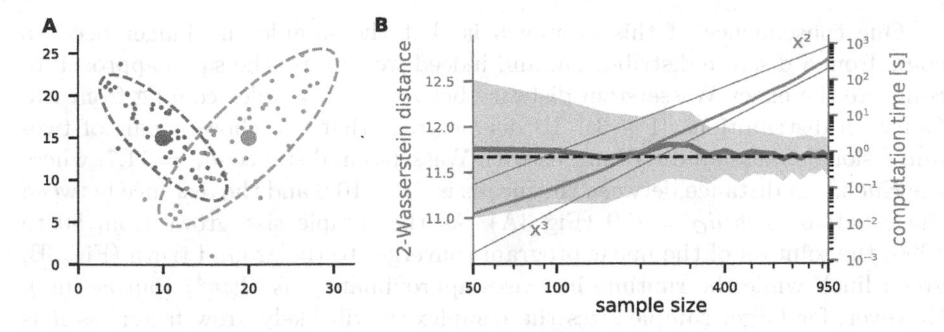

Fig. 3. Computation time and accuracy. (**A**) Two multivariate Gaussian distributions with 50 samples each. (**B**) The Wasserstein distance between the two probability distributions is computed using a different number of samples. The ground-truth distance is obtained by the closed-form solution and is shown with the dashed black line. The Wasserstein distance estimates using our linear program approach are shown in green (mean and standard deviation over 50 repetitions). The purple line shows the average runtime. (Color figure online)

discrete probability distributions. Figure 2A provides an example of two one-dimensional distributions, defined on the same support of size 10 (which could e.g. be a ten-point rating scale from 1 to 10). Both probability mass functions have 10 values so the resulting optimization variable γ has $10 \times 10 = 100$ entries. While this linear program is easily solvable, the problem becomes computationally hard if we add additional feature dimensions (for example, if we add another ten-point feature, each probability density will become a two-dimensional mass function over 100 values, so then γ has length 10,000). The number of variables in the transport map therefore grows exponentially with the number of features, thus this approach is intractable for datasets with many features.

Instead, we reduce the probability densities to the subspace where samples have actually been observed, rather than comparing distributions on the complete space M. That is, we consider both distributions uniformly distributed over their samples (Fig. 2B). The marginal distributions μ and ν in Fig. 2B therefore become uniform distributions with supports of size n and m respectively, where n and m are the two sample sizes. The size of the optimization variable γ now becomes upper bounded by nm regardless of the number of features. The cost matrix is given by the pairwise L_p-distance between all samples, which can be computed efficiently. We are not aware of a prior use of this shortcut, which however is only applicable when the number of samples is not large.

A way to see that both approaches are equivalent is to consider the constraints $\mathbf{x} \geq 0$ and $\mathbf{Ax} = \mathbf{b}$. When the number of features is large while the sample size is small, the sample density at most support values will have zero probability mass, because no sample has been observed at that point. Since the rows and the columns in the transport plan must sum to the marginal distributions, each entry with zero probability mass forces the corresponding row or column to be empty. Therefore these entries can be left out in the problem formulation and the size of the optimization variable is upper bounded by nm.

One consequence of this approach is that the samples no longer need to come from a discrete distribution, and indeed we can use the same approach to compute the exact Wasserstein distance between the samples coming from two Gaussian distributions (Fig. 3). To demonstrate that, we chose a pair of two-dimensional Gaussian distributions with Wasserstein distance $d_W = 11.7$, where the Euclidean distance between the means is $d_E = 10.0$ and the distance between the covariances is $d_C = 6.0$ (Fig. 3A). As the sample size grows from 50 to 1000, the solution of the linear program converges to the ground truth (Fig. 3B, green line), while the runtime increases approximately as $\mathcal{O}(m^3)$ (purple line). However, for larger sample sizes the complexity will likely grow faster, as it is known that integer linear programs have exponential complexity [10]. Note that the dimensionality of the feature space (in this example, it is two-dimensional) does not strongly influence runtime.

2.4 Data

Simulated Data. To demonstrate and validate our method, we simulated hierarchical datasets, i.e. we defined the *hierarchical Gaussian mixture model* (HGMM). Similar to a Gaussian mixture model, a HGMM has multiple classes from which units are drawn. But here, each unit defines a Gaussian distribution with a unit-specific mean and covariance matrix. In each class, the unit means come from a class-specific Gaussian distribution, while the unit covariance matrices come from a class-specific Wishart distribution.

Definition 2. *Let $\mathcal{N}$ and $\mathcal{W}$ denote Gaussian and Wishart distributions respectively. A hierarchical Gaussian mixture model is then defined by the number of classes (K), the number of units per class (N_i for $i = 1 \dots K$), the number of samples per unit (M_j for $j = 1 \dots \sum_{i=1}^{K} N_i$) and their feature dimensionality F, where*

- *each class i is characterized by a Gaussian distribution $\mathcal{N}(\mu_i, \Sigma_i)$ with $\mu_i \in \mathbb{R}^F$ and $\Sigma_i \in \mathbb{R}^{F \times F}$ and a Wishart distribution $\mathcal{W}(n_i, \Lambda_i)$ with $n_i \geq F$ and $\Lambda_i \in \mathbb{R}^{F \times F}$;*
- *each unit X_j belonging to a class i is characterized by a Gaussian distribution $\mathcal{N}(\nu_j, \Gamma_j)$. Unit means are samples from the class-specific Gaussian $\nu_j \sim \mathcal{N}(\mu_i, \Sigma_i)$ and unit covariance matrices are samples from the class-specific Wishart distribution $\Gamma_j \sim \mathcal{W}(n_i, \Lambda_i)$;*
- *the samples S_k of each unit j are iid distributed as $S_k \sim \mathcal{N}(\nu_j, \Gamma_j)$.*

A HGMM is specified by the set of class-specific parameters $\{\mu_i, \Sigma_i, n_i, \Lambda_i\}$. For example, Fig. 4 shows a two-dimensional ($F = 2$) HGMM with $K = 4$ classes, $N = 100$ units in each class, and $M = 15$ samples in each unit. The class-specific Gaussian distributions (dashed black lines) are far away from each other because their means μ_i are chosen to be sufficiently different. The Within-class similarity of the unit means can be adjusted by Σ_i (defining the shape of the dashed contours). Each class has its own Wishart scale matrix Λ_i; in this example, the peculiar Wishart scale of the green class makes green units easily

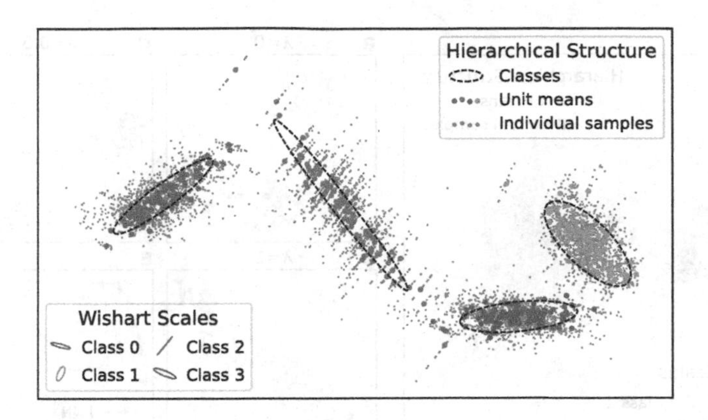

Fig. 4. Hierarchical Gaussian mixture model (HGMM). This two-dimensional example dataset has $K = 4$ classes with $N = 100$ units each. The gray points show the samples from all units ($M = 15$ samples per unit). Note that some of the units in the red and green classes have similar means, but their covariances are very different. (Color figure online)

distinguishable from the red units even when their means come close to the red class. Note, that the unit covariance matrices Γ_j in any given class are not all the same. Their sampling process (from a Wishart distribution) is equivalent to drawing n_i samples from a zero-centered Gaussian distribution with the Wishart scale as covariance matrix and estimating the sample covariance matrix. The larger the n_i, the closer all Γ_j are to the respective Λ_i. We used $n_i = 4$ in Fig. 4.

German Election Data. The German parliamentary election was held in September 2017 with six major parties making it to the parliament. Germany is divided into 299 voting districts (*Wahlkreise*). In each voting district, multiple polling stations (*Wahlbezirke*) are set up. In our analysis we consider each voting district to be a unit with its polling stations being its samples.

The election data were directly downloaded from the *Bundeswahlleiter* website (https://tinyurl.com/mpevp355). We removed results of all minor parties and normalized each polling station so that the percentages of the six major parties — CDU (including the Bavarian-only CSU), SPD, AfD, FDP, Grüne and Linke — sum to 1 (the feature dimension of each sample is therefore six). Voting by mail was counted in separate mail-only polling stations of the respective voting district.

For this dataset, we a priori defined four classes: *Cities* (all voting districts with population density of at least 1000 people per square kilometer; population densities (https://tinyurl.com/3262nf8b) also obtained from the *Bundeswahlleiter* website), *Southern Germany* (all districts in Bavaria and Baden-Würtemberg, excluding previously defined cities), *Eastern Germany* (former DDR, excluding cities) and *Western Germany* (the rest).

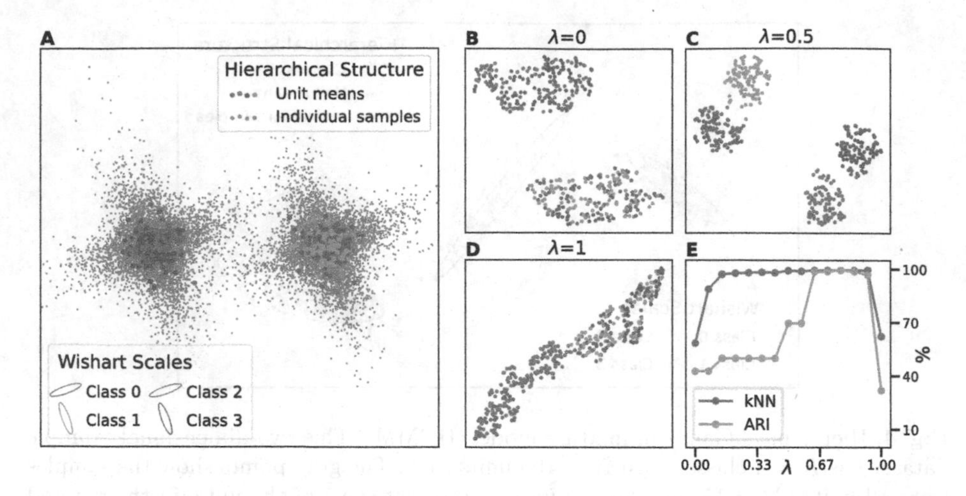

Fig. 5. Wasserstein t-SNE. (**A**) This two-dimensional ($F = 2$) HGMM was generated using $K = 4$ classes with $N = 100$ units each ($M = 30$ samples per unit). Two pairs of classes have the same distribution of unit means, while two other pairs of classes have the same distribution of unit covariance matrices. (**B**) The mean-based embedding ($\lambda = 0$) is not able to separate some of the classes. (**C**) The Wasserstein embedding ($\lambda = 0.5$) separates all four classes. (**D**) The covariance-based embedding ($\lambda = 1$) is not able to separate some of the classes. (**E**) The performance at different values of λ was assessed using the kNN accuracy ($k = 5$) in the 2D embedding and the adjusted Rand index (ARI) obtained from Leiden clustering of the original distance matrix (kNN graph with $k = 5$, resolution parameter $\gamma = 0.08$).

3 Results

3.1 Wasserstein t-SNE on Simulated Data

To perform *Wasserstein t-SNE*, we first compute the pairwise distance matrix between units in a dataset where each unit is considered to be a probability distribution over its samples. We then embed these units in 2D using the t-SNE algorithm.

Figure 5A shows a two-dimensional ($F = 2$) toy dataset that consists of $K = 4$ classes, with $N = 100$ units per class, and $M = 30$ samples per unit. The red and green classes have the same distribution of unit means; the same is true for the blue and the orange classes. Likewise, the red and orange classes have the same distribution of unit covariances; the same is true for the blue and the green classes. For a more detailed description see Sect. 2.4.

Depending on the value of λ in ($\star$), the resulting t-SNE embeddings show different structure. The mean-based embedding ($\lambda = 0$) only separates the dataset into two clusters (Fig. 5B). Since it only takes the means into account, the orange class cannot be separated from the blue one, and the red class cannot be separated from the green one. Similarly, the covariance-based embedding ($\lambda = 1$) only finds two clusters as well, mixing up blue with green and orange with red

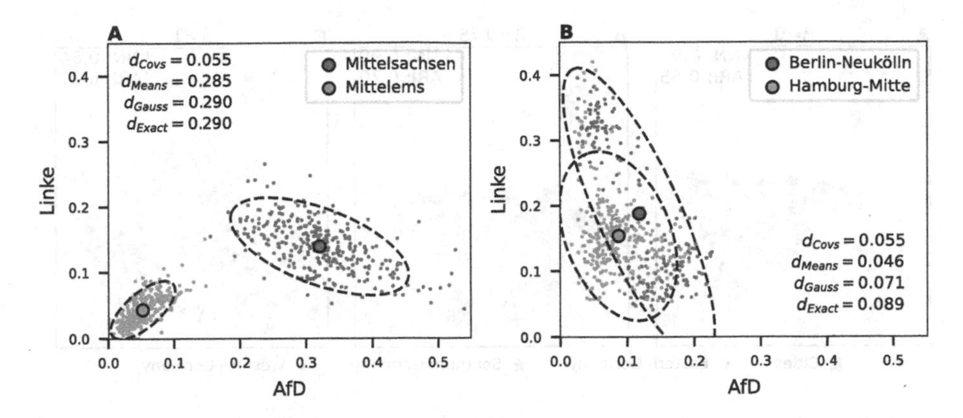

Fig. 6. Example voting districts in the 2017 German parliamentary election. Four voting districts (units) are shown together with their respective polling stations (samples). (**A**) Mittelems, which is located in the rural western Germany, exhibits positive correlation between the votes for AfD and for Linke, whereas Mittelsachsen in eastern Germany shows negative correlation. (**B**) The politically diverse district of Berlin-Neukölln has bimodal structure in the within-unit distribution of AfD and Linke votes, whereas Hamburg-Mitte does not show such bimodality.

(Fig. 5D). In contrast, the Wasserstein embedding ($\lambda = 0.5$) successfully separates all four classes from each other (Fig. 5C).

To measure the performance of the different λ values, we used two different metrics. One metric is the k-nearest-neighbor (kNN) classification accuracy that measures the probability that a unit is labeled correctly by the majority vote of its $k = 5$ nearest neighbors in the embedding (we used the sklearn implementation [5]). The second metric is the adjusted Rand index (ARI) [18] that evaluates the agreement between the ground truth classes and the results of unsupervised clustering. We used the Leiden clustering algorithm [19], applied to the Wasserstein distance matrix (here and below we used the leidenalg implementation [19] with resolution parameter $\gamma = 0.08$ on the kNN graph built with $k = 5$). Note that unlike the kNN accuracy, the ARI metric is independent of t-SNE.

Both metrics, kNN accuracy and ARI, peaked at $\lambda \in [0.7, 0.8]$ and showed markedly worse performance at both $\lambda = 0$ and $\lambda = 1$. Moreover, while the kNN accuracy was close to the peak at already $\lambda = 0.5$, the ARI achieved higher values only for $\lambda > 0.7$. The result indicates that putting more emphasis on the covariance structure helps the algorithm to cluster the data correctly. This shows the power of our generalized Wasserstein distance for Gaussian distributions, as in this case it outperforms the exact Wasserstein distance (corresponding to $\lambda = 0.5$).

3.2 German Parliamentary Election 2017

Gaussian Wasserstein *t*-SNE. The dataset of the 2017 German parliamentary election dataset contains 299 voting districts (units), each having about

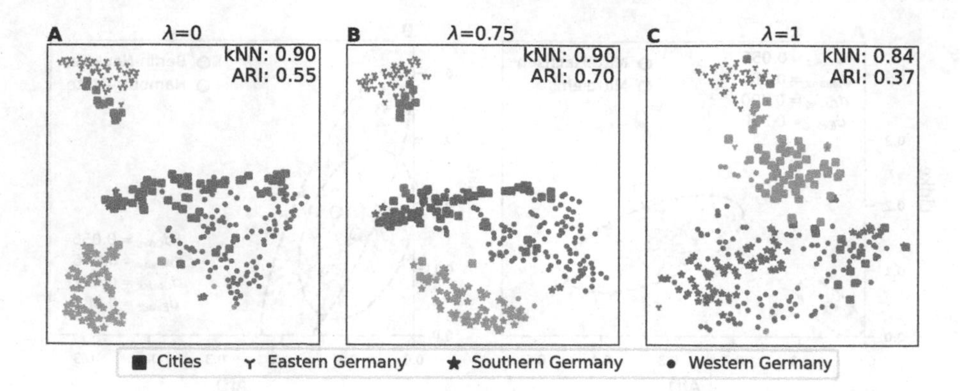

Fig. 7. Wasserstein t-SNE of the 2017 German parliamentary election. For clustering, we used the Leiden algorithm (on the original distance matrix) with $k = 5$ kNN graph and resolution parameter $\gamma = 0.08$. Colors correspond to the Leiden clusters; marker shape corresponds to the a priori classes. (**A**) The mean-based embedding with $\lambda = 0$ shows three clusters. (**B**) The Wasserstein embedding with $\lambda = 0.75$ shows four clusters. (**C**) The covariance-based embedding with $\lambda = 1$ shows three clusters.

150–850 polling stations (samples). The samples are represented as a points in six-dimensional space (corresponding to six political parties), as described in Sect. 2.4. For most units, the data could be reasonably well described by a multivariate Gaussian distribution (Fig. 6). For example, the Gaussian Wasserstein distance and the exact Wasserstein distance between Mittelsachsen and Mittelems districts, both equaled 0.290 (Fig. 6A). For some districts the approximation was less good: e.g. between Berlin-Neukölln and Hamburg-Mitte (Fig. 6B), the Gaussian Wasserstein distance was 0.071 while the exact Wasserstein distance was 0.089. This can be explained by the polarization of Berlin-Neukölln, which had a bimodal structure that could not be captured by a multivariate normal distribution. However, we found that most districts were well approximated by a Gaussian.

We computed the pairwise Wasserstein distances between all pairs of units for different values of λ, and embedded the resulting distance matrices with the t-SNE algorithm (Fig. 7). In parallel, we clustered each distance matrix by the Leiden algorithm (with resolution parameter $\gamma = 0.08$ and kNN graph with $k = 5$) and used the cluster assignments to color the t-SNE embeddings. The Wasserstein embedding with $\lambda = 0.75$ (Fig. 7B) outperformed the mean-based ($\lambda = 0$ and the covariance-based ($\lambda = 1$) embeddings, and achieved a kNN accuracy of 0.90 and an ARI of 0.70. While the difference in the kNN accuracy was not very different for other values of λ (0.90 for the mean-based and 0.82 for the covariance-based embeddings), the difference in the ARI was very pronounced (0.55 for the mean-based and 0.37 for the covariance-based embeddings).

The ARI is sensitive to the number of clusters, which is not pre-specified in the Leiden algorithm. While it automatically found three clusters with $\lambda = 0$ (Fig. 7A) and $\lambda = 1$ (Fig. 7C), it found four clusters with $\lambda = 0.75$, and these

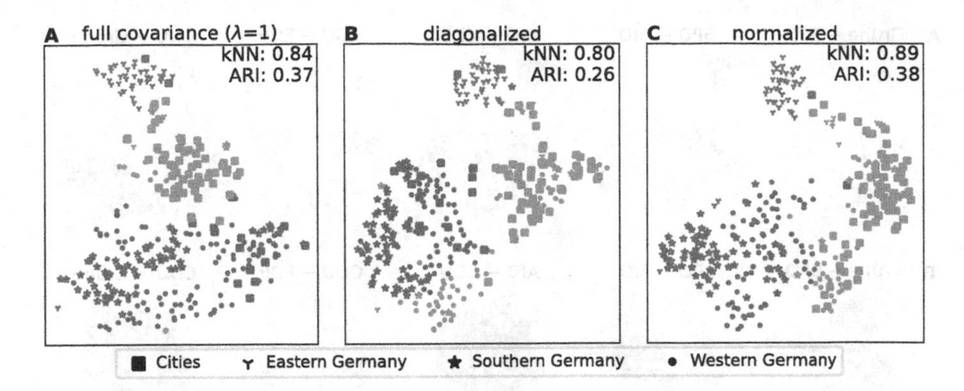

Fig. 8. Variants of the covariance-based ($\lambda = 1$) embedding. (**A**) The full covariance matrices were used to compute the pairwise distance matrix. The same embedding as in Fig. 7C. (**B**) Only the diagonal entries of the covariance matrices were used, i.e. the marginal variances. (**C**) Each covariance matrix was normalized to become a correlation matrix.

clusters corresponded well to the four classes (Cities, Western Germany, Eastern Germany, Southern Germany) that we defined a priori. In contrast, the mean-based distance matrix merged Western Germany with the Cities, and the covariance-based embedding merged Western Germany with Southern Germany. This decreased the ARI for these embeddings, which is also visible in Fig. 10C. While the exact ARI values depend on the choice of k and γ parameters, our results show that the Wasserstein distances with $0 < \lambda < 1$ can be an improvement compared to ignoring either the information about the means or about the covariance structure.

Covariance and Correlation Structure. The covariance-based embedding clearly showed three clusters (Fig. 7C), even though the means were completely ignored in this analysis. Where did this structure emerge from? Covariance is influenced by correlation and by marginal variances. To disentangle these two aspects of the data, we constructed a covariance-based embedding after zeroing out all off-diagonal values of all covariance matrices before computing the pairwise distance matrix (Fig. 8B). We also constructed a covariance-based embedding after normalizing all covariance matrices to be correlation matrices (Fig. 8C). Both embeddings were similar to the original covariance-based embedding (Fig. 8A), suggesting that there was meaningful information in marginal variances as well as in pairwise correlations.

Figure 8C demonstrates that class information was present in the within-unit correlations, and indeed we saw earlier that parties' results can correlate differently in different voting districts (Fig. 6). To visualize this effect, we overlayed pairwise correlation coefficients on the Gaussian Wasserstein embedding with $\lambda = 0.75$ (Fig. 9A). For several pairs of parties, correlation strongly depended on the class, e.g. AfD and SPD were positively correlated in the Cities, but negatively correlated in Eastern Germany. This indicates that people who vote

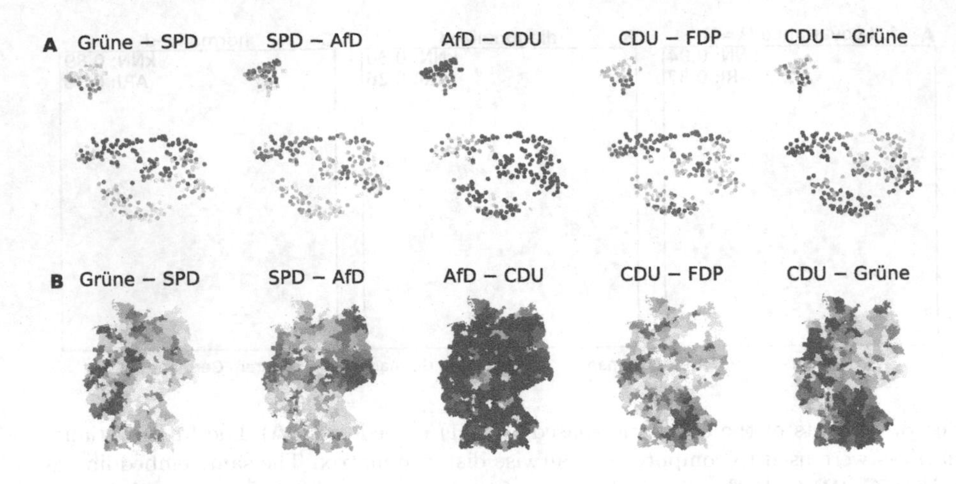

Fig. 9. Pairwise correlations between parties. Colors represent the Pearson correlation coefficient of the respective parties in each voting district, from blue (-1) to red (1). We chose five (out of 15) pairs of parties showing the most interesting patterns. (**A**) Party correlations overlayed over the Gaussian Wasserstein embedding with $\lambda = 0.75$ (as in Fig. 7B). (**B**) Party correlations overlayed over the geographical map of Germany. (Color figure online)

SPD in the cities tend to live in the same neighborhoods (i.e. same polling stations within a given district) as people who vote AfD, whereas in the rural east they tend to live in different neighborhoods. This suggests that these parties are perceived differently in different parts of the country, opposing each other in some of the regions but sharing sympathizers in others. Previous research has shown that many voters switched from SPD to AfD in the 2017 election [6]. Our analysis indicates that this effect may have happened mostly in the east.

Exact Wasserstein Embedding. To verify that the Gaussian approximation used above did not strongly distort the embedding, we also did an embedding based on the exact Wasserstein distance. As explained in Sect. 2.3, we calculated the exact Wasserstein distances using linear programming. While this approach is more faithful to the data, it requires much longer computation time; calculating all $299 \cdot 298/2 = 44,551$ pairwise exact Wasserstein distances between units took 43 h on a machine with 8 CPU cores at 3.0 GHz. The resulting embedding (Fig. 10A) was very similar to the Gaussian embedding with $\lambda = 0.5$ (Fig. 10B).

The Gaussian approximation has one important benefit beyond the faster runtime: namely, it allows to adjust the λ parameter. As a function of λ, the kNN accuracy was nearly flat (Fig. 10C) but the ARI of the Leiden clustering peaked for $\lambda > 0.6$, corresponding to the regime where the Leiden algorithm identifies four clusters in the data instead of three. At the same time, the performance of the exact Wasserstein embedding is similar to $\lambda = 0.5$ and hence is worse (Fig. 10C, black dashed lines). This shows that, compared to the exact

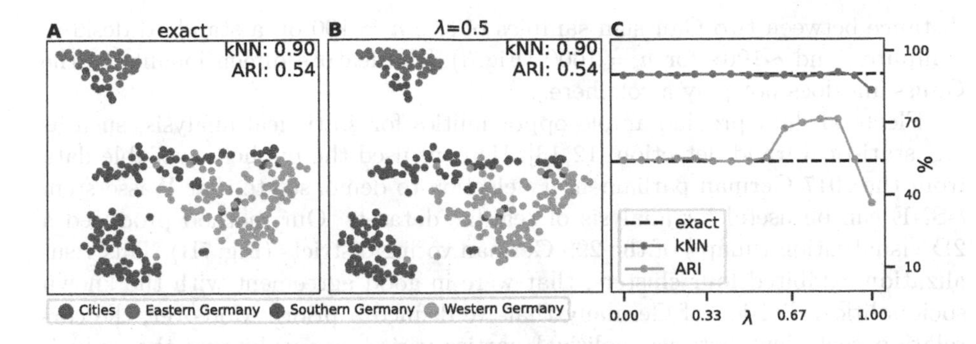

Fig. 10. Comparison of the Wasserstein t-SNE embeddings based on the Gaussian approximation and based on the exact Wasserstein distances. (**A**) The exact Wasserstein t-SNE embedding separates the classes. (**B**) The Gaussian Wasserstein embedding with $\lambda = 0.5$ shows similar structure. (**C**) The kNN classification accuracy ($k = 5$) and the ARI based on the Leiden clustering ($k = 5$ kNN graph, resolution parameter $\gamma = 0.08$) are shown for different values of λ. The black dashed lines show the kNN accuracy and the ARI of the exact Wasserstein embedding. (Color figure online)

Wasserstein distances, the Gaussian approximation with the flexible λ parameter can improve not only the runtime but also the final embedding.

4 Discussion

In this work we introduced *Wasserstein t-SNE* as a method to visualize hierarchical datasets. The main idea is to compute the pairwise distance matrix between the units of interest, using the Wasserstein metric to compare the distribution of their samples; then we use t-SNE to make a 2D embedding of the resulting distance matrix (Fig. 1). Using a simulated and a real-life dataset, we showed that our approach can outperform standard t-SNE based on unit averages (Figs. 5, 7).

There are two different ways to use Wasserstein t-SNE. One way is to approximate each unit by a multivariate Gaussian distribution which is fast but not scalable to high feature dimensionality. The second way is to compute the exact Wasserstein distances which is more accurate but substantially slower if there are many samples. Both ways require to compute and store the pairwise distance matrix which gets unfeasible for very large datasets. A benefit of using the Gaussian approximation is that it allows to put emphasis on either the means or the covariances of the units by adjusting the λ parameter in the distance definition ($\star$). We suggested this definition as a novel generalization of the closed-form solution for the 2-Wasserstein distance between Gaussian distributions. To compute the exact Wasserstein distances, we solve a linear program for each distance calculation. We developed an approach that scales with the number of samples and allows to compute Wasserstein distances between samples from continuous distributions. Using this approach, it took us $\sim$0.1 s to compute the Wasserstein distance between two Gaussian samples of size $n = 100$ on a standard desktop computer, and ∼100 s for $n = 1000$ (Fig. 3). The feature dimensionality of the Gaussians does not play a role here.

Electoral data provide ample opportunities for statistical analysis, such as e.g. statistical fraud detection [12,13]. Here we used the publicly available data from the 2017 German parliamentary election to demonstrate that Wasserstein t-SNE can be useful for analysis of real-life datasets. Our method produced a 2D visualization ('map') of the 299 German voting districts (Fig. 7B). This visualization exhibited four clusters, that were in good agreement with the known sociopolitical division of Germany that we defined a priori. Moreover, the correlation coefficient between political parties varied smoothly over the embedding (Fig. 9), and in some cases even changed the sign. This showed that the information about within-unit distributions can be valuable to provide concise visualizations of political landscape in an unsupervised way.

We are not aware of other methods specifically designed to visualize hierarchical data. The naive approach is to collapse units to their means. However, we showed that this can be suboptimal whenever there is meaningful information in the unit covariances (e.g. Fig. 7). An alternative is to append some of the covariance-based features to the unit means. This approach was, e.g., used in the Wisconsin Breast Cancer dataset [21] (popularized by its role as a UCI benchmark) where samples are cells and units are the respective patients. Here the variance of each feature (such as cell radius or cell smoothness) was appended to the dataset as an additional separate feature. While this allows to use some of the covariance information, it removes all information about feature correlation. Finally, it is possible to base all the analysis on the sample level, instead of the unit level. Such a 'sample-based' t-SNE embedding would show many more points than a 'unit-based' t-SNE embedding. However, for datasets like the one shown in Fig. 5 this would not result in a useful visualization, as it would yield only two clusters and not four (as there are two pairs of classes with strongly overlapping distribution of samples within the units).

In summary, we believe that Wasserstein t-SNE is a promising method to visualize hierarchical datasets. Our results on synthetic data and on the 2017 German election data demonstrate that Wasserstein t-SNE can outperform standard alternatives and uncover meaningful structure in the data. We hope that our method can be useful in various domains. For example, social science often deals with hierarchical datasets, such as the European Values Study [3] where geographical regions can be seen as units while individual participants of the survey can be seen as samples. We believe that Wasserstein t-SNE can also be useful beyond the social and political science, e.g. for biomedical data.

Acknowledgements. We thank Philipp Berens for comments and suggestions. This research was funded by the Deutsche Forschungsgemeinschaft (Excellence Cluster 2064 "Machine Learning: New Perspectives for Science", 390727645), and by the German Ministry of Education and Research through the Tübingen AI Center (01IS18039A).

References

1. Arjovsky, M., Chintala, S., Bottou, L.: Wasserstein generative adversarial networks. In: International Conference on Machine Learning, pp. 214–223. PMLR (2017)
2. Dowson, D., Landau, B.: The Fréchet distance between multivariate normal distributions. J. Multivar. Anal. **12**(3), 450–455 (1982)
3. European Values Study: Integrated Dataset (EVS 2017). GESIS Data Archive, Cologne. ZA7500 Data file Version 4.0.0 (2020)
4. Flamary, R., Cuturi, M., Courty, N., Rakotomamonjy, A.: Wasserstein discriminant analysis. Mach. Learn. **107**(12), 1923–1945 (2018). https://doi.org/10.1007/s10994-018-5717-1
5. Goldberger, J., Hinton, G.E., Roweis, S., Salakhutdinov, R.R.: Neighbourhood components analysis. In: Advances in Neural Information Processing Systems 17 (2004)
6. Görtz, A.: Sozialstrukturelle Ursachen für Wechselwahl: Warum wechseln WählerInnen von der SPD zur AfD? Social Science Open Access Repository (2020)
7. Huangfu, Q., Hall, J.J.: Parallelizing the dual revised simplex method. Math. Program. Comput. **10**(1), 119–142 (2018)
8. Jung, S., Keuper, M.: Internalized Biases in Fréchet inception distance. In: NeurIPS 2021 Workshop on Distribution Shifts: Connecting Methods and Applications (2021)
9. Kantorovich, L.V.: The mathematical method of production planning and organization. Manage. Sci. **6**(4), 363–422 (1939)
10. Karp, R.M.: Reducibility among combinatorial problems. In: Miller, R.E., Thatcher, J.W., Bohlinger, J.D. (eds.) Complexity of Computer Computations. The IBM Research Symposia Series, pp. 85–103. Springer, Boston (1972). https://doi.org/10.1007/978-1-4684-2001-2_9
11. Kobak, D., Linderman, G.C.: Initialization is critical for preserving global data structure in both t-SNE and UMAP. Nat. Biotechnol. **39**(2), 156–157 (2021)
12. Kobak, D., Shpilkin, S., Pshenichnikov, M.S.: Integer percentages as electoral falsification fingerprints. Ann. Appl. Stat. **10**(1), 54–73 (2016)
13. Kobak, D., Shpilkin, S., Pshenichnikov, M.S.: Statistical fingerprints of electoral fraud? Significance **13**(4), 20–23 (2016)
14. Van der Maaten, L., Hinton, G.: Visualizing data using t-SNE. J. Mach. Learn. Res. **9**(11), 2579–2605 (2008)
15. McInnes, L., John Healy, J.M.: UMAP: uniform manifold approximation and projection for dimension reduction. arXiv (2020)
16. Mémoli, F.: Gromov-Wasserstein distances and the metric approach to object matching. Found. Comput. Math. **11**(4), 417–487 (2011)
17. Poličar, P.G., Stražar, M., Zupan, B.: openTSNE: a modular Python library for t-SNE dimensionality reduction and embedding. BioRxiv (2019)
18. Rand, W.M.: Objective criteria for the evaluation of clustering methods. J. Am. Stat. Assoc. **66**(336), 846–850 (1971)
19. Traag, V.A., Waltman, L., Van Eck, N.J.: From Louvain to Leiden: guaranteeing well-connected communities. Sci. Rep. **9**(1), 1–12 (2019)
20. Villani, C.: The Wasserstein distances. In: Optimal Transport. Grundlehren der mathematischen Wissenschaften, vol. 338, pp. 93–111. Springer, Heidelberg (2009). https://doi.org/10.1007/978-3-540-71050-9_6
21. Wolberg, W.: Breast cancer Wisconsin (Original). UCI Machine Learning Repository (1992)

Nonparametric Bayesian Deep Visualization

Haruya Ishizuka[1(✉)] and Daichi Mochihashi[2]

[1] Bridgeston Corporation, Chuo City, Japan
haruya.ishizuka@bridgestone.com
[2] The Institute of Statistical Mathematics, Tachikawa, Japan
daichi@ism.ac.jp

Abstract. Visualization methods such as t-SNE [1] have helped in knowledge discovery from high-dimensional data; however, their performance may degrade when the intrinsic structure of observations is in low-dimensional space, and they cannot estimate clusters that are often useful to understand the internal structure of a dataset. A solution is to visualize the latent coordinates and clusters estimated using a neural clustering model. However, they require a long computational time since they have numerous weights to train and must tune the layer width, the number of latent dimensions and clusters to appropriately model the latent space. Additionally, the estimated coordinates may not be suitable for visualization since such a model and visualization method are applied independently. We utilize neural network Gaussian processes (NNGP) [2] equivalent to a neural network whose weights are marginalized to eliminate the necessity to optimize weights and layer widths. Additionally, to determine latent dimensions and the number of clusters without tuning, we propose a latent variable model that combines NNGP with automatic relevance determination [3] to extract necessary dimensions of latent space and infinite Gaussian mixture model [4] to infer the number of clusters. We integrate this model and visualization method into nonparametric Bayesian deep visualization (NPDV) that learns latent and visual coordinates jointly to render latent coordinates optimal for visualization. Experimental results on images and document datasets show that NPDV shows superior accuracy to existing methods, and it requires less training time than the neural clustering model because of its lower tuning cost. Furthermore, NPDV can reveal plausible latent clusters without labels.

Keywords: Data visualization · Gaussian processes · Nonparametric Bayesian models · Neural network

1 Introduction

Visualization methods such as t-SNE [1], which compress input to two- or three-dimensional visual coordinates to be mapped on a scatter plot, provide a useful

---

Supplementary Information The online version contains supplementary material available at https://doi.org/10.1007/978-3-031-26387-3_8.

M.-R. Amini et al. (Eds.): ECML PKDD 2022, LNAI 13713, pp. 121–137, 2023.
https://doi.org/10.1007/978-3-031-26387-3_8

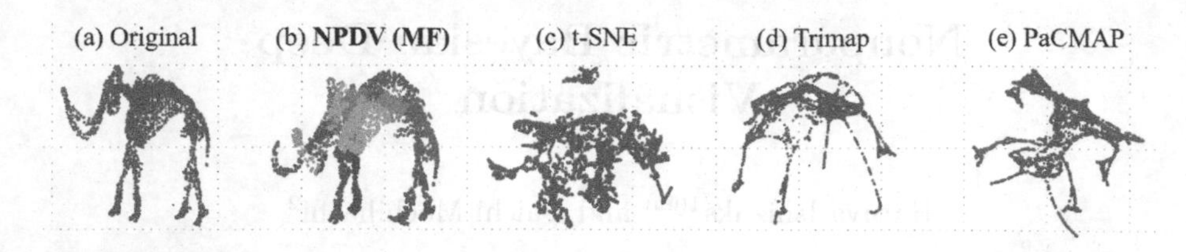

Fig. 1. 3D Visualization of 100-dimensional data generated by transforming the mammoth data in (a). Since NPDV(MF) estimates not only visual coordinates but also clusters, the associated plot in (b) can be colored by the cluster assignments, unlike existing methods in (c)–(e). See Sect. 5 for details. (Color figure online)

overview of high-dimensional data, and a number of methods have been proposed. These methods estimate visual coordinates based on the similarity of data points and fall into two categories. The first category is local methods to preserve neighbor structures in the original space [5–9]. These methods utilize nearest neighbor graph that represents pairwise similarity to retain distances between neighbors. Among them, t-SNE [1] and UMAP [10] are the most popular algorithms. The second is global methods to exploit relationships among three points, and preserve distances between data points distant from one another [11–13]. Particularly, Trimap [14] arguably shows comparable accuracy to t-SNE and UMAP. There is a trade-off between preserving the local or global structures. PaCMAP [15] exceptionally preserves both by combining the loss functions of local and global methods. They have improved knowledge discovery in various domains, such as bioinformatics [16] and audio processing [17].

However, they have several drawbacks to reveal hidden structures behind datasets. First, their performance may degrade when the observations are distributed on a low-dimensional manifold embedded in the observation space. In this case, the similarity between observations may differ from that in the manifold, resulting in an inaccurate visualization. This problem worsens when the manifold is embedded by a highly nonlinear function. Second, they cannot estimate clusters. Visualization together with clusters provides an intuitive understanding of the internal structure of a dataset; however, most of these methods only estimate visual coordinates. Supervised dimensionality-reduction [18–20] utilize labels to address this drawback; however, these are not always available. We present these problems through a simulation example. In this example, the 3D mammoth data shown in Fig. 1 (a) is embedded into a 100-dimensional space nonlinearly by a neural network; then, this data is visualized in 3D space using several methods[1]. Exiting methods in Fig. 1 (c)–(e) evidently fail to recover the original mammoth shape. In addition, the lack of estimating clusters make it difficult to interpret the internal structure of resulting plot.

A solution here is to visualize latent coordinates and clusters estimated using a neural clustering model [21–23]; however, this approach presents other issues. As these models can accurately model a low-dimensional manifold by leveraging a neural autoencoder and perform clustering simultaneously, this approach

[1] Rotatable plots are provided as an html file in the supplemental material.

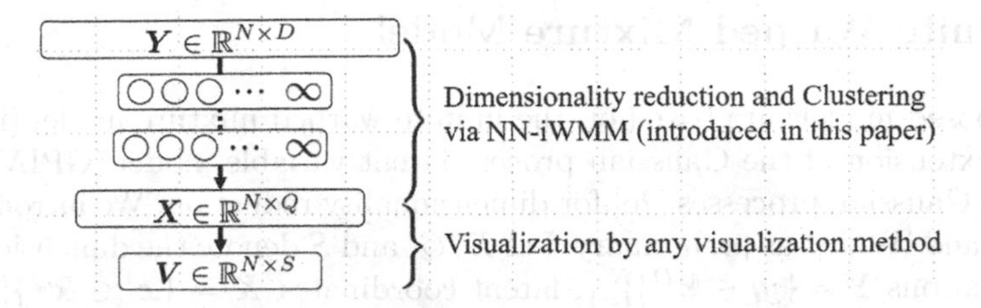

Fig. 2. Visualization flow of NPDV. NPDV integrates the estimation of latent coordinates $\boldsymbol{X}$ from observations $\boldsymbol{Y}$ using NN-iWMM and that of visual coordinates $\boldsymbol{V}$ from $\boldsymbol{X}$. Clustering is also performed by NN-iWMM.

addresses the aforementioned issues. However, they often require a long computational time to optimize model performance since they have numerous neural weights to train and need to search their appropriate hyperparameter settings, which is not suitable for scientific visualization. Particularly, the width of hidden layers, the number of latent dimensions, and clusters that are critical hyperparameters to reveal the latent structure. Furthermore, the estimated latent coordinates may not be suitable for visualization as such a model and visualization method are independently applied.

We utilize neural network Gaussian processes (NNGP) [2], which are equivalent to a neural network whose weights are marginalized, to address these issues. The marginalization greatly reduces the computational time by eliminating the necessity to optimize numerous weights and layer width while exploiting the power of neural networks. Additionally, to determine the number of latent dimensions and clusters without tuning, we propose the neural network infinite warped mixture model (NN-iWMM) by combining NNGP with automatic relevance determination [3] to extract the necessary dimensions of latent space and the infinite Gaussian mixture model [4] to infer the number of clusters. Finally, NN-iWMM and visualization methods are integrated as shown in Fig. 2, into nonparametric Bayesian deep visualization (NPDV) that jointly infers latent and visual coordinates to render latent coordinates optimal for visualization. Based on NPDV, we introduce NPDV(MF), which employs matrix factorization to linearly reduce the dimensionality, and NPDV(t-SNE) based on t-SNE; both methods enables to visualize the internal structure of dataset by utilizing the estimated clusters. As shown in Fig. 1 (b), we can observe from the simulation study that NPDV(MF) could accurately recover the mammoth shape. NPDV(t-SNE) achieves better accuracy than existing methods and NPDV(MF) for real-world data. Furthermore, NPDV(t-SNE) shows two preferable properties in unsupervised settings: (1) it takes considerably less training time than the neural clustering model and (2) has the ability to reveal plausible clusters without label information.

The remainder of this paper is organized as follows. We introduce the preliminaries in Sect. 2, and the proposed models and their training algorithm in Sects. 3 and 4, respectively. Subsequently, we demonstrate their advantages through simulation and real data experiments in Sects. 5 and 6.

2 Infinite Warped Mixture Model

The proposed models are based on the infinite warped mixture model (iWMM) [24], an extension of the Gaussian process latent variable model (GPLVM) [25] that uses Gaussian processes [26] for dimensionality reduction. We introduce the notation and iWMM as preliminary. Let D, Q, and S denote the dimensionalities of observations $\boldsymbol{Y} = \{\boldsymbol{y}_i \in \mathbb{R}^D\}_{i=1}^N$, latent coordinates $\boldsymbol{X} = \{\boldsymbol{x}_i \in \mathbb{R}^Q\}_{i=1}^N$, and visual coordinates $\boldsymbol{V} = \{\boldsymbol{v}_i \in \mathbb{R}^S\}_{i=1}^N$, respectively. Q is smaller than D and larger than S, and S is typically set to two or three. N represents the number of observations. $\mathcal{N}(\boldsymbol{m}, \boldsymbol{C})$ is a multivariate Gaussian distribution with a mean $\boldsymbol{m}$ and covariance $\boldsymbol{C}$. $\boldsymbol{I}_N$ represents an N-dimensional identity matrix.

GPLVM independently draws a latent coordinate $\boldsymbol{x}_i$ from $\mathcal{N}(\boldsymbol{0}, \boldsymbol{I}_Q)$ and draws the dth column of observations $\boldsymbol{y}_{\cdot d} = \{y_{id}\}_{i=1}^N$ from the distribution below:

$$p(\boldsymbol{y}_{\cdot d}|\boldsymbol{X}) = \mathcal{N}(\boldsymbol{y}_{\cdot d}|\boldsymbol{0}, K + \beta^{-1}\boldsymbol{I}_N), \quad (1)$$

where K is the Gram matrix, each component is given by the kernel function $k(\boldsymbol{x}, \boldsymbol{x}')$ evaluated at two coordinates $\boldsymbol{x}$ and $\boldsymbol{x}'$, and $\beta > 0$ is the precision.

Latent space sometimes has clusters; however, GPLVM fails to capture them as it assumes a unimodal Gaussian distribution as prior to $\boldsymbol{X}$. iWMM assumes the infinite Gaussian mixture model (∞-GMM) [4] defined under Dirichlet process theory [27] as prior to $\boldsymbol{X}$ to model latent clusters. Each coordinate $\boldsymbol{x}_i$ is drawn from the following distribution:

$$p(\boldsymbol{x}_i) = \sum_{k=1}^{\infty} \pi_k \mathcal{N}(\boldsymbol{x}_i|\boldsymbol{m}_k, \boldsymbol{R}_k^{-1}), \quad (2)$$

where π_k, $\boldsymbol{m}_k$, and $\boldsymbol{R}_k$ represent the mixing weight, mean, and precision matrix of the kth Gaussian distribution, respectively. For simplicity, $\boldsymbol{R}_k$ is assumed to be a diagonal matrix in this study.

3 Proposed Methods

We utilize neural network Gaussian processes (NNGP) [2], which are equivalent to a neural network whose weights and biases are marginalized, to eliminate the necessity to optimize neural weights, biases and the width of hidden layer while implicitly utilizing a neural network. Subsequently, we explain NNGP and introduce a latent variable model that combines iWMM and NNGP. Finally, we propose the NPDV.

3.1 Neural Network Gaussian Processes

In an L-layer fully connected neural network, let $\phi(\cdot)$ and N_ℓ denote the activation and width of the ℓth layer, respectively. It is assumed that the weights and biases of the ℓth layer, W_{ij}^ℓ and b_i^ℓ for $i = 1, 2 \cdots, N_\ell$ and $j = 1, 2, \cdots, N_{\ell-1}$, are

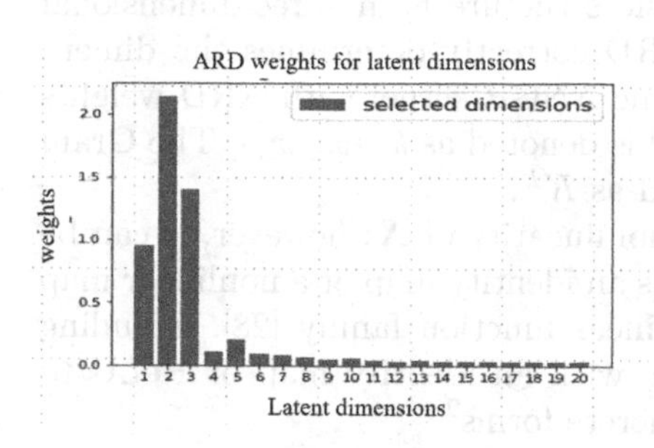

Fig. 3. ARD weights estimated by NPDV on simulation data. The three dimensions in red are selected as the necessary dimensions. (Color figure online)

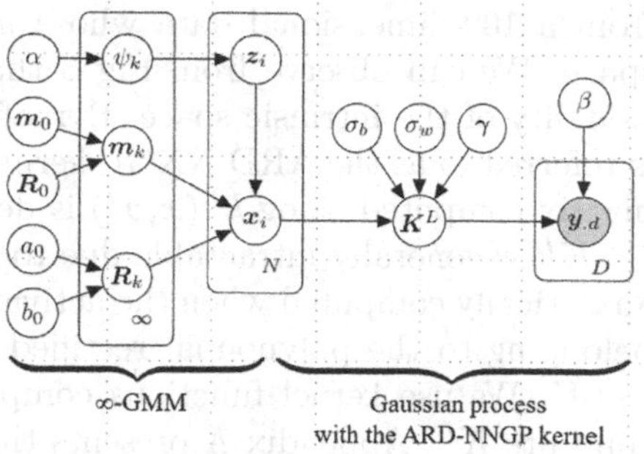

Fig. 4. Generative process of NN-iWMM. $\boldsymbol{X}$ is drawn from the ∞-GMM, and $\boldsymbol{Y}$ is then drawn from Gaussian processes with the ARD-NNGP kernel.

independently drawn from $\mathcal{N}(0, \sigma_w/N_\ell)$ and $\mathcal{N}(0, \sigma_b/N_\ell)$, respectively. A pre-activation of the ℓth layer, $a_i^\ell(\boldsymbol{x}_n)$, is then computed as a linear combination of the post-activations of the $(\ell-1)$th layer $\{\phi(a_i^{\ell-1}(\boldsymbol{x}_n))\}_{j=1}^{N_{\ell-1}}$:

$$a_i^\ell(\boldsymbol{x}_n) = b_i^\ell + \sum_{j=1}^{N_{\ell-1}} W_{ij}^\ell \phi(a_j^{\ell-1}(\boldsymbol{x}_n)). \tag{3}$$

$a_i^\ell(\boldsymbol{x}_n)$ is distributed with a Gaussian distribution from the central limit theorem when $N_\ell \to \infty$ because the summation in (3) is the sum of i.i.d. random variables. Because this result holds for any n, the output of the ℓth layer $\{a_i^\ell(\boldsymbol{x}_n)\}_{n=1}^N$ is jointly distributed with the Gaussian process with the Gram matrix K^ℓ. Each component of K^ℓ, $k^\ell(\boldsymbol{x}, \boldsymbol{x}')$, is computed as follows:

$$k^\ell(\boldsymbol{x}, \boldsymbol{x}') = \sigma_b^2 + \sigma_w^2 \mathbb{E}_{a_i^{\ell-1} \sim \mathcal{GP}(\boldsymbol{0}, K^{\ell-1})}[\phi(a_i^{\ell-1}(\boldsymbol{x}))\phi(a_i^{\ell-1}(\boldsymbol{x}'))]. \tag{4}$$

The NNGP kernel is computed by iterating the recursion (4) L times, where Gaussian processes with this kernel are equivalent to an L-layer ∞-width neural network whose weights and biases are marginalized. The first step of the original NNGP kernel is the inner product of the input: $k^0(\boldsymbol{x}, \boldsymbol{x}') = \sigma_b^2 + Q^{-1}\sigma_w^2 \boldsymbol{x}^{\mathrm{T}}\boldsymbol{x}'$. In contrast, we introduce automatic relevance determination (ARD) [3] weights $\gamma = \{\gamma_q\}_{q=1}^Q$ into $k^0(\boldsymbol{x}, \boldsymbol{x}')$ to estimate the importance of each dimension:

$$k^0(\boldsymbol{x}, \boldsymbol{x}') = \sigma_b^2 + Q^{-1}\sigma_w^2 \sum_{q=1}^{Q} \gamma_q x_q x_q'. \tag{5}$$

An ARD weight γ_q increases if the qth dimension is highly related to the observations, and becomes zero if it is totally irrelevant. Figure 3 shows the ARD weights estimated by NPDV and the selected dimensions when conducting the simulation study . In the simulation, 20 dimensional latent coordinates were estimated from a 100-dimensional data whose intrinsic structure is in three-dimensional space. We can observe from Fig. 3 that ARD correctly determines the dimensionality of the intrinsic space. Hereafter, the NNGP kernel with ARD weights is referred to as the ARD-NNGP kernel and is denoted as $k^L(\boldsymbol{x}_i, \boldsymbol{x}_j)$. The Gram matrix computed using $k^L(\boldsymbol{x}, \boldsymbol{x}')$ is denoted as K^L.

K^L is generally intractable due to the nonlinearity of $\boldsymbol{X}$; however, it can be analytically computed when the activation is an identity map or a nonlinear map belonging to the polynomial rectified nonlinear function family [28], including ReLU. We use kernel functions compatible with an identity map or ReLU to compute K^L. Appendix A presents the concrete forms[2].

3.2 NN-iWMM

NN-iWMM is formulated by substituting the ARD-NNGP kernel into iWMM and infers the latent coordinates, their cluster assignments and the number of latent clusters. Figure 4 shows its generative process consisting of generating latent coordinates $\boldsymbol{X}$ from ∞-GMM and mapping $\boldsymbol{X}$ to the observation space using Gaussian processes with the ARD-NNGP kernel.

In ∞-GMM, the mixing weights $\{\pi_k\}_{k=1}^{\infty}$ are drawn from the stick-breaking process GEM(α) [29], which defines a distribution equivalent to the Dirichlet process. Specifically, π_k is computed as $\pi_k = \psi_k \prod_{j=1}^{k-1}(1 - \psi_j)$, where each ψ_j is drawn from the beta distribution Beta$(1, \alpha)$. Subsequently, the mean $\boldsymbol{m}_k$ and diagonal components of the precision matrix $\{r_{kq}\}_{q=1}^{Q}$ of the kth Gaussian distribution are drawn from $\mathcal{N}(\boldsymbol{0}, \boldsymbol{I})$ and the gamma distribution Gam$(1, 1)$, respectively.

For $i = 1, 2, \cdots, N$, the cluster assignment z_i is drawn from the categorical distribution Cat$(\{\pi_k\}_{k=1}^{\infty})$ and $\boldsymbol{x}_i$ is generated from the z_ith Gaussian distribution $\mathcal{N}(\boldsymbol{m}_{z_i}, \boldsymbol{R}_{z_i})$. Then, the Gram matrix K^L is computed from $\boldsymbol{X}$ using (4) and (5). $\boldsymbol{y}_{\cdot d}$ is drawn from $\mathcal{N}(\boldsymbol{0}, K^L)$ for $d = 1, 2, \cdots, D$.

Due to the absence of weights and biases, it is unnecessary for NN-iWMM to optimize the weights and layer widths while leveraging a neural network. Additionally, unlike neural clustering models, using ∞-GMM and ARD weights allows to determine the number of latent dimensions and clusters, which are critical to model the latent space appropriately, without tuning.

3.3 Nonparametric Bayesian Deep Visualization

NPDV jointly estimates $\boldsymbol{X}$ and the visual coordinates $\boldsymbol{V}$ by integrating the two reduction steps, as shown in Fig. 2. We use weighted latent coordinates $\boldsymbol{X}\boldsymbol{\gamma}^{\mathrm{T}}$ as the input to prioritize the necessary dimensions of $\boldsymbol{X}$ in terms of ARD weights when estimating $\boldsymbol{V}$. Denoting $\mathcal{R}_{\mathrm{DR}}(\boldsymbol{X}\boldsymbol{\gamma}^{\mathrm{T}}, \boldsymbol{V})$ and $\boldsymbol{L}(\boldsymbol{Y}, \boldsymbol{X})$ as the visualization loss to estimate $\boldsymbol{V}$ and the loss of NN-iWMM to estimate $\boldsymbol{X}$, respectively, we introduce how to integrate these two losses into a Bayesian model using regularized Bayesian inference (regBayes) [30].

[2] All appendices are provided in the Supplemental Materials.

The regBayes framework enables the design of Bayesian models while considering the appropriate constraints on its posterior. This framework is built on a variational formulation of the Bayesian posterior $p(\boldsymbol{\theta}|\boldsymbol{Y}) \propto p(\boldsymbol{Y}|\boldsymbol{\theta})p(\boldsymbol{\theta})$, where $\boldsymbol{Y}$ and $\boldsymbol{\theta}$ are the observations and parameters, respectively. $p(\boldsymbol{\theta}|\boldsymbol{Y})$ can be viewed as a solution to the following variational optimization problem [31]:

$$\begin{cases} \min_{q(\boldsymbol{\theta})} & \mathrm{KL}[q(\boldsymbol{\theta})\|p(\boldsymbol{\theta})] - \int q(\boldsymbol{\theta}) \log p(\boldsymbol{Y}|\boldsymbol{\theta}) d\boldsymbol{\theta} \\ \text{s.t.} & q(\boldsymbol{\theta}) \in \mathcal{P} \end{cases} \tag{6}$$

where $\mathcal{P}$ is a set of probability distributions. In the regBayes framework, we consider the following optimization problem with constraints regarding the expectation of the regularizer $E_{q(\boldsymbol{\theta})}[\mathcal{R}(\boldsymbol{\theta}, \boldsymbol{Y})]$:

$$\begin{cases} \min_{q(\boldsymbol{\theta})} & \mathrm{KL}[q(\boldsymbol{\theta})\|p(\boldsymbol{\theta})] - \int q(\boldsymbol{\theta}) \log p(\boldsymbol{Y}|\boldsymbol{\theta}) d\boldsymbol{\theta} \\ \text{s.t.} & E_{q(\boldsymbol{\theta})}[\mathcal{R}(\boldsymbol{\theta}, \boldsymbol{Y})] \leq 0, \ q(\boldsymbol{\theta}) \in \mathcal{P}\,. \end{cases} \tag{7}$$

The optimal solution of (7) is obtained by

$$q^*(\boldsymbol{\theta}) \propto p(\boldsymbol{Y}|\boldsymbol{\theta})p(\boldsymbol{\theta}) \exp(-\lambda \mathcal{R}(\boldsymbol{\theta}, \boldsymbol{Y}))\,, \tag{8}$$

where λ is the Lagrange multiplier. From (8), the optimal posterior of $\boldsymbol{\theta}$ that satisfies the constraints is obtained by its right-hand term. For NPDV, $p(\boldsymbol{Y}|\boldsymbol{\theta})p(\theta)$ corresponds to the likelihood of NN-iWMM $\mathcal{L}(\boldsymbol{Y}, \boldsymbol{X})$:

$$\mathcal{L}(\boldsymbol{Y}, \boldsymbol{X}) = p(\boldsymbol{Y}|\boldsymbol{X})p(\boldsymbol{X}|\boldsymbol{z})p(\boldsymbol{z}|\{\boldsymbol{m}_k, \boldsymbol{r}_k, \psi_k\}_{k=1}^{\infty}))p(\{\boldsymbol{m}_k, \boldsymbol{r}_k, \psi_k\}_{k=1}^{\infty})\,. \tag{9}$$

$\mathcal{R}(\boldsymbol{Y}, \boldsymbol{\theta})$ is given by $\mathcal{R}_{\mathrm{DR}}(\boldsymbol{X}\gamma^{\mathrm{T}}, \boldsymbol{V})$. Therefore, the optimal posterior of the NPDV is obtained by

$$q^*(\boldsymbol{X}, \boldsymbol{V}) \propto \mathcal{L}(\boldsymbol{Y}, \boldsymbol{X}) \times \exp(-\lambda \mathcal{R}_{\mathrm{DR}}(\boldsymbol{X}\gamma^{\mathrm{T}}, \boldsymbol{V})\,. \tag{10}$$

Hence, the joint optimization of $\mathcal{L}(\boldsymbol{Y}, \boldsymbol{X})$ and $\mathcal{R}_{\mathrm{DR}}(\boldsymbol{X}\gamma^{\mathrm{T}}, \boldsymbol{V})$ is the posterior inference of $q^*(\boldsymbol{X}, \boldsymbol{V})$. NPDV makes the posterior of $\boldsymbol{X}$ suitable for visualization because $\mathcal{R}_{\mathrm{DR}}(\boldsymbol{X}\gamma^{\mathrm{T}}, \boldsymbol{V})$ in (10) serves as a regularizer to infer the posterior of NN-iWMM. λ is a hyperparameter that balances $\mathcal{L}(\boldsymbol{Y}, \boldsymbol{X})$ and $\mathcal{R}_{\mathrm{DR}}(\boldsymbol{X}\gamma^{\mathrm{T}}, \boldsymbol{V})$, and NPDV degenerates to NN-iWMM when $\lambda = 0$. $\boldsymbol{V}$ is treated as a deterministic parameter, as are the NNGP parameters $\{\sigma_w.\sigma_b, \gamma\}$.

The *any* dimensionality-reduction method can be used for $\mathcal{R}_{\mathrm{DR}}$. In this paper we combine matrix factorization (MF) and *t*-SNE, which are widely used in many different domains, with NPDV. We call these methods NPDV(MF) and NPDV(*t*-SNE), respectively. Notably, $\lambda = ND$ practically works well for both methods.

NPDV(MF): MF approximates a matrix using a product of two low-rank matrices, and is one of the widely used linear dimensionality-reduction methods. For NPDV(MF), the weighted coordinates $\boldsymbol{X}\gamma^{\mathrm{T}}$ are factorized as the rank S matrix $\boldsymbol{W}$ and visual coordinates $\boldsymbol{V}$: $\boldsymbol{X}\gamma^{\mathrm{T}} \approx \boldsymbol{W}\boldsymbol{V}^{\mathrm{T}}$. Denoting $\|\cdot\|_F^2$ as the

Frobenius norm, $\|\boldsymbol{A}\|_F^2 \equiv \sum_{i,j} a_{ij}^2$, $\boldsymbol{A} \in \mathbb{R}^{I \times J}$, and the associated visualization loss $\mathcal{R}_{\mathrm{MF}}(\boldsymbol{X}\boldsymbol{\gamma}^{\mathrm{T}}, \boldsymbol{V})$ can be computed as

$$\mathcal{R}_{\mathrm{MF}}(\boldsymbol{X}\boldsymbol{\gamma}^{\mathrm{T}}, \boldsymbol{V}) = \|\boldsymbol{X}\boldsymbol{\gamma}^{T} - \boldsymbol{W}\boldsymbol{V}^{\mathrm{T}}\|_F^2 . \tag{11}$$

NPDV(t-SNE): NPDV(MF) may fail to capture the nonlinear pattern as it linearly reduces the dimensionality of $\boldsymbol{X}\boldsymbol{\gamma}^{\mathrm{T}}$. We then introduce NPDV(t-SNE), which uses the t-SNE loss for $\mathcal{R}_{\mathrm{DR}}(\boldsymbol{X}\boldsymbol{\gamma}^{\mathrm{T}}, \boldsymbol{V})$ as a nonlinear counterpart. For NPDV(t-SNE), $\mathcal{R}_{\mathrm{DR}}(\boldsymbol{X}\boldsymbol{\gamma}^{\mathrm{T}}, \boldsymbol{V})$ is computed as the divergence of similarities in the latent and visual spaces. Denoting $\odot$ as the element-wise Hadamard product, the similarity of weighted latent coordinates, $\boldsymbol{x}_i \odot \boldsymbol{\gamma}$ and $\boldsymbol{x}_j \odot \boldsymbol{\gamma}$, is computed from the conditional probability based on the Gaussian kernel:

$$p_{j|i}^X = \frac{\exp(-\|\boldsymbol{\gamma} \odot \boldsymbol{x}_i - \boldsymbol{\gamma} \odot \boldsymbol{x}_j\|^2/2\tau_i^2)}{\sum_{\ell \neq i} \exp(-\|\boldsymbol{\gamma} \odot \boldsymbol{x}_i - \boldsymbol{\gamma} \odot \boldsymbol{x}_\ell\|^2/2\tau_i^2)}, \; p_{ij}^X = \frac{p_{i|j}^X + p_{j|i}^X}{2N}, \tag{12}$$

where τ_i^2 is the variance of the Gaussian distribution and is computed from the neighbors of $\gamma \odot \boldsymbol{x}_i$ using a binary search with perplexity ρ, which is a hyperparameter that controls the number of neighbors.

The similarity of two visual coordinates, $\boldsymbol{v}_i$ and $\boldsymbol{v}_j$, is evaluated using Student's t-distribution kernel as follows:

$$p_{ij}^V \equiv \frac{(1 + \|\boldsymbol{v}_j - \boldsymbol{v}_i\|^2)^{-1}}{\sum_k \sum_{\ell \neq k} (1 + \|\boldsymbol{v}_k - \boldsymbol{v}_\ell\|^2)^{-1}} . \tag{13}$$

$\mathcal{R}_{t\text{-SNE}}(\boldsymbol{X}\boldsymbol{\gamma}^{\mathrm{T}}, \boldsymbol{V})$ is computed as the KL divergence between $\{p_{ij}^X\}_{i,j}$ and $\{p_{ij}^V\}_{i,j}$:

$$\mathcal{R}_{t\text{-SNE}}(\boldsymbol{X}\boldsymbol{\gamma}^{\mathrm{T}}, \boldsymbol{V}) \equiv \sum_{i,j,i \neq j} p_{ij}^X \log \frac{p_{ij}^Y}{p_{ij}^V} . \tag{14}$$

4 Bayesian Training

We employ variational inference to train the NPDV. NN-iWMM is a special case of NPDV when $\lambda = 0$; hence, we focus on the training algorithm for NPDV. The parameters of NPDV are estimated by maximizing the evidence lower bound (ELBO) $\mathcal{L}$ of NPDV. $\mathcal{L}$ is derived from Jensen's inequality and the variational distribution $\mathcal{Q}$that is used to approximate the true posterior [32]:

$$\mathcal{L} = \mathbb{E}_{\mathcal{Q}}\left[\log \frac{\mathcal{L}(\boldsymbol{Y}, \boldsymbol{X}) \times \exp(-\lambda \mathcal{R}_{\mathrm{DR}}(\boldsymbol{X}\boldsymbol{\gamma}^{\mathrm{T}}, \boldsymbol{V}))}{\mathcal{Q}}\right] . \tag{15}$$

Following [33], the ∞-GMM is approximated by a finite Gaussian mixture model whose maximum number of mixtures is K. Additionally, we impose the following mean-field assumption on $\mathcal{Q}$.

$$\mathcal{Q} = \prod_{i=1}^{N} q(\boldsymbol{x}_i) q(z_i) \prod_{k=1}^{K} q(\psi_k) q(\boldsymbol{m}_k) q(\boldsymbol{r}_k) , \tag{16}$$

Algorithm 1. Variational inference algorithm for NPDV

Input observations $\boldsymbol{Y}$ and the number of layers L
1. Pre-training
Initialize $\boldsymbol{\Pi}_0 = [\{\boldsymbol{\mu}_i, \boldsymbol{S}_i\}_{i=1}^N, \boldsymbol{\zeta}, \sigma_b^2, \sigma_w^2, \boldsymbol{\gamma}, \beta]$
for i=1,2,... **do**
  Update $\boldsymbol{\Pi}_0$ with a gradient-based method
end for
Initialize $\boldsymbol{V}$
2. Training NPDV.
Initialize $\boldsymbol{\Pi}_1 = \{\boldsymbol{z}, \{\boldsymbol{m}_k, \boldsymbol{r}_k, \pi_k\}_{k=1}^K\}$
for i=1,2,... **do**
  Generate $\widetilde{\boldsymbol{X}}$ with (18)
  Approximate $\mathcal{L}_1$, $\mathcal{L}_2$, and $\mathcal{R}_{\rm DR}$ with $\widetilde{\boldsymbol{X}}$
  Update $\boldsymbol{\Pi}_1$ with the EM algorithm in Appendix B
  Update $\boldsymbol{\Pi}_2 = \{\boldsymbol{\mu}_i, \boldsymbol{S}_i\}_{i=1}^N, \boldsymbol{\zeta}, \boldsymbol{V}, \sigma_b^2, \sigma_w^2, \boldsymbol{\gamma}, \beta\}$ with a gradient-based method
end for

where $q(\boldsymbol{x}_i) = \mathcal{N}(\boldsymbol{\mu}_i, \boldsymbol{S}_i)$. The variational distributions for the mixture model $q(\psi_k)$, $q(\boldsymbol{m}_k)$, $q(\boldsymbol{r}_k)$, and $q(z_i)$ have the same form as in [33]. Using the mean-field assumption on $\mathcal{Q}$ and conditional independence of NN-iWMM, $\mathcal{L}$ is decomposed into the four terms:

$$\begin{aligned}\mathcal{L} &= \mathbb{E}_{q(\boldsymbol{X})}[\log p(\boldsymbol{Y}|\boldsymbol{X})] - \mathbb{E}_{q(\boldsymbol{X})}[\log q(\boldsymbol{X})].\\ &+ \mathbb{E}_{q(\boldsymbol{X},\boldsymbol{z},\boldsymbol{m},\boldsymbol{r},\boldsymbol{\phi})}\left[\log \frac{p(\boldsymbol{X}, \boldsymbol{z}, \{\boldsymbol{m}_k, \boldsymbol{r}_k, \pi_k\}_{k=1}^K)}{q(\boldsymbol{z}, \{\boldsymbol{m}_k, \boldsymbol{r}_k, \pi_k\}_{k=1}^K)}\right] - \lambda \mathbb{E}_{q(\boldsymbol{X})}[\mathcal{R}_{\rm DR}(\boldsymbol{X}\boldsymbol{\gamma}^{\rm T}, \boldsymbol{V})]\\ &= \mathcal{L}_1 + \textstyle\sum_{i=1}^N \mathcal{H}(q(\boldsymbol{x}_i)) + \mathcal{L}_2 - \lambda \mathcal{R}_{\rm DR}(\boldsymbol{X}\boldsymbol{\gamma}^{\rm T}, \boldsymbol{V}),\end{aligned} \tag{17}$$

Unlike Gaussian entropy $\mathcal{H}(q(\boldsymbol{x}_i))$, $\mathcal{L}_1$, $\mathcal{L}_2$, and $\mathcal{R}_{\rm DR}(\boldsymbol{X}\boldsymbol{\gamma}^{\rm T}, \boldsymbol{V})$ cannot be evaluated analytically. Therefore, we adapt the reparameterization trick [34] to approximate these quantities. Using this trick, Monte Carlo samples of $\boldsymbol{x}_i$, $\widetilde{\boldsymbol{x}}_i$, is generated by affine-transforming the standard Gaussian noise $\boldsymbol{\epsilon}$ with $\boldsymbol{\mu}_i$ and $\boldsymbol{S}_i$:

$$\widetilde{\boldsymbol{x}}_i = \boldsymbol{\mu}_i + \boldsymbol{S}_i \boldsymbol{\epsilon}_i\,; \boldsymbol{\epsilon}_i \sim \mathcal{N}(\boldsymbol{0}, \boldsymbol{I}_Q) \text{ for } i = 1, 2, \cdots, N\,. \tag{18}$$

Hereafter, we outline the evaluation of $\mathcal{L}_1$, $\mathcal{L}_2$, and $\mathcal{R}_{\rm DR}$ using $\widetilde{\boldsymbol{X}} = \{\widetilde{\boldsymbol{x}}_i\}_{i=1}^N$.

The naive approximation of $\mathcal{L}_1$ requires $\mathcal{O}(N^3)$ complexity because it inverts $K^L \in \mathbb{R}^{N\times N}$ and is difficult to train over a large dataset. We exploit the inducing-point approach [35] to reduce the complexity. Using this approach, $\mathcal{L}_1$ given $\widetilde{\boldsymbol{X}}$ is approximated by $\mathcal{O}(M^3)$, $(M \ll N)$ complexity based on M pseudo-inputs $\boldsymbol{\zeta} \in \mathbb{R}^{M\times Q}$ in the latent space, and the corresponding Gaussian process outputs $\boldsymbol{u}_d \in \mathbb{R}^M$. The cost of the variational mixture model $\mathcal{L}_2$, given $\widetilde{\boldsymbol{X}}$, is of the same form as the ELBO for ∞-GMM in [33]. Moreover, the parameters of the variational mixture model can be updated using the expectation-maximization (EM) algorithm. $\mathcal{R}_{\rm DR}(\boldsymbol{X}\boldsymbol{\gamma}^{\rm T}, \boldsymbol{V})$ is computed by substituting $\widetilde{\boldsymbol{X}}$

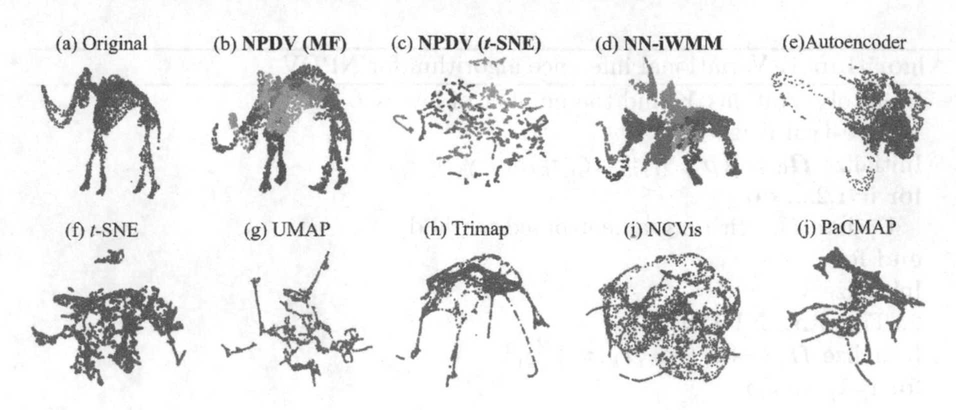

Fig. 5. 3D Visualization of a 100-dimensional data that are generated by transforming the mammoth data in (a) using a neural network. (b)–(d) are obtained by proposed methods and colored by the estimated clusters. (e)–(j) are obtained by existing methods (Color figure online)

and ARD weights with $\boldsymbol{\gamma}$. Appendix B provides the details of evaluating $\mathcal{L}_1$ and $\mathcal{L}_2$ and the update formulae of the parameters of the variational mixtures.

Algorithm 1 summarizes the NPDV training algorithm. First, we pretrain the NN-iWMM that assumes $\mathcal{N}(\mathbf{0}, \boldsymbol{I}_Q)$ as the prior of $\boldsymbol{X}$ to initialize $\boldsymbol{\Pi}_0 = [\{\boldsymbol{\mu}_i, \boldsymbol{S}_i\}_{i=1}^N, \boldsymbol{\zeta}, \sigma_w^2, \sigma_b^2, \boldsymbol{\gamma}, \beta]$. Subsequently, we initialize $\boldsymbol{V}$. After that, We generate Monte Carlo samples $\widetilde{\boldsymbol{X}}$ using (18) and approximate $\mathcal{L}_1$, $\mathcal{L}_2$, and $\mathcal{R}_{\text{DR}}$ to train the NPDV. Then, we update the parameters of the variational mixture model $\boldsymbol{\Pi}_1 = [\boldsymbol{z}, \{\boldsymbol{m}_k, \boldsymbol{r}_k, \psi_k\}_{k=1}^K]$ and the others $\boldsymbol{\Pi}_2 = [\{\boldsymbol{\mu}_i, \boldsymbol{S}_i\}_{i=1}^N, \boldsymbol{\zeta}, \boldsymbol{V}, \sigma_b^2, \sigma_w^2, \boldsymbol{\gamma}, \beta]$ using the EM algorithm and a gradient-based method, respectively.

Less Parametricity: Due to the absence of weights and biases, NPDV has significantly less parameters than autoencoder variants. A symmetric neural autoencoder has $2\sum_{\ell=1}^{L}(N_\ell + N_\ell N_{\ell-1})$ weights and biases, where N_ℓ and N_0 represent the width of ℓth hidden layer and dimensionality of the observations, respectively. It is not uncommon for them to have more than 10^7 parameters as N_ℓ often exceeds thousands. Conversely, because NNGP has no need to estimate weights and biases, NPDV has much less parameters than neural models.

Hyperparameter Settings: NPDV has several hyperparameters; however, not all of them are considered in practice. For the maximum latent dimensions and cluster, Q and K, if they are set to sufficiently large values, the necessary latent dimensions and number of clusters are estimated by the ARD mechanism and ∞-GMM. For the perplexity of NPDV(t-SNE), ρ, the number of inducing points M, learning rate η, and balance term λ, NPDV(t-SNE) achieves higher accuracy than the existing methods with $M=100$, $\rho=30$, $\eta=0.01$, and $\lambda=ND$. Therefore, we only focus on a single hyperparameter, the layer depth of the NN-iWMM L.

Table 1. Summary of the datasets used in Sect. 6. C and D are the number of labels and dimensionality of observations, respectively. For **20 news**, 20 labels are converted to 6 meta-labels according to the dataset guideline.

Dataset	Type	C	D
MNIST	Images	10	784
Fashion-MNIST	Images	10	784
20 news	Documents	6	1,000

5 Simulation Study

We present the qualitative properties of the NPDV through a simulation study. As mentioned in Sect. 1, the visualization accuracy may degrade when observations are distributed on a lower dimensional manifold. To imitate such situation, the 3D mammoth data in Fig. 5(a) is embedded into a 100-dimensional space by a neural network. Then, we visualize this 100-dimensional data in 3D space using several methods. Besides NN-iWMM, NPDV(MF) and NPDV(t-SNE), we apply six existing methods to the data as baselines: Autoencoder[3], t-SNE, UMAP [10], Trimap [14], NCvis [9], and PaCMAP [15]. Appendix C provides the details of the experimental settings.

Figure 5 shows the resulting plot of each method. In contrast to conventional visualization methods in Fig. 5(f)–(j), since NN-iWMM based methods infers clusters in addition to latent coordinates, we colored the associated plots by estimated clusters. Especially, Fig. 5 (b) shows that NPDV(MF) recovers the original mammoth shape more accurately than the existing methods in Fig. 5 (e)–(j) and other NN-iWMM based methods in Fig. 5 (c) and (d). This means NPDV(MF) accurately recovered the intrinsic manifold embedded in a high-dimensional space in this simulation. Furthermore, it enables to find the specific parts of the mammoth body, such as paws and horns, as clusters by coloring with the cluster assignments.

The plot of NPDV(t-SNE) in Fig. 5 (c) is blurred because the t-SNE occasionally fails to capture the global structure. However, as shown in the next section, NPDV(t-SNE) achieves superior performance to NPDV(MF) on real-world data.

6 Experiments on Real-World Data

We demonstrate several advantages of NPDV(t-SNE) through real-world data experiments on three datasets. **MNIST** contains hand-written digit images, where each image is labeled one of 0-9. **Fashion-MNIST** contains images of clothing, where each image is labeled with one of 10 categories, such as T-shirts or shoes. **20 news** corpus records English articles, where each article is classified into one of 20 labels. Table 1 summarises these datasets. We randomly

[3] The network architecture is the same as that of the data generation network.

Table 2. Average k-nearest neighbors classification accuracy with five different random seeds. The highest scores are in bold font.

Method	MNIST			Fashion-MNIST			20 news		
	k=10	k=20	k=30	k=10	k=20	k=30	k=10	k=20	k=30
t-SNE	0.930	**0.920**	0.915	0.819	0.794	0.783	0.726	0.700	0.690
UMAP	0.921	0.916	0.913	0.777	0.763	0.757	0.729	0.705	0.695
Trimap	0.902	0.897	0.891	0.774	0.760	0.757	0.740	0.720	0.713
NCVis	0.891	0.886	0.884	0.776	0.764	0.759	0.405	0.358	0.338
PaCMAP	0.902	0.896	0.894	0.778	0.766	0.759	0.741	0.724	0.717
VSB-DVM	**0.931**	0.920	**0.915**	**0.837**	0.819	0.806	0.778	0.757	0.749
NN-iWMM	0.893	0.884	0.881	0.765	0.748	0.741	0.725	0.706	0.698
NPDV(MF)	0.529	0.495	0.484	0.650	0.629	0.619	0.636	0.618	0.610
NPDV(t-SNE)	0.928	0.917	0.911	0.834	**0.820**	**0.808**	**0.786**	**0.761**	**0.750**

extracted 5,000 samples from each dataset for evaluation. For **MNIST** and **Fashion-MNIST**, all images are scaled within [0, 1]. For **20 news**, the original 20 labels were converted into six meta-labels, e.g., *rec* and *sci* for recreation and science, respectively, according to the dataset guideline[4], and documents were transformed into 1,000-dimensional tf.idf vectors after removing stopwords and performing lemmatization.

We used the k-nearest neighbor classification accuracy for $k = [10, 20, 30]$ as a metric. This metric increases when coordinates with the same label are close to one another and measures how accurately they can capture label differences. Furthermore, we qualitatively compared the resulting plots. For model comparison, in addition to the methods used in Sect. 5, we built VSB-DVM+t-SNE that applies t-SNE to latent coordinates estimated by the latest neural clustering model, VSB-DVM [23]. The hyperparameters of VSB-DVM are tuned by minimizing the loss of held-out 1,000 samples using Optuna [36] with 50 trials. Note that tuning VSB-DVM is time-consuming due to iterative model fitting. For NPDV(MF) and NPDV(t-SNE), the layer depth, L, the maximum number of latent dimensions and clusters, Q and K are set to $L = 6$, $Q = 100$ and $K = 50$, respectively. Appendix D provides The details of experimental settings and all visualization results.

Quantitative and Qualitative Comparison. Table 2 lists the k-nearest neighbor-classification accuracies. VSB-DVM+t-SNE and NPDV(t-SNE) outperform other methods for **Fashion-MNIST** and **20 news**. Notably, NPDV(t-SNE) shows comparable accuracies with well-tuned VSB-DVM+t-SNE. Figure 5 and 6 show the resulting plots on **Fashion-MNIST** and **20 news** obtained by the five methods. For NPDV(t-SNE), NN-iWMM and VSB-DVM+t-SNE, the

[4] http://qwone.com/~ jason/20Newsgroups/.

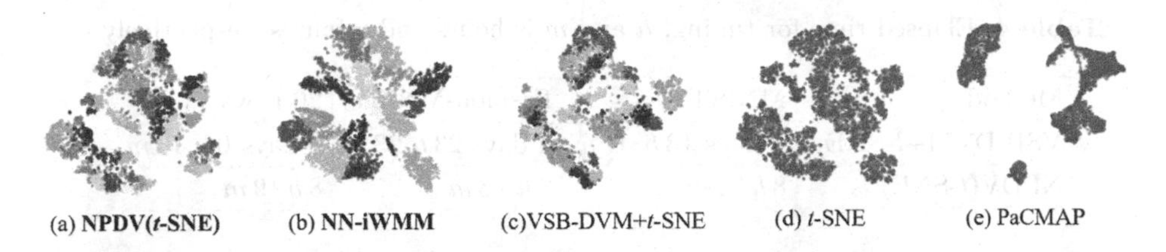

Fig. 6. Visualization of **Fashion-MNIST**. (a), (b) and (c) are colored by latent clusters that can be discovered by each method.

Fig. 7. Visualization of **20 news**. (a), (b) and (c) are colored by latent clusters that can be discovered by each method.

Table 3. Number of parameters of optimized models.

Method	MNIST	Fashion-MNIST	20 news
VSB-DVM+t-SNE	133.6×10^6	29.1×10^6	2.81×10^6
NPDV(t-SNE)	1.04×10^6	1.04×10^6	1.04×10^6

points are colored by the estimated clusters. The coloring helps to understand cluster structures more easily than t-SNE and PaCMAP, as in the simulation study. For **20 news**, NPDV(t-SNE) shows better cluster separation compared to VSB-DVM+t-SNE. We guess the reason why cluster structures are taken over to visual coordinates due to the joint training of NN-iWMM and t-SNE (Fig. 7).

Computational Cost Comparison with Neural Clustering Model. NPDV(t-SNE) incurs significantly less computational cost compared to VSB-DVM+t-SNE. VSB-DVM needs to estimate numerous weights. Consequently, it has 2.8–133 times more parameters than NPDV(t-SNE), as shown in Table 3. Additionally, VSB-DVM must tune several hyperparameters, which increases the computational time. Specifically, it took multiple days with 50 trials, as shown in Table 4. Conversely, we only focus on layer depth L to train NPDV(t-SNE). NPDV(t-SNE) finishes computation in a considerably shorter time than VSB-DVM even if we try all candidates $L = \{4, 5, 6, 7\}$.

Latent Cluster Discovery. We investigated the characteristics of the clusters estimated using the NPDV(t-SNE). In addition to the **20 news**, we used the

Table 4. Elapsed time for tuning. h and m is hours and minutes, respectively.

Method	MNIST	Fashion-MNIST	20 news
VSB-DVM+t-SNE	5 days 13 h 47 m	4 days 23 h 23 m	6 days 0 h 13 m
NPDV(t-SNE)	8 h 32 m	8 h 6 m	8 h 19 m

Brown corpus[5]. For the **Brown** corpus, we randomly extracted 5,000 sentences and converted them into sentence vectors using SIF weighting [37].

For **20 news**, we investigated the relationships between the four lower labels in *rec* and estimated clusters. The articles in *rec* belong to one of 23 clusters. Figure 8 is a cross table of these labels and clusters. Evidently, the same labeled articles belong to specific clusters, while these labels were not used during training. The link between labels and clusters indicates that a common topic exists within a cluster because each label represents a single topic. Figure 9 shows a scatter plot of the **Brown** corpus. We discovered that some clusters shared topics, such as religion, social thought, and science. Furthermore, similar themes (religion and social thought) were placed close, whereas different themes (science) were distant, based on the first two themes. Therefore, the distance between clusters reflects the similarity of themes.

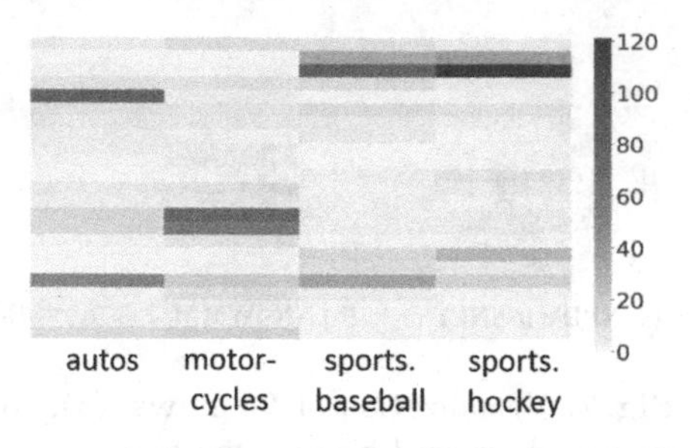

Fig. 8. Cross table of lower labels in *rec* (x-axis) and the clusters (y-axis).

From these visualizations, we found that a common topic exists within a cluster and the distance between clusters reflects the similarity of topics. As topics can be considered as intrinsic clusters in a dataset, NPDV(t-SNE) can help to reveal clusters and grasp their similarities.

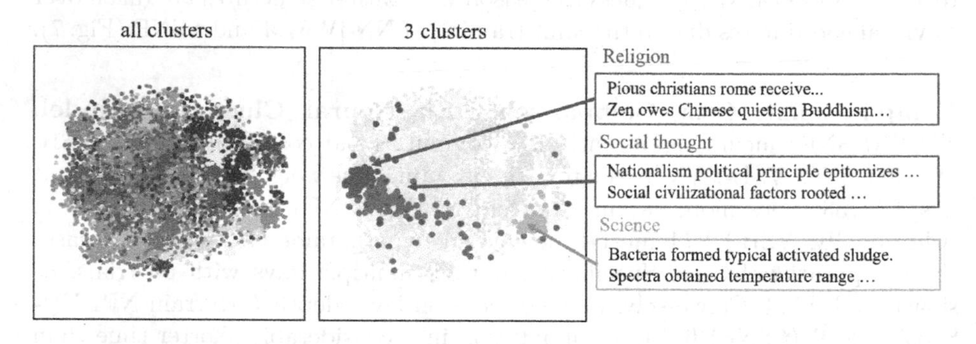

Fig. 9. Visualization of **Brown** corpus using NPDV(t-SNE). The points are colored by the estimated clusters. The right is an entire plot, and the left shows three clusters and their sample sentences. (Color figure online)

[5] http://korpus.uib.no/icame/manuals/BROWN/INDEX.HTML.

7 Conclusion

We proposed a nonparametric Bayesian latent variable model, NN-iWMM, and an associated visualization method, NPDV. NN-iWMM determines the layer widths, the dimensionality of latent space, and the number of clusters that are critical to model the latent space without tuning, while leveraging the power of neural networks implicitly. NPDV estimates the optimal latent coordinates to learn visual coordinates by integrating NN-iWMM and a visualization method. Additionally, we introduced NPDV(MF) and NPDV(t-SNE). Both methods enable to visualize the internal structure of dataset by utilizing the estimated clusters. Simulation studies demonstrated that NPDV(MF) infers the intrinsic latent manifold better than the existing methods. Real data experiments demonstrated that NPDV(t-SNE) outperforms conventional methods and shows comparable accuracy with a well-tuned neural clustering model. Furthermore, it shows two preferable properties in unsupervised settings: (1) NPDV(t-SNE) takes considerably less training time than the neural clustering model and (2) it has the ability to revealing plausible clusters without label information.

In this paper, we limit ourselves to study the properties of NPDV in the case of using matrix factorization or t-SNE. By considering combination with other methods, we expect to improve the accuracy and computing efficiency of NPDV.

References

1. van der Maaten, L., Hinton, G.: Visualizing data using t-SNE. J. Mach. Learn. Res. **9**(1), 2579–2605 (2008)
2. Lee, J.H., Bahri, Y., Novak, R., Schoenholz, S., Pennington, J., Sohl-Dickstein, J.: Deep neural networks as Gaussian processes. In: International Conference on Learning, Representation, vol.2018, no. 48, pp. 478–487 (2018)
3. Mackay, D.J.: Bayesian non-linear modeling for the prediction competition. ASHRAE Trans. **100**(2), 1053–1062 (1994)
4. Rassmusen, C.E.: The infinite Gaussian mixture model. In: Proceedings of the 12th International Conference on Neural Information Processing Systems, pp. 554–560
5. Kruscal, J.B.: Multidimensional scaling by optimizing goodness of git to a nonmetric hypothesis. Psychometrika **29**, 1–27 (1964)
6. Tenebaum, J.B., de Silva, J.C., Langford, A.: A global geometric framework for nonlinear dimensionality reduction. Science **290**(5500), 2319–2323 (2000)
7. Hinto, G.E., Roweis, S.: Stochastic neighbor embedding. In: Advances in Neural Information Processing Systems 15 (2002)
8. Tang, J., Liu, J., Zhang, M., Mei, Q.: Visualizing large-scale and high-dimensional data. In: The 25th International Conference on the World Wide Web, pp. 287–297 (2016)
9. Aleksandr, A., Maxim, P.: NCVis: noise contrastive approach for scalable visualization. Proceed. Web Conf. **2020**, 2941–2947 (2020)
10. McInnes, L., Healy, J., Saul, N., Großberge, L.: UMAP: uniform manifold approximation and projection. J. Open Source Softw. **3**(29), 2579–2605 (2018)
11. Hadsell, R., Chopra, S., LeCun, Y.: Dimensionality reduction by learning an invariant mapping. In: Proceedings of the IEEE Computer Society Conference on Computer Vision and Pattern Recognition, vol. 2, pp. 1735–1742 (2006)

12. van der Maaten, L., Weinberger, K.: Stochastic triplet embedding. In: Proceedings of the IEEE International Workshop on Machine Learning for Signal Processing, pp. 1–6 (2012)
13. Wilber, M.J., Kwak, I.S., Kriegman D.J., Belongie, S.: Learning concept embeddings with combined human-machine expertise. In: Proceedings of the IEEE International Conference on Computer Vision 2, pp. 981–989 (2015)
14. Ehsan, A., Manfred, K.W.: TriMap: large-scale dimensionality reduction using triplets. arXiv preprint arXiv:1910.00204 (2019)
15. Wang, Y., Huang, H.M., Rudin, C., Shaposhnik, Y.: Understanding how dimension reduction tools work: an empirical approach to deciphering t-SNE, UMAP, TriMap, and PaCMAP for Data Visualization. J. Mach. Learn. Res. **2021**(201), 1–73 (2021)
16. Wallach, I., Liliean, R.: The protein-small-molecule database, a non-redundant structural resource for the analysis of protein-ligand binding. Bioinformatics **25**(5), 615–620 (2010)
17. Hamel, P., Eck, D.: Learning features from music audio with deep belief networks. In: Proceedings of the International Society for Music Information Retrieval Conference, pp. 339–344 (2010)
18. Geng, X., Zhan, D.-C., Zhou, Z.-H.: Supervised nonlinear dimensionality reduction for visualization and classification. IEEE Trans. Syst. Man Cybern. **35**(6), 1098–1107 (2005)
19. Venna, A., Peltonen, J., Nybo, K., Aidos, H., Kaski, S.: Information retrieval perspective to nonlinear dimensionality reduction for data visualization. J. Mach. Learn. Res. **11**(13), 451–490 (2010)
20. Zheng, J., Zhang, H.H., Cattani, C., Wang, W.: Dimensionality reduction by supervised neighbor embedding using Laplacian search. Biomed. Sig. Process. Model. Complexity Living Syst. **2014**, 594379 (2014)
21. Xie, J., Girshick, R., Farhadi, A.: Unsupervised deep embedding for clustering analysis. In: Proceedings of the 33rd International Conference on International Conference on Machine Learning, vol. 48, pp. 478–487 (2016)
22. Fard, M.N., Thonet, T., Gaussier, E.: Deep k-means: jointly clustering with k-means and learning representations. ArXiv:1806.10069 (2018)
23. Yang, X., Yan, Y., Huang, K., Zhang, R.: VSB-DVM: an end-to-end Bayesian nonparametric generalization of deep variational Mixture Model. In: 2019 IEEE International Conference on Data Mining (2019)
24. Iwata, T., Duvenaud, D., Ghahramani, Z.: Warped mixtures for nonparametric cluster shapes. In: Proceedings of the Twenty-Ninth Conference on Uncertainty in Artificial Intelligence, pp. 311–320 (2013)
25. Lawrence, N.D.: Probabilistic non-linear principal component analysis with Gaussian process latent variable models. J. Mach. Learn. Res. **6**, 1783–1816 (2004)
26. Rassmusen, C.E., Williams, C.: Gaussian Processes for Machine Learning. The MIT Press (2006)
27. Ferguson, T.S.: A Bayesian analysis of some nonparametric problems. Ann. Statist. **1**(2), 209–230 (1973)
28. Cho, Y., Saul, L.K.: Kernel methods for deep learning. Adv. Neural. Inf. Process. Syst. **22**, 342–350 (2009)
29. Sethuraman, J.: Constructive definition of Dirichlet process. Statist. Sinica **4**(2), 639–650 (1994)
30. Zhu, J., Chen, N., Xing, E.P.: Bayesian inference with posterior regularization and applications to infinite latent SVMs. J. Mach. Learn. **15**(1), 1799–1847 (2014)
31. Zellner, A.: Optimal information processing and Bayes' theorem. Am. Stat. **42**(4), 278–280 (1988)

32. Bishop, C.M.: Pattern recognition and machine learning, 1st Edn. Springer (2006)
33. Blei, D., Jordan, M.: Variational inference for Dirichlet process mixtures. J. Bayesian Anal. **1**(1), 121–144 (2006)
34. Kingma, D.P., Welling, M.: Auto-encoding variational Bayes. In: Proceedings of the 2nd International Conference on Learning Representations (2013)
35. Titsias, M.K., Lawrence, N.D.: Bayesian Gaussian process latent variable model. In: Proceedings of the 13th International Workshop on Artificial Intelligence and Statistics, vol. 9, pp. 844–851 (2010)
36. Akiba, T., Sano, S., Yanase, T., Ohta, T., Koyama, M.: Optuna: a next-generation hyperparameter optimization framework. In: Proceedings of the 25th ACM SIGKDD International Conference on Knowledge Discovery & Data Mining, KDD 2019, pp. 2623–2631 (2019)
37. Arora, S., Liang, Y., Ma, T.: A simple but tough-to-beat baseline for sentence embeddings. In: International Conference on Learning Representations (2017)

FastDEC: Clustering by Fast Dominance Estimation

Geping Yang[1], Hongzhang Lv[1], Yiyang Yang[1(✉)], Zhiguo Gong[2(✉)], Xiang Chen[3], and Zhifeng Hao[4]

[1] Faculty of Computer, Guangdong University of Technology, Guangzhou, China
yyygou@gmail.com

[2] State Key Laboratory of Internet of Things for Smart City and Department of Computer and Information Science, University of Macau, Macau, China
fstzgg@um.edu.mo

[3] School of Electronics and Information Technology, Sun Yat-Sen University, Guangzhou, China

[4] College of Engineering, Shantou University, shantou, China

Abstract. k-Nearest Neighbors (k-NN) graph is essential for the various graph mining tasks. In this work, we study the density-based clustering on the k-NN graph and propose FastDEC, a clustering framework by fast dominance estimation. The nearest density higher (NDH) relation and dominance-component (DC), more specifically their integration with the k-NN graph, are formally defined and theoretically analyzed. FastDEC includes two extensions to satisfy different clustering scenarios: FastDEC$_D$ for partitioning data into clusters with arbitrary shapes, and FastDEC$_K$ for K-Way partition. Firstly, a set of DCs is detected as the results of FastDEC$_D$ by segmenting the given k-NN graph. Then, the K-Way partition is generated by selecting the top-K DCs in terms of the inter-dominance (ID) as the seeds, and assigning the remaining DCs to their nearest dominators.

FastDEC can be viewed as a much faster, more robust, and k-NN based variant of the classical density-based clustering algorithm: Density Peak Clustering (DPC). DPC estimates the significance of data points from the density and geometric distance factors, while FastDEC innovatively uses the global rank of the dominator as an additional factor in the significance estimation. FastDEC naturally holds several critical characteristics: (1) excellent clustering performance; (2) easy to interpret and implement; (3) efficiency and robustness. Experiments on both the artificial and real datasets demonstrate that FastDEC outperforms the state-of-the-art density methods including DPC.

Keywords: Density estimation · Clustering · k Nearest Neighbors

G. Yang and H. Lv—Equal Contribution.

Supplementary Information The online version contains supplementary material available at https://doi.org/10.1007/978-3-031-26387-3_9.

M.-R. Amini et al. (Eds.): ECML PKDD 2022, LNAI 13713, pp. 138–156, 2023.
https://doi.org/10.1007/978-3-031-26387-3_9

1 Introduction

Density-based clustering is widely used in various fields such as computer vision and outlier detection. As a fast pre-processing technology, it is used to partition the data into clusters with arbitrary shapes. Several classic and effective methods have been designed for such a task. This work aims to design an efficient, robust, and effective density-based clustering method. In Table 1, we list several perspectives that are significant to the density-based clustering, based on which, analyze the relevant methods including the classical and state-of-the-art.

Table 1. Characteristics of existing density-based methods. CC: Connected-Component detection. OS: over-segmentation. K-Way: supports K-Way Partition. Para: primary parameters. (DR: density reachable. DH: density-higher. NDH: nearest density-higher.)

Method	Connectivity	CC	Avoid OS	K-Way	Accuracy	Efficiency	Para	Robustness
DBSCAN	DR	×	×	×	Low	High	τ, min_{pt}	Low
Meanshift	DH	×	×	×	Media	Low	τ	Media
Quickshift	NDH	×	×	×	Media	Media	τ	Media
DPC	NDH	×	×	✓	High	Low	τ	Media
Quickshift++	Mutual k-NN	✓	✓	×	High	Low	k, β	Media
FINCH	1st-NN	✓	×	✓	Media	High	$k = 1$	High
QuickDSC	mutual k-NN	✓	✓	✓	High	High	k, β	Media
FastDEC (Ours)	k-NN dominance	✓	✓	✓	High	High	k	High

Density-based clustering algorithms conduct the connectivity between data points. Meanshift [8], a mode-seeking procedure, iteratively shifts data points to the location with local maximum density, while the density is estimated by Gaussian Kernel. Such a mechanism makes use of a density-higher (DH) connectivity as the cluster structure. DBSCAN [15] adopts a Flat Kernel [19] to identify the density-reachable (DR) connectivity. It significantly reduces computational cost but ignores the mode information. Rather than shifting to a location, Quickshift [29] only moves to the NDH data point if one exists within the τ-radius ball. Both methods hill-climb to the local modes of density and cluster upon these modes. Such processes asymptotically detect all modes individually, leading to **over-segmentation (OS)** [20]. That is the nearby modes are clustered as distinctive clusters, although it is better to group them as a single cluster.

Quickshift++ [20] uses the mutual k-Nearest Neighbors (mutual k-NN) graph to avoid the OS. Two points are merged into a connected-component (CC) if a mutual k-NN edge is detected. Based on the obtained CCs, QuickDSC [36] introduces a density-higher (DH) based way to generate a K-Way partition. However, both methods still require an additional parameter β to balance the segmentation. On the other hand, Density Peak Clustering (DPC) [26] prevents the OS by a new factor: geometric distance, that is the distance from a point to its nearest density-higher (NDH) point. The higher this value, the less probability the point to be a mode. Besides, DPC is extremely slow since it requires finding every possible NDH connection. Cluster by First Integer Neighbor Hierarchy (FINCH) [27] applies the first nearest neighbor (1st-NN) as the connectivity to form the CCs, based on which conduct the K-Way partition. Although FINCH is a parameter-free algorithm, it disregards the description of local modes, therefore may suffer from the OS as well.

To sum up (as Table 1), these perspectives are critical to our target. Connectivity is used to model the relationships between data points. Similar to the methods such as DBSCAN, and Meanshift, CCs can be returned directly as the final result, which leads to clusters with arbitrary shapes, but meanwhile makes the algorithm difficult to control, e.g., if K-Way partition is explicitly required. On the other hand, CC is relevant to the algorithm efficiency because the detected CCs as the densely connected points can be used to further form the clusters. From this point of view, DPC completely ignores such a density structure. Avoiding OS is key to the density-based methods as well. Finally, besides the efficiency and effectiveness, the algorithm robustness is discussed as well for some parameters are difficult to determine in practice [20,34]. For example, Quickshift++ and QuickDSC are outstanding for their performances, however, they need to tune two parameters, k and β, for various datasets, which diminishes the robustness of algorithms.

In this work, FastDEC, clustering by fast dominance estimation, is proposed to fulfill the requirements above. The advantages of FastDEC are highlighted as follows:

1. Efficient, effective, and robust. It can be viewed as a k-NN graph based variant of DPC. According to our analysis, all DPC needs is a k-NN graph. Thus, FastDEC requires only one parameter, that is the number of nearest neighbors k.
2. Easy to interpret. To further support K-Way partition, dominance-component (DC) and inter-dominance estimation are formally defined and theoretically analyzed. In the inter-dominance estimation, the global rank of the NDH dominator is innovatively involved, which further enhances the performance of the algorithm;
3. Experiments on distinctive datasets demonstrate that FastDEC[1] outperforms the state-of-the-art including Quickshift++ [20], FINCH [27], and QuickDSC [36].

2 Related Work

Clustering is a critical analysis tool in data mining and machine learning fields. Density-based clustering methods such as DBSCAN [15], MeanShift [8], QuickShift [29], and so on [19], effectively identify the clusters with arbitrary shapes. These methods use a τ-radius ball to estimate the density, which is difficult to choose in practice. The k-NN DE based methods [2,20,36] are proposed recently to compute the density efficiently.

However, the mode-seeking clustering suffers from over-segmentation [20]. The mutual k-NN graph [6,7,21] is proposed to overcome this problem. The methods

[1] FastDEC is released on https://github.com/gepingyang/FastDEC.

above cannot provide the K-Way partitions. BIRCH [35], Agglomerative clustering [11], and FINCH [27] are proposed to satisfy this requirement. However, they do not utilize the density structure. The K-Mode [4] and the LK-Mode [31] are able to provide a K-Way partition, but neither of them considers the connected-component (CC).

DPC [26], an effective clustering algorithm, is proposed to address this limitation. The idea is simple: the importance of each point is evaluated from two aspects: (1) density; (2) geometric distance: distance to the NDH point. Based on two aspects jointly, the most significant K points as selected as the seeds and based on which clusters the remaining points. DPC suffers from two issues: (1) parameter τ is hard to determine; (2) computational expensive. Based on DPC, several works [3,13,23,32] are designed to improve its effectiveness. Meanwhile, several methods are proposed to accelerate the DPC by space technology [25] or parallelization approach [1].

3 Preliminaries

Let $X = \{x_1, x_2, ..., x_n\}$ be the set of data points, where $x_i \in \mathbb{R}^f$ is an f-dimensional feature vector, and n is the number of points. The Euclidean distance is considered exclusively for clustering. A k Nearest Neighbors (k-NN) graph $G = (V, E)$ is conducted on X, where V denotes the vertex set, and E denotes the edge set. Distinguished from the K (capital letter), which corresponds to top-K significant points or K-Way partition, k-NN (lowercase letter) denotes the k Nearest Neighbors. For each $v \in V$, its k-NN is provided as $\mathcal{N}^k(v)$, where $\mathcal{N}^k(v)_{[j]}$ denotes the jth-NN of v.

Let $d(v)$ denote the density of vertex v, it is referred to as the degree in spectral methods [28]. Given a k-NN graph G, $d(v)$ can be computed through a density estimator. Gaussian Kernel Density Estimator (GKDE) [9] is the most robust but requires a predefined sphere (by τ). DBSCAN uses a Flat Kernel Density Estimator (FKDE), and Quickshift++ adopts a kth-NN Density Estimator (k^{th}-NN DE).

In FastDEC, given a vertex v, its density is computed by a typical k-NN Gaussian kernel density :

$$d(v) = \sum_{u \in \mathcal{N}^k(v)} \exp(-\frac{||v - u||^2}{2\sigma^2}) \tag{1}$$

where $||v - u||$ denotes the Euclidean distance between vertices v and u, and σ is the global bandwidth parameter. We straightly set σ as the mean value of all k-NN distances. The definitions of k-NN Flat kernel density estimator and k^{th} kernel density estimator can be found in Supplement Material.

Based on a density estimator, the density-higher relation is defined as follows:

Definition 1 *(density-higher). A vertex u is **density-higher (DH)** than v if $d(u) > d(v)$.*

Definition 2 *(nearest density-higher). A vertex u is the* ***nearest density-higher (NDH)*** *vertex of v if (1) u is DH than v and (2) there is no vertex w such that w is DH than v, and* $||v - w|| < ||v - u||$.

Both DH and NDH relations are asymmetric.

A vertex with the highest density in a local region is referred to as the mode [8]. However, depending on density only is insufficient to identify the modes correctly because of the over-segmentation (OS), that is the nearby vertices with local maximum densities are detected as distinctive modes [20]. From the connectivity perspective, they should be grouped as one component for their dense mutual relations. Quickshift++ [20] resolves this problem by introducing an additional tolerance parameter β to adjust the "closeness" between modes. Density Peak Clustering (DPC) addresses the OS by a new factor. For a vertex $v \in V$, $g(v)$, the geometric distance from v to its NDH vertex is defined as:

$$g(v) = \begin{cases} \min\limits_{u \in V: d(u) > d(v)} ||v - u||, & \text{if } u \text{ exists} \\ \max\limits_{u \in V} g(u), & \text{otherwise} \end{cases} \tag{2}$$

v is a local mode if its NDH vertex u exists; otherwise, v is a global mode.

The weight of a vertex is estimated by two factors: $w(v) = d(v) \cdot g(v)$. The idea is intuitive: the density value (e.g., $d(v)$) indicates the intensity of density, while geometric distance $g(v)$ reveals the scope of density. Thus, $w(v)$ jointly reflects the weight of v from two factors. However, DPC searches for every possible NDH relation with extremely high time-complexity $O(n^2)$. QuickDSC [36] uses a mutual k-NN graph [20] to reduce the search cost, but requires tuning a new parameter β. Our experiments demonstrate that the good settings of β are diverse in different datasets.

4 Proposed Framework

In this work, we propose FastDEC, a clustering framework by fast dominance estimation. Its general procedure is described in Fig. 1.

1. k-NN density estimation: build the space index on X for the k-NN graph retrieval. Based on the k-NN graph G, estimate the densities of vertices by Eq. 1;
2. Direct k-NN dominator (DkD) detection: identify the direct k-NN dominator (DkD) of all vertices from the G; Based on the DkD, a set of dominance-components can be explicitly identified, where the dominance-component (DC) is a special form of Connected-Component;
3. DC significance estimation: the significance of DCs is estimated from three factors. Based on which, the top-K DCs are selected as the seeds;
4. DC-based clustering: the remaining DCs are assigned to their Nearest Density Higher (NDH) DCs.

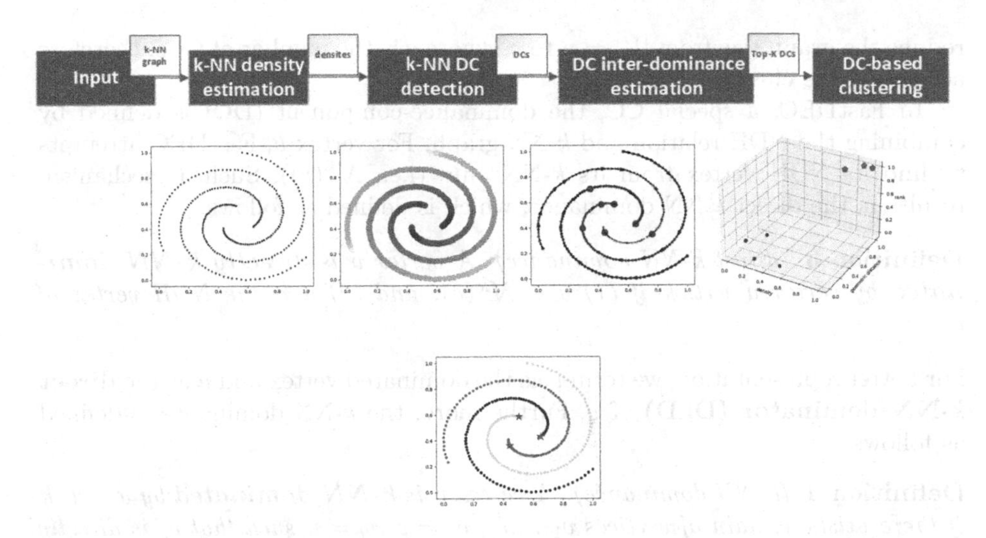

Fig. 1. FastDEC consists four primary stages: (1) k-NN density estimation; (2) Direct k-NN dominator (DkD) detection; (3) DC significance estimation; (4) DC-based clustering. Visualization is generated on Spiral. ($n = 312$, $k = 7$ and $K = 3$)

The first stage computes the vertex densities by accessing the k-NN graph G, which is essential for all density-based methods. **Note:** FastDEC can adopt an arbitrary density estimator. k-NN Gaussian Kernel DE (as Eq. 1) is used as the default for several critical characteristics: (1) it is as robust as the τ-ball based DE (as Table 1); (2) k is easy to control and the only parameter (so as the whole framework FastDEC); (3) based on the sophisticated index technologies [10,12], FastDEC can retrieve a large-scale k-NN graph efficiently. In the second stage, according to the given k, the DkDs of all vertices are captured as well. It is easy to show the equivalence between DkD and DC (to be introduced later). In the third stage, a significance estimation process is performed on the obtained DCs from three factors including density, geometric distance, and the global rank of the NDH vertex. The last stage follows the typical process of DPC [26] and QuickDSC [36]. The DkD detection and the DC significance estimation are the key components of FastDEC with k as the only control parameter.

4.1 Direct k-NN Dominator (DkD) Detection

Density alone is insufficient to correctly identify the modes/clusters because of the OS. To overcome it, Quickshift++ [20] is proposed to merge the nearby vertices as the connected-component (CC) by mutual k-NN edges. While Sarfraz et al. [27] utilize the symmetric 1st Nearest Neighbor (1st-NN) relation to segment the entire k-NN graph into CCs. The obtained CCs are used as the basic units for further clustering, therefore preventing the OS since the nearby vertices with high densities are grouped. Meanwhile, both methods enormously reduce the graph size from $|V| = n$ to c, where c is the number of CCs, therefore accelerate the clustering process.

In FastDEC, a special CC, the dominance-component (DC) is defined by combining the NDH relation and k-NN graph. For vertex v, FastDEC attempts to find its NDH vertex from its k-NN only (i.e., $\mathcal{N}^k(v)$). Such a mechanism results in the direct k-NN dominance, which is defined as follows.

Definition 3 *(direct k-NN dominance). A vertex v is* ***directly k-NN dominated*** *by vertex u wrt. k if (1) $u \in \mathcal{N}^k(v)$, and (2) u is the NDH vertex of v.*

For better representation, we term v as the dominated vertex and u as the **direct k-NN dominator (DkD)** of v. Furthermore, the k-NN dominance is defined as follows:

Definition 4 *(k-NN dominance). A vertex v is* ***k-NN dominated*** *by u wrt. k if there exists a chain of vertices $v_1,...,v_l$, $v_1 = v$, $v_l = u$ such that v_i is directly k-NN dominated by v_{i+1}.*

k-NN dominance is a canonical extension of direct k-NN dominance. It is transitive as the DH, meanwhile, it is not symmetric. The notion of k-NN dominance-connected is introduced to guarantee symmetry.

Definition 5 *(k-NN dominance-connected). A vertex v is* ***k-NN dominance-connected*** *to a vertex u wrt. k if there exists a vertex w such that both, u and v are k-NN dominated by w.*

k-NN dominance-connectivity is a symmetric relation. Through direct k-NN dominance, a vertex iteratively reaches a vertex **without DkD**. The latter is more significant and representative, because of the higher density value. We define such kind of vertices as the k-NN mode:

Definition 6 *(k-NN mode). A vertex without k-NN dominator wrt. k, is a k-NN mode.*

k-NN mode is a vertex with the highest density in the local region (i.e., k-NN). The region is further expanded by the direct k-NN dominance. Intuitively, the vertices dominated by a k-NN mode have strong affiliations. Now, we define the k-NN dominance-component (DC).

Definition 7 *(k-NN dominance-component). Let V be the vertex set, and $m \in V$ be a k-NN mode. A k-NN dominance-component DC of m wrt. k is a non-empty subset of V satisfying the following conditions:*

(1) $\forall u, v$: if $u \in DC$ and v is k-NN dominance-connected to u wrt. k, then $v \in DC$. (Maximality & Connectivity)
(2) $\forall v$ are k-NN dominated by m wrt. k, then $v \in DC$. (Dominance)

Note that the k-NN DC has several critical characteristics:

Algorithm 1. Direct k-NN Dominator (DkD) Detection

Input: X, DB, k
Output: DkD
1: $G \leftarrow$ Conduct k-NN graph of X by querying DB; ▷ k-NN graph construction
2: $d \leftarrow$ Estimate k-NN densities via Equation 1; ▷ k-NN density estimation
3: Initialize DkD(v) as *None* for $v \in V$; ▷ DkD Initialization
4: **for** $v \in V$ **do**
5: **for** $j := 1, ..., k$ **do** ▷ search from near to far by varying j in $[1, k]$
6: **if** $d(v) < d(\mathcal{N}^k(v)_{[j]})$ **then** ▷ $\mathcal{N}^k(v)_{[j]}$ denotes the jth nearest neighbor of v
7: DkD(v) $:= \mathcal{N}^k(v)_{[j]}$; ▷ v is dominated by its jth-NN
8: break;
9: **end if**
10: **end for**
11: **end for**
12: Return DkD

1. A *DC* has only a k-NN mode, that is the vertex with the highest density in the *DC*;
2. The density values along the path, that starts from a vertex to the k-NN mode, are guaranteed to be monotonically increasing;
3. The k-NN mode is an excellent representative of *DC*;
4. DCs are mutually disjoint;

The corresponding proofs are omitted because of the space limitation. **Note:** noise filtering [15] is compatible with FastDEC, although it is insignificant to the results of this work. It is disabled in all applicable methods [8,19,34] by setting $min_{pt} = 0$.

Our concern is to identify the DkD from X. The procedure is described as Algorithm 1, with the dataset X, the space index DB, and k as input. First, conduct the k-NN graph (Line 1), and then for each vertex $v \in V$, compute its density by an arbitrary DE (Line 2), and initialize its DkD as None (Line 3). Second, for each vertex v, we attempt to find its DkD (Lines 4–11), where $\mathcal{N}^k(v)_{[j]}$ denotes the jth-NN of v. Finally, the identified DkD information is returned. In this stage, the vertex without DkD is referred to as a k-NN mode. Based on the DkD, we can extract the dominance-components (DCs) explicitly by performing a Hill-Climbing search on all vertices (as Algorithm 2).

4.2 DC Dominance Estimation

Fundamentally dominator-component (DC) is a set of vertices that are compactly linked through the nearest density-higher (NDH) relations. In Fig. 2 (a),

Algorithm 2. Hill-Climbing: convert the DkD to DCs $\mathcal{DC}^k$ and the corresponding modes M^k explicitly

Input: DkD
Output: $\mathcal{DC}^k, M^k$

1: $M^k \leftarrow \emptyset$;
2: $c := 0$; ▷ Count the number of DCs
3: **for** $v \in V$ **do**
4:     **if** DkD(v) is *None* **then**
5:         $M^k \leftarrow M^k \cup v$;
6:         $c := c + 1$;
7:         $\mathcal{DC}^k_c \leftarrow \{v\}$; ▷ Initialize $\mathcal{DC}^k$
8:     **end if**
9: **end for**
10: **for** $v \in V$ **do**
11:     **for** $m := 1, ..., c$ **do**
12:         **if** v is k-NN dominated by $M^k(m)$ **then**;
13:             $\mathcal{DC}^k_m \leftarrow \mathcal{DC}^k_m \cup v$;
14:         **end if**
15:     **end for**
16: **end for**
17: Return $\mathcal{DC}^k$, M^k

Algorithm 3. FastDEC$_D$: Typical Density-based Method

Input: $X \in \mathbb{R}^{n \times m}$, DB, k
Output: clusters C

1: DkD $\leftarrow$ Algorithm 1(X, DB, k); ▷ DkD Detection
2: $\mathcal{DC}^k, M^k \leftarrow$ Algorithm 2(DkD); ▷ Extract DCs from DkD explicitly
3: Return $C \leftarrow \mathcal{DC}^k$. ▷ Return DCs as the Clusters

the change rates of NDH relations of three k-NN relevant density estimators[2] are shown, where k varies from 5 to 15. For instance, assume we obtain n NDH relations if set $k = 10$. If we set $k = 11$, n' NDH relations are changed, then the change rate is reported as n'/n for $k = 11$. Results on change rate demonstrate that the k-NN Gaussian Kernel based NDH relation is more stable, so it is set as the default DE.

Besides, NDH is adopted by Quickshift [29], DPC [26], and QuickDSC [36]. The former two methods explicitly search all possible NDH relations and lead to a huge amount of search cost; the latter uses an additional parameter β together with a mutual k-NN graph to "avoid" the unnecessary search. Our proposed k-NN dominance-component is conducted on NDH relation as well but with the restriction of k-NN. The idea is intuitive: obtain the majority of

[2] Density-Reachable (DR) in DBSCAN [15] is equivalent to τ based Flat Kernel. For the sake of comparison, we use a k-NN based one.

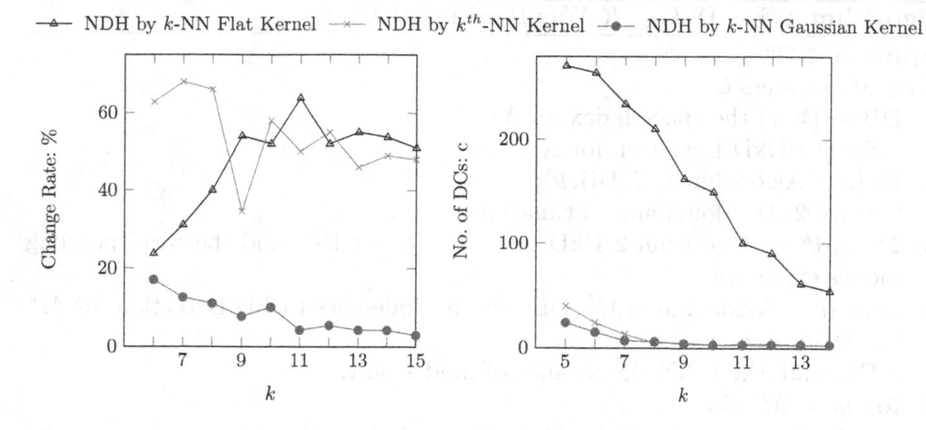

Fig. 2. The change rate of NDH relation and the No. of DCs (c) on Spiral by varying k.

such relations from the given k-NN graph, and perform the extra NDH search for the k-NN modes only. In Fig. 2 (b), the number of obtained DCs c is shown, with k varying from 5 to 15, where c is obtained by counting the number of k-NN modes. Again, NDH by k-NN Gaussian Kernel is the stablest. More interestingly and significantly, k can be used to control the number of DCs.

Based on the obtained components (e.g., CC/DC), the computational complexity is heavily reduced, and the algorithm is further enhanced with the capability of K-Way partition: select top-K significant DCs as the initial seeds/centers from the obtained c DCs. For CC-enabled methods, several strategies are proposed to find an ideal value of c. QuickDSC uses the parameter tuning in a brute-force manner, while FINCH applies the BIRTH [35] to identify the hierarchy structure of CCs, and according to which, generates K clusters. Regarding FastDEC, as k increases, the number of DCs $c \rightarrow c_{global}$, where c_{global} is the number of global modes. If k approaches 0, $c \rightarrow n$.

Algorithm 1 returns the DkD of all vertices. By revoking Algorithm 2, a set of DCs $\mathcal{DC}^k = \{DC_1, DC_2, ..., DC_c\}$ and the corresponding k-NN modes $M^k = \{m_1, m_2, ..., m_c\}$ are obtained for the given k. If $c \geq K$, the obtained DCs are sufficient for the subsequent process (e.g., DC dominance estimate). Otherwise, it is necessary to reduce the value of k to obtain at least K DCs. Fortunately, this step is extremely efficient by directly accessing the retained space index DB. Its detail is omitted for the space limitation.

Straightforwardly, similar to the typical density-based methods [8,15,19,20], the obtained DCs with arbitrary shapes can be reported as the clustering outcome. We term this method the FastDEC$_D$, its procedure is described as Algorithm 3. Except for the K-Way partition, FastDEC$_D$ satisfies most requirements listed in Table 1. To further support the K-Way partition, FastDEC$_K$ is designed.

Algorithm 4. FastDEC$_K$: K-Way Partition Method

Input: $X \in \mathbb{R}^{n \times m}$, k, K
Output: clusters C
1: DB ← Build the space index on X;
▷ Stage 1:DkD Detection for X
2: DkD ← Algorithm 1(X, DB, k);
▷ Stage 2: DC dominance estimation
3: $\mathcal{DC}^k, M^k$ ← Algorithm 2(DkD); ▷ Detect DCs and the corresponding modes explicitly
4: DkD_M ← Algorithm 1(M^k, DB, n); ▷ Mode-based DkD Detection for M^k by setting $k = n$
▷ Estimate the k-NN significance of modes only.
5: **for** $m_i \in M^k$ **do**
6: $s(m_i)$ ← estimate m_i by $DkD_M(m_i)$ via Equation 3;
7: **end for**
▷ Stage 3: DC-based clustering.
8: C ← top-K DCs;
9: **for** $DC_i \in \mathcal{DC}$ and DC_i is not top-K DCs **do**
10: Assign DC_i to the cluster $\hat{C} \in C$ that includes $DkD_M(m_i)$;
11: **end for**
12: Return C.

Similar to DPC and QuickDSC, the density and geometric distance are involved in estimation (as Eq. 3). Additionally, an interesting and novel factor: the rank of the NDH neighbor is considered as well. For a vertex $v \in V$, its significance is defined as:

$$s(v) = d(v) \cdot g(v) \cdot r(v) \tag{3}$$

where $r(v)$ denotes the rank of its NDH neighbor in $\mathcal{N}^k(v)$. Clearly, $r(v)$ ranges from 1 to n. The larger $r(v)$, the more significant v is. This factor can be viewed as **global rank** information for weighting the dominance of v.

The mode in existing density-based methods [8,26] is τ-based. In QuickDSC [36], the geometric distance (e.g., $g(v)$) works as a special form of τ that is highly adaptive to the data point (e.g., v). For the significance estimation, the global rank information in terms of the $r(v)$ is involved. This factor indicates the relative position of v in the whole dataset X. As a result, our decision map evolves from 2-dimensional to 3-dimensional (as Fig. 1).

On the other hand, it seems that the involvement of $r(v)$ heavily aggravates the search cost of estimation. Fortunately, the significance ($s(v)$) is consistent to the vertex weight ($w(v)$) adopted in DPC and QuickDSC.

Theorem 1. *Given a k-NN dominance component DC with its k-NN mode m, its significance value $s(m) > s(v)$ for all vertices $v \in DC$ if $w(m) > w(v)$.*

Proof. Rewrite $s(v) = w(v) \cdot r(v)$. It is easy to infer that $r(m) > r(v)$ definitely for all $v \in DC$, because $r(v) \leq k$ and $r(m) > k$. The theorem holds since $r(m) > r(v)$ and $w(m) > w(v)$ for all $v \in DC$.

Regarding the vertex weight $w(v)$, it is easy to infer that $d(m) > d(v)$ for all $v \in DC$ since m is the k-NN mode. For the geometric distance, $g(m) > g(v)$ in the majority of the cases [36]. Therefore, once the DCs and the corresponding modes are obtained, **we only need to estimate the significance of the modes.**

The general procedure is described as Algorithm 4. First, a space index DB is built on X (Line 1). Based on this, the DkD detection is executed for X (Line 2). Second, the DCs and the corresponding modes are explicitly found (Line 3). For each mode m_i, since its density is known, its geometric distance and the global rank are computed by querying the DB (Line 4). Meanwhile, the DkD information of the modes, denoted as DkD_M, is detected as well. Therefore, it is possible to **estimate the "n-NN significance" of modes** (Lines 5–7). The last stage of FastDEC follows the general process of DPC and QuickDSC: top-K significant modes are selected as the seeds (Line 8). The remaining DCs are assigned to the clusters according to the DkD_M (Lines 9–11).

4.3 Complexity Analysis and Implementation

FastDEC_K consists of an initial stage, and three major stages. FastDEC_D includes the initial stage, DkD detection, and a hill-climbing process. The initial stage requires building a space index (e.g., DB), and the DkD detection stage utilizes it to construct a k-NN graph. We use KD-Tree as the default space index, and Random Project Tree [10] + Nearest Neighbor Descend [12] are used to obtain the approximate k-NN by considering the scalability. Both ways require a complexity of $O(nlog(n)k)$. The DkD detection requires a complexity of $O(nk)$. DC significance estimation costs a complexity of $O(clog(n)k + cn)$ where c is the number of DCs. The last stage needs a complexity of cK. The most consuming step is the k-NN graph retrieval, the whole framework is still near-linear to n and k. FastDEC is implemented by Python, NumPy[3] is used for acceleration. We might further speed up our framework with parallel technologies.

5 Evaluation

To evaluate the performance of FastDEC, experiments are conducted on several artificial and real-world datasets of different sizes.

Datasets. The artificial and real-world datasets are selected from various sources. Details of the employed datasets are described in Table 2, and Principal Component Analysis (PCA), as a typical preprocessing of step dimension reduction, is applied on MNIST to reduce the number of feature dimensions from 784 to 20.

[3] https://numpy.org/.

Table 2. Datasets in evaluation.

Name	# instances	# features	# classes	Type
Banana-ball [24]	2,000	2	3	Artificial
Flame [17]	240	2	2	Artificial
R15 [30]	600	2	15	Artificial
Spiral [5]	312	2	3	Artificial
S2 [16]	5,000	2	15	Artificial
Seeds [14]	210	7	3	Real-World
Banknote [14]	1,372	4	2	Real-World
Segmentation [14]	2,310	19	7	Real-World
Phonemes [18]	4,509	258	5	Real-World
MFCCs [14]	7291	22	10	Real-World
MNIST [22]	70,000	20	10	Real-World

Table 3. ARI Comparison (%) on Artificial and Real-world datasets. The best and second best results are highlighted by **bold** and underline respectively.

	Typical Density-based Methods					K-Way Partition Methods				
Datasets	DBSCAN	Meanshift	Quickshift	Quickshift++	FastDEC$_D$	FINCH	DPC	QuickDSC	SNN-DPC	FastDEC$_K$
Banana-ball	47.70	62.70	90.57	99.67	**99.83**	74.96	69.66	99.50	**99.83**	**99.83**
Flame	19.93	7.41	98.81	**100.00**	**100.00**	0.00	72.78	**100.00**	95.02	**100.00**
R15	26.37	98.21	92.78	**99.28**	**99.28**	**99.28**	73.94	**99.28**	**99.28**	**99.28**
Spiral	1.00	10.22	**100.00**	58.77	**100.00**	0.01	98.56	58.76	**100.00**	**100.00**
S2	0.00	**93.83**	92.64	92.80	93.67	88.01	48.15	93.47	92.89	**93.67**
Seeds	38.39	**72.37**	61.66	70.76	70.76	39.10	68.23	77.27	**78.36**	76.64
Banknote	82.60	17.81	28.33	95.39	**95.67**	10.86	89.25	95.39	83.53	**96.53**
Segmentation	40.38	51.27	47.17	49.23	**59.66**	42.99	57.80	60.59	60.26	**60.94**
Phonemes	42.36	40.82	71.10	74.95	**76.51**	65.59	79.59	74.60	73.37	**76.51**
MFCCs	13.65	35.74	17.17	43.69	**47.02**	24.35	15.77	23.99	27.04	**31.74**
MNIST	6.74	21.98	N/A	46.41	**46.84**	N/A	N/A	35.96	N/A	**45.05**

Baselines. FastDEC$_D$ is compared with typical density-based methods: DBSCAN [15], Meanshift [8], Quickshift [29] and Quickshift++ [20]. FastDEC$_K$ is compared with the methods that support K-Way partition: FINCH [27], DPC [26], SNN-DPC [23], and QuickDSC [33]. More specifically, some recent DPC-based methods [1,3,13,25,32] are not involved, for the unavailability of their implementations.

Evaluation Metric. We use the Adjusted Rand Index (ARI), Normalized Mutual Information (NMI) and Adjusted Mutual Information (AMI) to evaluate the clustering results. For all metrics, the higher value indicates better performance. AMI-based results can be found in Supplement Material.

Experiment Setup. All experiments are executed on a Win-10 64-bits machine with Intel(R) Core $i5-9400F$ CPU(2.90 GHz), and 16 GB of main memory. For

Table 4. NMI (%) on Artificial and Real-world datasets. The best and second best results are highlighted by **bold** and underline respectively.

Datasets	Typical Density-based Methods					K-Way Partition Methods				
	DBSCAN	Meanshift	Qucikshift	Quickshift++	FastDEC$_D$	FINCH	DPC	QuickDSC	SNN-DPC	FastDEC$_K$
Banana-ball	67.98	61.72	88.53	99.30	**99.65**	80.17	68.73	99.10	**99.65**	**99.65**
Flame	37.45	32.88	97.06	**100.00**	**100.00**	15.25	79.76	**100.00**	89.94	**100.00**
R15	74.25	98.65	97.97	**99.42**	**99.42**	**99.42**	92.02	**99.42**	**99.42**	**99.42**
Spiral	**100.00**	45.44	**100.00**	59.57	**100.00**	40.10	98.05	59.57	**100.00**	**100.00**
S2	0.00	**94.62**	93.85	94.08	94.52	93.20	77.18	94.38	94.09	**94.52**
Seeds	46.17	**69.00**	64.33	67.97	67.97	50.79	65.03	73.23	**74.10**	73.29
Banknote	73.58	26.63	44.63	91.94	**92.35**	13.19	83.00	91.94	78.49	**93.59**
Segmentation	60.16	62.37	65.44	71.13	**73.03**	59.71	70.78	72.45	68.34	**73.82**
Phonemes	60.28	60.75	73.12	84.00	**85.41**	72.10	**88.65**	83.53	82.74	85.41
MFCCs	52.30	68.00	66.06	**74.83**	74.22	61.51	49.35	58.27	61.27	**63.10**
MNIST	23.52	39.40	N/A	64.80	**66.56**	N/A	N/A	54.74	N/A	**66.25**

Table 5. Overall runtime (secs) on Artificial and Real-world datasets. The best and second best results are highlighted by **bold** and underline respectively.

Datasets	Typical Density-based Methods					K-Way Partition Methods				
	DBSCAN	Meanshift	Quickshift	Quickshift++	FastDEC$_D$	FINCH	DPC	QuickDSC	SNN-DPC	FastDEC$_K$
Banana-ball	**0.02**	9.17	0.40	0.05	0.04	0.15	2.63	0.15	25.55	**0.04**
Flame	**0.00**	0.54	0.01	**0.00**	**0.00**	0.03	0.01	**0.00**	0.34	**0.00**
R15	**0.00**	1.01	0.03	0.01	**0.00**	0.03	0.07	0.01	1.94	**0.00**
Spiral	**0.00**	0.38	0.04	**0.00**	**0.00**	0.11	0.02	**0.00**	0.49	**0.00**
S2	0.05	12.44	1.66	0.05	**0.04**	0.57	4.56	0.07	207.52	**0.04**
Seeds	**0.00**	0.16	0.01	**0.00**	**0.00**	0.04	0.02	**0.00**	0.43	**0.00**
Banknote	**0.01**	6.49	0.15	0.08	0.03	0.13	0.39	0.05	19.95	**0.02**
Segmentation	**0.09**	8.68	0.30	0.21	0.14	**0.14**	1.08	0.17	41.38	**0.14**
Phonemes	**0.27**	180.57	1.08	7.71	7.12	**0.59**	4.17	7.46	123.40	7.02
MFCCs	**0.74**	30.36	2.96	1.02	0.96	1.11	10.74	1.28	413.45	**1.08**
MNIST	**31.00**	8306.19	N/A	237.38	200.43	N/A	N/A	210.05	N/A	**178.44**

the k-NN based methods: Quickshift++, QuickDSC, FastDEC$_D$, SNN-DPC and FastDEC$_K$, k is varied from 5 to 200 with step size 5 in all datasets. Regarding τ-ball based methods: DBSCAN, Meanshift, Quickshift, and DPC, τ varies from 0.05 to 2 with step size 0.05 in all datasets. DBSCAN requires an additional parameter min_{pt} that corresponds to the noise filtering. Regarding PCA, we also use the implementation provided by Scikit-Learn. Besides, Quickshift++ and QuickDSC need a parameter β, we vary it from 0.1 to 0.3 with step size 0.1.

5.1 Comparison on Artificial and Real-World Datasets

The evaluation of the clustering results on real-world datasets is shown in Table 3, Table 4, and Table 5. The bold represents the best result, and the second-best result is highlighted by underline, "N/A" denotes we cannot obtain the clustering result on the experimental host (e.g., Out-of-Memory). By varying the control parameters, for each method, the best clustering result and the corresponding execution time is reported. The runtime 0.00 indicates that the runtime is less than 0.005 seconds. Note: rather than the Nearest Density Higher (NDH) relation, FastDEC adopts the Nearest Density Higher **or Equal** (NDHE) relation in the Segmentation.

Based on the tables above, several findings are drawn:

1. FastDEC (two versions) always achieves the best or the second-best results among all datasets. The performance of the τ-ball based methods (DBSCAN, DPC, Meanshift, Quickshift) is relatively poor. The performance of Quickshift++ and QuickDSC are competitive. The former is a typical density-based method, while the latter is a K-Way partition method;
2. Both Quickshift++ and QuickDSC need an additional parameter β which requires extra effort in tuning. FINCH is parameter-free by ignoring the mode information, thus its performance is relatively poor. For our method, only FastDEC$_D$ needs to tune the k, but still, the overall performance by varying k is outstanding and robust;
3. Regarding the algorithm runtime, there is no doubt that DBSCAN is the fastest method. FINCH is quite fast as well by utilizing the 1st-NN graph. On the contrary, SNN-DPC is rather slow for extracting shared-nearest-neighbor relations. However, the methods above ignore the DH information and lead to a substantial reduction in clustering quality. FastDEC obtains the best or the second-best result in terms of the runtime on almost all datasets, which demonstrates its efficiency.

To sum up, FastDEC provides a balanced yet excellent solution that considers effectiveness, efficiency, and parameter-tuning.

5.2 Robustness Testing

Now, we test the robustness of the algorithm parameters. We only demonstrate the experiments on two datasets MFCCs and MNIST, more results are attached as Supplement Material. Again, comparisons are conducted for typical density-based methods and K-Way partition methods separately. For the former, Meanshift, Quickshift, DPC and DBSCAN use the primary parameter τ, but the remaining use the k. We convert them into the same granularity: for example, setting $k = 1$ is identical to setting $\tau = 0.01$. For parameter β, we use the suggested setting 0.3 [20]. Specifically, Fig. 3 (a) and Fig. 3 (b) show the NMI results of the typical density-based methods in the MFCCs and MNIST respectively. Figure 3 (c) and Fig. 3 (d) are that of the K-Way partition methods respectively. According to Fig. 3 (a) and Fig. 3 (b), τ-based algorithms are parameter sensitive, so it is difficult to capture the best setting. On the other hand, the performance of k-based algorithms, especially the FastDEC$_D$, is quite stable. It again demonstrates our method is robust. Regarding the K-Way partition, the performance of FastDEC$_K$ is compromised as well. It completely outperforms the other algorithms including the parameter-free algorithm FINCH.

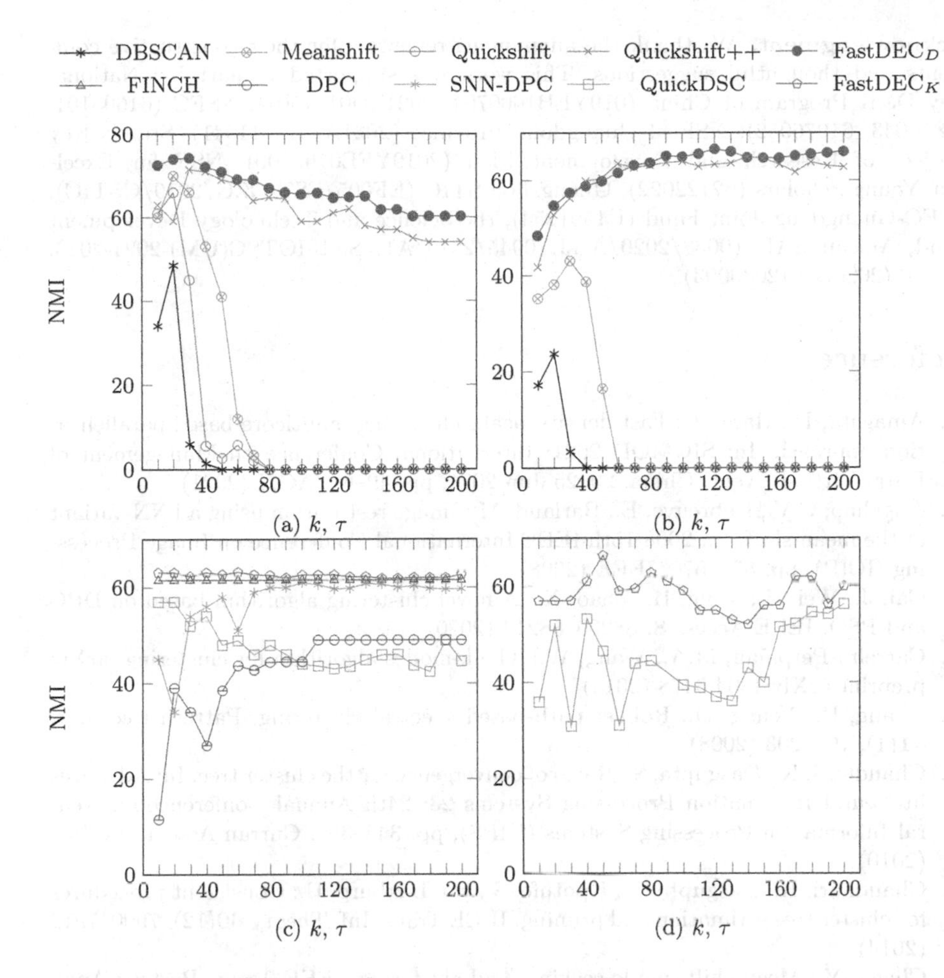

Fig. 3. Robustness Testing of parameters k and τ on MFCCs, MNIST.

6 Conclusion

In this paper, we propose FastDEC, a novel density-based clustering framework. It provides two versions to satisfy different requirements: FastDEC$_D$ for identifying the clusters with arbitrary shapes, and FastDEC$_K$ for K-Way partition. FastDEC detects the nearest density higher (NDH) relation from a k-NN graph and forms the dominance-component (DC), based on which, conduct the clusters. The experiments on real-world datasets demonstrate that our method has excellent performance, meanwhile, it is efficient and easy to tune. In the future, we consider a distributed version of FastDEC to handle the extremely large datasets.

Acknowledgment. We thank the anonymous reviewers for their constructive comments and thoughtful suggestions. This work was supported in part by: National Key D&R Program of China (019YFB1600704, 2021ZD0111501), NSFC (61603101, 61876043, 61976052), NSF of Guangdong Province (2021A1515011941), State's Key Project of Research and Development Plan (2019YFE0196400), NSF for Excellent Young Scholars (62122022), Guangzhou STIC (EF005/FST-GZG/2019/GSTIC), NSFC-Guangdong Joint Fund (U1501254), the Science and Technology Development Fund, Macau SAR (0068/2020/AGJ, 0045/2019/A1, SKL-IOTSC(UM)-2021-2023, GDST (2020B1212030003).

References

1. Amagata, D., Hara, T.: Fast density-peaks clustering: multicore-based parallelization approach. In: SIGMOD 2021: International Conference on Management of Data, Virtual Event, China, 20–25 Jun 2021, pp. 49–61. ACM (2021)
2. Angelino, C.V., Debreuve, E., Barlaud, M.: Image restoration using a kNN-variant of the mean-shift. In: 2008 15th IEEE International Conference on Image Processing (ICIP), pp. 573–576. IEEE (2008)
3. Cai, J., Wei, H., Yang, H., Zhao, X.: A novel clustering algorithm based on DPC and PSO. IEEE Access **8**, 88200–88214 (2020)
4. Carreira-Perpiñán, M.Á., Wang, W.: The k-modes algorithm for clustering. arXiv preprint arXiv:1304.6478 (2013)
5. Chang, H., Yeung, D.: Robust path-based spectral clustering. Pattern Recognit. **41**(1), 191–203 (2008)
6. Chaudhuri, K., Dasgupta, S.: Rates of convergence for the cluster tree. In: Advances in Neural Information Processing Systems 23: 24th Annual Conference on Neural Information Processing Systems (NIPS), pp. 343–351. Curran Associates, Inc. (2010)
7. Chaudhuri, K., Dasgupta, S., Kpotufe, S., von Luxburg, U.: Consistent procedures for cluster tree estimation and pruning. IEEE Trans. Inf. Theory **60**(12), 7900–7912 (2014)
8. Cheng, Y.: Mean shift, mode seeking, and clustering. IEEE Trans. Pattern Anal. Mach. Intell. **17**(8), 790–799 (1995)
9. Comaniciu, D., Meer, P.: Mean shift: a robust approach toward feature space analysis. IEEE Trans. Pattern Anal. Mach. Intell. **24**(5), 603–619 (2002)
10. Dasgupta, S., Freund, Y.: Random projection trees and low dimensional manifolds. In: Proceedings of the Annual ACM Symposium on Theory of Computing (STOC), pp. 537–546 (2008)
11. Davidson, I., Ravi, S.S.: Agglomerative hierarchical clustering with constraints: theoretical and empirical results. In: Jorge, A.M., Torgo, L., Brazdil, P., Camacho, R., Gama, J. (eds.) PKDD 2005. LNCS (LNAI), vol. 3721, pp. 59–70. Springer, Heidelberg (2005). https://doi.org/10.1007/11564126_11
12. Dong, W., Charikar, M., Li, K.: Efficient k-nearest neighbor graph construction for generic similarity measures. In: Proceedings of the 20th International Conference on World Wide Web (WWW), pp. 577–586. ACM (2011)
13. Du, M., Ding, S., Jia, H.: Study on density peaks clustering based on k-nearest neighbors and principal component analysis. Knowl. Based Syst. **99**, 135–145 (2016)

14. Dua, D., Graff, C.: UCI machine learning repository (2017). http://archive.ics.uci.edu/ml
15. Ester, M., Kriegel, H.P., Sander, J., Xu, X.: A density-based algorithm for discovering clusters in large spatial databases with noise. In: Knowledge Discovery and Data Mining (KDD), pp. 226–231 (1996)
16. Fränti, P., Virmajoki, O.: Iterative shrinking method for clustering problems. Pattern Recognit. **39**(5), 761–775 (2006)
17. Fu, L., Medico, E.: FLAME, a novel fuzzy clustering method for the analysis of DNA microarray data. BMC Bioinform. **8**, 3 (2007)
18. Hastie, T., Tibshirani, R., Friedman, J.: The Elements of Statistical Learning. SSS, Springer, New York (2009). https://doi.org/10.1007/978-0-387-84858-7
19. Hinneburg, A., Keim, D.A.: An efficient approach to clustering in large multimedia databases with noise. In: Knowledge Discovery and Data Mining (KDD), pp. 58–65 (1998)
20. Jiang, H., Jang, J., Kpotufe, S.: Quickshift++: Provably good initializations for sample-based mean shift. In: International Conference on Machine Learning (ICML), vol. 80, pp. 2299–2308. PMLR (2018)
21. Jiang, H., Kpotufe, S.: Modal-set estimation with an application to clustering. In: Proceedings of the 20th International Conference on Artificial Intelligence and Statistics (AISTATS), vol. 54, pp. 1197–1206. PMLR (2017)
22. LeCun, Y., Bottou, L., Bengio, Y., Haffner, P.: Gradient-based learning applied to document recognition. Proc. IEEE **86**(11), 2278–2324 (1998)
23. Liu, R., Wang, H., Yu, X.: Shared-nearest-neighbor-based clustering by fast search and find of density peaks. Inf. Sci. **450**, 200–226 (2018)
24. Myhre, J.N., Mikalsen, K.Ø., Løkse, S., Jenssen, R.: Robust clustering using a kNN mode seeking ensemble. Pattern Recognit. **76**, 491–505 (2018)
25. Rasool, Z., Zhou, R., Chen, L., Liu, C., Xu, J.: Index-based solutions for efficient density peak clustering (extended abstract). In: 37th IEEE International Conference on Data Engineering, ICDE 2021, Chania, Greece, 19–22 Apr 2021, pp. 2342–2343. IEEE (2021)
26. Rodriguez, A., Laio, A.: Clustering by fast search and find of density peaks. Science **344**(6191), 1492–1496 (2014)
27. Sarfraz, M.S., Sharma, V., Stiefelhagen, R.: Efficient parameter-free clustering using first neighbor relations. In: 2019 IEEE/CVF Conference on Computer Vision and Pattern Recognition (CVPR), pp. 8934–8943 (2019)
28. Shi, J., Malik, J.: Normalized cuts and image segmentation. IEEE Trans. Pattern Anal. Mach. Intell. **22**(8), 888–905 (2000)
29. Vedaldi, A., Soatto, S.: Quick shift and kernel methods for mode seeking. In: Forsyth, D., Torr, P., Zisserman, A. (eds.) ECCV 2008. LNCS, vol. 5305, pp. 705–718. Springer, Heidelberg (2008). https://doi.org/10.1007/978-3-540-88693-8_52
30. Veenman, C.J., Reinders, M.J.T., Backer, E.: A maximum variance cluster algorithm. IEEE Trans. Pattern Anal. Mach. Intell. **24**(9), 1273–1280 (2002)
31. Wang, W., Carreira-Perpiñán, M.Á.: The laplacian k-modes algorithm for clustering. arXiv preprint arXiv:1406.3895 (2014)
32. Xie, J., Gao, H., Xie, W., Liu, X., Grant, P.W.: Robust clustering by detecting density peaks and assigning points based on fuzzy weighted k-nearest neighbors. Inf. Sci. **354**, 19–40 (2016)
33. Yang, Y., et al.: GraphLSHC: towards large scale spectral hypergraph clustering. Inf. Sci. **544**, 117–134 (2021)

34. Yang, Y., Gong, Z., Li, Q., U, L.H., Cai, R., Hao, Z.: A robust noise resistant algorithm for POI identification from flickr data. In: Twenty-Sixth International Joint Conference on Artificial Intelligence (IJCAI), pp. 3294–3300. ijcai.org (2017)
35. Zhang, T., Ramakrishnan, R., Livny, M.: SIGMOD, pp. 103–114. ACM Press, New York (1996)
36. Zheng, X., Ren, C., Yang, Y., Gong, Z., Chen, X., Hao, Z.: QuickDSC: clustering by quick density subgraph estimation. Inf. Sci. **581**, 403–427 (2021)

SECLEDS: Sequence Clustering in Evolving Data Streams via Multiple Medoids and Medoid Voting

Azqa Nadeem(✉) and Sicco Verwer

Delft University of Technology, Delft, The Netherlands
{azqa.nadeem,s.e.verwer}@tudelft.nl

Abstract. Sequence clustering in a streaming environment is challenging because it is computationally expensive, and the sequences may evolve over time. K-medoids or Partitioning Around Medoids (PAM) is commonly used to cluster sequences since it supports alignment-based distances, and the k-centers being actual data items helps with cluster interpretability. However, offline k-medoids has no support for concept drift, while also being prohibitively expensive for clustering data streams. We therefore propose SECLEDS, a streaming variant of the k-medoids algorithm *with constant memory footprint*. SECLEDS has two unique properties: i) it uses multiple medoids per cluster, producing stable high-quality clusters, and ii) it handles concept drift using an intuitive *Medoid Voting scheme* for approximating cluster distances. Unlike existing adaptive algorithms that create new clusters for new concepts, SECLEDS follows a fundamentally different approach, where the clusters themselves evolve with an evolving stream. Using real and synthetic datasets, we empirically demonstrate that SECLEDS produces high-quality clusters regardless of drift, stream size, data dimensionality, and number of clusters. We compare against three popular stream and batch clustering algorithms. The state-of-the-art BanditPAM is used as an offline benchmark. SECLEDS achieves comparable F1 score to BanditPAM while reducing the number of required distance computations by 83.7%. Importantly, SECLEDS outperforms all baselines by 138.7% when the stream contains drift. We also cluster real network traffic, and provide evidence that SECLEDS can support network bandwidths of up to 1.08 Gbps while using the (expensive) dynamic time warping distance.

Keywords: Sequence clustering · K-medoids · Data streams · Concept drift · Network traffic

1 Introduction

Stream clustering is the problem of clustering a potentially unbounded stream of items in a single pass, where the items arrive sequentially without any particular

Supplementary Information The online version contains supplementary material available at https://doi.org/10.1007/978-3-031-26387-3_10.

M.-R. Amini et al. (Eds.): ECML PKDD 2022, LNAI 13713, pp. 157–173, 2023.
https://doi.org/10.1007/978-3-031-26387-3_10

order, *e.g.,* network traffic, financial transactions, and sensor data. Stream clustering algorithms must have low memory overhead, be computationally efficient, and robust to concept drift, *i.e.,* evolving data distributions [25]. Maintaining high cluster quality in a fully online setting is extremely difficult. Therefore, hybrid online-offline algorithms are popular among existing approaches, *e.g.,* CluStream [2], StreamKM++ [1], DenStream [6], BIRCH [31]. These algorithms have an online component that summarizes the data stream, and an offline component that periodically uses that information to create the final clusters. There also exist algorithms that store part of the stream for handling outliers, *e.g.,* BOCEDS [15], MDSC [9]. Existing stream clustering algorithms handle concept drift by having variable number of clusters: they add new clusters for newly observed behavior and discard clusters that contain too many old data items. This leads to higher memory requirements for managing buffers and intermediate solutions. Batch clustering algorithms can also be used in a streaming setting by considering a batch size of one, *e.g.,* Minibatch k-means [24]. However, they start to under-perform when the stream contains drift.

In recent years, sequential data has increasingly become popular because of the powerful insights that it provides regarding behavior analytics [5], *e.g.,* for attacker strategy profiling [21], fraud detection [13], human activity recognition [7]. Clustering sequences in an offline setting is challenging in itself because sequences are often out-of-sync, requiring expensive alignment-based distance measures, which are often not supported by many clustering algorithms. K-medoids or Partitioning Around Medoids (PAM) has often been used to cluster sequences because the k-centers are represented by actual data items, called medoids or prototypes [27,28]. This has multiple benefits: i) it makes the cluster interpretation simpler; ii) it enables the use of non-metric distances such as dynamic time warping (DTW); and iii) it allows to estimate exact storage requirements based on the k-fixed clusters. Although the state-of-the-art offline k-medoids algorithms, *i.e.,* FastPAM1 [23] and BanditPAM [26] have reduced the runtime complexity to $\mathcal{O}(nlogn)$, they are still not efficient enough to be used in streaming settings, and the cluster quality will degrade over time as the stream evolves. To the best of our knowledge, there exists no streaming version of the k-medoids algorithm that can efficiently cluster sequential data.

Contributions. In this paper, we propose *SECLEDS*, a lightweight streaming version of the k-medoids algorithm with *constant memory footprint.* SECLEDS has two unique properties: Firstly, it uses p-medoids per cluster to maintain stable high-quality clusters. Note the difference from IMMFC [30], which uses the information of multiple medoids in independent sub-solutions to select the final medoids. We initialize the p-medoids using a non-uniform sampling strategy similar to k-means++. Secondly, a Medoid Voting scheme is used to estimate a cluster's center of mass. The offline k-medoids has a SWAP step that tests each point in a cluster to determine the next medoid. SECLEDS cannot do this because it does not store any part of the stream. Instead, it maintains votes for each medoid that estimate how representative (valuable) it is given the data seen so far. A user-supplied decay factor enables SECLEDS to slowly forget the votes

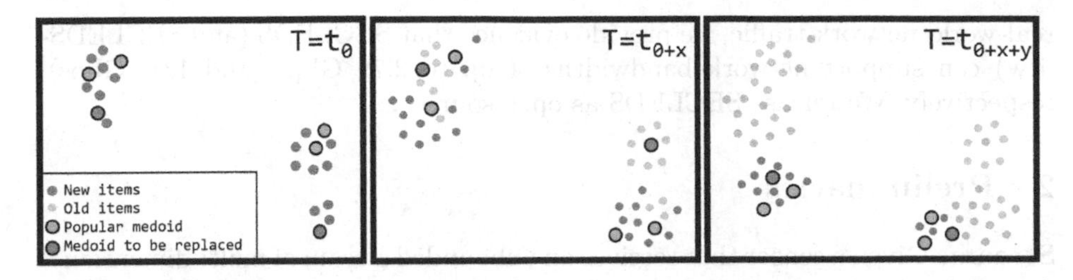

Fig. 1. An illustration of SECLEDS' clusters following an evolving data stream. The medoids close to recent data gain more votes, while the medoids with the least votes are replaced with new data items from the stream. In effect, the k-clusters handle concept drift by capturing different concepts in the stream at different time steps.

regarding past data. The least representative medoids are then replaced with new data items. This way, rather than creating new clusters for new concepts, the k-clusters themselves evolve with the data stream. Figure 1 shows how the clusters follow a data stream as it evolves. Thus, the k-clusters represent different concepts in the stream at different time steps. SECLEDS addresses the following real-world constraints:

I. A runtime efficient medoid-based clustering algorithm with a fixed memory footprint that can handle high-bandwidth data streams,
II. An algorithm that produces high-quality clusters in a streaming environment while being able to deal with concept drift,
III. Accurate sequence clusters using alignment-based distances, and
IV. Minimal parameter settings to support ease-of-use.

Empirical Results. We experiment on several real and synthetic data streams that contain 2D points and univariate sequences. We empirically demonstrate that SECLEDS produces high-quality clusters regardless of drift, stream size, and number of clusters. We use the following state-of-the-art and popular clustering algorithms as baselines: a) Streaming: CluStream, StreamKM++; b) Batch: Minibatch k-means; c) Offline: BanditPAM. Particularly, BanditPAM is used as a benchmark for the best achievable clustering on a static dataset. The results show that i) SECLEDS achieves comparable F1 score to BanditPAM, while reducing the required number of distance computations by 83.7%; ii) SECLEDS outperforms all baselines by 138.7% when the stream contains drift; iii) SECLEDS is faster than BanditPAM and CluStream on most clustering tasks.

We also discuss a use-case from the network security domain where network traffic is often randomly sampled to keep the storage requirements within a predefined budget. Consequently, temporal patterns in the network traffic are lost that could have been useful for downstream tasks, *e.g.*, behavior analytics. We propose a smarter sampling technique that uses medoid-based stream clustering (SECLEDS) to summarize the network traffic: SECLEDS clusters sequences of network traffic and periodically stores the medoids of each cluster, thus reducing the storage needs while preserving temporal patterns in the data. By clustering real-world network traffic, we provide evidence that SECLEDS (and SECLEDS-dtw) can support network bandwidths of up to 2.79 Gbps (and 1.08 Gbps), respectively. We release SECLEDS as open-source[1].

2 Preliminaries

Stream. Given a sensor that receives an unbounded stream of multi-dimensional data points $X = \{x_1, x_2, \dots\}$ with dimensionality d, arriving at time steps $T = \{t_1, t_2, \dots\}$, a sequential data stream is defined as $S = \{s_1, \dots s_n, \dots\}$, where s_i is a time window w over X such that $s_i = \{x_i, x_{i+1}, \dots x_{i+w}\}$, and y_i is its associated class label. Traditional point clustering considers $w = 1$, while for sequence clustering, we consider $w > 1$. We use two configurations, *i.e.*, $d = 2, w = 1$ (2D point clustering) and $d = 1, w = 100$ (univariate sequence clustering). A case of bivariate variable length sequences is given in appendix.

Concept Drift. Real-world data streams often change unexpectedly over time. This shift alters the statistical properties of their underlying distribution. In machine learning, this is called concept drift [18,32]. Concept drift is typically categorized into four types [17]: (i) *Sudden drift* where a new concept arises abruptly; (ii) *Gradual drift* where an old concept is slowly replaced by a new one; (iii) *Incremental drift* where a concept incrementally turns into another one; and (iv) *Recurring concepts* are old concepts that reappear from time to time. Years of research has gone into developing *concept drift detectors* that either monitor the underlying data distribution, error rate or perform hypothesis testing to trigger model retraining [4,17]. Typical stream clustering algorithms handle concept drift by introducing new clusters for new concepts, and discarding old irrelevant clusters [14]. Although intuitively appealing, this requires user-supplied parameters that define what 'new' means.

3 SECLEDS: Sequence Clustering in Evolving Streams

SECLEDS is a lightweight streaming variant of the classical k-medoids (PAM) algorithm. To support high bandwidth data streams, SECLEDS does not store any part of the stream in memory—it receives an item, assigns it to one of the k-clusters, and then discards it. This way, SECLEDS has a guaranteed constant memory footprint [**I**]. However, this requirement cannot be achieved using the offline BUILD and SWAP steps of PAM. Instead, SECLEDS performs a non-uniform sampling (similar to k-means++) on an initial batch of the stream to initialize the medoids. It also makes use of *multiple medoids per cluster* to provide a stable cluster definition in a streaming setting where noise and concept drift are common properties [**II**].

The efficiency of a chosen distance measure is usually the primary performance bottleneck in sequence clustering. Thus, minimizing the number of distance computations is key to scaling SECLEDS to large data streams. This is

[1] SECLEDS: https://github.com/tudelft-cda-lab/SECLEDS.

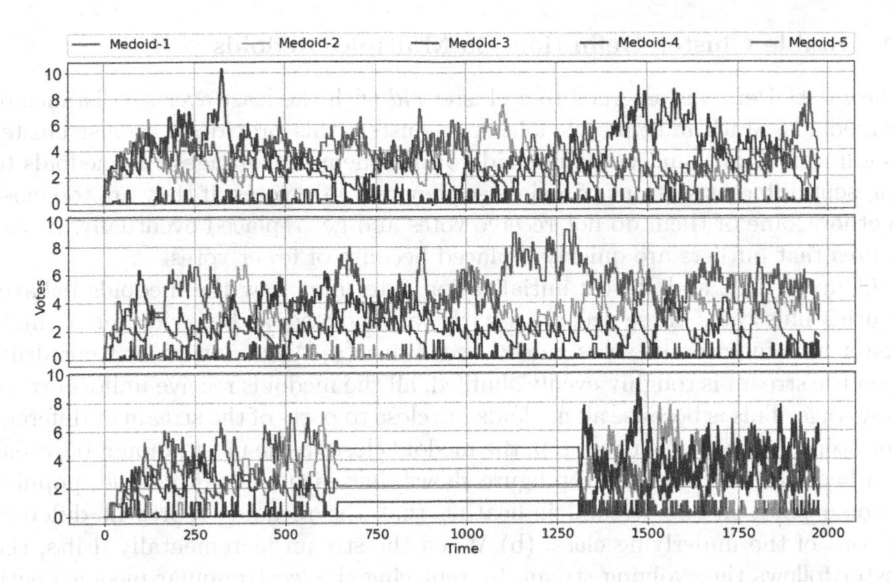

Fig. 2. The effect of cluster initialization and concept drift on medoid selection of a cluster, where $p = 5$. (Top): Given a uniformly distributed stream, every medoid becomes popular at some point. (Middle): For an incrementally drifted stream, the medoid close to drifted data becomes popular. (Bottom): For a class-ordered stream, all clusters start from one class, until one medoid migrates to the correct class. In this case, the correct class is observed from t_{1300}. Medoid-3 migrates first and becomes popular, while medoid-1 migrates last.

achieved by introducing a *Medoid Voting scheme* whose purpose is twofold: (i) it determines the center of mass of a cluster, thus is able to estimate how representative a medoid is given the data seen so far; and (ii) by using this information, old irrelevant medoids are replaced by new ones that are located near recent data. Hence, better medoids can be found without having to perform additional distance computations [**I**]. This also allows SECLEDS to support robust but computationally expensive distance measures specifically meant for sequential data, *e.g.*, dynamic time warping [**III**]. Finally, SECLEDS handles concept drift by regularly *forgetting* past data and occupying newer regions/concepts in the data stream. This is achieved by applying exponential decay λ to the medoid votes at each time step [**II**].

SECLEDS has a modular implementation in Python. The k-clusters, p-medoids per cluster, and decay rate λ are the only three user-supplied parameters needed for the algorithm, making it useful for exploratory data analysis [**IV**]. We believe these parameters are easier to tune compared to many radius- or density-based hyperparameters in existing clustering algorithms, which require a deeper understanding of the data distribution in advance.

3.1 Stable Cluster Definition via Multiple Medoids

A new data item s is assigned to a cluster *cid* with the least average distance to its medoids. With multiple medoids per cluster, this provides a robust cluster assignment. Additionally, the medoid voting scheme encourages the medoids to represent different sections of a class so they can gain votes: if they are too close together, some of them do not receive votes and get replaced eventually. It also ensures that outliers are quickly replaced because of fewer votes.

Concept drift and cluster initialization determine how the medoids behave. Figure 2 illustrates three scenarios with varying medoid behavior for a single cluster as a function of votes gained over time: (a) Assuming no concept drift, when the stream is roughly evenly shuffled, all the medoids receive uniform votes on average. This is because all medoids are close to parts of the stream at different time steps. At a specific time step, the medoid close to the most amount of recent data becomes popular. The top figure shows that each medoid becomes popular at some point in the stream, indicating that the medoids represent different sections of the underlying class. (b) When the stream incrementally drifts, the cluster follows the evolving stream by replacing the least popular medoids with recent data items from the stream. Since the new medoids are now closer to new data, they gain more votes and become popular. This has roughly the same effect as the first case. (c) When the data arrives one class at a time, all clusters are initialized in a single class. As data from a new class appears, one medoid from the closest cluster migrates to it and starts gaining votes. Over time, the older popular medoids lose their votes because of exponential decay, and eventually migrate to the new class. This is shown from t_{1300} onward in the bottom figure, highlighting the importance of multiple medoids in noisy streams.

3.2 Center of Mass Estimation

The voting scheme provides an estimate of a cluster's center of mass by assigning more votes to recently observed data in the stream S, while exponential decay helps to forget votes regarding older data. Without decay, older clusters with popular medoids never evolve. Thus, these properties help to replace irrelevant medoids, *e.g.*, those that are located i) close to the least amount of recent data, or ii) in a region where new data no longer arrives. Note that we only apply exponential decay to the *most recently updated cluster*, so that we do not forget valuable information about other clusters while the data from this class arrives.

4 The SECLEDS Algorithm

SECLEDS has three modules: an initialization module (INIT), an assignment module (ASSIGN), and an update module (UPDATE). The task is to assign each item in S to one of the k-clusters. SECLEDS maintains and updates a model of the stream seen so far in the form of k-clusters, $\mathcal{C} = \{C_1, \dots C_k\}$. For clarity, we use t to denote the clusters at time t. These superscripts are removed from

Algorithm 1. Each cluster is represented by a set of p-medoids and their votes, *i.e.*, for $1 \leq i \leq k$, $C_i^t = \{(m_{i,1}^t, v_{i,1}^t) \ldots (m_{i,p}^t, v_{i,p}^t)\}$, where for $1 \leq j \leq p$: $m_{i,j}^t \in S$ is the jth medoid of the ith cluster at time t having $v_{i,j}^t \in \mathbb{R}$ votes.

Init. A batch $\mathbb{B}$ from the start of S is used to initialize the clusters. The batch can be small but enough to select $k \cdot p$ medoids. In the experiments, we used a batch size of $(1.5 \cdot k \cdot p)$. We use a non-uniform sampling strategy, similar to the Lloyd's algorithm [16], to select the primary medoid of each cluster: SECLEDS selects the first medoid of the first cluster $(m_{1,1}^1)$ arbitrarily from the batch. Another $k-1$ medoids are sampled with a probability proportional to the squared distance between $m_{1,1}^1$ and other items in $\mathbb{B}$. This initializes the primary medoid of each cluster. The other p-1 medoids for each cluster C_i^1 are the items in $\mathbb{B}$ that are closest to its primary medoid $m_{i,1}^1$. This way, the medoids maintain cluster separation by reducing the risk of medoids from multiple clusters overlapping each other. All medoids start with 0 votes.

Assign and Update. With the clusters initialized, the stream processing begins. ASSIGN and UPDATE are called for each item in S. ASSIGN has 3 steps: (i) An incoming item s at time t is assigned to the cluster C_{cid}^t for which its previous medoids C_{cid}^{t-1} have the least average distance to s, formally defined in Eq. (1) for any given distance function $d(.,.)$. (ii) The closest medoid to s receives a vote, while exponential decay λ is applied to the other medoids *i.e.*, for all $1 \leq j \leq p$ and $j' \neq j$: $v_{cid,j}^t = (v_{cid,j}^{t-1} + 1)$ if $j = \arg\min_j d(s, m_{cid,j}^{t-1})$, otherwise $v_{cid,j'}^t = v_{cid,j'}^{t-1} \cdot (1-\lambda)$. This way, the medoids maintain an estimate of their centers of mass without storing any part of the stream. (iii) The votes of all other clusters remain the same, *i.e.*, $v_{i,j}^t = v_{i,j}^{t-1}$ for all $i \neq cid$ and $1 \leq j \leq p$.

$$C_{cid}^t = \arg\min_{1 \leq cid \leq k} \frac{\sum_{j=1}^{p} d(s, m_{cid,j}^{t-1})}{p} \tag{1}$$

At every time step t, the new data item s is promoted to be a medoid of C_{cid}^t: the medoid having the least votes which is not the newest medoid is replaced by s, *i.e.*, $\{m_{cid,j}^t = s,\ v_{cid,j}^t = 0\}$ where $j = \arg\min_j v_{cid,j}^{t-1}$ and $m_{cid,j}^{t-1} \neq \eta_{cid}^t$, where η_{cid}^t keeps track of the newest medoid of cluster cid at time t. Inspired by Tabu search [11], including η_{cid}^t ensures that the most-recently updated medoid is not selected to be replaced each time. Tabu search is a local search meta-heuristic that selects which values to change except for the last δ ones ($\delta = 1$ in this case).

Time Complexity. Given k clusters, p medoids, b batch size, and n items in the stream, SECLEDS has a time complexity of $\mathcal{O}(\texttt{n})$: SECLEDS selects the first medoid at random from the initial batch, and then performs b distance computations to find the other k-1 primary medoids. The rest of the k(p-1) medoids are also selected using the same distance information. In total, this requires $\mathcal{O}(\texttt{kb})$ distance computations. For every $s \in S$, SECLEDS computes the average distance to each cluster, which requires kp distance computations. Over an entire run, this gives nkp distance computations. In the UPDATE module, SECLEDS reallocates medoid votes without any distance computations, making

Algorithm 1: SECLEDS for clustering sequences in evolving streams

Input: Data stream, nclusters, nprototypes: S, k, p

1 **function** SECLEDS(S, k, p)
2 $b \leftarrow 1.5 \cdot k \cdot p$
3 $\mathbb{B} \leftarrow$ Collect b items from S
4 $\mathcal{C} \leftarrow$ INIT($\mathbb{B}$, k, p) // INIT
5 **forall** s *in* $S[b:]$ **do**
6 $cid \leftarrow \arg\min_{1 \leq cid \leq k} \frac{1}{p} \cdot \sum_{j=1}^{p} d(s, m_{cid,j})$ // ASSIGN
7 $j \leftarrow \arg\min_j d(s, m_{cid,j})$ for all $1 \leq j \leq p$
8 $v_{cid,j} \leftarrow (v_{cid,j} + 1)$, $v_{cid,j'} \leftarrow v_{cid,j'} \cdot (1 - \lambda)$ for $j' \neq j$
9 $j \leftarrow \arg\min_j v_{cid,j}$ where $m_{cid,j} \neq \eta_{cid}$ for all $1 \leq j \leq p$ // UPDATE
10 $m_{cid,j} \leftarrow s$, $v_{cid,j} \leftarrow 0$
11 **yield** *cid*

12 **function** INIT($\mathbb{B}, k, p$)
13 Choose $m_{1,1} \in \mathbb{B}$ arbitrarily. Let $C_1 \leftarrow \{(m_{1,1}, 0)\}$
14 **for** $i \leftarrow 2 \ldots k$ **do**
15 Choose $m_{i,1} \in \mathbb{B}$ with probability $d(m_{i,1}, m_{1,1})^2$, $m_{i,1} \neq m_{1,1}$
16 Let $C_i \leftarrow \{(m_{i,1}, 0)\}$
17 **for** $i \leftarrow 1 \ldots k$ **do**
18 $dist \leftarrow d(b, m_{i,1})$ for all $b \in \mathbb{B}$ and $b \neq m_{i,1}$
19 Choose $\{m_{i,2} \ldots m_{i,p}\}$ having smallest values in *dist*
20 Update $C_i \leftarrow \{(m_{i,1}, 0) \ldots (m_{i,p}, 0)\}$
21 **return** $\{C_1, \ldots, C_k\}$

the runtime negligible. Since k and p are small user-supplied parameters, the overall runtime complexity is $\mathcal{O}(\mathtt{n})$.

Space Complexity. After initialization, SECLEDS only stores the p medoids and their votes for the k clusters. Since these are (small) user-defined parameters, the space complexity of SECLEDS is $\mathcal{O}(1)$.

5 Experimental Setup

Datasets. We use three synthetic and a real dataset containing 2D points and univariate sequences, see Table 1. The data generation process is given in appendix. The synthetic datasets are released in the SECLEDS code repository.

Blobs: The blob dataset was created using `scikit-learn` [22]. The dataset contains $n = 100,000$ two-dimensional points ($d = 2$), equally distributed in $k = 10$ classes, with varying standard deviations.

Table 1. Summary of experimental datasets.

Dataset	Type	Drift	Stream size (n)	Clusters (k)	Dimensions (d,w)
Blobs	Synthetic	No	100,000	10	(2,1)
Sine-curve	Synthetic	No	50,000	4	(1,100)
Sine-curve-drifted	Synthetic	Yes	50,000	4	(1,100)
CTU13-9	Real-world	Yes	213,386	2	(1,100)

Sine-Curve: A sine-curve generator was used to create $k = 4$ synthetic univariate sine curves of length $1,250,000$ each, using varying *frequency*, *phase* and *error* (see appendix). Each curve is partitioned using a non-overlapping window of length $w = 100$ to obtain the experimental dataset. In total, $n = 50,000$ curves are obtained, equally divided across $k = 4$ classes.

Sine-Curve-Drifted: Incremental concept drift is added to the Sine-curve dataset by shifting the phase of each curve by a factor of $(drift \cdot c\_id)$, where c_id is the curve index in the stream, and $drift = 0.05$. Note that adding drift to the frequency of the sine curves produces similar results.

CTU13-9: CTU13 [10] is an open source dataset composed of network traffic (netflows) coming from real botnet-infected hosts and normal hosts. We use scenario-9, containing 10 bot-infected hosts and 6 benign hosts. A total of $2,087,509$ (normal and botnet) netflows were captured over 5 h and 37 min. We obtain $n = 213,386$ univariate sequences of length $w = 100$ using a sliding window model [33] with $step\_size = 1$ (see appendix).

We use Euclidean distance for all datasets. For the Sine-curve and network traffic datasets, we additionally use Dynamic time warping (DTW). Note that Euclidean distance can only be used with fixed-length sequences and often produces less accurate results compared to DTW [12,29].

Stream Configuration. A data stream S of size n is constructed from a chosen experimental dataset. For each experiment, the clustering task is executed $trials$-times, randomly shuffling the stream each time, to make the results data-order-invariant. A clustering task invokes SECLEDS and the baselines such that each algorithm observes the exact same order of data arrival. In this paper, we set $trials = 10$, unless otherwise reported. All experiments are run on Intel Xeon E5620 quad-core processor with 74 GB RAM.

Evaluation. We use two metrics for performance evaluation: i) *runtime* to cluster a stream size of n; ii) *F1 score* computed from the pairwise co-occurrences of items in the stream using Eq. (2), as originally defined in [19].

$$eval(a,b) = \begin{cases} y_a = y_b \wedge C_x = C_y, & \text{true positive} \\ y_a = y_b \wedge C_x \neq C_y, & \text{false negative} \\ y_a \neq y_b \wedge C_x = C_y, & \text{false positive} \\ y_a \neq y_b \wedge C_x \neq C_y, & \text{true negative} \end{cases} \tag{2}$$

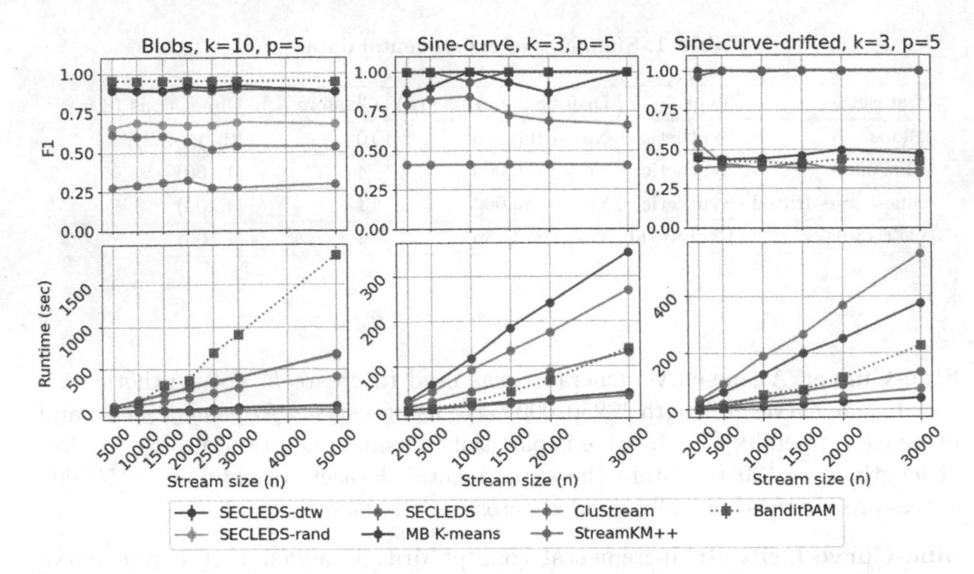

Fig. 3. Clustering Blobs and Sine-curves: SECLEDS's runtime grows approximately linearly with stream size, while maintaining competitive F1 score with the best-performing baselines, *i.e.*, BanditPAM and Minibatch k-means. SECLEDS consistently performs better than all baselines in the presence of concept drift.

where y_a and y_b are labels of items a and b that are placed in clusters C_x and C_y. Since clusters do not have pre-defined labels, data from one class may be assigned to arbitrary clusters in different runs. Thus, instead of looking at the predicted label, we measure F1 using the pairwise co-occurrences of true labels.

Baselines. We compare SECLEDS with state-of-the-art open-source partition-based clustering algorithms with k-fixed clusters: a) Streaming: CluStream, StreamKM++; b) Batching: MiniBatch k-means; c) Offline: BanditPAM. Mini-Batch k-means and StreamKM++ are online versions of k-means, while CluStream is an adaptive, online-offline algorithm. BanditPAM (v1.0.5) is used as a benchmark for the best achievable clustering on a static dataset. We set time_window=1, max_micro_clusters=$k \cdot p$, halflife=0.5 for CluStream; chunk_size=1, halflife=0.5 for StreamKM++; batch_size=1, max_iter=1 for Minibatch k-means.

6 Empirical Results

Key Findings. In this section, we empirically demonstrate the following results:

1. SECLEDS produces high-quality clusters, regardless of concept drift, stream size n, data dimensionality (d, w), and number of clusters k. SECLEDS shows competitive F1 compared to the best performing baseline (BanditPAM), while reducing the number of required distance computations by 83.7%.

2. SECLEDS outperforms *all baselines* by 138.7% when the stream contains concept drift. SECLEDS outperforms the *best-performing streaming baseline* by 58.2% on Blobs, 33.3% on Sine-curve, and 143.7% on Sine-curve-drifted.
3. SECLEDS-dtw clusters ~5.5 h of network traffic in just 8% of the time. Thus, it can handle networks with bandwidths of up to 1.08 Gbps, which is significantly higher than the requirements of a typical enterprise network.

Point vs. Sequence Clustering. We use Blobs with $k = 10$ on stream sizes $n = (5000, \ldots 50,000)$, and Sine-curve with $k = 3$ on stream sizes $n = (2000, \ldots 30,000)$. For both, we set $p = 5$, $\lambda = 0.1$, $trials = 10$. The mean and standard deviation of the F1 scores and runtimes are given in Fig. 3. The benchmark (BanditPAM) achieves a mean F1 of 0.95 and 1.0 for Blobs and Sine-curve, respectively.

SECLEDS outperforms both CluStream and StreamKM++ on the Blobs dataset, and additionally outperforms Minibatch k-means on the Sine-curve dataset. Minibatch k-means performs exceptionally well on point clustering, but loses its edge on sequence clustering. This is because the centroids are computed by collapsing temporally-linked dimensions into single values that do not adequately represent the sequences. An improvement in F1 score is observed for CluStream and StreamKM++ on the higher dimensional Sine-curve dataset because of fewer clusters ($k = 10$ vs. $k = 3$).

We also compare the effect of euclidean and dynamic time warping distance on the Sine-curve dataset. Although, they both produce equivalent results, it must be noted that euclidean distance only works with fixed-length sequences. An example of SECLEDS-dtw on clustering bivariate sequences $d = 2, w =$(min:15, max:121) from *UJI Pen Characters* [8] is given in the appendix.

Initialization Quality. Stream clustering algorithms are greatly impacted by the quality of cluster initialization. To test this, we compare SECLEDS against SECLEDS-rand (initialized with randomly selected medoids from the initial batch $\mathbb{B}$). Evidently, the clusters take a long time to converge, regardless of the stream size. The cumulative F1 score over time for these configurations is given in the appendix, showing that although the impact of poor initialization is reduced over time, SECLEDS-rand does not completely recover from it. Thus, the distance-based non-uniform sampling strategy proves to be extremely helpful in initializing good clusters.

Clustering with Concept Drift. We use Sine-curve-drifted with $k = 3$, $p = 5$, $\lambda = 0.1$, $trials = 10$ on stream sizes $n = (2000, \ldots 30,000)$. SECLEDS outperforms *all baselines* by 138.7%, and outperforms the *best-performing streaming baseline* by 143.7%, on average. BanditPAM no longer serves as a benchmark because it only has a static view of the data, *i.e.,* it does not distinguish between class distributions at $T = t_x$ and $T = t_{x+y}$. Both SECLEDS and CluStream maintain their F1 scores with concept drift, but SECLEDS is 161.8% better than CluStream. StreamKM++ and Minibatch k-means observe a significant reduction in their performance. We hypothesize that it might be due to the lack

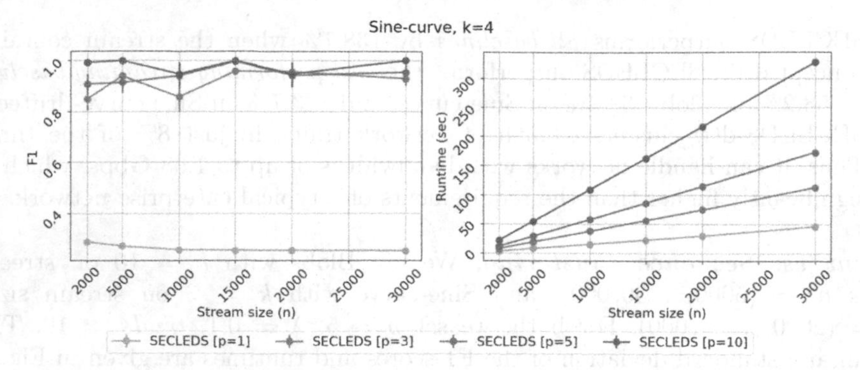

(a) $p = \{3, 5\}$ provide the best trade-off between runtime and F1 for Sine-curve.

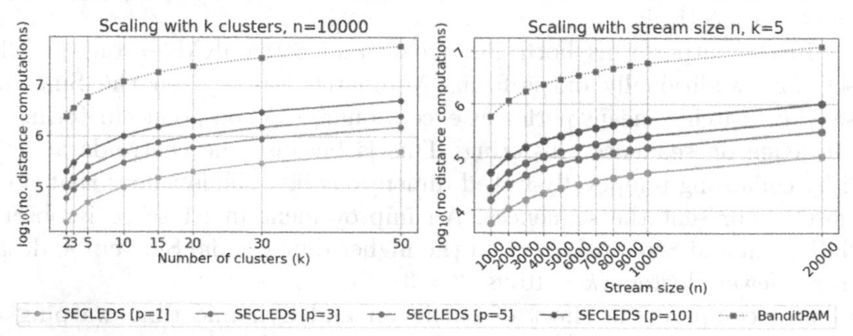

(b) SECLEDS requires 83.7% fewer distance computations compared to BanditPAM.

Fig. 4. Scaling with n, k, and p: (a) Empirical results; (b) Theoretical estimate.

of exponential decay in k-means, which limits the movement of the centroids towards newer data. This experiment provides strong evidence for SECLEDS' ability to handle concept drift with only k-fixed clusters.

Runtime Analysis. StreamKM++ and Minibatch k-means are among the fastest clustering algorithms on all datasets, which is expected since they are based on k-means. CluStream does not scale well for high-dimensional datasets, and is much slower than SECLEDS on sequence clustering. As the stream size n grows, SECLEDS also becomes faster than the high-performance implementation of BanditPAM on both point and sequence clustering. Interestingly, the runtimes of BanditPAM, CluStream and StreamKM++ seem to be affected by concept drift: given the same dataset and constant parameters, their runtimes increase approximately twofold when there is drift in the data. We hypothesize that this is a side effect of the sampling strategy used to speed up these algorithms.

Scaling with n***,*** k***, and*** p***.*** We use Sine-curve with $k = 4$, $p = \{1, 3, 5, 10\}$, $\lambda = 0.1$, $trials = 10$ on stream sizes $n = (2,000, \ldots 30,000)$. The mean and standard deviation of the F1 and runtime of SECLEDS is reported in Fig. 4a.

Table 2. Clustering real network traffic: Compared to BanditPAM, SECLEDS requires fewer distance computations, is faster, and has a better cluster quality. SECLEDS-dtw is slower but produces better clusters than SECLEDS. Overall, ∼5.5 h of network traffic is clustered in under 27 min (Bold = best scores).

	Stream config.	# Distances ($k = 2$)	Run time ($k = 2$)	**F1** ($k = 2$)	F1 ($k = 5$)
BanditPAM	Time-ordered	10.3×10^6	978.03 s	0.64	0.38
	Cross-validated		984.8 s	0.64	0.38
SECLEDS	Time-ordered	$\mathbf{2.1 \times 10^6}$	**629.39 s**	**0.85**	0.82
	Cross-validated		**631.84 s**	0.79	0.76
SECLEDS-dtw	Time-ordered	$\mathbf{2.1 \times 10^6}$	1623.05 s	**0.85**	**0.88**
	Cross-validated		1626.89 s	**0.81**	**0.80**

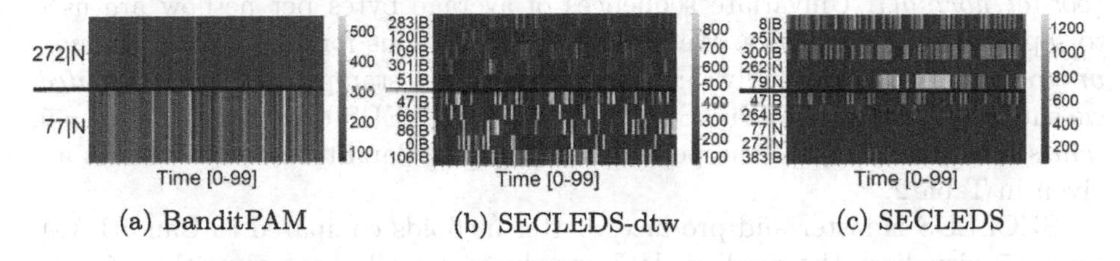

(a) BanditPAM (b) SECLEDS-dtw (c) SECLEDS

Fig. 5. Visualizing the medoids of BanditPAM, SECLEDS & SECLEDS-dtw on $k = 2$, $p = 5$. Each row is a medoid. The label denotes curve identifier and y_i.

A single medoid per cluster, which is standard for PAM-based algorithms, does poorly in a streaming setting. Intuitively, more medoids help to improve the stability of the clusters, but the relationship is not linear. If p is set too low, the medoids keep jumping to various regions in the dataset, and if it is set too high, the medoids slow down the evolution of the clusters, having an equally detrimental effect on the performance. The optimal value of p with respect to performance and runtime is dataset-dependent. For Sine-curve, $p = \{3, 5\}$ are good alternatives. Additionally, although SECLEDS has multiple medoids per cluster, it performs significantly fewer distance computations compared to the (almost linear) BanditPAM. Figure 4b shows this for increasing stream size n, number of clusters k, and number of medoids p, with BanditPAM as reference.

6.1 Use Case: Intelligent Network Traffic Sampling via SECLEDS

A typical enterprise network has a bandwidth of 25 Mbps[2], which produces about 17,000 packets *per second*, consuming 2 terabytes of storage space *each day*! To conserve space, the packets are aggregated into network flows (netflows) at the router level, and only a fraction of them are stored for analysis *i.e.*, 1

[2] https://mosaicnetworx.com/it-challenges/bits-bytes-understanding-enterprise-network-speeds/.

in N netflows are stored. Naturally, randomly sampled network traffic does not preserve the temporal patterns of the data, thus limiting the efficacy of traffic profiling and behavior analytics.

We propose to cluster sequences of netflows using SECLEDS, and to periodically store a medoid snapshot of each cluster, since they are representative of the network traffic seen so far. This way, each snapshot stores an overview of temporally-linked netflows. The number of clusters k can be chosen depending on the required granularity of behaviors captured by the clusters. It can also be approximated from an initial batch using, *e.g.,* [3]. The number of medoids p can be configured according to available storage space, network bandwidth, and the intervals at which to store the medoids.

We demonstrate this use case by generating a stream from the CTU13-9 netflows. The construction and feature engineering processes are given in the appendix. In short, the ground truth provides two classes, *i.e.,* $y_i \in \{botnet, normal\}$. Univariate sequences of average bytes per netflow are used to separate the two classes. We use two configurations for the stream: i) *Time-ordered:* the sequences arrive in order of their timestamps; ii) *Cross-validated:* we shuffle the stream. We run SECLEDS and SECLEDS-dtw with $k = 2$, $p = 5$, $trials = 5$, and compare the performance against BanditPAM. The results are given in Table 2.

SECLEDS is faster and produces better medoids compared to BanditPAM. Figure 5 visualizes the final medoids produced by all three algorithms in the form of temporal heatmaps. Temporal heatmaps have previously been used to visualize temporal similarities in [20]. Each row shows a sequence (medoid), and the colors indicate the magnitude of the curve at each time step. Both medoids of BanditPAM are from the normal class. Although the medoids of SECLEDS-dtw are all from the botnet class, it is evident that they capture distinct behaviors of the malicious hosts. SECLEDS finds medoids from both classes, but the clusters are impure, *i.e.*, more medoids on average are from different classes. As such, the cluster quality of SECLEDS-dtw is significantly better than that of SECLEDS.

The results indicate that there are many smaller classes in the network stream, reflecting the various behaviors of benign and infected hosts. When k is set to a larger number, the clustering algorithms find smaller, purer data regions, *e.g.,* for $k = 5$, SECLEDS-dtw produces 4 pure clusters (2 normal and 2 botnet), while SECLEDS only produces 1 pure (normal) cluster, see appendix for their temporal heatmaps. Table 2 shows the F1 scores for $k = 5$. Note that although the clustering results for $k = 5$ are better than $k = 2$, the former obtains a lower F1 score as a side-effect of the metric: it penalizes higher number of clusters when less class labels are available by lowering the recall. As such, we recommend to over-estimate k in order to sample many regions from the network traffic.

Finally, SECLEDS is faster than SECLEDS-dtw, as expected. SECLEDS clusters the entire stream in 3.1% of the traffic collection time, and SECLEDS-dtw in 8% of the collection time. This experiment provides evidence that SECLEDS can handle much larger network bandwidths.

Network Bandwidth Support. SECLEDS-dtw can handle networks with a bandwidth of up to 1.08 Gbps, which is more than sufficient for small to medium enterprises. The experiments in Table 2 show that SECLEDS spends $\frac{1626.89}{213386} = 0.0076$ seconds on average to cluster a single sequence of length $w = 100$. Thus, SECLEDS can cluster 131.58 sequences per second. Assuming that the sequence windows w are non-overlapping over the traffic stream, SECLEDS can process 13,158 individual netflows per second. The CTU13-9 dataset is composed of 115,415,321 packets aggregated into 2,087,509 netflows. Assuming uniform distribution, each netflow contains about 55.2 packets. SECLEDS can, thus, process 726,315.79 packets per second. Given that each network packet is about 1500 bytes, this makes a total of 1.089 Gigabytes per second. Similarly, SECLEDS can handle network bandwidths of up to 2.79 Gbps.

7 Conclusions

We propose SECLEDS, a streaming version of k-medoids with constant memory footprint. SECLEDS uses a combination of multiple medoids per cluster and a medoid voting scheme to create k-clusters that evolve with evolving data streams. Testing on several real and synthetic datasets and comparing against state-of-the-art baselines, we demonstrate that i) SECLEDS achieves competitive F1 score compared to the benchmark (BanditPAM) on streams without concept drift; ii) SECLEDS outperforms all baselines by 138.7% on streams with concept drift; iii) SECLEDS reduces the number of required distance computations by 83.7% compared to the benchmark, making it faster than BanditPAM and CluStream for several clustering tasks, iv) SECLEDS can support high-bandwidth network streams of up to 1.08 Gbps using the expensive dynamic time warping distance. These results reinforce the importance of designing lightweight medoid-based stream clustering algorithms.

Acknowledgements. We thank Ruben te Wierik, Silviu Fucarev, and Rami Al-Obaidi for their contributions to the SECLEDS algorithm.

References

1. Ackermann, M.R., Märtens, M., Raupach, C., Swierkot, K., Lammersen, C., Sohler, C.: Streamkm++ a clustering algorithm for data streams. JEA **17**, 2–1 (2012)
2. Aggarwal, C.C., Philip, S.Y., Han, J., Wang, J.: A framework for clustering evolving data streams. In: VLDB, pp. 81–92. Elsevier (2003)
3. de Andrade Silva, J., Hruschka, E.R.: Extending k-means-based algorithms for evolving data streams with variable number of clusters. In: ICMLA, vol. 2, pp. 14–19. IEEE (2011)
4. Barros, R.S.M., Santos, S.G.T.C.: A large-scale comparison of concept drift detectors. Inf. Sci. **451**, 348–370 (2018)
5. Boeva, V., Nordahl, C.: Modeling evolving user behavior via sequential clustering. In: Cellier, P., Driessens, K. (eds.) ECML PKDD 2019. CCIS, vol. 1168, pp. 12–20. Springer, Cham (2020). https://doi.org/10.1007/978-3-030-43887-6_2

6. Cao, F., Estert, M., Qian, W., Zhou, A.: Density-based clustering over an evolving data stream with noise. In: SDM, pp. 328–339. SIAM (2006)
7. Cook, D.J., Krishnan, N.C., Rashidi, P.: Activity discovery and activity recognition: a new partnership. IEEE Trans. Cybern. **43**(3), 820–828 (2013)
8. Dua, D., Graff, C.: UCI machine learning repository (2017)
9. Fahy, C., Yang, S.: Finding and tracking multi-density clusters in online dynamic data streams. IEEE Trans. Big Data **8**, 178–192 (2019)
10. Garcia, S., Grill, M., Stiborek, J., Zunino, A.: An empirical comparison of botnet detection methods. Comput. Secur. **45**, 100–123 (2014)
11. Glover, F.: Future paths for integer programming and links to artificial intelligence. Comput. Oper. Res. **13**(5), 533–549 (1986)
12. Guijo-Rubio, D., Durán-Rosal, A.M., Gutiérrez, P.A., Troncoso, A., Hervás-Martínez, C.: Time-series clustering based on the characterization of segment typologies. IEEE Trans. Cybern. **51**(11), 5409–5422 (2020)
13. Guo, J., Liu, G., Zuo, Y., Wu, J.: Learning sequential behavior representations for fraud detection. In: ICDM, pp. 127–136. IEEE (2018)
14. Hyde, R., Angelov, P., MacKenzie, A.R.: Fully online clustering of evolving data streams into arbitrarily shaped clusters. Inf. Sci. **382**, 96–114 (2017)
15. Islam, M.K., Ahmed, M.M., Zamli, K.Z.: A buffer-based online clustering for evolving data stream. Inf. Sci. **489**, 113–135 (2019)
16. Lloyd, S.: Least squares quantization in PCM. IEEE Trans. Inf. Theory **28**(2), 129–137 (1982)
17. Lu, J., Liu, A., Dong, F., Gu, F., Gama, J., Zhang, G.: Learning under concept drift: a review. TKDE **31**(12), 2346–2363 (2018)
18. Lu, N., Zhang, G., Lu, J.: Concept drift detection via competence models. Artif. Intell. **209**, 11–28 (2014)
19. Manning, C., Raghavan, P., Schütze, H.: Introduction to information retrieval. Nat. Lang. Eng. **16**(1), 100–103 (2010)
20. Nadeem, A., Hammerschmidt, C., Gañán, C.H., Verwer, S.: Beyond labeling: using clustering to build network behavioral profiles of malware families. In: Stamp, M., Alazab, M., Shalaginov, A. (eds.) Malware Analysis Using Artificial Intelligence and Deep Learning, pp. 381–409. Springer, Cham (2021). https://doi.org/10.1007/978-3-030-62582-5_15
21. Nadeem, A., Verwer, S., Moskal, S., Yang, S.J.: Alert-driven attack graph generation using s-PDFA. IEEE Trans. Dependable Sec. Comput. **19**(2), 731–746 (2021)
22. Pedregosa, F., et al.: Scikit-learn: Machine learning in python. J. Mach. Learn. Res. **12**, 2825–2830 (2011)
23. Schubert, E., Rousseeuw, P.J.: Faster k-medoids clustering: improving the PAM, CLARA, and CLARANS algorithms. In: Amato, G., Gennaro, C., Oria, V., Radovanović, M. (eds.) SISAP 2019. LNCS, vol. 11807, pp. 171–187. Springer, Cham (2019). https://doi.org/10.1007/978-3-030-32047-8_16
24. Sculley, D.: Web-scale k-means clustering. In: WWW, pp. 1177–1178 (2010)
25. Silva, J.A., Faria, E.R., Barros, R.C., Hruschka, E.R., de Carvalho, A.C., Gama, J.: Data stream clustering: a survey. CSUR **46**(1), 1–31 (2013)
26. Tiwari, M., Zhang, M.J., Mayclin, J., Thrun, S., Piech, C., Shomorony, I.: Banditpam: almost linear time k-medoids clustering via multi-armed bandits. NeurIPS **33**, 10211–10222 (2020)
27. Ushakov, A.V., Vasilyev, I.: Near-optimal large-scale k-medoids clustering. Inf. Sci. **545**, 344–362 (2021)

28. Wang, T., Li, Q., Bucci, D.J., Liang, Y., Chen, B., Varshney, P.K.: K-medoids clustering of data sequences with composite distributions. IEEE Trans. Signal Process. **67**(8), 2093–2106 (2019)
29. Wang, X., Mueen, A., Ding, H., Trajcevski, G., Scheuermann, P., Keogh, E.: Experimental comparison of representation methods and distance measures for time series data. Data Min. Knowl. Disc. **26**(2), 275–309 (2013)
30. Wang, Y., Chen, L., Mei, J.P.: Incremental fuzzy clustering with multiple medoids for large data. IEEE Trans. Fuzzy Syst. **22**(6), 1557–1568 (2014)
31. Zhang, T., Ramakrishnan, R., Livny, M.: Birch: a new data clustering algorithm and its applications. Data Min. Knowl. Disc. **1**(2), 141–182 (1997)
32. Žliobaitė, I., Pechenizkiy, M., Gama, J.: An overview of concept drift applications. In: Japkowicz, N., Stefanowski, J. (eds.) Big Data Analysis: New Algorithms for a New Society. SBD, vol. 16, pp. 91–114. Springer, Cham (2016). https://doi.org/10.1007/978-3-319-26989-4_4
33. Zubaroğlu, A., Atalay, V.: Data stream clustering: a review. Artif. Intell. Rev. **54**(2), 1201–1236 (2021)

Knowledge Integration in Deep Clustering

Nguyen-Viet-Dung Nghiem(✉), Christel Vrain, and Thi-Bich-Hanh Dao

Univ. Orléans, INSA Centre Val de Loire, LIFO EA 4022, 45067 Orléans, France
nguyen-viet-dung.nghiem@etu.univ-orleans.fr,
{christel.vrain,thi-bich-hanh.dao}@univ-orleans.fr

Abstract. Constrained clustering that integrates knowledge in the form of constraints in a clustering process has been studied for more than two decades. Popular clustering algorithms such as K-means, spectral clustering and recent deep clustering already have their constrained versions, but they usually lack of expressiveness in the form of constraints. In this paper we consider prior knowledge expressing relations between some data points and their assignments to clusters in propositional logic and we show how a deep clustering framework can be extended to integrate this knowledge. To achieve this, we define an expert loss based on the weighted models of the logical formulas; the weights depend on the soft assignment of points to clusters dynamically computed by the deep learner. This loss is integrated in the deep clustering method. We show how it can be computed efficiently using Weighted Model Counting and decomposition techniques. This method has the advantages of both integrating general knowledge and being independent of the neural architecture. Indeed, we have integrated the expert loss into two well-known deep clustering algorithms (IDEC and SCAN). Experiments have been conducted to compare our systems IDEC-LK and SCAN-LK to state-of-the-art methods for pairwise and triplet constraints in terms of computational cost, clustering quality and constraint satisfaction. We show that IDEC-LK can achieve comparable results with these systems, which are tailored for these specific constraints. To show the flexibility of our approach to learn from high-level domain constraints, we have integrated implication constraints, and a new constraint, called span-limited constraint that limits the number of clusters a set of points can belong to. Some experiments are also performed showing that constraints on some points can be extrapolated to other similar points.

Keywords: Deep clustering · Knowledge integration · Constrained clustering

1 Introduction

Clustering is an important task in Data Mining, which aims at partitioning data instances into groups (clusters) such that instances in the same cluster

---

Supplementary Information The online version contains supplementary material available at https://doi.org/10.1007/978-3-031-26387-3_11.

M.-R. Amini et al. (Eds.): ECML PKDD 2022, LNAI 13713, pp. 174–190, 2023.
https://doi.org/10.1007/978-3-031-26387-3_11

are similar and instances in different clusters are dissimilar. Prior knowledge has been integrated into the clustering process by means of constraints, leading to a new field called Constrained Clustering. Constraints can be instance-level constraints, mainly must-link, resp. cannot-link constraints, which state that two instances must be, resp. cannot be in the same cluster. Constraints can also specify requirements that the clusters must satisfy. Many works have been developed to integrate constraints: by enforcing constraints [25], by balancing clustering quality and constraint satisfaction [8], or by learning a metric taking into account the constraints [3,28]. Most clustering approaches are based on a distance between objects leading to the difficulty of choosing the right representation of data. The emergence of deep learning and its ability to learn new data representation in a lower dimension space have led to deep clustering approaches [6,12,26]. The integration of constraints into deep clustering has been studied in [14,30] but most of the work focused only on instance-level constraints. To the best of our knowledge few work consider the integration of different types of constraints, such as [30], where a loss is defined for each type of constraints and the loss criterion is therefore a combination of the loss for each kind of constraints. Such a framework has two drawbacks: the integration of a new family of constraints requires the design of the corresponding loss, and defining the global loss needs to set the parameters to combine the different losses for the constraints.

In our work, we take another point of view which is to define a general constraint satisfaction score. Then, an expert loss, based on this constraint satisfaction score, is introduced into a deep clustering framework for backpropagation. We show that the constraint satisfaction score can be computed through its transformation into a Weighted Model Counting problem [21]. Based on logical formulas, our approach has the advantage of being flexible, so that different types of knowledge can be integrated without designing specific losses. The framework is summarized in Fig. 1.

Our contributions are:

- We propose a logical formulation of the constrained clustering problem and a unified definition of expert loss to integrate constraints into a deep clustering framework.
- Given a constraint set, the expert loss is based on a constraint satisfaction score, defined thanks to the notion of semantic models of a logical formulae, thus making it independent from the type of constraints. Moreover, this can be computed by Weighted Model Counting.
- We show that our framework can be integrated into different clustering frameworks by considering two well-known deep clustering methods, namely IDEC and SCAN, and extending them to integrate knowledge.
- Experiments on five datasets with randomly generated constraint sets show that our framework is competitive with state-of-the-art deep constrained clustering systems on pairwise and triplet constraints.
- To illustrate the genericity of our approach, we introduce a new type of constraints, called a span-limited constraint.

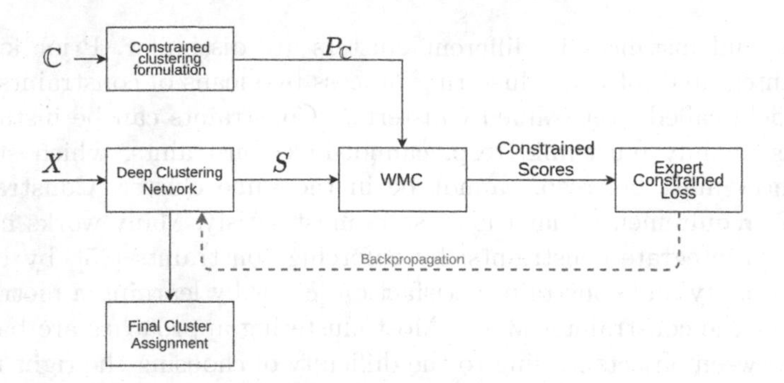

Fig. 1. Overview of our DC-LK framework. The constrained clustering problem is formulated in a logical form $P_{\mathbb{C}}$. A deep clustering framework is used to compute a soft assignment S of data points $\mathbf{X}$ to clusters. The constrained score is computed based on S and the constraint problem $P_{\mathbb{C}}$ and is used to define the expert loss. It is backpropagated to the deep clustering network.

- We analyze the efficiency of our framework both on runtime and on constraint satisfaction for complex constraints.
- We show that satisfying constraints when training the model allows to improve the satisfaction of unseen constraints on test data.

The rest of the paper is organized as follows. Related work is reviewed in Sect. 2. The formulation of the constrained clustering problem and of the expert loss is presented in Sect. 3. Section 4 presents knowledge integration into two deep clustering frameworks IDEC and SCAN using the expert loss. Section 5 describes the experiments and analyzes the results and Sect. 6 concludes and discusses future work.

2 Related Work

Constrained Clustering. Many approaches have now been developed for constrained clustering. Most of them focus on pairwise (must-link/cannot-link) constraints and the early work consisted in adapting classic methods such as k-means or spectral clustering to enforce them [5,18]. Other constraints have been introduced as for instance cardinality constraints [22]. Nevertheless all these approaches are usually designed for one kind of constraints whereas the expert knowledge is often multiform including both pairwise constraints, cardinality constraints and much more complex constraints as given for instance in [9], thus requiring new frameworks for constrained clustering. It has been shown that declarative frameworks such as ILP [1,19], SAT [11] or Constraint Programming [8] allow to integrate a large variety of constraints, while satisfying all the constraints.

Deep Clustering. Recently, deep clustering approaches have been extensively proposed following the success of deep neural networks in supervised learning.

Several research directions have been considered: adapting to clustering well-known supervised learning architectures such as convolutional neural network [6], changing data representation through an autoencoder and then enforcing the clustering structure on the latent space [12,26]. Another approach is to mine the nearest neighbor based on pretext features (an embedding for a specific task such as inpainting patches, predicting noise, instance discrimination). It helps to promote similar predictions of the neighbors, thus, improving the clustering quality [23]. More ambitious approaches have been proposed as for instance generative models that both cluster data and generate samples for a given clustering [15,20], but they usually suffer from relatively low performances.

Deep Learning with Knowledge. Knowledge integration can be seen as a generalization of semi-supervised learning. While label information is easy to represent, expert knowledge can be various and thus expressed in many different ways. To tackle this problem, several work [27,29] have studied the integration of knowledge expressed in logic in Deep Supervised Learning: knowledge is then enforced on each individual input instance. [29] gives a precise formulation of the loss regardless of the logical form (whether it is represented in CNF, DNF or in a arbitrary form) at the price of a high computational complexity. [27] learns a knowledge loss (using a logic graph embedder) for a specific logical form (d-DNNF), which is much faster but requires a substantial amount of constraints.

In a clustering setting, even if each point receives a label, the output on all the data is expected to represent a partition (or another structure). This means that the constraints are not put on the output of a single point, but they can link several outputs, which is much more challenging.

[14] integrates triplet constraints in a deep clustering framework. DCC [30] has proposed a deep clustering framework, which can integrate several types of constraints such as pairwise, triplet or cardinality. However for each type of constraint, a specific loss is designed. This differs from our approach, where the same definition of expert loss is given for any type of constraints, as soon as the constraint can be expressed using a logical formulation.

3 Expert Loss for Knowledge Integration

3.1 Expert Knowledge Representation

In this paper, we are interested in integrating knowledge in a deep clustering system. We suppose that we have n points $x_1, \ldots, x_n$ and that we want to learn a partition of them into k clusters. We also have expert knowledge on the desired partition, which is expressed by expert constraints and written in propositional logic. Let β_{ij}, $i \in \{1, ..., n\}, j \in \{1, ..., k\}$ be formulas, such that β_{ij} is $True$ when point i is assigned to cluster j. Using β, the predicate meaning that two points u, v are in the same cluster can be expressed by

$$Together(u, v) \stackrel{\text{def}}{=} \wedge_{i \in [1,k]}((\beta_{ui} \wedge \beta_{vi}) \vee (\neg\beta_{ui} \wedge \neg\beta_{vi}))$$

and the predicate meaning that two points i, j are in different clusters:

$$Apart(u, v) \stackrel{\text{def}}{=} \wedge_{i \in [1,k]} (\neg \beta_{ui} \vee \neg \beta_{vi})$$

The constraints we consider are:

- must-link (cannot-link) constraints stating that two points u, v must be (resp. cannot be) in the same cluster are expressed by the statement $Together(u, v)$ (resp. $Apart(u, v)$).
- triplet constraints (a, p, n) expressing that a is closer to p than to n:

$$Together(a, n) \implies Together(a, p)$$

- more complex implication constraints, as for instance:

$$Together(a, b) \wedge Together(c, d) \wedge Apart(a, e) \Rightarrow Together(e, f) \wedge Apart(c, e)$$

- a new type of constraints, called *span-limited constraint* that expresses that a given group I of points cannot be dispatched in more than a given number m of clusters.

$$\vee_{J \subset \{1,\ldots,k\}, |J|=m} \wedge_{i \in I} \vee_{j \in J} \beta_{ij}$$

3.2 Constraint-Satisfaction Score

We consider a deep clustering algorithm that produces a soft assignment S of points to clusters, where S_{ij} denotes the soft assignment value of point i to cluster j. We aim at integrating in this system a new loss, called **expert loss** that takes into account the satisfaction of constraints. This loss has to be generic so as to integrate different kinds of constraints, this explains why we rely on propositional logic. This expert loss is based on the weighted models of the logical formulas, where the weights depend on the soft assignment of points to clusters dynamically computed by the deep learner.

Given a partition $\mathbf{p}$ and the soft assignment matrix S, we define the partition score between the partition and S by:

$$Score(\mathbf{p}, S) = \prod_{i \in [1,n]} S_{i\mathbf{p}_i} \tag{1}$$

where $\mathbf{p}_i$ denotes the assignment of i in the partition $\mathbf{p}$.

Given a set of constraints $\mathbb{C}$, we denote by $\mathbb{P}_\mathbb{C}$ the set of partitions that satisfy all the constraints in $\mathbb{C}$. To measure how likely the soft assignment S is with respect to the constraint set $\mathbb{C}$, we define a constraint-satisfaction score as follows:

$$Score(\mathbb{C}, S) = \sum_{p \in \mathbb{P}_\mathbb{C}} Score(\mathbf{p}, S) \tag{2}$$

To illustrate this, let us suppose that we have only 2 points and 2 clusters. Then the score of the partition that assigns each point to cluster 1 is $S_{11} * S_{21}$. If we add a must-link constraint betwwen the two points, then two partitions satisfy this constraint, and $Score(\mathbb{C}, S) = S_{11} * S_{21} + S_{12} * S_{22}$.

3.3 Constraint-Satisfaction Score Computed by a WMC Problem

The constraint-satisfaction score (2) can be computed directly. However, enumerating all the partitions in $\mathbb{P}_{\mathbb{C}}$ is expensive ($\mathcal{O}(k^n)$). We show that this problem can be converted into a Weighted Model Counting problem with an appropriate choice of formulas and of weights.

Let us recall that if α is a logical formula defined on a set of variables $\mathbf{Y}$, where each variable Y_i is associated with a weight w_i, then the weighted model counting (WMC) of α is defined by [21]:

$$WMC(\alpha, w) = \sum_{\mathbf{y} \models \alpha} \prod_{i:\mathbf{y} \models Y_i} w_i \prod_{i:\mathbf{y} \models \neg Y_i} (1 - w_i) \tag{3}$$

In (3), the weights w_i and $(1-w_i)$ can occur several times. Sentential Decision Diagrams (SDD) [10] can be used for a more efficient representation to compress the computation into an arithmetic tree.

Theorem 1. *Let* $\mathbf{B}$ *be a set of logical variables* $\{B_{ij} : i \in [1,n], j \in [1,k]\}$*. We define* β_{ij} *as follows:*

$$\begin{aligned} \beta_{ij} &\stackrel{def}{=} B_{ij} \wedge \textstyle\bigwedge_{t \in [1,j-1]} \neg B_{it} \text{ for all } j \in [1, k-1], \\ \beta_{ik} &\stackrel{def}{=} \textstyle\bigwedge_{t \in [1,k-1]} \neg B_{it} \end{aligned} \tag{4}$$

Let w_B *the weight for the variables defined by:*

$$w_B(B_{ij}) = \begin{cases} S_{ij}/(1 - \sum_{t \in [1,j-1]} S_{it}) & \text{if } \sum_{t \in [1,j-1]} S_{it} < 1 \\ 1 & \text{otherwise} \end{cases} \tag{5}$$

Given a set of constraints $\mathbb{C}$ *expressed using* β*. Then we have:*

$$Score(\mathbb{C}, S) = WMC(\beta \wedge \mathbb{C}, w_B) \tag{6}$$

Theorem 2. *With the definition of* β *by (4), the formula stating that each point belongs to a single cluster is expressed by (7) and is a tautology, i.e. (7)* $\equiv \top$.

$$\bigwedge_i (\beta_{i1} \wedge \ldots \wedge \beta_{ik} \bigwedge_{j,l \in [1,k]: j \neq l} (\neg \beta_{ij} \vee \neg \beta_{il})) \tag{7}$$

In (4), β_{ij} true means the point i is assigned to cluster j and not to any other cluster j' such that $j' < j$. In Theorem 2, the fact that (7) is a tautology means using β we ensure the result is a partition, which is required in a clustering problem. The proofs of the two theorems are given in the supplementary material[1].

Translation of expert constraints in terms of B The constraints are expressed in terms of β. For sake of efficiency, expert knowledge is expressed with B. For instance, a cannot-link constraint is written:

$$\wedge_{i \in [1,k]} (\neg \beta_{ui} \vee \neg \beta_{vi}) \iff \wedge_{i \in [1,k]} \left[\neg B_{ui} \vee \neg B_{vi} \vee_{t \in [1,i-1]} (B_{ut} \vee B_{vi}) \right]$$

[1] https://github.com/dung321046/Knowledge-Integration-in-Deep-Clustering.

3.4 Decomposition of the Problem

The score that we have defined in (2) takes into account all the constraints of $\mathbb{C}$ together in a single loss term. In order to have a greater impact when learning but also for complexity reasons, the whole problem is decomposed into a set of sub-problems c, one for each expert constraint $c \in \mathbb{C}$. Given a constraint $c \in \mathbb{C}$, we define $Score(c, S)$ in the same way as in (2):

$$Score(c, S) = \sum_{p \in \mathbb{P}_c} Score(\mathbf{p}, S)$$

where $\mathbb{P}_c$ is the set of partitions satisfying c, and we define $Score(\mathbb{C}, S)$ by:

$$Score(\mathbb{C}, S) = \prod_{c \in \mathbb{C}} Score(c, S) \tag{8}$$

3.5 Expert Loss

The expert loss L_{expert} is defined as

$$L_{expert} = -\log Score(\mathbb{C}, S) = -\sum_{c \in \mathbb{C}} \log Score(c, S) \tag{9}$$

4 Integrating Knowledge in Deep Clustering Frameworks

We present here our framework called DC-LK (Deep Clustering with Logical Knowledge), whose general scheme is shown in Fig. 1. Given $\mathbf{X}$ a set of n points, k a number of clusters and $\mathbb{C}$ a set of constraints expressing expert knowledge, it computes a cluster assignment $p = \{p_1, p_2, ..., p_n\}$, $p_i \in \{1, \ldots, k\}$, expressing that point i belongs to cluster p_i, guided by the expert constraints. First, expert constraints are formulated in logic and represented as SDD structures. Data $\mathbf{X}$ is processed through a deep clustering network thus computing a soft cluster assignment $S = (S_{ij})$ that represents the likelihood of point i to belong to cluster j. The expert constraint loss depending on S is computed by Weight Model Counting and this loss is integrated with the deep learner loss for back-propagation. Any deep clustering learner [4,12,23] that computes a soft cluster assignment S could be used.

We consider two methods for integrating the expert loss in a deep clustering learner. Since at each epoch there are two objectives, one is for the clustering task, the other one is to satisfy the constraint, we propose two main methods to combine them: separated back-propagation and joined back-propagation. The separated back-propagation calculates and backpropagates the clustering loss (which depending on the architecture can be composed of several losses) and the expert loss separately. In contrast, the joined method combines all the losses into a weighted sum for back-propagation.

In this work, we have integrated our expert loss into two deep clustering frameworks IDEC [12] and SCAN [23].

4.1 IDEC-LK

IDEC The neural structure of IDEC is an autoencoder, which allows to learn new representation of data $\mathbf{Z} = encode(\mathbf{X})$. K-means is applied to find the cluster centers $(\mu_1, \ldots, \mu_k)$ in the embedding space. The soft cluster assignments of all points (to all clusters) is computed based on Student's t-distribution:

$$S_{ij} = \frac{(1+ \parallel z_i - \mu_j \parallel^2)^{-1}}{\sum_{j'}(1+ \parallel z_i - \mu_{j'} \parallel^2)^{-1}} \tag{10}$$

For learning, IDEC uses the clustering loss and the reconstruction loss. The reconstruction loss is the mean square distance between the original data $\mathbf{X}$ and the reconstructed output $\tilde{\mathbf{X}} = decode(encode(\mathbf{X}))$.

$$L_{recon} = \sum_{i=1}^{n} \|x_i - \hat{x}_i\|^2 \tag{11}$$

The clustering loss is based on the Kullback-Leibler difference between the soft-assignment S_{ij} and an "ideal" target assignment P, which amplifies the separation between the clusters, defined by: $P_{ij} = (S_{ij}^2/f_j)/(\sum_{j'} S_{ij'}^2/f_{j'})$ where $f_j = \sum_{i=1}^{n} S_{ij}, j = 1, \ldots, k$ are the soft cluster frequencies. Then, the clustering loss is:

$$L_{clustering} = \sum_i \sum_j P_{ij} \log \frac{P_{ij}}{S_{ij}} \tag{12}$$

Expert Integration. We have implemented separated back-propagation for IDEC-LK, in order for constraints to have a stronger impact on learning. It may be less efficient in run time, since forward and backward are done twice, but this is not a problem, given the simple architecture of IDEC. Moreover, for efficiency reasons, we use mini-batch learning for constraint sets. So, the loss of IDEC-LK model is defined as:

$$\begin{aligned} L_1 &= \lambda_r \times L_{recon} + \lambda_c \times L_{clustering} \\ L_2 &= \lambda_e \times L_{expert} \end{aligned} \tag{13}$$

where λ_r, λ_c and λ_e are coefficients controlling each loss, L_{recon} is the reconstruction loss for the autoencoder, $L_{clustering}$ is the IDEC clustering loss based on KL divergence, L_{expert} is our expert loss. The detailed algorithm is given in 1.

4.2 SCAN-LK

SCAN. The neural structure of SCAN is a convolutional neural network (CNN), which is pretrained by SimCLR [7] using contrastive learning. With a suitable K value, it is observed that the K nearest neighbors of a point in the pretext embedding are instances of the same cluster. Let us denote $S_i \in \mathbb{R}^k_{[0,1]}$ the soft

Algorithm 1. Training process of IDEC-LK

Input: Input data: $\mathbf{X}$, Number of clusters: k, Constraint set: $\mathbb{C}$, Maximum iterations: $MaxIter$; Coefficients: λ_r, λ_e, λ_c
Output: Cluster assignment $\mathbf{p}$
1: Initialize parameters with pre-trained autoencoder
2: Initialize μ with K-means on the representations learned by pre-trained autoencoder
3: Generate $\mathbb{T}$ - a set of SDD structures from all $c \in \mathbb{C}$
4: **for** $iter := 1$ to $MaxIter$ **do**
5: **for** $batch := 1$ to $NumConstrainedBatches$ **do**
6: $X_{batch} = \{x : x \in C_{batch}\}$
7: Calculate $Z_{batch} = encode(X_{batch})$
8: Forward distribution S via t-distribution with Z, μ; (Eq. (10)) from the set of points $x \in C_{batch}$
9: Feed S to SDD structures $\mathbb{T}$ to calculate L^{batch}_{expert}
10: Backpropagate L_2 and update parameters
11: **end for**
12: **for** $batch := 1$ to $BatchAllInputs$ **do**
13: Calculate $Z_{batch} = encode(X_{batch})$
14: Forward distribution S_{batch} via t-distribution with Z_{batch}, μ (Eq. (10))
15: Calculate target distribution P_{batch}
16: Feed Z_{batch} to the decoder to obtain the reconstruction $\tilde{X}_{batch}$
17: Calculate L^{batch}_{recon}, $L^{batch}_{clustering}$, (Eq. (11), (12), respectively)
18: Backpropagate L_1 and update parameters
19: **end for**
20: **end for**
21: $\mathbf{p} = \{\arg\max_{j \in [1,k]} Q_{ij} : i \in [1, n]\}$
22: Return $\mathbf{p}$

assignment vector of x_i and $\mathcal{N}_{x_i}$ the neighborhood of x_i. The loss function is defined by:

$$L = \lambda_{nn} \times \frac{1}{n} \sum_{i=1}^{n} \sum_{x_j \in \mathcal{N}_{x_i}} \log\langle S_i \cdot S_j \rangle + \lambda_{entropy} \times \sum_{h=1}^{k} S^*_h \log S^*_h \qquad (14)$$

where $\lambda_{nn}, \lambda_{entropy}$ are the coefficients, $\langle \cdot \rangle$ is the dot product operator, $S^*_h = \frac{1}{n} \sum_{i=1}^{n} S_{ih}$. The first term enforces the similarity in predictions of x_i with its neighbors while the second term enforces the even distribution of points to all k clusters.

Expert Integration. In this case, we chose joined back-propagation because forward and backward operations are too expensive in SCAN architecture. This leads to a more complicated handling of constraints involved in batches (which are randomly generated in SCAN architecture). For each batch, the algorithm searches for expert constraints containing points in the batch and completes the batch by the points involved in these constraints. To limit the size of the batch, for each point at most one constraint involving it is chosen at random. The total loss of SCAN-LK model is defined as:

$$L = \lambda_{nn} \times L_{nearest} + \lambda_{entropy} \times L_{entropy} + \lambda_{expert} \times L_{expert} \tag{15}$$

where $\lambda_{nn}, \lambda_{entropy}$ and λ_{expert} are coefficients controlling each loss. When the stopping condition is reached (for instance the change of the loss is under a given threshold), a final assignment is computed for all i by taking $p_i = \arg\max_j S_{ij}$.

Algorithm 2. Training process of SCAN-LK

Input: Input data: $\mathbf{X}$, Number of clusters: k, Constraint set: $\mathbb{C}$, Maximum iterations: $MaxIter$; Coefficients: $\lambda_{nn}, \lambda_{entropy}, \lambda_{expert}$
Output: Cluster assignment $\mathbf{p}$
1: Initialize parameters with SimCLR
2: Generate $\mathbb{T}$ - a set of SDD structures from all $c \in \mathbb{C}$
3: **for** $iter := 1$ to $MaxIter$ **do**
4: **for** $batch := 1$ to $BatchAllInputs$ **do**
5: Load X_{batch} - points in the batch
6: Load C_{batch} - expert constraints containing X_{batch} and $X_{batch}^{constrained}$ - all points in C_{batch}
7: Forward $X_{batch}^{constrained}$ to obtain $S_{batch}^{constrained}$
8: Feed $S_{batch}^{constrained}$ to SDD structures $\mathbb{T}$ to calculate L_{expert}^{batch}
9: Calculate $L_{nearest}^{batch}$, $L_{entropy}^{batch}$
10: Backpropagate $L = \lambda_{nn} \times L_{nearest}^{batch} + \lambda_{entropy} \times L_{entropy}^{batch} + \lambda_{expert} \times L_{expert}^{batch}$ and update parameters
11: **end for**
12: **end for**
13: Forward all data $\mathbf{X}$ to obtain S.
14: $\mathbf{p} = \{\arg\max_{j \in [1,k]} S_{ij} : i \in [1, n]\}$
15: Return $\mathbf{p}$

---

5 Experiments

Experiments are conducted to address the following aims: (i) to evaluate IDEC-LK and SCAN-LK on clustering quality and to compare them with other systems on pairwise and triplet constraints; (ii) to evaluate our system on constraints that have never been used in deep clustering experiments, namely implication constraints and span-limited constraints. Our experiment could be reproduced using the avaiable source[2].

[2] https://github.com/dung321046/Knowledge-Integration-in-Deep-Clustering.

5.1 Experiment Settings

Datasets. We use five datasets, which are challenging and also used in many recent deep constrained clustering methods [14,30]. **MNIST** contains 70,000 handwritten single-digits from 10 classes. Among them 60,000 images are used to perform clustering, the remaining ones are used to evaluate the interest of span-limited constraints. Similarly, **STL10** has 5,000 color images for clustering and expert learning and 8,000 images for testing. **Fashion** has 60,000 images associated to a label from 10 classes. **CIFAR10** consists of 50,000 color images in 10 classes. **Reuters** contains around 810,000 English news stories labeled with a category tree [17].

Experiment Setting. For all the experiments, we first run SDAE [24] and SimCLR [7] for learning a new representation space. The same pre-trained model is given as input to the clustering algorithms. We compare our system to IDEC [12] (unconstrained), PCK-means [2] (pairwise constraints), MPCK-means [3] (pairwise constraints), and DCC [30] (pairwise and triplet constraints).

No supervised information is used for setting the parameters, therefore for SDAE, IDEC, SCAN and DCC, we use the default parameters. Our hyper-parameters are detailed in the supplementary. We put two stopping conditions: either when the maximum number of epochs is reached or when the percentage of assignments that differ from the previous epoch is less than 0.01%.

Experiments are run on a 2.6 GHz Intel Core i7 processor and a NVIDIA GeForce RTX 2060 graphics card.

5.2 Experiments and Analysis for Clustering Quality

In this section, we study the impact of pairwise and triplet constraints on the clustering quality. For all datasets, the true class of objects is available and we use it to evaluate the accuracy of the clustering. We consider two measures: Normalized Mutual Information (NMI) and clustering accuracy (ACC), with a one-to-one mapping between clusters and labels, computed by the Hungarian algorithm [16].

For testing the influence of pairwise constraints, we consider 4 numbers of constraints (10, 100, 500, 1000). For each test case, we randomly generate five sets of constraints, we run the system only once for each set of constraints and we report the mean and the standard deviation in Table 1 and Table 2. We do the same for triplet constraints (see Fig. 2). All results are given in the supplementary materials.

Pairwise Constraints. In MNIST, Fashion and Reuters datasets, IDEC-LK has competitive performances to the state-of-the-art methods. In terms of complexity, the times to run each epoch of DCC and of IDEC-LK are quite the same. However, the convergence of IDEC-LK is slower than for DCC (See Table 1).

In CIFAR10 and STL10, all the systems based on SDAE (IDEC, DCC, IDEC-LK) have a poor performance. The performance of IDEC with CIFAR10 data is reported in the supplementary material. With 1000 pairwise constraints, SCAN-LK helps to improve the clustering performance of CIFAR10 and STL10, with a ratio of 3% and 6% respectively.

Table 1. Comparison on clustering quality with 1000 pairwise constraints. Green (blue) numbers are for the best (second-best) values, respectively.

Data	Models	NMI	ACC	Time (s)
MNIST	DCC	0.8689 ± 0.0008	0.8815 ± 0.0007	277 ± 9
MNIST	MPCK-means	0.7589 ± 0.0171	0.7788 ± 0.0413	211 ± 3
MNIST	PCK-means	0.7463 ± 0.0228	0.7698 ± 0.0543	32.97 ± 15.90
MNIST	IDEC-LK	0.8680 ± 0.0017	0.8826 ± 0.0012	388 ± 27
Fashion	DCC	0.6000 ± 0.0019	0.5241 ± 0.0039	140 ± 16
Fashion	MPCK-means	0.5749 ± 0.0138	0.5312 ± 0.0292	205 ± 4
Fashion	PCK-means	0.5714 ± 0.0212	0.5314 ± 0.0293	37.02 ± 13.19
Fashion	IDEC-LK	0.6009 ± 0.0019	0.5230 ± 0.0034	358 ± 17
Reuters	DCC	0.5655 ± 0.0086	0.7477 ± 0.0030	3.46 ± 0.41
Reuters	MPCK-means	0.5262 ± 0.0330	0.7251 ± 0.0412	167 ± 2
Reuters	PCK-means	0.5174 ± 0.0288	0.7343 ± 0.0377	14.80 ± 2.34
Reuters	IDEC-LK	0.5927 ± 0.0105	0.7563 ± 0.0079	27.52 ± 9.88

Table 2. Comparison on clustering quality between the baselines and SCAN-LK with 1000 pairwise constraints.

Data	Models	NMI	ACC	#Unsat
CIFAR10	SCAN	68.30	79.39	183 ± 17
CIFAR10	SCAN-LK	71.81 ± 0.19	82.11 ± 0.27	55 ± 12.77
STL10	SCAN	65.11	75.58	194.67 ± 2.52
STL10	SCAN-LK	72.48 ± 0.79	83.57 ± 0.95	4.33 ± 0.58

Triplet Constraints. The triplet constraints have less impact on the clustering quality than the pairwise ones because they convey conditional information on the points. Relying on a Kolmogorov-Smirnov test with $p = 0.05$ [13], IDEC-LK has better clustering quality in Reuters while it has similar performances with DCC in the other datasets.

5.3 Experiments and Analysis for Constraint Satisfaction

In this section, we aim at illustrating two points: first our method can leverage complex domain knowledge and second, it can learn from it, that is, it does not only aim at satisfying the constraints but it is also able to satisfy unseen constraints of the same type. The second point is crucial because acquiring constraints is expensive, and the set of training constraints is only a minute fraction of all possible interpretations of the domain knowledge.

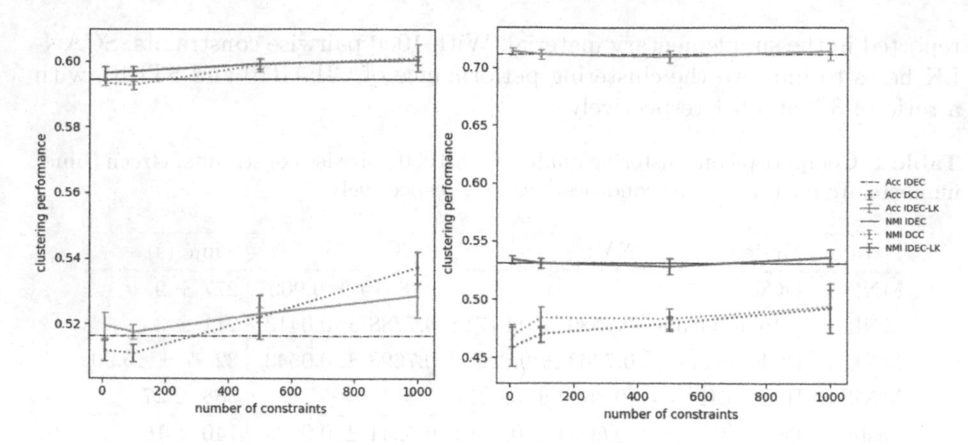

Fig. 2. The clustering performances with triplet constraints on Fashion (left) and Reuters (right) dataset

Implication Constraint. Introduced in Sect. 3.1, the first part (if-clause) is denoted as P and the second part (then-clause) as Q. To study the interest of such constraints, we generate 5 sets of 100 constraints at random based on the ground truth. For each constraint, the number of Together/Apart constraints in P is 3, the number of Together/Apart in Q is 1 and we define the notion of P-Q distribution: around 20% constraints satisfy $P = \bot$, the remaining 80% is $(P = \top, Q = \top)$.

Table 3. Comparison between IDEC and IDEC-LK on the satisfaction of implication constraints.

Data	Models	$\overline{Score}$	#Unsat
MNIST	IDEC	0.8777 ± 0.0118	13.6 ± 1.5
MNIST	IDEC-LK	0.8856 ± 0.0130	12.4 ± 1.7
Fashion	IDEC	0.7620 ± 0.0442	24.8 ± 4.6
Fashion	IDEC-LK	0.7743 ± 0.0449	23.2 ± 4.5
Reuters	IDEC	0.8290 ± 0.0376	17.8 ± 4.7
Reuters	IDEC-LK	0.8357 ± 0.0381	17.0 ± 5.0

In Table 3, IDEC-LK shows the improvement of the average constrained scores, denoted by $\overline{Score}$, compared to the value of IDEC. In all three datasets, the constraint satisfaction score has been improved, and the number of unsatisfied constrained is reduced.

Span-limited Constraint. In this experiment, we run IDEC-LK algorithm with MNIST and SCAN-LK with STL10. We aim at testing the interest of span-limited constraints that state that a group of points can be dispatched on a fixed number of clusters. For generating such constraints for MNIST dataset, we have selected four groups from the labels: $\mathcal{G} = \{G(3,9), G(6,8), G(1,7), G(2,5)\}$ (pairs of digits that share some similar shapes) and stated that elements in each group must be dispatched into two clusters. To create a group $G(u,v)$, we have selected randomly 100 images of either digit u or digit v. To test whether such constraints have a true impact, we chose them so that a quarter of them have been assigned to a wrong cluster by SDAE+K-Means (without constraints). For generating span-limited constraints for STL10 dataset, we have applied the same process with $\mathcal{G} = \{G(airplane, bird), G(cat, dog), G(deer, horse)\}$ and we have randomly selected 1,000 images for each group. Because the matching between labels and clusters is unknown, for each group $G(u,v)$, we set the two clusters to be the ones with the highest number of points in $G(u,v)$.

Figure 3 shows the changes in satisfaction of span-limited constraints. A quarter of the images in each spanning group initially does not belong to the two major clusters obtained with IDEC. After training with IDEC-LK with *all* points in all the four groups, the points that were already in these clusters remain, while the other points have changed to belong to one of the two clusters.

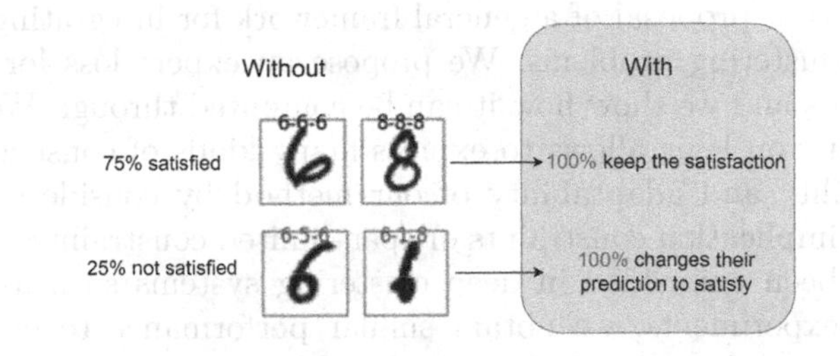

Fig. 3. Training results with span-limited constraints of cluster 6 and 8 in MNIST. IDEC (Without) vs IDEC-LK (With)

Let us notice that the model is learned using constraints on the train set. In order to analyse the capacity of generalizing constraints, we study the ability of the model to satisfy constraints on the test set. Figure 4 presents the satisfaction of span-limited constraints from a test set, which are unknown when training the model: we consider two datasets MNIST (10,000 test points, IDEC) and STL10 (8,000 test points, SCAN) and for each we compare the behavior of two models, respectively learned without/with the constraints on the train set. All data in the test set are neither used in the clustering process, nor in knowledge integration. We can observe that the number of unsatisfied constraints in the test set is reduced by 85.7% for MNIST and 46.79% for STL10.

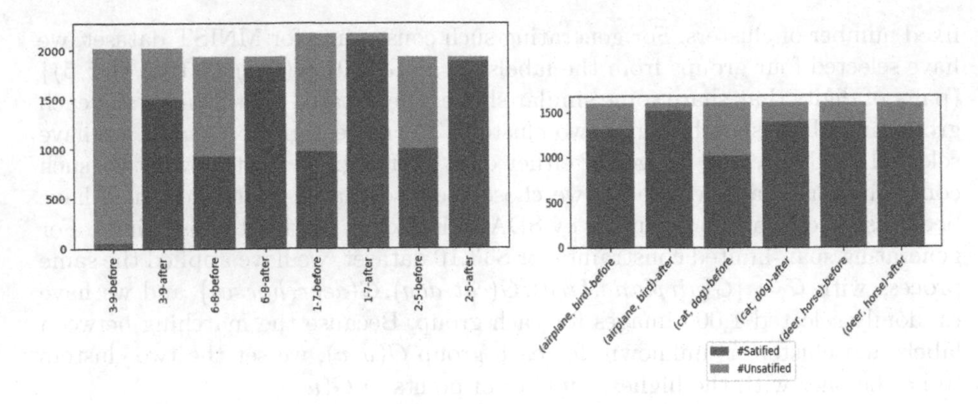

Fig. 4. Study on the effect of learning on constraints on test data: MNIST with IDEC-LK and STL10 with SCAN-LK, span-limited constraints

6 Conclusion

Our work is the first proposal of a general framework for integrating knowledge in constrained clustering problems. We propose an expert loss for integrating expert constraints and we show how it can be computed through Weight Model Counting. Relying on logic allows to express many kinds of constraints and we show the flexibility and adaptability of our method by considering new constraints such as implication constraints or span-limited constraints. This general framework has been embedded in deep clustering systems such as IDEC and SCAN. In our experiments, we obtain similar performance to other systems with well-known constraint types, but we also show the ability to integrate new constraints and even to generalize the constraints to unseen points. We plan to embed our proposal in other deep clustering architectures to show the generality of our approach.

The main limitation of this work is the complexity for computing the SDD trees, so that it prevents from incorporating cluster-level constraints or constructing a single loss for the whole constraint set. So we aim at reducing complexity by introducing new formulations or approximation schemes. However, let us notice that SDD trees are computed only once at the beginning of the learning process.

References

1. Babaki, B., Guns, T., Nijssen, S.: Constrained clustering using column generation. In: CPAIOR 2014, pp. 438–454 (2014)
2. Basu, S., Banjeree, A., Mooney, E., Banerjee, A., Mooney, R.J.: Active semi-supervision for pairwise constrained clustering. In: SDM, pp. 333–344 (2004)
3. Bilenko, M., Basu, S., Mooney, R.J.: Integrating constraints and metric learning in semi-supervised clustering. In: ICML 2004. pp. 11–18 (2004)

4. Bo, D., Wang, X., Shi, C., Zhu, M., Lu, E., Cui, P.: Structural deep clustering network. In: Proceedings of The Web Conference 2020, pp. 1400–1410 (2020)
5. Bradley, P., Bennett, K., Demiriz, A.: Constrained k-means clustering. Technical report MSR-TR-2000-65, Microsoft Research (2000)
6. Caron, M., Bojanowski, P., Joulin, A., Douze, M.: Deep clustering for unsupervised learning of visual features. In: Ferrari, V., Hebert, M., Sminchisescu, C., Weiss, Y. (eds.) Computer Vision – ECCV 2018. LNCS, vol. 11218, pp. 139–156. Springer, Cham (2018). https://doi.org/10.1007/978-3-030-01264-9_9
7. Chen, T., Kornblith, S., Norouzi, M., Hinton, G.: A simple framework for contrastive learning of visual representations. In: ICML, pp. 1597–1607. PMLR (2020)
8. Dao, T.B.H., Duong, K.C., Vrain, C.: Constrained clustering by constraint programming. Artif. Intell. **244**, 70–94 (2017)
9. Dao, T.B.H., Vrain, C., Duong, K.C., Davidson, I.: A framework for actionable clustering using constraint programming. In: ECAI 2016, pp. 453–461 (2016)
10. Darwiche, A.: SDD: a new canonical representation of propositional knowledge bases. In: IJCAI (2011)
11. Davidson, I., Ravi, S.S., Shamis, L.: A SAT-based framework for efficient constrained clustering. In: ICDM 2010, pp. 94–105 (2010)
12. Guo, X., Gao, L., Liu, X., Yin, J.: Improved deep embedded clustering with local structure preservation. In: IJCAI 2017, pp. 1753–1759 (2017)
13. Hodges, J.L.: The significance probability of the SMIRNOV two-sample test. Ark. Mat. **3**(5), 469–486 (1958)
14. Ienco, D., Pensa, R.G.: Deep triplet-driven semi-supervised embedding clustering. In: Kralj Novak, P., Šmuc, T., Džeroski, S. (eds.) DS 2019. LNCS (LNAI), vol. 11828, pp. 220–234. Springer, Cham (2019). https://doi.org/10.1007/978-3-030-33778-0_18
15. Jiang, Z., Zheng, Y., Tan, H., Tang, B., Zhou, H.: Variational deep embedding: An unsupervised and generative approach to clustering (2016)
16. Kuhn, H.W.: The Hungarian method for the assignment problem. Naval Res. Logist. Quart. **2**(1–2), 83–97 (1955)
17. Lewis, D.D., Yang, Y., Rose, T.G., Li, F.: Rcv1: a new benchmark collection for text categorization research. JMLR **5**, 361–397 (2004)
18. Lu, Z., Carreira-Perpinan, M.A.: Constrained spectral clustering through affinity propagation. In: IEEE CVPR, pp. 1–8. IEEE (2008)
19. Mueller, M., Kramer, S.: Integer linear programming models for constrained clustering. In: DS 2010, pp. 159–173 (2010)
20. Mukherjee, S., Asnani, H., Lin, E., Kannan, S.: Clustergan: latent space clustering in generative adversarial networks. In: Proceedings of the AAAI Conference on Artificial Intelligence, vol. 33, pp. 4610–4617 (2019)
21. Sang, T., Beame, P., Kautz, H.A.: Performing Bayesian inference by weighted model counting. In: AAAI, vol. 5, pp. 475–481 (2005)
22. Tang, W., Yang, Y., Zeng, L., Zhan, Y.: Optimizing MSE for clustering with balanced size constraints. Symmetry **11**(3), 338 (2019)
23. Van Gansbeke, W., Vandenhende, S., Georgoulis, S., Proesmans, M., Van Gool, L.: SCAN: learning to classify images without labels. In: Vedaldi, A., Bischof, H., Brox, T., Frahm, J.-M. (eds.) ECCV 2020. LNCS, vol. 12355, pp. 268–285. Springer, Cham (2020). https://doi.org/10.1007/978-3-030-58607-2_16
24. Vincent, P., Larochelle, H., Lajoie, I., Bengio, Y., Manzagol, P.A., Bottou, L.: Stacked denoising autoencoders: Learning useful representations in a deep network with a local denoising criterion. J. Mach. Learn. Res. **11**(12), 3371–3408 (2010)

25. Wagstaff, K., Cardie, C., Rogers, S., Schrödl, S.: Constrained K-means Clustering with Background Knowledge. In: ICML 2001, pp. 577–584 (2001)
26. Xie, J., Girshick, R., Farhadi, A.: Unsupervised deep embedding for clustering analysis. In: ICML 2016, pp. 478–487 (2016)
27. Xie, Y., Xu, Z., Kankanhalli, M.S., Meel, K.S., Soh, H.: Embedding symbolic knowledge into deep networks. In: NIPS, pp. 4233–4243 (2019)
28. Xing, E.P., Ng, A.Y., Jordan, M.I., Russell, S.: Distance metric learning with application to clustering with side-information. In: NIPS, vol. 15, p. 12 (2002)
29. Xu, J., Zhang, Z., Friedman, T., Liang, Y., Broeck, G.: A semantic loss function for deep learning with symbolic knowledge. In: ICML, pp. 5502–5511 (2018)
30. Zhang, H., Zhan, T., Basu, S., Davidson, I.: A framework for deep constrained clustering. Data Min. Knowl. Disc. **35**(2), 593–620 (2021). https://doi.org/10.1007/s10618-020-00734-4

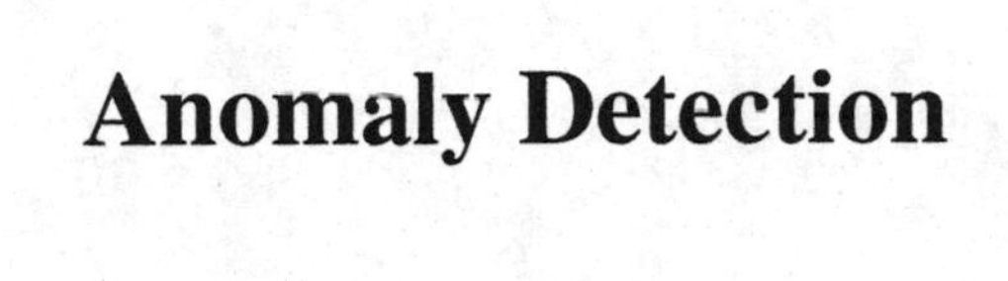
Anomaly Detection

ARES: Locally Adaptive Reconstruction-Based Anomaly Scoring

Adam Goodge[1,3](✉), Bryan Hooi[1,2], See Kiong Ng[1,2], and Wee Siong Ng[3]

[1] School of Computing, National University of Singapore, Singapore, Singapore
adam.goodge@u.nus.edu, {dcsbhk,seekiong}@nus.edu.sg
[2] Institute of Data Science, National University of Singapore, Singapore, Singapore
[3] Institute for Infocomm Research, A*STAR, Singapore, Singapore
wsng@i2r.a-star.edu.sg

Abstract. How can we detect anomalies: that is, samples that significantly differ from a given set of high-dimensional data, such as images or sensor data? This is a practical problem with numerous applications and is also relevant to the goal of making learning algorithms more robust to unexpected inputs. Autoencoders are a popular approach, partly due to their simplicity and their ability to perform dimension reduction. However, the anomaly scoring function is not adaptive to the natural variation in reconstruction error across the range of normal samples, which hinders their ability to detect real anomalies. In this paper, we empirically demonstrate the importance of local adaptivity for anomaly scoring in experiments with real data. We then propose our novel Adaptive Reconstruction Error-based Scoring approach, which adapts its scoring based on the local behaviour of reconstruction error over the latent space. We show that this improves anomaly detection performance over relevant baselines in a wide variety of benchmark datasets.

Keywords: Anomaly detection · Machine learning · Unsupervised learning

1 Introduction

The detection of anomalous data is important in a wide variety of applications, such as detecting fraudulent financial transactions or malignant tumours. Recently, deep learning methods have enabled significant improvements in performance on ever larger and higher-dimensional datasets. Despite this, anomaly detection remains a challenging task, most notably due to the difficulty of obtaining accurate labels of anomalous data. For this reason, supervised classification methods are unsuitable for anomaly detection. Instead, unsupervised methods are used to learn the distribution of an all-normal training set. Anomalies

Supplementary Information The online version contains supplementary material available at https://doi.org/10.1007/978-3-031-26387-3_12.

M.-R. Amini et al. (Eds.): ECML PKDD 2022, LNAI 13713, pp. 193–208, 2023.
https://doi.org/10.1007/978-3-031-26387-3_12

are then detected amongst unseen samples by measuring their closeness to the normal-data distribution.

Autoencoders are an extremely popular approach to learn the behaviour of normal data. The reconstruction error of a sample is directly used as its anomaly score; anomalies are assumed to have a higher reconstruction error than normal samples due to their difference in distribution. However, this anomaly score fails to account for the fact that reconstruction error can vary greatly even amongst different types of normal samples. For example, consider a sensor system in a factory with various activities on weekdays but zero activity on weekends. The weekday samples are diverse and complex, therefore we could expect the reconstruction error of these samples to vary greatly. Meanwhile, even a small amount of activity in a weekend sample would be anomalous, even if the effect on the reconstruction error is minimal. The anomaly detector would likely detect false positives (high reconstruction error) among weekday samples and false negatives among weekend samples. Although encoded within the input data attributes, the context of each individual sample (i.e. day of the week) is neglected at the detection stage by the standard scoring approach. Instead, all samples are assessed according to the same error threshold or standard. This invokes the implicit assumption that the reconstruction errors of all samples are identically distributed, regardless of any individual characteristics and context which could potentially influence the reconstruction error significantly.

In this work, we aim to address this problem by proposing **Adaptive Reconstruction Error-based Scoring (ARES)**. Our **locally-adaptive** scoring method is able to automatically account for any contextual information which affects the reconstruction error, resulting in more accurate anomaly detection. We use a flexible, **neighbourhood-based** approach to define the context, based on the location of its latent representation learnt by the model.

Our scoring approach is applied at test time, so it can be retrofitted to pre-trained models of any size and architecture or used to complement existing anomaly detection techniques. Our score is simple and efficient to compute, requiring very little additional computational time, and the code is available online[1]. In summary, we make the following contributions:

1. We **empirically show** with **real, multi-class data** the variation in reconstruction error among different samples in the normal set and why this justifies the need for local adaptivity in anomaly scoring.
2. We propose a novel anomaly scoring method which adapts to this variation by evaluating anomaly status based on the local context of a given sample in the latent space.
3. We evaluate our method against a wide range of baselines with various benchmark datasets. We also study the effect of different components and formulations of our scoring method in an ablation study.

[1] https://github.com/agoodge/ARES.

2 Background

2.1 Autoencoders

Autoencoders are neural networks that output a reconstruction of the input data [26]. They are comprised of two components: an encoder and decoder. The encoder compresses data from the input level into lower-dimensional latent representations through its hidden layers. The encoder output is typically known as the bottleneck, from which the reconstruction of the original data is found through the decoder hidden layers. The network is trained to minimise the reconstruction error over the training set:

$$\min_{\theta,\phi} \|\mathbf{x} - (f_\theta \circ g_\phi)(\mathbf{x})\|_2^2, \tag{1}$$

where g_ϕ is the encoder, f_θ the decoder. The assumption is that data lies on a lower-dimensional manifold within the high-dimensional input space. The autoencoder learns to reconstruct data on this manifold by performing dimension reduction. By training the model on only normal data, the reconstruction error of a normal sample should be low as it close to the learnt manifold on which it has been reconstructed by the autoencoder, whereas anomalies are far away and are reconstructed with higher error.

2.2 Local Outlier Factor

The Local Outlier Factor (LOF) method is a neighbourhood-based approach to anomaly detection; it measures the density of a given sample relative to its k nearest neighbours' [8]. Anomalies are assumed to be in sparse regions far away from the one or more high-density clusters of normal data. A lower density therefore suggests the sample is anomalous. The original method uses the 'reachability distance'; defined for a point A from point B as:

$$\max\{k\text{-distance}(B), d(A,B)\} \tag{2}$$

where $d(\cdot,\cdot)$ is a chosen distance metric and k-distance(B) is the distance of B to its kth nearest neighbour. The local reachability density and subsequently the local outlier factor of A based on its set of neighbours $\mathsf{N}_k(A)$, is:

$$\text{lrd}_k(A) := \left(\frac{\sum_{B \in \mathsf{N}_k(A)} \text{reachability-distance}_k(A,B)}{|\mathsf{N}_k(A)|} \right)^{-1}. \tag{3}$$

$$\text{LOF}_k(A) = \frac{\sum_{B \in \mathsf{N}_k(A)} \text{lrd}_k(B)}{|\mathsf{N}_k(A)| \cdot \text{lrd}_k(A)}, \tag{4}$$

3 Related Work

Anomalies are often assumed to occupy sparse regions far away from high-density clusters of normal data. As such, a great variety of methods detect anomalies by measuring the distances to their nearest neighbours, most notably KNN [35], which directly uses the distance of a point to its kth nearest neighbour as the anomaly score. LOF [8] measures the density of a point relative to the density of points in its local neighbourhood, as seen in Sect. 2. This is beneficial as a given density may be anomalous in one region but normal in another region. This local view accommodates the natural variation in density and therefore allows for the detection of more meaningful anomalies. More recently, deep models such as graph neural networks have been used to learn anomaly scores from neighbour-distances in a data-driven way [15]. Naturally, this relies on an effective distance metric. This is often itself a difficult problem; even the most established metrics have been shown to lose significance in high-dimensional spaces [6] due to the 'curse of dimensionality'.

Reconstruction-based methods rely on deep models, such as autoencoders, to learn to reconstruct samples from the normal set accurately. Samples are then flagged as anomalous if their reconstruction error higher than some pre-defined threshold, as it is assumed that the model will reconstruct samples outside of the normal set with a higher reconstruction error [4,9,13,14,26].

More recent works have used autoencoders or other models are deep feature extractors, with the learnt features then used in another anomaly detection module downstream, such as Gaussian mixture models [7], DBSCAN [3], KNN [5] and auto-regressive models [1]. [11] calibrates different types of autoencoders against the effects of varying hardness within normal samples.

Other methods measure the distance of a sample to a normal set-enclosing hypersphere [25,30]. or its likelihood under a learnt model [12,24,34]. Generative adversarial networks have also been proposed for anomaly detection, mostly relying either on the discriminator score or the accuracy of the generator for a given sample to determine anomalousness [2,33].

In all of these methods, the normal set is often restricted to a subset of the available data; e.g. just one class from a multi-class dataset in experiments. In practice, normal data could belong to multiple classes or modes of behaviour which all need to be modelled adequately. As such, developing anomaly detection methods that can model a diverse range of normal behaviours and adapt their scoring appropriately are important.

4 Methodology

4.1 Problem Definition

We assume to have m normal training samples $\mathbf{x}_1^{(\text{train})}, \cdots, \mathbf{x}_m^{(\text{train})} \in \mathbb{R}^d$ and n testing samples, $\mathbf{x}_1^{(\text{test})}, \cdots, \mathbf{x}_n^{(\text{test})} \in \mathbb{R}^d$, each of which may be normal or anomalous. For each test sample $\mathbf{x}$, our algorithm should indicate how anomalous it is through computing its **anomaly score** $s(\mathbf{x})$. The goal is for anomalies to be given higher anomaly scores than normal points. In this work, the fundamental question is:

Given an autoencoder with encoder g_ϕ *and decoder* f_θ*, how can we use the latent encoding* $\mathbf{z} = g_\phi(\mathbf{x})$ *and reconstruction* $\hat{\mathbf{x}} = (f_\theta \circ g_\phi)(\mathbf{x})$ *of a sample* $\mathbf{x}$ *to score its anomalousness?*

Our approach is not limited to standard autoencoders, and can be applied to other models that involve a latent encoding $\mathbf{z}$ and a reconstruction $\hat{\mathbf{x}}$.

In practice, the anomaly score is compared with a user-defined anomaly threshold; samples which exceed this threshold are flagged as anomalies. Different approaches can be employed to set this threshold, such as extreme value theory [28]. Our focus is instead on the approach to anomaly scoring itself, which allows for any choice of thresholding scheme to be used alongside it.

4.2 Statistical Interpretation of Reconstruction-Based Anomaly Detection

In the standard approach, with the residual as $\boldsymbol{\varepsilon} := \mathbf{x} - \hat{\mathbf{x}}$, the anomaly score is:

$$R(\boldsymbol{\varepsilon}) := \|\boldsymbol{\varepsilon}\|_2^2 = \|\mathbf{x} - \hat{\mathbf{x}}\|_2^2. \tag{5}$$

This can be seen as a negative log-likelihood-based score that assumes $\boldsymbol{\varepsilon}$ follows a Gaussian distribution with zero mean and unit variance: $\boldsymbol{\varepsilon} \sim \mathcal{N}(0, \mathbf{I})$:

$$-\log P(\boldsymbol{\varepsilon}) = \frac{d}{2}\log(2\pi) + \frac{R(\boldsymbol{\varepsilon})}{2} \tag{6}$$

The negative log-likelihood is an intuitive anomaly score because anomalous samples should have lower likelihood, thus higher negative log-likelihood.

The key conclusion, ignoring the constant additive and multiplicative factors, is that $R(\boldsymbol{\varepsilon})$ can be equivalently seen as a negative log-likelihood-based score, based on a model with the implicit assumption of constant mean as well as *constant variance residuals across all samples*, known in the statistical literature as **homoscedasticity**. Most crucially, this assumption applies to all $\mathbf{x}$ regardless of its individual latent representation, $\mathbf{z}$, which implies that the reconstruction error of $\mathbf{x}$ does not depend on the location of $\mathbf{z}$ in the latent space. We question the validity of this assumption in real datasets.

In our earlier example of a factory sensor system, the variance of a sample $\mathbf{x}$ depends greatly on its characteristics (i.e. day of the week), with high variance during weekdays and low variance on weekends. This is instead a prime example of **heteroscedasticity**. Furthermore, since the latent representation $\mathbf{z}$ typically encodes these important characteristics of $\mathbf{x}$, there is likely to be a clear dependence between $\mathbf{z}$ and the reconstruction error $R(\boldsymbol{\varepsilon})$, which will be examined in Sect. 4.3. This relationship could be exploited to improve the modelling of reconstruction errors and thus detection accuracy by adapting the anomaly score to this relationship at the sample-level.

4.3 Motivation and Empirical Results

In this section, we train an autoencoder on all 10 classes of `MNIST` to empirically examine the variation in the reconstruction error of real data to scrutinise the homoscedastic assumption. In doing so, we exemplify the shortcomings of the standard approach which implicitly assumes all reconstruction errors from a fixed Gaussian distribution. The autoencoder architecture and training procedure is the same as those described in Sect. 5. We make the following observations:

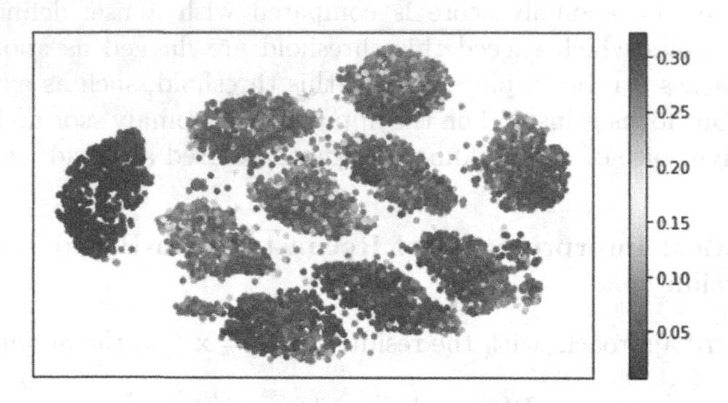

Fig. 1. t-SNE plot of the latent encodings with colour determined by reconstruction error of the associated sample.

1. **Inter-class variation:** In Fig. 1, the t-SNE [21] projections of the latent representations of training points are coloured according to their reconstruction error. We see that the autoencoder learns to separate each class approximately into distinct clusters. In this context, the class label can be seen as a variable characteristic between different samples within the normal set (similar to the weekday vs. weekend example).
2. **Intra-class variation:** Also in Fig. 1, we can see that there is significant variation even within any single cluster. In other words, there are noticeably distinct regions of high (and low) reconstruction errors within each individual cluster. This shows that there are additional characteristics that influence whether a given normal sample has a high or low reconstruction error besides its class label.
   We observe that there is significant variation in reconstruction error between the different clusters. Most notably, samples in the leftmost cluster (corresponding to class 1) has significantly lower reconstruction errors than most others. Indeed, as shown in Fig. 2, the distribution of reconstruction errors associated with class 1 samples is very different to those of the other classes. This shows that there is significant variation in reconstruction errors between classes, and that it is inappropriate to assume that the reconstruction errors of all samples can be modelled by a single, fixed Gaussian distribution. For example, a reconstruction error of 0.06 would be high for a class 1 or class 9 sample, but low for a class 2 or class 8 sample.

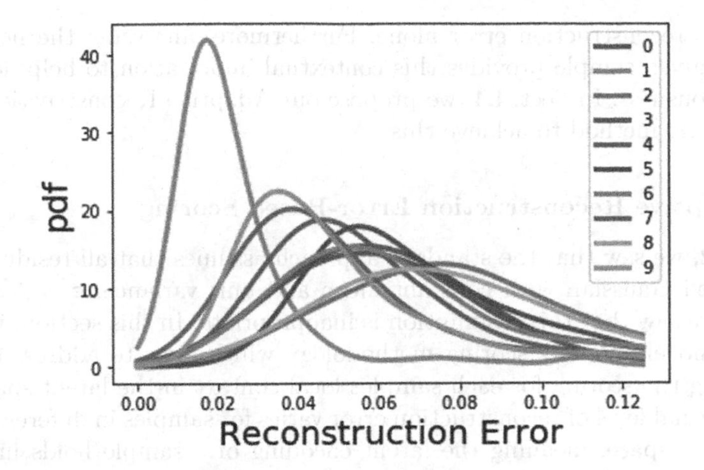

Fig. 2. Probability density function of reconstruction errors associated with different classes of training samples, estimated via kernel density estimation.

3. **Neighbourhood correlation:** In Fig. 3, we plot the reconstruction error of test samples against the average reconstruction error of its nearest training set neighbours in the latent space (neighbourhood error). We see that the reconstruction error of a test point increases as its neighbourhood error increases. Furthermore, the variance in test errors increases for larger neighbourhood errors: a clear heteroscedastic relationship. This information is useful in determining anomalousness: a test error of 0.1 would be anomalously high if its neighbourhood error is 0.02, but normal if it is 0.06.

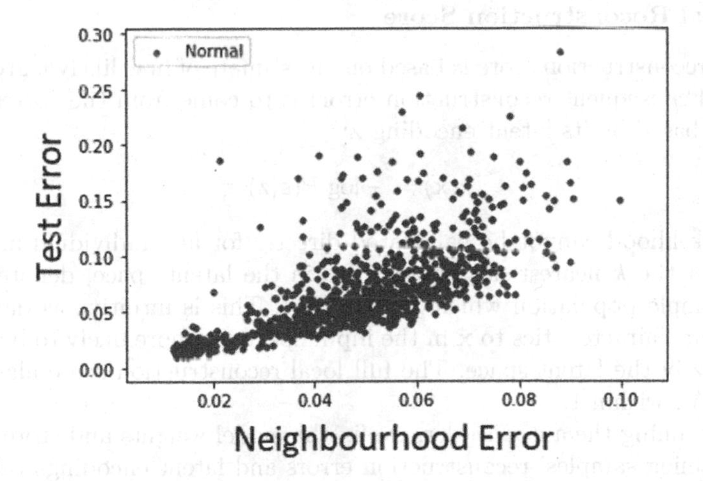

Fig. 3. Average reconstruction error of the training set nearest neighbours of a point versus its own reconstruction error for normal test samples.

Given these observations, we propose that anomalies could be more accurately detected by incorporating contextual information into the anomaly scoring beyond reconstruction error alone. Furthermore, analysing the neighbourhood of a given sample provides this contextual information to help determine its anomalousness. In Sect. 4.4, we propose our Adaptive Reconstruction Error-based Scoring method to achieve this.

4.4 Adaptive Reconstruction Error-Based Scoring

In Sect. 4.2, we saw that the standard approach assumes that all residuals come from a fixed Gaussian with constant mean and unit variance: $\varepsilon \sim \mathcal{N}(0, \mathbf{I})$. In Sect. 4.3, we saw that this assumption is inappropriate. In this section, we detail ARES, a novel anomaly scoring methodology which aims to address this flaw by adapting the scoring for each samples local context in the latent space.

The normal level of reconstruction error varies for samples in different regions of the latent space, meaning the latent encoding of a sample holds important information regarding anomalousness. As such, ARES is inspired by the **joint-likelihood** of a samples residual ε with its latent encoding $\mathbf{z}$, defined as follows:

$$-\log P(\varepsilon, \mathbf{z}) = -\log P(\varepsilon|\mathbf{z}) - \log P(\mathbf{z}). \tag{7}$$

The first term, $-\log P(\varepsilon|\mathbf{z})$, is the conditional (negative log-) likelihood of the points residual conditioned on its latent encoding. The second term, $-\log P(\mathbf{z})$, is the likelihood of observing the latent encoding from the normal set.

We now detail our approach to interpret these terms into tractable, efficient scores which we name the **local reconstruction score** and **local density score** respectively. These scores are combined to give the overall ARES anomaly score.

4.5 Local Reconstruction Score

The local reconstruction score is based on an estimate of how likely a given residual ε (and consequent reconstruction error) is to come from the corresponding sample $\mathbf{x}$, based on its latent encoding $\mathbf{z}$:

$$r(\mathbf{x}) = -\log P(\varepsilon|\mathbf{z}) \tag{8}$$

This likelihood cannot be calculated directly for any individual $\mathbf{z}$. Instead, we consider the k nearest neighbours of $\mathbf{z}$ in the latent space, denoted $\mathsf{N}_k(\mathbf{z})$, to be a sample population which z belongs to. This is intuitive as data points with similar characteristics to $\mathbf{x}$ in the input space are more likely to be encoded nearer to $\mathbf{z}$ in the latent space. The full local reconstruction score algorithm is shown in Algorithm 1.

After training the autoencoder, we fix the model weights and store in memory all training samples' reconstruction errors and latent encodings taken from the bottleneck layer. For a test sample $\mathbf{x}$, we find its own latent encoding $\mathbf{z}$ and reconstruction $\hat{\mathbf{x}}$. We then find $\mathsf{N}_k(\mathbf{z})$, the set of k nearest neighbours to $\mathbf{z}$ amongst the training set encodings. We only find nearest neighbours among the training samples as they are all assumed to be normal and therefore should have reconstruction errors within the normal range.

We want to obtain an estimate of the reconstruction error that could be expected of $\mathbf{x}$, assuming its a normal point, based on the reconstruction errors of its (normal) neighbours. We can then compare this expected value, conditioned on its unique location in the latent space, with the points true reconstruction error to determine its anomalousness.

In a fully probabilistic approach, we could do this by measuring the likelihood of the test points reconstruction error under a probability distribution, e.g. a Gaussian, fit to the neighbours' reconstruction errors. However, it is unnecessarily restrictive to assume any closed-form probability distribution to adequately model this population for the unique neighbourhood of each and every test sample. Instead, we opt for a non-parametric approach; we measure the difference between the test points reconstruction error and the median reconstruction error of its neighbouring samples. The larger the difference between them, the more outlying the test point is in comparison to its local neighbours, therefore the more likely it is to be anomalous. In practice, using the median was found to perform better than the mean as it is more robust to extrema. With this, we obtain the local reconstruction score as:

$$r(\mathbf{x}) = \|\mathbf{x} - \hat{\mathbf{x}}\|_2^2 - \underset{\mathbf{n} \in \mathsf{N}_k(\mathbf{z})}{\mathsf{median}}(\|\mathbf{n} - \hat{\mathbf{n}}\|_2^2). \tag{9}$$

Nearest neighbour search in the latent space is preferable over the input space as dimension reduction helps to alleviates the curse of dimensionality (see Sect. 3), resulting in more semantically-meaningful neighbors. Secondly, from a practical perspective, the neighbour search is less time-consuming and computationally intensive in lower dimensional spaces.

Algorithm 1. Local Reconstruction Score

Input: Autoencoder $A(\cdot) = (f_\theta \circ g_\phi)(\cdot)$, training set $\mathbf{X}_{\text{train}}$, test sample $\mathbf{x}_{\text{test}}$
Parameters: neighbour count k
Output: Local reconstruction score $r(\mathbf{x}_{\text{test}})$

1: Train autoencoder A on training set according to: $\min_{\theta,\phi} \|\mathbf{X}_{\text{train}} - \hat{\mathbf{X}}_{\text{train}}\|_2^2$ where $\hat{\mathbf{x}}^i_{\text{train}} = (f_\theta \circ g_\phi)(\mathbf{x}^i_{\text{train}})$ for $\mathbf{x}^i_{\text{train}} \in \mathbf{X}_{\text{train}}$.
2: Find latent encoding of $\mathbf{x}_{\text{test}}$: $\mathbf{z}_{\text{test}} := g_\phi(\mathbf{x}_{\text{test}})$
3: Find the set of k nearest neighbours to $\mathbf{z}_{\text{test}}$ among latent encodings of the training data: $\mathsf{N}_k(\mathbf{z}_{\text{test}}) := \{\mathbf{z}^1_{\text{train}}, ..., \mathbf{z}^k_{\text{train}}\}$ where $\mathbf{z}^i_{\text{train}} = g_\phi(\mathbf{x}^i_{\text{train}}) \in$ for $i = \{1, ..., k\}$.
4: **return** $r(\mathbf{x}_{\text{test}}) = \|\mathbf{x}_{\text{test}} - \hat{\mathbf{x}}_{\text{test}}\|_2^2 - \underset{\mathbf{n} \in \mathsf{N}_k(\mathbf{z})}{\mathsf{median}}(\|\mathbf{x}^i_{\text{train}} - \hat{\mathbf{x}}^i_{\text{train}}\|_2^2)$

4.6 Local Density Score

The local reconstruction score corresponds to the conditional term of the joint likelihood. We now introduce the local density score, which corresponds to the likelihood of observing the given encoding in the latent space. This is a density estimation task, which concerns the relative distance of $\mathbf{z}$ to its nearest neighbours, unlike the local reconstruction score which focuses on the reconstruction error of neighbours. Anomalies are assumed to exist in sparse regions, where normal samples are unlikely to be found in significant numbers. Any multivariate distribution P, with trainable parameters Θ, could be used to estimate this density:

$$d(\mathbf{x}) := -\log P(\mathbf{z}; \Theta) \tag{10}$$

Note that it is common to ignore constant factor shifts in the anomaly score. Thus, the distribution P need not be normalized; even unnormalized density estimation techniques can be used as scoring functions. We note that LOF is an example of an unnormalized score which is similarly locally adaptive like the local reconstruction score, so LOF is used for the local density score in our main experiments. Other methods are also tested and their performance is shown in the ablation study.

The overall anomaly score for sample $\mathbf{x}$ is:

$$s(\mathbf{x}) := r(\mathbf{x}) + \alpha d(\mathbf{x}), \tag{11}$$

where $r(\mathbf{x})$ is its local reconstruction score and $d(\mathbf{x})$ the local density score. These two scores are unnormalized, so we use a scaling factor α to balance the relative magnitudes of the two scores. We heuristically set it equal to 0.5 for all datasets and settings for simplicity, as this was found to balance the two scores sufficiently fairly in most cases. We choose not to treat α as a hyper-parameter to be tuned to optimise performance, although different values could give better performance for different datasets. The effect of changing α is shown in the supplementary material.

Computational Runtime: The average runtimes of experiments with the `MNIST` dataset can be found in the supplementary material. We see that, despite taking longer than the standard reconstruction error approach, the additional computational runtime of ARES anomaly scoring is insignificant in relation to the model training time. Anomaly scoring with ARES is just 1.2% of the overall time taken (including training) in the one-class case (0.017 min for scoring versus 1.474 min for training), and 3.5% in the multi-class case. The additional run-time is a result of the k nearest neighbour search. An exact search algorithms would be $\mathcal{O}(nm)$ with n train and m test samples, however approximate methods can achieve near-exact accuracy much more efficiently.

5 Experiments

In our experiments, we aim to answer the following research questions:

RQ1 (Accuracy): Does ARES perform better than existing anomaly detection methods?

RQ2 (Ablation Study): How do different components and design choices of ARES contribute to its performance?

Table 1. Name and descriptions of the datasets used in experiments, including the number of samples and dimensions.

Dataset	#Dim	#Classes	#Samples	Description
SNSR [31]	48	Multi-class	58,509	Electric current signals
MNIST [19]	784	Multi-class	70,000	0-9 digit images
FMNIST [32]	784	Multi-class	70,000	Fashion article images
OTTO [22]	93	Multi-class	61,878	E-commerce types
MI-F [29]	58	Single-class	25,286	CNC milling defects
MI-V [29]	58	Single-class	23,125	CNC milling defects
EOPT [16]	20	Single-class	90,515	Storage system failures

5.1 Datasets and Experimental Setup

Table 1 shows the datasets we use in experiments. The single-class datasets consist of ground truth normal-vs-anomaly labels, as opposed to the multiple class labels in multi-class datasets. In the latter, we distinguish between the 'one-class normality' setup, in which one class label is used as the normal class and all other classes are anomalous. Alternatively, in the 'multi-class normality' setup, one class is anomalous and all other classes are normal. For a dataset with N classes, there are N possible arrangements of normal and anomaly classes. We train separate models for each arrangement and find their average score for the final result. We use the Area-Under-Curve (AUC) metric to measure performance as it does not require an anomaly score threshold to be set.

As anomaly scores are calculated for each test sample independently of each other, the proportion of anomalies in the test set has no impact on the anomaly score assigned to any given sample. Therefore, we are able to use a normal:anomaly ratio of 50:50 in our experiments for the sake of simplicity and an unbiased AUC metric. Besides the normal sampels in the test set, the remaining normal samples are split 80:20 into training and validation sets for all models. Full implementation details can be found in the supplementary material.

5.2 Baselines

We test the performance of ARES against a range of baselines. We use the scikit learn implementations of LOF (in the input space) [8], IForest [20], PCA [27] and OC-SVM [10]. Publicly available codes are used for DAGMM [7], RAPP-SAP and RAPP-NAP [17], and we use Pytorch to build the autoencoder (AE and ARES) and variational autoencoder (VAE). All experiments were conducted in Windows OS using an Nvidia GeForce RTX 2080 Ti GPU.

We do not tune hyper-parameters relating to the model architectures or training procedures for any method. The effect of variation in hyper-parameters is studied in the ablation study in Sect. 5.4 and the supplementary material instead. We set the number of neighbours $k = 10$ for both the local density and local reconstruction score in our main experiments.

5.3 RQ1 (Accuracy)

Table 2 shows the average AUC scores as a percentage (i.e. multiplied by 100). In the one-class normality setting, ARES significantly improves performance over the baselines in all multi-class datasets besides FMNIST. In the single-class datasets, this improvement is even greater, e.g. +8% lift for MI-F and EOPT. Compared with AE, we see that local adaptivity helps to detect true anomalies by correcting for the natural variation in reconstruction error in the latent space. This effect is even more pronounced in the multi-class normality setting, where ARES gives the best performance on all datasets. Here, the normal set is much more diverse and therefore local adaptivity is even more important. We show the standard deviations and additional significance test scores in the supplementary material, which also show statistically significant ($p < 0.01$) improvement.

Table 2. Mean AUC scores for each datasets and normality setting. The best scores are highlighted in bold and we mark the most significant improvements over AE ($p < 0.01$) with **. Further tests are in the supplementary material.

Dataset	LOF	IFOREST	PCA	OC-SVM	SAP	NAP	DAGMM	VAE	AE	ARES
One-class Normality										
SNSR	97.98	89.16	92.01	95.85	98.79	98.74	88.08	89.49	98.30	**98.83***
MNIST	96.85	85.44	95.68	90.35	95.35	97.25	89.60	91.73	96.96	**97.89****
FMNIST	91.35	91.39	90.13	90.74	89.66	**93.08**	87.97	77.59	92.33	91.63
OTTO	84.76	70.34	80.09	81.43	81.61	82.77	68.02	82.39	85.26	**87.86****
MI-F	59.79	81.53	55.07	76.69	81.78	80.61	82.23	76.93	71.19	**89.52**
MI-V	83.97	84.35	87.32	83.58	88.24	89.35	75.45	89.03	90.75	**93.94**
EOPT	55.01	61.61	54.72	59.66	59.87	61.69	60.63	68.08	59.85	**68.43**
Multi-class Normality										
SNSR	60.74	52.70	52.94	52.79	57.52	58.32	4.77	61.36	57.28	**69.78****
MNIST	77.40	56.49	70.41	58.56	84.76	86.38	54.24	84.12	80.04	**93.25****
FMNIST	71.50	64.75	66.92	60.27	68.50	72.09	57.56	71.18	71.03	**72.49**
OTTO	63.01	54.14	58.27	62.96	57.47	63.44	58.96	61.88	59.59	**63.54**

The neighbours of a normal sample with high reconstruction error tend to be have high reconstruction errors themselves. By basing the anomaly scoring on their relative difference, ARES uses this to better detect truly anomalous samples. Furthermore, ARES also uses the local density of the point, which depends purely on its distance to the training samples in the latent space. This is important as there may be some anomalies with such low reconstruction error that comparison with neighbours does not alone indicate anomalousness (for example the weekend samples mentioned earlier). These samples could be expected to be occupy very sparse regions of the latent space due to their significant deviation from the normal set, which means they can be better detected through the local density score. By combining these two scores, we are able to detect a wider range of anomalous data than either could individually.

5.4 RQ2 (Ablation Study)

Density Estimation Method. Table 3 shows the performance of ARES with other density estimation methods. KNN is the distance to the kth nearest neighbour in the latent space with $k = 20$. GD is the distance of a point to the closest of N Gaussian distributions fit to the latent encodings of samples for each of the classes in the training set ($N = 1$ in the one-class normality case). NF is the likelihood under a RealNVP normalizing flow [23].

Table 3. Mean AUC scores for each choice of local density score. The best scores are highlighted in bold.

Dataset	LOF	KNN	GD	NF
One-class Normality				
SNSR	**98.83**	98.66	95.68	98.29
MNIST	**97.89**	95.24	87.29	97.30
FMNIST	91.03	**92.94**	90.26	92.14
OTTO	**87.76**	83.14	81.27	86.74
MI-F	**89.52**	76.12	80.41	84.47
MI-V	**93.94**	91.60	79.61	92.89
EOPT	**68.43**	67.63	63.35	61.07
Multi-class Normality				
SNSR	**69.78**	67.42	61.86	60.63
MNIST	**93.25**	92.92	91.14	86.25
FMNIST	72.49	72.77	**73.02**	70.48
OTTO	63.54	61.71	**66.30**	63.67

LOF performs best overall, closely followed by KNN. GD performs poorly in one-class experiments, however it is better in multi-class normality experiments and even the best for FMNIST and OTTO. This could be as the use of multiple distributions provides more flexbility. NF is noticeably worse; previous studies have found that normalizing flows are not well-suited to detect out-of-distribution data [18].

In the supplementary material, we further test these density estimation methods by varying their hyper-parameters and find LOF to still come out best. Further supplementary experiments show that the local density score generally performs better than the local reconstruction score in the multi-class normality case and vice versa in the one-class case. Combining them, as in ARES, generally gives the best performance overall across different latent embedding sizes.

Robustness to Training Contamination. In practice, it is likely that a small proportion of anomalies 'contaminate' the training set. In this section, we study the effect of different levels of contamination on ARES, defined as $n\%$ of the total number of samples in the training set. Table 4 shows that increasing n worsens performance overall. With more anomalies in the training set, it is more likely that anomalies are found in the nearest neighbour set of more test points, which skews both the average neighbourhood reconstruction error as well as their density in the latent space and degrades performance.

Table 4. Mean AUC scores in MNIST with training set anomaly contamination (n%) and different neighbourhood sizes. The best scores are highlighted in bold.

	Neighbourhood size				
n	10	50	100	200	500
One-class Normality					
0	**97.89**	97.85	97.80	97.73	97.56
0.5	95.77	**96.08**	96.06	95.95	95.72
1	82.50	94.45	**95.97**	95.81	95.42
2	90.65	92.88	93.29	**93.33**	92.99
3	88.41	91.35	91.92	**92.07**	91.86
5	84.91	87.85	89.20	90.26	**90.65**
10	79.69	81.51	83.16	84.88	**86.82**
Multi-class Normality					
0	**93.25**	92.71	92.08	91.31	89.68
0.5	68.38	76.07	78.36	80.75	**81.42**
1	52.38	58.42	67.62	65.75	**68.91**
2	59.50	63.13	65.71	67.98	**69.68**
3	57.09	59.00	60.78	62.89	**65.35**
5	50.86	51.48	52.40	53.47	**55.78**
10	**52.50**	50.80	50.87	51.14	52.11

We find that ARES is more robust to training set contamination with a higher setting of the number of neighbours (k). In the one-class setup, we see that ARES performs better with higher values of k as the proportion of anomalies increases. By using more neighbours, the effect of any individual anomalies on the overall neighbourhood error is reduced, which helps to maintain better performance. This effect is even more stronger in the multi-class normality setup. The highest neighbour count of $k = 500$ gives the best performance in all cases except for $n = 0\%$ and 10%. As the multi-class training sets are much larger than their one-class normality counterparts, $n\%$ corresponds to a much larger number of anomalous contaminants, which explains their greater effect for a given k.

6 Conclusion

Autoencoders are extremely popular deep learning models used for anomaly detection through their reconstruction error. We have shown that the assumption made by the standard reconstruction error score, that reconstruction errors are identically distributed for all normal samples, is unsuitable for real datasets. We empirically show that there is a heteroscedastic relationship between latent space characteristics and reconstruction error, which demonstrates why adaptivity to local latent information is important for anomaly scoring. As such, we have developed a novel approach to anomaly scoring which adaptively evaluates the anomalousness of a samples reconstruction error, as well as its density in the latent space, relative to those of its nearest neighbours. We show that our approach results in significant performance improvements over the standard approach, as well as other prominent baselines, across a range of real datasets.

Acknowledgements. This work was supported in part by NUS ODPRT Grant R252-000-A81-133.

References

1. Abati, D., Porrello, A., Calderara, S., Cucchiara, R.: Latent space autoregression for novelty detection. In: ICCV, pp. 481–490 (2019)
2. Akcay, S., Atapour-Abarghouei, A., Breckon, T.P.: GANomaly: semi-supervised anomaly detection via adversarial training. In: Jawahar, C.V., Li, H., Mori, G., Schindler, K. (eds.) ACCV 2018. LNCS, vol. 11363, pp. 622–637. Springer, Cham (2019). https://doi.org/10.1007/978-3-030-20893-6_39
3. Amarbayasgalan, T., Jargalsaikhan, B., Ryu, K.H.: Unsupervised novelty detection using deep autoencoders with density based clustering. Appl. Sci. **8**(9), 1468 (2018)
4. An, J.: Variational autoencoder based anomaly detection using reconstruction probability. In: SNU Data Mining Center 2015-2 Special Lecture on IE (2015)
5. Bergman, L., Cohen, N., Hoshen, Y.: Deep nearest neighbor anomaly detection. arXiv preprint arXiv:2002.10445 (2020)
6. Beyer, K., Goldstein, J., Ramakrishnan, R., Shaft, U.: When is "nearest neighbor" meaningful? In: Beeri, C., Buneman, P. (eds.) ICDT 1999. LNCS, vol. 1540, pp. 217–235. Springer, Heidelberg (1999). https://doi.org/10.1007/3-540-49257-7_15
7. Bo, Z., Song, Q., Chen, H.: Deep autoencoding gaussian mixture model for unsupervised anomaly detection (2018)
8. Breunig, M.M., Kriegel, H.P., Ng, R.T., Sander, J.: LOF: identifying density-based local outliers. In: SIGMOD, vol. 29, pp. 93–104. ACM (2000)
9. Chen, J., Sathe, S., Aggarwal, C., Turaga, D.: Outlier detection with autoencoder ensembles. In: SDM, pp. 90–98. SIAM (2017)
10. Chen, Y., Zhou, X.S., Huang, T.S.: One-class SVM for learning in image retrieval. In: ICIP pp. 34–37. Citeseer (2001)
11. Deng, A., Goodge, A., Lang, Y.A., Hooi, B.: CADET: calibrated anomaly detection for mitigating hardness bias. In: IJCAI (2022)
12. Dinh, L., Sohl-Dickstein, J., Bengio, S.: Density estimation using real NVP. arXiv preprint arXiv:1605.08803 (2016)
13. Feng, W., Han, C.: A novel approach for trajectory feature representation and anomalous trajectory detection. In: ISIF, pp. 1093–1099 (2015)
14. Goodge, A., Hooi, B., Ng, S.K., Ng, W.S.: Robustness of autoencoders for anomaly detection under adversarial impact. In: IJCAI (2020)

15. Goodge, A., Hooi, B., Ng, S.K., Ng, W.S.: Lunar: Unifying local outlier detection methods via graph neural networks. In: Proceedings of the AAAI Conference on Artificial Intelligence (2022)
16. inIT: Tool wear detection in CNC mill (2018). https://www.kaggle.com/init-owl/high-storage-system-data-for-energy-optimization
17. Kim, K.H., et al.: RaPP: novelty detection with reconstruction along projection pathway. In: ICLR (2019)
18. Kirichenko, P., Izmailov, P., Wilson, A.G.: Why normalizing flows fail to detect out-of-distribution data. arXiv preprint arXiv:2006.08545 (2020)
19. Lecun, Y.: Mnist (2012). http://yann.lecun.com/exdb/mnist/
20. Liu, F.T., Ting, K.M., Zhou, Z.H.: Isolation-based anomaly detection. ACM Trans. Knowl. Discov. Data (TKDD) **6**(1), 1–39 (2012)
21. Maaten, L.V.D., Hinton, G.: Visualizing data using t-SNE. J. Mach. Learn. Res. **9**, 2579–2605 (2008)
22. Otto, G.: Otto group product classification challenge (2015). https://www.kaggle.com/c/otto-group-product-classification-challenge
23. Papamakarios, G., Pavlakou, T., Murray, I.: Masked autoregressive flow for density estimation. In: NeurIPS, pp. 2338–2347 (2017)
24. Rezende, D.J., Mohamed, S.: Variational inference with normalizing flows. arXiv preprint arXiv:1505.05770 (2015)
25. Ruff, L., et al.: Deep one-class classification. In: ICML, pp. 4393–4402 (2018)
26. Sakurada, M., Yairi, T.: Anomaly detection using autoencoders with nonlinear dimensionality reduction. In: MLSDA, p. 4. ACM (2014)
27. Shyu, M.L., Chen, S.C., Sarinnapakorn, K., Chang, L.: A novel anomaly detection scheme based on principal component classifier. Technical report, Miami Univ Coral Gables FL Dept of Electric and Computer Engineering (2003)
28. Siffer, A., Fouque, P.A., Termier, A., Largouet, C.: Anomaly detection in streams with extreme value theory. In: SIGKDD, pp. 1067–1075 (2017)
29. SMART: Tool wear detection in CNC mill (2018). https://www.kaggle.com/shasun/tool-wear-detection-in-cnc-mill
30. Tax, D.M., Duin, R.P.: Support vector data description. Mach. Learn. **54**(1), 45–66 (2004)
31. UCI: Sensorless drive diagnosis (2015)
32. Xiao, H., Rasul, K., Vollgraf, R.: Fashion-mnist: a novel image dataset for benchmarking machine learning algorithms (2017)
33. Zenati, H., Foo, C.S., Lecouat, B., Manek, G., Chandrasekha, V.R.: Efficient GAN-based anomaly detection (2019)
34. Zhai, S., Cheng, Y., Lu, W., Zhang, Z.: Deep structured energy based models for anomaly detection. ICML **48**, 1100–1109 (2016)
35. Zimek, A., Gaudet, M., Campello, R.J.G.B., Sander, J.: Subsampling for efficient and effective unsupervised outlier detection ensembles. In: SIGKDD, pp. 428–436 (2013)

R2-AD2: Detecting Anomalies by Analysing the Raw Gradient

Jan-Philipp Schulze[1,3], Philip Sperl[1,3](✉), Ana Răduţoiu[1], Carla Sagebiel[2], and Konstantin Böttinger[3]

[1] Technical University of Munich, Munich, Germany
[2] Heidelberg University, Heidelberg, Germany
[3] Fraunhofer Institute for Applied and Integrated Security, Garching bei München, Germany
{jan-philipp.schulze,philip.sperl, konstantin.boettinger}@aisec.fraunhofer.de

Abstract. Neural networks follow a gradient-based learning scheme, adapting their mapping parameters by back-propagating the output loss. Samples unlike the ones seen during training cause a different gradient distribution. Based on this intuition, we design a novel semi-supervised anomaly detection method called R2-AD2. By analysing the temporal distribution of the gradient over multiple training steps, we reliably detect point anomalies in strict semi-supervised settings. Instead of domain dependent features, we input the raw gradient caused by the sample under test to an end-to-end recurrent neural network architecture. R2-AD2 works in a purely data-driven way, thus is readily applicable in a variety of important use cases of anomaly detection.

Keywords: Anomaly detection · Semi-supervised learning · Deep learning · Data mining · IT security

1 Introduction

Anomalies are inputs that significantly deviate from the given notion of normal. Depending on the use case, anomalies may lead to attacks on the infrastructure, fraudulent transactions or points of interest in general. In recent years, research on semi-supervised anomaly detection (AD) gained traction (e.g. [28,31,37]), where we leverage prior knowledge about the anomalous distribution to boost the overall detection performance. This setting is often found in real-world settings, where a few anomalies have already been detected manually while the rest are unknown. Unlike classification tasks, a semi-supervised AD method should not just differentiate between normal inputs and known anomalies, but also reveal yet unseen types of anomalies.

The lack of absolute training data modelling all types of anomalies complicates the use of machine learning algorithms with an automatic feature selection,

A. Răduţoiu, C. Sagebiel—The research was done while working at Fraunhofer AISEC.

M.-R. Amini et al. (Eds.): ECML PKDD 2022, LNAI 13713, pp. 209–224, 2023.
https://doi.org/10.1007/978-3-031-26387-3_13

e.g. deep learning (DL) methods. In our research, we alleviate this problem by analysing an abstract representation of the input: its temporal gradient distribution. Intuitively, a neural network (NN) trained only on the known normal data will fail to process anomalies in the same manner. We analyse this discrepancy with the help of an auxiliary NN. To reduce the manual work and domain expert knowledge required, we designed our AD method to be purely data-driven. Instead of hand-crafted features, we analyse the raw gradient caused by individual inputs for anomalous patterns. In our thorough empirical study, we show that our method generalises to several use cases and data types. Based on this principle, we call our novel AD method R2-AD2, raw gradient anomaly detection. In summary, our contributions to AD research are:

- We introduce a novel data-driven end-to-end neural architecture to analyse the temporal distribution of the gradient to detect point anomalies.
- To the best of our knowledge, R2-AD2 is the first semi-supervised AD method based on the analysis of gradients.
- We thoroughly analyse the performance gain by R2-AD2 on ten data sets against five baseline methods.
- To support future research, we open-sourced[1] our code.

1.1 Related Work

R2-AD2 is a DL-based, semi-supervised AD method building on the analysis of the input's gradient space. In the following, we discuss related work from all of the three categories. For a broader overview on AD, we recommend the surveys of Pang et al. [27] and Ruff et al. [30].

Anomaly Detection Based on Deep Learning Methods. DL methods deliver high performance even on complex inputs, but are data-demanding. Due to the inherent class imbalance of AD, it is challenging to apply DL methods. Over the past years, a variety of solutions arose, which we loosely group in three categories: methods based on 1) the reconstruction error, 2) the distance to the training data and 3) end-to-end architectures. Reconstruction-based methods use a representation or distribution estimation method, e.g. autoencoders (AEs) [2,5,40] or generative adversarial nets (GANs) [1,22,33]. Intuitively, when the network is fitted on the normal data, there is a measurable difference between the reconstructed and the input sample when an anomaly is processed. The main problems are noisy data sets, causing a low reconstruction error for some anomalies, and anomalies close to normal samples, which are easy to reconstruct. Distance-based methods, e.g. one-class classifiers [6,31,36], introduce a transformer network. Using a suitable metric, the transformer network maps normal samples close to each other, but anomalies far away. Problems may arise when the data set contains multiple notions of normal, which cannot be mapped to the very same centre of normality. R2-AD2 uses an end-to-end neural architecture, directly mapping the input to an anomaly score. Usually, end-to-end

[1] https://github.com/Fraunhofer-AISEC/R2-AD2.

Table 1. Work across different detection domains analysing the gradient space.

	Unsupervised	Semi-supervised	Supervised
Anomaly det.	[17,18]	R2-AD2	–
Out-of-Distr. det.	[15,38]	[20]	–
Adversarial det.	–	–	[7,21,23,34]

architectures [26,28,37] require normal as well as anomalous training samples. However, manually finding anomalies is a time-consuming and error-prone process. Research on substituting real anomalies by artificially created ones, e.g. geometric transformations [3,8] or out-of-distribution (OOD) samples [12], tries to solve this issue. These methods need careful adaptations to the respective data set. In R2-AD2, we mitigate the problem by using a simple source for trivial anomalies: a Gaussian distribution as done in A^3 [37]. Our evaluation motivates that our analysis in the gradient space of NNs allows to find a suitable boundary between real normal and real anomalous samples even with this simple source for counterexamples.

Semi-supervised Anomaly Detection. In the past years, research about semi-supervised AD has gained traction. In real-world scenarios, a few known anomalies – much less than the normal samples – may already be available. These known anomalies may have been found manually or by an unsupervised AD method. Semi-supervised AD methods use this kind of prior knowledge to boost the overall detection performance. DeepSAD [31] is a semi-supervised extension of one-class classifiers. Deviation Networks (DevNet) [28] is based on distance metrics. The authors of A^3 [37] analyse the hidden activations of NNs for anomalous patterns. Reconstruction errors are evaluated in ABC [44] and ESAD [14]. Expanding the view to OOD detection, DROCC [9] uses generated counterexamples based on the prior knowledge about real anomalies. For semi-supervised AD methods, the distribution of the known anomalies may severely impact the generalisation performance [45]. Thus, a main challenge is the detection of unknown anomalies, i.e. anomalies, which have not yet been detected manually.

Gradient-Based Detection of Anomalous Instances. R2-AD2 analyses the gradient space of NNs. Despite the variety of AD research, this idea has barely been covered by previous work. We give an overview in Table 1. Kwon et al. [18] propose using the l_2-norm of an AE's gradient. The same authors refine the idea in their AD method GradCon [17]. Here, they measure the cosine similarity between past normal gradients and the current input. Expanding the view to research topics related to AD, we see applications in OOD and adversarial detection. In OOD detection, multi-class data and thus known class labels are assumed, which is not applicable to AD, where we merely distinguish between monolithic sets of normal and anomalous data. Sun et al. detect OOD samples by measuring the Mahalanobis distance of the gradient. In GradNorm [15], the authors used the Kullback-Leibler divergence on the l_1-norm of the gradient. Similarly, Lee et al. [20] use the l_1-norm, but also incorporate some known OOD samples. In adversarial detection, samples, which have been specifically generated to alter the decision of a NN, are detected. In contrast to AD and OOD detection, adversarial detection is usually considered a supervised problem because counterexamples can be easily generated. In GraN [23], the authors used the l_1-norm of the gradient, whereas in Gradient Similarity [7] the authors took the l_2-norm of the gradient along the cosine similarity to distinguish between benign and adversarial samples. Lee et al. [21] train a classifier on the layer-wise l_2-norm. In DA3G [34], the authors analyse the raw gradient of the last two layers of classifiers. In R2-AD2, we refrain from using hand-crafted features or manually selecting points of interest as each choice incorporates prior knowledge from the algorithm designer, which may not be backed by the training data. Instead, we analyse the temporal distribution of the entire raw gradient by our end-to-end DL-based architecture. Our evaluation shows that R2-AD2 outperforms past AD methods on a variety of use cases and data types.

2 Prerequisites

In AD, we discover samples that deviate from the training data set $\mathcal{X}_{\text{norm}}$. Implicitly, we assume all samples in $\mathcal{X}_{\text{norm}}$ to be normal, even when polluted by unknown anomalies. In literature, there is some ambiguity in the definition of semi-supervised AD, which is sometimes referred to as supervised AD. In this regard, we follow the notation of Ruff et al. [31]. In our semi-supervised scenario, further we have access to a small data set $\mathcal{X}_{\text{anom}}$, containing a few known anomalies, i.e. $|\mathcal{X}_{\text{norm}}| \gg |\mathcal{X}_{\text{anom}}|$. Note that AD differs from related topics as OOD detection. In OOD detection, we do have access to an underlying classifier and its multi-class training data set. Instead, in AD, we consider the entire normal data set as one class and detect deviations from it. We refer to the survey of Salehi et al. [32] for an in-depth discussion of AD and its related research topics.

2.1 Activation Anomaly Analysis

Parts of R2-AD2 are inspired by the semi-supervised AD method A^3 [37]. Sperl et al. introduced their so-called target-alarm architecture. The target network, e.g. an AE, learns the distribution of the normal data. An auxiliary NN, called the alarm network, analyses the hidden activations of the target network while processing normal as well as anomalous inputs. As additional source of anomalous patterns, they input synthetic anomalies generated from a Gaussian prior.

In R2-AD2, we extend the target-alarm architecture to analyse the temporal gradient distribution of AEs. We use a recurrent alarm network to concurrently analyse the gradient of multiple AEs for anomalous patterns. Each AE reflects a different training state of the very same architecture. Our evaluation shows that the temporal gradient distribution allows a more reliable anomaly detection performance even under severe data pollution and unknown anomalies.

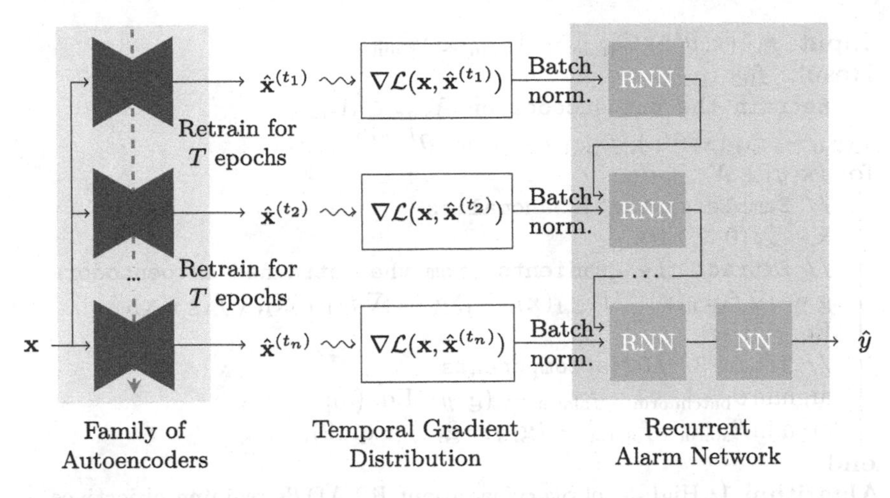

Fig. 1. Data flow of R2-AD2: we map the input sample $\mathbf{x}$ to an anomaly score $\hat{y} \in [0, 1]$, where 1 is highly anomalous. The input is processed by a family of AEs, each yielded by successive training on the normal samples. We measure the discrepancy between the predicted and the original input by calculating the respective gradient. An auxiliary network, called the alarm network, analyses this sequence of gradients for anomalous patterns.

3 R2-AD2

R2-AD2 builds upon our main intuition:

> Let $f_{\text{AE}}(\mathbf{x}; \boldsymbol{\theta})$ be an AE trained on the data set $\mathcal{X}_{\text{norm}}$ containing normal samples. The evolution of the gradient $\nabla f_{\text{AE}}(\mathbf{x}) = \nabla_{\boldsymbol{\theta}} \mathcal{L}(\mathbf{x}, f_{\text{AE}}(\mathbf{x}; \boldsymbol{\theta}))$ is useful to decide if the current input $\mathbf{x}$ is normal or anomalous.

Our intuition is a natural extension of the manual analysis of the gradient as done in past research [17,18]. Instead of considering certain features, e.g. magnitudes or directions, we analyse the gradient in its entirety. Let $f_{\text{AE}}^{(i)}(\mathbf{x}) = f_{\text{AE}}(\mathbf{x}; \boldsymbol{\theta}^{(i)})$ be the target AE after the i-th training epoch. With each training step, the mapping parameters $\boldsymbol{\theta}$ adapt more to the training data and hence to the normal samples. Thus, we embed the temporal distribution of the gradient in R2-AD2. Let $g_{f_{\text{AE}}}(\mathbf{x})$ denote the function that extracts the gradients over time given the target AE $f_{\text{AE}}(\cdot)$:

$$g_{f_{\text{AE}}}(\mathbf{x}) = [\nabla f_{\text{AE}}^{(i)}(\mathbf{x})]_{i=T_0+jT, j\in\mathbb{N}} = [\nabla f_{\text{AE}}^{(T_0)}(\mathbf{x}), \nabla f_{\text{AE}}^{(T_0+T)}(\mathbf{x}), \ldots], \quad (1)$$

where T is some sampling frequency and T_0 an offset.

Given the temporal gradient distribution, an auxiliary NN, called the *alarm network* $f_{\text{alarm}}(\cdot)$, analyses it for anomalous patterns. The alarm network is a binary classifier outputting an anomaly score, where 1 is highly anomalous. Both networks are combined to the overall end-to-end architecture of R2-AD2

Input: $f_{\mathrm{AE}}(\mathbf{x}; \boldsymbol{\theta}^{T_0}), \mathcal{D}_{\mathrm{train}} = \mathcal{X}_{\mathrm{train}} \times \mathcal{Y}_{\mathrm{train}}$
Result: $f_{\mathrm{R2\text{-}AD2}}$
`// Retrain the autoencoder on` $\mathcal{X}_{\mathrm{norm}} \subset \mathcal{X}_{\mathrm{train}}$
$f_{\mathrm{AE},0} \leftarrow f_{\mathrm{AE}}(\mathbf{x}; \boldsymbol{\theta}^{T_0}),\ f_{\mathrm{AE},1} \leftarrow f_{\mathrm{AE}}(\mathbf{x}; \boldsymbol{\theta}^{T_0+T}), \ldots;$
for $(\mathbf{x}, y) \in \mathcal{X}_{train}$ **do**
`// Sample synthetic anomalies`
$\tilde{\mathbf{x}} \leftarrow \mathcal{N}(\mathbf{0.5}, \mathbf{1.0});$
`// Extract the gradients from the retrained autoencoders`
$\mathbf{g} \leftarrow [\nabla f_{\mathrm{AE},0}(\mathbf{x}), \nabla f_{\mathrm{AE},1}(\mathbf{x}), \ldots],\ \tilde{\mathbf{g}} \leftarrow [\nabla f_{\mathrm{AE},0}(\tilde{\mathbf{x}}), \nabla f_{\mathrm{AE},1}(\tilde{\mathbf{x}}), \ldots],$
Eq. (1);
`// Train R2-AD2's components`
$\mathrm{argmin}\, \boldsymbol{\theta}_{\mathrm{batchnorm.}} : f_{\mathrm{alarm}} \leftarrow (\mathbf{g}, y)$, Eq. (3);
$\mathrm{argmin}\, \boldsymbol{\theta}_{\mathrm{alarm}} : f_{\mathrm{alarm}} \leftarrow (\mathbf{g}, y), (\tilde{\mathbf{g}}, 1)$, Eq. (2);
end
Algorithm 1: High-level overview about R2-AD2's training objectives.

depicted in Fig. 1 and formally defined as: $f_{\mathrm{R2\text{-}AD2}}(\mathbf{x}) = f_{\mathrm{alarm}}(g_{f_{\mathrm{AE}}}(\mathbf{x})) \in [0, 1]$. Due to the sequential nature of the gradient, the alarm network is a recurrent neural network (RNN). We combine the RNN with a time-distributed batch normalisation [16] layer and fully-connected output layers. In our research, we found the batch normalisation layer to be essential to scale small gradients, especially after several training epochs of the target network.

Training Objectives. AD is characterised by its inherent class imbalance, where known anomalies are rare and might not cover the entire anomaly distribution. In R2-AD2, we solve this problem by sampling trivial counterexamples from a Gaussian prior, i.e. $\tilde{\mathbf{x}} \sim \mathcal{N}(\mu, \sigma^2)$. Even though these synthetic anomalies do not resemble real ones, our analysis in the gradient space results in a meaningful decision barrier between real normal and real anomalous inputs. As result, the training objective of R2-AD2 becomes a simple classification using the binary cross entropy (BXE) as loss:

$$\underset{\theta_{\mathrm{alarm}}}{\mathrm{argmin}}\, \mathbb{E}[\mathcal{L}_{\mathrm{BXE}}\left(y, f_{\mathrm{R2\text{-}AD2}}\left(\mathbf{x}\right)\right) + \mathcal{L}_{\mathrm{BXE}}\left(1, f_{\mathrm{R2\text{-}AD2}}\left(\tilde{\mathbf{x}}\right)\right)], \tag{2}$$

where $(\mathbf{x}, y) \sim P_{\mathcal{D}}, \tilde{\mathbf{x}} \sim \mathcal{N}(\mathbf{0.5}, \mathbf{1.0})$. Our input data is scaled to $\mathbf{x} \in [0, 1]^N$, thus the synthetic anomalies are likely outside this interval, i.e. clearly anomalous. Due to the random nature of the counterexamples, we adapt the batch normalisation layer on the training data only, i.e.:

$$\underset{\theta_{\mathrm{batchnorm.}}}{\mathrm{argmin}}\ \mathbb{E}[\mathcal{L}_{\mathrm{BXE}}\left(y, f_{\mathrm{R2\text{-}AD2}}\left(\mathbf{x}\right)\right)]. \tag{3}$$

In Algorithm 1, we summarise R2-AD2's training process.

Table 2. Data sets under evaluation. If multiple anomaly types were available, we tested R2-AD2 on classes unknown during training in our transfer experiments.

Data	Normal	Train ano. ⊆	Test ano.	Encoder
CC [29]	Normal	Anomalous	Anomalous	20, 10, 5
CoverType [4]	1–3	4–5	4–7	40, 20, 10
DarkNet [10]	Non-Tor/-VPN	Tor	Tor, VPN	60, 30, 15
DoH [25]	Benign	Mal	Mal	20, 10, 5
FMNIST [43]	0–3	4–6	4–9	8C3-8C3-8
IDS [35]	Benign	Bot, BF	Bot, BF, Infil., Web	60, 40, 20
KDD [39]	Normal	DoS, Probe	DoS, Probe, R2L, U2R	40, 20, 10
MNIST [19]	0–3	4–6	4–9	8C3-8C3-8
Mam. [42]	Normal	Malignant	Malignant	5, 3, 2
URL [24]	Benign	Def., Mal	Def., Mal., Phi., Spam	60, 30, 15

4 Experimental Setup

We evaluated R2-AD2 in challenging experiments mimicking real-world scenarios. In Table 2, we show the ten data sets under evaluation, ranging from commonly used baseline data sets to important applications of AD, e.g. intrusion or fraud detection. We scaled all numerical values to $[0, 1]$ and 1-Hot encoded categorical entries. If not given by the data set, 75% were used for the training split, 5% for validation and 20% for testing. While training R2-AD2's AE, 25% of the training data were held back to evaluate the gradient distribution of some fresh normal samples while training the alarm network.

Baseline Methods. R2-AD2 is a deep semi-supervised AD method based on the analysis of the gradient space of AEs. AEs themselves can be used as AD method by measuring the reconstruction error, when only trained on the normal data. We used the mean squared error as anomaly score, i.e. $\hat{y} = \|f_{\mathrm{AE}}(\mathbf{x}) - \mathbf{x}\|_2^2$. GradCon [17] is a AD method based on the analysis of the gradient space of NNs. We favoured GradCon over the authors' initial AD method based on l_2-norms [18] as it generally performed better according to their evaluation. Both aforementioned baseline methods are unsupervised, thus do not profit from known anomalies. Expanding our view to deep semi-supervised AD, DeepSAD [31] is a commonly used baseline. In the same category, DevNet [28] and A^3 [37] are currently the best performing methods.

Parameter Choices. We designed R2-AD2 as a data-driven method, which readily applies to a diverse set of use cases and data types. Thus, we chose one common set of hyperparameters for the entire evaluation. Across all data sets, we analysed a target network trained for $T_0 = 10$ epochs across 2 retraining steps, each with $T = 5$ epochs resulting in three models. The alarm network had the dimensions $1000, 500, 200, 75$ except for the small Mammography data set, where we used $100, 50, 25, 10$. LSTM [13] elements were used for the first two dimensions, ReLU-activated dense layers else. R2-AD2 was trained for 100 epochs at a learning rate of 0.001 using Adam as optimiser. For a fair comparison, we chose the same hyperparameters for the other baseline methods if applicable.

5 Evaluation

We carefully followed the best practices introduced by Hendrycks & Gimpel [11] and report the performance as area under the ROC curve (AUC) and average precision (AP). Both metrics measure the performance independently of a chosen detection threshold. An ideal AD method scores an AUC and AP of 1. To measure the significance of our results, we report the p-value of the Wilcoxon signed-rank test [41]. It evaluates the null hypothesis that a ranked list of measurements was derived from the same distribution.

5.1 Known Anomalies

In our first experiment, we evaluated the performance gain in an ideal semi-supervised AD setting. We limited the number of known anomalies to 100 randomly chosen samples, i.e. far less than normal samples available. In Table 3, we summarise the results. R2-AD2 took the lead across all baseline methods, scoring the best on 7 out of 10 data sets.

Table 3. Detection of known anomalies, i.e. the training and test data set contained the same anomaly classes. We limited the number of known anomalies to 100 and show the results after five detection runs.

	Ours		Unsupervised baselines				Semi-supervised baselines					
	R2-AD2		AE		GradCon		DeepSAD		DevNet		A^3	
	AUC	AP	AUC	AP	AUC	AP	AUC	AP	AUC	AP	AUC	AP
CC	**.98 ± .01**	**.81 ± .03**	.95 ± .00	.43 ± .00	.80 ± .16	.34 ± .08	.88 ± .03	.34 ± .26	**.98 ± .00**	.74 ± .00	.88 ± .06	.51 ± .21
CT	**.84 ± .02**	**.43 ± .05**	.76 ± .02	.25 ± .03	.70 ± .05	.21 ± .06	.57 ± .07	.16 ± .05	.83 ± .01	.33 ± .02	.46 ± .06	.08 ± .01
DN	**.92 ± .01**	**.75 ± .02**	.54 ± .01	.23 ± .01	.62 ± .09	.24 ± .04	.68 ± .15	.36 ± .12	.90 ± .01	.69 ± .02	.84 ± .02	.54 ± .05
DoH	**.98 ± .00**	**1.00 ± .00**	.85 ± .01	.98 ± .00	.74 ± .06	.97 ± .01	.73 ± .08	.97 ± .01	.91 ± .01	.99 ± .00	.83 ± .04	.98 ± .01
FMN.	.92 ± .00	.95 ± .00	.86 ± .00	.92 ± .00	.82 ± .02	.88 ± .03	.69 ± .02	.77 ± .02	.93 ± .02	.96 ± .01	**.95 ± .01**	**.97 ± .00**
IDS	**.93 ± .01**	**.89 ± .01**	.84 ± .01	.52 ± .04	.45 ± .10	.19 ± .06	.67 ± .08	.38 ± .12	.87 ± .01	.67 ± .07	.88 ± .02	.67 ± .11
KDD	.88 ± .02	.90 ± .03	**.95 ± .00**	**.95 ± .00**	.74 ± .04	.83 ± .01	.85 ± .07	.88 ± .06	.92 ± .03	.94 ± .01	.93 ± .04	**.95 ± .02**
MN.	.97 ± .01	**.98 ± .00**	.75 ± .01	.79 ± .01	.82 ± .04	.84 ± .04	.70 ± .02	.75 ± .02	.97 ± .00	**.98 ± .00**	**.98 ± .00**	**.98 ± .01**
Mam.	**.94 ± .01**	**.69 ± .03**	.90 ± .01	.27 ± .01	.89 ± .01	.26 ± .02	.69 ± .23	.16 ± .12	**.94 ± .01**	.65 ± .04	.88 ± .04	.43 ± .06
URL	**.95 ± .01**	**.99 ± .00**	.92 ± .00	.98 ± .00	.90 ± .01	.97 ± .00	.94 ± .01	**.99 ± .00**	**.95 ± .01**	**.99 ± .00**	.94 ± .01	**.99 ± .00**
mean	**.93**	**.84**	.83	.63	.75	.57	.74	.58	.92	.79	.86	.71
p-val	–	–	.02	.01	.00	.00	.00	.00	.38	.05	.06	.05

As expected, the unsupervised baseline methods could not match the performance of the semi-supervised methods as they do not profit from the known anomalies. Looking at the AUC, R2-AD2 was 24% better than the other gradient-based AD method, GradCon. KDD was the only data set, where the unsupervised methods took the lead. Here, some unknown anomalies are within the test data set. Similar to the discussion of Ye et al. [45], we believe the semi-supervised methods overfitted to the known anomalies. Comparing our performance to GradCon, we see strong evidence that the analysis of the raw gradient is favourable over a hand-crafted feature set: GradCon's analysis of the cosine similarity works well on some data sets (e.g. MNIST and DarkNet), but does not generalise to all ten data sets. R2-AD2 had the more consistent performance.

Considering the semi-supervised baselines, the largest margin was on DoH, where R2-AD2 performed 8% better than DevNet, and 6% better on IDS compared to A^3. A^3 has a similar architecture as R2-AD2, but analyses the hidden activations of a single AE instead of the temporal gradient distribution. Overall R2-AD2 performed 8% better than A^3. Only on the image data sets, A^3 was the preferable method. Summarising this section, R2-AD2 clearly profited from the prior knowledge available in semi-supervised AD and allowed a more reliable detection performance compared to other state-of-the-art methods.

5.2 Noise Resistance

In real-world settings, it is usually infeasible to guarantee a clean training data set. We evaluated this scenario by polluting the data with anomalous training samples labelled as normal. All semi-supervised methods still had access to 100 known anomalies. We summarise the performance in Fig. 2 for DoH and DarkNet, where R2-AD2 took the lead in our first experiment, and MNIST and KDD,

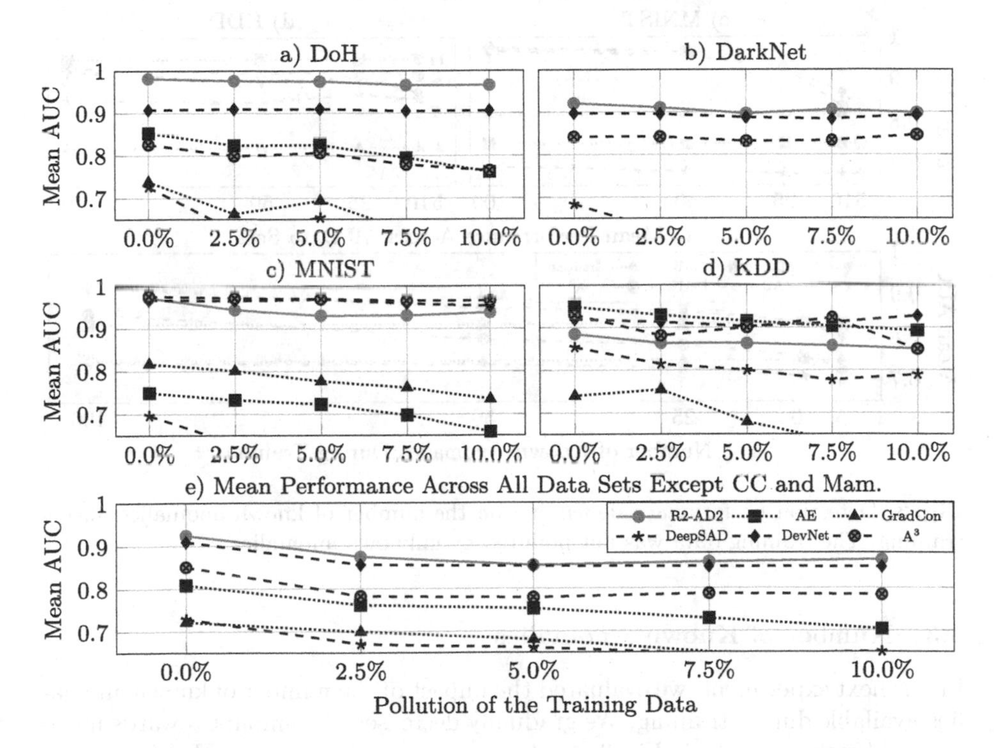

Fig. 2. Detection performance depending on the training data pollution. All semi-supervised methods had access to 100 known anomalies. Note that CC and Mammography did not contain enough anomalies, thus were excluded.

where the baseline methods performed better. Additionally, we show the mean performance across all data sets, which scaled to this experiment.

Looking at the mean performance, R2-AD2 took the lead across all pollution levels. The performance dropped only by –6%, when every tenth training sample was an anomaly labelled as normal. For the unsupervised baselines, the performance drop was considerably larger, e.g. –13% for the AE. The known anomalies seemed to stabilise the performance. Across the data sets, the general ranking between the baseline methods did not change: AD methods that performed well on cleaner data sets also performed well on polluted data sets. Our evaluation showed that R2-AD2 is resistant to noisy training data sets as often found in real-world settings.

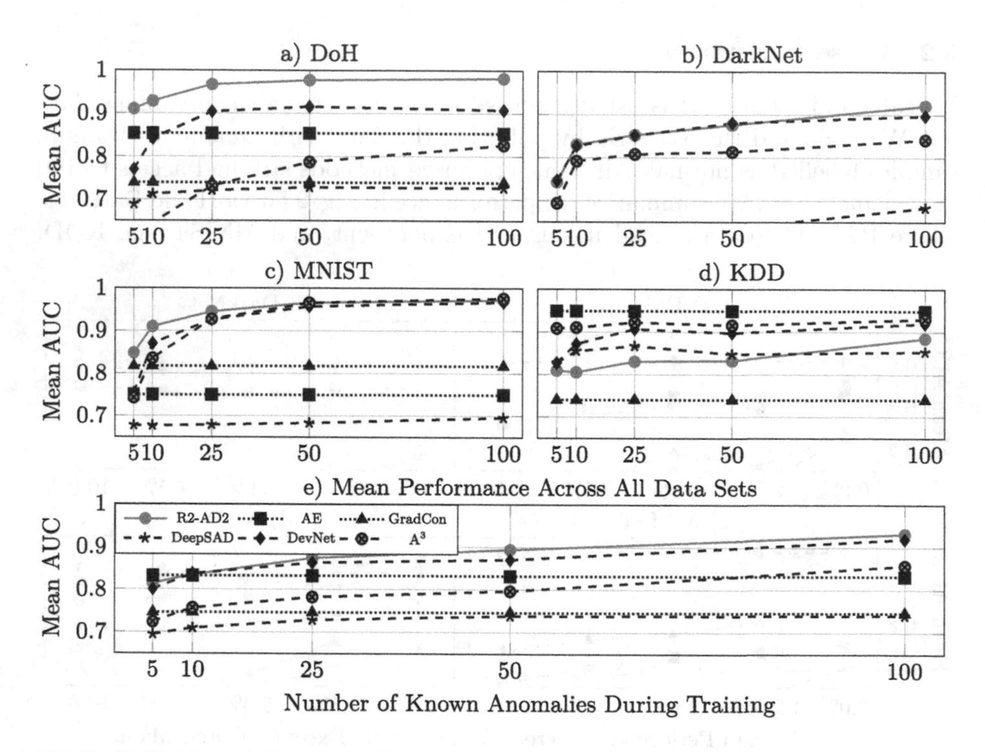

Fig. 3. Detection performance depending on the number of known anomalies during training. The training data was not polluted by unknown anomalies.

5.3 Number of Known Anomalies

In our next experiment, we evaluated the impact of the number of known anomalies available during training. We gradually decreased the amount towards unsupervised regimes shown in Fig. 3. As the known anomalies were randomly selected among all anomaly classes, some classes might have been excluded during training.

As expected, the unsupervised methods remained at their initial performance as they do not incorporate known anomalies. R2-AD2 exceeded the performance of the AE with as little as ten known anomalies. Looking at the mean performance, R2-AD2 was better than the semi-supervised methods across all anomaly counts. Interestingly, R2-AD2 took the lead on MNIST for small amounts of prior knowledge, i.e. less than 50 anomalies. In this experiment, we saw R2-AD2 to perform well even with little prior knowledge about known anomalies.

Table 4. Detection of unknown anomalies, i.e. there are more anomaly classes in the test set than there were in the training data. We limited the number of known anomalies to 100 and show the results after five detection runs.

	Ours		Unsupervised baselines				Semi-supervised baselines					
	R2-AD2		AE		GradCon		DeepSAD		DevNet		A^3	
	AUC	AP	AUC	AP	AUC	AP	AUC	AP	AUC	AP	AUC	AP
CT	.64 ± .02	.21 ± .01	**.76 ± .02**	**.25 ± .03**	.70 ± .05	.21 ± .06	.51 ± .07	.12 ± .05	.63 ± .05	.16 ± .02	.38 ± .07	.07 ± .02
DN	**.86 ± .02**	**.65 ± .02**	.54 ± .01	.23 ± .01	.62 ± .09	.24 ± .04	.51 ± .05	.29 ± .03	.62 ± .05	.28 ± .03	.56 ± .07	.23 ± .03
FMN.	.92 ± .02	.95 ± .01	.86 ± .00	.92 ± .00	.82 ± .02	.88 ± .03	.69 ± .02	.77 ± .03	**.94 ± .01**	**.96 ± .01**	.92 ± .02	.95 ± .01
IDS	**.92 ± .01**	**.80 ± .01**	.84 ± .01	.52 ± .04	.45 ± .10	.19 ± .06	.62 ± .25	.42 ± .25	.87 ± .01	.66 ± .05	.86 ± .03	.64 ± .13
KDD	.87 ± .04	.90 ± .03	**.95 ± .00**	**.95 ± .00**	.74 ± .04	.83 ± .01	.86 ± .03	.90 ± .01	.91 ± .01	.93 ± .01	.91 ± .03	.93 ± .02
MN.	.93 ± .01	**.96 ± .01**	.75 ± .01	.79 ± .01	.82 ± .04	.84 ± .04	.69 ± .02	.75 ± .01	.93 ± .01	.95 ± .00	**.94 ± .01**	**.96 ± .01**
URL	**.94 ± .01**	**.99 ± .00**	.92 ± .00	.98 ± .00	.90 ± .01	.97 ± .00	.92 ± .01	.98 ± .00	.92 ± .02	.98 ± .00	.92 ± .02	.98 ± .01
mean	**.87**	**.79**	.80	.66	.72	.59	.69	.60	.83	.70	.78	.68
p-val	–	–	.47	.30	.05	.03	.02	.03	.47	.30	.30	.30

Table 5. Mean performance depending on the number of retraining steps of the target network. We evaluated the detection of known anomalies given 100 anomalous training samples and show the results after five detection runs.

Number of AEs	1		2		3		4	
Metric	AUC	AP	AUC	AP	AUC	AP	AUC	AP
Mean performance	.87	.77	**.93**	.83	**.93**	**.84**	.92	**.84**

5.4 Unknown Anomalies

In our final experiment, we evaluated the transfer performance, i.e. how well the knowledge about known anomalies transfers to unknown ones. In real-world setting, the known anomalies often cover only a small part of all possible anomalies. Semi-supervised AD methods are expected to use the prior knowledge about known anomalies to detect unknown ones. We simulate this setting by limiting the training anomaly classes. Also in this experiment, R2-AD2 took the lead with a mean AUC of 87% as shown in Table 4. R2-AD2 was 5% better compared to the next best baseline, DevNet. Except for CoverType and KDD, R2-AD2 was better than the unsupervised methods. This experiment suggested that the raw gradient contains features that generalise across anomaly types.

Throughout our evaluation, we have seen R2-AD2 to deliver superior anomaly detection performance under common limitations in AD. R2-AD2 reliably detected known anomalies, yet generalised to unknown ones. Moreover, the detection performance remained at a high level even under polluted data sets and little prior knowledge about potential anomalies. With R2-AD2 we provide a reliable AD method applicable to a variety of important applications of AD.

5.5 Ablation Study

In our ablation study, we took a critical look on the temporal component of R2-AD2. We analysed the gradient of multiple training states of the target AE for anomalous patterns. Would it have been sufficient to consider a single time step only? In Table 5, we evaluated the gradient of 1, 2, 3 and 4 time steps. For the single time step case, we replaced the LSTM elements by dense layers. The mean performance considerably decreased when only evaluating a single time step. In comparison to three time steps, the AUC dropped by –6%. A single extra time step improved the performance. Expanding the number of time steps did not result in further improvements. We conclude that the temporal gradient distribution contains features important to AD, which are not present in a static one-step analysis.

6 Discussion and Future Work

In R2-AD2, we expanded the analysis of the gradient space of NNs – to the best of our knowledge – the first time to semi-supervised AD. Based on our evaluation, we have seen that the temporal gradient distribution allows to reliably detect anomalous inputs under diverse extents of prior knowledge on several important fields of application. Due to the end-to-end nature of R2-AD2, it readily integrates in other application areas. We hope to spark interest in porting our framework to sequential inputs like sensor measurements or video streams. Moreover, as we have seen related work in other detection areas, e.g. OOD or adversarial detection, we see potential to apply a gradient-based analysis to other important data mining and IT security applications e.g. deepfake detection.

7 Summary

In this paper, we introduced R2-AD2: a semi-supervised AD method based on the analysis of the temporal gradient distribution of NNs. R2-AD2 showed superior performance in a purely data-driven way, generalising to several important applications of AD. Our evaluation motivated that R2-AD2 is less susceptible to noisy training data than other state-of-the-art AD methods and requires less known anomalies for reliable detection performance. With R2-AD2, we extend the analysis of the NN's gradient the first time to semi-supervised AD, providing a reliable AD method to researchers and practitioners.

Acknowledgement. This research was supported by the Bavarian Ministry of Economic Affairs, Regional Development and Energy in the project "Cognitive Security".

Ethical Implications. Data-driven AD reveals data points that differ from the training data distribution. Underrepresented groups in the training data may cause a bias in the detection results. In example of the census data set, which we analysed during our evaluation, e.g. the origin of the citizens could be used for the anomaly decision leading to ethical implications. To this end, we encourage users of R2-AD2 and AD in general to thoroughly evaluate potential biases in the data.

References

1. Akcay, S., Atapour-Abarghouei, A., Breckon, T.P.: GANomaly: semi-supervised anomaly detection via adversarial training. In: Jawahar, C.V., Li, H., Mori, G., Schindler, K. (eds.) ACCV 2018. LNCS, vol. 11363, pp. 622–637. Springer, Cham (2019). https://doi.org/10.1007/978-3-030-20893-6_39
2. Beggel, L., Pfeiffer, M., Bischl, B.: Robust anomaly detection in images using adversarial autoencoders. In: Brefeld, U., Fromont, E., Hotho, A., Knobbe, A., Maathuis, M., Robardet, C. (eds.) ECML PKDD 2019. LNCS (LNAI), vol. 11906, pp. 206–222. Springer, Cham (2020). https://doi.org/10.1007/978-3-030-46150-8_13
3. Bergman, L., Hoshen, Y.: Classification-based anomaly detection for general data. In: International Conference on Learning Representations (2020). https://openreview.net/forum?id=H1lK_lBtvS
4. Blackard, J.A., Dean, D.J.: Comparative accuracies of artificial neural networks and discriminant analysis in predicting forest cover types from cartographic variables. Comput. Electron. Agric. **24**(3), 131–151 (1999). https://doi.org/10.1016/S0168-1699(99)00046-0
5. Borghesi, A., Bartolini, A., Lombardi, M., Milano, M., Benini, L.: Anomaly detection using autoencoders in high performance computing systems. In: Proceedings of the AAAI Conference on Artificial Intelligence, vol. 33, no. 01, pp. 9428–9433 (2019). https://doi.org/10.1609/aaai.v33i01.33019428
6. Chalapathy, R., Menon, A.K., Chawla, S.: Anomaly detection using one-class neural networks. arXiv:1802.06360 [cs, stat] (2019). http://arxiv.org/abs/1802.06360, arXiv: 1802.06360
7. Dhaliwal, J., Shintre, S.: Gradient similarity: an explainable approach to detect adversarial attacks against deep learning. arXiv:1806.10707 [cs] (2018). http://arxiv.org/abs/1806.10707, arXiv: 1806.10707
8. Golan, I., El-Yaniv, R.: Deep anomaly detection using geometric transformations. In: Bengio, S., Wallach, H., Larochelle, H., Grauman, K., Cesa-Bianchi, N., Garnett, R. (eds.) Advances in Neural Information Processing Systems, vol. 31. Curran Associates, Inc. (2018), https://proceedings.neurips.cc/paper/2018/file/5e62d03aec0d17facfc5355dd90d441c-Paper.pdf
9. Goyal, S., Raghunathan, A., Jain, M., Simhadri, H.V., Jain, P.: DROCC: deep robust one-class classification. In: Proceedings of the 37th International Conference on Machine Learning, pp. 3711–3721. PMLR (2020).https://proceedings.mlr.press/v119/goyal20c.html, iSSN: 2640-3498
10. Habibi Lashkari, A., Kaur, G., Rahali, A.: DIDarknet: a contemporary approach to detect and characterize the darknet traffic using deep image learning. In: 2020 the 10th International Conference on Communication and Network Security, ICCNS 2020, pp. 1–13. Association for Computing Machinery, New York (2020). https://doi.org/10.1145/3442520.3442521

11. Hendrycks, D., Gimpel, K.: A baseline for detecting misclassified and out-of-distribution examples in neural networks. In: International Conference on Learning Representations (2017). https://openreview.net/forum?id=Hkg4TI9xl
12. Hendrycks, D., Mazeika, M., Dietterich, T.: Deep anomaly detection with outlier exposure. In: International Conference on Learning Representations (2019). https://openreview.net/forum?id=HyxCxhRcY7
13. Hochreiter, S., Schmidhuber, J.: Long short-term memory. Neural Comput. **9**(8), 1735–1780 (1997). https://doi.org/10.1162/neco.1997.9.8.1735
14. Huang, C., Ye, F., Zhao, P., Zhang, Y., Wang, Y.F., Tian, Q.: ESAD: end-to-end deep semi-supervised anomaly detection. In: The 32nd British Machine Vision Conference (2021). https://www.bmvc2021-virtualconference.com/conference/papers/paper_0329.html
15. Huang, R., Geng, A., Li, Y.: On the importance of gradients for detecting distributional shifts in the wild. arXiv:2110.00218 [cs] (2021). http://arxiv.org/abs/2110.00218, arXiv: 2110.00218
16. Ioffe, S., Szegedy, C.: Batch normalization: accelerating deep network training by reducing internal covariate shift. In: Proceedings of the 32nd International Conference on Machine Learning, pp. 448–456. PMLR (2015). https://proceedings.mlr.press/v37/ioffe15.html, iSSN: 1938-7228
17. Kwon, G., Prabhushankar, M., Temel, D., AlRegib, G.: Backpropagated gradient representations for anomaly detection. In: Vedaldi, A., Bischof, H., Brox, T., Frahm, J.-M. (eds.) ECCV 2020. LNCS, vol. 12366, pp. 206–226. Springer, Cham (2020). https://doi.org/10.1007/978-3-030-58589-1_13
18. Kwon, G., Prabhushankar, M., Temel, D., AlRegib, G.: Novelty detection through model-based characterization of neural networks. In: 2020 IEEE International Conference on Image Processing (ICIP), pp. 3179–3183 (2020). https://doi.org/10.1109/ICIP40778.2020.9190706, iSSN: 2381-8549
19. Lecun, Y., Bottou, L., Bengio, Y., Haffner, P.: Gradient-based learning applied to document recognition. Proc. IEEE **86**(11), 2278–2324 (1998). https://doi.org/10.1109/5.726791
20. Lee, J., AlRegib, G.: Open-set recognition with gradient-based representations. In: 2021 IEEE International Conference on Image Processing (ICIP), pp. 469–473 (2021). https://doi.org/10.1109/ICIP42928.2021.9506430, iSSN: 2381-8549
21. Lee, J., Prabhushankar, M., AlRegib, G.: Gradient-based adversarial and out-of-distribution detection. In: International Conference on Machine Learning (ICML) Workshop on New Frontiers in Adversarial Machine Learning (2022)
22. Li, D., Chen, D., Jin, B., Shi, L., Goh, J., Ng, S.-K.: MAD-GAN: multivariate anomaly detection for time series data with generative adversarial networks. In: Tetko, I.V., Kůrková, V., Karpov, P., Theis, F. (eds.) ICANN 2019. LNCS, vol. 11730, pp. 703–716. Springer, Cham (2019). https://doi.org/10.1007/978-3-030-30490-4_56
23. Lust, J., Condurache, A.P.: GraN: an efficient gradient-norm based detector for adversarial and misclassified examples. In: ESANN 2020, p. 6 (2020)
24. Mamun, M.S.I., Rathore, M.A., Lashkari, A.H., Stakhanova, N., Ghorbani, A.A.: Detecting malicious URLs using lexical analysis. In: Chen, J., Piuri, V., Su, C., Yung, M. (eds.) NSS 2016. LNCS, vol. 9955, pp. 467–482. Springer, Cham (2016). https://doi.org/10.1007/978-3-319-46298-1_30
25. MontazeriShatoori, M., Davidson, L., Kaur, G., Lashkari, A.H.: Detection of DoH tunnels using time-series classification of encrypted traffic. In: The 5th IEEE Cyber Science and Technology Congress, pp. 63–70 (2020). https://doi.org/10.1109/DASC-PICom-CBDCom-CyberSciTech49142.2020.00026

26. Pang, G., Cao, L., Chen, L., Liu, H.: Learning Representations of ultrahigh-dimensional data for random distance-based outlier detection. In: Proceedings of the 24th ACM SIGKDD International Conference on Knowledge Discovery & Data Mining, KDD 2018, pp. 2041–2050. Association for Computing Machinery, New York (2018). https://doi.org/10.1145/3219819.3220042
27. Pang, G., Shen, C., Cao, L., Hengel, A.V.D.: Deep learning for anomaly detection: a review. ACM Comput. Surv. **54**(2), 38:1–38:38 (2021). https://doi.org/10.1145/3439950
28. Pang, G., Shen, C., van den Hengel, A.: Deep anomaly detection with deviation networks. In: Proceedings of the 25th ACM SIGKDD International Conference on Knowledge Discovery & Data Mining, KDD 2019, pp. 353–362. Association for Computing Machinery, New York (2019). https://doi.org/10.1145/3292500.3330871
29. Pozzolo, A.D., Caelen, O., Johnson, R.A., Bontempi, G.: Calibrating probability with undersampling for unbalanced classification. In: 2015 IEEE Symposium Series on Computational Intelligence, pp. 159–166 (2015). https://doi.org/10.1109/SSCI.2015.33
30. Ruff, L., et al.: A unifying review of deep and shallow anomaly detection. In: Proceedings of the IEEE, pp. 1–40 (2021). https://doi.org/10.1109/JPROC.2021.3052449
31. Ruff, L., et al.: Deep semi-supervised anomaly detection. In: International Conference on Learning Representations (2020). https://openreview.net/forum?id=HkgH0TEYwH
32. Salehi, M., Mirzaei, H., Hendrycks, D., Li, Y., Rohban, M.H., Sabokrou, M.: A unified survey on anomaly, novelty, open-set, and out-of-distribution detection: solutions and future challenges. arXiv:2110.14051 [cs] (2021). http://arxiv.org/abs/2110.14051, arXiv: 2110.14051
33. Schlegl, T., Seeböck, P., Waldstein, S.M., Langs, G., Schmidt-Erfurth, U.: f-AnoGAN: fast unsupervised anomaly detection with generative adversarial networks. Med. Image Anal. **54**, 30–44 (2019). https://doi.org/10.1016/j.media.2019.01.010
34. Schulze, J.-P., Sperl, P., Böttinger, K.: DA3G: detecting adversarial attacks by analysing gradients. In: Bertino, E., Shulman, H., Waidner, M. (eds.) ESORICS 2021. LNCS, vol. 12972, pp. 563–583. Springer, Cham (2021). https://doi.org/10.1007/978-3-030-88418-5_27
35. Sharafaldin, I., Lashkari, A.H., Ghorbani, A.A.: Toward generating a new intrusion detection dataset and intrusion traffic characterization. In: ICISSP, pp. 108–116 (2018)
36. Sohn, K., Li, C.L., Yoon, J., Jin, M., Pfister, T.: Learning and evaluating representations for deep one-class classification. In: International Conference on Learning Representations (2021). https://openreview.net/forum?id=HCSgyPUfeDj
37. Sperl, P., Schulze, J.-P., Böttinger, K.: Activation anomaly analysis. In: Hutter, F., Kersting, K., Lijffijt, J., Valera, I. (eds.) ECML PKDD 2020. LNCS (LNAI), vol. 12458, pp. 69–84. Springer, Cham (2021). https://doi.org/10.1007/978-3-030-67661-2_5
38. Sun, J., et al.: Gradient-based novelty detection boosted by self-supervised binary classification. arXiv:2112.09815 [cs] (2021). http://arxiv.org/abs/2112.09815, arXiv: 2112.09815

39. Tavallaee, M., Bagheri, E., Lu, W., Ghorbani, A.A.: A detailed analysis of the KDD CUP 99 data set. In: 2009 IEEE Symposium on Computational Intelligence for Security and Defense Applications, pp. 1–6 (2009). https://doi.org/10.1109/CISDA.2009.5356528, iSSN: 2329-6275
40. Vu, H.S., Ueta, D., Hashimoto, K., Maeno, K., Pranata, S., Shen, S.M.: Anomaly detection with adversarial dual autoencoders. arXiv:1902.06924 [cs] (2019). http://arxiv.org/abs/1902.06924, arXiv: 1902.06924
41. Wilcoxon, F.: Individual comparisons by ranking methods. In: Kotz, S., Johnson, N.L. (eds.) Breakthroughs in Statistics: Methodology and Distribution, Springer Series in Statistics, pp. 196–202. Springer, New York (1992). https://doi.org/10.1007/978-1-4612-4380-9_16
42. Woods, K.S.: Comparative evaluation of pattern recognition techniques for detection of microcalcifications in mammography. Int. J. Pattern Recogn. Artif. Intell. **07**(06), 1417–1436 (1993). https://doi.org/10.1142/S0218001493000698
43. Xiao, H., Rasul, K., Vollgraf, R.: Fashion-MNIST: a novel image dataset for benchmarking machine learning algorithms. arXiv:1708.07747 [cs, stat] (2017). http://arxiv.org/abs/1708.07747, arXiv: 1708.07747
44. Yamanaka, Y., Iwata, T., Takahashi, H., Yamada, M., Kanai, S.: Autoencoding binary classifiers for supervised anomaly detection. In: Nayak, A.C., Sharma, A. (eds.) PRICAI 2019. LNCS (LNAI), vol. 11671, pp. 647–659. Springer, Cham (2019). https://doi.org/10.1007/978-3-030-29911-8_50
45. Ye, Z., Chen, Y., Zheng, H.: Understanding the effect of bias in deep anomaly detection. In: Zhou, Z.H. (ed.) Proceedings of the Thirtieth International Joint Conference on Artificial Intelligence, International Joint Conferences on Artificial Intelligence Organization , IJCAI-2021, pp. 3314–3320 (2021). https://doi.org/10.24963/ijcai.2021/456

Hop-Count Based Self-supervised Anomaly Detection on Attributed Networks

Tianjin Huang(✉), Yulong Pei, Vlado Menkovski, and Mykola Pechenizkiy

Department of Mathematics and Computer Science, Eindhoven University of Technology, 5600 MB Eindhoven, The Netherlands
{t.huang,y.pei.1,v.menkovski,m.pechenizkiy}@tue.nl

Abstract. A number of approaches for anomaly detection on attributed networks have been proposed. However, most of them suffer from two major limitations: (1) they rely on unsupervised approaches which are intrinsically less effective due to the lack of supervisory signals of what information is relevant for capturing anomalies, and (2) they rely only on using local, e.g., one- or two-hop away node neighbourhood information, but ignore the more global context. Since anomalous nodes differ from normal nodes in structures and attributes, it is intuitive that the distance between anomalous nodes and their neighbors should be larger than that between normal nodes and their (also normal) neighbors if we remove the edges connecting anomalous and normal nodes. Thus, estimating hop counts based on both global and local contextual information can help us to construct an anomaly indicator. Following this intuition, we propose a hop-count based model (HCM) that achieves that. Our approach includes two important learning components: (1) Self-supervised learning task of predicting the shortest path length between a pair of nodes, and (2) Bayesian learning to train HCM for capturing uncertainty in learned parameters and avoiding overfitting. Extensive experiments on real-world attributed networks demonstrate that HCM consistently outperforms state-of-the-art approaches.

Keywords: Self-supervised anomaly detection · Attributed networks

1 Introduction

Attributed networks are ubiquitous in a variety of real-world applications. Attributed networks can be utilized to represent data from different domains. For example, in a social network, each node can represent a user, an edge denotes the friend relation between users, and user profiles are the attributes to describe users. A citation network consists of papers as the nodes, citation relations as the edges, and words in paper abstracts can be the attributes of papers. Unlike plain networks where only structural information exists, attributed networks also contain rich features to provide more details to describe (elements of) networks. Due

T. Huang and Y. Pei—Both authors contributed equally to this research.

M.-R. Amini et al. (Eds.): ECML PKDD 2022, LNAI 13713, pp. 225–241, 2023.
https://doi.org/10.1007/978-3-031-26387-3_14

to the ubiquity of attributed networks, various tasks on attributed networks have been widely studied such as community detection [5,21], link prediction [2,15] and network embedding [10,18].

Among these tasks on attributed networks, anomaly detection is perhaps one of the most important ones in the current analytics tasks – it can shed light on a wide range of real-world applications such as fraud detection in finance and spammers discovery in social media.

Unlike in anomaly detection on plain networks, detecting anomalies on attributed networks should rely on two sources of information: (1) the structural patterns of how nodes interconnect or interact with each other, which are reflected by the topological structures, (2) the distributions of nodal features. Therefore, it is more challenging to detect anomalies on attributed networks. Figure 1 shows a toy example of anomalies on an attributed network. In Fig. 1, nodes represent the individuals and links represent the connections. Node 10 can be identified as an attribute anomaly since it connects to the community with US cities while its location attribute is a China City. Node 11 can be identified as a structural anomaly since it connects to almost all other nodes where some connections are unreasonable.

To solve this challenging problem, various supervised and unsupervised approaches have been proposed recently. Unsupervised anomaly detection methods are preferable in practice because of the prohibitive cost for accessing the ground-truth anomalies [3]. Hence, in this study, we focus on unsupervised fashion to detect anomalies in the sense that we do not require access to ground truth anomalies in the training data. Intrinsically, unsupervised learning approaches demonstrate worse performance than supervised learning approaches (when they are applicable) due to the lack of supervisory signals. Recently, various self-supervised techniques have been proposed to learn a better feature representation via designing supervisory signals directly from input data. It has been shown, e.g., in the computer vision domain, that self-supervised learning can greatly benefit model's robustness and out-of-distribution detection [9].

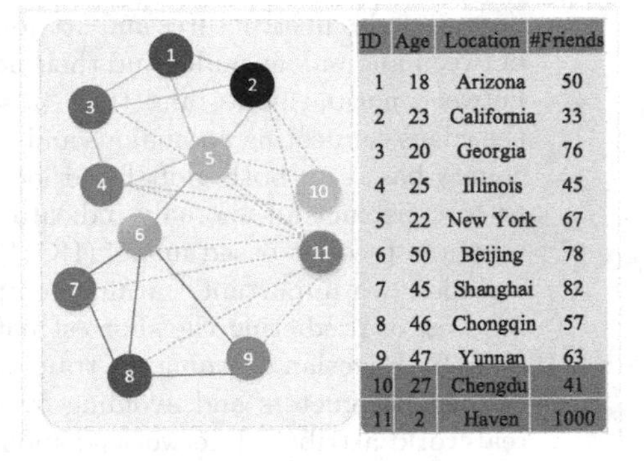

Fig. 1. A toy example for illustrating anomalies on attributed networks.

In this paper, we employ the idea of self-supervised learning by forcing the prediction of the shortest path length between pairs of nodes (denoted as hop counts for convenience) to detect anomalies on attributed networks. This idea allows to address another limitation of existing approaches that utilize only the local context of nodes to detect anomalous nodes while neglecting the global context information that provides a complementary view of the network structure patterns including anomalous structures. It has been shown that the global context of a node is helpful to learn node representation that can finely characterize the similarity and differentiation between nodes [23]. Therefore, it is reasonable to assume that effectively anomaly detection methods should consider local and global contextual information.

1.1 Our HCM Approach in a Nutshell

We use hop count prediction as a self-supervised learning task to capture local and global contextual information for node representation. We use GCN to learn node representations for refining the graph. Then the learned node representation is used to predict hop counts for arbitrary node pairs. We design two anomaly scores based on the intuition that the true hop counts between anomalous nodes and their neighbors shall be larger than that between normal nodes and their neighbors, e.g. in our toy example in Fig. 1, the hop counts between node 11 and their neighbors should be larger than 1 since node 11 is probably from the community with reddish nodes.

We adopt Bayesian learning to train our HCM model because Bayesian methods are appealing in their ability to capture uncertainty in learned parameters and avoid overfitting [28]. Specifically, we exploit Stochastic Gradient Langevin Dynamics (SGLD) [28] to optimize our model and conduct Bayesian inference.

1.2 Summary of the Contributions

The contributions of this paper are summarized as follows:

- We employ self-supervised learning ideas in the graph anomaly detection domain and make use of both global and local contextual information of nodes, i.e., hop counts, to detect anomalies on attributed networks.
- With the help of the self-supervised learning technique, we propose our HCM model to learn node representations capturing local and global contextual information of nodes. And based on the HCM model, we design two anomaly scores to detect anomalies. Experimental results demonstrate the effectiveness of our approach.
- We exploit SGLD to optimize HCM model for capturing uncertainty in learned parameters and avoid overfitting. We experimentally demonstrate that our model optimized by SGLD performs better than SGD in anomaly detection. Besides, SGLD achieves a steadier behaviour during the training process.

2 Related Work

Anomaly detection is an important task in data mining and machine learning. Previous studies roughly can be categorized into four types [3]: community analysis, subspace selection, residual analysis and deep learning methods. Community analysis methods [1,6,7] detect anomalies by identifying current node's abnormality with other nodes within the same community. Subspace selection approaches [25,26] learn a subspace for features and then discover anomalies in the learned subspace. Residual analysis methods [16,24] explicitly model the residual information by reconstructing the input attributed network based on matrix factorization. With the popularity of deep learning techniques, methods using deep neural networks such as graph convolutional networks (GCN) and network embedding to detect anomalies have been proposed [3,17,22]. Among these methods, seven popular methods are chosen as baselines in this paper including LOF [1], AMEN [25], Radar [16], ANOMALOUS [24], DOMINANT [3], MADAN [7] and ResGCN [22]. LOF [1] measures how isolated the object is with respect to the surrounding neighborhood and detects anomalies at the contextual level. LOF only considers nodal attributes. AMEN [25] analyzes the abnormality of each node from the ego-network point of view. Radar [16] detects anomalies by characterizing the residuals of attribute information and its coherence with network information. ANOMALOUS [24] is a joint anomaly detection framework to select attributes and detect anomalies using CUR decomposition of a matrix. DOMINANT [3] selects anomalies by ranking the reconstruction errors where the errors are learned by GCN. MADAN [7] uses the heat kernel as filtering operator to exploit the link with the Markov stability to find the context for multi-scale anomalous nodes. ResGCN [22] learns the residual information using a deep neural network, and reduces the adverse effect from anomalous nodes using the residual-based attention mechanism.

3 Problem Definition

Following the commonly used notations, we use bold uppercase characters for matrices, e.g., $\boldsymbol{X}$, bold lowercase characters for vectors, e.g., $\boldsymbol{b}$, and normal lowercase characters for scalars, e.g., c. The ith row of a matrix $\boldsymbol{X}$ is denoted by $\boldsymbol{X}_{i,:}$ and $(i,j)^{th}$ element of matrix $\boldsymbol{X}$ is denoted as $\boldsymbol{X}_{i,j}$. The Frobenius and L_2 norm of a matrix are represented as $\|\cdot\|_F$ and $\|\cdot\|_2$ respectively. The number of elements of a set is denoted by $|\cdot|$.

Definition 1 ***Attributed Networks***. *An attributed network $\mathcal{G} = \{V, E, \boldsymbol{X}\}$ consists of: (1) a set of nodes $V = \{v_1, v_2, ..., v_n\}$, where $|V| = n$ is the number of nodes; (2) a set of edges E, where $|E| = m$ is the number of edges; and (3) the node attribute matrix $\boldsymbol{X} \in \mathbb{R}^{n \times d}$, the ith row vector $\boldsymbol{X}_{i,:} \in \mathbb{R}^d$ is the attribute of node v_i where d is the number of attributes.*

The topological structure of attributed network $\mathcal{G}$ can be represented by an adjacency matrix $\boldsymbol{A}$, where $\boldsymbol{A}_{i,j} = 1$ if there is an edge between node v_i and node v_j. Otherwise, $\boldsymbol{A}_{i,j} = 0$. We focus on the undirected networks in this study and it is trivial to extend it to directed networks. The attribute of $\mathcal{G}$ can be represented by an attribute matrix $\boldsymbol{X}$. Thus, the attributed network can be represented as $\mathcal{G} = \{\boldsymbol{A}, \boldsymbol{X}\}$. With these notations and definitions, same to previous studies [3,16,22,24], we formulate the task of anomaly detection on attributed networks:

Problem 1 ***Anomaly Detection on Attributed Networks.*** *Given an attributed network $\mathcal{G} = \{\boldsymbol{A}, \boldsymbol{X}\}$, which is represented by the adjacency matrix $\boldsymbol{A}$ and attribute matrix $\boldsymbol{X}$, the task of anomaly detection is to find a set of nodes that are rare and differ singularly from the majority reference nodes of the input network.*

4 Hop-Count Based Model

In this section we first present a detailed description for our proposed HCM model. Then the SGLD optimization strategy will be introduced. Finally, we discuss the anomaly scores based on different strategies.

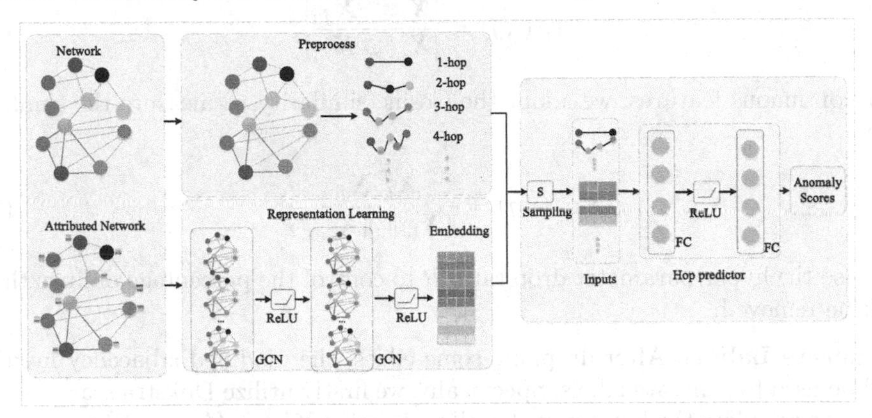

Fig. 2. The framework of our proposed HCM model. The dotted lines in Preprocess component denote the dropped edges.

4.1 Model Framework

In this section, we introduce the framework of our proposed HCM model in detail. HCM model is based on self-supervised task designed by some innate characteristics of the data. Recently, several self-supervised tasks have been proposed for graph data. For example, node clustering task where a clustering operation will be used to provide pseudo labels; Graph partition task where graph partitioning will provide pseudo labels [31]; PairDistance task where hop counts of node pairs are used as label [11]; PairwiseAttrSim where the similarities of node pairs are taken as label [11]. Here we adopt PairDistance as our self-supervised task. There are three reasons accounting for using PairDistance as our self-supervised task: 1) It is cheap to get the true label during the training process; 2) The hop counts capture both global and local information of a network; 3) hop counts can be utilized to directly construct anomaly scores of nodes. Our HCM model is composed of three components (Fig. 2): Preprocess components, Graph Convolutional network (GCN) and Multilayer perception (MLP). Preprocess is to generate labels during the training process; GCN is to learn the representation of the attributes networks; MLP is the classifier to predict hop counts for node pairs.

Preprocess. There are two operations in Preprocess component. The first is to drop edges with small similarities in the adjacency matrix. The second is to generate labels, namely the hop counts of all node pairs.

Drop Edges Considering that the anomalies on the attribute network could potentially perturb links, we remove some edges in the adjacency matrix before utilizing it to generate labels. Specifically, we remove the potential perturbed links according to the similarities of node pairs. [30] has demonstrated that removing links with small similarities can effectively purify perturbed links, especially for adversarial perturbed links. Similar to [30], for binary features, we adopt the Jaccard similarity to measure the similarities:

$$S_{Jar}(i,j) = \frac{|\boldsymbol{X}_{i,:} \cap \boldsymbol{X}_{j,:}|}{|\boldsymbol{X}_{i,:} \cup \boldsymbol{X}_{j,:}|}. \tag{1}$$

For continuous features, we adopt the Cosine similarity to measure the similarities:

$$S_{Sin}(i,j) = \frac{\boldsymbol{X}_{i,:}^T \boldsymbol{X}_{j,:}}{\|\boldsymbol{X}_{i,:}\|_2 \|\boldsymbol{X}_{j,:}\|_2}. \tag{2}$$

We use the hyper-parameter drop ratio R to control the percentage of links that will be removed.

Generate Labels. After dropping some edges, the modified adjacency matrix will be used to generate labels. Specifically, we firstly utilize Dijkstra search algorithm to calculate the hop counts for all node pairs: $Y_{hop}^1 = \{(v_m, v_n)|hop(v_m, v_n) = 1, v_m, v_n \in V\}$, $Y_{hop}^2 = \{(v_m, v_n)|hop(v_m, v_n) = 2, v_m, v_n \in V\}$, $Y_{hop}^3 = \{(v_m, v_n)\ |hop(v_m, v_n) = 3, v_m, v_n \in V\}$..., $Y_{hop}^i = \{(v_m, v_n)\ |hop(v_m, v_n) = i, v_m, v_n \in V\}$, where $hop(v_m, v_n)$ denotes the hop counts between node v_m and v_n. Then these hop counts will be grouped into C classes: $hop(v_m, v_n) = 1, hop(v_m, v_n) = 2, ..., hop(v_m, v_n) >= C$. Besides, Considering that the $|Y_{hop}^i|$ are different for each set Y_{hop}^i, which will lead to imbalanced training set. In order to avoid the bias induced by imbalanced training set, we sample a same amount of node pairs from each set Y_{hop}^i during the training, the amount of node pairs is decided by the smallest $|Y_{hop}^i| \cdot S$ where S is sampling ratio.

GCN. GCN learns node representations by passing and aggregating messages between neighboring nodes. Different types of GCN have been proposed recently [8,14], and we focus on one of the most widely used versions proposed in [14]. Formally, a GCN is defined as

$$\boldsymbol{h}_i^{(l+1)} = f\Big(\sum_{j \in Ne(i)} \frac{1}{\sqrt{\widetilde{\boldsymbol{D}}_{i,i}\widetilde{\boldsymbol{D}}_{j,j}}} \boldsymbol{h}_j^{(l)} \boldsymbol{W}^{(l)}\Big), \tag{3}$$

where $\boldsymbol{h}_i^{(l)}$ is the latent representation of node v_i in layer l, $Ne(i)$ is the set of neighbors of node v_i, and $\boldsymbol{W}^l$ is the layer-specific trainable weight matrix. $f(\cdot)$

is a non-linear activation function and we select ReLU as the activation function following previous studies [14] (written as $f_{ReLU}(\cdot)$ below). $\widetilde{\boldsymbol{D}}$ is the diagonal degree matrix of $\widetilde{\boldsymbol{A}}$ defined as $\widetilde{D}_{i,i} = \sum_j \widetilde{A}_{i,j}$ where $\widetilde{\boldsymbol{A}} = \boldsymbol{A} + \boldsymbol{I}$ is the adjacency matrix of the input attributed network $\boldsymbol{G}$ with self connections $\boldsymbol{I}$. Equivalently, we can rewrite GCN in a matrix form:

$$\boldsymbol{H}^{(l+1)} = f_{ReLU}\left(\widetilde{\boldsymbol{D}}^{-\frac{1}{2}}\widetilde{\boldsymbol{A}}\widetilde{\boldsymbol{D}}^{-\frac{1}{2}}\boldsymbol{H}^{(l)}\boldsymbol{W}^{(l)}\right). \tag{4}$$

For the first layer, $\boldsymbol{H}^{(0)} = \boldsymbol{X}$ is the attribute matrix of the input network. Therefore, we have

$$\boldsymbol{H}^{(1)} = f_{ReLU}\left(\widetilde{\boldsymbol{A}}\boldsymbol{X}\boldsymbol{W}^{(0)}\right). \tag{5}$$

In this paper, we denote the representation learned by the GCN as $\boldsymbol{Z}$.

MLP. The target of MLP component is to construct hop counts prediction task. MLP is composed of multiple fully connected layers. During training phase, CrossEntropy loss is used to construct supervisory signals. During inference, it predicts hop counts for any arbitrary node pair representations. Given any node pairs $\boldsymbol{Z}_{m,:}$ and $\boldsymbol{Z}_{n,:}$, it can be formally expressed as following:

$$\boldsymbol{z}_d = |\boldsymbol{Z}_{m,:} - \boldsymbol{Z}_{n,:}| \tag{6}$$

$$F_{mlp}(v_m, v_n) = f^k_{ReLU}(\boldsymbol{z}_d \cdot \boldsymbol{W}), \tag{7}$$

where $F_{mlp}(v_m, v_n)$ is the hop counts prediction of node v_m and v_n, f^k_{ReLU} is k fully connected layers, $\boldsymbol{W}$ is total trainable weight matrix of MLP and $|\cdot|$ here is the absolute function. The reason for using absolution function is based on the consideration that the hop counts from node $\boldsymbol{Z}_{m,:}$ to $\boldsymbol{Z}_{n,:}$ and from node $\boldsymbol{Z}_{n,:}$ to $\boldsymbol{Z}_{m,:}$ are the same on undirected networks. For directed networks, absolution function can be removed.

For the convenience of formulation, we denote the whole model as follows:

$$F_{\boldsymbol{w}}(\boldsymbol{X}, \boldsymbol{A}) = F_{mlp}(v_m, v_n) \circ H^l, \tag{8}$$

where $\boldsymbol{w} \in \Omega$ are the parameters of the HCM model and it contains parameters from GCN and MLP components.

The Self-supervised Learning Loss. As described above, we adopt PairDistance as the self-supervised task. PairDistance task is to predict hop counts for any given node pairs. And it can be formulated as a multi-class classification problem. Therefore, we formulate the self-supervised learning loss as follows:

$$\mathcal{L}_{self}(\boldsymbol{w}, \boldsymbol{A}, \boldsymbol{X}) = \frac{1}{|\mathcal{S}|} \sum_{(v_m, v_j) \in \mathcal{S}} L(F_{mlp}(v_m, v_n), hop(v_m, v_n)), \tag{9}$$

where $\mathcal{S}$ is a set of randomly sampled node pairs and $L(\cdot)$ is the CrossEntropy loss.

4.2 Model Training

The Bayesian framework suggests that the parameters of a model are random variables instead of determined. Therefore, it is crucial to estimate the distribution of model parameters. According to Bayes' theorem, the posterior distribution of the parameters can be defined as follows:

$$p(\boldsymbol{w}|\boldsymbol{X}, \boldsymbol{A}, \boldsymbol{Y_{hop}}) = \frac{p(\boldsymbol{Y_{hop}}|\boldsymbol{X}, \boldsymbol{A}, \boldsymbol{w})p(\boldsymbol{w})}{p(\boldsymbol{Y_{hop}}|\boldsymbol{X}, \boldsymbol{A})}, \tag{10}$$

where $p(\boldsymbol{w})$ is a prior distribution for the model's parameters $\boldsymbol{w}$.

According to Eq.(10), there are two different approaches to do the inference:(1) Maximum posterior probability (MAP) estimation finds the mode of the posterior distribution. (2) Bayesian inference computes the posterior distribution itself. Considering that the MAP-estimation can not capture the model's uncertainty, we choose the second approach as our solution. The Bayesian inference of outputs is given as followings:

$$p(y^*_{hop}|x^*, \boldsymbol{X}, \boldsymbol{A}, \boldsymbol{Y_{hop}}) = \int_{\Omega} p(y^*|x^*, \boldsymbol{w})p(\boldsymbol{w}|\boldsymbol{X}, \boldsymbol{A}, \boldsymbol{Y_{hop}})d\boldsymbol{w}, \tag{11}$$

Due to the integration, it is impossible to achieve the prediction by computing the Eq.(11) directly. Fortunately, Stochastic gradient Langevin dynamics (SGLD) [28] provides a general framework to solve Eq. (11). The SGLD update is defined as follows:

$$\delta_{\boldsymbol{w}_t} = \frac{\epsilon_t}{2}(\nabla_{\boldsymbol{w}_t} logp(y_{hop}|\boldsymbol{w}_t) + \nabla_{\boldsymbol{w}} logp(\boldsymbol{w}_t)) + \eta_t \tag{12}$$

$$\eta_t \in N(0, \epsilon) \tag{13}$$

$$\boldsymbol{w}_{t+1} = \boldsymbol{w}_t - \delta_{\boldsymbol{w}_t}, \tag{14}$$

where ϵ_t is the step size. The log prior term will be implemented as weight decay. [28] has shown that under suitable conditions, e.g., $\sum \epsilon_t = \infty$ and $\sum \epsilon_t^2 < \infty$ and others, $\boldsymbol{w}$ converges to the posterior distribution. The pseudocode of HCM model training are summarized in Algorithm 1 (In Appendix A).

4.3 Model Inference for Anomaly Detection

Inference. Based on SGLD update, the posterior inference can be achieved by adding Gaussian noise to the gradients at each step and the prediction can be estimated by the posterior sample averages after a "burn in" phase [28]:

$$\widetilde{Y}_{hop}(v_i, v_j) = \frac{1}{T}\sum_{t=1}^{T} F_{\boldsymbol{w}_t}(v_i, v_j|\boldsymbol{X}, \boldsymbol{A}) \tag{15}$$

Anomaly Scores. Based on the HCM model, we design two anomaly criteria for catching anomalies. The first anomaly criterion is the averaged predicted hop counts between a node and its neighbors. Since anomalous nodes are different from their local neighbors, the predicted hop counts are expected to be larger than normal nodes. Therefore, the higher averaged predicted hop counts, the higher probability of being anomalous nodes. The second anomaly criterion combines the averaged predicted hop counts and the inferred variance of posterior samples. [13] shows the evidence that the lack of data could increase the model's uncertainty. Since anomalies are rare on the network, the inferred variance on anomalous nodes tends to be higher.

Average Hop Prediction (AHP) which is the averaged hop predictions between a node and its neighbors. The expression is defined as followings:

$$S_{AHP}(v_i) = \frac{\sum_{v_j} \widetilde{Y}_{hop}(v_i, v_j)}{|\mathcal{N}(v_i)|}, v_j \in \mathcal{N}(v_i), \tag{16}$$

where $\mathcal{N}(v_i)$ is the neighbors of node v_i. $S_{AHP}(v_i)$ represents the averaged hop prediction for node v_i. For convenience, we use $S_{AHP}(V)$ denote the averaged hop predictions for all nodes:

$$S_{AHP}(V) = \{S_{AHP}(v_1), S_{AHP}(v_2), ..., S_{AHP}(v_i)\}, v_i \in \boldsymbol{V} \tag{17}$$

Average Hop Prediction+Inferred Variance (HAV) which integrates the averaged hop predictions and inferred variance. The variance of hop prediction between node v_i and node v_j is defined as followings:

$$\delta(v_i, v_j) = \frac{\sum_t^T (F_{w_t}(v_i, v_j|\boldsymbol{X}, \boldsymbol{A}) - \widetilde{Y}_{hop}(v_i, v_j))^2}{T}, \tag{18}$$

where $v_i, v_j \in \boldsymbol{V}$. For one node, we use the averaged variance of hop predictions between the node and its neighbors to present the anomaly scores of this node.

$$S_{IV}(v_i) = \frac{\sum_{v_j} \delta(v_i, v_j)}{|\mathcal{N}(v_i)|}, v_j \in \mathcal{N}(v_i). \tag{19}$$

Similarly, We use $S_{IV}(V)$ to denote the variance for all nodes.

$$S_{IV}(V) = \{S_{IV}(v_1), S_{IV}(v_2), ..., S_{IV}(v_i)\}, v_i \in \boldsymbol{V}. \tag{20}$$

Finally, we integrate the anomaly scores from the predicted hop counts and the inferred variance together. The expression is defined as follows:

$$S_{HAV}(V) = \frac{S_{AHP}(V)}{\max(S_{AHP}(V))} + \frac{S_{IV}(V)}{\max(S_{IV}(V))}, \tag{21}$$

where $\max(\cdot)$ represents the maximum value of the set. Considering that the predicted hop counts and inferred variance have different scales, we use the maximum to normalize each score. The pseudocode of anomaly scores are summarized in Algorithm 2 (In Appendix A).

5 Experiments

In this section, we evaluate our proposed HCM model empirically. We aim to answer the following three research questions:

- RQ1: Is the HCM model effective in detecting anomalies on attributed networks?
- RQ2: Does SGLD optimization framework benefit the anomaly detection performance of our approach on attributed networks?
- RQ3: How do the hyper-parameters in the HCM model affect the anomaly detection performance?

We first introduce the datasets and experimental settings. Then we report and analyze the experimental results.

5.1 Datasets

We use five real-world attributed networks to evaluate the effectiveness of our proposed method. These networks have been widely used in previous anomaly detection studies [3,7,16,22,24]. These networks can be categorized into two types: networks with ground-truth anomaly labels and networks with injected anomaly labels.

- For networks with ground-truth anomaly labels, Amazon and Enron[1] have been used. Amazon is a co-purchase network [20]. The anomalous nodes are defined as nodes having the tag *amazonfail*. Enron is an email network [19]. Attributes represent content length, number of recipients, etc., and each edge indicates the email transmission between sender and receiver. Spammers are the anomalies.
- For networks with injected anomaly labels, we select BlogCatalog, Flickr and ACM[2]. BlogCatalog is a blog sharing website. A list of tags associated with each user is used as the attributes. Flickr is an image hosting and sharing website. Similarly, tags are the attributes. ACM is a citation network. Words in abstracts are used as attributes.

A brief statistics of these attributed networks are showed in Table 2 (In Appendix B). Note that for the data with injected labels, to make a fair comparison, we follow previous studies for anomaly injection [3,22]. In specific, two anomaly injection methods have been used to inject anomalies by perturbing topological structure and nodal attributes, respectively:

- **Structural anomalies**: we perturb the topological structure of the input network to generate structural anomalies. We randomly select s nodes from the network and then make those nodes fully connected, and then all the s nodes forming the clique are labeled as anomalies. t cliques are generated repeatedly and totally there are $s \times t$ structural anomalies.

[1] https://www.ipd.kit.edu/mitarbeiter/muellere/consub/.
[2] http://people.tamu.edu/~xhuang/Code.html.

- **Attribute anomalies**: we perturb the nodal attributes of the input network to generate attribute anomalies. Same to [3,22,27], we first randomly select $s \times t$ nodes as the attribute perturbation candidates. For each selected node v_i, we randomly select another k nodes from the network and calculate the Euclidean distance between v_i and all the k nodes. Then the node with largest distance is selected as v_j and the attributes $\boldsymbol{X}_{j,:}$ of node v_j is changed to $\boldsymbol{X}_{i,:}$ of node v_i. The selected node v_j is regarded as the generated attribute anomaly.

In the experiments, we set $s = 15$ and set t to 10, 15, and 20 for BlogCatalog, Flickr and ACM, respectively which are the same to [3,22] in order to make a fair comparison. To facilitate the learning process, in our experiments, we use Principal Component Analysis (PCA) [29] to reduce the dimensionality of attributes to 20 following [4].

5.2 Experimental Settings

In the experiments, we use the HCM model consisting of four convolution layers and two fully connected layers for Amazon and Enron networks, two convolution layers and two fully connected layers for BlogCatalog, Flickr and ACM networks. The units of each convolution layer are 128. The units of the fully connected layer are set to 256 and the classes C respectively. The learning rate is set to 0.01. The weight decay is set to 5e−8. The default drop ratio R, classes C and sampling ratio S are set to 0.2, 4 and 0.3 respectively.

In the experiments, following the previous studies [3,7,16,22,24], we use the area under the receiver operating characteristic curve (ROC-AUC) as the evaluation metric for anomaly detection. We compare the proposed HCM model with the following anomaly detection methods:**LOF** [1], **AMEN** [25], **Radar** [16], **ANOMALOUS** [24], **DOMINANT** [3], **MADAN** [7] and **ResGCN** [22].

Table 1. Performance of different anomaly detection methods w.r.t. ROC-AUC on Flickr, BlogCatalog, ACM, Amazon and Enron networks.-:No results.

	Flickr	BlogCatalog	ACM	Amazon	Enron
LOF [1]	0.488	0.491	0.473	0.490	0.440
AMEN [25]	0.604	0.665	0.533	0.470	0.470
Radar [16]	0.728	0.725	0.693	0.580	0.650
ANOMALOUS [24]	0.716	0.728	0.718	0.602	0.695
DOMINANT [3]	0.781	0.749	0.749	0.625	0.685
MADAN [7]	–	–	–	0.680	0.660
ResGCN [22]	0.780	0.785	0.768	**0.710**	0.660
HCM-AHP (Ours)	0.791	**0.808**	**0.806**	0.62	0.670
HCM-HAV (Ours)	**0.792**	0.798	0.761	0.708	**0.715**

5.3 Experimental Results

We conduct experiments to evaluate the performance of our proposed method and to compare it with several state-of-the-art approaches, to which we refer as baselines, on two different types of networks: networks with ground-truth anomalies and injected anomalies respectively.

Results on the Networks with Known Ground-Truth Anomalies Results on the network with ground-truth anomalies are showed in Table 1. It shows that the HCM-HAV score achieves the best among the baselines on Enron network and comparable results on Amazon. The HCM-AHP score leads to comparable results achieved with DOMINANT and ANOMALOUS, demonstrating the effectiveness of utilizing both global and local information to detect anomalies. The better performance of the HCM-HAV score than the HCM-AHP score indicates that the inferred variance plays a role in finding anomalies in the Amazon and Enron networks.

Results on the Networks with Injected Anomalies. Results on the network with injected anomalies are showed in Table 1. As can be seen, HCM-HAV and HCM-AHP scores outperform the baseline methods in all cases, which further validates the effectiveness of our method by utilizing both global and local information to detect anomalies.

The insight of HCM-AHP is that the hop counts between the anomalous nodes and their neighbors are expected to be larger than that between normal nodes and their neighbors. The insight of HCM-HAV score integrates the influence of inferred variance into HCM-AHP. Figure 3 shows that the average HCM-AHP and HCM-HAV among anomalous nodes are bigger than among non-anomalous nodes, which further validates the rationality of the insight. Interestingly, we note that the HCM-HAV score doesn't perform better than HCM-AHP score on BlogCatalog and ACM networks, which indicates that the integration of inferred variance and HCM-AHP score does not constantly improve the performance. However, HCM-HAV can consistently achieve good results on all networks compared to HCM-AHP.

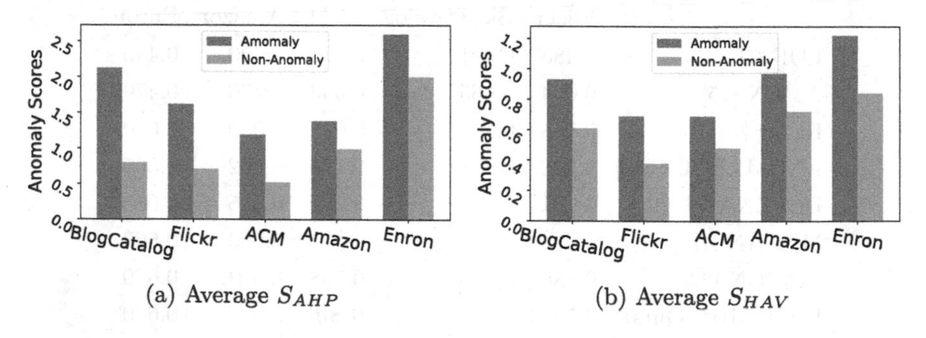

(a) Average S_{AHP} (b) Average S_{HAV}

Fig. 3. Average scores of anomalous nodes and non-anomalous nodes.

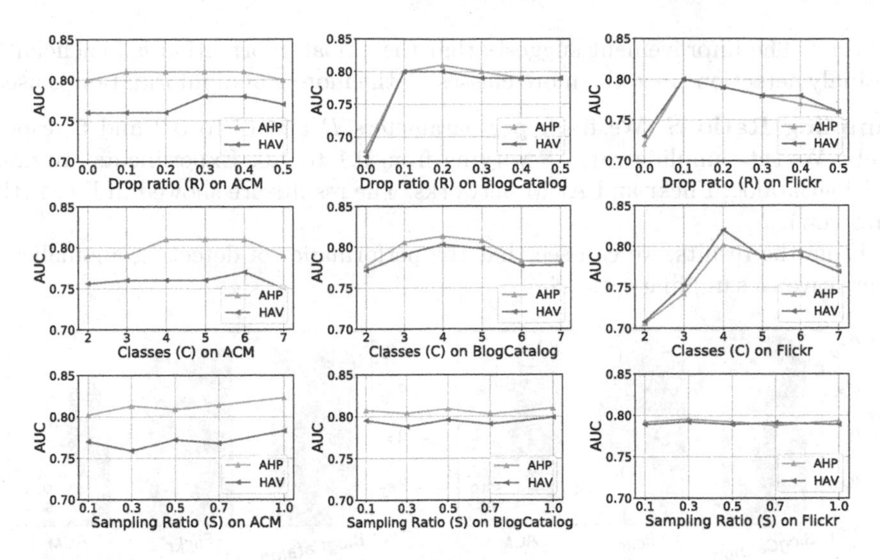

Fig. 4. Influence of the hyper-parameters: Drop ratio (R) (ranging from 0 to 0.5), Classes (C) (ranging from 2 to 7), Sampling Ratio (S) (ranging from 0.1 to 1.0). Experiments are conducted on ACM, BlogCatalog and Flickr respectively.

5.4 Parameter Analysis

We conducted a series of experiments to study the impact of hyper-parameters on the performance of anomaly detection. There are three hyper-parameters introduced in our proposed method: (1) drop ratio R in the Preprocess component, which controls the percentages of dropped edges, (2) classes C, which controls the max hop counts that the training set contains, (3) sampling ratio S, that controls the amount of node pairs engaging in the training process of each iteration.

Drop Ratio R**.** We fix classes C and sampling ratio S to 5 and 0.3 respectively. Then we monitor the anomaly performance by ranging drop ratio R from 0 t 0.5 on BlogCatalog, Flickr and ACM datasets. The results are showed in Fig. 4 (the first row).

From the results, we can see that removing a few edges with low similarity is beneficial in detecting anomalies. Specifically, the experiments show that the performance of anomaly detection increases greatly on BlogCatalog and Flickr datasets when a few edges are removed. However, it is not encouraged to remove a large amount of edges since the performance of anomaly detection has slightly decreased when a large amount of edges are removed.

Classes C**.** We set drop ratio R and sampling ratio S to 0.2 and 0.3 respectively. The classes C varies from 2 to 7. Experiments are done on BlogCatalog, Flickr and ACM networks and the results are showed in Fig. 4 (the second row).

From the figures, we can see that the performance of anomaly detection is improved with the increase of the classes C, especially on Flickr and BlogCatalog datasets. The improvement suggests that the global information is beneficial to anomaly detection since the more classes C, the more global information is used.

Sampling Ratio S**.** We fix hyper-parameters R and C to 0.2 and 5 respectively. We set sampling ratio S varying from 0.1 to 1.0. Experiments are done on BlogCatalog, Flickr and ACM networks. The results are showed in Fig. 4 (the third row).

From the results, we can see that the performance of detecting anomalies is insensitive to sampling ratio S.

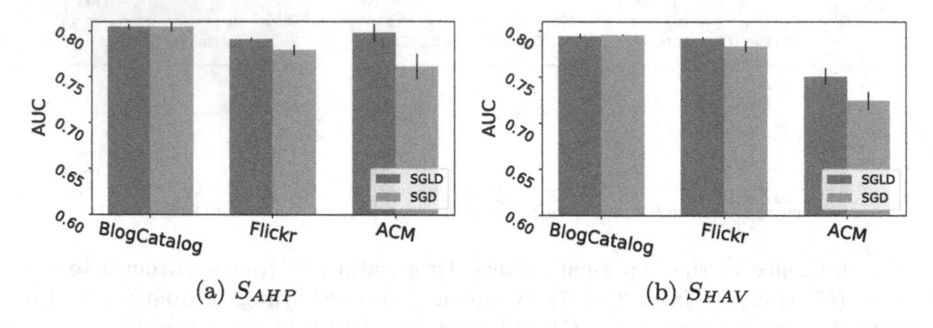

(a) S_{AHP} (b) S_{HAV}

Fig. 5. ROC-AUC achieved by S_{HAV} and S_{AHP} under SGLD and SGD training strategy respectively.

5.5 Ablation Study

For the proposed HCM model, SGLD is used instead of stochastic gradient descent (SGD) to optimize the model since we adopt Bayesian learning. Here we conduct experiments to show the benefits of utilizing SGLD to train our proposed model. Specifically, we firstly test the anomaly detection performances of the model optimized by SGD and SGLD respectively. Then we monitor the trend of ROC-AUC during the training process under SGD and SGLD training strategy respectively. The results are showed in Fig. 5a, Fig. 5b and Fig. 6. We can observe the following from these figures:

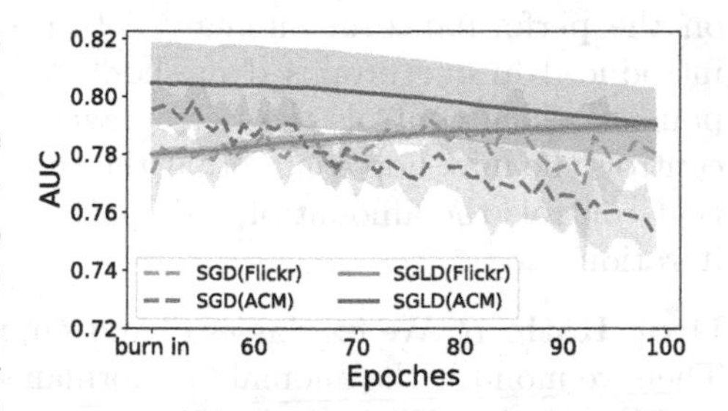

Fig. 6. ROC-AUC trends during the training process under SGLD and SGD training strategy respectively. The ROC-AUC value is the mean of 10 times repeated experiments. The filled region along trends is the standard deviation of the 10 times repeated experiments.

- From Fig. 5a and Fig. 5b, we see that the ROC-AUC performance achieved by SGLD is better than that by SGD on Flickr and ACM networks, which shows that SGLD based Bayesian learning benefits the anomaly detection on attributed networks.

- From Fig. 6, we see that the AUC trend under SGLD is steadier than that under SGD.

6 Conclusion

In this paper, we proposed a HCM model based on self-supervised technique for anomaly detection. To the best of our knowledge, it is the first model to take both local and global contextual information into account for anomaly detection. Specifically, we utilize a hop-count based self-supervised task to learn node representation with capturing local and global contextual information. The learned node representation will be used to predict hop counts for arbitrary node pairs as well. Besides, we designed two new anomaly scores for detecting anomalies based on the hop counts prediction via the HCM model. Finally, we introduce SGLD to train the model for capturing uncertainty in learned parameters and avoiding overfitting. The extensive experiments demonstrate: 1) the consistent effectiveness of the HCM model in anomaly detection on attributed networks that we proposed; 2) SGLD-based Bayesian learning strategy is beneficial to achieve a better and stable performance in detecting anomalies.

References

1. Breunig, M.M., Kriegel, H.P., Ng, R.T., Sander, J.: LOF: identifying density-based local outliers. In: Proceedings of the 2000 ACM SIGMOD International Conference on Management of Data, pp. 93–104 (2000)
2. Brochier, R., Guille, A., Velcin, J.: Link prediction with mutual attention for text-attributed networks. In: Companion Proceedings of The 2019 World Wide Web Conference, pp. 283–284 (2019)
3. Ding, K., Li, J., Bhanushali, R., Liu, H.: Deep anomaly detection on attributed networks. In: Proceedings of the 2019 SIAM International Conference on Data Mining, pp. 594–602. SIAM (2019)
4. Ding, K., Li, J., Liu, H.: Interactive anomaly detection on attributed networks. In: Proceedings of the Twelfth ACM International Conference on Web Search and Data Mining, pp. 357–365 (2019)
5. Falih, I., Grozavu, N., Kanawati, R., Bennani, Y.: Community detection in attributed network. In: Companion Proceedings of the The Web Conference 2018, pp. 1299–1306 (2018)
6. Gao, J., Liang, F., Fan, W., Wang, C., Sun, Y., Han, J.: On community outliers and their efficient detection in information networks. In: Proceedings of the 16th ACM SIGKDD International Conference on Knowledge Discovery and Data Mining, pp. 813–822 (2010)
7. Gutiérrez-Gómez, L., Bovet, A., Delvenne, J.C.: Multi-scale anomaly detection on attributed networks. arXiv preprint arXiv:1912.04144 (2019)
8. Hamilton, W., Ying, Z., Leskovec, J.: Inductive representation learning on large graphs. In: Advances in Neural Information Processing Systems, pp. 1024–1034 (2017)
9. Hendrycks, D., Mazeika, M., Kadavath, S., Song, D.: Using self-supervised learning can improve model robustness and uncertainty. In: Advances in Neural Information Processing Systems, pp. 15663–15674 (2019)

10. Huang, X., Li, J., Hu, X.: Accelerated attributed network embedding. In: Proceedings of the 2017 SIAM International Conference on Data Mining, pp. 633–641. SIAM (2017)
11. Jin, W., Derr, T., Liu, H., Wang, Y., Wang, S., Liu, Z., Tang, J.: Self-supervised learning on graphs: Deep insights and new direction. arXiv preprint arXiv:2006.10141 (2020)
12. Johnson, D.B.: Efficient algorithms for shortest paths in sparse networks. J. ACM (JACM) **24**(1), 1–13 (1977)
13. Kendall, A., Gal, Y.: What uncertainties do we need in bayesian deep learning for computer vision? In: Advances in Neural Information Processing Systems, pp. 5574–5584 (2017)
14. Kipf, T.N., Welling, M.: Semi-supervised classification with graph convolutional networks. arXiv preprint arXiv:1609.02907 (2016)
15. Li, J., Cheng, K., Wu, L., Liu, H.: Streaming link prediction on dynamic attributed networks. In: Proceedings of the Eleventh ACM International Conference on Web Search and Data Mining, pp. 369–377 (2018)
16. Li, J., Dani, H., Hu, X., Liu, H.: Radar: residual analysis for anomaly detection in attributed networks. In: IJCAI, pp. 2152–2158 (2017)
17. Liang, J., Jacobs, P., Sun, J., Parthasarathy, S.: Semi-supervised embedding in attributed networks with outliers. In: Proceedings of the 2018 SIAM International Conference on Data Mining, pp. 153–161. SIAM (2018)
18. Meng, Z., Liang, S., Bao, H., Zhang, X.: Co-embedding attributed networks. In: Proceedings of the Twelfth ACM International Conference on Web Search and Data Mining, pp. 393–401 (2019)
19. Metsis, V., Androutsopoulos, I., Paliouras, G.: Spam filtering with Naive Bayes-which naive bayes? In: CEAS, vol. 17, pp. 28–69. Mountain View, CA (2006)
20. Müller, E., Sánchez, P.I., Mülle, Y., Böhm, K.: Ranking outlier nodes in subspaces of attributed graphs. In: 2013 IEEE 29th International Conference on Data Engineering Workshops (ICDEW), pp. 216–222. IEEE (2013)
21. Pei, Y., Chakraborty, N., Sycara, K.: Nonnegative matrix tri-factorization with graph regularization for community detection in social networks. In: Twenty-Fourth International Joint Conference on Artificial Intelligence (2015)
22. Pei, Y., Huang, T., van Ipenburg, W., Pechenizkiy, M.: RESGCN: attention-based deep residual modeling for anomaly detection on attributed networks. arXiv preprint arXiv:2009.14738 (2020)
23. Peng, Z., Dong, Y., Luo, M., Wu, X.M., Zheng, Q.: Self-supervised graph representation learning via global context prediction. arXiv preprint arXiv:2003.01604 (2020)
24. Peng, Z., Luo, M., Li, J., Liu, H., Zheng, Q.: Anomalous: a joint modeling approach for anomaly detection on attributed networks. In: IJCAI, pp. 3513–3519 (2018)
25. Perozzi, B., Akoglu, L.: Scalable anomaly ranking of attributed neighborhoods. In: Proceedings of the 2016 SIAM International Conference on Data Mining, pp. 207–215. SIAM (2016)
26. Perozzi, B., Akoglu, L., Iglesias Sánchez, P., Müller, E.: Focused clustering and outlier detection in large attributed graphs. In: Proceedings of the 20th ACM SIGKDD International Conference on Knowledge Discovery and Data Mining, pp. 1346–1355 (2014)
27. Song, X., Wu, M., Jermaine, C., Ranka, S.: Conditional anomaly detection. IEEE Trans. Knowl. Data Eng. **19**(5), 631–645 (2007)

28. Welling, M., Teh, Y.W.: Bayesian learning via stochastic gradient Langevin dynamics. In: Proceedings of the 28th International Conference on Machine Learning (ICML-11), pp. 681–688 (2011)
29. Wold, S., Esbensen, K., Geladi, P.: Principal component analysis. Chemom. Intell. Lab. Syst. **2**(1–3), 37–52 (1987)
30. Wu, H., Wang, C., Tyshetskiy, Y., Docherty, A., Lu, K., Zhu, L.: Adversarial examples for graph data: deep insights into attack and defense. In: Proceedings of the 28th International Joint Conference on Artificial Intelligence, pp. 4816–4823. AAAI Press (2019)
31. You, Y., Chen, T., Wang, Z., Shen, Y.: When does self-supervision help graph convolutional networks? arXiv preprint arXiv:2006.09136 (2020)

Deep Learning Based Urban Anomaly Prediction from Spatiotemporal Data

Bhumika(✉) and Debasis Das

Indian Institue of Technology Jodhpur, Jodhpur, India
{bhumika.1,debasis}@iitj.ac.in

Abstract. Urban anomalies are unusual occurrences like congestion, crowd gathering, road accidents, natural disasters, crime, etc., that cause disturbance in society and, in worst cases, may cause loss to property or life. Prediction of these anomalies at the early stages may prevent significant loss and help the government to maintain urban sustainability. However, predicting different kinds of urban anomaly is difficult because of its dynamic nature (e.g., holiday versus weekday, office versus shopping mall) and presence in various forms (e.g., road congestion may be caused by blocked driveway or accident). This work proposes a novel integrated framework *UrbanAnom* that utilizes a data fusion approach to predict urban anomaly data using gated graph convolution and recurrent unit. To evaluate our urban anomaly prediction framework, we utilize multi-stream datasets of New York City's urban anomalies, points of interest (POI), roads, calendar, and weather that were collected via smart devices in the city. The extensive experiments show that our proposed framework outperforms baseline and state-of-the-art models.

Keywords: Urban anomaly · Deep learning · Data fusion · Spatio-temporal data · Gated graph convolution network · Gated recurrent unit

1 Introduction

The term *anomaly* refers to a deviation from the normal or expected pattern, and examples of anomalies include fraud, real-world events, criminal activity, traffic congestion, crowding, etc. One such type of anomaly is the *urban anomaly*, which we see around us in the form of traffic congestion, fairs, market promotions, fire incidents, criminality, etc., and may pose hazards to the general public's safety or result in financial losses. Statistics show that the annual cost of traffic congestion in four major Indian cities-Delhi, Mumbai, Bengaluru, and Kolkata is Rs. 1.47 lakh crore [4]. Therefore, reducing life or economic losses might be possible with early and accurate urban anomaly prediction. The local government, for instance, can organize transportation and mobility management during the festival season to avoid an unneeded stampede. With the help of this study, we hope to promote sustainable urbanization by foreseeing various types of urban anomalies.

Predicting urban anomalies traditionally involves a lot of effort. For instance, feature-based techniques rely on extracted features [20], which necessitate

M.-R. Amini et al. (Eds.): ECML PKDD 2022, LNAI 13713, pp. 242–257, 2023.
https://doi.org/10.1007/978-3-031-26387-3_15

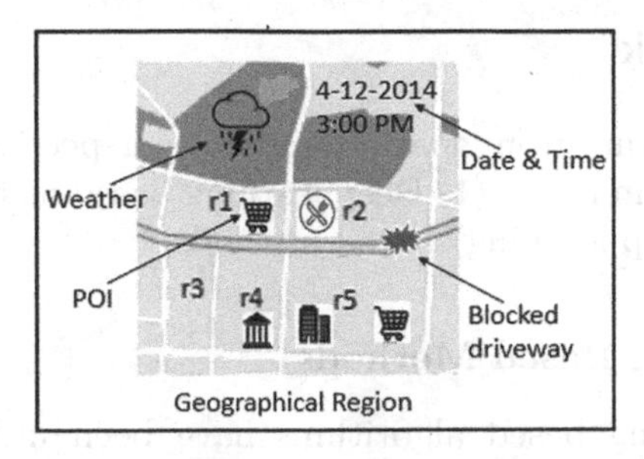

Fig. 1. Blocked driveway anomaly in urban areas

domain expertise to accurately capture the intricate dynamics of the urban anomaly. While other research [9,10] attempts to forecast anomalies but are only capable of coarse-grained prediction with low accuracy, we are aiming to predict different types of anomaly in a region bordered by roadways. An example of anomalies in urban areas is shown in Fig. 1, where blocked driveway anomaly affects region $r2$ and $r5$ directly and inferred from different contextual information like weather, point of interest (POI), and date & time. Difficulties in modelling anomaly prediction from several viewpoints are dynamic nature, rare occurrences, area dependency, and direct-indirect influencing factors; these difficulties drive a novel framework design.

We present the integrated framework where multiple deep neural networks capture different aspects of data and fuse their output to achieve common objective. To address the challenge of the dynamic nature of the urban anomaly, we extract spatial and temporal insights. Next, to solve the challenge of region dependency, we form regions with the help of road network with the intuition that it may pay attention towards illegal parking and blocked driveway. To incorporate the influence factors of urban anomaly, we use weather-related features. Spatial and temporal aspects are learned by the framework separately. Later, to join these modules, we fuse the output as input to global attention layer. To get the relevance of different anomalous events, the attention layer is used to predict future events more accurately. Finally, the hidden states from the attention layer are passed as input to multi-layer perceptron to predict the anomaly category in a region. We utilize a number of real-world datasets, including those gathered from New York City's 311 complaints, POIs, and weather stations. Our contributions are summarized as below:

- A novel framework, namely UrbanAnom is proposed to predict urban anomaly of specific categories in particular region.
- We propose a GatedGCN based method to capture inter-region relationships in the city. A Stacked GRU based modeling approach is chosen to take advantage of long-term and short-term temporal dependency.
- The extensive experiments on real-world urban anomaly dataset shows that UrbanAnom outperforms in terms of different metrics like F-measure, macro-F1, and micro-F1 of 83%, 85%, and 83%, respectively, from baseline as well as state-of-the-art models.

2 Related Work

A review of the literature from spatio-temporal perspective is presented in this section with two aims in mind: (1) Deep learning based methods, and (2) Hybrid learning (graph + deep learning) based methods.

2.1 Deep Learning Based Methods

Recently, deep learning based algorithms have been utilised with promising results in a variety of anomaly detection and prediction tasks, including crowd gathering, traffic accidents, criminal prediction, etc. Among deep learning techniques, recurrent neural networks have demonstrated superior performance in a variety of spatial-temporal tasks, including weather forecasting, stock market forecasting, accident forecasting, etc. In particular, Jiang et al. [12] predicted crowd dynamics from video data. They made predictions about future crowd density and flow using a multi-task convLSTM encoder-decoder. Another work done by Zhou et al. [29] have suggested utilising deep learning to predict crimes like robberies and burglary by combining spatial, temporal, and semantic data into latent space. Huang et al. [11] suggested a multi-view multi-model spatial-temporal learning (MiST) framework used a recurrent neural network and pattern fusion module to forecast city-wide anomalous events. For predicting traffic accidents, they suggested a dynamic fused network framework that makes advantage of hierarchical deep learning. Additionally, Shimosaka et al. [18] proposed mixed-order poisson regression from GPS data to find nationwide abnormal events. In order to predict urban anomalies, Huang et al. [10] created a hierarchical deep neural network that combined geographical, temporal, and category aspects. We also use the concept of integrating different deep learning models since an integrated framework can better capture the dynamic behavior of events. The performance of deep learning-based approaches is usually improved by the use of attention mechanisms in current trends. We also employ this concept to enhance the performance of the entire framework.

2.2 Hybrid Learning (Graph + Deep Learning) Based Methods

In general, several factors influence urban anomalies, and data analysis using a single dimension does not reveal any underlying correlations. Therefore, Zhang et al. [26] suggested employing graph embedding and neural network to detect anomaly from spatio-temporal data. A multi-modal fusion model for urban anomaly prediction from a spatial and temporal perspective was put out by Liu et al. [16]. They obtained spatial information using a graph convolution network, and temporal features using a gated recurrent unit. On the other hand, we change the general architecture and take into account extra contextual information like the calendar and weather data. Zhao et al. [27] solution to the traffic prediction problem, which incorporates both spatial and temporal relationships, used a temporal graph convolution network. In a different article, authors Liu et al. [15] also suggested a system that used adaptive graph convolution and temporal convolution to solve the challenge of urban anomaly detection. To efficiently capture various inherent information, the integrated architecture has been applied to numerous prediction tasks. Urban anomalies can be predicted using spatial and temporal clues, but we also need to take the context into account. Therefore, we employ an integrated framework to capture different aspects.

3 Preliminaries

In this section, we first define the terms and then formally define the problem statement for urban anomaly prediction. Particularly, we consider R geographical regions of an urban area and A anomaly categories with T time window.

3.1 Notation

Definition 1. ***Region Graph*** *In this study, we use the map segmentation method [24] on the road network, such as highway and arterial roads, to split the city into regions* $R = r_1, r_2, \ldots, r_n$. *In the proposed architecture, we take inter-region graph formulation into account. Each region* $r_i \in R$ *functions as a node* $v \in V$ *of the graph* $G = (V, E)$ *in the inter-region case, where* V *indicates the set of all disjoint regions and* E *is the set of all connecting pathways of the regions. If two regions are close to one another, an edge* $e_k \in E$ *exists between* v_i *and* v_j *such that* $i \neq j$ *where* $(u, v) \in V$. *The region graph's adjacency matrix can be defined as* $\mathcal{RG} \in \mathbb{R}^{V*V}$. *In* $\mathcal{RG}$, *we specifically set the element* $\mathcal{RG}_{ij} = 1$ *if a connecting path exists between two regions and* $\mathcal{RG}_{ij} = 0$ *if there isn't a connecting path.*

Definition 2. ***Point of interest*** *The point of interest (POI) dataset includes latitude and longitude positions for hospitals, businesses, educational institutions, retail locations, etc., that serve as a feature of graph nodes. The rationale behind taking POI into account is to identify correlations between various places depending on how they function (such as a hospital or commercial area). For instance, a similar functioning region in an urban location would have a similar anomalous pattern, according to Yuan et al. [23] study. If* F *is the number of POI categories, then the adjacency matrix for POI can be written as* $\mathcal{PI} \in \mathbb{R}^{V*F}$. *The element* $\mathcal{PI}_{ij}$ *is set to* 1 *in* $\mathcal{PI}$ *if a specific POI category is present in a region and* $\mathcal{PI}_{ij}$ *to 0 in all other cases.*

Definition 3. ***Temporal Anomaly Stream*** *Data in the prediction of urban anomalies shows a temporal stream that varies over time. This temporal stream is represented for a region* r_i *at time step* k *as* $\mathcal{TS} = (Y_i^{1k}, Y_i^{2k}, \ldots, Y_i^{lk})$, *where* $\mathcal{TS} \in \mathbb{R}^{n*l*k}$ *is the record of an anomaly of* l *category in* k *time slots at* r_i *region. When an anomaly of category* a_l *occurs at the* k^{th} *time step at region* r_i *the adjacency matrix for the temporal stream has the value* $\mathcal{TS}_i^{lk} = 1$ *and* $\mathcal{TS}_i^{lk} = 0$ *otherwise.*

Definition 4. ***Weather and Calendar Context*** *It stands to reason that weather has an impact on anomalous occurrence because obstructed driveway reports are more often in adverse weather. As a result, we incorporate the weather as a crucial component, which is denoted as* $\mathcal{W} \in \mathbb{R}^{1 \times f_w}$*. The urban anomaly changes with time as well. For instance, because most individuals tended to sleep at night, there are fewer complaints at night. Additionally, there is a different pattern of complaints during the week and on vacations. We therefore divide the given day, which consists of 24 h, into six-hour periods and represent this as a one-hot encoding vector,* $\mathcal{CL} \in \mathbb{R}^{1 \times f_{cl}}$.

3.2 Problem Statement

Solutions for predicting urban anomalies typically focus on extracting spatio-temporal data without taking context into consideration. On the other hand, we take into account semantic, spatial, and temporal data. The goal is to learn a predictive function that predicts l anomaly categories across n regions in s future time steps given historical data with l anomaly categories $A = (a_1, a_2, \ldots, a_l)$ over n region $R = (r_1, r_2, \ldots, r_n)$ and k time step $T = (t_1, t_2, \ldots, t_k)$. The formal representation of problem is given as:

$$y_n^{l,(k+s)} = \Psi(\mathcal{RG}, \mathcal{PI}, \mathcal{TS}, \mathcal{W}, \mathcal{CL}); \quad (1)$$

where $\Psi(\cdot)$ is a approximation function that we want to learn with input arguments region graph ($\mathcal{RG}$), point of interests ($\mathcal{PI}$), temporal anomaly ($\mathcal{TS}$), weather information ($\mathcal{W}$), and calendar data ($\mathcal{CL}$). The outcome is $y_n^{l,(k+s)}$, which is a prediction of all anomaly categories l in every region n over the next s time steps.

4 Framework: UrbanAnom

In this section, UrbanAnom framework is described in detail with introduction of the model input and the motivation for proposed framework. As Fig. 2 shows the architecture of UrbanAnom that consist four major modules: Semantic Spatial Module, Context Aware Temporal Module, Global Attention, and Multi-Layer Perceptron.

Definition 5. ***Anomaly context tensor*** *The input for the model is adjacency region matrix (*$\mathcal{RG}$*), point of interest matrix (*$\mathcal{PI}$*), temporal anomaly stream (*$\mathcal{TS}$*), calendar (*$\mathcal{CL}$*), and weather embedding (*$\mathcal{W}$*). As shown in Fig. 2, context aware temporal module have extracted* $\mathcal{RG}$*,* $\mathcal{PI}$ *and* $\mathcal{CL}$*,* $\mathcal{W}$ *data along with* $\mathcal{TS}$*. In case of semantic spatial dependency adjacency matrix* $\mathcal{RG}$ *and* $\mathcal{PI}$ *are fed into GatedGCN.*

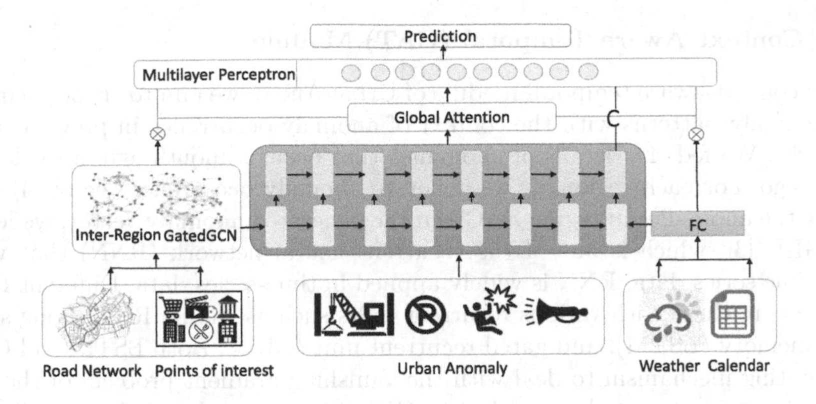

Fig. 2. Graphical representation of proposed architecture

4.1 Semantic Spatial (SS) Module

To consider spatial information from the geographical region along with semantic signals, we also include POI details. The intuition behind including POI is that similar functioning regions may have similar kinds of anomalies. Traditionally, convolution neural network (CNN) is used to capture the local spatial information, but it does not work well with non-euclidean space such as graphs. We divide the city into regions according to road network, and considering each region as node of the graph, CNN is unable to capture complex topological relationships from the graph network. Recently, CNN variant that works over graph has been come into existence, such as graph convolution network (GCN). A benchmark over graph neural network given by [5], shows gatedGCN works better in the node classification, graph classification, and link prediction, which is anisotropic variant of GCN. Therefore, we decide to use gatedGCN [1] in the task of extracting spatial information of geographical region.

Let h_i^l is a hidden unit at layer l attached with node i. The updated unit h_i^{l+1} at next layer $l+1$ in GatedGCN uses bath normalization, edge gates and residual connections which is represented by the equations:

$$h_i^{l+1} = h_i^l + ReLU(BN(U^l h_i^l + \sum_{j \in N_i} e_{ij}^l \odot V_l h_j^l)), \tag{2}$$

where $\odot$ is a Hadamard product, $U^l, V^l \in \mathbb{R}^{d*d}$ and edge gates e_{ij}^l is used as soft attention represented in equation as:

$$e_{ij}^l = \frac{\sigma(\hat{e}_{ij}^l)}{\sum_{j' \in N_i} \sigma(\hat{e}_{ij'}^l) + \epsilon}, \tag{3}$$

$$\hat{e}_{ij}^l = \hat{e}_{ij}^{l-1} + (BN(A^l h_i^{l-1} + B^l h_j^{l-1} + C^l \hat{e}_{ij}^{l-1})), \tag{4}$$

where ϵ is hyperparameter for numeric stability, σ represent activation function, $A^l, B^l, C^l \in \mathbb{R}^{d*d}$. The difference between GCN and gatedGCN is that the later one take cares edge feature at each layer. In summary, we use GatedGCN [1] to learn semantic spatial relationship of anomalies.

4.2 Context Aware Temporal (CAT) Module

In the context-aware temporal module of UrbanAnom, we aim to encode sequential anomaly patterns with the context of anomaly occurrence in previous days or weeks. We fed the details of anomalies that occur a month ago, a week ago, a day ago. For each region r_i, we generate anomaly occurrence vector A_l^t that reflect the anomalies in region r_i. Given the generated anomaly vector, we leverage GRU [3], which is one of the recurrent neural network (RNN) that work with time-series data. RNN is widely applied in time-series data. Different RNN variants are there with various recurrent units such as vanilla RNN, Long short term memory (LSTM), and gated recurrent unit (GRU). Both LSTM and GRU have gating mechanism to deal with the vanishing gradient problem of the traditional recurrent neural network, but GRU is less complex and more efficient in small training data. Our framework is flexible to change recurrent units; the effect of change is also explored.

In GRU, on each timestep t, we have input x^t, i.e., features and hidden state h^t. The hidden state also acts as a memory block, and operations of the memory block are controlled by two gates, namely the update gate and reset gate. Update gate controls what part of the hidden state is updated versus preserved, and reset gate controls what part of the previous state are used to compute new content. The operation on memory block performed by gates using following equations at each timestep:

$$u^t = \sigma(W_u h^{t-1} + U_u x^t + b_u) \tag{5}$$

Update gate (u^t) controls what part of hidden state are updated versus preserved.

$$r^t = \sigma(W_r h^{t-1} + U_r x^t + b_r) \tag{6}$$

Reset gate (r^t) controls what part of previous state are used to compute new content.

$$h'^t = tanh(W_h(r^t \cdot h^{t-1}) + U_h x^t + b_h) \tag{7}$$

(h'^t) represent the next hidden state.

$$h^t = (1 - u^t) \cdot h^{t-1} + u^t \cdot h'^t \tag{8}$$

(h^t) is the current hidden state. Here W_u, W_r, W_h are weights and b_u, b_r, b_h are biases. While σ and tanh are activation functions, Dot (.) represent element wise product.

4.3 Global Attention Module

Limitation of neural network based architectures is that they represent fixed length internal representation, which is not good for representing long dependencies. In our case, for a specific region r_i, complex dependencies exist among spatial and temporal anomaly occurrences. In our UrbanAnom architecture, we use attention mechanisms that pick the most essential signals to capture short and long distance dependencies, in order to avoid the situation where only the last hidden vector is used to represent spatial and temporal patterns [17]. The ability of the attention mechanism to selectively focus on a portion of crucial information has been demonstrated in machine translation and image analysis tasks. This inspires us to prefer global attention above representations of hidden spatial and temporal states. Attention mechanism is given by the equations as:

$$attn_n = tanh(W_{attn} h_n + b_{attn}) \tag{9}$$

$$\Lambda_n = \frac{exp(attn_n^T W_m)}{\sum_{n'} exp(attn_n^T W_m)} \tag{10}$$

$$\hat{a} = \sum_{n=1}^{N} \Lambda_n W_{attn} h_n \tag{11}$$

where W_{attn}, b_{attn}, W_m are training parameters, h_n shows the hidden state learnt from lower layer and the number of input vectors represented by N. Learned importance weight represented by α_n and $\hat{a}$ represent new hidden representation called attention vector. In our case we utilize the attention over hidden states of both semantic spatial module and context aware temporal module.

$$\Upsilon = \Delta(\hat{a}(SS_{h^t}), \hat{a}(CAT_{h^t})) \tag{12}$$

where Υ represents the global attention, SS_{h^t} and CAT_{h^t} are the hidden states of semantic spatial module and context aware temporal module. The symbol Δ is the fusion function between SS and CAT attention output.

4.4 Multi-layer Perceptron Based Prediction Module

The multi-Layer perceptron is used as a last layer in the prediction phase of the proposed UrbanAnom architecture to generate the presence of anomalies in various categories of each unique region r_i. The MLP is able to describe anomaly occurrence probabilities using a softmax function. The output of SS, CAT, and Υ is dynamically fused into a multilayer perceptron network to generate the final anomaly prediction represented as an equation below:

$$y_n^{l,k} = fc(W_{ss} * SS + W_{CAT} * CAT + W_{at} * \Upsilon_t) \tag{13}$$

where W_{ss}, W_{CAT}, W_{at} are learnable parameters and $fc(\cdot)$ represent the fully connected layer of perceptrons. The cross entropy loss function is defined as:

$$L = \sum_{n,l,k \in A} y_n^{l,k} log \hat{y}_n^{l,k} + (1 - y_n^{l,k}) log(1 - \hat{y}_n^{l,k}) + \lambda R_{reg} \tag{14}$$

where $\hat{y}_n^{l,k}$ denotes the predicted anomalous event of the l category in region r_i in k^{th} time slot. We use L_2 norm as regularization R_{reg} function and λ is adjustable hyperparameter. The model parameters are learned during minimization of loss function.

5 Evaluation

We conduct experiments to determine the efficiency of the proposed framework using datasets from New York City, including NYC-Urban Anomaly, NYC-POI, NYC-Road Network, NYC-Weather, and NYC-calendar. In this section, we provide a description datasets used, parameter settings, performance validation, parameter sensitivity, and evaluation of variants.

5.1 Dataset

In developed nations, the anomaly reporting system also emerged along with the rise in urbanisation. As a result, we run independent experiments to predict anomalies of different categories while validating our proposed framework using various real-world datasets from New York City[1]. The city of New York has a 311 emergency service platform[2] that lets residents file complaints by phone call, text message, or mobile app. Traffic congestion, crime, fire events, and other anomalies can occur in metropolitan areas, but we have chosen the data given by the 311 emergency service in New York City as our dataset for anomalies.

Brief explanation of the datasets is given as: 1) NYC-Urban Anomaly: Dataset contains latitude, longitude, complaint type, and timestamp information. Four types of anomalies, including blocked driveways, noise, illegal parking, and building use, have been the subjects of our experiments. The reason for selecting only these categories is because they are common occurrences and simple to compare with prior research. The distribution of urban anomalies is depicted in Fig. 3, with darker colours denoting more anomalies in a given area. 2) NYC-POI: The dataset containts geo-coordinated information on different categories are grouped into six main categories, including education, food & dinning, health & beauty etc. The POI data is extracted from OpenStreetMap API[3]of year 2017 and assume there is no major change in POI information. Table 1 shows the statistics of the dataset. 3) NYC-Road Network: The main component of the transportation system is the road network. To segment the road network dataset of New York City into regions, a map segmentation method. The road network information is given in the website[4]. 4) NYC-Weather: We acquired meteorological information for New York City from WunderGround[5], which included temperature information as well as 18 characteristics of various weather conditions, such as sunny, rain, and haze, etc. 5) NYC-Calender: The calendar information, including the days of the week, the weekend, and the holidays, was retrieved from the Holiday library[6].

[1] https://opendata.cityofnewyork.us/.
[2] https://portal.311.nyc.gov/.
[3] https://www.openstreetmap.org/.
[4] https://figshare.com/articles/dataset/Urban_Road_Network_Data/2061897.
[5] https://www.wunderground.com/.
[6] https://pypi.org/project/holidays/.

Table 1. Dataset statistics of urban anomaly and POI from NYC

Urban Anomaly from NYC	
Category	**Instances**
Blocked Driveway	74,698
Noise	134,690
Illegal Parking	57,374
Building Use	24,319

Point-of-interest (POI) from NYC	
Category	**Instances**
Business to Business	3717
Education	1062
Government & Community	3116
Food & Dinning	3385
Health & Beauty	4336
Real Estate & Construction	4675

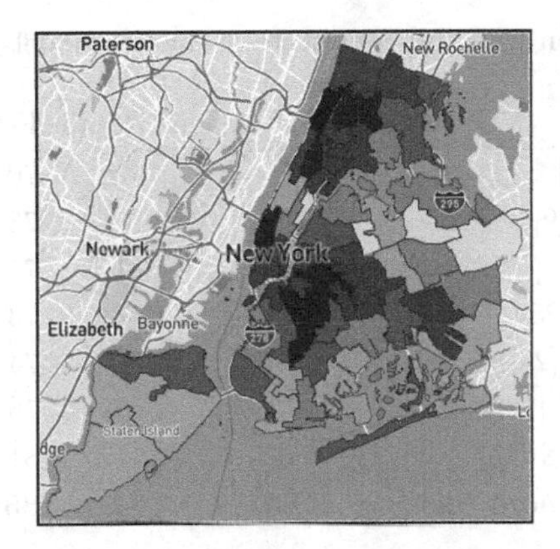

Fig. 3. Anomaly distribution in NYC

5.2 Parameter Settings

All experiments were conducted on Google Colab using a GPU specification of Tesla P100-PCIE-16 GB and an Intel Xeon 2.20 GHz processor. The Colab environment allotted 12 GB RAM and 34 GB of hard drive space for memory. We use the PyTorch toolkit to create the model in Python. We divided the dataset into three sets: a training set (8 months), a validation set (2 months), and a test set (4 months) (2 months). The hyper-parameters are fine-tuned using the validation set, and test data is used for the final performance assessment. Adam optimizer to train it with a $1e-3$ learning rate. In stacked GRU, we set the hidden dimension size to 32, and 4 GRU layers are used with the topmost layer as attention. In GateGCN, 2 layers are used, the global attention dimension is set to 32, and MLP layers are 3. Batch size for the experiment is set to 64, and regularization parameter λ is set to 0.01.

5.3 Performance Validation

We compare the performance of UrbanAnom with various baselines; Table 2 describes the results of urban anomaly prediction in terms of accuracy, precision, recall, and F-measure. In the urban anomaly prediction, UrbanAnom outperforms the existing baseline approaches, according to the evaluation results.

Table 2. Comparison with different baselines in percentage (%)

Methods	Accuracy	Precision	Recall	F-Measure
SVR [2]	69	68	65	66
LR [8]	67	66	67	66
ST-RNN [14]	72	69	70	69
LSTM [7])	75	71	73	72
GRU [3]	74	73	72	72
ARM [6]	79	77	76	76
UrbanAnom	**85**	**83**	**84**	**83**

Table 3. Comparison with different state-of-the-art models in percentage (%)

Methods	Accuracy	Precision	Recall	F-Measure
DCA [28]	–	75	62	70
CUAPS [9]	66	70	76	73
UAPD [21]	66	69	74	71
ind+int [25]	69	68	77	73
DAUAD [26]	–	70	75	74
ST-MFM [16]	74	73	80	79
DST-MFN [15]	–	77	84	81
UrbanAnom	**85**	**83**	**84**	**83**

This illustrates the advantage of taking into account the integration of semantic spatial, context-aware temporal modules. Second, methods based on neural networks perform better than standard machine learning because neural networks are better at learning hidden and non-linear correlations. Third, attention-based techniques capture long-term dependencies more effectively than a simple recurrent neural network. Finally, just using time-series data is inefficient compared to recurrent neural networks; in order to accurately forecast anomalies, we must include spatial, temporal, and semantic information.

Additionally, we look into how well UrbanAnom predicts various categories of anomalies; the results are given in Figs. 4a, 4b, 4c, and 4d. Noise, illegal parking, blocked driveways, and building use are some of the different categories we include for evaluation. We find that UrbanAnom performed better than baseline methods in predicting individual category anomalies and is capable of modelling region, time, and category data efficiently. Last but not least, we compare our model to current state-of-the-art urban anomaly prediction models. The results are reported in Table 3. As shown in Table 3, UrbanAnom predicts urban anomaly more accurately in terms of different performance metrics.

5.4 Parameter Sensitivity

In this subsection, we examine the robustness of UrbanAnom, we examine the effect of different hyperparameter settings (i.e., attention dimension, hidden state

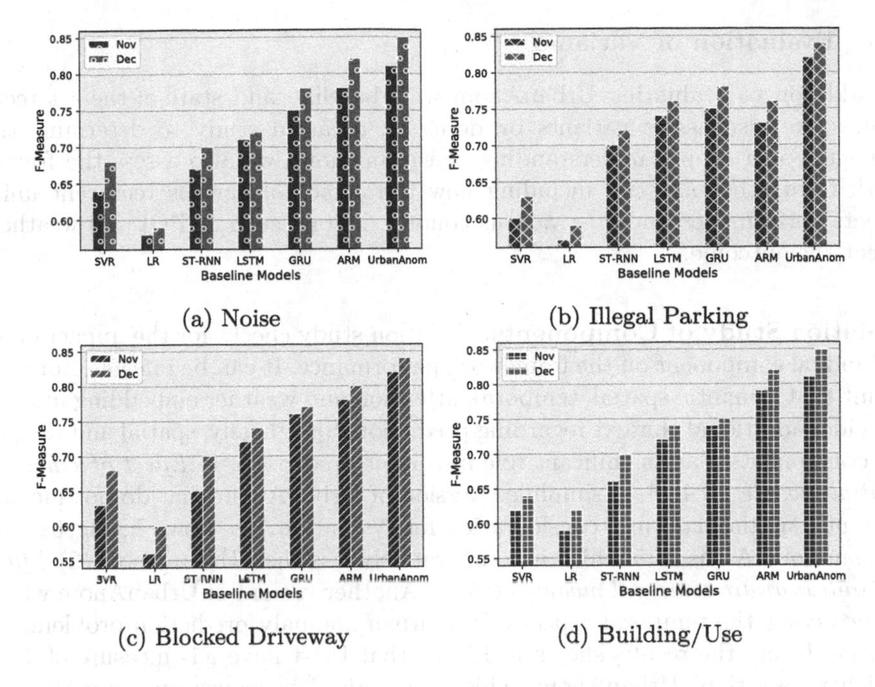

(a) Noise (b) Illegal Parking

(c) Blocked Driveway (d) Building/Use

Fig. 4. Predicting results for individual anomaly category

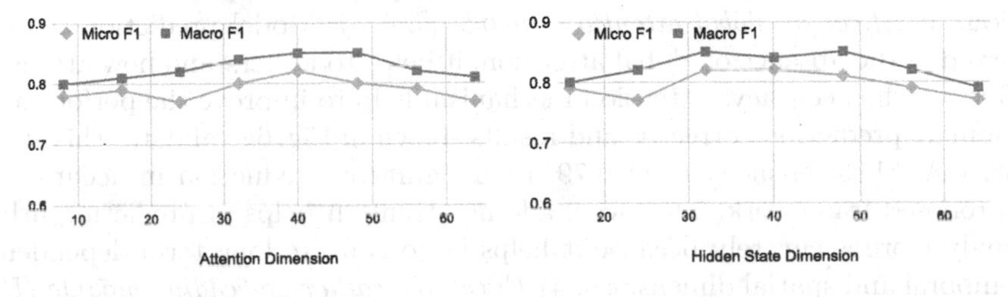

(a) Attention dimension effect (b) Hidden state dimension effect

Fig. 5. Hyper-parameter sensitivity results

dimension), and the results are shown in Fig. 5a, and Fig. 5b. Other parameter are set to the default value, except the parameter being tested. We observe from Fig. 5a that increasing the dimension size increases the performance initially, but it also occurs with extra computation cost. Therefore, we set the dimension size as 32 to provide a balance between efficiency and computation cost. From Fig. 5b, it is also observed that we got peak performance when hidden state dimension set to 32. We also observe that both hyperparameters, i.e., attention dimension and hidden state dimension have less effect on the performance, which shows the robustness of the proposed framework.

5.5 Evaluation of Variants

In addition to evaluating UrbanAnom with baseline and state-of-the-art techniques, we also assess variants by doing an ablation study to determine the impact. For a deeper understanding of UrbanAnom, we also assess the framework from other aspects, including how the choice of various recurrent units affects performance and how various context factors, such as POI and weather, affect the outcome.

Ablation Study of Components. Ablation study checks for the impact of an individual component on the framework performance. It can be inferred from the result that semantic spatial, temporal, attention and weather embedding module provides additional context regarding prediction. Specifically, spatial and temporal components plays significant role in correct prediction. 1) *Effect of semantic spatial module (UA-s)*: A simplified version of UrbanAnom that do not include semantic spatial data into consideration for evaluation. As shown in Fig. 6a, F1-measure of UA-s is 0.75 which is comparatively less than UrbanAnom. 2) *Effect of context aware temporal module (UA-c)*: Another variant of UrbanAnom which do not cover the temporal aspect of the urban anomaly prediction problem. As observed from the results shown in Fig. 6a that UA-t have F1- measure of 0.78, which is less than UrbanAnom. This effect raised beacause anomaly changes with respect to time and temporal component plays an important role in prediction. 3) *Effect of global attention module (UA-a)*: Model prediction score is observed in the absence of global attention; it helps to understand how attention is affecting the accuracy. Attention mechanism help to improve the performance of anomaly prediction correctly, and results shown in Fig. 6a validates this statement. UA-a has F1-measure of 0.79, its a significant reduction in accuracy of the proposed framework. Therefore, adding attention helps in predicting urban anomaly more accurately because it helps us to capture long term dependency in temporal and spatial dimensions. 4) *Effect of weather embedding module (UA-w)*: Weather information can be an important aspect on anomalies, so to check its effect, we evaluate our model without using it. As the intuition that weather may affect urban lives, results show its applicability. It is clear from Fig. 6a, UA-w performs less than UrbanAnom with F1-measure of 0.81.

Context, Recurrent Unit, and Graph Model Selection Effect. The effect of context information, recurrent unit, and graph model selection is shown in Fig. 6. Insights of the Fig. 6 are discussed as: 1) *Context information effect:* To improve the accuracy of predictions, we incorporate context information such as POI (P), weather (W), and calendar (R) into our framework. The impact of contexts on the model's performance is important to know the effect of individual context contribution in total prediction accuracy. In 6b, the F-measure is shown for the emphcontext-W (without $POI/W/R$) information. For instance, $P-W$ stands for model performance without POI data. As shown, extra information improves the model's accuracy by 10%, and weather information has a large

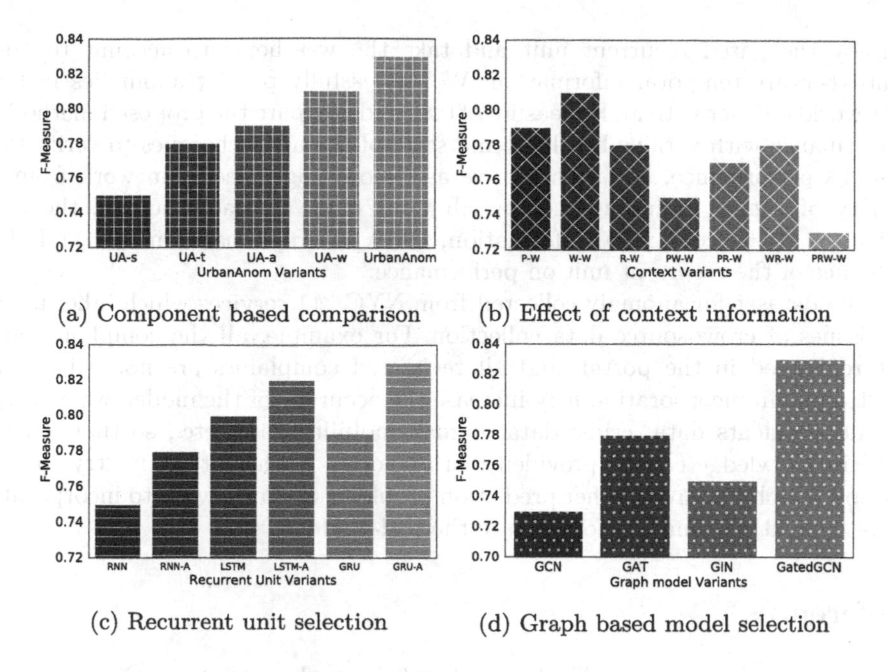

(a) Component based comparison

(b) Effect of context information

(c) Recurrent unit selection

(d) Graph based model selection

Fig. 6. Evaluation of variants

influence across all situations. The removal of all the context categories $PRW-W$ decrease the prediction accuracy significantly. 2) *Recurrent unit selection effect:* We utilize different recurrent units, i.e., RNN, RNN-A, LSTM, LSTM-A, GRU, GRU-A for temporal insights, where A denotes local attention. It is clear from the results shown in Fig. 6c that selection of the recurrent unit makes a huge difference in the accuracy of the model. The first observation that we have drawn is that RNN has lower performance among all the variants and GRU-A has the best one. Second, local attention also increases the F1-measure, as all the variants RNN-A, LSTM-A, and GRU-A perfoms better than their base model. 3) *Graph based model selection effect:* We experimentd with different variants of GCN, i.e., GCN [13], graph attention network (GAT) [19], graph isomorphic network (GIN) [22], and GatedGCN for spatial relation extraction. It is clear from the results shown in Fig. 6d that GatedGCN have shown better performance than other graph based models. We can also infer that GCN has lower performance among all the variants.

6 Conclusion

To minimise the economic loss to our society, urban anomaly prediction is a crucial endeavour. In this research, we proposed a solution to the problem of urban anomaly prediction. In order to represent semantic spatial relations, the metropolitan area is separated into regions based on the road network. We employ the gated recurrent unit and take the weather into account to add context-aware temporal information. We successfully predict anomalies in the real-world dataset with an F-measure of 0.83 and compare the proposed method's performance with various baselines and state-of-the-art techniques to verify the model's performance. Additionally, we assess our suggested framework from a variety of angles, including the contributions made by each module, the significance of the contextual information, effect of graph based model and the influence of the recurrent unit on performance.

The dataset for anomaly collected from NYC 311 services which inherits all the issues of crowdsource data collection. For example, all the complaints are not registered in the portal, and all registered complaints are not validated. Multi-domain incorporation may increase the accuracy of the model; we will try to add accidents data, crime data, human mobility data, etc., so that better domain knowledge can be provided to the model. In the future, we try to use proposed architecture in other prediction problems also and want to incorporate more context information to simulate the real-world.

References

1. Bresson, X., Laurent, T.: Residual gated graph convnets. arXiv preprint arXiv:1711.07553 (2017)
2. Chang, C.C., Lin, C.J.: LIBSVM: a library for support vector machines. ACM Trans. Intell. Syst. Technol. (TIST) **2**(3), 1–27 (2011)
3. Chung, J., Gulcehre, C., Cho, K., Bengio, Y.: Empirical evaluation of gated recurrent neural networks on sequence modeling. arXiv preprint arXiv:1412.3555 (2014)
4. Dash, D.K.: Traffic congestion costs four major Indian cities rs. 1.5 lakh crore a year. https://timesofindia.indiatimes.com/india/traffic-congestion-costs-four-major-indian-cities-rs-1-5-lakh-crore-a-year/articleshow/63918040.cms (2018)
5. Dwivedi, V.P., Joshi, C.K., Laurent, T., Bengio, Y., Bresson, X.: Benchmarking graph neural networks. arXiv preprint arXiv:2003.00982 (2020)
6. Feng, J., et al.: Deepmove: predicting human mobility with attentional recurrent networks. In: Proceedings of the 2018 World Wide Web Conference, pp. 1459–1468 (2018)
7. Hochreiter, S., Schmidhuber, J.: Long short-term memory. Neural Comput. **9**(8), 1735–1780 (1997)
8. Hosmer, D.W., Lemeshow, S.: Applied Logistic Regression. Wiley, New York (2000)
9. Huang, C., Wu, X., Wang, D.: Crowdsourcing-based urban anomaly prediction system for smart cities. In: Proceedings of the 25th ACM International on Conference on Information and Knowledge Management, pp. 1969–1972 (2016)
10. Huang, C., Zhang, C., Dai, P., Bo, L.: Cross-interaction hierarchical attention networks for urban anomaly prediction. In: IJCAI (2020)
11. Huang, C., Zhang, C., Zhao, J., Wu, X., Yin, D., Chawla, N.: Mist: a multiview and multimodal spatial-temporal learning framework for citywide abnormal event forecasting. In: The World Wide Web Conference, pp. 717–728 (2019)
12. Jiang, R., et al.: Deepurbanevent: a system for predicting citywide crowd dynamics at big events. In: Proceedings of the 25th ACM SIGKDD International Conference on Knowledge Discovery & Data Mining, pp. 2114–2122 (2019)

13. Kipf, T.N., Welling, M.: Semi-supervised classification with graph convolutional networks. arXiv preprint arXiv:1609.02907 (2016)
14. Liu, Q., Wu, S., Wang, L., Tan, T.: Predicting the next location: A recurrent model with spatial and temporal contexts. In: Thirtieth AAAI Conference on Artificial Intelligence (2016)
15. Liu, R., Zhao, S., Cheng, B., Yang, H., Tang, H., Li, T.: Deep spatio-temporal multiple domain fusion network for urban anomalies detection. In: Proceedings of the 29th ACM International Conference on Information & Knowledge Management, pp. 905–914 (2020)
16. Liu, R., Zhao, S., Cheng, B., Yang, H., Tang, H., Yang, F.: St-MFM: a spatiotemporal multi-modal fusion model for urban anomalies prediction. In: Proceedings of the Twenty-fourth European Conference on Artificial Intelligence (2020)
17. Qin, Y., Song, D., Chen, H., Cheng, W., Jiang, G., Cottrell, G.: A dual-stage attention-based recurrent neural network for time series prediction. arXiv preprint arXiv:1704.02971 (2017)
18. Shimosaka, M., Tsubouchi, K., Chen, Y., Ishihara, Y., Sato, J.: Moire: mixed-order poisson regression towards fine-grained urban anomaly detection at nationwide scale. In: 2020 IEEE International Conference on Big Data (Big Data), pp. 963–970. IEEE (2020)
19. Veličković, P., Cucurull, G., Casanova, A., Romero, A., Lio, P., Bengio, Y.: Graph attention networks. arXiv preprint arXiv:1710.10903 (2017)
20. Wang, Y., Xu, J., Xu, M., Zheng, N., Jiang, J., Kong, K.: A feature-based method for traffic anomaly detection. In: Proceedings of the 2Nd ACM SIGSPATIAL Workshop on Smart Cities and Urban Analytics, pp. 1–8 (2016)
21. Wu, X., Dong, Y., Huang, C., Xu, J., Wang, D., Chawla, N.V.: UAPD: predicting urban anomalies from spatial-temporal data. In: Ceci, M., Hollmén, J., Todorovski, L., Vens, C., Džeroski, S. (eds.) ECML PKDD 2017. LNCS (LNAI), vol. 10535, pp. 622–638. Springer, Cham (2017). https://doi.org/10.1007/978-3-319-71246-8_38
22. Xu, K., Hu, W., Leskovec, J., Jegelka, S.: How powerful are graph neural networks? arXiv preprint arXiv:1810.00826 (2018)
23. Yuan, J., Zheng, Y., Xie, X.: Discovering regions of different functions in a city using human mobility and pois. In: Proceedings of the 18th ACM SIGKDD International Conference on Knowledge Discovery and Data Mining, pp. 186–194 (2012)
24. Yuan, N.J., Zheng, Y., Xie, X.: Segmentation of urban areas using road networks. Microsoft Corp., Redmond, WA, USA, Technical reports MSR-TR-2012-65 (2012)
25. Zhang, H., Zheng, Y., Yu, Y.: Detecting urban anomalies using multiple spatio-temporal data sources. In: Proceedings of the ACM on Interactive, Mobile, Wearable and Ubiquitous Technologies, vol. 2, no. 1, pp. 1–18 (2018)
26. Zhang, M., Li, T., Shi, H., Li, Y., Hui, P.: A decomposition approach for urban anomaly detection across spatiotemporal data. In: IJCAI, pp. 6043–6049 (2019)
27. Zhao, L., et al.: T-GCN: a temporal graph convolutional network for traffic prediction. IEEE Trans. Intell. Transp. Syst. (2019)
28. Zheng, Y., Zhang, H., Yu, Y.: Detecting collective anomalies from multiple spatio-temporal datasets across different domains. In: Proceedings of the 23rd SIGSPATIAL International Conference on Advances in Geographic Information Systems, pp. 1–10 (2015)
29. Zhou, B., et al.: Escort: fine-grained urban crime risk inference leveraging heterogeneous open data. IEEE Syst. J. **15**, 4656–4667 (2020)

Detecting Anomalies with Autoencoders on Data Streams

Lucas Cazzonelli[(✉)] and Cedric Kulbach

FZI Research Center for Information Technology, Haid-und-Neu-Str. 10-14, 76131 Karlsruhe, Germany
{cazzonelli,kulbach}@fzi.de

Abstract. Autoencoders have achieved impressive results in anomaly detection tasks by identifying anomalous data as instances that do not match their learned representation of normality. To this end, autoencoders are typically trained on large amounts of previously collected data before being deployed. However, in an online learning scenario, where a predictor has to operate on an evolving data stream and therefore continuously adapt to new instances, this approach is inadequate. Despite their success in offline anomaly detection, there has been little research leveraging autoencoders as anomaly detectors in such a setting. Therefore, in this work, we propose an approach for online anomaly detection with autoencoders and demonstrate its competitiveness against established online anomaly detection algorithms on multiple real-world datasets. We further address the issue of autoencoders gradually adapting to anomalies and thereby reducing their sensitivity to such data by introducing a simple modification to the models' training approach. Our experimental results indicate that our solution achieves a larger gap between the losses on anomalous and normal instances than a conventional training procedure.

Keywords: Anomaly detection · Autoencoders · Data streams · Unsupervised learning

1 Introduction

Autoencoders (*AEs*) were found to enable the unsupervised learning of useful non-linear data representations. Consequently, they sparked significant progress in many areas of machine learning, including the area of unsupervised anomaly detection that deals with the identification of unusual events.

For this purpose, *AE*s are typically trained on large amounts of normal, non-anomalous data instances, reducing the networks' reconstruction error for this kind of data. Due to the absence of anomalies in the training data as well as the constraints imposed on *AE*s, losses for anomalies remain at a higher level and can therefore be distinguished from losses of normal instances. However, collecting and fitting large amounts of data can be an obstacle in many real-world applications. Data might for example not be available from the get-go but rather

M.-R. Amini et al. (Eds.): ECML PKDD 2022, LNAI 13713, pp. 258–274, 2023.
https://doi.org/10.1007/978-3-031-26387-3_16

arrive little by little in small chunks or even in continuous streams of individual samples. In many cases, a model may also be subject to distributional changes over time in the form of concept drift [27]. This is especially true for anomaly detection, as the definition of what constitutes an anomaly is not definitive [31] and may for instance depend heavily on the temporal context in which the data appears. As a result, incremental model updates can often be required to adapt the model to such variations after the initial training.

Based on this observation, an anomaly detector operating on a continuous and potentially infinite stream of individual samples should fulfill the requirements proposed by [4] to

R1: be able to process a single instance at a time,
R2: be able to process each instance in a limited amount of time,
R3: use a limited amount of memory,
R4: be ready to predict at any time,
R5: be able to adapt to changes in the data distribution.

To establish a unified view, we additionally formalize the task of online *Anomaly Detection* (*AD*) in Sect. 2.

As feed-forward neural networks, *AE*s inherently feature a fixed time (**R2**) complexity and, due to their fixed structure, a constant memory (**R3**) consumption with respect to the number of processed examples. Since they are conventionally optimized with gradient-based methods, *AE*s also allow processing individual examples (**R1**) to adapt to the most recent data instance (**R5**) and predicting at any time (**R4**). Even their tendency to forget previously learned data patterns [19] when being exposed to new tasks does not necessarily inhibit the performance of *AE* for *AD*. This can even be beneficial that forgetting old concepts or distributions of normality may often even be desired, as the sudden reappearance of data that would have previously been considered normal could be deemed as abnormal.

In conclusion, *AE*s seem well suited for the task of online *AD*. Nevertheless, only a handful of studies attempted to leverage the remarkable representation learning capabilities of *AE*s to detect anomalies in data streams. Rather, previous contributions focused on adapting conventional offline learning *AD* algorithms to fulfil the requirements of online learning (see e.g. [15,24,26]). In this work, we present an *AE*-based anomaly detector for the above requirements and demonstrate its ability to outperform previous state-of-the-art anomaly detectors on multiple real-world datasets. Further, we address the problem of contaminated training data as studied by [5]. Unlike in traditional offline learning, the training data of an online *AD* will inevitably be contaminated with anomalous instances, since the ability to remove such data from the data stream would make the model redundant in the first place. Thus, depending on the proportion and arrival of anomalies and other external circumstances, reconstruction losses for abnormal data can be expected to decrease to a greater extent than when training on non-contaminated data, degrading the model's detection accuracy [32]. To mitigate this effect on our approach, we propose an improved optimization objective for *AE*-based *AD*, that leads to increases in discrimination performance. To the best of our knowledge, we are the first to address the issue of training data contamination in the context of AE-based online AD.

To enable to enable the usage of our results in future applications, we implemented our approach as an extension[1] to the online learning framework River [17][2]. In conclusion, we provide the following contributions:

C1: A formalization of streaming AD.
C2: A competitive approach for AE-based anomaly detection on data streams.
C3: An alternative optimization technique to improve the performance of AEs anomaly detectors under the influence of anomalous training examples.
C4: An in-depth evaluation of our approaches with established real-world datasets.
C5: A Python package (See footnote 1) extending River [17] that facilitates the reproducibility and reuse of our work.

2 Problem Statement

When using an offline learning approach AEs infringe most of the defined requirements. Predictions for new data instances have to be calculated immediately at the time of their arrival to conform with **R1**. The requirements **R4** and **R5** pose the problem that online AD models also need to continuously adapt to their stream of inputs. In addition, data streams often have high frequency, requiring efficient resource utilization, as required by **R2** - **R4**. Therefore, the separation of training and testing stages as performed in most offline scenarios is not applicable when evaluating data streams. Accordingly, we define the problem of anomaly detection as follows:

Definition 1. *Incremental Anomaly Detection*
Let $\mathcal{S}^+ = \{(\boldsymbol{x}^{(i)}, y^{(i)}) \mid \forall i \in \mathbb{N}\}$ *be a potentially infinite sequence of tuples each consisting of a feature-vector* $\boldsymbol{x}^{(i)} \in \mathbb{R}^d$, $d \in \mathbb{N}$ *and a corresponding anomaly label* $y^{(i)} \in \{0, 1\}$. *Further, let* $\mathcal{S} = \{(\boldsymbol{x}^{(i)}, y^{(i)}) \mid \forall i \in \{1, \ldots, I\}\}$ *be a sub-sequence of previously observed instances. Let* $\mathcal{A} = \{\boldsymbol{x}^{(i)} \mid y^{(i)} = 1\, i \in \{1, \ldots, I\}\}$ *be the set of anomalies in* $\mathcal{S}$, *where* $\frac{|\mathcal{A}|}{I} \ll 0.5$.
Also, let $\mathcal{S}_{train}$ *and* $\mathcal{S}_{valid}$ *be disjoint subsets of* $\mathcal{S}$. *Given a validation protocol* $\mathcal{V}$ *and an arbitrary anomaly detection function* $g : \mathbb{R}^d \to \{0, 1\}$ *that was trained on* $\mathcal{S}_{train}$, *the objective of incremental anomaly detection can then be defined as*

$$\min_g \mathcal{V}(\mathcal{L}_c, \mathcal{S}_{valid}, g(\,\cdot\,; \mathcal{S}_{train})), \tag{1}$$

where $\mathcal{L}_c : \{0, 1\} \times \mathbb{R} \to \mathbb{R}$ *denotes a classification loss that quantifies the dissimilarity of the true label* $y^{(i)}$ *and the prediction* $\hat{y}^{(i)} = g(\boldsymbol{x}^{(i)})$.

[1] Available at https://github.com/lucasczz/DAADS.
[2] Available at https://github.com/online-ml/river.

As a validation protocol $\mathcal{V}$ we employ a *test-then-train* [4] protocol that is sensitive to *concept drifts.* In a *test-then-train* evaluation, models are first validated based on a performance metric $\mathcal{L}_c$ and then trained on each instance in $\mathcal{S}$ at arrival-time. Although our definition of incremental *AD* includes supervised approaches, we limit our analysis to unsupervised *AD*, where only the feature instances $\boldsymbol{x}^{(i)}$ are available for training.

3 Related Work

In this section, we depict the work related to our approach by presenting online learning approaches for the *AD* task as well as offline learning *AD* approaches that are based on *AE*. Finally, we summarize previous work on online learning and *AE*-based *AD*.

3.1 Offline Anomaly Detection

Anomaly Detection is a very active area of research that has produced a variety of different approaches for identifying anomalous data examples. Early studies focused on statistical measures in this context (see [21]). Machine learning approaches have relied on the assumption that anomalies are found in low-density areas. Highly successful examples of such approaches include *One Class Support Vector Machines* (*OC-SVM*s) [25], which learn a hyperplane that separates high-density normal data from low-density anomalous data, and *Local Outlier Factor* (*LOF*) [6], which scores data points based on the relative densities of their neighborhoods. Rather than directly assessing differences in density, *Isolation Forests* [25] separate anomalies with an ensemble of trees that iteratively splits the data along random attribute thresholds. Since the algorithm assumes that anomalies are easily separable from the rest of the data, they can be identified as instances isolated by a small number of splits. According to [14], *Isolation Forests* provide better scalability compared to previous techniques such as *OC-SVM*s and *LOF*.

In more recent years, significant progress in the area of *AD* has been achieved with deep learning models. Some of the most successful model types in this regard are *AE*s. Inspired by the functioning principle of *PCA*-based *AD* , Sakurada and Yairi [23] proposed to exploit the limited generalization of *AE* feature mappings, which lead to higher reconstruction losses when facing previously unobserved data patterns. Numerous studies subsequently improved the original *AD* algorithm's performance by introducing advanced model variations. Zhou and Paffenroth [29] for instance introduced a robust *AE* that iteratively separates noise from the underlying clean data whereas Gong et al. [11] mapped the *AE*'s representations to a fixed number of memorized vectors to avoid learning representations that generalize to anomalous data.

3.2 Online Anomaly Detection

Despite the success of *AE*s in offline learning, only a few studies investigated their usage as online anomaly detectors for data streams.

Mirsky et al. [16] adapted the concept of *AE* ensembles [7] to the task of online learning and added a secondary downstream *AE* which reconstructs the loss values of the ensemble, forming the *Kit-Net* model. To assign data features to individual *AE*s in the ensemble, *Kit-Net* requires initial training data.

In the realm of online *AD* previous research has been focused on adaptations of well-established conventional machine learning techniques. Particularly popular in this context are streaming variants of *LOF* and *IsolationForest.* Based on *Isolation Forests*, Tan et al. [26] for instance proposed the *Half Space Tree* (*HST*) *AD* approach that builds an ensemble of trees by iteratively halving random attribute subspaces of an initial data subset. Other tree-based methods include *RS-Forests* [28] and *Robust Random Cut Forest* (*RRCF*) [12]. The *xStream* anomaly detector introduced by [15] elaborates on the concept of iterative subspace cuts by applying random matrix projections before dividing examples based on randomly selected features that were created in the process. *LOF*-based streaming *AD* techniques include *Incremental Local Outlier Factor* (*iLOF*) [20] and its advancements, *MiLOF* [24] and *DILOF* [18], both of which aim to overcome the loss of density data that would result from simply removing the oldest instances to maintain fixed memory and run time usage.

There are several studies on training *AE*s and other neural networks in a more general online learning setting [1,22,30]. However, none have investigated the online learning of *AE*s in the specific context of *AD*, which differs considerably from most conventional scenarios, since the often targeted minimization of the reconstruction loss may even be disadvantageous in the case of *AD* if losses for anomalies are affected by it.

To overcome the high sensitivity of neural networks to the learning rate, Baydin et al. [3] proposed *Hypergradient Descent* (*HD*) optimization algorithms, which simultaneously optimize network parameters as well as the learning rate using gradient descent. We evaluate the *HD* modification of *Stochastic Gradient Descent* (*SGD*) for our approach in Sect. 6.

4 Streaming Anomaly Detection with Autoencoders

According to Definition 1, we define an online *AD* system as depicted in Fig. 1. We define an anomaly detector as $g = \tau \circ f$, where $f : \mathbb{R}^d \to \mathbb{R}$ is an anomaly scoring function that judges the abnormality of its input in the form of an anomaly score $z^{(i)}$, and $\tau : \mathbb{R} \to \{0, 1\}$ is a *thresholding* function that decides whether to label the current input as an anomaly given $z^{(i)}$. Since the choice of an appropriate threshold mostly depends on the characteristics of the particular application, e.g., the relative cost of false-positive versus false-negative classifications, we omit the thresholding problem and focus on the computation of anomaly scores in this work.

While arbitrary *AD* models can implement the scoring function f, we aim to improve upon the performance of existing techniques by introducing basic-*AE* and *Denoising AE* (*DAE*) online anomaly scorers as well as a *Probability Weighted AE* (*PW-AE*) which we specifically develop to address the issue of contamination. In the following, we briefly explain the functional concept of these models. In accordance with **R1**, we begin the score calculation for the latest sample with calling the reconstruction function $r_{\boldsymbol{\theta}}$ on the model input $\boldsymbol{x}^{(i)}$ by performing a forward pass through the *AE*, generating a reconstruction $\hat{\boldsymbol{x}}^{(i)}$. Subsequently, we calculate the reconstruction loss $l^{(i)} = \mathcal{L}(\boldsymbol{x}^{(i)}, \hat{\boldsymbol{x}}^{(i)})$. To accommodate for fluctuations of average reconstruction losses, we scale $l^{(i)}$ by applying the post-processing function π, yielding an anomaly score $z^{(i)}$. For simplicity, we scale the losses by the average of a sliding window μ_ω to obtain the anomaly score $z^{(i)}$. We further use the loss value $l^{(i)}$ to optimize the *AE*'s parameters $\boldsymbol{\theta}$ for every streaming instance by calculating a parameter update $\Delta^{(i)}\boldsymbol{\theta}$ with an optimization function *opt*, which we subsequently subtract from the current model parameters $\boldsymbol{\theta}$. For the optimization function *opt* any gradient-based optimization technique can be used, although we will subsequently assume standard *SGD* for the sake of simplicity. In Algorithm 1, we describe our approach towards online *AD* with the basic *AE* variant. For this type of model, we calculate the weight updates $\Delta^{(i)}\boldsymbol{\theta}$ according to the conventional *SGD* approach as

$$\Delta^{(i)}\boldsymbol{\theta} = \gamma \nabla_{\boldsymbol{\theta}}\, l^{(i)}, \tag{2}$$

where γ represents the learning rate and $\nabla_{\boldsymbol{\theta}}\, l^{(i)}$ the gradient of $l^{(i)}$ with respect to the model parameters $\boldsymbol{\theta}$. Apart from calculating a new reconstruction and training loss using dropout, we use the same update rule for the *DAE* as for the base variant. Through training with corrupted input data, *DAE*s were shown to

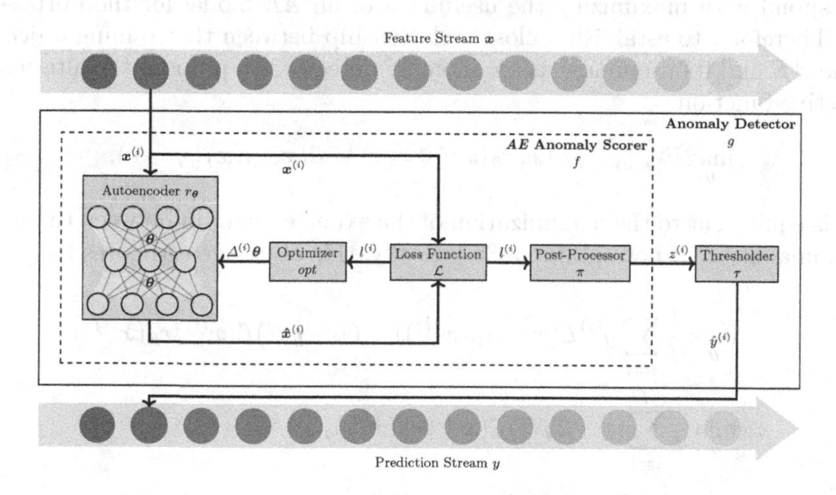

Fig. 1. Conceptual overview of proposed *AE*-anomaly scorer f within an incremental anomaly detector g.

Algorithm 1. Basic anomaly detection with *AE*

1: **Input:**
2: stream of inputs X, window size w_size, learning rate γ,
3: thresholding function τ, architecture *arch*
4: **Output:** stream of anomaly classifications $\hat{Y}$
5: $\boldsymbol{\theta} \leftarrow \text{init}(arch)$ ▷ Perform Glorot weight initialization [10]
6: $\omega \leftarrow \text{Window}(w\_size)$ ▷ Initialize sliding window
7: **for all** $\boldsymbol{x}^{(i)} \in X$ **do** ▷ Start Stream X
8: $l^{(i)} \leftarrow \mathcal{L}(\boldsymbol{x}^{(i)}, r_{\boldsymbol{\theta}}(\boldsymbol{x}^{(i)}))$ ▷ Calculate reconstruction loss
9: $z^{(i)} \leftarrow \frac{l^{(i)}}{\mu_\omega}$ ▷ Apply post-processing
10: $\hat{y}^{(i)} \leftarrow \tau(z^{(i)})$ ▷ Apply thresholding
11: $\Delta^{(i)}\boldsymbol{\theta} \leftarrow opt(l^{(i)})$ ▷ Calculate parameter update
12: $\boldsymbol{\theta} \leftarrow \boldsymbol{\theta} - \Delta^{(i)}\boldsymbol{\theta}$ ▷ Apply parameter update
13: $\omega.append(l^{(i)})$ ▷ Add loss to sliding window
14: **yield:** $\hat{y}^{(i)}$
15: **end for**

be able to achieve more robust representations and thus better *AD* accuracy in previous studies [23].

From a probabilistic point of view, *SGD* optimizes the objective function

$$\min_{\boldsymbol{\theta}} \mathbb{E}_{(\boldsymbol{x},y)\sim\hat{p}_S} \mathcal{L}(\boldsymbol{x}, r_{\boldsymbol{\theta}}(\boldsymbol{x})), \tag{3}$$

where $\boldsymbol{x}$ follows the empirical distribution defined by previously observed data $\hat{p}_S$. Due to the fact, that Equation (3) rewards decreasing loss values for both normal and anomalous instances, minimizing the above objective does not directly correspond with maximizing the usefulness of an *AE*-model for the purpose of *AD*. Therefore, to establish a closer relationship between the training objective of the *AE* and its usefulness as an anomaly detector, we propose the alternative objective function

$$\max_{\boldsymbol{\theta}} \mathbb{E}_{(\boldsymbol{x},y)\sim\hat{p}_S} \mathcal{L}(\boldsymbol{x}, r_{\boldsymbol{\theta}}(\boldsymbol{x}) \mid y=1) - \mathcal{L}(\boldsymbol{x}, r_{\boldsymbol{\theta}}(\boldsymbol{x}) \mid y=0), \tag{4}$$

that is equivalent to the maximization of the expected margin between the scores of anomalous- and normal data. Empirically, this objective estimates to

$$\begin{aligned} &\max_{\boldsymbol{\theta}} \frac{1}{I} \sum_{i=1}^{I} y^{(i)} \mathcal{L}(\boldsymbol{x}^{(i)}, r_{\boldsymbol{\theta}}(\boldsymbol{x}^{(i)})) - (1-y^{(i)}) \mathcal{L}(\boldsymbol{x}^{(i)}, r_{\boldsymbol{\theta}}(\boldsymbol{x}^{(i)})) \\ \Leftrightarrow &\min_{\boldsymbol{\theta}} \frac{1}{I} \sum_{i=1}^{I} (1-2y^{(i)}) \mathcal{L}(\boldsymbol{x}^{(i)}, r_{\boldsymbol{\theta}}(\boldsymbol{x}^{(i)})). \end{aligned} \tag{5}$$

Most *AE*-based *AD* applications assume that training is performed on data, where the number of anomalous instances ($y^{(i)} = 1$) is close to zero. Under this circumstance Equation (5) approximates the standard *AE* optimization goal.

However, this assumption is not necessarily valid for every phase of a data stream. Anomalies can, for example, concentrate in small time windows and therefore negatively affect the performance of a model trained to optimize the objective in Equation (3), as losses decrease for such instances. In the following, we derive an improved update rule from the objective defined in Equation (5). We refer to *DAE*s trained with this update rule as *PW-AE*s.

Since the true value of $y^{(i)}$ is unknown at the time of training, we substitute $y^{(i)}$ with an estimate of the probability $\hat{p}(y^{(i)} = 1)$, that we calculate by assuming $l^{(i)} \sim \mathcal{N}(\mu_\omega, \sigma_\omega)$, with μ_ω and σ_ω being the average and standard deviation of a sliding window ω of previous losses. Under these assumptions, we estimate

$$\hat{p}_{\mathcal{N}}(y^{(i)} = 1) = \Phi(\frac{l^{(i)} - \mu_\omega}{\sigma_\omega}), \tag{6}$$

where Φ denotes the cumulative standard-normal distribution function. Although reconstruction losses cannot be accurately represented by a normal distribution, they could improve the sensitivity of *AD* to anomalies. To incorporate the prior knowledge that on average $p(y^{(i)} = 1) \ll 0.5$ (see Definition 1), we calculate the final probability estimate $\hat{p}(y^{(i)} = 1)$ as

$$\hat{p}(y^{(i)} = 1) = \hat{p}_{\mathcal{N}}(y^{(i)} = 1) * \frac{1}{2p_0}, \tag{7}$$

where p_0 is a hyperparameter that determines the value of $\hat{p}_{\mathcal{N}}(y^{(i)} = 1)$ for which the weight update $\Delta^{(i)}$ assumes a value of 0. Using this probability estimate, the weight update for optimizing the objective in Equation (5) is given by

$$\begin{aligned} \Delta^{(i)}\boldsymbol{\theta} &= (1 - 2\hat{p}_{\mathcal{N}}(y^{(i)} = 1) * \frac{1}{2p_0})\gamma\nabla_{\boldsymbol{\theta}} l^{(i)} \\ &= (1 - \frac{\hat{p}_{\mathcal{N}}(y^{(i)} = 1)}{p_0})\gamma\nabla_{\boldsymbol{\theta}} l^{(i)} \end{aligned} \tag{8}$$

5 Experiments

We evaluated our approaches (*AE*, *DAE*, and *PW-AE*) based on 3 commonly used real-world datasets in comparison with state-of-the-art approaches for online *AD* in the following experiments:

Performance. We compare all approaches based on the area-under-the-curve measures of their *Receiver Operating Characteristic* (*ROC-AUC*) and *Precision Recall* (*PR-AUC*) curves as well as their runtimes.

Contamination Robustness. To evaluate the robustness against the degree of contamination, we present performance measures for different proportions of anomalies.

Parameters. Finally, we present an evaluation based on different design choices for the latent space, the optimizer, and the learning rate of *AD*.

In the following, we describe the used datasets and the experimental setting in which we evaluated our approach along with established online *AD* algorithms.

5.1 Data Streams

We conducted our experiments on incremental data streams which we emulated using the Covertype, Shuttle[3] and Creditcard [8] real-world data streams. The key characteristics of the selected streams are presented in the following.

Covertype originally consists of 7 classes representing different types of forest cover. Due to its non-stationary data distribution, Covertype is frequently used to emulate data streams in online learning (see e.g. [24,26]. We modified the dataset according to the standard procedure from previous work on *AD* (e.g. [26]), where the most frequent class is selected as normal and the rarest class as anomalous, resulting in an anomaly share of 0.96%.[4] We removed the remaining classes and categorical attributes, leaving a total of 10 numerical attributes. We then grouped the anomalous samples in short sequences of 2 to 9 samples and randomly distributed them throughout the data stream.

Shuttle originally contains 8 classes. According to [26], we used the first class as normal data and samples of the remaining classes, except for the excluded fourth class, as anomalous. The share of anomalies in this modified dataset is 7.15%. Originally in time order, the dataset donor later randomized the order of data examples causing the dataset to exhibit a largely stationary data distribution [26].

Creditcard [8] consists of credit card transactions performed in September 2013 out of which only 0.17% were marked as fraudulent, causing a highly imbalanced class distribution. Due to the confidentiality of the original transactions, the data is provided in the form of the first 28 principal components as well as the timestamp and monetary amount of each transaction. Since Creditcard features stream-typical characteristics such as concept drift, as well as a close relation to real-world fraud detection, the dataset was intensively investigated regarding online *AD* (see [8]).

Due to their different properties regarding the occurrence of concept drifts as well as their share of anomalous data, the selected data streams cover the necessary range of possible streaming-*AD* scenarios.

5.2 Setup

In this section, we present the experimental setup, necessary to obtain the results presented in Sect. 6. While the proposed *AD* algorithms can, in theory, be executed using any *AE* architecture and optimization algorithm, we used shallow

[3] Both published at the UCI ML Repository https://archive.ics.uci.edu/ml (last accessed February 18, 2023).

[4] All modified datasets are available at https://github.com/lucasczz/DAADS (last accessed February 18, 2023).

networks with coupled encoder- and decoder weights and *Scaled Exponential Linear Unit* (*SELU*) activations along with basic *SGD* optimization for the sake of simplicity and computational efficiency. For calculating the models' reconstruction errors, we used a smooth *Mean Absolute Error* (*MAE*) function, which we selected due to its advantage of being less prone to outliers compared to *Mean Squared Errors* (*MSE*s) [9].

For the performance evaluation, we used a learning rate of 0.02 for *AE* and *DAE*, and a higher value of 0.1 for the *PW-AE* to account for the adaptive reduction of its learning rate. Like [16], we determined the network width by using a *latent layer ratio* specifying the number of units in the hidden layer relative to the number of input features to account for the varying dimensionality between data streams. We evaluated the basic *AE* with a latent layer ratio of 10% and, due to the regularization induced by applying 10% dropout, we chose 100% for the *DAE* and *PW-AE* models. For the p_0 parameter of the latter variant, we used a value of 0.9.

We evaluate the proposed incremental *AE* anomaly detectors along with several well-established online anomaly detectors such as *iLOF* [20], *HST* [26], *RRCF* [12], *xStream* [15] and *Kit-Net* [16]. Except for *xStream*, we executed all reference algorithms using the authors' suggested parameter values and restricted the sizes of all sliding windows to a maximum of 250 data points. We ran *xStream* with an improved configuration due to poor performance with its default setting. For *Kit-Net*, we used the first 100 examples in each data stream to map the input features to the individual *AE*s.

6 Results

In this section, we present the results for the experiments depicted in Sect. 5. Figure 2 shows an example of the distribution of individual anomaly scores generated by different *AD* models for Shuttle. Denote that the value range of the anomaly scores differs significantly between the individual models, which can be remedied by adjusting the value range of the anomaly threshold. While all models yield higher average scores for anomalous examples, it can be seen that the *AE* models provided the most significant gap between scores of anomalous and normal samples throughout most of the data stream segment. The *ROC-AUC* and *PR-AUC* scores shown in Table 1 reflect this observation. While the tree-based *RRCF* and *HST* come close to the *AE* models in terms of *ROC-AUC*, the latter clearly outperformed all other approaches in terms of *PR-AUC*. For Covertype and Creditcard, we found similar results in that the *DAE* and *PW-AE* models yielded significantly higher *PR-AUC* values than the baseline approaches while being among the highest performing models in terms of *ROC-AUC*. Their high performance on all three of the investigated datasets highlights the models' ability to accurately identify anomalies for both data with and without concept drift and different degrees of class imbalance. While providing overall slightly less accurate anomaly scores, the basic *AE* also managed to exceed the baseline models' *PR-AUC* for all datasets except for Covertype. The *AE*s were among

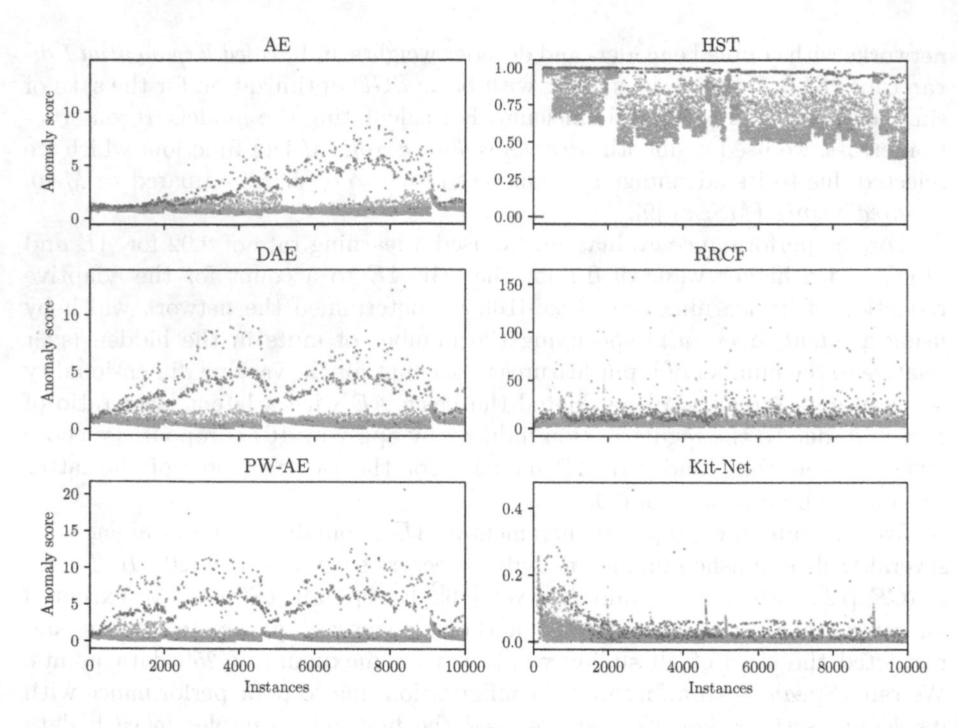

Fig. 2. Distribution of anomaly scores as a function of the number of processed instances on Shuttle. Anomalies are drawn in red. (Color figure online)

the most computationally efficient models, using only a fraction of the runtime (**R2** and **R4**) of most competing algorithms. Although being slower than *Kit-Net* on Covertype and Shuttle, the *AE* models' runtimes were still within the same order of magnitude. For Creditcard, the runtimes were, on the other hand, significantly lower, which made the *AE*s on average faster than the competing algorithms when considering all datasets. Since the memory requirement of an *AE* is influenced by the choice of architecture rather than by data instances occurring over time, the memory consumption is constant and can be intentionally limited/selected by adjusting the network's size. When comparing individual *AE* models, the *DAE* and *PW-AE* demonstrated a distinct advantage in terms of accuracy metrics. Especially for Covertype, the alternative model architecture led to significantly higher *PR-AUC* scores, which came at the cost of longer runtimes due to the additional forward pass needed to reconstruct the corrupted input for each training step. This advantage is also apparent by the distribution of scores displayed in Fig. 3, where the scores of anomalous data examples separate from the scores of normal data at a much higher rate at the start of the data stream. We achieved further improvements in discriminative performance by using the *PW-AE* update rule for Creditcard and - likely due to its higher grade of contamination - especially for Shuttle. With regard to **R5**, Fig. 3 shows, that the *PW-AE* adapted more quickly to normal samples in Shuttle than the

Table 1. Average benchmarking results for 10 random seeds. Best results are displayed in bold. Runtimes are given in core-minutes of an Intel(R) Xeon(R) Platinum 8180M CPU.

	Covertype			Creditcard			Shuttle		
Model	ROC-AUC	PR-AUC	Runtime	ROC-AUC	PR-AUC	Runtime	ROC-AUC	PR-AUC	Runtime
iLOF [20]	0.934	0.339	7.44	0.922	0.127	8.96	0.551	0.193	1.52
HST [26]	0.905	0.141	21.07	0.931	0.171	21.47	0.975	0.78	3.92
RRCF [2]	0.968	0.301	240.82	**0.950**	0.103	354.46	0.947	0.461	43.54
xStream [15]	0.754	0.033	13.85	0.711	0.005	15.38	0.724	0.118	2.35
Kit-Net [16]	0.905	0.205	**0.95**	0.943	0.141	6.02	0.803	0.259	**0.17**
AE	0.956	0.266	1.56	0.940	0.234	**1.67**	0.973	0.886	0.26
DAE	**0.984**	**0.501**	2.60	0.943	0.247	2.66	0.981	0.922	0.42
PW-AE	0.982	0.451	2.87	0.945	**0.258**	2.95	**0.986**	**0.955**	0.47

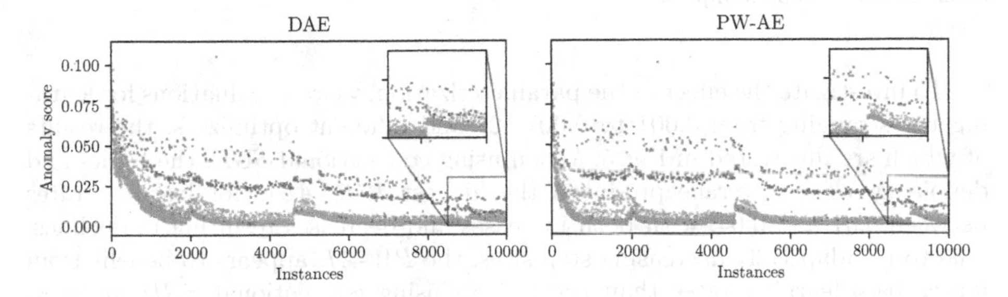

Fig. 3. Distribution of unscaled *AE* anomaly scores concerning the number of processed instances on Shuttle. Anomalies are drawn in red.

DAE due to its higher default learning rate. While the *DAE*'s losses briefly collapse to a single range of values, the *PW-AE*'s losses remain almost perfectly linearly separable throughout the whole stream segment and separate at a much faster rate after the sudden increase of normal losses due to a faster adaptation to normal data and even a slight increase in losses for anomalies. Since the *PW-AE*'s training procedure reinforces the model's concept of abnormality, its benefit likely depends on the accuracy of the underlying base model, which is supported by the performance gains on Shuttle for which the *DAE* and *AE* models are already able to label the majority of data accurately. Nevertheless, the distribution of anomaly scores demonstrates that our alternative *PW-AE* training approach can significantly widen the gap between the scores of anomalous and normal instances and therefore improve the separability of anomalies under the influence of contaminated training data as reflected by its increases in *ROC-AUC*, and *PR-AUC*.

Regarding the robustness of different contamination shares, Fig. 4 shows that the *PW-AE* appears to benefit from higher levels of contamination. As the proportion of anomalies randomly distributed in the Covertype or Shuttle data increases, the benefit of the *PW-AE* training procedure increases, leading to an increase in the *ROC-AUC* of the model compared to the architecturally identical *DAE* trained at a fixed learning rate. This observation supports the premise that our modified procedure can mitigate the effects of training on anomalous data.

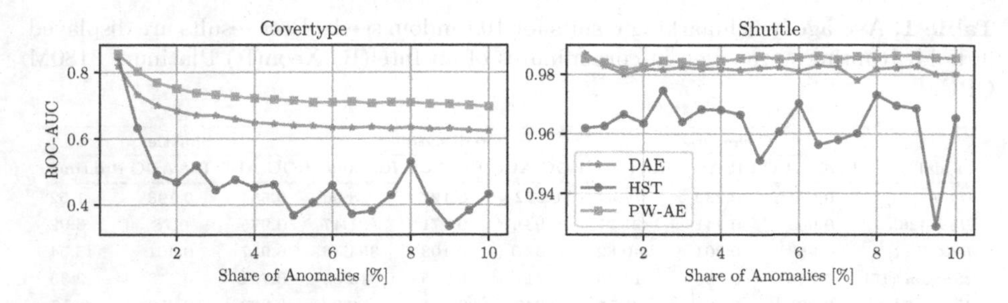

Fig. 4. *ROC-AUC* scores for varying anomaly percentages, which we generated by randomly sampling and inserting anomalies into the data. To allow for larger anomaly percentages, we used sampling with replacement. Each dataset created in this manner consisted of 50,000 examples.

To investigate the effect of the parameterization, we ran evaluations for learning rates ranging from 0.001 to $0.001 \cdot 2^8$ and different optimizers, the results of which are illustrated in Fig. 5. When using conventional *SGD*, the basic- and denoising-*AE* on average produced the highest *ROC-AUC* for learning rates between 0.01 and 0.04, with both increases and decreases resulting in declines. Due to its adaptively decreasing step sizes, the *PW-AE* appears to benefit from larger base learning rates than the models using conventional *SGD* updates. While yielding marginally lower *ROC-AUC* at learning rates below 0.05, the *PW-AE*'s average *ROC-AUC* for higher learning rates exceeded those of the best-performing *AE* and *DAE*, indicating that the *PW-AE*'s performance advantage must be caused by its alternative training approach rather than a difference in learning rate. The optimal learning rates are relatively consistent across datasets, supporting the transferability of our results. Overall, the *Adam* [13] and *HD-SGD* [3] optimization algorithms did not provide significant advantages over basic *SGD*, since neither approach was able to improve the *ROC-AUC* or contribute to the robustness with respect to the base learning rate. Considering the additional computational complexity of *Adam* and *HD-SGD* compared to plain *SGD*, the latter is therefore likely the best suited out of the investigated optimization algorithms for online *AD*.

To investigate the impact of the network width on *AD* performance, we performed test-then-train evaluations with latent layer ratios between 10% and 200%. As displayed in Fig. 6 all model variants achieved *ROC-AUC* scores close to or greater than 0.96 for all investigated latent layer ratios, indicating that the performance of the proposed approach is rather robust to the choice of network width. In terms of differences between model variants, it appears that the basic *AE* favors lower latent layer ratios in some cases, which its lack of regularization could explain.

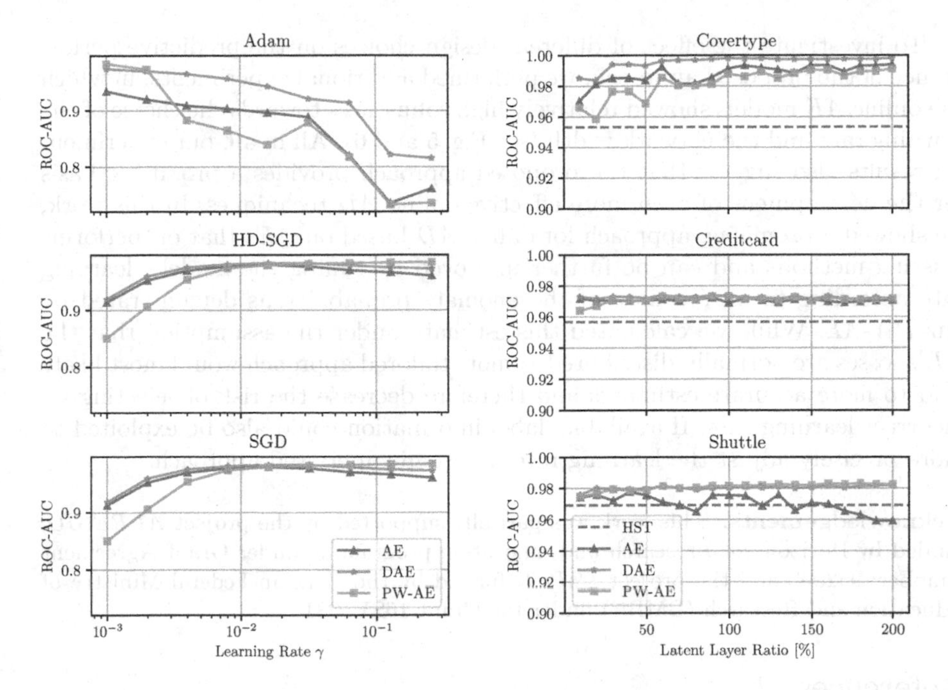

Fig. 5. *ROC-AUC* scores for variable learning rates γ averaged over the first 50,000 samples of each dataset.

Fig. 6. *ROC-AUC* scores for *AE* anomaly detectors with varying latent layer ratios on the first 50,000 samples of each dataset.

7 Conclusion and Future Work

In this work, we formalized the task of online *AD* based on predefined requirements for *AD* on streaming data, for which we introduced an AE-based framework. Our experimental results on three real-world datasets showed the proposed approach to yield significantly more accurate predictions than established baseline models, as is evident by their at least competitive or often superior *ROC-AUC* scores and their *PR-AUC* scores, which in the case of the most performant *PW-AE* and *DAE* regularly exceeded the scores of the next best non-*AE* model by more than 20%. Using only a fraction of the processing time required by most baseline models, the proposed models were also among the most computationally efficient models.

To mitigate the drawback of the conventional *AE* training objective in the case of contaminated training data, we further proposed an alternative *AE* objective for *AD*, which we used to derive a modified simple yet effective modification of the *SGD* update rule that reduces the learning rate for data instances that the *AE* deems to have a high chance of being an outlier. Our experiments indicate that the *PW-AE* models using this technique can reduce losses on normal data while maintaining higher losses on anomalous data compared to the models.

To investigate the effect of different design choices on the predictive performance of the proposed approach, we performed additional experiments, in which the online AE models showed relatively high robustness towards the choice of the learning rate and the network width (see Fig. 5 and 6). All in all, our experimental results also suggest that the proposed approach provides a promising basis for the development of even more effective online AD techniques. In this work, we showed a promising approach for online AD based on AEs that outperforms existing methods and can be further improved by tuning the model's learning rate according to an estimate of the anomaly probability as demonstrated by our *PW-AE*. While we calculated this estimate under the assumption that the AE's losses are normally distributed, a more tailored approach would most likely lead to more accurate estimates and therefore decrease the risk of selecting an incorrect learning rate. If available, label information could also be exploited to more precisely adjust the learning rate in a semi-supervised approach.

Acknowledgements. This work was partially supported by the project *AI REGIO*, funded by Horizon 2020 research and innovation programme under Grant Agreement Number: 952003 and the project *CoPoT*, funded by the German Federal Ministry of Education and Research (BMBF) under the Grant 16SV8281.

References

1. Ashfahani, A., Pratama, M.: Autonomous deep learning: continual learning approach for dynamic environments. In: Proceedings of the 2019 SIAM International Conference on Data Mining (SDM) (2019). https://doi.org/10.1137/1.9781611975673.75
2. Bartos, M.D., Mullapudi, A., Troutman, S.C.: rrcf: implementation of the robust random cut forest algorithm for anomaly detection on streams. J. Open Source Softw. (2019). https://doi.org/10.21105/joss.01336
3. Baydin, A.G., et al.: Online learning rate adaptation with hypergradient descent. In: International Conference on Learning Representations (ICLR) (2018)
4. Bifet, A., et al.: Machine Learning for Data Streams: With Practical Examples in MOA. MIT Press, Cambridge (2018)
5. Borges, N., Meyer, G.G.L.: Coping with training contamination in unsupervised distributional anomaly detection. In: 2009 43rd Annual Conference on Information Sciences and Systems (CISS) (2009). https://doi.org/10.1109/CISS.2009.5054728
6. Breunig, M., et al.: LOF: identifying density-based local outliers. In: Proceedings of the 2000 ACM SIGMOD International Conference on Management of Data (ACM SIGMOD) (2000). https://doi.org/10.1145/342009.335388
7. Chen, J., et al.: Outlier detection with autoencoder ensembles. In: Proceedings of the 2017 SIAM International Conference on Data Mining (SDM) (2017). https://doi.org/10.1137/1.9781611974973.11
8. Dal Pozzolo, A., et al.: Learned lessons in credit card fraud detection from a practitioner perspective. Expert Syst. Appl. (2014). https://doi.org/10.1016/j.eswa.2014.02.026
9. Girshick, R.: Fast R-CNN. arXiv preprint arXiv:1504.08083 (2015)

10. Glorot, X., Bengio, Y.: Understanding the difficulty of training deep feedforward neural networks. In: Proceedings of the Thirteenth International Conference on Artificial Intelligence and Statistics (AISTATS) (2010)
11. Gong, D., et al.: Memorizing normality to detect anomaly: memory-augmented deep autoencoder for unsupervised anomaly detection. In: 2019 IEEE/CVF International Conference on Computer Vision (ICCV) (2019). https://doi.org/10.1109/ICCV.2019.00179
12. Guha, S., et al.: Robust random cut forest based anomaly detection on streams. In: International Conference on Machine Learning (ICML) (2016)
13. Kingma, D.P., Ba, J.: Adam: a method for stochastic optimization. In: International Conference for Learning Representations (ICLR) (2015)
14. Liu, F.T., Ting, K.M., Zhou, Z.-H.: Isolation Forest. In: 2008 Eighth IEEE International Conference on Data Mining (ICDM) (2008). https://doi.org/10.1109/ICDM.2008.17
15. Manzoor, E., Lamba, H., Akoglu, L.: xStream: Outlier Detection in Feature-Evolving Data Streams. In: Proceedings of the 24th ACM SIGKDD International Conference on Knowledge Discovery & Data Mining (ACM SIGKDD) (2018). https://doi.org/10.1145/3219819.3220107
16. Mirsky, Y., et al.: Kitsune: An Ensemble of Autoencoders for Online Network Intrusion Detection. arXiv preprint arXiv:1802.09089 (2018)
17. Montiel, J., et al.: River: Machine Learning for Streaming Data in Python. arXiv preprint arXiv:2012.04740 (2020)
18. Na, G.S., Kim, D., Yu, H.: DILOF: effective and memory efficient local outlier detection in data streams. In: Proceedings of the 24th ACM SIGKDD International Conference on Knowledge Discovery & Data Mining (ACM SIGKDD) (2018). https://doi.org/10.1145/3219819.3220022
19. Parisi, G.I., et al.: Continual lifelong learning with neural networks: a review. Neural Netw. (2019). https://doi.org/10.1016/j.neunet.2019.01.012
20. Pokrajac, D., Lazarevic, A., Latecki, L.J.: Incremental local outlier detection for data streams. In: 2007 IEEE Symposium on Computational Intelligence and Data Mining (IEEE CIDM) (2007). https://doi.org/10.1109/CIDM.2007.368917
21. Rousseeuw, P.J., Leroy, A.M.: Robust Regression and Outlier Detection. Wiley, New York (1987)
22. Sahoo, D., et al.: Online deep learning: learning deep neural networks on the fly. In: Twenty-Seventh International Joint Conference on Artificial Intelligence (IJCAI) (2018). https://doi.org/10.24963/ijcai.2018/369
23. Sakurada, M., Yairi, T.: Anomaly detection using autoencoders with nonlinear dimensionality reduction. In: MLSDA Workshop (2014). https://doi.org/10.1145/2689746.2689747
24. Salehi, M., et al.: Fast memory efficient local outlier detection in data streams. IEEE TKDE **28**, 3246–3260 (2016). https://doi.org/10.1109/TKDE.2016.2597833
25. Schölkopf, B., et al.: Support vector method for novelty detection. In: Advances in Neural Information Processing Systems (NeurIPS) (2000)
26. Tan, S.C., Ting, K.M., Liu, T.F.: Fast anomaly detection for streaming data. In: Proceedings of the 22nd International Joint Conference on Artificial Intelligence (IJCAI) (2011)
27. Widmer, G., Kubat, M.: Learning in the presence of concept drift and hidden contexts. Mach. Learn. (1996). https://doi.org/10.1007/BF00116900
28. Wu, K., et al.: RS-Forest: a rapid density estimator for streaming anomaly detection. In: 2014 IEEE International Conference on Data Mining (ICDM) (2014). https://doi.org/10.1109/ICDM.2014.45

29. Zhou, C., Paffenroth, R.C.: Anomaly detection with robust deep autoencoders. In: Proceedings of the 23rd ACM SIGKDD International Conference on Knowledge Discovery and Data Mining (ACM SIGKDD) (2017). https://doi.org/10.1145/3097983.3098052
30. Zhou, G., Sohn, K., Lee, H.: Online incremental feature learning with denoising autoencoders. In: Proceedings of the Fifteenth International Conference on Artificial Intelligence and Statistics (AISTATS) (2012)
31. Japkowicz, N., Stefanowski, J. (eds.): Big Data Analysis: New Algorithms for a New Society. SBD, vol. 16. Springer, Cham (2016). https://doi.org/10.1007/978-3-319-26989-4
32. Zong, B., et al.: Deep Autoencoding Gaussian Mixture Model for Unsupervised Anomaly Detection. In: International Conference on Learning Representations (ICLR) (2018)

Anomaly Detection via Few-Shot Learning on Normality

Shin Ando(✉) and Ayaka Yamamoto

School of Management, Tokyo University of Science, Shinjuku City, Japan
ando@rs.tus.ac.jp, 8620510@ed.tus.ac.jp

Abstract. One of the basic ideas for anomaly detection is to describe an enclosing boundary of normal data in order to identify cases outside as anomalies. In practice, however, normal data can consist of multiple classes, in which case the anomalies may appear not only outside such an enclosure but also in-between 'normal' classes. This paper addresses deep anomaly detection aimed at embedding 'normal' classes to individually close but mutually distant proximities. We introduce a problem setting where a limited number of labeled examples from each 'normal' class is available for training. Preparing such examples is much more feasible in practice than collecting examples of anomalies or labeling large-scale, normal data. We utilize the labeled examples in a margin-based loss reflecting the inter-class and the intra-class distances among the embedded labeled data. The two terms and their relations are derived from an information-theoretic principle. In an empirical study using image benchmark datasets, we show the advantage of the proposed method over existing deep anomaly detection models. We also show case studies using low-dimensional mappings to analyze the behavior of the proposed method.

Keywords: Deep anomaly detection · Generative adversarial networks · Deep one-class classification · Data description · Few-shot learning

1 Introduction

Deep anomaly detection (DAD) [8,11] has received strong interests in recent years, but remains to be among the challenging tasks for deep learning. A basic goal in DAD is to find a compact representation of the data observed under 'normality' such that unobserved 'anomalies' are more likely to be distant or exhibit strong discrepancy from them.

GAN-based anomaly detection is a category of DAD, which learns the manifold of normal data distribution and identifies anomalies primarily based on the error between the original and an image reconstructed through a generator network [1,14,18]. The deep data description models [9,12,13] are extensions of one-class classification and support vector data description [15]. They form a category of DAD which learns an embedding function and a data-enclosing hypersphere with the minimum volume in the embedded space, with an implicit

M.-R. Amini et al. (Eds.): ECML PKDD 2022, LNAI 13713, pp. 275–290, 2023.
https://doi.org/10.1007/978-3-031-26387-3_17

assumption that normal data comes from a single class or source. At testing, the anomaly score is determined by the distance from the center of the hypersphere.

In practice, however, normal data can consist of multiple classes, and the anomalies may appear not only outside its boundary but also in-between 'normal' classes. In such cases, the conventional approach to find a single enclosure of normal data may increase the possibility of detecting anomalies outside, but it can also increase the possibility of overlooking anomalies between classes. In this paper, we alternatively attempt to find an embedding where each class is condensed to a proximity, but at the same time mutually distant and dispersed. It allows for a unified approach to detecting anomalies, as cases which appear far from the nearby normal classes.

We propose a framework utilizing a small number of labeled examples, or prototypes, from each 'normal' class. Practically, preparing a limited number of labeled data is far less expensive than collecting examples of anomalies or labeling large-scale normal data. The prototypes are used in a tune-up training, after a pre-training using large-scale unlabeled data by generative adversarial networks. This input setting differs from semi-supervised anomaly detection [13], which takes few examples of anomalies for calibrating anomaly scores, and also from few-shot learning [7] and out-of-distribution detection [9] which exploit a large-scale, labeled dataset from related tasks.

The training in the proposed framework is driven by an information-theoretic principle, which can formalize deep representation learning as a reduction of intra-class distances and an expansion of the inter-class distances at a trade-off. We propose a margin-based loss, which penalizes prototype pairs which increase intra-class margins or reduce inter-class margins. We conduct an empirical study to evaluate the proposed framework in comparison to existing DAD models and to analyze its embedding of the normal classes.

The main contribution of this paper is two-fold: (1) an anomaly detection framework under a new setting, utilizing small-scale, labeled normal data which are not practically expensive, (2) a margin-based loss derived from an information-theoretic principle to integrate small-scale labeled data into deep representation learning.

The rest of this paper is organized as follows. Section 2 describes the previous studies on deep anomaly detection and the relation between the information bottleneck and deep learning. Section 4 describes the technical details of the proposed framework. Section 5 presents the empirical results and the analyses from our experiments using public image datasets. We state our conclusion in Sect. 6.

2 Related Work

2.1 Deep Anomaly Detection

Two primary purposes of deep learning models in anomaly detection frameworks are: (1) reconstructing test samples and (2) providing distance metrics in the embedded space. The examples of (1) include deep autoencoders and GANs

[6]. In AnoGAN [14], the generator network G learns the normal data distribution manifold from which a test image X is reconstructed. The anomaly score is given by the residual difference after minimizing the absolute difference with the generated image X'. GANomaly [1], and Efficient-GAN based Anomaly Detection [20] similarly uses the reconstruction loss, after mapping the test image to and from the embedded space using the encoder and the generator networks, as anomaly scores.

The basis of measuring anomalousness by reconstruction error is that the trained generator acquires a mapping from a uniform distribution to the normal data distribution manifold [1,14]. In cases that the test sample is an anomaly, the image reconstructed by the generator should naturally deviate from the original and towards the normal data distribution.

Deep-SVDD [12] is an extension of the support vector data description [15] and an example of (2). It aims to learn an embedding in which the normal data can be enclosed by a hyper-spherical boundary. The boundary defines a one-class classifier, which identifies outliers based on the distance from its center. Deep multi-sphere SVDD (DMSVDD) extended the idea to learn multiple hyper-spheres, to addressed anomalies among multiple classes of normal data [5].

Multi-class Data Description (MCDD) [9] exploits the Deep SVDD model for out-of-distribution detection (OOD). It trains a DNN such that the embedding function f maps the labeled data onto the proximity of the centers of corresponding classes. The in-distribution classes are modeled as Gaussian components in the embedded space. Deep SVDD employs a max-margin loss for training, while in MCDD, implementations with a max-margin loss and a GDA-based MAP loss were introduced.

In our proposed model, we use GANs for pre-training an initial embedding, and a margin-based loss for fine-tuning the embedding for anomaly detection.

2.2 Information Bottleneck

The information bottleneck (IB) [2,17] is a principle for signal encoding to achieve a larger compression rate and a smaller distortion. It was adopted to machine learning for finding a sparse representation of an input variable X, which maintains the predictive power over an output variable Y. The sparseness and the predictive power of the representation Z are measured by its statistical dependence, i.e., mutual information, with respect to X and Y, respectively.

The IB principle is formalized as a minimization problem over a Lagrangian

$$\mathcal{L} = I(X;Z) - \beta I(Y;Z) \tag{1}$$

where $I(Y;Z)$ quantifies the amount of relevant information on Y. Since Z is generated from X, $I(X;Z)$ decreases as the rate of compression increases. The multiplier β represents the trade-off between the two terms.

In [16], the IB principle was introduced to analyse the layer-wise compression efficiency in DNNs. It was also employed in [13] for deriving a semi-supervised training loss.

[3] presented an analysis of the IB problem in a case where Y is the class variable and Z is the d-dimensional deep representation from the embedding function $f : \mathcal{X} \to \mathcal{Z} \subset \mathbb{R}^d$, with several modeling assumptions on Z. The first assumption is that the conditional distribution $p(z|y)$ is an isotropic Gaussian component for each class y, i.e.,

$$\begin{aligned} p(z|y) &= \mathcal{N}(z; \mu_y, \sigma_y I) \\ &= \frac{1}{(2\pi\sigma_y^2)^{d/2}} \exp\left(-\frac{\|z-\mu_y\|^2}{2\sigma_y^2}\right) \end{aligned}$$

where μ_y and σ_y denotes the class mean and standard deviation, respectively. The marginal distribution $p(z)$ is empirically approximated as an average of the Dirac delta functions

$$p(z) = \frac{1}{N}\sum_{i=1}^{N} \delta\left(z - f(x_i)\right)$$

The mutual information $I(Y;Z)$, which is equivalent to the expected Kullback-Leibler divergence between $p(y|z)$ and $p(z)$, is then rewritten as

$$\begin{aligned} I(Y;Z) &= E_{y,z}\left[\log \frac{p(z|y)}{p(z)}\right] \\ &= \sum_{y=1}^{K} \frac{1}{n_y} \sum_{z} \log \frac{\exp\left(-\frac{\|z-\mu_y\|^2}{2\sigma_y}\right) - \log \sigma_{y'}^d}{\sum_{y'} \exp\left(-\frac{\|z-\mu_{y'}\|^2}{2\sigma_{y'}}\right) - \log \sigma_{y'}^d} + \text{const.} \end{aligned} \tag{2}$$

$p(z|x)$ was defined as a probability that x is mapped to z, given the randomness of DNN such as batch normalization and dropouts. It was also modeled by an isotropic Gaussian component centered at $f(x)$ with a common standard deviation $\hat{\sigma}$.

$$\begin{aligned} p(z|x) &= \mathcal{N}(z|f(x), \hat{\sigma}^2 I) \\ &= \frac{1}{(2\pi\hat{\sigma}^2)^{d/2}} \exp \frac{\|z - f(x)\|^2}{2\hat{\sigma}^2} \end{aligned}$$

The mutual information $I(X;Z)$ then was rewritten as

$$\begin{aligned} I(X;Z) &= E_{x,z}\left[\log \frac{p(z|x)}{p(z)}\right] \\ &= \frac{1}{N^2} \sum_{z \in \mathcal{Z}} \sum_{i=1}^{N} \left(-\frac{\|z - f(x_i)\|^2}{2\hat{\sigma}^2} + \log \hat{\sigma}^d\right) + \text{const.} \end{aligned} \tag{3}$$

Based on (3), $I(X;Z)$ was approximated as the sum of mutual distances in the embedded space. From (2), $I(Y;Z)$ increases as the class distribution concentrates to its center, which broadly interprets as the reduction of the intra-class distances. It was argued that by jointly minimizing $I(X;Z)$ and $I(Y;Z)$,

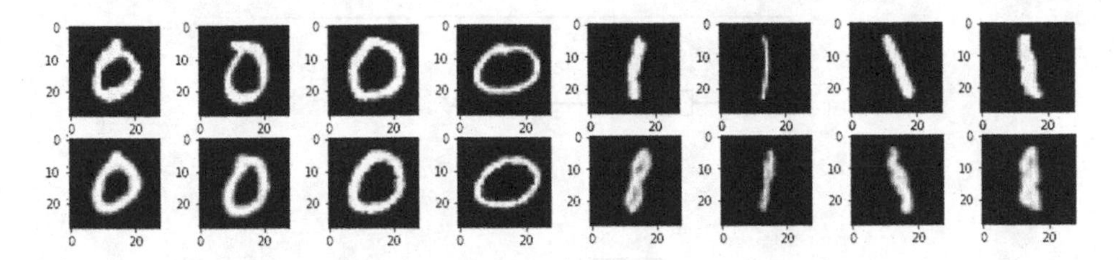

Fig. 1. Original and reconstructed images

the inter-class mutual distances will increase while the intra-class distances decrease, by virtue of substantially larger class deviations σ_y compared to the deviation of the randomness, $\hat{\sigma}$, after training.

Motivated by the above analysis from [3], we implement a margin-based loss with a focus on the inter-class and intra-class properties for deep anomaly detection.

3 Motivating Example

In this section, we examine the behavior of GAN-based anomaly detection to motivate the proposed framework. Adversarial training has several attractive properties for anomaly detection. For example, GANs can learn deep representation from 'normal' data in an unsupervised manner. A trained generator network can be used to 'reconstruct' a test case, and its 'error' from the original case can provide a natural anomaly score as mentioned in Sect. 2.1.

In the following, we describe a DAD process using the BiGAN [4] framework, comprised of a generator, discriminator, and an encoder, used in EGBAD [19] and GANomaly [1]. We conducted an unsupervised, adversarial training of BiGAN in a standard setup for anomaly detection using the MNIST benchmark, which is to remove one class designated as an 'anomaly' class from training and compile the 'normal' data from the remaining classes.

After training, the test cases were initially mapped to a Euclidean space and reconstructed back to an image using the encoder and the generator networks. The examples of the original and the reconstructed images, from the setup that the digit '1' was designated as anomalies, are shown in Fig. 1.

The original images are shown in the first row while the reconstructed images are shown in the second row. The images shown on the left half is those of digit '0', a normal class, and the images on the right half are those of digit '1', the anomaly class, respectively.

Graphically, the reconstructed normal images resemble natural handwriting while the reconstructed anomaly images exhibit unnatural forms, with subtle resemblance of other digits. In terms of the pixel-wise comparison, however, the reconstructed 'anomalous' images, are not substantially different from their originals. Meanwhile, the reconstructed normal images exhibit slight modifications from their originals.

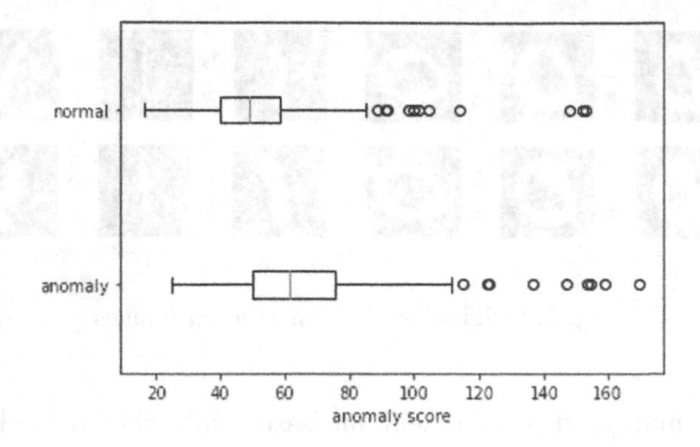

Fig. 2. Anomaly score distributions

Figure 2 compares the distribution of mean absolute errors between the original and the reconstructed test images belonging to the normal and the anomalous classes. There is a notable overlap of interquartiles between the two distributions. In this case, it is therefore unlikely that the reconstruction errors as anomaly scores produce a good detection performance.

The graphical results suggests that the learned manifold can include intermediate patterns of different classes in the training data, since GANs learn a mapping between a continuous, Euclidean unit space and a distribution manifold. With a large variety in normal data, intermediate patterns may allow a reconstruction of anomalous cases without significant error. We should also beware of the class-wise bias over the reconstruction error, producing relatively higher anomaly scores for classes with larger and more complex patterns.

Based on these preliminary analyses, we avoid the reconstruction and pixel-wise comparison process but instead were motivated to find an embedding in which the anomalies can be detected by the distances from the nearby 'normal' classes, thus consider a setting where typical examples of normal classes can be utilized.

4 Prototype Data Description

In the previous section, we introduced our motivation to utilize labeled examples of normal data into training and exploit distances in the embedded space in testing. This section describes the framework which integrates these examples into training based on an information-theoretic principle.

Generally, a set of normal data for training is large-scale and unlabeled as the cost of observation under normality is small, but the reward for labeling such data is also small to none. However, it can be feasible to collect a limited number of examples from each class. Here, we assume such a small-scale dataset comprised of K samples from each of N classes, much like the setting of N-way-K-shot learning, is available. The problem input thus consists of a large-scale, unlabeled normal dataset available for pre-training, and a set of labeled examples for the tune-up training.

Let us denote the unlabeled dataset by $\mathcal{X} = \{x_i\}_{i=1}^{M}$ and the set of labeled data, or prototypes, by $\mathcal{P} = \{(p_j, y_j)\}_{j=1}^{K \times N}$. N and K denotes the number of classes and the number of prototypes for each class, respectively.

We represent by random variables X and Z, the structured input data, e.g., images, and their the embedding, respectively. The class variable Y takes a value from $\mathcal{Y} = \{1, \ldots, N\}$. We denote the embedding function of a DNN with parameters W by $f : \mathcal{X} \to \mathcal{Z}$.

As referenced in Sect. 2.2, minimizing the information bottleneck loss interprets to expanding inter-class distances and reducing intra-class distances at a trade-off. The intra-class distances, represented as (2), are measured with regards to class means and variances, reflecting the modeling assumption that the class distributions are Gaussian components. For the task at hand, however, the estimated parameters may not be robust given the small scale of the labeled data.

Alternatively, we attempt to reduce the diameter of the class-enclosing convex. Let R_c^* denote the largest intra-class distance among samples of class c,

$$R_c^* = \max_{k,j:y_k=y_j=c} \|f(p_j; W) - f(p_k; W)\| \tag{4}$$

As R_c^* is equivalent to the diameter of the convex hull of the samples of c, we can minimize its volume and subsequently the intra-class distances, by descending along the gradients of R_c^* with regards to W. Note that the small scale of the labeled data allows for computing the mutual distance matrix in a feasible time. Still, it is inefficient to iterate the descent and the update for the single largest distance, R_c^*. We, instead, take the sum of intra-class, pair-wise distances over a threshold R_{intra}, as a loss $\mathcal{L}_{\text{intra}}(W)$.

$$\mathcal{L}_{\text{intra}}(W) = \sum_{j,k:y_j=y_k} \max\{0, \|f(p_j; W) - f(p_k; W)\| - R_{\text{intra}}\} \tag{5}$$

By setting R_{intra} to a q_{intra}-quantile over the intra-class distances, (5) takes a summation over the largest $q \times N$ intra-class pairs. Using the gradients of (5) achieves a substantially faster descent compared to using that of R_c^*.

With regards to the inter-class distances, we look at (3), which can be approximated by the sum of all pair-wise distances in the embedded space. The constants and the terms related to the variance $\hat{\sigma}^2$ are irrelevant to the optimization and can be ignored. Further, we take the homogeneous-class pairs out of consideration, as a joint minimization with (2) should reduce the distances between them as argued in Sect. 2.2.

The inter-class loss $\mathcal{L}_{\text{inter}}$ is defined to take a summation over the heterogeneous-class pairs, which falls short of the inter-class margin threshold R_{inter}.

$$\mathcal{L}_{\text{inter}}(W) = \sum_{j,k:y_j \neq y_k} \max\{0, R_{\text{inter}} - \|f(p_j; W) - f(p_k; W)\|\} \tag{6}$$

R_{inter} is set to a q_{inter}-quantile among the inter-class distances. (6), thus, is focused on penalizing pairs which correspond to relatively smaller inter-class margins.

We minimize $\mathcal{L} = \mathcal{L}_{\text{intra}} + \beta\mathcal{L}_{\text{inter}}$ to increase the overall margins between classes. For anomaly detection, it is intuitive to employ a margin-based loss, as data near the boundaries of normal data or classes are more critical to the detection performance compared to those near the class center. We will refer to this framework as Prototype Data Description (PDD), as it models the enclosing hull of normal-class prototypes.

At testing, we compute the anomaly score using kernel density estimation (KDE) in the embedded space. Let $\hat{\mathcal{P}}_n = \{f(p_j) : y_j = n\}$ denote the set of embedded prototypes with class label n, and $D^{(n)}(z)$ the kernel density estimation of z given $\hat{\mathcal{P}}_n$. A large density indicates a closeness to the prototypes of n.

We define a scaled density function a_n such that

$$a_n(x) = \frac{D^{(n)}(f(x)) - D^{(n)}_{\text{max}}}{D^{(n)}_{\text{max}} - D^{(n)}_{\text{min}}}$$

with scaling parameters

$$D^{(n)}_{\text{max}} = \max_{z \in \bigcup_{\mathcal{Y} \setminus n} \hat{\mathcal{P}}_i} D_n(z),\ D^{(n)}_{\text{min}} = \min_{z \in \bigcup_{\mathcal{Y}} \hat{\mathcal{P}}_i} D_n(z)$$

In [24,25], the test cases were scored for open-set recognition using the distance to the closest class-prototypes. In our study, we similarly use the inverse of the largest scaled density of the test case x as the anomaly score[1].

$$A(x) = \exp\left(-\max_n a_n(x)\right) \tag{7}$$

5 Empirical Results

5.1 Setup

This section presents an empirical study for comparative and graphical evaluation of PDD. We set up anomaly detection tasks using benchmark image datasets following previous studies, by excluding one class designated as anomalies from training. The training set is thus compiled from the remaining classes and the performance is measured for the detection of the anomaly class samples in the test set. The following experiments are conducted with three public datasets: MNIST [21], Fashion-MNIST [22], and CIFAR10 [23]. Their properties are summarized in Table 1.

As baselines for comparison, we conducted the same experiments using EGBAD [19], GANomaly [1], and MCDD [9]. EGBAD and GANomaly are baselines of unsupervised GAN-based DAD, given only the 'normal' class images

[1] A sample code of the PDD is provided in https://github.com/ProtoDD/pdd.

Table 1. Datasets

Dataset	#Image size × Channels	#Instances	# Classes
MNIST	28 × 28 × 1	70,000	10
Fashion-MNIST	28 × 28 × 1	60,000	10
CIFAR	32 × 32 × 3	60,000	10

without labels in training. MCDD is a baseline of data description and OOD, given the same dataset but with complete labels of 'normal' classes in training. PDD is given a 9-way, 20-shot labeled prototypes in addition to the same unlabeled dataset. The prototypes were chosen randomly and removed from the default training set. Note that PDD is at a disadvantage compared to the OOD model, while at an advantage compared to the unsupervised DAD with regards to the amount of supervising information used in training. For performance measures, we compute the Area Under the ROC curve (AUROC) and the Area Under the Precision-Recall Curve (AUPRC).

The baseline models were implemented based on their publicly available codes[2,3,4]. The hyperparameters of the baselines were determined by grid search around the suggested values from their respective papers. The summary of the main training parameters are shown in the appendix (Table 2).

The hyperparameters related to IB and KDE were empirically determined: IB trade-off $\beta = 3$, KDE bandwith $w = 10$, distance quantiles $q_{\text{intra}} = 0.5$, $q_{\text{inter}} = 0.25$. The GAN-architectures for pre-training are shown in the appendix (Tables 3 and 4). The training were conducted in a single-GPU environment with a Tesla P100 with 16GB memory. The optimizer was ADAM with learning rate 0.002.

5.2 Comparative Analysis

This section presents comparisons between performances of PDD and the baseline models. We report the AUROC measures due to the low AUPRCs of the baselines. The AUPRCs of the proposed model are reported in the next section.

Figure 3 shows the comparisons with unsupervised DAD baselines on ten anomaly detection tasks based on MNIST datasets. The markers indicate the mean over ten repetitions, while the error bars indicate the standard deviation. Similarly, Figs. 4 and 5 show comparisons over the tasks based on Fashion-MNIST and CIFAR-10 datasets, respectively.

From Figs. 3, 4 and 5, PDD has substantial advantage over GAN-based DADs in these thirty tasks. Overall, EGBAD may not be adequate for handling multi-class normal data. It is omitted from Fig. 5 due to its low measures. Additionally,

[2] https://github.com/houssamzenati/Efficient-GAN-Anomaly-Detection.
[3] https://github.com/samet-akcay/ganomaly.
[4] https://github.com/donalee/DeepMCDD.

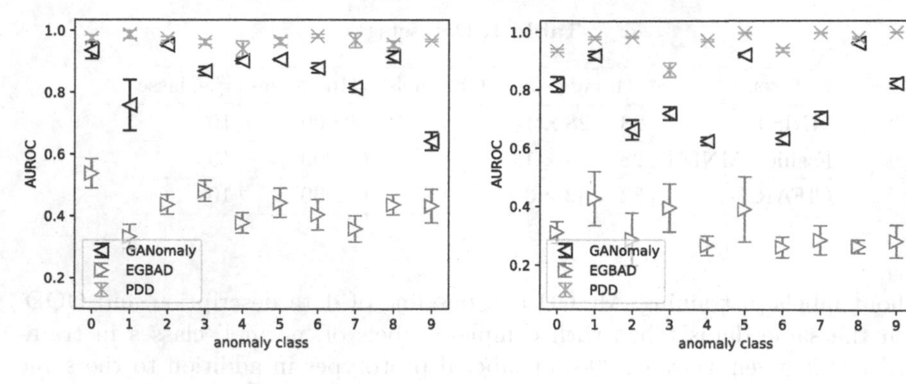

Fig. 3. vs Unsupervised-DAD (MNIST)

Fig. 4. vs Unsupervised DAD (Fashion)

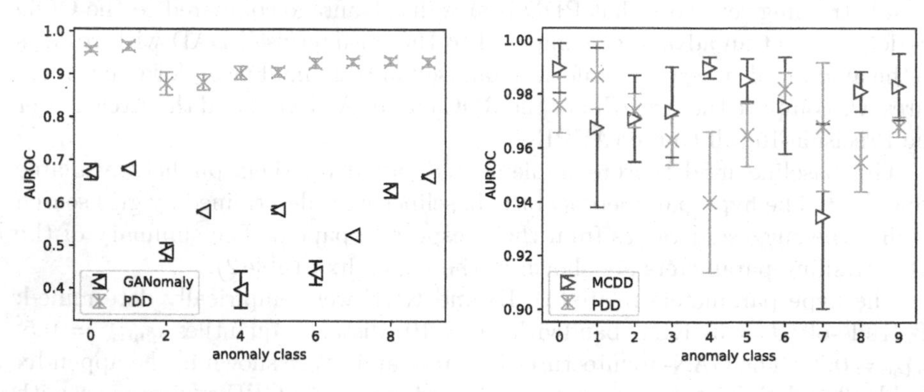

Fig. 5. vs Unsupervised DAD (CIFAR-10)

Fig. 6. vs OOD (MNIST)

PDD showed relatively small variances over the class designated as anomaly while GANomaly showed high variance depending on the class.

The comparison between PDD and the OOD baseline are shown separately in Figs. 6, 7 and 8. The markers and the error bars respectively indicate the mean and the standard deviation over ten repetitions in each task.

Over the thirty tasks, PDD and MCDD averaged AUC higher than 0.85. MCDD generally showed larger variances than PDD, and neither outperformed the other overall. We note that PDD shows comparable performances while exploiting a limited amount of supervising information compared to MCDD.

The run time of PDD were 1.5 min per epoch on average. The run time of the baselines in proportion to that of PDD were as follows: PDD:1.0, EGBAD: 0.56, GANomaly: 0.84, MCDD: 1.4.

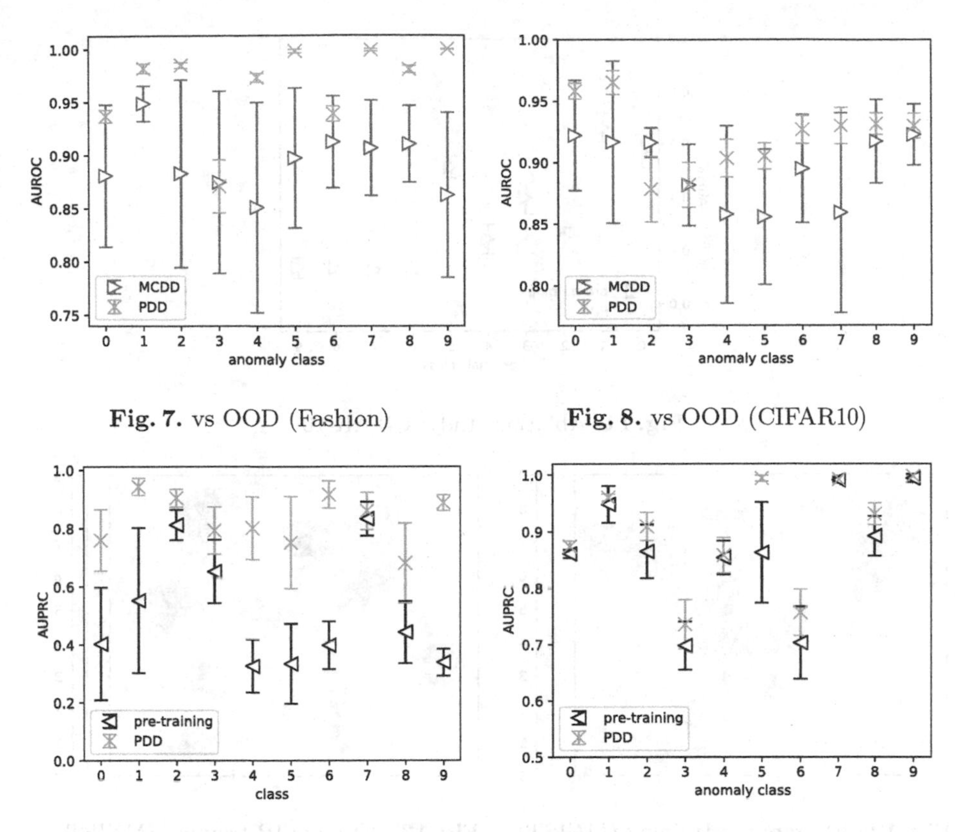

Fig. 7. vs OOD (Fashion)

Fig. 8. vs OOD (CIFAR10)

Fig. 9. Ablation study: MNIST

Fig. 10. Ablation study: Fashion-MNIST

5.3 Ablation Study

In this and the following sections, we evaluate the impact of the IB-training by quantitative and graphical comparison of the proposed model after pre-training and after IB-loss training.

Figures 9, 10 and 11 show the comparisons of AUPRCs in the same thirty tasks as the previous experiment. The markers indicate the means over ten repetitions, while the error bars reflect the standard deviations.

The initial embedding by GANs produced comparative performances in several tasks, but in many tasks it yields substantially low AUPRC measures. Over the thirty tasks, IB-training achieves substantial improvements on top of the initially acquired embeddings.

5.4 Graphical Analysis

This section presents graphical analyses on low-dimensional mappings of the embedded images from typical cases. Figures 12 and 13 show 2-D mapping of the

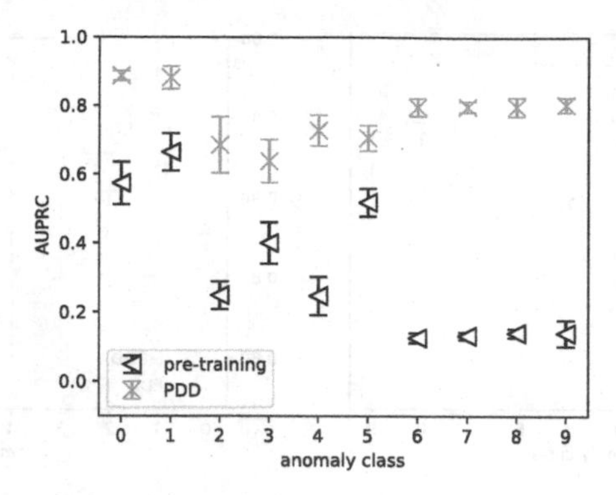

Fig. 11. Ablation study: CIFAR-10

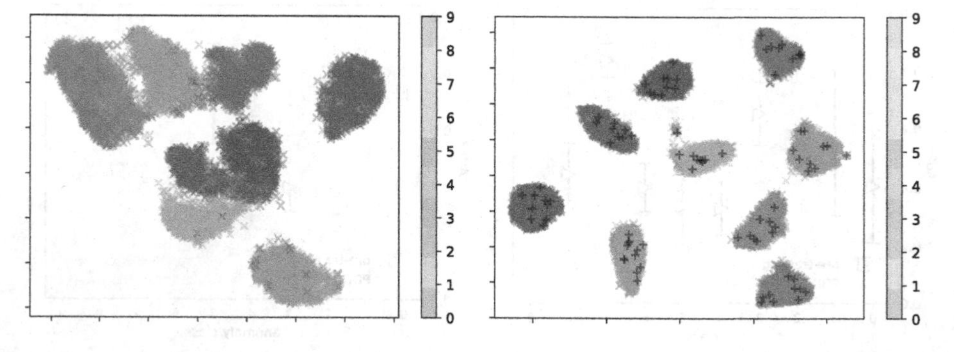

Fig. 12. 2D-map: pre-training (MNIST) **Fig. 13.** 2D-map: IB-training (MNIST)

MNIST test images after GAN pre-training and IB training, respectively. The 2-D mappings were generated by UMAP [10], in the setup where the anomaly class is digit '1', The '×' markers of different colors indicate the embeddings of respective classes. The black '+' markers indicate the prototypes.

Figure 13 shows that the class-wise distributions from MNIST can be identified after pre-training, but they are mostly unseparated and few are overlapping at that point. Meanwhile, from Fig. 13, the class distributions are evenly separated and individually enclosed in different proximities after IB training.

The 2-D mappings of the Fashion-MNIST testset images after pre-training and IB training are shown in Figs. 14 and 15, respectively. The black and colored markers indicate the classes and the prototypes, and the anomaly class is digit '6'. Figure 14 shows that most of the classes are overlapping with one or more other class distributions after pre-training. In comparison, the classes in Fig. 15 shows either some reduction in overlaps or increased separation. Note that groups of classes with inherently similar patterns, e.g., {5: Sandal, 7: Sneaker, 9: Boots} and {2: Pullover, 4: Coat}, are unseparated in 2-D, but their overlaps are reduced substantially after IB-training.

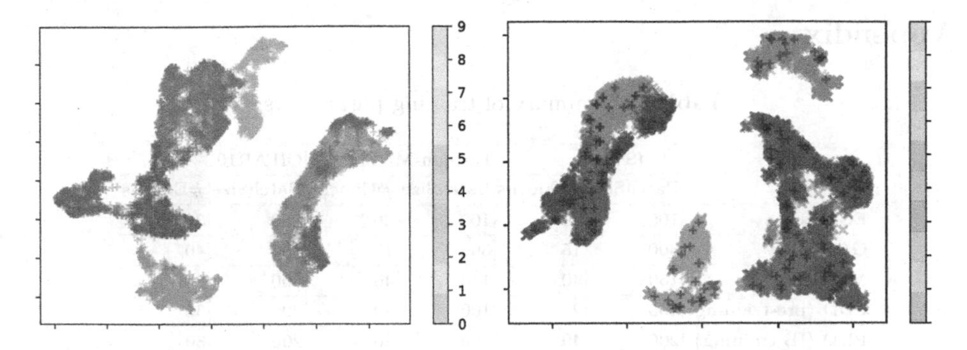

Fig. 14. 2D-map: pre-training (Fashion) **Fig. 15.** 2D-map: IB-training (Fashion)

From many cases similar to those above, we can expect PDD to expand the inter-class margins while gathering each class to a different proximity. The graphical analysis also indicates that the KDE-based anomaly scores can be effective for detecting anomalies that appear between separated normal classes.

Generally, not all normal classes can be separated. As seen in the Fashion-MNIST experiment, when classes are intrinsically similar, they are embedded to adjoint regions. We note that it is not the goal to cluster all independent classes, and the capacity for anomaly detection will not necessarily be impaired when similar classes are unseparated. Practically, if when classes are similar, e.g., shoes and boots, there may be instances which are hard to discern, but they are unlikely to be considered anomalies.

6 Conclusion

In this paper, we addressed the task of anomaly detection in the presence of multi-class, normal data, with a new, practically feasible input setting, utilizing a small set of class prototypes. We implemented a deep neural network training driven by an information-theoretic principle, with a loss based on intra-class and inter-class distances among the prototypes. Our empirical evaluation showed that the proposed method holds substantial advantages over unsupervised DAD models and also is comparable to an OOD data description model in terms of the performance measures. From the graphical analyses, the proposed model can typically learn mutually distant and dispersed embedding of the class prototypes, which enables its density-based anomaly scores.

Given that proposed model addresses a setting slightly different from those of existing tasks, e.g., ODD and Unsupervised Anomaly Detection, building a benchmark which is also practically relevant is an important future work.

Futhermore, we aim to understand its more practical characteristics such its sensitivity to noise and the number of prototypes, the means and its impact of selecting "good" prototypes.

Appendix

Table 2. Summary of training parameters

	MNIST		Fashion-MNIST		CIFAR10	
	BatchSize	#Epochs	BatchSize	#Epochs	BatchSize	#Epochs
EGBAD	100	20	100	30	150	30
GANomaly	300	15	300	15	300	40
MCDD	150	40	150	40	200	80
PDD (pre-training)	200	20	100	40	200	30
PDD (IB training)	200	40	100	40	200	30

Table 3. GAN architecture (MNIST/Fashion-MNIST)

(a) Generator

Layer unit	D_{out}
Input	100
Lin + BN + ReLU	7 × 7 × 512
CnvTr+BN+LkReLU	14 × 14 × 256
CnvTr+BN+LkReLU	14 × 14 × 128
CnvTr+BN+LkReLU	28 × 28 × 64
CnvTr+Tanh	28 × 28 × 1

(b) Discriminator

Layer unit	D_{out}
Input	28 × 28 × 1
Cnv+BN+LkReLU	28 × 28 × 8
Cnv+BN+LkReLU	14 × 14 × 16
Cnv+BN+LkReLU	7 × 7 × 32
Cnv+BN+LkReLU	7 × 7 × 64
Fltn+Lin+Sgmd	1

Table 4. GAN architecture (CIFAR-10)

(a) Generator

Layer unit	D_{out}
Input	100
ConvTr+BN+ReLU	4 × 4 × 512
CnvTr+BN+ReLU	8 × 8 × 256
CnvTr+BN+ReLU	16 × 16 × 128
CnvTr+BN+ReLU	32 × 32 × 64
CnvTr+Tanh	64 × 64 × 3

(b) Discriminator

Layer unit	D_{out}
Input	64 × 64 × 3
Cnv+LkReLU	32 × 32 × 64
Cnv+BN+LkReLU	16 × 16 × 128
Cnv+BN+LkReLU	8 × 8 × 256
Cnv+BN+LkReLU	4 × 4 × 512
Conv+Sgmd	1

References

1. Akcay, S., Atapour-Abarghouei, A., Breckon, T.P.: GANomaly: semi-supervised anomaly detection via adversarial training. In: Jawahar, C.V., Li, H., Mori, G., Schindler, K. (eds.) ACCV 2018. LNCS, vol. 11363, pp. 622–637. Springer, Cham (2019). https://doi.org/10.1007/978-3-030-20893-6_39
2. Alemi, A.A., Fischer, I., Dillon, J.V., Murphy, K.: Deep variational information bottleneck. In: 5th International Conference on Learning Representations, ICLR 2017, Toulon, France, 24–26 April 2017, Conference Track Proceedings. OpenReview.net (2017)

3. Ando, S.: Deep representation learning with an information-theoretic loss. CoRR abs/2111.12950 (2021)
4. Ding, R., Guo, G., Yang, X., Chen, B., Liu, Z., He, X.: BiGAN: collaborative filtering with bidirectional generative adversarial networks. In: Proceedings of the 2020 SIAM International Conference on Data Mining, SDM 2020, pp. 82–90. SIAM (2020)
5. Ghafoori, Z., Leckie, C.: Deep multi-sphere support vector data description. In: Proceedings of the 2020 SIAM International Conference on Data Mining, SDM 2020, pp. 109–117. SIAM (2020)
6. Goodfellow, I.J., et al.: Generative adversarial nets. In: Proceedings of the 27th International Conference on Neural Information Processing Systems. NIPS'14, vol. 2, pp. 2672–2680. MIT Press, Cambridge (2014)
7. Jeong, T., Kim, H.: OOD-MAML: meta-learning for few-shot out-of-distribution detection and classification. In: Advances in Neural Information Processing Systems, vol. 33, pp. 3907–3916. Curran Associates, Inc. (2020)
8. Kwon, D., Kim, H., Kim, J., Suh, S.C., Kim, I., Kim, K.J.: A survey of deep learning-based network anomaly detection. Clust. Comput. (2017)
9. Lee, D., Yu, S., Yu, H.: Multi-class data description for out-of-distribution detection. In: Proceedings of the 26th ACM SIGKDD International Conference on Knowledge Discovery and Data Mining. KDD '20, pp. 1362–1370. Association for Computing Machinery, New York (2020)
10. McInnes, L., Healy, J., Saul, N., Großberger, L.: UMAP: uniform manifold approximation and projection. J. Open Source Softw. **3**(29), 861 (2018)
11. Pang, G., Shen, C., Cao, L., Hengel, A.V.D.: Deep learning for anomaly detection: a review. ACM Comput. Surv. 54(2) (Mar 2021)
12. Ruff, L., et al.: Deep one-class classification. In: Proceedings of the 35th International Conference on Machine Learning. Proceedings of Machine Learning Research, vol. 80, pp. 4393–4402. PMLR (2018)
13. Ruff, L., et al.: Deep semi-supervised anomaly detection. In: 8th International Conference on Learning Representations, ICLR 2020. OpenReview.net (2020)
14. Schlegl, T., Seeböck, P., Waldstein, S.M., Schmidt-Erfurth, U., Langs, G.: Unsupervised anomaly detection with generative adversarial networks to guide marker discovery. In: Niethammer, M., et al. (eds.) IPMI 2017. LNCS, vol. 10265, pp. 146–157. Springer, Cham (2017). https://doi.org/10.1007/978-3-319-59050-9_12
15. Tax, D.M.J., Duin, R.P.W.: Support vector data description. Mach. Learn. **54**, 45–66 (2004)
16. Tishby, N., Zaslavsky, N.: Deep learning and the information bottleneck principle. In: 2015 IEEE Information Theory Workshop (ITW), pp. 1–5 (2015)
17. Tishby, N., Pereira, F.C., Bialek, W.: The information bottleneck method. Comput. Res. Repos. (CoRR) physics/0004057 (2000)
18. Zenati, H., Romain, M., Foo, C., Lecouat, B., Chandrasekhar, V.: Adversarially learned anomaly detection. In: 2018 IEEE International Conference on Data Mining (ICDM), pp. 727–736 (2018)
19. Zenati, H., Foo, C.S., Lecouat, B., Manek, G., Chandrasekhar, V.R.: Efficient GAN-based anomaly detection. CoRR abs/1802.06222 (2018),
20. Zenati, H., Foo, C.S., Lecouat, B., Manek, G., Chandrasekhar, V.R.: Efficient GAN-based anomaly detection (2019)
21. Lecun, Y., Bottou, L., Bengio, Y., Haffner, P.: Gradient-based learning applied to document recognition. Proc. IEEE **86**(11), 2278–2324 (1998). https://doi.org/10.1109/5.726791

22. Xiao, H., Rasul, K., Vollgraf, R.: Fashion-MNIST: a novel image dataset for benchmarking machine learning algorithms, August 2017
23. Krizhevsky, A.: Learning multiple layers of features from tiny images. Master's thesis (2009)
24. Liu, B., Kang, H., Li, H., Hua, G., Vasconcelos, N.: Few-shot open-set recognition using meta-learning. In: Proceedings of the IEEE/CVF Conference on Computer Vision and Pattern Recognition (CVPR), June 2020
25. Jeong, M., Choi, S., Kim, C.: Few-shot open-set recognition by transformation consistency. In: Proceedings of the IEEE/CVF Conference on Computer Vision and Pattern Recognition (CVPR). pp. 12566–12575, June 2021

Interpretability and Explainability

Interpretations of Predictive Models for Lifestyle-related Diseases at Multiple Time Intervals

Yuki Oba[1(✉)], Taro Tezuka[2], Masaru Sanuki[3], and Yukiko Wagatsuma[3]

[1] Graduate School of Science and Technology, University of Tsukuba, Tsukuba, Japan
s2230178@s.tsukuba.ac.jp

[2] Faculty of Engineering, Information and Systems, University of Tsukuba, Tsukuba, Japan
tezuka@iit.tsukuba.ac.jp

[3] Faculty of Medicine, University of Tsukuba, Tsukuba, Japan
{sanuki,ywagats}@md.tsukuba.ac.jp

Abstract. Health screening is practiced in many countries to find asymptotic patients of diseases. There is a possibility that applying machine learning to health screening datasets enables predicting future medical conditions. We extend this approach by introducing interpretable machine learning and determining health screening items (attributes) that contribute to detecting lifestyle-related diseases in their early stages. Furthermore, we determine how contributing attributes change within one to four years of time. We target diabetes and chronic kidney disease (CKD), which are among the most common lifestyle-related diseases. We trained predictive models using XGBoost and estimated each attribute's contribution levels using SHapley Additive exPlanations (SHAP). The results indicated that numerous attributes drastically change their levels of contribution over time. Many of the results matched our medical knowledge, but we also obtained unexpected outcomes. For example, we found that for predicting HbA1c and creatinine, which are indicators of diabetes and CKD, respectively, the contribution from alanine transaminase goes up as the time interval lengthens. Such findings can provide insights into the underlying mechanisms of how lifestyle-related diseases aggravate.

Keywords: Interpretable machine learning · Data-driven medicine · Health screening · Disease prediction · Tabular data

1 Introduction

Health screening is practiced in many countries to find asymptotic patients of diseases. It has become increasingly important to detect early symptoms of lifestyle-related diseases through health screening due to their increasing rate of patients among diseases. Machine learning provides a promising approach for making such predictions. In addition to finding asymptotic patients, there is high potential in

M.-R. Amini et al. (Eds.): ECML PKDD 2022, LNAI 13713, pp. 293–308, 2023.
https://doi.org/10.1007/978-3-031-26387-3_18

health screening data. Medical researchers can obtains insights from analyzing health screening data. Such an approach is often called data-driven medicine.

Data-driven medicine has been gaining much popularity due to its potentially high impact on medical practices and drug discovery. In the field of artificial intelligence (AI) research, interpretable machine learning provides a good measure for data-driven medicine. Medical researchers can gain an understanding from the insights provided by the interpretations of machine learning models into the underlying mechanisms of diseases. Clinically, an interpretation can suggest testing other examination items not included in the original health screening records to further understand the patient's condition.

In this paper, we introduce the aspect of time into interpretable machine learning to health screening. Existing works on predictions and interpretations have conducted analyses only at a single time interval, that is, using a specific time interval between features and the target attribute. However, as a lifestyle-related disease develops over time, different test items are affected. The change can be observed by comparing attributes contributing to making predictions at different points in time. A number of attributes may contribute to making long-term predictions, while others are useful for short-term predictions. Such a difference may have been known to clinicians, but it was difficult to observe it quantitatively. Interpretable machine learning on health screening records can now provide a suitable means for such an analysis.

To see the dynamics of contributing attributes, we observed the differences in attributes that help make predictions at time intervals between one and four years. Specifically, we trained models using XGBoost and ranked the attributes in accordance with their contributions using SHapley Additive exPlanations (SHAP) [11].

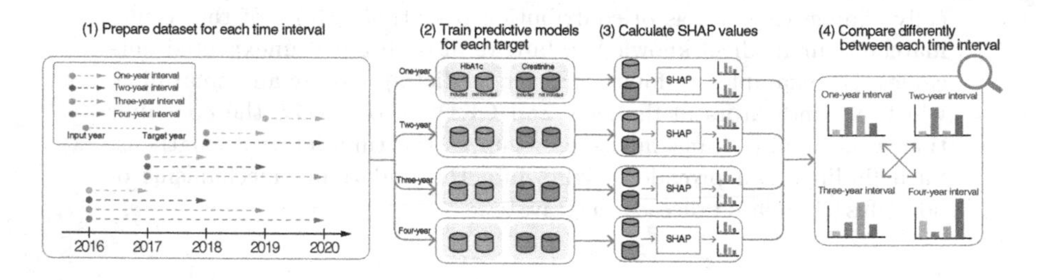

Fig. 1. The framework for interpreting predictive models presented in this paper.

We chose type 2 diabetes and chronic kidney disease (CKD) as the target diseases because they are the most common among lifestyle-related diseases. From the experiments, we found that the contributing attributes significantly differ depending on time intervals. Many attributes that contributed to predicting the disease indicators one year later were different from those four years later. Our results show that the interpretation of machine learning results can effectively uncover such dynamics and provide more insights into how lifestyle-related diseases aggravate. This can also help patients and clinicians by enabling them to diagnose diseases at early stages. In addition, our approach provides a framework for applying interpretable machine learning to data-driven medicine in general and can contribute to the development of evidence-based medicine for lifestyle-related diseases. Figure 1 shows the overview of our experiments. The main contributions of the paper are as follows.

- We created a high-performance predictive model of lifestyle-related diseases for asymptomatic healthy patients based on a large-scale health screening dataset.
- We implemented a SHAP-based interpretation system for finding attributes that contribute to making predictions for the aggravation of diabetes and CKD.
- We found that different attributes contribute to the aggravation of diabetes and CKD depending on the time span being considered.

Our code is available at https://github.com/itumizu/interpretation_at_multiple_time_intervals.

2 Related Work

2.1 Prediction of Diabetes Stages Using Medical Records

Many studies proposed models to predict the onset of diabetes using medical records. Model building methods range from classical statistical approaches to modern machine learning-based techniques. For predictive models and methodologies, Kavakiotis *et al.* is a good survey on machine learning and data mining for diabetes research.

For classical statistical approaches, Sisodia *et al.* trained models such as decision trees and support vector machines (SVMs) to predict whether a patient would develop diabetes [19]. Choi *et al.* proposed a predictive model for type 2 diabetes using electronic medical records [4]. They used logistic regression, linear discriminant analysis, quadratic discriminant analysis, and k-nearest neighbor. Furthermore, Dagliati *et al.* applied a random forest and logistic regression on electronic health records to predict the onset of diabetes complications [5]. As modern machine learning techniques, Lai *et al.* used logistic regression and a gradient boosting machine to predict type 2 diabetes [10]. Most studies only focused on making predictions. However, several studies analyzed the implications of trained predictive models. For example, Manini *et al.* used a Bayesian network to investigate a causal structure for clinical complications in type 1 diabetes [13].

In most existing work, models were trained by medical data obtained from electronic medical records in hospitals. In contrast, our dataset comes from annual medical checkups. Additionally, since our dataset contains data from healthy subjects, it has an advantage in investigating the early stages of diabetes.

2.2 Prediction of Chronic Kidney Diseases Stages Using Medical Records

CKD is known to cause and be caused by other diseases. Many studies aim to determine whether hospitalized patients having CKD later develop other diseases as a complication. Tangri *et al.* predicted whether patients with CKD would deteriorate to renal failure [21]. They used the Cox proportional hazards model to analyze the factors associated with CKD progression. Kunwar *et al.* built a model to classify whether a patient has CKD using a naive Bayesian method and a neural network and compared the performance [9]. Wang *et al.* used health checkup data to predict the risk of CKD using random forest, XGBoost, and ResNet by using the regression of creatinine levels [23]. Moreno-Sanchez proposed a model for early detection of CKD by combining AdaBoost and decision trees [15]. They analyzed the relationship between the feature importance obtained from the constructed model and the attributes entered into the model. For dealing with CKD complications, Ravizza *et al.* used logistic regression with electronic medical record data to predict the risk of developing CKD in patients with diabetes [17]. Belur Nagaraj *et al.* proposed a model to identify patients with type 2 diabetes who will develop end-stage renal failure in the future [2].

CKD progresses slowly due to lifestyle-related effects. Therefore, there may be no subjective symptoms until the disease becomes severe. Prediction of the long-term progression of the disease and detection of signs will help prevent the onset of the disease. However, existing studies almost entirely aimed at predicting the onset of CKD for patients already having certain diseases rather than targeting healthy subjects.

2.3 Interpretable Prediction of Diseases

Some studies have introduced interpretability into their prediction. Xie *et al.* identified two new risk factors for type 2 diabetes by training predictive models [24]. Their analysis included an SVM, random forest, and neural network. Using time-series data, Park *et al.* proposed to use deep attention networks to make interpretable predictions for vascular diseases [16]. Their model was based on an RNN, which is appropriate for long-time sequence data, but it may not suit the health screening records that we target. For interpretation at different time intervals, Shakeri *et al.* compared attributes that contribute to the prediction of sepsis onset at two and six hours using SHAP [18].

Broome *et al.* reviewed the status of machine learning and AI in making decisions regarding diabetes care. They covered numerous use cases of machine learning in identifying pre-diabetes patients, automated insulin dosing systems, and customized meal and lifestyle recommendations. Finally, Dankwa-Mullan *et al.* surveyed various ways AI can be used for diabetes care, discussing how interpretable models are critical for many applications [6].

Existing studies have added interpretability to conventional methods. On the other hand, few studies have attempted to use interpretability to find new relationships. Interpretable predictions lead to understanding the model's behavior and why the predicted results were obtained. Moreover, interpretation of machine learning models can find abstract connections between the data learned by the model and the prediction target. Our method uses modern machine learning and interpretation techniques to analyze the relationship between the data and the predictor, focusing on the difference in the time intervals.

3 Dataset

3.1 Structure and Attributes

We used a medical checkup dataset collected by a regional health care center in Mito Kyodo General Hospital in Japan. The dataset consists of three annual medical checkup records. The number of samples (participants) for 2016, 2017, 2018, 2019, and 2020 was 4,133, 4,261, 4,270, 4,015, and 4,367, respectively.

We conducted a prediction task with time intervals ranging from one to four years. We used records from participants who took medical checkups in both input and target years.

The number of participants in each combination were 2,396 for 2016-2017, 2,527 for 2017-2018, 2,511 for 2018-2019, 2,701 for 2019-2020, 2,140 for 2016-2018, 2,179 for 2017-2019, 2,531 for 2018-2020, 1,896 for 2016-2019, 2,274 for 2017-2020 and 1,975 for 2016-2020. We removed attributes that were missing in over 95% of the participants.

We trained our model using the 38 remaining attributes: age, sex, height, weight, waist circumference, body mass index (BMI), systolic blood pressure, diastolic blood pressure, total cholesterol, high-density lipoprotein (HDL) cholesterol, low-density lipoprotein (LDL) cholesterol, fasting blood sugar (FBS), hemoglobin A1c (HbA1c), status of diabetes mellitus, hemoglobin, red blood cell count, hematocrit, white blood cell count, uric acid, hematuria (blood in urine), urine protein, urine sugar, fecal occult blood for day 1, fecal occult blood for day 2, neutral fat, cholinesterase, creatinine, albumin, alanine transaminase, aspartate transaminase, γ-glutamyl transpeptidase, C-reactive protein, electrocardiogram, abdominal echo, chest X-ray, status of gastric intestinal series, ophthalmology, and serum abnormalities. In addition, there are answers to the 20 self-administered questions [14] shown in Table 1. We did not include Q13 and Q16 in the attributes we used because the questions have changed since 2018, and responses to the same questions are no longer available.

Sex was selected from male or female. Electrocardiogram, abdominal echo, chest X-ray, status of gastric intestinal series, ophthalmology, serum abnormalities were selected from "nothing particular," "mild abnormality, "follow-up," "requires treatment," "requires further testing," and "under medical treatment." Fecal occult blood for day 1 and day 2 were selected from "negative," "positive," and "missing." Urine protein and urine sugar were selected from (-), (+-), (+), (2+), (3+), (4+), (5+). Q1–Q12, Q14–Q15, Q17, Q20, and Q22 were yes/no questions. Q18 was selected from "every day," "sometimes," and "none." Q19 was selected from "less than 180 ml," "180–360 ml," "360–540 ml," and "more than 540 ml," where arbitrary alcoholic drinks was quantified by converting to

Table 1. Self-administered questions in our medical checkup dataset

Number	Question
Q1	Using anti-hypertensive drug
Q2	Using insulin injection or antidiabetic (hypoglycemic) drug
Q3	Using anti-cholesteremic agent
Q4	Have you ever been diagnosed as having a stroke (cerebral hemorrhage or infarction) by a physician or had medical treatment?
Q5	Have you ever been diagnosed as having heart disease (angina pectoris or myocardial infarction) by a physician or had medical treatment?
Q6	Have you ever been diagnosed as having chronic renal failure by a physician or got artificial dialysis?
Q7	Have you ever been diagnosed as having anemia?
Q8	Have you smoked in the last month?
Q9	Have you put on weight by 10 kg since your 20s?
Q10	Have you exercised for more than 30 min each time, for more than two times per week, and for more than one year?
Q11	Do you walk daily or do other physical activity equal to walking for more than 1 h per day?
Q12	Do you walk faster than those in the same age group as you?
Q14	Do you eat faster than others?
Q15	Do you have dinner within two hours before going to bed more than three times a week?
Q17	Do you skip breakfast more than three times a week?
Q18	How often do you drink alcohol (such as sake, shochu, beer, whisky, etc.)?
Q19	When drinking, how much alcohol do you consume?
Q20	Do you sleep enough?
Q21	Do you want to improve your life style (life habit) such as exercise or eating?
Q22	If you have any chance to get health guidance on improving your life style (life habit), will you use it?

sake containing the same amount of alcohol. Q21 was selected from "I am not planning on improving," "I would like to try," "I am starting," "I am improving (less than six months)", and "I am improving (more than six months)."

3.2 Ethical Considerations

Annual medical examinations are conducted along with the Japanese Industry Safety and Health Act and are performed to facilitate lifestyle change and early disease diagnosis, which in turn would lower health expenditure and improve quality of life. This study was reviewed and approved by the ethics review committee of the authors' institution and conducted in accordance with the principles of the Declaration of Helsinki. Written informed consent was obtained from each participant.

4 Method

4.1 Target Attributes

We used HbA1c and creatinine as target attributes because they are commonly-used indicators of diabetes and CKD, respectively. We refer to attributes used for predicting as "features." They are the inputs to predictive models. In statistical terms, features are independent variables, and target attributes are dependent variables.

HbA1c is widely used as a criterion for conducting a diabetes diagnosis. Creatinine is a metabolite created by energy production in muscles, and high amounts of it are found in patients with CKD. The estimated glomerular filtration rate (eGFR) is also widely used for the diagnosis of CKD. Because the eGFR can be computed deterministically from creatinine, age, sex, and race, we aimed at predicting the amount of creatinine.

4.2 Prediction Tasks

The models were trained using features from a single year. Because our dataset contains health screening records from 2016 to 2020, we selected two years from 2016 to 2020 and used the latter half of the years chosen as the prediction target. For example, in one experiment, features from 2016 were used to predict the target attribute in 2020. In another experiment, features from 2017 were used to predict the target attribute in 2020. For each pair, features from the earlier year were used to predict the target attribute in the latter year.

To find attributes that contribute in making predictions in different ways, we conducted two types of experiments: (1) Training with the target attribute, together with strongly relevant attributes, from earlier years included as features, and (2) training without the target attribute and strongly relevant attributes from earlier years removed from features. In the latter type of experiments, HbA1c, FBS, and status of diabetes mellitus were removed from the attributes when predicting HbA1c. For predicting creatinine, only creatinine from earlier years was removed from features.

4.3 Preprocessing

In our dataset, each participant is labeled with one of six possible stages of diabetes, namely *nothing particular*, *mild abnormality*, *follow-up*, *requires treatment*, *requires further testing*, and *under medical treatment*. These stages were defined by the Japan Society of Ningen Dock[1]. For predicting HbA1c, we only used data

[1] https://www.ningen-dock.jp/wp/wp-content/uploads/2018/06/Criteria-category.pdf.

from participants whose stage in the input year is in *nothing particular*, *mild abnormality*, or *follow-up*.

When predicting creatinine, we used the eGFR to filter out a number of participants. To measure the kidneys' filtering capacity, the eGFR, which is calculated from creatinine and age, is commonly used. We calculated the eGFR using the following formula 1 defined by the Japanese Society of Nephrology [8].

$$eGFR = 194 \times Creatinine^{-1.094} \times Age^{-0.287} (\times 0.739 \text{ if female}) \quad (1)$$

We used six categories defined by the Kidney Disease: Improving Global Outcomes (KDIGO) organization [20]. In the six categories, we only used data from participants whose category in the input year is G1, G2, and G3a (*i.e.* eGFR $\geqq 45\,\mathrm{mL/min/1.73\,m^2}$).

For attributes taking continuous values, we replaced missing values with the average value. For attributes taking discrete values, a missing value is treated as an additional category. These methods are commonly used to handle missing values in data used for training models.

Contradictory samples were removed from the original dataset. Specifically, we removed the participants who answered *no* to the question *Q2: Using insulin injection or antidiabetic (hypoglycemic) drug* in the output year, despite their stage being in *under medical treatment* that year. We assumed they did not answer the questions correctly and removed them from the dataset.

For HbA1c prediction using preprocessed data, the numbers of participants by time intervals were 1,410 for four years, 3,159 for three years, 5,281 for two years, and 7,811 for one year. For creatinine prediction, the numbers of participants by time intervals were 1,544 for four years, 3,466 for three years, 5,753 for two years, and 8,540 for one year.

We conducted five-times-five nested cross-validation to compare machine learning techniques. Namely, we divided the dataset into five folds. For each training session, we used one fold as a test dataset and the rest for training and validation. The folds not used for testing were split into five further folds. Four of them were used for training, and one was used for validation, that is, hyperparameter optimization. The dataset was split into folds participant-wise. In other words, no participant is contained in two or more folds.

4.4 Training and Interpretation

We trained XGBoost [3] to predict the target attributes. It is known that XGBoost shows high performance for tabular data.

For interpretation, we used SHAP, which is based on game theory, to measure how each attribute contributed as a part of a coalition with other features in making the prediction correct. We used TreeExplainer [12] to calculate the SHAP values. There are model-independent methods for calculating SHAP values, such as Kernel SHAP. However, these methods compute SHAP values by making many predictions using perturbed input data. Therefore, the combinations become vast as the number of attributes increases and the computation time becomes longer. In TreeExplainer, SHAP values are calculated on the basis of the branching information used for prediction by the tree-based model, enabling fast and accurate calculation.

We optimized model parameters using the Optuna framework [1]. For each condition, we repeated training for 100 trials to optimize hyperparameters. In optimization, we changed the learning rate, max depth, min child weight, gamma, colsample by tree, and subsample as hyperparameter in the specified range. the learning rate is selected from {0.1, 0.01, 0.001}. max depth is selected from {1, 2, 3, 4, 5, 6, 7, 8, 9, 10}. min child weight is selected from {1, 2, 3, 4, 5}. gamma is selected from {0.0, 0.1, 0.2, 0.3, 0.4}. colsample by tree is selected from { 0.6, 0.7, 0.8, 0.9, 1.0} subsample is selected from {0.6, 0.7, 0.8, 0.9, 1.0}. We trained XGBoost for 1,000 rounds in each trial. For each method, if the validation root mean squared error (RMSE) did not improve for 20 rounds, we stopped training. After training, we selected the model having the highest validation RMSE and compared the results.

5 Evaluation

5.1 Prediction Accuracy

Table 2 shows the RMSE, mean absolute error (MAE), and R^2 score for conditions where (1) the target attribute in an earlier year is used as one of the features and (2) the target attribute in an earlier year is not used as a feature. When the time interval was longer, all error measures increased. This indicates that the longer the time interval, the more difficult it is to make predictions. The graphs also show that errors increase when the target attribute is not used as a feature.

Table 2. Prediction performance of each condition

Target	Target attribute	Time interval (year)	RMSE	MAE	R^2 score
HbA1c	Included	1	0.188 ± 0.006	0.140 ± 0.003	0.709 ± 0.016
		2	0.200 ± 0.014	0.144 ± 0.004	0.685 ± 0.030
		3	0.233 ± 0.014	0.166 ± 0.005	0.614 ± 0.025
		4	0.234 ± 0.027	0.159 ± 0.011	0.603 ± 0.063
	Not included	1	0.294 ± 0.007	0.226 ± 0.005	0.284 ± 0.029
		2	0.304 ± 0.015	0.230 ± 0.007	0.268 ± 0.039
		3	0.328 ± 0.019	0.245 ± 0.010	0.232 ± 0.020
		4	0.335 ± 0.023	0.245 ± 0.010	0.188 ± 0.052
Creatinine	Included	1	0.055 ± 0.001	0.041 ± 0.001	0.876 ± 0.005
		2	0.059 ± 0.001	0.044 ± 0.001	0.858 ± 0.011
		3	0.062 ± 0.003	0.046 ± 0.002	0.847 ± 0.013
		4	0.065 ± 0.003	0.048 ± 0.001	0.838 ± 0.005
	Not included	1	0.103 ± 0.003	0.080 ± 0.001	0.566 ± 0.007
		2	0.104 ± 0.003	0.081 ± 0.002	0.560 ± 0.027
		3	0.106 ± 0.003	0.082 ± 0.002	0.551 ± 0.013
		4	0.107 ± 0.005	0.082 ± 0.002	0.559 ± 0.024

5.2 HbA1c Included as a Feature

We first present the results for when HbA1c from an earlier year is included as a feature. For example, the amount of HbA1c in 2016 was used to predict HbA1c in 2020. Because there is a strong correlation between the amounts of HbA1c measured at two different years, the precision of prediction tends to be much higher than when not including them as a feature. Such high precision is useful for practical applications. However, because many predictions are explained by the amount of HbA1c in the earlier year, it is more difficult to see how other attributes contribute to making predictions. For this reason, including HbA1c as a feature might not be an ideal approach for scientific investigation on clarifying how various attributes affect the aggravation of the disease. Therefore, we also trained models in which HbA1c from an earlier year was not included as a feature. Such a model results in a lower prediction accuracy but enables to see contributing attributes other than HbA1c.

Figure 2a indicates how highly-ranked attributes change over a four-year period. The attributes are sorted in decreasing order of SHAP. The lines connect the same attributes across the years. The top three attributes (HbA1c, status of diabetes mellitus, and age) only slightly changed. However, attributes with lower ranks changed drastically over time. For example, waist circumference contributes largely to making predictions for the one-year interval (ranked 8th), but not much for the four-year interval (ranked 21st). In addition, alanine transaminase does not contribute for a short time interval (ranked 29th for the one-year interval) but contributes more in a longer time interval (ranked 6th for the four-year interval).

Figures 3a and b indicate the waist circumference and alanine transaminase for each sample, and the SHAP values corresponding to these values.

Alanine transaminase is one of the leading indicators of liver conditions. When the liver is in a normal condition, it works as an enzyme. When the liver's condition deteriorates, alanine transaminase, working inside the cells, leaks into the bloodstream, and its amount in the blood increases. Changes in the amount of alanine transaminase in the blood have been reported to be associated with diabetes [7,22].

Among highly contributing attributes are uric acid, Q18, and Q19. Uric acid is one of the commonly used indicators of kidney function. The amount of uric acid in the blood increases when the kidneys are unable to filter it out. Causes for such deficiency are alcohol consumption or decreased kidney function. Q18 and Q19 are questions about the frequency and amount of alcohol consumption. Excessive drinking places an undue burden on the liver's ability to break down alcohol, resulting in a decline in liver function. As the time interval increases, the rank of uric acid goes up, while those of Q18 and Q19 go down. The fact that alanine transaminase and uric acid are more contributing in long-term predictions than in short-term ones may indicate that liver and kidney conditions have a long-term effect on diabetes. At the time of writing, we are unsure as to why the rankings of Q18 and Q19 go down over time. It could be because drinking habits may change over four years while uric acid stays the same.

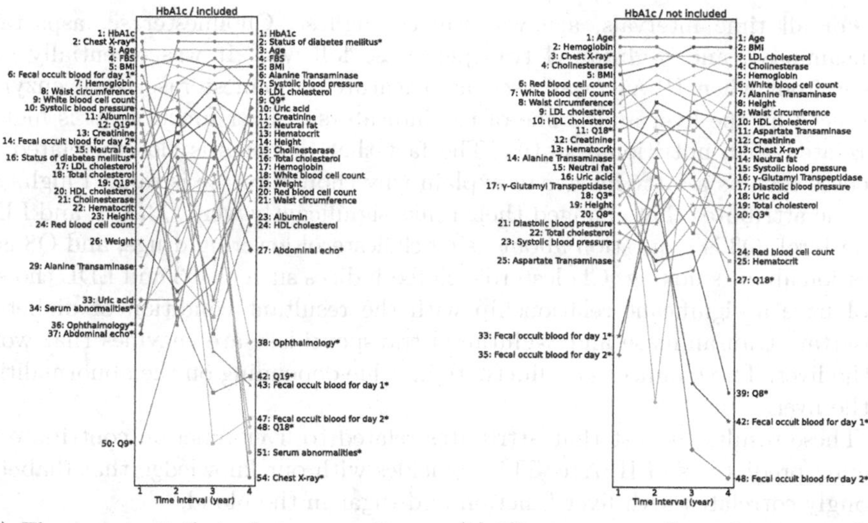

(a) The target attribute from an earlier year is included as a feature.

(b) The target attribute from an earlier year is **not included** as a feature.

Fig. 2. Ranking of attributes by SHAP values when predicting HbA1c.

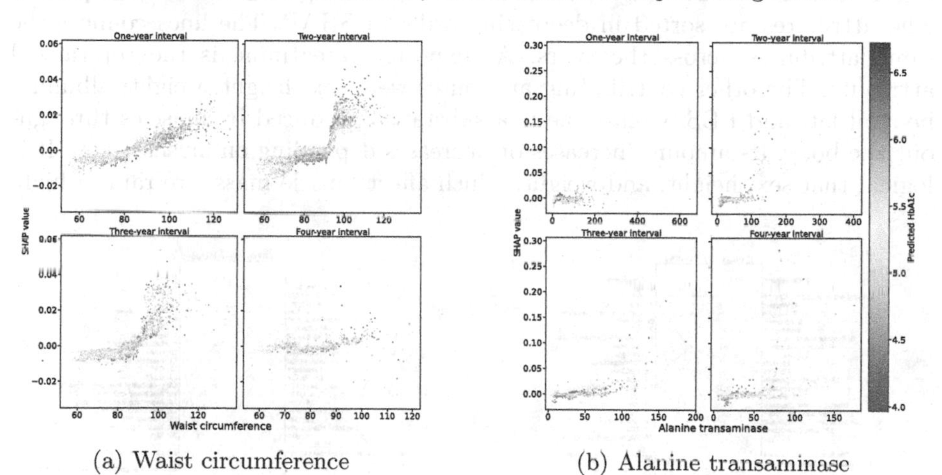

(a) Waist circumference

(b) Alanine transaminase

Fig. 3. Waist circumference, alanine transaminase and SHAP values for each sample when predicting HbA1c using related attribute from earlier years.

5.3 HbA1c Not Included as a Feature

Figure 2b indicates the changes in high-ranked attributes. The change in ranking was nearly the same as the previous results when HbA1c was included as a feature. One difference is that hemoglobin appears as a high-ranked attribute, possibly due to its correlation to HbA1c.

For all time intervals, age was ranked highest. Cholinesterase, aspartate transaminase, and γ-glutamyl transpeptidase followed. It was essentially different from when HbA1c was used as a feature. Cholinesterase is an enzyme produced by the liver and is one of the indicators of liver function. It is highly correlated with nutritional status. The fact that diabetes is closely related to diseases such as a fatty liver may explain why cholinesterase is ranked high.

The attributes that changed their ranks significantly were Q3, Q8, and LDL cholesterol. Q3 is a question about using cholesterol-lowering drugs, and Q8 is a question about smoking. Cholesterol-related indices such as Q3 and LDL cholesterol have a significant relationship with the resultant condition of diabetes. Aspartate transaminase and γ-glutamyl transpeptidase are enzymes that work in the liver. These indices also fluctuate in value depending on the abnormalities of the liver.

These results suggest that attributes related to liver function contribute to making predictions of HbA1c. This coincides with our knowledge that diabetes strongly correlates with liver function and sugar in the blood.

5.4 Creatinine Included as a Feature

Figure 4a indicates how highly-ranked attributes change over a four-year period. The attributes are sorted in decreasing order of SHAP. The lines connect the same attributes across the years. As expected, creatinine is the top-ranked attribute. The other contributing attributes were sex, height, weight, albumin, neutral fat, and FBS. Creatinine is a substance produced by muscles throughout the body. Its amount increases or decreases depending on muscle mass. It is logical that sex, height, and weight, which affect muscle mass, are ranked high.

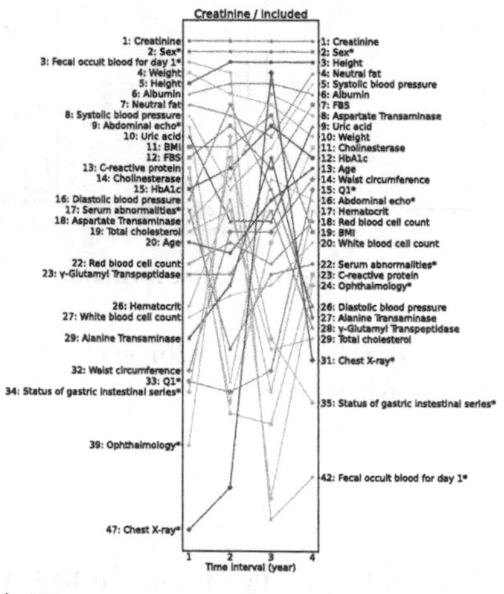

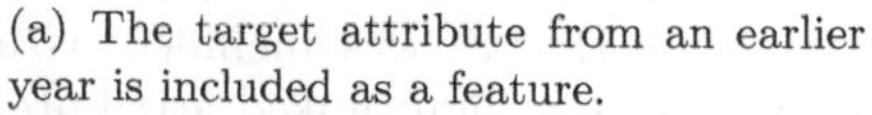
(a) The target attribute from an earlier year is included as a feature.

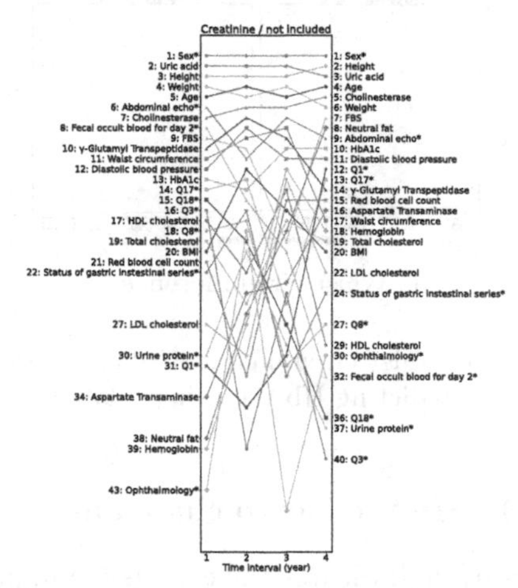

(b) The target attribute from an earlier year is **not included** as a feature.

Fig. 4. Ranking of attributes by SHAP values when predicting creatinine.

Albumin is a type of protein found in the blood. Because it is produced in the liver, it represents how well the liver is functioning. If the liver is not working correctly, the production of albumin decreases. The kidneys filter out albumin, but if they are not functioning right, they may not be filtered out and run off into the urine.

There are attributes with significant changes. For example, fecal occult blood for the day is ranked 3rd in the one-year interval and 4th in the two-year interval. However, it is ranked below 40th for the three-year and four-year intervals. In addition, waist circumference goes up from being ranked 32nd (one-year interval) to 14th (four-year interval), suggesting its long-term effect on CKD.

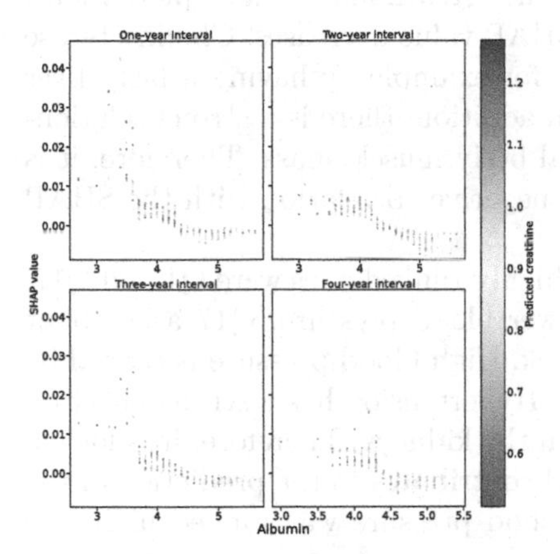

Fig. 5. Albumin and SHAP values for each sample when predicting creatinine using creatinine from earlier years.

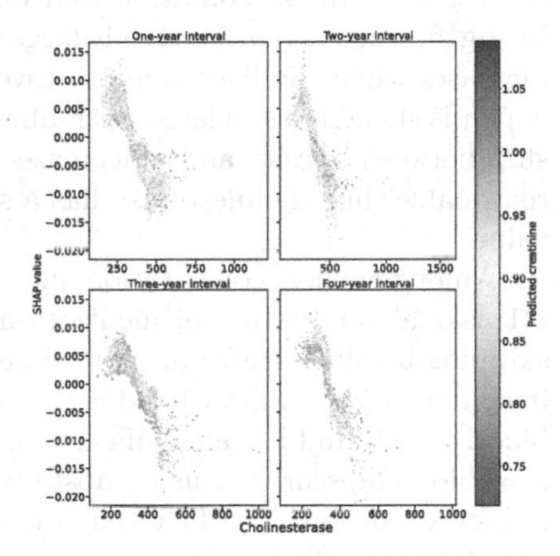

Fig. 6. Cholinesterase and SHAP values for each sample when predicting creatinine **without** using creatinine from earlier years.

Figure 5 shows that the SHAP value is higher in the positive direction when the albumin level is small. In particular, when kidney function deteriorates, nephrotic syndrome develops, which is a disease in which albumin in the blood flows out into the urine, decreasing in the amount in the blood. Therefore, the amount of albumin in the blood is as essential as creatinine when detecting the changes in kidney function at an early stage.

Waist circumference is also known as abdomen circumference, which increases due to the accumulation of fat in the gut and under the stomach's skin as the time interval between features and the target attribute lengthens. The ranks of weight and BMI lower while that of waist circumference rises. Therefore, although one can estimate the degree of obesity from other indices, waist circumference is considered a clear indicator of fat accumulation and can represent the degree of obesity. In contrast, it cannot be defined only by height and weight.

5.5 Creatinine Not Included as a Feature

Figure 4b shows the change of high-ranked attributes. When creatinine was not included as a feature, gender, height, and weight were ranked similarly as when creatinine was included. The other highly-ranked attributes were uric acid, cholinesterase, and abdominal echo. Uric acid and Cholinesterase are indicators of the function of the liver. Uric acid is also used to represent the level of kidney healthiness. Like HbA1c, they are also closely related to kidney function. Abdominal echo is an ultrasound examination of the abdominal organs such as the kidneys, liver, and pancreas. Therefore, it is reasonable that these attributes indicate the status of organs related to renal function and are listed as an essential attribute in the condition that does not use creatinine to make predictions. In Fig. 6, when cholinesterase is large, the SHAP value decreases. Cholinesterase increases when the liver condition worsens, for example by having a fatty liver which is strongly associated with obesity. In addition, there is a strong relationship between obesity and a decrease in total body muscle mass. Therefore, it is reasonable that cholinesterase has a strong negative correlation with the SHAP value.

Among medical consultation questions, highly-ranked ones were Q1 and Q17. Q1 asks about the use of medication to lower blood pressure. Q17 asks about skipping breakfast three or more times a week. High blood pressure is caused by irregular sleep and disordered eating habits. Hypertension has a strong effect on blood vessels and has a significant impact on the kidneys. Therefore, it is logical that blood pressure status is substantial and contributes to the prediction of the kidney condition. The fact that diastolic blood pressure was ranked high also supports this effect.

A healthy diet is one of the essential factors in maintaining good health, regardless of kidney condition. In addition, breakfast provides the energy needed for daytime activities and moderates blood sugar fluctuations during the day. It also indicates the relationship between lifestyle and diet in maintaining kidney function.

6 Conclusion

We analyzed the time dynamics of relationships between the predictions in our predictive model and the attributes in health screening data. Overall, the combination of XGBoost and SHAP turned out to be extremely powerful for finding contributing attributes from health screening data.

The experiments showed that as the time interval between features and the target attribute changes, many attributes change their degree of significance. A number of them matched with our existing knowledge on the mechanism of diabetes and CKD, but there were also interesting, unexpected observations that may provide insight to medical researchers. For example, the rank of alanine transaminase rises as the time interval lengthens, both for predicting HbA1c and creatinine. This suggests that alanine transaminase is a good early indicator for the target diseases.

The investigation of time dynamics of interpretations can lead to finding new relationships between health screening and the progression of diabetes and CKD. In many cases, the results matched our knowledge regarding diabetes and CKD, suggesting the effectiveness of using interpretable machine learning to investigate the underlying mechanisms of diseases.

In this work, we interpreted models that predict future medical states using health records from a single year as an input. Our current approach cannot capture how the dynamics over several years affect the medical condition in the future. For example, our predictive model cannot take into account whether the patient's medical test result is deteriorating rapidly in a few years or not. If we can train predictive models that take health records from several years as input, we can see the effect of such dynamics on the future outcome. We, therefore, plan to develop predictive models that take time-series data as input and obtain interpretations for those models.

Acknowledgements. This work was supported by JST COI Grant Number JPMJCE1301 to Y.W.; by G-7 Scholarship Foundation, Uehara Memorial Foundation, and JSPS KAKENHI Grant Number 18KK0308 to T.T. We are grateful for the staff of Mito Kyodo General Hospital for data preparation and research support.

References

1. Akiba, T., et al.: Optuna: a next-generation hyperparameter optimization framework. In: Proceedings of the 25th ACM SIGKDD International Conference on Knowledge Discovery & Data Mining, pp. 2623–2631 (2019)
2. Belur Nagaraj, S., et al.: Machine-learning-based early prediction of end-stage renal disease in patients with diabetic kidney disease using clinical trials data. Diabetes Obes. Metab. **22**(12), 2479–2486 (2020)
3. Chen, T., Guestrin, C.: XGBoost: a scalable tree boosting system. In: Proceedings of the ACM International Conference on Knowledge Discovery and Data Mining, pp. 785–794 (2016)
4. Choi, B.G., et al.: Machine learning for the prediction of new-onset diabetes mellitus during 5-year follow-up in non-diabetic patients with cardiovascular risks. Yonsei Med. J. **60**(2), 191–199 (2019)
5. Dagliati, A., et al.: Machine learning methods to predict diabetes complications. J. Diabetes Sci. Technol. **12**, 295–302 (2018)
6. Dankwa-Mullan, I., et al.: Transforming diabetes care through artificial intelligence: the future is here. Population Health Man. **22**(3), 229–242 (2019)
7. Itabashi, F., et al.: Combined associations of liver enzymes and obesity with diabetes mellitus prevalence: the Tohoku medical megabank community-based cohort study. J. Epidemiol. **32**(5), 221–227 (2020)
8. Japanese Society of Nephrology: Clinical practice guidebook for diagnosis and treatment of chronic kidney disease 2012. Japan. J. Nephrol. **54**(8), 1031–1191 (2012)
9. Kunwar, V., et al.: Chronic kidney disease analysis using data mining classification techniques. In: 2016 6th International Conference-Cloud System and Big Data Engineering (Confluence), pp. 300–305. IEEE (2016)

10. Lai, H., et al.: Predictive models for diabetes mellitus using machine learning techniques. BMC Endocr. Disord. **101**, 1–9 (2019)
11. Lundberg, S.M., Lee, S.I.: A unified approach to interpreting model predictions. In: Proceedings of the 31st International Conference on Neural Information Processing Systems, pp. 4768–4777 (2017)
12. Lundberg, S.M., et al.: From local explanations to global understanding with explainable AI for trees. Nature Machine Intell. **2**(1), 56–67 (2020)
13. Marini, S., et al.: A Dynamic Bayesian Network model for long-term simulation of clinical complications in type 1 diabetes. J. Biomed. Inform. **57**, 369–376 (2015)
14. Ministry of Health: Labour and Welfare. Standardized questionnaire for health checkups, Japan (2013)
15. Moreno-Sanchez, P.A.: Features importance to improve interpretability of chronic kidney disease early diagnosis. In: 2020 IEEE International Conference on Big Data (Big Data), pp. 3786–3792 (2020)
16. Park, S., et al.: Interpretable prediction of vascular diseases from electronic health records via deep attention networks. In: Proceedings of IEEE 18th International Conference on Bioinformatics and Bioengineering (2018)
17. Ravizza, S., et al.: Predicting the early risk of chronic kidney disease in patients with diabetes using real-world data. Nat. Med. **25**(1), 57–59 (2019)
18. Shakeri, E., et al.: Exploring features contributing to the early prediction of sepsis using machine learning. In: 2021 43rd Annual International Conference of the IEEE Engineering in Medicine & Biology Society (EMBC), pp. 2472–2475 (2021)
19. Sisodia, D., Sisodia, D.S.: Prediction of diabetes using classification algorithms. Procedia Comput. Sci. **132**, 1578–1585 (2018)
20. Stevens, P.E., Levin, A.: Evaluation and management of chronic kidney disease: synopsis of the kidney disease: improving global outcomes 2012 clinical practice guideline. Ann. Intern. Med. **158**(11), 825–830 (2013)
21. Tangri, N., et al.: A predictive model for progression of chronic kidney disease to kidney failure. JAMA **305**(15), 1553–1559 (2011)
22. Vozarova, B., et al.: High alanine aminotransferase is associated with decreased hepatic insulin sensitivity and predicts the development of type 2 diabetes. Diabetes **51**(6), 1889–1895 (2002)
23. Wang, W., Chakraborty, G., Chakraborty, B.: Predicting the risk of chronic kidney disease (ckd) using machine learning algorithm. Appl. Sci. **11**(1), 202 (2021)
24. Xie, Z., et al.: Building risk prediction models for type 2 diabetes using machine learning techniques. Prev. Chronic Dis. **16**, 1–9 (2019)

Fair and Efficient Alternatives to Shapley-based Attribution Methods

Charles Condevaux[1(✉)], Sébastien Harispe[2], and Stéphane Mussard[1]

[1] Univ. Nîmes CHROME, Nîmes Cedex 1, France
{charles.condevaux,stephane.mussard}@unimes.fr
[2] EuroMov Digital Health in Motion, Univ Montpellier, IMT Mines Ales, Montpellier, France
sebastien.harispe@mines-ales.fr

Abstract. Interpretability of predictive machine learning models is critical for numerous application contexts that require decisions to be understood by end-users. It can be studied through the lens of local explainability and attribution methods that focus on explaining a specific decision made by a model for a given input, by evaluating the contribution of input features to the results, e.g. probability assigned to a class. Many attribution methods rely on a game-theoretic formulation of the attribution problem based on an approximation of the popular Shapley value, even if the underlying rationale motivating the use of this specific value is today questioned. In this paper we introduce the FESP - Fair-Efficient-Symmetric-Perturbation - attribution method as an alternative approach sharing relevant axiomatic properties with the Shapley value, and the Equal Surplus value (ES) commonly applied in cooperative games. Our results show that FESP and ES produce better attribution maps compared to state-of-the-art approaches in image and text classification settings.

Keywords: Machine learning interpretability · XAI · Local interpretability · Attribution method

1 Introduction

Deep learning models are today state-of-the-art to tackle a large variety of machine learning problems in image or natural language processing (NLP) to cite a few. The use of these efficient models is however still limited due to their intrinsic black-box nature, i.e. deciphering the complex input-output mapping performed by trained deep learning models -sometimes involving billions of parameters- is still an open problem [21]. Indeed, many application contexts require not only models with good average performance, but also significant explanations allowing to fully understand and interpret predictor outputs. This is not only true for obvious critical use cases, e.g. in the medical field, in which sensitive decisions have to be supported by evidence [10,21,29]. More generally, legitimate concerns about potential harmful bias of inscrutable models are more expressed. Due to those issues, regulators introduce more and more legal requirements imposing life-impacting automated decision making to be explainable [15,28].

Supplementary Information The online version contains supplementary material available at https://doi.org/10.1007/978-3-031-26387-3_19.

M.-R. Amini et al. (Eds.): ECML PKDD 2022, LNAI 13713, pp. 309–324, 2023.
https://doi.org/10.1007/978-3-031-26387-3_19

In this context, numerous works analyze approaches contributing to deep learning model explainability, in particular through the notions of global and local interpretability while dealing with predictive tasks [4,16,21]. Global interpretability sheds some light on the general model behavior, e.g. global decision rules, while local interpretability focuses on explaining a specific decision (output) for a given input instance. This paper is concerned with local interpretability, and in particular with *attribution methods* (AMs). These methods aim at explaining the prediction made by a predictor for an input instance by assigning a scalar *attribution value* to each input feature. The purpose is therefore, for a given instance, to distinguish the features best explaining a model output prediction. The core problem is thus to define an AM that assigns a relevant attribution value to each feature in a specific predictive context.

Assigning attribution values in a meaningful way is an open question that has been studied through different angles. A large body of works focuses on AMs backed on axiomatic motivations defining supposedly intuitive properties that these methods should respect [2,22,24,36,38]. A wide range of contributions are in particular considering the Shapley value [32] as ground truth since it defines the unique way to solve the game-theoretic formulation of the attribution problem considering admitted axioms in coalition games - attribution is made considering a cooperative game between model features; attribution among the features is then made based on the Shapley value [2,22,38]. In that context, several approaches have been proposed to approximate the prohibitive computation of the Shapley value which requires evaluating 2^N feature subsets considering inputs of N features (NP-hard). Nevertheless, even if contributions stress that AMs based on the Shapley value seem to agree with human intuitive expectations [2], no clear agreement on that matter has been reached and the ground truth status of the Shapley value is today questioned [12,20]. Axiomatically grounded and algorithmically efficient, AMs have still to be investigated.
The contributions proposed in this paper are threefold:

1. We introduce FESP (Fair-Efficient-Symmetric-Perturbation value), an axiomatically grounded and algorithmically efficient AM that shares some properties with the Shapley value.
2. We propose the use of the equal surplus (ES) value, an $O(N)$ AM employed in cooperative games, which is linear, efficient and symmetric.
3. We show that FESP and ES achieve good accuracy on image and text classification compared with usual AMs.[1][2] The results outline their benefits with different benchmarks, e.g. issued from SHAP [22] and gradients [33,38].

The paper is structured as follows: Sect. 2 introduces existing AMs and discussions about the Shapley value. Section 3 presents FESP and ES. Section 4 evaluates ES, FESP and existing AMs on image and text classification tasks, with discussions on performances with respect to different protocols. Section 5 discusses our findings before mentioning perspectives they open.

[1] Our experiments: https://github.com/ccdv-ai/fesp_es.git. This work used HPC resources of IDRIS (allocation 2022-AD011011309R2) made by GENCI.

[2] This work has benefited from LAWBOT (ANR-20-CE38-0013) grant.

2 State of the Art

In this section we present the attribution problem focusing on a multiclass classification setting, as well as state-of-the art AMs proposed to solve it.

2.1 The Attribution Problem

Considering a predictor, the attribution problem consists in attributing a scalar value to each input feature characterizing an instance with respect to (w.r.t.) a predicted value (e.g. class probability in a classification setting or real value in a regression setting). This value represents the contribution of a specific feature to the prediction, e.g. in a classification setting, this value may be useful to understand which input features support a given class.

Without loss of generality, a multiclass classification setting is considered with a set of classes $\mathcal{C} := [\![1, C]\!]$, with $[\![a, b]\!]$ denoting the interval of all integers between a and b included. In that context, a predictor f takes an N-dimensional feature input $\mathbf{x} := [x_1, \ldots, x_N] \in \mathbb{R}^N$ and produces a probability distribution $f(\mathbf{x}) := [f_1(\mathbf{x}), \ldots, f_C(\mathbf{x})] \in [0, 1]^C$, with $f_i(\mathbf{x})$ the probability assigned to class $i \in \mathcal{C}$ by f for $\mathbf{x}$. $\mathcal{N} := [\![1, N]\!]$ is the set of feature indices.

Considering this setting, given predictor f and an input $\mathbf{x} \in \mathbb{R}^N$, an AM φ aims at computing a contribution vector $\varphi(\mathbf{x}, f_i)$ for any class $i \in \mathcal{C}$ such as $\varphi(\mathbf{x}, f_i) = [\varphi_1(\mathbf{x}, f_i), \ldots, \varphi_N(\mathbf{x}, f_i)] \in \mathbb{R}^N$, with $\varphi_j(\mathbf{x}, f_i)$ the attribution value of feature $j \in \mathcal{N}$ w.r.t. $f_i(\mathbf{x})$. Otherwise stated, considering the AM φ, $\varphi_j(\mathbf{x}, f_i)$ is the contribution of feature j to the probability assigned by the predictor f to class i for the input $\mathbf{x}$.

The two main classes of approaches studied in the literature to solve the attribution problem are introduced hereafter. They are both based on the evaluation of a perturbation of the input features on the predictive value under study.

2.2 Attribution Using Feature Coalisation Analysis

In the local interpretability setting, numerous perturbation-based approaches define an AM φ by evaluating the contribution $\varphi_j(\mathbf{x}, f_i)$ of a specific feature $j \in \mathcal{N}$ (to $f_i(\mathbf{x})$) as its contribution to coalitions of features. Considering a coalition including all features except j (i.e. $\mathcal{N} \setminus \{j\}$), the contribution of j to that coalition is assessed by evaluating the impact of a perturbation of x_j on $f_i(\mathbf{x})$. Such a perturbation aims at mimicking the removal of the studied feature, e.g. by naively setting its value to zero or a baseline value.

For any $\mathcal{S} \subseteq \mathcal{N}$, $\mathbf{x}(\mathcal{S})$ refers to the vector $\mathbf{x}$ in which all feature values x_k, $k \in \mathcal{N} \setminus \mathcal{S}$ have been substituted by a baseline value. As the input $\mathbf{x}$ is implicitly fixed in our discussions, $f_i(\mathcal{S})$ is used to denote $f_i(\mathbf{x}(\mathcal{S}))$, which is the probability assigned by f to class $i \in \mathcal{C}$ w.r.t. $\mathbf{x}(\mathcal{S})$.

The *marginal contribution* of a feature $j \in \mathcal{N}$ to a coalition $\mathcal{S}$ ($j \notin \mathcal{S}$) is thus defined by $f_i(\mathcal{S} \cup \{j\}) - f_i(\mathcal{S})$. Numerous AMs based on this notion of marginal contribution have been studied [6,13,39,41]. Game theory allows us to obtain such contributions, through the (least) core [18] but also through the Shapley value often considered as the ground truth value to explain the role of a given variable [2].

Attribution Value as the Shapley Value: The Shapley value averages marginal contributions over all possible feature coalitions:

$$\varphi_j^{Sh}(\mathbf{x}, f_i) := \sum_{\mathcal{S} \subseteq \mathcal{N} \setminus \{j\}} P(\mathcal{S}) \Big(f_i(\mathcal{S} \cup \{j\}) - f_i(\mathcal{S}) \Big),$$

for all $j \in \mathcal{N}$; $f_i(\emptyset) := 0$ for all $i \in \mathcal{C}$ by convention, and $P(\mathcal{S}) := (N - S - 1)!S!/N!$ $(S := |\mathcal{S}|)$.

The Shapley value implies (and is implied by) four axioms: *efficiency, additivity, symmetry* and the *null player axiom*, see [32].[3] These axioms make the Shapley value appealing from a theoretical point of view, and have motivated the *de facto* ground truth status given to this value.

However, considering N features, 2^N coalitions have to be evaluated which makes the Shapley value prohibitively expensive to compute. A natural way to reduce computation complexity is to rely either on coalition sampling to compute the marginal contributions [8], on local coalitions [9] or on Boolean circuits [3]. The first approach can however be slow to converge when the number of features is large. Instead of directly modifying original inputs, DASP [2] relies on distribution propagation using an auxiliary network based on Lightweight Probabilistic Deep Networks [14]. This model sequentially produces an estimate for each coalition size, thus allowing to greatly reduce the complexity from $O(2^N)$ to $O(N^2)$. Although this approximation is accurate, building an additional network is cumbersome, especially when fine tuning a pretrained model (as it requires rewriting each layer and activation function).

Attribution Based on Occlusion: In order to determine whether a feature or a group of features impacts a prediction, occlusion models measure the effect of removing them from the input (marginal contribution). In computer vision, these feature coalitions generally take the form of a sliding block [42], of a predefined size, inside which pixels are disturbed or replaced by a specific value (e.g. 0). Although such perturbation and occlusion models can accurately measure the marginal contribution of a variable, they tend to be slower than other AMs since they require multiple forward passes to fully cover the input and are thus dependent on the number of features. The size of the block is also an additional hyperparameter which can have a significant impact on overall performances.

2.3 Attribution Based on Gradient Analysis

Gradient-based approaches rely on various gradient computations through back-propagation evaluations. They compute the attribution value of a feature evaluating the partial derivative of the studied predicted value with regard to the feature value, e.g., $\varphi_j(\mathbf{x}, f_i)$ is defined as a function of $\partial f_i(\mathbf{x})/\partial x_j$. In this context $\varphi_j(\mathbf{x}, f_i)$ is then evaluated based on the impacts on $f_i(\mathbf{x})$ induced by a local change of x_j. The function φ_j should be carefully chosen to respect some properties or specific behaviors. For

[3] It is noteworthy that *additivity* implies *linearity* but the converse does not hold. Invoking *linearity* enlarges the class of admissible AMs, see Theorem 1 below.

instance, multiplying the gradient by the input [34] increases the sharpness of the attribution map but fails to handle specific functions like ReLU, which can produce zero values. More sophisticated models like DeepLift [33] and Integrated Gradient [38] satisfy a desirable axiom called completeness which is closely related to the efficiency axiom in cooperative game theory: for a baseline $\mathbf{x}'$ we have $\sum_{j \in \mathcal{N}} \varphi_j(\mathbf{x}, f) = f(\mathbf{x}) - f(\mathbf{x}')$.

To compute the contribution map, DeepLift takes all neurons and compares their activations after feeding a true sample and a reference input which can depend on the task and on the dataset. This model is inspired by Layer-wise Relevance Propagation which relies on a similar idea without the use of a reference [5]. Integrated Gradient averages different gradients: the input is modified multiple times along a linear path between itself and a baseline often set to zero. This continuous setting has been connected to another branch of the literature based on coalisation analysis, such as the Aumann-Shapley value [37].

3 Fair-Efficient-Symmetric Perturbations-based AMs

3.1 The Equal Surplus Value

It is well established that the Shapley value is easy to interpret since it displays the average of all marginal contributions of each feature; in this respect, it is a marginalist value. It shares some common properties with other marginalist values which form the Linear-Efficient-Symmetric values family (LES values) [31]. To our knowledge, this family has not been studied in the context of the attribution problem. The axioms respected by LES values are introduced hereafter.

Axiom 1 Linearity: *For all predictors f, g, an AM φ satisfies linearity if, $\varphi(\mathbf{x}, \alpha_1 f_i + \alpha_2 g_i) = \alpha_1 \varphi(\mathbf{x}, f_i) + \alpha_2 \varphi(\mathbf{x}, g_i)$, for all $\alpha_1, \alpha_2 \in \mathbb{R}$ and for all classes $i \in \mathcal{C}$.*

Axiom 2 Efficiency: *For all predictors f, an AM φ satisfies efficiency if, $\sum_{j \in \mathcal{N}} \varphi_j(\mathbf{x}, f_i) = f_i(\mathcal{N})$, for all classes $i \in \mathcal{C}$.*

Axiom 3 Symmetry: *For all predictors f, an AM φ satisfies symmetry if, for all features $j \in \mathcal{N}$, $\varphi_j(\mathbf{x}, f_i) = \varphi_{\pi(j)}(\mathbf{x}_\pi, f_i)$ for all permutations π over the set of $N!$ permutations on $\mathcal{N}$ and for all classes $i \in \mathcal{C}$.*

LES values have been extensively characterized outside the machine learning literature first by [31], then by [17,25,27] through the following theorem:

Theorem 1. *For all predictors f and all classes $i \in \mathcal{C}$, an AM φ satisfies linearity, efficiency and symmetry if and only if there exists a unique sequence of $N-1$ real numbers $\{b_s\}_{s=1}^{N-1}$ such that for each $j \in \mathcal{N}$ with $b_0 = 0$ and $b_N = 1$:*

$$\varphi_j(\mathbf{x}, f_i) = \sum_{\mathcal{S} \subseteq \mathcal{N} \setminus \{j\}} P(\mathcal{S}) \Big(b_{s+1} f_i(\mathcal{S} \cup \{j\}) - b_s f_i(\mathcal{S}) \Big).$$

LES values are all based on marginal contributions, therefore they provide feature contributions and interpretations very close to the usual Shapley value. The Shapley value φ^{Sh} is indeed a particular case of the LES family considering all marginal contributions equally weighted ($b_s = 1$ for all $s = 1, \ldots, N-1$). Other well-known LES values, studied in the cooperative game literature are: the Equal Surplus value (ES) [11], the Solidarity value [26], the Prenucleolus [30], and the Consensus value [19]. The ES value φ_j^{ES} ($b_s = 0$ if $1 < s < N$, $b_s = 1$ if $s = N$, $b_s = N-1$ if $s = 1$) is a peculiar member of the LES family since it is of complexity $O(N)$ whereas the others are $O(2^N)$:

$$\varphi_j^{ES}(\mathbf{x}, f_i) = f_i(\{j\}) + \frac{f_i(\mathcal{N}) - \sum_{k=1}^{N} f_i(\{k\})}{N}. \tag{1}$$

The first term of the right-hand side of Equation (1) is the contribution of feature x_j alone: its individual marginal contribution compared to a model composed of all features with baseline values $f_i(\{j\}) - f_i(\emptyset)$. The second term is the equal surplus: $f_i(\mathcal{N}) - \sum_{j=1}^{N} f_i(\{j\})$, i.e. the additional gain produced by the grand coalition in excess of the sum of the individual marginal contributions of features x_j, which evolve independently of the others.[4]

3.2 FESP

An AM grounded on the individual marginal contributions of each feature $f_i(\{j\}) - f_i(\emptyset)$ as in the ES is welcome since it outlines the role of each feature independently of the others. However, the equal surplus term is a constant for all features, consequently it cannot display the interaction of each feature with the grand coalition. In order to capture this specific effect, the exclusion of one feature from the whole set of features is employed, which consists in the occlusion technique. Occlusion related to feature x_j over class i may be simply characterized by $f_i(\mathcal{N} \setminus \{j\})$ instead of the equal surplus $f_i(\mathcal{N}) - \sum_{j=1}^{N} f_i(\{j\})$.

Then, two extreme feature coalitions could be considered for an AM: the one with the feature itself (such as Fig. 1 on the right-hand side - considering features as super-pixels), and the one associated with occlusion, i.e. the entire image minus a given feature (center of Fig. 1). On this basis, we propose the following family of AMs based on extreme feature coalitions:

Definition 1. Family of AMs based on extreme feature coalitions:

$$\varphi_j(\mathbf{x}, f_i) = w_i f_i(\{j\}) + (1 - w_i)(-f_i(\mathcal{N} \setminus \{j\})), \tag{2}$$

with $w_i \in [0, 1]$ a weight associated with class $i \in \mathcal{C}$.

The first component of the family, $w_i f_i(\{j\})$, is grounded on the individual marginal feature contribution, which is always positive. Then, as far as the feature is discriminant, its contribution to the classification in class i increases. The second component,

[4] The study of the independence is of importance for the tractability of the Shapley value, this is the case with fully factorized data distributions [7].

Fig. 1. Extreme feature coalitions

$(1 - w_i)(-f_i(\mathcal{N} \setminus \{j\}))$, is the contribution of occlusion, it is always negative. Occlusion of a discriminant feature x_j for class i entails that the probability f_i collapses, implying that the second component tends to zero. If an AM does not lie in the family of extreme feature coalitions, anything guaranties that bad features would be penalized by occlusion. Indeed, whenever a feature x_j is not discriminant for the classification in class i, the second component becomes negative, and the attribution value $\varphi(\mathbf{x}, f_i)$ can also become negative so that feature x_j is considered non-explanatory for the task. Furthermore, in order to gauge whether a feature is more *relevant* than another, the *fair treatment* axiom must be respected.

A feature x_k is said to be more relevant compared to feature x_ℓ when the association of x_k with all feature coalitions $\mathcal{S} \setminus \{k, \ell\}$ provides a greater attribution value compared to that of x_ℓ [27]. This property is welcome for all classification tasks such as image and text classifications. For instance, in an image classification setting, if a pixel x_k is more relevant compared to another one, because it allows some important shapes to be outlined, then the AM provides a higher contribution for x_k.

Axiom 4 Fair Treatment: *For all models f, and two given features x_k, x_ℓ, an AM φ satisfies fair treatment if, whenever feature x_k is more relevant compared to feature x_ℓ, i.e. $f_i(\mathcal{S} \cup \{k\}) \geq f_i(\mathcal{S} \cup \{\ell\})$ for all $\mathcal{S} \setminus \{k, \ell\}$, then $\varphi_k(\mathbf{x}, f_i) \geq \varphi_\ell(\mathbf{x}, f_i)$, for any given class $i \in \mathcal{C}$.*

FESP is an $O(N)$ complexity AM that shares a common structure with members of the LES family: it respects *efficiency*, *symmetry* and *fair treatment* (see Appendix A and B).

Proposition 1. *If an AM φ lies in the family of AMs based on extreme feature coalitions, and if it satisfies efficiency, then it is the FESP (Fair-Efficient-Symmetric-Perturbation) value given by, for $j \in \mathcal{N}$ and for $i \in \mathcal{C}$,*

$$\varphi_j^{FESP}(\mathbf{x}, f_i) = w_i f_i(\{j\}) + (1 - w_i)(-f_i(\mathcal{N} \setminus \{j\})), \tag{3}$$

$$w_i = \frac{f_i(\mathcal{N}) + \sum_{k=1}^{N} f_i(\mathcal{N} \setminus \{k\})}{\sum_{k=1}^{N} f_i(\{k\}) + \sum_{k=1}^{N} f_i(\mathcal{N} \setminus \{k\})}. \tag{4}$$

4 Experiments

This section presents results and evaluation protocols defined for comparing AMs on image and text classification tasks.We report experiments running ES and FESP alongside Integrated Gradients [38], DeepLIFT rescale [33], GradientShap [22] and Occlusion model. We also compare to the SHAP library using the DeepExplainer model [22] for vision tasks and the NLP pipeline named ShapExplainer for language tasks.

Local Explainability. A model is first trained to solve the predictive task under consideration. In order to focus on AM evaluation and to avoid any interpretative bias, we consider *simple* predictive tasks for which good performances are today easily achieved. Based on the predictor obtained, an AM is then evaluated regarding the features it brings out as important to explain the prediction obtained for a given input (only predictions are performed, no training phase is involved while evaluating AMs).

Top-k Model Accuracy. This metric consists in evaluating how the predictor accuracy evolves only using top-k input-dependent contributing features according to an AM φ. If φ identifies the features that best explain an input classification, the predictor should keep achieving good performances only considering those features, i.e., the more φ performs correctly, the better should be the predictor accuracy only considering a subset of features provided by φ. Unselected features for a given input are simply masked during prediction; the shape of the predictor input is not modified. For a given task, the same predictor is therefore employed independently of the features considered during prediction.

4.1 Image Classification: Protocols and Results

A pretrained VGG16 model [35] is fine tuned on a binary classification task related to Oxford-IIIT Pet Dataset[5] (`dog` vs `cat`, fine-tuning: 3 epochs over 6325 images). It achieves 99% accuracy (1024 images, features are pixels).

Masking Strategy. An image segmentation dataset gives for an image, the pixels of the shape of interest (segmentation mask), as well as its label, e.g. for an image labeled `dog`, the pixels of the dog are known. As focus is put on a simple classification task, it is assumed that pixels inside the mask should be relevant and have high attribution values (Appendix C presents a similar experiment modifying the pixels outside the segmentation mask with a random value). Considering an AM φ and a given image $\mathbf{x}$, the top-k contributing pixels of $\mathbf{x}$ are computed w.r.t. φ, with k a fixed number of pixels set based on the size of the segmentation mask of $\mathbf{x}$. AM-$Precision@k$ is computed: AM-$Precision@20(\varphi, \mathbf{x}, i)$ is the precision of φ on $\mathbf{x}$ over class i, only considering the top-k pixels, k being here equal to 20% of the size of the segmentation mask for $\mathbf{x}$ on class i. The precision of an AM is set to the average of the precision obtained for each image.

[5] https://www.robots.ox.ac.uk/~vgg/data/pets/.

Averaging ES and FESP. Despite their $O(N)$ complexity, computing ES and FESP is slower than gradient based attribution methods since a forward pass is required for each feature. For a 224 × 224 RGB image, 50,176 passes would be in theory necessary to compute attribution values. In practice, removing or inferring a class by modifying a single pixel has little to no impact on the prediction of the VGG16. 56 × 56 superpixels are considered for ES, FESP and Occlusion, so that the image becomes a grid of 16 superpixels. These methods are run on the superpixels, and the process is repeated by moving the grid with a stride of 8. All pixels inside a given superpixel get the same attribution score φ_j for the current pass; these scores are then averaged resulting in an overlapping process (each pixel gets masked the same number of times in order to get a balanced average).[6] This approach is similar to Occlusion and DeepExplain implementation.[7]

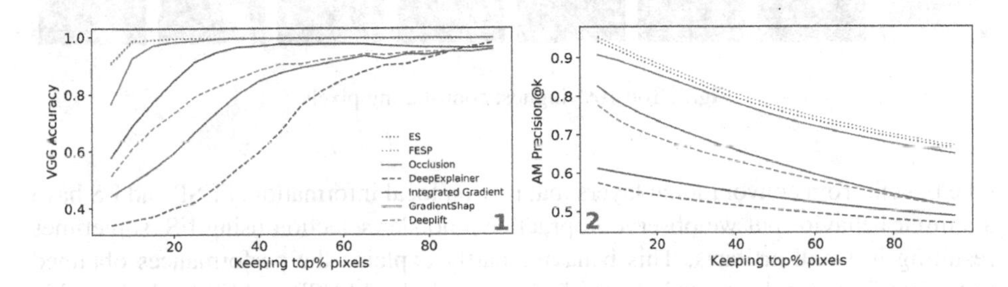

Fig. 2. Effect of feature selection on the predictor accuracy/precision.

Plot 1 (Fig. 2) shows the accuracy of the pretrained model (VGG16) while only considering the top-k features (pixels) evaluated as important by each AM. Considering the top 10% of the features, ES and FESP provide good predictive performances (90% accuracy). Except Occlusion, the other AMs must consider almost 90% of the selected features in order to reach the full input predictor accuracy (99%). This accuracy is reached only using 15% to 20% of the features identified as important by FESP and ES.

Plot 2 (Fig. 2) presents the AM-$Precision@k$, i.e. the capacity of each AM to outline expected informative pixels (pixels of the segmentation mask). AMs generally tend to consider the most important pixels to be inside the segmentation mask at first. FESP, ES and Occlusion achieves very good performances compared to other methods according to that test.

Figure 3 shows which image parts are recognized as relevant by the different AMs to explain the network prediction (top-10%).

We observe very different behaviors. AMs based on backpropagation independently treat pixels and therefore may return a noisy representation that is difficult to understand even for a relatively simple task where the discriminating criteria are fairly high level. The other AMs choose very localized areas, ignoring the rest of the image since they

[6] Good tradeoff between performance and time complexity since large superpixels lead to higher performances while small ones tend to be noisier.

[7] https://github.com/slundberg/shap.

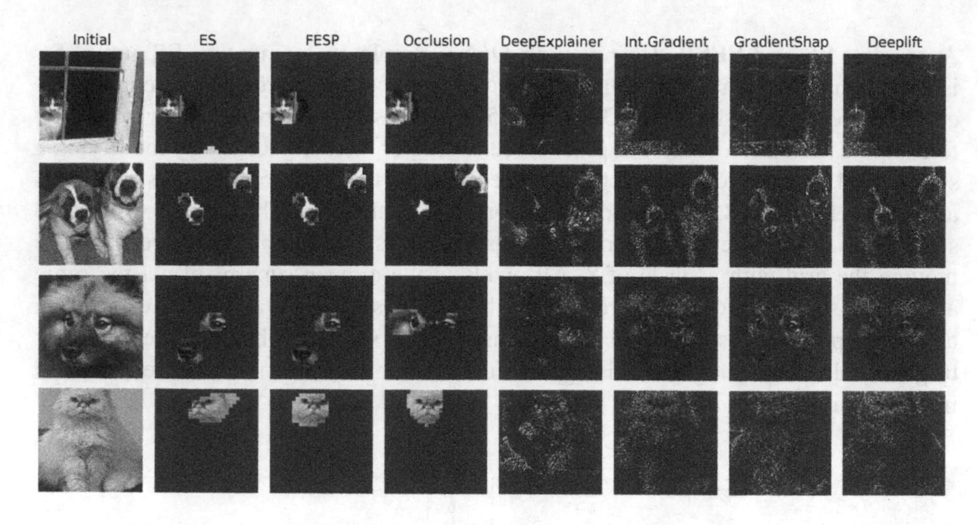

Fig. 3. Top 10% highest contributing pixels.

take benefit from convolutional layers that rely on local information. FESP and ES have a similar behavior but we observe in practice a noisier selection using ES, sometimes resulting in small artifacts. This behavior partly explains the performances obtained in the prediction task on partially masked inputs. Indeed FESP and ES tend to quickly identify very discriminant groups of features enabling to achieve good predictive performances even with a very limited set of features (Fig. 2). Thus, normalizing and merging the best performing AMs can be a good solution to improve the overall selection as shown in Appendix D.

Robustness of ES and FESP. An additional evaluation protocol is conducted based on recent contributions on AM evaluations [40] (refining [1]). It relies on a binary classification setting involving fictive composite images, each one being composed of 2×2 images from the Oxford-IIIT Pet Dataset. Each composite image is labeled `cat` or `dog` and only contains a single image among four corresponding to its label. Considering a specific composite image and a good predictive model, we assume that an efficient AM should make it easy to distinguish the single image corresponding to its label. Each composite image is a random mix of: (i) a labeled image (`cat` or `dog`), (ii) 3 unrelated additional images, (iii) the locations of the 4 images in the 2×2 grid. The train and test sets are generated using the same approach compared to disjoint subsets of images. The same pretrained VGG16 architecture is fine-tuned on 6325 images over 4 epochs (96% accuracy on the test set).

Figure 4 shows that ES and FESP display the highest top-k model accuracy, indeed an accuracy greater than 90% is reached with 3% of top pixels (plot 1). The AM-$Precision@k$ (plot 2) gives the percentage of pixels located within the labeled subimage (among the four) given the selection of the top-k contributing pixels. According to this metric, FESP is between Occlusion and ES with 95% for only 2% of top pixels.

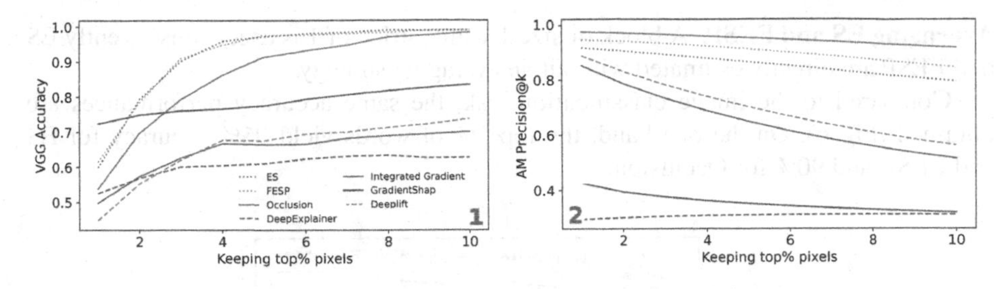

Fig. 4. Effect of feature selection on accuracy and precision.

As shown in Fig. 5 with 2 × 2 images, Occlusion, FESP and ES bring out relevant areas of the classes `dog` and `cat` and tend to be less noisy than gradient-based techniques (more images are available in Appendix D).

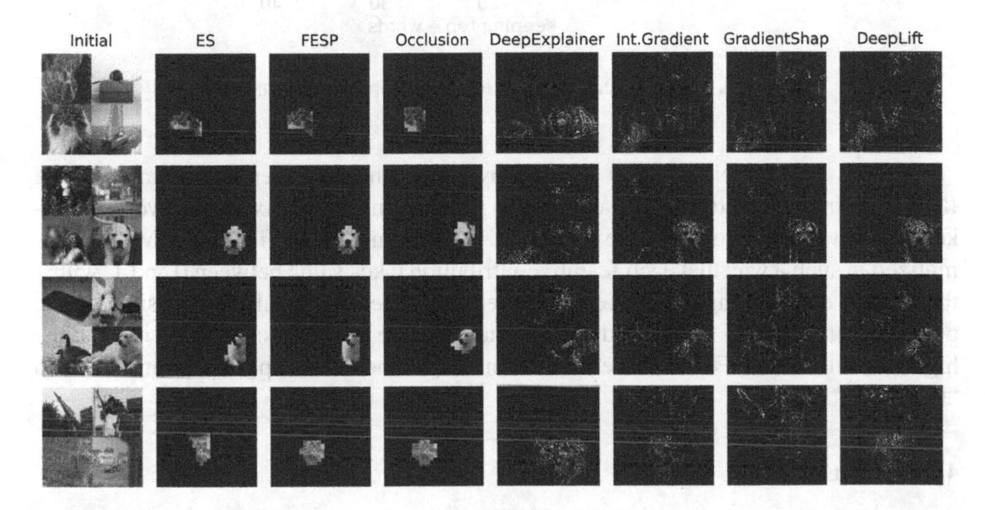

Fig. 5. Top 5% highest contributing pixels.

4.2 Text Classification: Protocols and Results

A binary text classification task is performed using IMDB dataset [23]. The model is a pretrained RoBERTa fine tuned on IMDB dataset, for which features are words (95.5% accuracy).[8] Testing is made on a subset of 1024 samples of the official testing set.

Masking Strategy. The masking strategy is task dependent. Transformers can take benefit from the softmax function inside the self-attention mechanism to fully mask a token and avoid all connections. This is not possible with convolution layers used in vision tasks.

[8] https://huggingface.co/textattack/roberta-base-imdb.

Averaging ES and FESP. A block of size 1 with a stride of 1 is used, consequently ES and FESP are directly estimated without an averaging strategy.

Compared to the image classification task, the same accuracy performances are obtained (Fig. 6). On the one hand, the top 5% of words yield 95% accuracy for ES and FESP, and 90% for Occlusion.

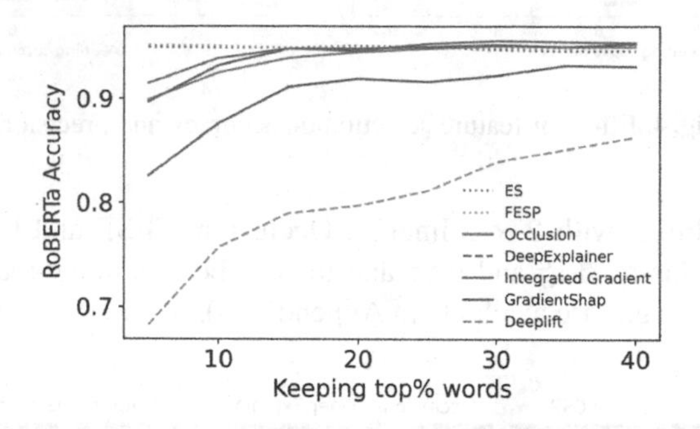

Fig. 6. Effect of feature selection on the predictor accuracy.

Finally, Fig. 7 depicts words selected by the seven AMs over one example of the IMDB testing set. Additional examples are provided in Appendix D. For words chunked into several subwords by the tokenizer, the maximum score is used. AMs are normalized in such a way that each feature contribution takes value between 0 and 1, with 1 the highest contributing tokens being colored red. As expected, all AMs easily capture positive words such as "love" and "sexy", but these are not necessarily associated to the highest contribution. For instance Occlusion assigns the most important contribution to "this" and "I".

4.3 Discussions

Occlusion, ES and FESP. FESP and ES behave similarly most of the time although ES being slightly noisier since each feature evolves independently (of the grand coalition). In order to remedy this problem, FESP straddles the line between ES and Occlusion since it can be considered as a weighted mean of the two methods. In terms of interpretability, these models differ greatly. Occlusion is unable to determine the sign of the feature contributions unlike FESP & ES. This difference makes interpretability difficult in many cases, especially for word importance tasks as can be shown in Fig. 7. In the case of image classifications a feature with a high contribution does not mean that it contributes positively to the prediction in the case of Occlusion.

From Local to Global Explainability. Although we focused on local explainability, ES and FESP can also be used in a global explainability context. This can be achieved by using specific metrics (e.g. accuracy, coefficient of determination) to measure the average impact of a feature on the predictions of a given predictor (see Appendix E for examples).

ES

In this film I prefer Deacon Frost. He's so sexy! I love his glacial eyes! I like Stephen Dorff and the vampires, so I went to see it. I hope to see a gothic film with him. "Blade" it was very "about the future". If vampires had been real, I would be turned by Frost!

FESP

In this film I prefer Deacon Frost. He's so sexy! I love his glacial eyes! I like Stephen Dorff and the vampires, so I went to see it. I hope to see a gothic film with him. "Blade" it was very "about the future". If vampires had been real, I would be turned by Frost!

Occlusion

In this film I prefer Deacon Frost. He's so sexy! I love his glacial eyes! I like Stephen Dorff and the vampires, so I went to see it. I hope to see a gothic film with him. "Blade" it was very "about the future". If vampires had been real, I would be turned by Frost!

Shap Explainer

In this film I prefer Deacon Frost. He's so sexy! I love his glacial eyes! I like Stephen Dorff and the vampires, so I went to see it. I hope to see a gothic film with him. "Blade" it was very "about the future". If vampires had been real, I would be turned by Frost!

Integrated gradient

In this film I prefer Deacon Frost. He's so sexy! I love his glacial eyes! I like Stephen Dorff and the vampires, so I went to see it. I hope to see a gothic film with him. "Blade" it was very "about the future". If vampires had been real, I would be turned by Frost!

GradientShap

In this film I prefer Deacon Frost. He's so sexy! I love his glacial eyes! I like Stephen Dorff and the vampires, so I went to see it. I hope to see a gothic film with him. "Blade" it was very "about the future". If vampires had been real, I would be turned by Frost!

DeepLift

In this film I prefer Deacon Frost. He's so sexy! I love his glacial eyes! I like Stephen Dorff and the vampires, so I went to see it. I hope to see a gothic film with him. "Blade" it was very "about the future". If vampires had been real, I would be turned by Frost!

Normalized importance

0 1

Fig. 7. Word importance normalized scores.

Fig. 8. Merged top 30% highest contributing pixels relative to mask size.

5 Conclusion

We have presented Equal Surplus (ES) and FESP (Fair-Efficient-Symmetric-Perturbation), two AMs based on marginalist values that can be used for local explainability of deep supervised learning models. These AMs compute attribution values that share relevant axiomatic properties with the Shapley value while ensuring an $O(N)$ time complexity for N-dimensional inputs.

According to the proposed evaluations based on two image and text classification tasks, FESP, ES and Occlusion seem to be more suited for tasks with spatial or temporal dependencies such as computer vision and NLP. Indeed, in these contexts, backpropagation and gradient-based approaches tend to be noisy and generally more difficult to interpret for humans. Additionally, our results also corroborate literature findings highlighting that backpropagation gradient-based approaches tend to act like shape detectors, and therefore achieve good results in distinguishing the global shape of an object of interest in a local attribution setting [1]. This paves the way to the study of various AMs mixing both approaches highlighting different but often complementary features (see Fig. 8).

Finally, the quantitative and qualitative results achieved by ES and FESP motivate the study of fast and axiomatically grounded AMs derived from LES values, which could reveal more, for example with the employ of AMs issued from the least square prenucleolus [30,31], or in the global attribution setting.

References

1. Adebayo, J., Gilmer, J., Muelly, M., Goodfellow, I., Hardt, M., Kim, B.: Sanity checks for saliency maps. arXiv preprint arXiv:1810.03292 (2018)
2. Ancona, M., Öztireli, C., Gross, M.: Explaining deep neural networks with a polynomial time algorithm for shapley value approximation. In: International Conference on Machine Learning, pp. 272–281. PMLR (2019)
3. Arenas, M., Barceló, P., Bertossi, L.E., Monet, M.: The tractability of shap-score-based explanations for classification over deterministic and decomposable boolean circuits. In: Thirty-Fifth AAAI Conference on Artificial Intelligence, AAAI 2021, Thirty-Third Conference on Innovative Applications of Artificial Intelligence, IAAI 2021, The Eleventh Symposium on Educational Advances in Artificial Intelligence, EAAI 2021, Virtual Event, 2–9 Feb 2021, pp. 6670–6678. AAAI Press (2021)
4. Arrieta, A.B., et al.: Explainable artificial intelligence (XAI): Concepts, taxonomies, opportunities and challenges toward responsible AI. Inform. Fusion **58**, 82–115 (2020)
5. Bach, S., Binder, A., Montavon, G., Klauschen, F., Müller, K.R., Samek, W.: On pixel-wise explanations for non-linear classifier decisions by layer-wiserelevance propagation. PloS One **10**, e0130140 (2015)
6. Brink, R., Funaki, Y., Ju, Y.: Reconciling marginalism with egalitarianism: consistency, monotonicity, and implementation of egalitarian shapley values. Soc. Choice Welfare **40**, 693–714 (2013)
7. den Broeck, G.V., Lykov, A., Schleich, M., Suciu, D.: On the tractability of shap explanations. Proceed. AAAI Conf. Artif. Intell. **35**(07), 6505–6513 (2021)
8. Castro, J., Gómez, D., Tejada, J.: Polynomial calculation of the shapley value based on sampling. Comput. Operat. Res. **36**(5), 1726–1730 (2009)
9. Chen, J., Song, L., Wainwright, M.J., Jordan, M.I.: L-shapley and c-shapley: efficient model interpretation for structured data. In: International Conference on Learning Representations (2019). https://openreview.net/forum?id=S1E3Ko09F7
10. Ching, T., et al.: Opportunities and obstacles for deep learning in biology and medicine. J. R. Soc. Interface **15**(141), 20170387 (2018)

11. Driessen, T.S.H., Funaki, Y.: Coincidence of and collinearity between game theoretic solutions. Oper.-Res.-Spektrum **13**(1), 15–30 (1991)
12. Frye, C., Rowat, C., Feige, I.: Asymmetric shapley values: incorporating causal knowledge into model-agnostic explainability. In: Advances in Neural Information Processing Systems 33 (2020)
13. Funaki, Y., Hoede, K., Aarts, H.: A marginalistic value for monotonic set games. Internat. J. Game Theory **26**, 97–111 (1997)
14. Gast, J., Roth, S.: Lightweight probabilistic deep networks. In: Proceedings of the IEEE Conference on Computer Vision and Pattern Recognition (2018)
15. Goodman, B., Flaxman, S.: European union regulations on algorithmic decision-making and a "right to explanation". AI magazine **38**(3), 50–57 (2017)
16. Guidotti, R., Monreale, A., Ruggieri, S., Turini, F., Giannotti, F., Pedreschi, D.: A survey of methods for explaining black box models. ACM Comput. Surv. (CSUR) **51**(5), 1–42 (2018)
17. Hernández-Lamoneda, L., Juárez, R., Sánchez-Sánchez, F.: Dissection of solutions in cooperative game theory using representation techniques. Internat. J. Game Theory **35**, 395–426 (2007)
18. Heskes, T., Sijben, E., Bucur, I.G., Claassen, T.: Causal shapley values: exploiting causal knowledge to explain individual predictions of complex models. In: Advances in Neural Information Processing Systems, vol. 33, pp. 4778–4789. Curran Associates, Inc. (2020)
19. Ju, Y., Borm, P., Ruys, P.: The consensus value: a new solution concept for cooperative games. Soc. Choice Welfare **28**, 685–703 (2007)
20. Kumar, I.E., Venkatasubramanian, S., Scheidegger, C., Friedler, S.: Problems with shapley-value-based explanations as feature importance measures. In: International Conference on Machine Learning, pp. 5491–5500. PMLR (2020)
21. Linardatos, P., Papastefanopoulos, V., Kotsiantis, S.: Explainable AI: a review of machine learning interpretability methods. Entropy **23**(1), 18 (2021)
22. Lundberg, S.M., Lee, S.I.: A unified approach to interpreting model predictions. In: Guyon, I., et al. (eds.) Advances in Neural Information Processing Systems 30, pp. 4765–4774. Curran Associates, Inc. (2017)
23. Maas, A.L., Daly, R.E., Pham, P.T., Huang, D., Ng, A.Y., Potts, C.: Learning word vectors for sentiment analysis. In: Proceedings of the 49th Annual Meeting of the Association for Computational Linguistics: Human Language Technologies, pp. 142–150. Association for Computational Linguistics, Portland, Oregon, USA (June 2011). http://www.aclweb.org/anthology/P11-1015
24. Montavon, G., Lapuschkin, S., Binder, A., Samek, W., Müller, K.R.: Explaining nonlinear classification decisions with deep Taylor decomposition. Pattern Recogn. **65**, 211–222 (2017)
25. Nembua, C.C., Andjiga, N.G.: Linear, efficient and symmetric values for TU-games. Econ. Bullet. **3**, 1–10 (2008)
26. Nowak, A.S., Radzik, T.: A solidarity value for n-person transferable utility games. Internat. J. Game Theory **23**, 43–48 (1994)
27. Radzik, T., Driessen, T.: On a family of values for TU-games generalizing the shapley value. Math. Soc. Sci. **65**, 105–111 (2013)
28. Ras, G., van Gerven, M., Haselager, P.: Explanation methods in deep learning: users, values, concerns and challenges. In: Escalante, H.J., et al. (eds.) Explainable and Interpretable Models in Computer Vision and Machine Learning. TSSCML, pp. 19–36. Springer, Cham (2018). https://doi.org/10.1007/978-3-319-98131-4_2
29. Rudin, C.: Stop explaining black box machine learning models for high stakes decisions and use interpretable models instead. Nat. Mach. Intell. **1**(5), 206–215 (2019)

30. Ruiz, L.M., Valenciano, F., Zarzuelo, J.M.: The least square prenucleolus and the least square nucleolus. two values for TU games based on the excess vector. Int. J. Game Theory **25**, 113–134 (1996)
31. Ruiz, L.M., Valenciano, F., Zarzuelo, J.M.: The family of least square values for transferable utility games. Games Econom. Behav. **24**, 109–130 (1998)
32. Shapley, L.S.: A value for n-person games. Contrib. Theory Games **2**(28), 307–317 (1953)
33. Shrikumar, A., Greenside, P., Kundaje, A.: Learning important features through propagating activation differences. In: Precup, D., Teh, Y.W. (eds.) Proceedings of the 34th International Conference on Machine Learning Research, 06–11 Aug 2017, vol. 70, pp. 3145–3153. PMLR (2017)
34. Shrikumar, A., Greenside, P., Shcherbina, A., Kundaje, A.: Not just a black box: Llearning important features through propagating activation differences. CoRR abs/1605.01713 (2016)
35. Simonyan, K., Zisserman, A.: Very deep convolutional networks for large-scale image recognition (2015)
36. Sun, Y., Sundararajan, M.: Axiomatic attribution for multilinear functions. In: Proceedings of the 12th ACM conference on Electronic commerce, pp. 177–178 (2011)
37. Sundararajan, M., Najmi, A.: The many shapley values for model explanation. In: III, H.D., Singh, A. (eds.) Proceedings of the 37th International Conference on Machine Learning Research, 13–18 Jul 2020, vol. 119, pp. 9269–9278. PMLR (2020)
38. Sundararajan, M., Taly, A., Yan, Q.: Axiomatic attribution for deep networks. In: International Conference on Machine Learning, pp. 3319–3328. PMLR (2017)
39. Wang, J., Zhang, Y., Kim, T.K., Gu, Y.: Shapley q-value: a local reward approach to solve global reward games. Proceed. AAAI Conf. Artif. Intell. **34**(05), 7285–7292 (2020)
40. Yona, G., Greenfeld, D.: Revisiting sanity checks for saliency maps (2021). https://doi.org/10.48550/ARXIV.2110.14297. https://arxiv.org/abs/2110.14297
41. Young, P.: Monotonic solutions of cooperative games. Internat. J. Game Theory **29**, 65–72 (1985)
42. Zeiler, M.D., Fergus, R.: Visualizing and understanding convolutional networks. CoRR abs/1311.2901 (2013)

SMACE: A New Method for the Interpretability of Composite Decision Systems

Gianluigi Lopardo[1(✉)], Damien Garreau[1], Frédéric Precioso[2], and Greger Ottosson[3]

[1] Université Côte d'Azur, Inria, CNRS, LJAD, Nice, France
glopardo@unice.fr

[2] Université Côte d'Azur, Inria, CNRS, I3S, Nice, France

[3] IBM France, Sophia Antipolis, France

Abstract. Interpretability is a pressing issue for decision systems. Many *post hoc* methods have been proposed to explain the predictions of a single machine learning model. However, business processes and decision systems are rarely centered around a unique model. These systems combine multiple models that produce key predictions, and then apply rules to generate the final decision. To explain such decisions, we propose the Semi-Model-Agnostic Contextual Explainer (SMACE), a new interpretability method that combines a geometric approach for decision rules with existing interpretability methods for machine learning models to generate an intuitive feature ranking tailored to the end user. We show that established model-agnostic approaches produce poor results on tabular data in this setting, in particular giving the same importance to several features, whereas SMACE can rank them in a meaningful way.

Keywords: Interpretability · Composite AI · Decision-making

1 Introduction

Machine learning is increasingly being leveraged in systems that make automated decisions. However, the massive adoption of artificial intelligence in many industries is hindered by mistrust, due to the lack of explanations to support specific decisions (Jan et al. 2020). Interpretability is deeply linked to trust and, as a result of public concern, has become a regulatory issue. For example, the European guidelines for trustworthy AI[1] recommend that "AI systems and their decisions should be explained in a manner adapted to the stakeholder concerned."

While numerous interpretability methods for single machine learning models exist (Linardatos et al. 2021), in many practical applications, a decision is rarely made by a unique model. In fact, composite AI systems, combining machine learning models together with explicit rules, are very popular, particularly in business settings. Incorporating decision rules is important, for two main reasons. Firstly, *decision rules are*

[1] https://digital-strategy.ec.europa.eu/en/library/ethics-guidelines-trustworthy-ai.

Supplementary Information The online version contains supplementary material available at https://doi.org/10.1007/978-3-031-26387-3_20.

M.-R. Amini et al. (Eds.): ECML PKDD 2022, LNAI 13713, pp. 325–339, 2023.
https://doi.org/10.1007/978-3-031-26387-3_20

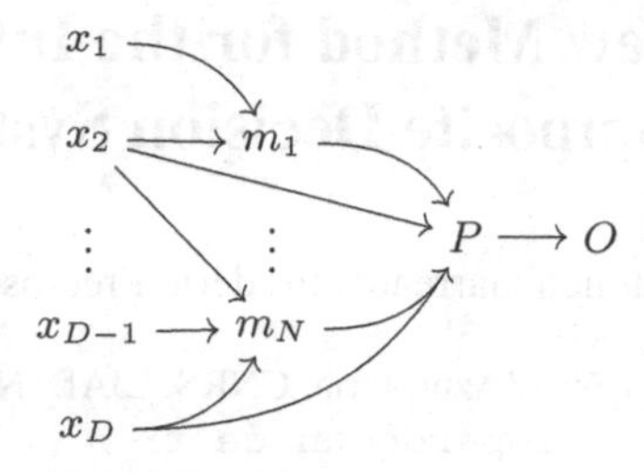

Fig. 1. Structure of a composite decision system with D input features $x_1, \ldots, x_D$, and N models $m_1, \ldots, m_N$. A decision policy P (*i.e.*, a set of decision rules) is finally applied to produce an outcome O. Note that in general both the models and the rules take a subset of input features as input, tough not necessarily the same.

crucial for expressing policies that can change (even very quickly) over time. For example, depending on last quarter's financial results, a company might be more or less risk-averse and therefore have a more or less conservative policy. Using an individual machine learning model would require to retrain it with new data each time the policy changes. In contrast, with a rule-based system, risk aversion can be managed by changing only a rule. Secondly, *machine learning models are not suitable for incorporating strict rules*. Indeed, while often a policy may represent a soft preference, in many cases we may have strict rules, due to domain needs or regulation. For example, we may have to require that clients' age be over 21 in order to offer them a service. It is typically difficult to account for such strict rules in a machine learning setting.

We focus our study on tabular data, most commonly used in businesses' day-to-day operations, often corresponding to customer records. Our interest in this paper is the interpretability of composite decision-making systems that include multiple machine learning models aggregated through decision rules in the form

```
if {premise}, then {consequence}.
```

Here, `premise` is a logical conjunction of conditions on input attributes (*e.g.*, age of a customer) and outputs of machine learning models (*e.g.*, the churn risk of a customer); `consequence` is a decision concerning a user (*e.g.*, propose a new offer to a customer). A phone company's policy for proposing a new offer can be

$$\texttt{if age} \leq 45 \texttt{ and churn\_risk} \geq 0.5\texttt{, then offer } 10\% \texttt{ discount.}$$

On the one hand, a number of additional challenges arise in this framework (see Sect. 3). On the other hand, there is knowledge we can leverage: we know the decision policy and how the models are aggregated. It is worth exploiting this information instead of considering the whole system as a black-box and being completely model-agnostic. In contrast, we want to be agnostic about the nature of individual models: we call this situation "semi-model-agnostic."

In this setting, we present the *Semi-Model-Agnostic Contextual Explainer* (SMACE), a novel interpretability method for composite decision systems that combines a geometric approach (for decision rules) with existing interpretability solutions (for machine learning models) to generate explanations based on feature importance. The key idea of SMACE is to agglomerate individual models explanations in a manner similar to that used by the whole decision system. By making the appropriate assumptions (see Sect. 4.2), we can see a decision system as a decision tree where some nodes refer to machine learning models. In a nutshell, we agglomerate the explanations for each model in a linear fashion, following the structure of this tree. We therefore combine an *ad hoc* method for the interpretability of decision trees, with *post hoc* methods for the models.

Contributions. The main contributions of this paper are

- The description of a new method, SMACE, for the interpretability of composite decision-making systems;
- The Python implementation of SMACE, available as an open source package at https://github.com/gianluigilopardo/smace;
- The evaluation of SMACE *vs* some popular methods showing that the latter perform poorly in our setting.

The rest of the paper is organized as follows. In Sect. 2, we briefly present some related work on both decision trees and *post hoc* methods for machine learning. Section 3 outlines the main challenges we want to address. In Sect. 4 the mechanisms behind SMACE are explained step by step; an overview is given in Sect. 4.3. Finally, we provide an evaluation of our method compared to established *post hoc* solutions in Sect. 5, before concluding in Sect. 6.

2 Related Work

A decision policy can be embedded in a decision tree. Small CART trees (Breiman et al. 1984) are intrinsically interpretable, thanks to their simple structure. However, as the number of nodes grows, interpretability becomes more challenging. Alvarez (2004) and Alvarez and Martin (2009) propose to study the partition generated by the tree in the feature space to rank features by importance. A similar approach has been used to build interpretable random forests (Bénard et al. 2021). We develop a solution inspired by this idea based on the distance between a point and the decision boundaries generated by the tree. The main difference in our setting is that each node can be a machine learning model.

Indeed, we also need to deal with machine learning interpretability. LIME (Ribeiro et al. 2016) explains the prediction of any model by locally approximating it with a simpler, intrinsically interpretable linear surrogate. Upadhyay et al. (2021) extends LIME to business processes, by modifying the sampling. Anchors (Ribeiro et al. 2018) extracts sufficient conditions for a certain prediction, in the form of rules. SHAP (Lundberg and Lee 2017) addresses this problem from a Game Theory perspective, where each input feature is a player, by estimating Shapley values (Shapley 1953). Despite the solid theoretical foundation, there is concern (Kumar et al. 2020) about its suitability for explanations. Labreuche and Fossier (2018) leverages Shapley values to explain the result of aggregation models for Multi-Criteria Decision Aiding. However, their solution requires full knowledge of the models involved, whereas we want to be agnostic about individual models. SMACE requires feature importance measures, provided for instance by LIME and SHAP (or different approaches as proposed by Främling (2022)).

Overall, perturbation-based methods have some drawbacks and are not always reliable (Slack et al. 2020). In addition, methods using linear surrogates are not suitable to deal with step functions (*e.g.*, the ones encoded by strict decision rules), which often leads to attributing the same contribution to multiple features. In the case of LIME for tabular data this behavior was pointed out by Garreau and von Luxburg (2020) and Garreau and Luxburg (2020a).

3 Challenges

As mentioned in the previous section, the field of interpretable machine learning has many unresolved issues. When trying to explain a decision that relies on multiple machine learning models, a number of additional problems arise:

- *Rule-induced nonlinearities*: decision rules will cause sharp borders in the decision space. For example: a car rental rule might state "`age` of renter must be above 21". Explanations for a machine learning based risk assessment close to the decision boundary `age` = 21, *e.g.*, must accurately indicate `age` as an important feature.
- *Out-of-distribution sampling*: the decision rules surrounding a machine learning model will eliminate a portion of the decision space. Explanatory methods based on sampling like LIME and SHAP are known to distort explanations because of this (see Sect. 2).
- *Combinations of decision rules and machine learning*: for a specific decision, a subset of rules triggered and a machine learning-based prediction was generated. How do we compose a prediction based on both sources?
- *Multiple machine learning models*: when multiple models are involved in a decision, we must also be able to aggregate multiple feature contributions. These may be (partially) overlapping and conflicting.

In addition, we want to have two desirable properties: *(1) the contribution associated with a feature must be positive if it satisfies the condition, negative otherwise*; *(2) the magnitude of the contribution associated with a feature must be greater the closer its value is to the decision boundary*.

4 SMACE

We now present SMACE in more details, starting with a thorough description of our setting in Sect. 4.1 and a discussion of our assumptions in Sect. 4.2. Section 4.3 contains the overview of the method, with additional details in Sect. 4.4, 4.5, and 4.6.

4.1 Setting

Let $x \in \mathbb{R}^{Q\times D}$ be the input data, where each row $x^{(i)} = (x_1, \dots, x_D)^\top \in \mathbb{R}^D$ is an instance and D is the cardinality of the *input features set* F. Let the set $M = \{m_1, \dots, m_N\}$ be the set of models. We will refer to their outputs $m_1(x), \dots, m_N(x)$ as the *internal features*, whose values we also denote as $y^{(1)}, \dots, y^{(N)}$ when there is no ambiguity. The union of input and internal features is the set of $D + N$ *features* to which the decision policy can be applied.

We define $\tilde{x} := (x_1, \dots, x_D, m_1(x), \dots, m_N(x))^\top$ as the completion of x with the outputs of the N models. Likewise, we call $\xi = (\xi_1, \dots, \xi_D)^\top$ the example to be explained and $\tilde{\xi} = (\xi_1, \dots, \xi_D, m_1(\xi), \dots, m_N(\xi))^\top$ its completion. A decision rule R is formally defined by a set of conditions on the features in the form $\tilde{x}_j \geq \tau$, for some cutoff $\tau \in \mathbb{R}$. Figure 1 illustrates a generic composite decision system.

4.2 Assumptions

The definition of SMACE is based on three assumptions required to frame the setting. Ideas for solving some of their limitations are discussed in Sect. 6.

Assumption 1. *Decision rules only refer to numerical values.*

This assumption allows us to take a simple geometric approach for the explainability of the decision tree. Note that this does not imply any restriction on the input of the machine learning models, that can still be categorical.

Assumption 2. *Each decision rule is related to a single feature, without taking into account feature interactions.*

For instance, this assumption excludes conditions like `if` $\tilde{x}_1 \geq \tilde{x}_2$. Geometrically, this implies decision trees with splits parallel to the axes, such as CART (Breiman et al. 1984), C4.5 (Quinlan 1993), and ID3 (Quinlan 1986).

Assumption 3. *The machine learning models only use input features to make predictions.*

We disregard the cases in which a machine learning model takes as input the output of other machine learning models. We remark that this is a very reasonable assumption that covers most real-world applications. Note that Assumptions 1 and 2 refer to the decision rules, while Assumption 3 is the only referring to the machine learning models and does not concern their nature.

4.3 Overview

For each example ξ whose decision we want to explain, we first perform two parallel steps:

– **Explain the results of the models**: for each machine learning model m, we derive the (normalized) contribution $\hat{\phi}_j^{(m)}$ for each input feature j. By default, SMACE relies on KernelSHAP to allocate these importance values;
– **Explain the rule-based decision**: measure the contribution r_j of each feature (that is, each input feature and each internal feature directly involved in the decision policy), through Algorithm 2.

Then, to get the **overall explanations** (see Algorithm 1), we combine these partial explanations. The total contribution of the input feature $j \in F$ to the decision for a given instance is

$$e_j = r_j + \sum_{m \in M} r_m \hat{\phi}_j^{(m)} . \tag{1}$$

That is, we weight the contribution of input features to each model with the contribution of that model in the decision rule, and we add the direct contribution of feature j to the decision rule (if a feature is not directly involved in a decision rule, its contribution is zero).

4.4 Explaining the Results of the Models

We need to attribute the output of each machine learning model to its input values. For instance, this is what KernelSHAP does, and by default SMACE relies on it. In any case, SMACE requires a measure of feature importance for the input features, but not necessarily based on SHAP. Any other measure of feature importance is possible. Given the contribution $\phi_j^{(m)}$ of each input feature j for each machine learning model m we define the normalized contribution as

$$\hat{\phi}_j^{(m)} = \begin{cases} \dfrac{\left|\phi_j^{(m)}\right|}{\sum_{i \in F}\left|\phi_i^{(m)}\right|} , & \text{if } \max_{i \in F}\left|\phi_i^{(m)}\right| \neq 0 , \\ 0 , & \text{otherwise.} \end{cases} \tag{2}$$

Indeed, two models m_k and m_h might give results $y^{(k)}$ and $y^{(h)}$ on very different scales, for instance because they do not have the same unit. In the example above, we may have models computing the churn risk and the life time value. The first value estimates a probability, so it belongs to $[0, 1]$, while the second is the expected economic return that the company may get from a customer, and it could be a quantity scaling as thousands of euros. In general, if m_k predicts the churn risk and m_h predicts the life time value, for a feature j in input to both models, we might expect $\left|\phi_j^{(h)}\right| \gg \left|\phi_j^{(k)}\right|$. In order to have a meaningful comparison between the models, we therefore need to scale the ϕ values and we use as scale factor the sum of the ϕ values for each model. The quantities $\hat{\phi}$ defined by means of Eq. (2) are of the same order of magnitude and dimensionless, so can be aggregated. In addition, $\hat{\phi}$ is defined such that

$$\forall j \in F , \forall m \in M , \quad 0 \leq \hat{\phi}_j^{(m)} \leq 1 .$$

Note that the second part of Eq. (2) is equivalent to taking the convention $\frac{0}{0} = 0$: the denominator is zero if and only if each contribution is zero. The definition implies that if the model m relies on a single feature j, the latter will have

$$\hat{\phi}_j^{(m)} = 1 \implies r_m \hat{\phi}_j^{(m)} = r_m \,,$$

i.e., the whole contribution of the model m to the decision is attributed to the input feature j, which in fact is the only one responsible for its output.

4.5 Explaining the Rule-Based Decision

In Sect. 2 we stated that the set of conditions used by a decision system can be interpreted as a CART tree, such as the one in Fig. 2, where each split represents a condition on a feature. A first approach to explain the decision of such a tree can be to show the trace followed by the point within the tree to the user. However, the trace does not contain enough information to understand the situation: a large change in some conditions may have no impact on the result, whereas a very small increase in one value may lead to a completely different classification, if we are close to a split value.

In addition, there may be many conditions within a decision rule, and simply listing them all would make it difficult to understand the decision. In fact, each condition is a split in the decision tree and each split produces a decision boundary. The collection of decision boundaries generated by the tree induces a partition of the input space and we call decision surface the union of the boundaries of the different areas corresponding to the different classes. Because of Assumption 2, at each point $z \in S$, the decision surface is piecewise-affine, consisting of a list of hyperplanes, each referring to one feature. By projecting an example point $\tilde{x}$ onto each component j of the surface S, we obtain the point $\pi_j^{(S)}(\tilde{x})$ (see Eq. (3)) at minimum distance that satisfies the condition on the j th feature (see Fig. 2). This distance is a measure of the robustness of the decision with respect to changes along feature j. Conversely, the smaller the distance, the more *sensitive* the decision.

As mentioned in Sect. 3, we want the method to assign a greater contribution to features with higher sensitivity. In this way, values close to the decision boundary are highlighted to the end user and the domain expert, who will be able to draw the appropriate conclusions. The explainability problem is therefore addressed by studying the decision surfaces generated by the decision tree.

However, to properly compare these contributions, we must first normalize the features. We must then query the models on the training set in order to obtain the values $y^{(1)}, \dots, y^{(N)}$. We thus apply a min-max normalization on both input features

$$\forall i \in \{1, \dots, Q\}, \quad x_j'^{(i)} = \frac{x_j^{(i)} - \min x_j}{\max x_j - \min x_j} \,,$$

and internal features, likewise. In this way, the values of each feature is scaled in $[0, 1]$. For the sake of convenience, we continue to denote the features x_i' and $y'^{(k)}$ as x_i and $y^{(k)}$, but from now on we consider them as scaled.

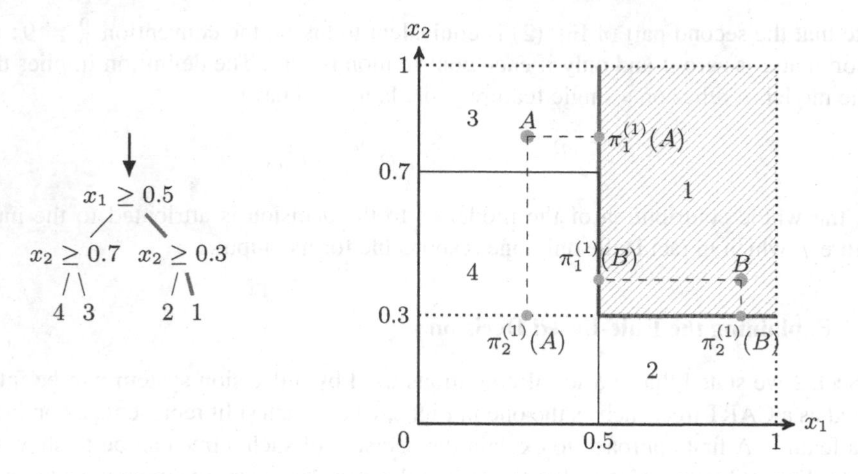

Fig. 2. On the left, a decision tree classifier based on x_1 and x_2. In **blue and bold**, the trace for leaf 1. On the right, the partition it generates. A and B are instance points, classified respectively as 3 and 1. The decision surface for leaf 1 is in **blue and bold**. The dashed lines indicate the distance between the points and the decision boundaries. (Color figure online)

Each decision surface S has as many components (hyperplanes) as there are features defining it. For instance, the decision surface for leaf 1 of Fig. 2 has two components: h_1 and h_2, along x_1 and x_2, respectively. The projection $\pi_j^{(S)}(x)$ of point x onto h_j is

$$\pi_j^{(S)}(\tilde{x}) \in \arg\min_{z \in h_j} \|\tilde{x} - z\|_2 \,. \tag{3}$$

For instance, let us consider the decision tree of Fig. 2 and the partition it generates. Let us say we are interested in leaf 1 (the grid subspace shown in Fig. 2) generated by the trace in blue. Example B satisfies both conditions, while A only satisfies the condition on x_2. We also note that the decision for B is very sensitive with respect to changes along axis x_2, while it is more robust with respect to x_1. We compute the contribution r_j of a feature j for the classification of point $\tilde{x}$ in leaf ℓ by means of Algorithm 2 as

$$r_j(\tilde{x}) = \begin{cases} \left|\tilde{x}_j - \pi_j^{(\ell)}(\tilde{x})\right| - 1 \,, & \text{if } \tilde{x}_j < h_j \,, \\ 1 - \left|\tilde{x}_j - \pi_j^{(\ell)}(\tilde{x})\right| \,, & \text{if } \tilde{x}_j \geq h_j \,. \end{cases} \tag{4}$$

We can see that for point A, the feature x_1 has a high negative contribution, since it does not satisfy the condition on it, while x_2 has a positive contribution. Point B satisfies both conditions: both features have positive contributions, but $r_2(B) > r_1(B)$, since the decision is more sensitive with respect to x_2.

4.6 Overall Explanations

Finally, once the partial explanations have been obtained, we agglomerate them via Eq. (1). We thus obtain a measure of the importance of features for a specific decision made

Algorithm 1. Overview of smace.

function SMACE_EXPLAIN(rule R (set of conditions), list of models M, example to explain $\xi \in \mathbb{R}^{\mathbb{D}}$)
  $\tilde{\xi} \leftarrow \xi$, $\phi \leftarrow \{0\}^N$, $r \leftarrow \{0\}^{D+N}$, $e \leftarrow \{0\}^D$
  for $m \in M$ **do**
    $\hat{\phi}^{(m)} \leftarrow$ EXPLAIN_MODEL(ξ, m) ▷ explain the result of model m (Sect. 4.4)
    $\tilde{\xi} \leftarrow (\xi_1, \ldots, \xi_D, \ldots, m(\xi))$
  end for
  for $j = 1, \ldots, D + N$ **do**
    $r_j \leftarrow$ RULE_CONTRIBUTION$(R, j, \tilde{\xi})$ ▷ explain the rule-based decision
  end for
  for $j = 1, \ldots, D$ **do**
    $e_j \leftarrow r_j + \sum_{m \in M} r_m \hat{\phi}_j^{(m)}$ ▷ aggregate
  end for
  return e
end function

by a system combining rules and machine learning models. Our measure of importance highlights the most critical features, those therefore most involved in the decision. In this way, a domain expert can analyse a decision by focusing on these features to make her or his own qualitative assessment.

Computational Cost. The most computationally expensive step of SMACE is to get explanations for the underlying models. It basically consists in N calls to the explainer on (at most) D input features. For instance, in the case of KernelSHAP, this would be $N \times 1000 \times D$.

5 Evaluation

What makes interpretability even more challenging is the lack of adequate metrics to appropriately assess the quality of explanations. In this section we compare the results obtained with SMACE and those obtained by applying the default implementations of SHAP[2] and LIME[3] on the whole decision system. We first perform a qualitative analysis on simple use cases, where we can get a complete understanding of the decision provided by the system. We show that SHAP and LIME do not satisfy the properties stated in Sect. 4.5 and we therefore argue that they are not suitable in this context. Finally, we perform a sanity check on aggregate explanations on three different realistic use cases.

[2] https://github.com/slundberg/shap.
[3] https://github.com/marcotcr/lime.

Algorithm 2. Computing RULE_CONTRIBUTION.

function RULE_CONTRIBUTION(rule R, variable j, example to explain $\tilde{\xi}$)
  $S \leftarrow R$ ▷ projection to the decision surface S generated by R
  $\pi_j^{(S)}(\tilde{\xi}) \leftarrow \arg\min_{z \in h_j} \|\tilde{\xi} - z\|_2$
  if $\tilde{\xi}$ satisfies condition on j **then**
    $r_j \leftarrow 1 - \left|\tilde{\xi}_j - \pi_j^{(S)}(\tilde{\xi})\right|$
  else
    $r_j \leftarrow \left|\tilde{\xi}_j - \pi_j^{(S)}(\tilde{\xi})\right| - 1$
  end if
  return r_j
end function

5.1 Qualitative Analysis

The input data consists of 1000 instances, each with three randomly generated components as uniform in $[0, 1]^3$. Note that decision rules on these data generate partitions analogous to those in Fig. 2, but in dimension 3.

Rules Only. Let us first evaluate the case of a decision system consisting of only three simple conditions applied to only three input features. The decision policy contains rule R_1:

$$\texttt{if } x_1 \leq 0.5 \texttt{ and } x_2 \geq 0.6 \texttt{ and } x_3 \geq 0.2 \texttt{ then } 1, \texttt{ else } 0.$$

Note that there are no models, R_1 is based solely on the input data. The method then reduces to the application of Eq. (4), discussed in Sect. 4.5.

Example with Two Violated Attributes. Take the example to be explained in an arbitrary position with respect to the boundaries: $\xi^{(1)} = (0.6, 0.1, 0.4)^\top$. The decision is 0, since the rule R_1 is not satisfied, indeed the conditions $\xi_1^{(1)} \leq 0.5$ and $\xi_2^{(1)} \geq 0.6$ are violated. We want to know why $\xi^{(1)}$ is not classified as 1 and the contributions of the three features to that decision. The comparison is shown in Table 1. The results of SMACE are computed (Eq. (4)) as

$$\begin{cases} r_1 = |0.6 - 0.5| - 1 = -0.9\,, \\ r_2 = |0.1 - 0.6| - 1 = -0.5\,, \\ r_3 = 1 - |0.4 - 0.2| = 0.8\,. \end{cases}$$

In this case, we see that all the three methods agree in their signs, satisfying property *(1)*. However, SHAP and LIME attribute the same contribution to x_1 and x_2 even though the sensitivities of the values are different. They do not satisfy property *(2)*: the contribution of x_1 should be higher in magnitude than that of x_2, since it is closer to the boundary. This behavior is due to the nonlinearities brought by the decision rules, as mentioned in Sect. 2. The point is that the sampling is performed in a space away from the boundary, and so by perturbing the example in a small neighborhood, the output does not change.

Table 1. Example in generic position, three conditions on three input features. LIME and SHAP are producing flat explanations on the variables x_1 and x_2, even if their sensitivities for the decision are very different. SMACE captures this information.

Condition	Example ($\xi^{(1)}$)	**SMACE**	SHAP	LIME
$x_1 \leq 0.5$	0.6	−0.9	−0.08	−0.21
$x_2 \geq 0.6$	0.1	−0.5	−0.08	−0.21
$x_3 \geq 0.2$	0.4	0.8	0.02	0.04

Table 2. Slight violation on one attribute, conditions on three input features. LIME and SHAP do not highlight the high sensitivities for x_2 and x_3, which are exactly on their respective decision boundary.

Condition	Example ($\xi^{(2)}$)	**SMACE**	SHAP	LIME
$x_1 \leq 0.5$	0.51	−0.99	−0.29	−0.22
$x_2 \geq 0.6$	0.60	1.00	0.12	0.14
$x_3 \geq 0.2$	0.20	1.00	0.03	−0.20

Slight Violation on One Attribute. We now consider the specific case where two features are exactly on the decision boundary, while one condition is slightly violated. Let us consider the example $\xi^{(2)} = (0.51, 0.6, 0.2)^\top$. The decision-making system classifies $\xi^{(2)}$ as 0 for a slight violation of the rule on the first attribute.

In Table 2 we see that SMACE highlights the slight violation of the rule on x_1.

Simple Hybrid System. Let us add two simple linear models m_1 and m_2. The models are defined as

$$\begin{cases} m_1(x) = 1x_2 + 2x_3 \,, \\ m_2(x) = 700x_1 - 500x_2 + 1000x_3 \,. \end{cases}$$

We are interested in rule R_3:

$$\texttt{if}\; x_1 \leq 0.5 \;\texttt{and}\; x_2 \geq 0.6 \;\texttt{and}\; m_1 \geq 1 \;\texttt{and}\; m_2 \leq 600 \;\texttt{then}\; 1\,,\; \texttt{else}\; 0\,,$$

and we want to explain the decision for $\xi^{(1)}$. The comparison on the whole system is in Table 3. Again, LIME and SHAP are producing identical results on x_1 and x_2, missing useful information. SMACE disagrees with the other methods on the sign of x_3, correctly giving a negative sign (Property *(1)*). Indeed, the input feature x_3 has a high contribution for the model m_2 and m_2 is not satisfying the condition ($m_2(\xi^{(1)}) = 770 > 600$), so it has a negative contribution.

By analyzing individual explanations, we have shown that SMACE produces meaningful results by assigning each feature a contribution proportional to its distance from the boundary. On the contrary, SHAP and LIME often assign the same contribution to different features, not providing useful information about the relative importance of each feature.

Table 3. Simple hybrid system, comparison on the whole decision system. LIME and SHAP both produce the same explanations for features 1 and 2.

Example ($\xi^{(1)}$)	**SMACE**	SHAP	LIME
$\xi_1^{(1)} = 0.6$	−1.03	−0.08	−0.19
$\xi_2^{(1)} = 0.1$	−1.73	−0.08	−0.19
$\xi_3^{(1)} = 0.4$	−0.54	0.02	0.09

5.2 Sanity Check

In the previous section, we showed that SMACE is able to produce meaningful feature attributions. We now demonstrate that SMACE also retains an ability to identify the set of features contributing negatively to a decision, regardless of individual attribution. If a feature contributes negatively, it means it must be moved to meet its condition. Correctly identifying negative features is a desirable property: to change the decision, each of them must be moved.

We consider 100 random instances which do not satisfy the rules (described in the supplementary), from three different datasets, and we apply SMACE, SHAP, and LIME. For each method, we extract the set of negative features. Note that to be sure that the rule will be satisfied, each negative feature should be shifted to a specific value: none of the three methods is giving this information. We then generate 1000 samples by shifting negative features with a local perturbation. The average decision made on these perturbed samples is an indicator of the quality of the explanations provided by each of the three methods.

Cancer Treatment. A machine learning model is trained to predict whether a breast cancer is benign or malignant from information about its size and structure. An automated decision system is then applied to decide on treatment: if the risk of the tumor being malignant is too high, it proceeds in full reliance on the model. If, on the other hand, the probability is low, but the size and composition of the tumor are suspicious, further investigation is carried out. The decision system consists of 30 continuous *input features* and 1 *internal feature* (coming from the model). We use the *Breast Cancer Wisconsin Data Set*.[4]

In this example, we want to explain *why* the treatment was not proposed, *i.e.*, which input features are negatively contributing to the decision. Given the large number of parameters to be analyzed, it is useful to order them by importance, in order to speed up the investigation by giving the right priorities. The graph at the top left of the Fig. 3 shows the comparison. SMACE curve is always above the others: it is better at detecting negative features.

Fraud Detection. A financial authority must track mobile money transactions, promptly halting anomalous transactions suspected of fraud. The authority uses a decision-making system to approve or block transactions, according to a *fraud score*, computed through a machine learning classifier, and the amount and balanced involved

[4] https://www.kaggle.com/datasets/uciml/breast-cancer-wisconsin-data.

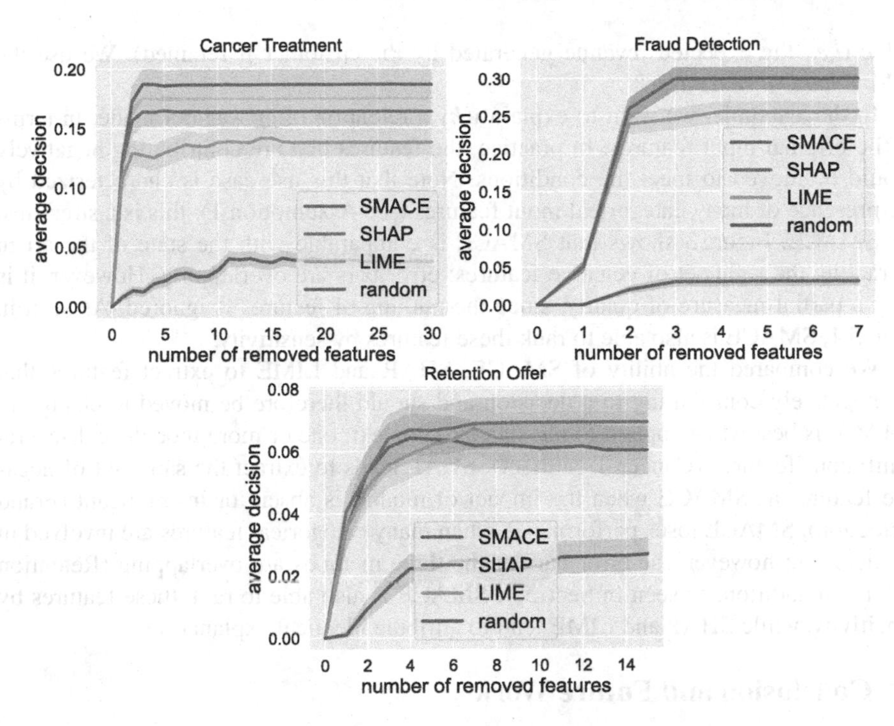

Fig. 3. Comparison of SMACE, SHAP, and LIME on the ability to identify the set of features contributing negatively to a decision, regardless of individual attribution. Correctly identifying negative features is a desirable property: to change the decision, each of them must be moved. When the conditions are not met, the three methods are used to extract the negative features, and we generate perturbed samples around the original values. We then compare the average decision made on the samples.

in the transaction. We use the *Synthetic Financial Datasets For Fraud Detection*[5] As before, we extract and perturb the negative features set for each method.

The graph at the top right of Fig. 3 shows that SMACE and SHAP are on par. In this decision system, the conditions based on the input features matter significantly more than the one on the model. This means that SMACE explanations are almost entirely based on Eq. (4) and, consistently with what we saw in Sect. 5.1, SMACE and SHAP are able to extract the correct set of negative features. However, we remark that SHAP is likely to assign them the same (negative) contribution: SMACE carries more information.

Retention Offer. Let us consider a mobile phone company which wants to predict if a customer is going to leave for a competitor, and to decide if a retention offer should be made, while not spending more on retention than the value of retaining the customer. The decision policy is based on information about the customer and their subscription (input features), and two models (producing internal features) predicting the *churn risk* (*i.e.*, the likelihood that the customer will cancel their subscription) and the *lifetime*

[5] https://www.kaggle.com/ealaxi/paysim1.

value (*i.e.*, the expected revenue generated by the customer if retained). We use the IBM *Telco Churn* dataset.[6]

In this example, we want to explain *why* a retention offer was not made, in terms of the original input features. In practice, the features that are contributing negatively should be moved to meet the conditions. Note that this use case is characterized by the presence of many categorical input features (see Assumption 1): this is a stress test for SMACE. Figure 3 shows that SMACE is comparable with the state of the art in extracting the right set of negative features: error bars are overlapping. However, it is only a partial measure of quality, since the ranking of features is ignored. As seen in Sect. 5.1, SMACE is also able to rank these features by sensitivity.

We compared the ability of SMACE, SHAP, and LIME to extract features that are negatively contributing to a decision and should therefore be moved to change it. SMACE is best when applied to the standard context: one or more models and several continuous features (Cancer Treatment). SHAP tends to extract the same set of negative features as SMACE when the impact of models is absent or insignificant (Fraud Detection). SMACE loses performance when many categorical features are involved in the decision: however, the error bars of the three methods are overlapping (Retention Offer). In addition, as seen in Sect. 5.1, SMACE is also able to rank these features by sensitivity, while SHAP and LIME tend to attribute identical explanations.

6 Conclusion and Future Work

We addressed the problem of explaining decisions produced by a decision-making system composed of both machine learning models and decision rules. We proposed SMACE, to generate feature importance based explanations. Up to the best of our knowledge, it is the first method specifically designed for these systems. SMACE approaches the problem with a projection-based solution to explain the rule-based decision and by aggregating it with models explanations. We finally showed that model-agnostic approaches designed to explain machine learning models are not well-suited for this problem, due to the complications coming with the rules. In contrast, SMACE provides meaningful results by meeting our requirements, *i.e.*, adapting to the needs of the end user.

In future work, we plan to extend SMACE, making it usable in a wider range of applications. A particularly interesting approach to include categorical features in the rules is implemented in CatBoost (Prokhorenkova et al. 2018), a gradient boosting toolkit. The idea is to group categories by *target statistics*, which can replace them. SMACE could also be generalized to more complex model configurations, where some models take as input the output of other models. One natural extension would be to recursively weight the importance of each model with the contribution it brings for other models.

Acknowledgments. This work has been supported by the French government, through the NIM-ML project (ANR-21-CE23-0005-01) and through the 3IA Côte d'Azur (ANR-19-P3IA-0002), and by EU Horizon 2020 project AI4Media (contract no. 951911, https://ai4media.eu/).

[6] https://github.com/IBMDataScience/DSX-DemoCenter/tree/master/DSX-Local-Telco-Churn-master.

References

Jan, S.T.K., Ishakian, V., Muthusamy, V., the need for process-aware explanations: AI trust in business processes. In: Proceedings of the AAAI Conference on Artificial Intelligence **34**, pp. 13403–13404 (2020)

Linardatos, P., Papastefanopoulos, V., Kotsiantis, S.: Explainable AI: a review of machine learning interpretability methods. Entropy **23**(1), 18 (2021)

Breiman, L., Friedman, J.H., Olshen, R.A., Stone, C.J.: Classification and regression trees. Routledge (1984)

Alvarez, I.: Explaining the result of a decision tree to the end-user. In: ECAI, vol. 16, page 411 (2004)

Alvarez, I., Martin, S.: Explaining a result to the end-user: a geometric approach for classification problems. In: Exact 09, IJCAI 2009 Workshop on explanation aware computing (International Joint Conferences on Artificial Intelligence), pages 102 (2009)

Bénard, C., Biau, G., Da Veiga, S., Scornet, E.: Sirus: stable and interpretable rule set for classification. Electron. J. Stat. **15**(1), 427–505 (2021)

Ribeiro, M.T., Singh, S., Guestrin, C.: "Why should I trust you?" explaining the predictions of any classifier. In: Proceedings of the 22nd ACM SIGKDD International Conference on Knowledge Discovery and Data Mining, pp. 1135–1144 (2016)

Upadhyay, S., Isahagian, V., Muthusamy, V., Rizk, Y.: Extending LIME for business process automation. arXiv preprint arXiv:2108.04371 (2021)

Ribeiro, M.T., Singh, S., Guestrin, C.: Anchors: high-precision model-agnostic explanations. In: Proceedings of the AAAI Conference on Artificial Intelligence, vol. 32 (2018)

Lundberg, S.M., Lee, S.-I.: A unified approach to interpreting model predictions. Adv. Neural. Inf. Process. Syst. **30**, 4765–4774 (2017)

Shapley, L.S.: A value for n-person games. Contributions to the Theory of Games, number 28 in Annals of Mathematics Studies, pp. 307–317, II (1953)

Kumar, E.I., Venkatasubramanian, S., Scheidegger, C., Friedler, S.: Problems with Shapley-value-based explanations as feature importance measures. In: International Conference on Machine Learning, pp. 5491–5500. PMLR (2020)

Labreuche, C., Fossier, S.: Explaining multi-criteria decision aiding models with an extended Shapley Value. In: IJCAI, pp. 331–339 (2018)

Främling, K.: Contextual importance and utility: a theoretical foundation. In: Long, G., Yu, X., Wang, S. (eds.) AI 2022. LNCS (LNAI), vol. 13151, pp. 117–128. Springer, Cham (2022). https://doi.org/10.1007/978-3-030-97546-3_10

Slack, D., Hilgard, S., Jia, E., Singh, S., Lakkaraju, H.: Fooling LIME and SHAP: adversarial attacks on post hoc explanation methods. In: Proceedings of the AAAI/ACM Conference on AI, Ethics, and Society, pp. 180–186 (2020)

Garreau, D., von Luxburg, U.: Looking deeper into tabular LIME. arXiv preprint arXiv:2008.11092, v. 1 (2020)

Garreau, D., Luxburg, U.: Explaining the explainer: a first theoretical analysis of LIME. In: International Conference on Artificial Intelligence and Statistics, pp. 1287–1296. PMLR (2020a)

Quinlan, R.J.: C4.5: programs for machine learning. Morgan Kaufmann Publishers Inc., San Francisco (1993). ISBN 1-55860-238-0. http://portal.acm.org/citation.cfm?id=152181

Quinlan, R.J.: Induction of decision trees. Mach. Learn. **1**(1), 81–106 (1986)

Prokhorenkova, L.O., Gusev, G., Vorobev, A., Dorogush, A.V., Gulin, A.: Catboost: unbiased boosting with categorical features. In: NeurIPS (2018)

Calibrate to Interpret

Gregory Scafarto(✉), Nicolas Posocco(✉), and Antoine Bonnefoy(✉)

EURA NOVA, Marseille, France
{gregory.scafarto,nicolas.posocco,antoine.bonnefoy}@euranova.eu

Abstract. Trustworthy Machine learning (ML) is driving a large number of ML community works in order to improve ML acceptance and adoption. The main aspect of trustworthy ML are the followings: fairness, uncertainty, robustness, explainability and formal guaranties. Each of these individual domains gains the ML community interest, visible by the number of related publications. However few works tackle the interconnection between these fields. In this paper we show a first link between uncertainty and explainability, by studying the relation between calibration and interpretation. As the calibration of a given model changes the way it scores samples, and interpretation approaches often rely on these scores, it seems safe to assume that the confidence-calibration of a model interacts with our ability to interpret such model. In this paper, we show, in the context of networks trained on image classification tasks, to what extent interpretations are sensitive to confidence-calibration. It leads us to suggest a simple practice to improve the interpretation outcomes: *Calibrate to Interpret*.

Keywords: Interpretability · Calibration · Classification · Trustworthy machine learning

1 Introduction

Despite being state of the art on many tasks, deep neural networks (DNNs) are still considered as black boxes which is problematic in contexts where decision making is critical. In order to ensure that a model can be safely used, one needs to have access to the uncertainty over its predictions, and to understand what drives those predictions. Both aspects, namely uncertainty and explainability, are frequently tackled via the use of post-hoc calibration and local interpretation respectively. The current work studies the interaction between these two central aspects of trustworthy ML.

There are numerous ways to interpret a model's behaviour, depending on what one has access to (internal model structure, training phase, ...). This work focuses on methods which interpret the decisions of already trained image classifiers, considering model-aware as well as model-agnostic local interpretation methods, following the definitions given in [34], the former have access to the internal structure of the model while the latter only relies on the predictions scores. These methods provide a saliency map highlighting important pixels for each prediction. Regarding predictions uncertainty,

---

Supplementary Information The online version contains supplementary material available at https://doi.org/10.1007/978-3-031-26387-3_21.

M.-R. Amini et al. (Eds.): ECML PKDD 2022, LNAI 13713, pp. 340–355, 2023.
https://doi.org/10.1007/978-3-031-26387-3_21

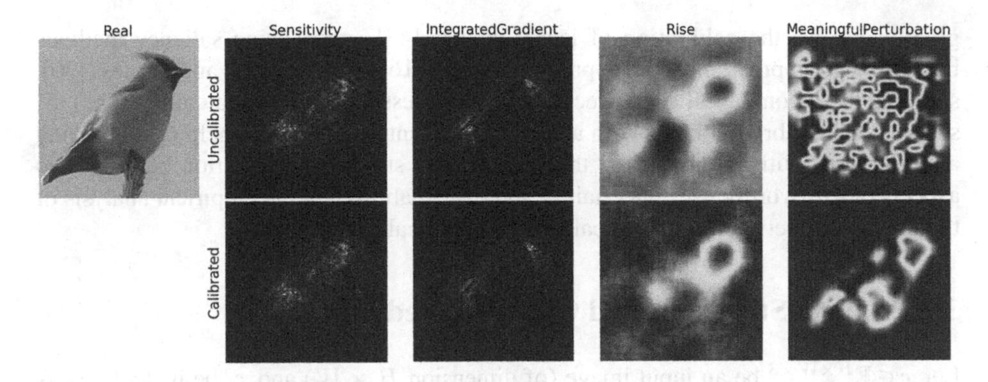

Fig. 1. Visual comparison of saliency changes due to calibration

post-hoc calibration methods adapt models so that their scores are consistent with conditional probabilities related to the predicted class [21]. It is a convenient way to tackle the overconfidence of modern neural networks [11,25].

This focus was made to fit the flexible context in which calibration and interpretation are handled after training the model. Since both post-hoc calibration and local interpretation are linked to the scores associated with the predicted class, we address their interdependence via the following questions: Does calibration impact the saliency maps obtained as interpretations? Do modifications, if any, improve the faithfulness of interpretation methods? Are saliency maps with calibrated models more human-friendly?

Our empirical evaluations highlights that there is indeed a positive interaction between the calibration of a model and its interpretability by enabling some widely used interpretation methods to work more efficiently on it. This impact is visible in terms of faithfulness, visual-coherence of the saliency maps, and their robustness most notably for model-agnostic approaches like Meaningful Perturbation (MP) [9]. Examples of saliency maps produced by various methods before and after the calibrations are presented in Fig. 1.

After positioning our work in Sect. 2, we introduce the problems of calibration and local interpretation, as well as the methods used in the literature to tackle them in Sect. 3. Section 4 then describes the conducted experiments and reports their outcomes. Finally results are discussed and we mention some future works before concluding in Sect. 5[1]

2 Related Works

Trustworthy ML has recently gained interest, with the development of modern uncertainty quantification, explainability, robustness, fairness and formal guaranties. However few publications have investigated links between these aspects. Among these we can mention attempts to link: robustness and calibration [33,35,44] showing that models which are robust to adversarial attacks are more interpretable; data augmentation, calibration and interpretation showing that the MixUp data augmentation procedure

[1] Code available at https://github.com/euranova/calibrate_to_interpret.

greatly impacts the calibration of learnt models [43] and existing saliency methods being used to improve the MixUp procedure itself [16]; or calibration and fairness [30], showing the incompatibility between most of fairness definitions and calibration. [15] showed that calibration helps with attention based interpretation by helping stabilizing attention distributions. Following these works, we study the interaction between two aspects of trustworthy ML: uncertainty and explainability via the empirical analysis of the interaction between post-hoc calibration and local interpretation.

3 Problem Statement and Other Related Works

Let $x \in \mathbb{R}^{H \times W \times 3}$ be an input image (of dimension $H \times W$) and x_i be its features. A model F is trained to classify sample images among C classes. It maps each input x to a logit vector L which is then converted, generally through a softmax function, into an output vector $F(x) \in [0,1]^C$ so that $\sum_{c=1}^{C} F(x)_c = 1$. The decision associated with such prediction is $y = \text{argmax}_c F(x)_c$.

3.1 Calibration

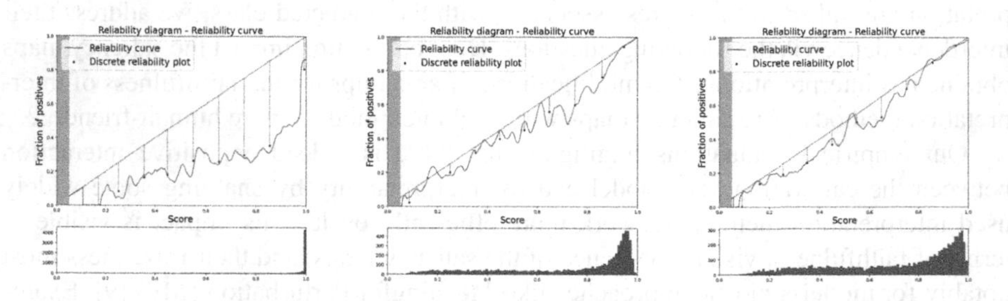

Fig. 2. Confidence reliability plots and curves for VGG16 trained on CIFAR100, when uncalibrated (left, $ECE_{conf} = 0.2$), calibrated using Temperature Scaling (center, $ECE_{conf} = 0.04$) and calibrated with Dirichlet calibration (right, $ECE_{conf} = 0.05$)

Many critical applications motivate the evaluation of confidence-calibration, which indicates how confident a model is in the class it predicts. It is well-known that modern neural networks tend to be miscalibrated, generally overconfident on their predictions [11,25]. Thus, some works have been done on the evaluation of calibration [27] and the improvement of such calibration for a given classifier, would it be during training [22] or as a post-processing step [20,21,29].

As local interpretation allows to interpret each specific predictions, we focused our work on its interaction with confidence-calibration, i.e. the calibration of scores for the predicted class $argmax(F(x))$.

A model is said to be well confidence-calibrated if:

$$\forall s \in [0,1], P\left(y = z \mid F(x)_z = s\right) = s \tag{1}$$

with $z = argmax(F(x))$

This notion can be quantified with estimates of the Expected Calibration Error (ECE), defined in the confidence-calibration setting as:

$$ECE^{conf} = \mathbb{E}_{max(F(X))}\left[|\mathbb{P}(y = z \mid F(x)_z = s) - s|\right] \tag{2}$$

In order to calibrate our models, we focus on post-hoc calibration, which can be applied to already trained models and relies exclusively on the scores given by the model. These very nice properties make these approaches plug-and-play, which justifies their popularity. We use in this paper the following techniques, which impact is illustrated in Fig. 2.

Temperature Scaling. If L is the logits vector output of the classifier before softmax activation σ, calibrated scores are given by $F(x)_c^{cal} = \sigma(LT^{-1})_c$ [29]. The temperature scaler T is fit to maximize the likelihood on a holdout set. Its role is to smooth or sharpen predicted scores. It allows DNNs to output more conservative scores, which are generally are over-confident.

Dirichlet Calibration. Dirichlet calibration [21] considers that score vectors follow a Dirichlet distribution. It transforms $log(F(x))$ instead of impacting the logits, considering that scores result from a softmax: $F(x)_c^{cal} = \sigma\Big(W \ln\left(F(x)\right) + b\Big)_c$, $W \in \mathbb{R}^{c \times c}$ and $b \in \mathbb{R}^c$ being fit on a holdout set.

3.2 Interpretation Methods

We use various methods in our experiments to interpret model's predictions. They cover a large range of approaches used for interpretation, from model-aware to model-agnostic ones.

Model-Aware Interpretation. While some approaches only use gradient information like the Sensitivity method (S) [38] and Guided Backpropagation [40], others combine such information with a latent representation to understand which input features led to each decision, like Grad-Cam [36] and FullGrad [41]. These gradient-based methods suffer from a few short-comings, such as neurons saturation [37], which have been overcome by methods averaging the gradients over a linear interpolation between the input image and a reference, like Integrated Gradient [42]. Some other methods avoid using gradient and directly use the predicted scores given by the model, to backpropagate it to the input of the model like DeepLIFT [37], a fast algorithm approximating Shapley values. Finally, other methods rank the importance of each latent representation like Score-Cam [46], which relies on a model-aware occlusion based approach.

Model-Agnostic Interpretation. We also use methods that consider interpreted models as black-boxes, such as occlusion-based approaches which degrade the input and analyse predicted score variations to define the importance of each part of the input, like Deletion [5,18], MP [9] or RISE [28]. Other methods use surrogate models in order to locally emulate the complex model behaviour in an interpretable fashion, such as LIME [34] and SHAP [23].

We selected a subset of these interpretation methods. To represent model-agnostic methods, we used RISE and MP, which do not rely on any other information than the output scores of the model for any given input, and Sensitivity, Integrated Gradient for model-aware methods. Although these approaches are very popular, we did not evaluate LIME and SHAP methods as they are not totally suited for images and are computationally expensive.

Evaluating the validity and limitations of these methods has been the aim of several works [2,17], for example by measuring their sensitivity to adversarial effects [10], their alignment with human perception [24], their faithfulness to the model being explained [13], or their stability [3]. We can rely on these works to build our experimental assessment of the calibration's impact on interpretations.

4 Evaluation of Calibration's Impact on Interpretation

4.1 Objectives

Although local interpretation approaches differ, they all rely on the output score vector $F(x)$, which is modified by the calibration process. Our aim is to assess whether or not these modifications have an impact on interpretations, and if this potential impact is rather positive or negative.

Assessing the quality of feature importance - here provided by saliency maps - is challenging, yet we argue that a good interpretation should at least: be *faithful* to the model it explains, meaning that removing pixels defined as salient should have an impact on the model outputs [14], interpretations should be *robust* [3], so that similar inputs should lead to similar interpretations, and be composed of structured and smoothly-varying components [39], in order to respect human expectations in terms of *visual coherence*.

To evaluate the impact of calibration on interpretations, we first assess if the calibration process actually impacts the resulting saliency maps by making pair-wise comparisons between them. We then quantify the impact of these changes by evaluating how the classifier's confidence drops when progressively removing important features. Third we assess the visual coherence of the produced saliency, by qualifying their structure and smoothly-varying properties, and we finally evaluate the gain in stability of interpretations when calibrating a model.

4.2 Experimental Setup

We designed cautiously an experimental setup to ensure the comparison of diverse models, image classification datasets and methods. Our goal was to isolate carefully the

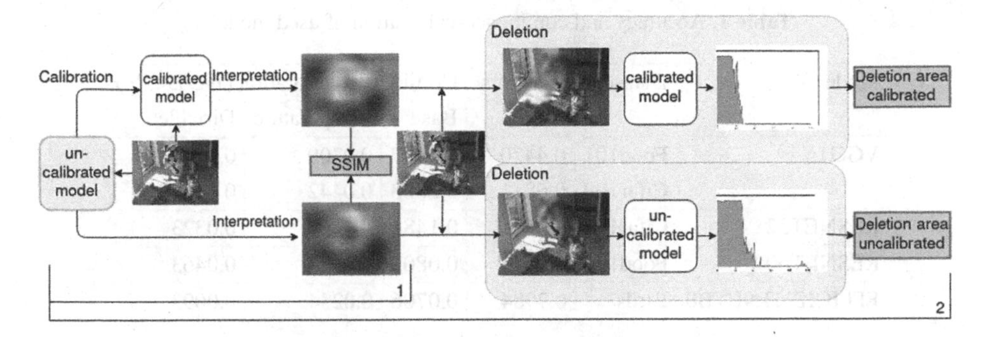

Fig. 3. Protocols to evaluate calibration's impact on interpretation methods: 1 - Comparison of interpretations using SSIM, and 2 - Progressive deletion impact

impact of the calibration on the interpretation procedures. Post-hoc calibration does not impact the models deeply, hence the observable modifications in interpretations is caused by the calibration step, as we compare the interpretation outcomes on uncalibrated models and their calibrated counterparts, all other things being equal.

Models: The following experiments were conducted with VGG, RESNET and EfficientNet models, classical and diverse architectures for image classification. As DNNs are known to be overconfident [11,25], in practice most pre-trained models available in model zoos are not calibrated, and when applying standard learning algorithm we directly obtain uncalibrated models. Yet, to be able to observe the effect of calibration, the tasks or datasets, should be sufficiently complex for the model, so that the accuracy of the model is not perfect, so that there is no room for calibration improvements.

Datasets: We chose three datasets (of various resolutions) to run our experiments: CIFAR-100 [19], Food101 [4] and Birds (CUB-200) [45]. These datasets allow a proper use of calibration in its rigorous context, since each image contains exactly one instance of known classes, and they present different visual properties and classification complexities. To ensure reproducible research, we used pretrained models on CIFAR-100 (VGG16 and RESNET50[2]) and Food-101 (ResNet50[3]). As no pre-trained EfficientNet on Birds dataset were available, we fine-tuned (100 epochs with a learning rate of 1e-4 using Adam optimizer) an EfficientNet pretrained on ImageNet (from torchvision) onto the Birds dataset.

Calibration: The calibration step has been performed using the calibration methods described previously (Temperature Scaling and Dirichlet calibration). These methods have been chosen as they are, respectively, baseline and state of the art post-hoc calibration techniques. For both methods and every model, we used a calibration set composed of 3000, 2500 and 2500 samples taken from the test set for CIFAR-100, Food-101 and

[2] https://github.com/chenyaofo/pytorch-cifar-models.
[3] https://github.com/Herick-Asmani/Food-101-classification-using-ResNet-50.

Table 1. Accuracy and confidence-calibration of used models

Model	Dataset	Accuracy	Confidence calibration (ECE^{conf})		
			Base	Temperature	Dirichlet
VGG16	Food101	0.4470	0.1997	0.0300	0.063
	Cifar100	0.6811	0.2003	0.0442	0.0477
RESNET32	Cifar100	0.6371	0.1484	0.0463	0.0323
RESNET50	Food101	0.8173	0.0803	0.0323	0.0463
EFFICIENTNETB0	Birds	0.7984	0.0706	0.0234	0.0093

Birds respectively. We evaluated the ECE before and after the calibration using the continuous estimator $ECE^{conf}_{density}$ introduced in [31] using the bandwidth automatically set with Silverman's rule, on 2500 samples. ECEs and accuracies of the different models, given in Table 1, confirm the initial mis-calibration of the raw models, and the effectiveness of the calibration step.

Interpretation: For Integrated Gradients (IG), black and white references combined are used with 30 equidistant points on the convex path from the reference to the input. For RISE, we randomly sample 4000 8×8 binary masks (higher dimensions would require more sampling), upscale them using bicubic interpolation, and values of the mask are drawn from a 0.6 Bernoulli law. For MP, optimization problems are solved using the Adam optimizer ($\alpha = 0.1, \beta = 0.4, lr = 0.1$) for 600 steps (these optimization parameters have not been fine-tuned).

All computed saliency maps are min-max-normalized. Some of these, resulting from each of the method applied on calibrated and uncalibrated model, can be observed in Fig. 1. The saliency maps analyzed in the following experiments were obtained from 2500 images randomly sampled from the remaining test set of each of the datasets (which have not been used for calibration).

4.3 Does Calibration Impact Interpretations?

We start by comparing each pair of interpretations coming from uncalibrated and calibrated models using SSIM [47], the structural similarity index, which is based on human perception and thus more relevant than MAE (mean absolute error) or MSE (mean squared error). This is synthesised in Fig. 3 part 1.

The results of this experiment, presented in Fig. 4, show that for all studied interpretation methods, there is a significant difference between saliency maps produced from the calibrated and uncalibrated models. This impact is particularly important on model-agnostic interpretation methods like MP and RISE.

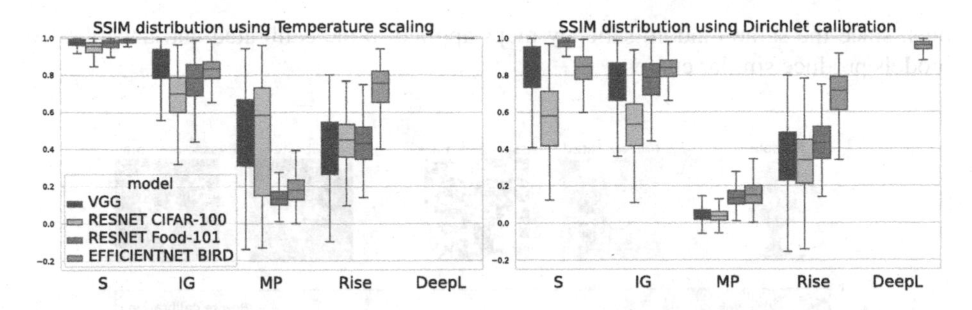

Fig. 4. Distribution of the SSIM between interpretations from calibrated and uncalibrated models. A score close to 1 means no change while a score at -1 means totally opposite images. (Note, Deeplift is not impacted using temperature: distribution at 1)

4.4 Does Calibration Improve the Faithfulness of Interpretation Methods?

We further investigate this impact to determine if detected visual changes in saliency maps improve their quality in terms of *faithfulness* [14]. For each pair of interpretations built by the previous experiment, we apply a deletion procedure [28]: input features are ranked according to their importance given by the interpretation, and are progressively neutralized while we observe the impact on the score of the predicted class from the degraded input image. We compute the *deletion area* – area under the score curve wrt to the percentage of neutralized pixels taking a hundred steps, from 1 to 100. In practice, to preserve the input's distribution, neutralized pixels are replaced by an 11×11 gaussian blurred patch, with $\sigma = 10$, centered on this pixel. The whole procedure is summed up in Fig. 3 part 2 and deletion curves are shown in Fig. 6.

Additionally to the four interpretation methods we use a random interpretation baseline as reference for the comparison. We remove a given percentage of 100 randomly sorted superpixels computed with the SLIC algorithm [1]. We averaged the deletion curve obtained with five different random orders.

We compute the deletion area on the random baseline, uncalibrated models and calibrated ones. Uncalibrated models are more confident, hence to ensure a fair comparison, we normalize the deletion curves in order to set the score of the predicted class from the initial clean image to one. The normalized curve then shows how much the confidence of the model drops, with respect to its initial confidence, when we neutralize pixels in decreasing importance.

The different interpretation methods show consistent results over the various datasets. Before any calibration considerations, they show very different level of faithfulness, MP being the most faithful with a predicted score quickly dropping as the percentage of neutralized pixels grows, followed by RISE, the other model-agnostic method. MP and RISE are also the most positively impacted by the calibration of the model. These two observations are confirmed by the measurement of the deletion areas shown in the second line of the Fig. 7. For the other interpretation methods, the calibration procedure shows little impact, as expected from the SSIM experiment. One can also confirm a posteriori that the normalization of the deletion curves does not introduce any bias, since the same random saliency maps applied to the calibrated and uncalibrated models produce similar curves.

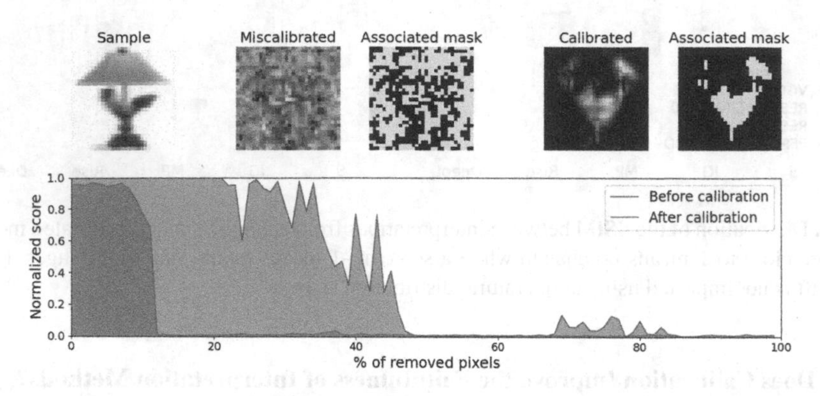

Fig. 5. Saliency maps, Otsu-binarized masks and deletion curves, for the calibrated model and the uncalibrated one, using MP, on a given sample. Map explaining the calibrated model is consistent with human expectation and more faithful to the model.

We conduct another analysis to assess the gain that calibration brings in an element-wise comparison of the deletion area. We compare each interpretation with the random baseline and consider that a prediction is well explained if the deletion area is lower for the method than the one obtained with the random saliency. The proportion of well explained images are reported in Fig. 7 (third line) as Better Than Random ratio (BTR). This BTR is always improved, except for one model/method case. The improvement varies from limited to important depending on the approach and the dataset. We read the greater impact of the Dirichlet calibration wrt the Temperature scaling, on both the BTR and deletion area, besides that their respective ECE values are cases comparable in most cases, by the fact that Dirichlet calibration can change the predicted class while temperature scaling does not.

The great improvement brought by calibration to MP is promising for model-agnostic interpretation. Notably, it is known that without computationally expensive hyperparameters tuning, MP is sensitive to visual artifacts [8]. As Fig. 5 suggests, calibration seems to help dealing with those.

To sum up, we measure a positive impact on the faithfulness as measured by the deletion area and the BTR ratio presented in Sect. 4.4 for model-agnostic methods, in worst case the interpretability is not impacted by the calibration. The most faithful method without calibration, namely MP, is also the most improved method. This suggests that the interpretability of the model in itself could depend on its calibration.

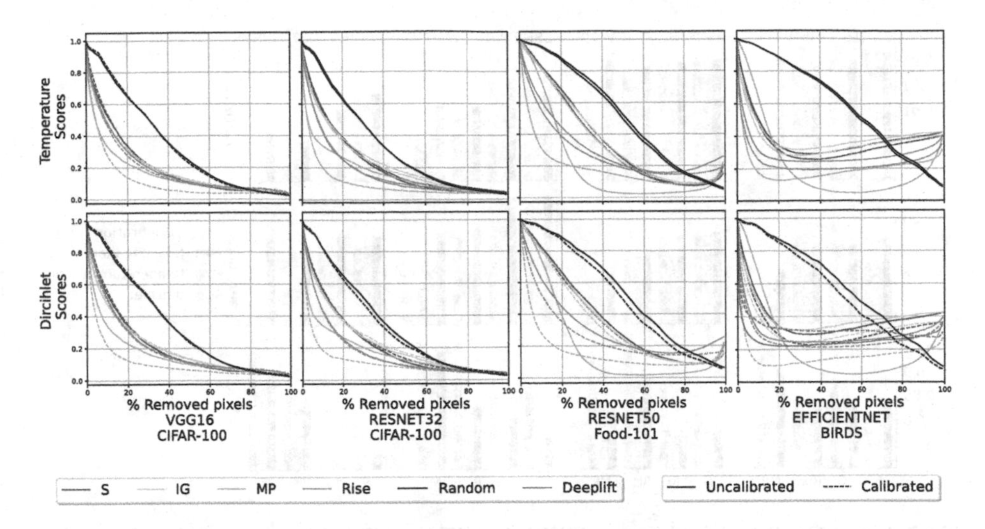

Fig. 6. Deletion curves before (plain line) and after calibration (dashed line) for three different models using Temperature Scaling (upper) and Dirichlet calibration (lower)

4.5 Are Saliency Maps with Calibration More Human-Friendly?

Another important aspect of interpretation is the readability for users, smoothly varying saliency maps are easier to comprehend, as they tend to highlight structures that we, as humans, recognize. Therefore, to quantify the complexity of a produced saliency map, we first distinguish activated pixels (foreground of saliency maps) from non-activated ones (background) using Otsu binarization [26], and compute the total variation (TV) of obtained binary images. A higher total variation suggests a more noisy interpretation.

Figure 7 shows that Otsu TV is always improved (lowered) by calibration for model-agnostic methods, or at worst unaffected for other methods, which means that interpretations exhibit smoother variations and fewer highlighted regions, while the mean deletion area is constant or improved. Hence the interpretations of the calibrated models are more human readable while being equally or more faithful to the model.

4.6 In Depth Analysis of Meaningful Perturbation

To better understand our findings, we focus now on MP, the most faithful method, which appears to be the most impacted by the calibration.

Effect of the Mis-Calibration. In order to evaluate if, for MP, the faithfulness improvement correlates with the mis-calibration level, we apply the deletion experiment using Temperature Scaling for different fixed temperatures. Figure 8 highlights that the minimum of the deletion area is obtained when the model is calibrated, showing a clear non linear correlation between the scaler's temperature value and the deletion area. Interestingly the second plot highlights a positive correlation between the mean deletion area

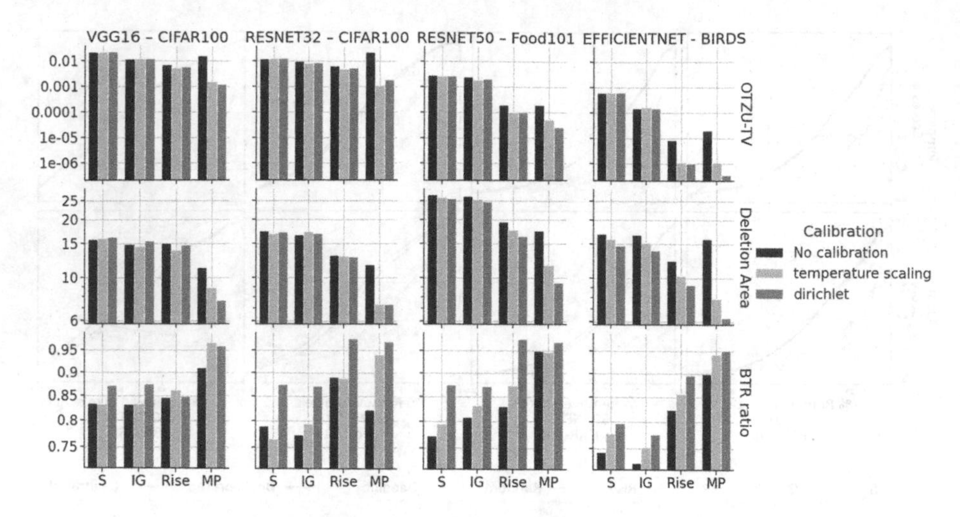

Fig. 7. Ratio of well explained images (BTR), Otsu TV and Deletion Area of various interpretation methods on multiple models and datasets before and after calibration.

and the ECE. The correlation differs whether the mis-calibration is due to an overconfident model or an underconfident model for the same range of ECE. The overconfident models are those with a temperature smaller than the best obtained.

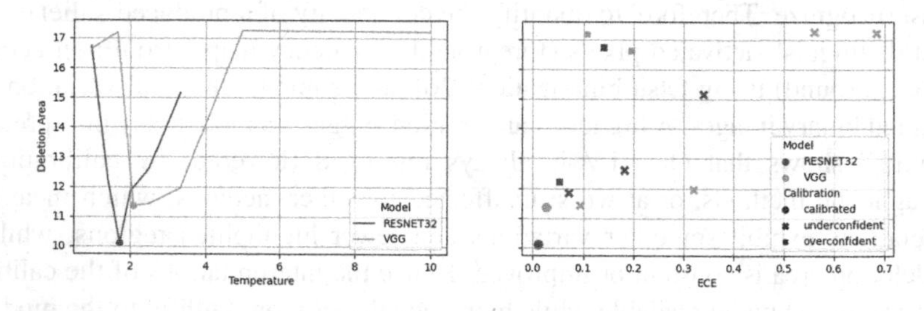

Fig. 8. Mean deletion area on CIFAR-100 for VGG16 and RESNET32 conditioned on the temperature. Marked points indicate calibrated models.

Interpretation Stability. As shown previously, calibration improves the deletion area and the total variation of saliency maps from MP, the best method in terms of deletion area, meaning that produced interpretations are both more faithful (according to [14] definition) and more spatially coherent. One possible explanation for this improvement is that calibration improves the stability of the method. It is known that MP is sensitive to visual artifacts [8] without proper hyperparameters tuning. Calibration seems to be a really adequate solution to prevent the apparition of such artifacts, especially when parameter tuning is not feasible. Indeed, as underlined in [3], interpretation methods tend to be unstable when the input is slightly modified. As calibration is related

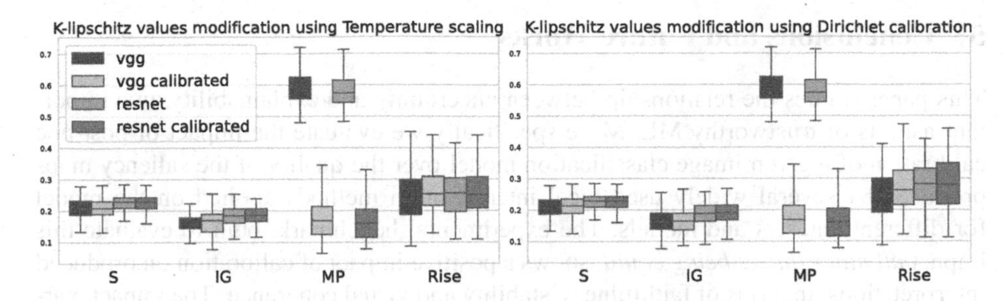

Fig. 9. K-lipschitz distribution obtained with calibrated and uncalibrated model using Temperature Scaling (left) and Dirichlet calibration (right).

to the model's robustness [33] it is natural to wonder whether or not the calibration of a given model improves the robustness of the interpretations made on its predictions. To that end, we compute an approximation of the Lipschitz constant of the saliency function, following the experiment introduced in [3], by adding small amount of gaussian noise to the input, so that the behaviour of the model does not change, and compute the normalized l_2-norm of the interpretations obtained with and without calibration. This experiment was applied on 150 randomly sampled points from CIFAR-100. For each point, we sample 40 neighbour points (inside the ball of radius 0.05 and with the point of interest as center). We only apply this experiment on 150 points in order to keep it feasible in a reasonable amount of time (for MP and Rise processing each point takes around an hour with an NVIDIA 1650).

Figure 9 reveals that calibration, while having no robustness impact on most methods, considerably improves the stability of MP. This is consistent with the $10\times$ decrease of TV for MP revealed in the previous section. Also, RISE, which is the least stable even after calibration, could be improved by increasing the number of sampled masks at the expense to an higher computational cost.

4.7 Discussions

While these results shed light on non-trivial interaction between interpretation methods and calibration, it is hard to express from the gain in term of deletion area a definitive statement about interpretation methods validity. This come from the difficulty of evaluating explanation, therefore to the best of our knowledge no consensus has been attained in terms of the best method to validate interpretation methods. The deletion experiments, while truly evaluating interpretations quality, seems to favor methods which highlight regions over methods which highlight high frequency details, which would account for the poor faithfulness performance as measured by the deletion area. Moreover, we are convinced that a good interpretation method should be able to reflect the uncertainty of the model and therefore should be impacted by the calibration. We argue that this could be a clue to setup a new sanity check for interpretation methods.

5 Conclusions and Future Works

This paper studies the relationship between uncertainty and explainability, two important aspects of trustworthy ML. More specifically we evaluate the impact of post-hoc calibration of a given image classification model over the quality of the saliency maps produced by several widely used local interpretation methods applied on the model for different datasets and models. The experimental benchmark, built to evaluate this impact *all other things being equal*, shows a positive impact of calibration on produced interpretations, in terms of faithfulness, stability and visual coherence. The impact, particularly beneficial on model-agnostic interpretation methods such as Meaning Perturbation (MP) [9], is in the worst case neutral. For these reasons, we suggest a simple practice to improve interpretation outcomes: *Calibrate to Interpret.*

A side benefit of the study is to rank the competing interpretation approaches, which shows that model-agnostic ones perform better with regard to faithfulness, measured by the deletion area, and visual coherence, measured by the Otsu-TV. Interestingly the stronger impact of calibration appears on the best interpretation method namely MP, and resolves one of its main drawbacks: its sensitivity to artifacts. We highlight for this method that there is a clear correlation between the calibration level of a model and the faithfulness of the MP method applied to its predictions.

This work opens the road to deeper analyses. A direct extension would be to analyse if *in-training* solutions to enforce calibration would impact similarly the interpretations. This would require a different benchmark setup, where modifications induced to the model by *in-training* calibration are properly framed. Additionally, although computational cost considerations prevented us from realizing experiments with the ROAR procedure [12] to measure the faithfulness of interpretation methods, we think the theoretical properties of this approach make it a great option to deepen our experimental evaluation. Furthermore, conclusions obtained through the analysis of the TV could be strengthen using a human evaluation. Finally, we wonder if other kinds of explanation paradigm, e.g. concept based [6], sample based [32] or even attention based [7] can also benefit from calibration.

References

1. Achanta, R., Shaji, A., Smith, K., Lucchi, A., Fua, P., Süsstrunk, S.: Slic superpixels compared to state-of-the-art superpixel methods. IEEE Trans. Pattern Anal. Mach. Intell. **34**(11), 2274–2282 (2012). https://doi.org/10.1109/TPAMI.2012.120
2. Adebayo, J., Gilmer, J., Muelly, M., Goodfellow, I., Hardt, M., Kim, B.: Sanity checks for saliency maps. In: Advances in Neural Information Processing Systems, NIPS 2018, pp. 9525–9536. Curran Associates Inc., Red Hook 2018)
3. Alvarez-Melis, D., Jaakkola, T.S.: On the robustness of interpretability methods. arXiv preprint arXiv:1806.08049 (2018)
4. Bossard, L., Guillaumin, M., Van Gool, L.: Food-101 – mining discriminative components with random forests. In: Fleet, D., Pajdla, T., Schiele, B., Tuytelaars, T. (eds.) ECCV 2014. LNCS, vol. 8694, pp. 446–461. Springer, Cham (2014). https://doi.org/10.1007/978-3-319-10599-4_29
5. Chang, C.H., Creager, E., Goldenberg, A., Duvenaud, D.: Explaining image classifiers by counterfactual generation. In: International Conference on Learning Representations (2019)

6. Chen, C., Li, O., Tao, C., Barnett, A.J., Su, J., Rudin, C.: This looks like that: deep learning for interpretable image recognition. In: Advances in Neural Information Processing Systems, pp. 8928–8939 (2019)
7. Chen, M., Radford, A., Child, R., Wu, J., Jun, H., Luan, D., Sutskever, I.: Generative pretraining from pixels. In: International Conference on Machine Learning, pp. 1691–1703. PMLR (2020)
8. Fong, R., Vedaldi, A.: Interpretable explanations of black boxes by meaningful perturbation, pp. 3449–3457, October 2017. https://doi.org/10.1109/ICCV.2017.371
9. Fong, R.C., Vedaldi, A.: Interpretable explanations of black boxes by meaningful perturbation. In: 2017 IEEE International Conference on Computer Vision (ICCV), pp. 3449–3457 (2017). https://doi.org/10.1109/ICCV.2017.371
10. Ghorbani, A., Abid, A., Zou, J.: Interpretation of neural networks is fragile. Proceedings of the AAAI Conference on Artificial Intelligence 33(01), pp. 3681–3688, July 2019. https://ojs.aaai.org/index.php/AAAI/article/view/4252
11. Guo, C., Pleiss, G., Sun, Y., Weinberger, K.Q.: On calibration of modern neural networks. In: Precup, D., Teh, Y.W. (eds.) Proceedings of the 34th International Conference on Machine Learning. Proceedings of Machine Learning Research, vol. 70, 06–11 Aug 2017, pp. 1321–1330. PMLR. http://proceedings.mlr.press/v70/guo17a.html
12. Hooker, S., Erhan, D., Kindermans, P.J., Kim, B.: Evaluating feature importance estimates (2018)
13. Hooker, S., Erhan, D., Kindermans, P.J., Kim, B.: A benchmark for interpretability methods in deep neural networks. In: Wallach, H., Larochelle, H., Beygelzimer, A., d'Alché-Buc, F., Fox, E., Garnett, R. (eds.) Advances in Neural Information Processing Systems 32, pp. 9737–9748. Curran Associates, Inc. (2019)
14. Jacovi, A., Goldberg, Y.: Towards faithfully interpretable nlp systems: how should we define and evaluate faithfulness? arXiv preprint arXiv:2004.03685 (2020)
15. Jain, R., Madhyastha, P.: Model explanations under calibration. CoRR arXiv preprint arXiv:1906.07622 (2019)
16. Kim, J.H., Choo, W., Song, H.O.: Puzzle mix: exploiting saliency and local statistics for optimal mixup. In: International Conference on Machine Learning, pp. 5275–5285. PMLR (2020)
17. Kindermans, P.-J., Hooker, S., Adebayo, J., Alber, M., Schütt, K.T., Dähne, S., Erhan, D., Kim, B.: The (Un)reliability of saliency methods. In: Samek, W., Montavon, G., Vedaldi, A., Hansen, L.K., Müller, K.-R. (eds.) Explainable AI: Interpreting, Explaining and Visualizing Deep Learning. LNCS (LNAI), vol. 11700, pp. 267–280. Springer, Cham (2019). https://doi.org/10.1007/978-3-030-28954-6_14
18. Kononenko, I., Robnik-Sikonja, M.: Explaining classifications for individual instances. IEEE Trans. Knowl. Data Eng. **20**(05), 589–600 (2008). https://doi.org/10.1109/TKDE.2007.190734
19. Krizhevsky, A., Hinton, G., et al.: Learning multiple layers of features from tiny images (2009)
20. Kull, M., Filho, T.S., Flach, P.: Beta calibration: a well-founded and easily implemented improvement on logistic calibration for binary classifiers. In: Singh, A., Zhu, J. (eds.) Proceedings of the 20th International Conference on Artificial Intelligence and Statistics. Proceedings of Machine Learning Research, vol. 54, 20–22 Apr 2017, pp. 623–631. PMLR. https://proceedings.mlr.press/v54/kull17a.html
21. Kull, M., Nieto, M.P., Kängsepp, M., Silva Filho, T., Song, H., Flach, P.: Beyond temperature scaling: obtaining well-calibrated multi-class probabilities with Dirichlet calibration. In: Advances in Neural Information Processing Systems, pp. 12295–12305 (2019)

22. Kumar, A., Sarawagi, S., Jain, U.: Trainable calibration measures for neural networks from kernel mean embeddings. In: Dy, J., Krause, A. (eds.) Proceedings of the 35th International Conference on Machine Learning. Proceedings of Machine Learning Research, vol. 80, 10–15 July 2018, pp. 2805–2814. PMLR. http://proceedings.mlr.press/v80/kumar18a.html
23. Lundberg, S.M., Lee, S.I.: A unified approach to interpreting model predictions. In: Guyon, I., Luxburg, U.V., Bengio, S., Wallach, H., Fergus, R., Vishwanathan, S., Garnett, R. (eds.) Advances in Neural Information Processing System, vol. 30. Curran Associates, Inc. (2017)
24. Mohseni, S., Block, J.E., Ragan, E.: Quantitative evaluation of machine learning explanations: a human-grounded benchmark. In: 26th International Conference on Intelligent User Interfaces, IUI 2021, pp. 22–31. Association for Computing Machinery, New York (2021). https://doi.org/10.1145/3397481.3450689
25. Nguyen, A., Yosinski, J., Clune, J.: Deep neural networks are easily fooled: High confidence predictions for unrecognizable images. In: 2015 IEEE Conference on Computer Vision and Pattern Recognition (CVPR), pp. 427–436 (2015). https://doi.org/10.1109/CVPR.2015.7298640
26. Otsu, N.: A threshold selection method from gray-level histograms. IEEE Trans. Syst. Man Cybern. **9**(1), 62–66 (1979). https://doi.org/10.1109/TSMC.1979.4310076
27. Pakdaman Naeini, M., Cooper, G., Hauskrecht, M.: Obtaining well calibrated probabilities using bayesian binning. In: Proceedings of the AAAI Conference on Artificial Intelligence 29(1), February 2015, https://ojs.aaai.org/index.php/AAAI/article/view/9602
28. Petsiuk, V., Das, A., Saenko, K.: Rise: Randomized input sampling for explanation of black-box models. In: BMVC (2018)
29. Platt, J.C.: Probabilistic outputs for support vector machines and comparisons to regularized likelihood methods. In: Advances in Large Margin Classifiers, pp. 61–74. MIT Press (1999)
30. Pleiss, G., Raghavan, M., Wu, F., Kleinberg, J., Weinberger, K.Q.: On fairness and calibration. Advances in Neural Information Processing Systems 30 (2017)
31. Posocco, N., Bonnefoy, A.: Estimating expected calibration errors. In: Farkaš, I., Masulli, P., Otte, S., Wermter, S. (eds.) ICANN 2021. LNCS, vol. 12894, pp. 139–150. Springer, Cham (2021). https://doi.org/10.1007/978-3-030-86380-7_12
32. Pruthi, G., Liu, F., Kale, S., Sundararajan, M.: Estimating training data influence by tracing gradient descent. In: Larochelle, H., Ranzato, M., Hadsell, R., Balcan, M.F., Lin, H. (eds.) Advances in Neural Information Processing Systems, vol. 33, pp. 19920–19930. Curran Associates, Inc. (2020). https://proceedings.neurips.cc/paper/2020/file/e6385d39ec9394f2f3a354d9d2b88eec-Paper.pdf
33. Qin, Y., Wang, X., Beutel, A., Chi, E.: Improving uncertainty estimates through the relationship with adversarial robustness, June 2020
34. Ribeiro, M.T., Singh, S., Guestrin, C.: "Why should i trust you?": explaining the predictions of any classifier. In: Proceedings of the 22nd ACM SIGKDD International Conference on Knowledge Discovery and Data Mining, KDD 2016, pp. 1135–1144. Association for Computing Machinery (2016). https://doi.org/10.1145/2939672.2939778
35. Ross, A.S., Doshi-Velez, F.: Improving the adversarial robustness and interpretability of deep neural networks by regularizing their input gradients. In: AAAI Conference on Artificial Intelligence (2018)
36. Selvaraju, R.R., Cogswell, M., Das, A., Vedantam, R., Parikh, D., Batra, D.: Grad-cam: visual explanations from deep networks via gradient-based localization. In: 2017 IEEE International Conference on Computer Vision (ICCV), pp. 618–626 (2017)
37. Shrikumar, A., Greenside, P., Kundaje, A.: Learning important features through propagating activation differences. In: Precup, D., Teh, Y.W. (eds.) Proceedings of the 34th International Conference on Machine Learning. Proceedings of Machine Learning Research, vol. 70, pp. 3145–3153. PMLR, 06–11 August 2017. http://proceedings.mlr.press/v70/shrikumar17a.html

38. Simonyan, K., Vedaldi, A., Zisserman, A.: Deep inside convolutional networks: Visualising image classification models and saliency maps. CoRR abs/1312.6034 (2014)
39. Smilkov, D., Thorat, N., Kim, B., Viégas, F., Wattenberg, M.: Smoothgrad: removing noise by adding noise. arXiv preprint arXiv:1706.03825 (2017)
40. Springenberg, J.T., Dosovitskiy, A., Brox, T., Riedmiller, M.A.: Striving for simplicity: The all convolutional net. CoRR abs/1412.6806 (2015)
41. Srinivas, S., Fleuret, F.: Full-gradient representation for neural network visualization. In: Advances in Neural Information Processing Systems (2019)
42. Sundararajan, M., Taly, A., Yan, Q.: Axiomatic attribution for deep networks. JMLR.org (2017)
43. Thulasidasan, S., Chennupati, G., Bilmes, J., Bhattacharya, T., Michalak, S.: On mixup training: improved calibration and predictive uncertainty for deep neural networks. arXiv preprint arXiv:1905.11001 (2019)
44. Tsipras, D., Santurkar, S., Engstrom, L., Turner, A., Madry, A.: Robustness may be at odds with accuracy (2018). http://arxiv.org/abs/1805.12152, cite arxiv:1805.12152
45. Wah, C., Branson, S., Welinder, P., Perona, P., Belongie, S.: The Caltech-UCSD Birds-200-2011 Dataset. Technical report CNS-TR-2011-001, California Institute of Technology (2011)
46. Wang, H., Wang, Z., Du, M., Yang, F., Zhang, Z., Ding, S., Mardziel, P., Hu, X.: Score-cam: score-weighted visual explanations for convolutional neural networks. In: 2020 IEEE/CVF Conference on Computer Vision and Pattern Recognition Workshops (CVPRW), pp. 111–119 (2020)
47. Wang, Z., Bovik, A.C., Sheikh, H.R., Simoncelli, E.P.: Image quality assessment: from error visibility to structural similarity. IEEE Trans. Image Process. **13**(4), 600–612 (2004)

Knowledge-Driven Interpretation of Convolutional Neural Networks

Riccardo Massidda(✉) and Davide Bacciu

Università di Pisa, Pisa, Italy

riccardo.massidda@phd.unipi.it, davide.bacciu@unipi.it

Abstract. Since the widespread adoption of deep learning solutions in critical environments, the interpretation of artificial neural networks has become a significant issue. To this end, numerous approaches currently try to align human-level concepts with the activation patterns of artificial neurons. Nonetheless, they often understate two related aspects: the distributed nature of neural representations and the semantic relations between concepts. We explicitly tackled this interrelatedness by defining a novel semantic alignment framework to align distributed activation patterns and structured knowledge. In particular, we detailed a solution to assign to both neurons and their linear combinations one or more concepts from the WordNet semantic network. Acknowledging semantic links also enabled the clustering of neurons into semantically rich and meaningful neural circuits. Our empirical analysis of popular convolutional networks for image classification found evidence of the emergence of such neural circuits. Finally, we discovered neurons in neural circuits to be pivotal for the network to perform effectively on semantically related tasks. We also contribute by releasing the code that implements our alignment framework.

Keywords: Interpretability · Convolutional neural network

1 Introduction

Neural representations offer limited insights in terms of human-level interpretation. Overcoming this limitation is one of the most compelling challenges in deep learning research and is crucial when considering artificial neural networks deployed for safety- and privacy-critical tasks. Because of the opacity of their internal behavior, the literature tends to define neural networks as black boxes [10]. Nonetheless, recent research highlights how, in particular domains, some components of a neural network might instead be characterized by clear-cutting intepretations [8,20]. For this reason, both theoretical research and practical interpretability approaches require sound methods to reliably and accurately identify associations between high-level concepts and neural components.

Supplementary Information The online version contains supplementary material available at https://doi.org/10.1007/978-3-031-26387-3_22.

M.-R. Amini et al. (Eds.): ECML PKDD 2022, LNAI 13713, pp. 356–371, 2023.
https://doi.org/10.1007/978-3-031-26387-3_22

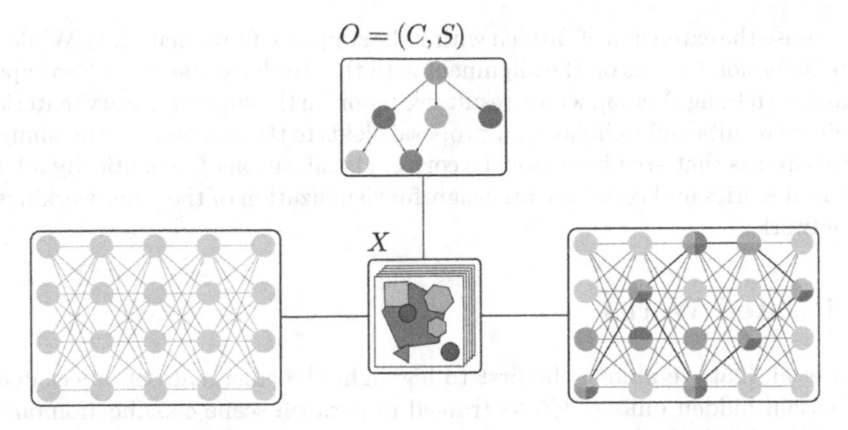

Fig. 1. Overview of the proposed methodology. A set of neural directions D is semantically aligned with an ontology O through a pixel-level annotated dataset X, whose labels are in a two-way relationship with the ontology concepts C. Semantic relations S enable the retrieval of subgraphs composed of architecturally connected and semantically related directions.

Early works tackled the alignment of human-level concepts with either distinct units [2] or directions within the output space of hidden layers [14]. Nonetheless, they considered concepts as independent entities, without adopting structured knowledge representation. In this context, we present a unified approach for the semantic alignment of neural components and visual concepts, applied to Convolutional Neural Networks (CNNs) and computer vision scenarios. Our approach considers concepts as members of a computational ontology and actively exploits their semantic relations (Fig. 1). Firstly, we improve the expressiveness of the alignment by acknowledging specialization between concepts. For instance, if an artificial neuron responds to the human notion of "feline", the framework can propagate the partial alignment with the concepts of "cat" or "tiger" without the need for explicit "feline" annotations. Consequently, we tackle the identification of semantically aligned components in two complementary scenarios: by selecting concepts aligned to a given direction, and by retrieving directions aligned to a given concept. Lastly, our main original contribution leverages semantic alignment to identify meaningful subgraphs composed of architecturally connected and semantically related components within the network. We refer to these subgraphs as circuits, following the term "neural circuit" and its widespread use in neuroscience. These circuits offer new insights into the content of distributed neural representations and provide a novel instrument for network inspection and interpretation.

We validate our approach by inspecting several renowned CNN architectures for scene classification by exploiting the Broden segmented dataset [2]. As a side contribution of our empirical validation, we extend the original Broden dataset by associating its labels with WordNet synsets [17]. We consider WordNet as a simple ontology that contains a taxonomy of concepts [18]. Furthermore, we publicly release the extension of Broden within the supplementary materials. While the main discussion focuses on the alignment with the Broden dataset, we also experimented with ImageNet [3], whose results we report in the supplementary materials. Empirical results highlight how our proposal yields to the emergence of meaningful neural circuits that are pivotal for the correct classification of semantically related visual categories and consitute an insightful visualization of the inner workings of the network.

2 Related Works

Zhou et al. [25] are among the first to highlight the emergence of object detectors within hidden units of CNNs trained to perform scene classification on the Places dataset [27]. Their work manually annotated such detectors by visualizing manipulated examples that maximized units activations. Olah et al. [20] approached the problem similarly by employing feature visualization techniques [21] to manually assign specific roles to individual neurons. Their contribution also highlights the role of neural units to fulfill complex tasks throughout the network. Bau et al. [2] introduced Network Dissection to automatically analyze neural activations and identify meaningful neurons in CNNs trained on the Places-365 dataset [26]. Their work introduced a pixel-level annotated image dataset called Broden, which marks portrayed objects and patterns. Zhou et al. [29] studied the role of semantically aligned units by measuring the accuracy drop when removing units aligned to a given concept. On top of Network Dissection, Mu et al. [19] discussed the consequences of analyzing compositions of visual concepts by applying logical operations to the annotations. Despite the different methodological approaches, the works discussed above analyze neural models by considering single units as meaningful artifacts as in localist networks [23]. Instead, our proposal acknowledges and investigates single units, their linear combinations, and knowledge-driven generated clusters.

More generally, other techniques aim to fulfill concept-based analysis of neural activations without restraining meaningful information to single neurons. Firstly, Fong et al. [6] expanded Network Dissection with linear combinations of hidden neurons in CNNs to identify distributed concept detectors. Similarly, Kim et al. [14] defined concept activation vectors (CAVs) as linear classifiers of visual concepts over the activations of an hidden layer. Furthermore, they proposed a measure, called TCAV, of the influence of these classifiers on specific outcomes of the network. Always using linear classifiers, Zhou et al. [28] proposed an interpretative framework based on the decomposition of hidden representations into a meaningful basis composed by such classifiers. While exploiting the expressiveness of hidden layers, these techniques considered concepts as independent entities, missing to acknowledge their semantic relations. On the contrary, we explicitly use ontological information to obtain interpretative results.

Finally, our approach might be understood in terms of ontology matching, i.e. the task of meaningfully aligning different ontologies to reduce the gap between overlapping representations [22]. Our work can be associated with extensional based techniques, where the semantic distance between concepts from two different ontologies is estimated according to a measure of the overlapping of their extensions [4]. In comparison, our approach exploits the portrayal of visual concepts to mediate their extension and estimates the difference between an explicit ontology and concepts implicitly expressed by neural representations.

3 Ontology-Driven Semantic Alignment

Given a pre-trained CNN for computer vision, our framework estimates semantic alignment between a set of visual concepts C and a set of neural directions D. We consider directions within the output space of neural layers as a useful instrument to inquire which concepts the network is able to effectively represent and discriminate. Formally, we define a neural direction $d \in D$ as a pair

$$d = (l, v), \tag{1}$$

where $v \in \mathbb{R}^{N_l}$ is a vector weighting the N_l units at the l-th layer of the network. Given an input image x, the output of a convolutional layer l is a tensor $f^l(x) \in \mathbb{R}^{N_l \times H_l \times W_l}$, where each unit corresponds to a channel. Furthermore, we treat fully connected layers as a specialization where $H_l = 1, W_l = 1$. For an input image x, the output of a neural direction d is the activation map

$$A_d(x) = f^l(x) \cdot v, \tag{2}$$

where $A_d(x) \in \mathbb{R}^{H_l \times W_l}$. Notably, when v corresponds to a vector $e^{(i)}$ from the canonical basis of $\mathbb{R}^{N_l}$, the activation map coincides with the output of the i-th neuron at layer l, i.e., the i-th channel. Furthermore, to simplify the notation, we always include a bias term β within v.

Given a segmented dataset X, for each example image $x \in X$, we require the existence of a binary mask $L_c(x)$, known as the concept mask, that marks the locations portraying the visual concept c. This requirement can be fulfilled by any dataset for object detection or semantic segmentation that provides semantic labeling of pixels. Furthermore, we require an ontology $O = (C, S)$ formalized as an extensional relational structure, where C is a set of concepts and S is a set of truth valued binary relations [9].

The presence of the specialization relation in the ontology enables the retrieval of masks for concepts which were not directly annotated in the dataset (Sect. 3.1). Consequently, by relating activations and concept masks, our approach computes an estimate of the alignment for direction-concept pairs (Sect. 3.2) and enables the retrieval of directions aligned to a given concept (Sect. 3.3). Finally, we exploit semantic relations between aligned concepts to cluster consecutive directions into meaningful neural circuits (Sect. 3.4).

3.1 High-Level Concept Masks

We consider each concept as an ideal function whose argument is an object of the world and whose value is a truth-value [7]. Thus, the extension E_c of a concept c,

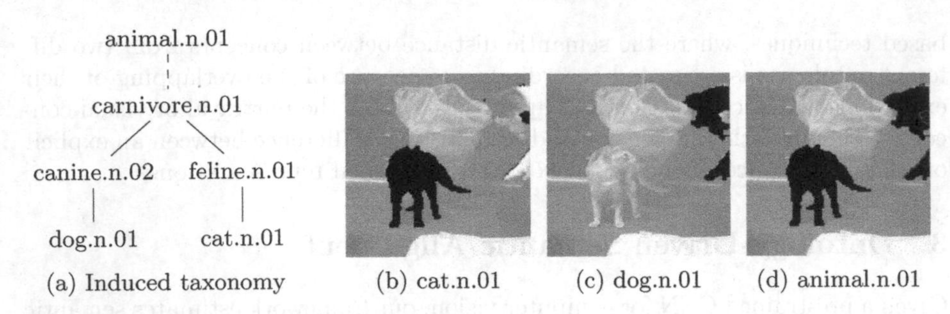

(a) Induced taxonomy (b) cat.n.01 (c) dog.n.01 (d) animal.n.01

Fig. 2. Example of mask generation for the higher-level concept of "animal" using taxonomical information. The induced taxonomy, built over the WordNet hypernymy (*is-a*) relation, enables the retrieval of the mask by exploiting masks annotated for "dog" and "cat" concepts from the Broden dataset, without having access to explicit annotations of the concept "animal".

is the set of all the objects of the world satisfying it. The specialization relation, also known as "is-a", is the semantic relation that expresses the inclusion between the extensions of concepts in an ontology [5]. The concept c is a specialization of c' if and only if the extension $E_{c'}$ contains E_c. Formally, $c \sqsubseteq c' \iff E_c \subseteq E_{c'}$.

Given a pixel position p in an image x, we define $Q(x_p)$ as the set of objects portrayed by that location. Consequently, a boolean concept mask $L_c(x)$ annotates for each possible pixel position p whether one of the portrayed objects by x_p pertains to the extension E_c. Consequently, if a concept c specializes c', then each location in the concept mask $L_c(x)$ implies the same location in the concept mask $L_{c'}(x)$. Formally,

$$\begin{aligned} L_c(x)_p &\iff (Q(x_p) \cap E_c) \neq \emptyset \\ &\implies (Q(x_p) \cap E_{c'}) \neq \emptyset \quad \{c \sqsubseteq c' \iff E_c \subseteq E_{c'}\} \\ &\iff L_{c'}(x)_p. \end{aligned} \tag{3}$$

Specialization induces a hierarchical taxonomy represented by a Directed Acyclic Graph (DAG). In the DAG, the node corresponding to the concept c is a child of the one corresponding to c' if and only if $c \sqsubseteq c'$. Hence, for each image, the mask of a concept at a certain level of the taxonomy can be obtained indirectly as the union of the masks of its children. The proposed approach is thus able to align higher-level concepts without explicit annotations by analyzing the concept masks of its descendants in the DAG (Fig. 2).

3.2 Alignment Measure

Firstly, we address how to estimate semantic alignment for a given neural direction $d \in D$ and a concept $c \in C$. For this purpose, we define a binary classifier over the activations of the neural direction to discriminate a visual concept. Therefore, for each example x in the segmented dataset X, we threshold an activation map $A_d(x)$ into a boolean activation mask

$$M_d(x) = A_d(x) > 0. \tag{4}$$

Typically, the activation mask $M_d(x)$ has a different shape than the input x and the concept mask $L_c(x)$. To be comparable, either the concept mask or the activation mask should be scaled to respectively match their shapes. This operation approximates the relation between a neural direction and its receptive field. For any pixel location p, the outcome of the binary classifier $M_d(x)$ in p should depend solely on x_p and ideally on each $L_c(x)_p$. This approximation discards the effects of striding and padding over the receptive field of convolutional units [1], which will be subject of future research. To ease the notation, in the following we assume that either $L_c(x)$ or $M_d(x)$ have been adequately scaled to an arbitrary shape (H, W). Furthermore, we define the operator $|K|$ to count the number of true values in an arbitrary boolean mask K.

Given this formulation, semantic alignment can be estimated by adopting an arbitrary classification performance metric. Therefore, the Jaccard similarity, also known as Intersection over Union (IoU),

$$\sigma_{\mathrm{IoU}}(d, c) = \frac{\sum_{x \in X} |M_d(x) \wedge L_c(x)|}{\sum_{x \in X} |M_d(x) \vee L_c(x)|}, \tag{5}$$

or the Sørensen-Dice coefficient, also known as F1 score,

$$\sigma_{\mathrm{F1}}(d, c) = \frac{\sum_{x \in X} 2|M_u(x) \wedge L_c(x)|}{\sum_{x \in X} |M_u(x)| + |L_c(x)|}, \tag{6}$$

consitute insightful measures of semantic alignment.

Furthermore, we provide an original probabilistic model of the influence of visual concepts on the output of hidden directions. We model each visual concept $c \in C$ and each direction $d \in D$ as a pair of Bernoulli random variables Y_c, Z_d. We assume that directions are conditionally independent given the concepts. Consequently, we propose a measure in terms of the maximum likelihood estimate

$$\sigma_{\mathcal{L}}(d, c) = \mathcal{L}(Y_c = 1 \mid Z_d = 1) = \frac{\sum_x |L_c(x) \wedge M_d(x)|}{\sum_x |L_c(x)|}, \tag{7}$$

of a concept being in the receptive field of a direction. This measure corresponds to the recall of the classifier, and offers different interpretative insights than other more restraining measures adopted in earlier works such as σ_{IoU} or equivalently σ_{F1}. In particular, $\sigma_{\mathcal{L}}$ is of use when the vector v of a direction $d = (l, v)$ pertains to the canonical basis of $\mathbb{R}^{N_l}$, thus it represents the output of a specific unit. Ideally, in a localist scenario, each unit of the network would activate only when stimulated by a specific visual concept. In practice, for a human observer, most units are polysemantic, i.e. they respond to multiple and possibly unrelated visual concepts. By trading off precision and recall, measures such as σ_{IoU} and σ_{F1} would ignore such concepts. On the contrary, $\sigma_{\mathcal{L}}$ can effectively highlight the partial alignment of concepts in polysemantic neurons.

3.3 Direction Learning

Other than aligning existing directions, we also address the issue of learning a vector v to determine a direction $d = (l, v)$ semantically aligned to a given concept c within the l-th layer of the network. Firstly, we consider two independents splits $X_{\text{train}}, X_{\text{val}}$ of a segmented dataset X. By solving a minimization problem, we determine the vector direction as

$$\begin{aligned} v &= \underset{v}{\operatorname{argmin}} \sum_{x \in X_{\text{train}}} \sum_{p} \ell(A_d(x)_p, L_c(x)_p) \\ &= \underset{v}{\operatorname{argmin}} \sum_{x \in X_{\text{train}}} \sum_{p} \ell((f^l(x) \cdot v)_p, L_c(x)_p) \end{aligned} \tag{8}$$

where p iterates over the locations of the activation map and of the concept mask, while ℓ is an arbitrary loss function for binary classification. Consequently, we estimate semantic alignment $\sigma(d, c)$ by computing one of the previously detailed measures on the X_{val} split of the dataset.

3.4 Neural Circuits

Given a threshold τ on the estimate $\sigma(d, c)$, we retrieve a set

$$\Psi = \{(d, c) \mid \sigma(d, c) > \tau\} \subseteq D \times C, \tag{9}$$

containing sufficiently aligned direction-concept pairs. The set Ψ offers a useful interpretative instrument by collecting the human-concepts that the network is able to sufficiently discriminate within the analyzed layers. Since we are also interested in the relation between aligned concepts, we connect alignment pairs within a directed graph $G = (\Psi, E)$ such that

$$((d, c), (d', c')) \in E \iff s(c, c') \wedge a(d, d'), \tag{10}$$

where s is a binary predicate stating the similarity of ontological concepts and a is a truth-valued function ensuring that d precedes d' in the network architecture. We detail the definition of the predicate s in the experimental setup.

Furthermore, we propose to weight the edges of the graph by estimating the influence between concept-aligned directions. The TCAV measure estimates the influence of a direction in an hidden layer towards a logit within the last layer of a classifier [14]. To compute the weight w_e of an edge $e = ((d, c), (d', c'))$, we generalize the TCAV measure to estimate the influence between two hidden directions. Firstly, we consider a function

$$h(f^l(x)) = A_{d'}(x) = f^{l'}(x) \cdot v' \tag{11}$$

that given the output of the l-th layer produces the activation map of the direction d'. Then, we are able to redefine the "conceptual sensitivity" as the directional derivative

$$g_{d,d'}(x) = \lim_{\epsilon \to 0} \frac{h(f^l(x) + \epsilon\bar{v}) - h(f^l(x))}{\epsilon} \quad (12)$$

$$= \nabla_v h(f^l(x)) \quad (13)$$

$$= v \cdot \nabla h(f^l(x)), \quad (14)$$

where $\bar{v} \in \mathbb{R}^{N_l \times H_l \times W_l}$ is a tensor obtained by repeating the vector v for each possible location of the activation map $f^l(x)$. We are interested in measuring if the direction aligned with the concept c positively influences the direction aligned with c' when a portrayal of c' is in the receptive field. To do so, we measure the fraction of inputs portraying c' that were positively influenced by the direction aligned to c. Formally,

$$w_e = \frac{\sum_{x \in X} |L_{c'}(x) \wedge (g_{d,d'}(x) > 0)|}{\sum_{x \in X} |L_{c'}(x)|} - 0.5, \quad (15)$$

where the estimate is adjusted to be either positive or negative whether the count is above or below half the inputs. Consequently, a positive value of w_e signifies a positive contribution of direction d towards d', while a negative value represents a negative contribution. As in the semantic alignment, either the concept or the sensitivity masks are scaled to match the same shape and approximate the receptive field.

Typically, because of the constraint enforced by the semantic relation s, the graph G will not be a connected graph. By extracting each non-trivial connected component, we obtain a set

$$T = \{t \mid t \subseteq \Psi, |t| > 1, G[t] \text{ is connected}\}, \quad (16)$$

where each $t \in T$ is a semantically related and architecturally connected neural circuit. Since weight estimation is a costly operation, we propose to limit the analysis to edges within neural circuits.

4 Results

We introduce an alpha version of *Bisturi*[1], a free and open source PyTorch-based library for the semantic alignment of CNNs for computer vision. *Bisturi* implements our unified framework and some of its specializations such as Network Dissection [2] and TCAV [14]. The experimental analysis focuses on the semantic alignment of neural directions with visual concepts representing concrete objects. To obtain an ontologically annotated segmented dataset, we associated each object label of the Broden dataset [2] to a member of the WordNet semantic network [17]. Since WordNet contains a taxonomy of concepts and various semantic relations, we considered it as a simple ontology [18]. As speculated,

[1] https://github.com/rmassidda/bisturi.

the specialization relation automatically increased the number of alignable concepts, from the original 672 object Broden labels to 1177 distinct visual concepts. We also extensively studied semantic alignment exploiting the ILSVRC11 ImageNet dataset [3,24], by generating approximated concept masks from existing bounding-box annotations. Nonetheless, while reporting results on ImageNet in the supplementary materials, the current section focuses on Broden to directly compare with previous literature.

Overall, we report an analysis of the last layers of three popular CNN architectures trained to classify the 365 different scenes and views from the Places-365 dataset [26]. In detail, we considered:

- The last three fully connected layers and the last two convolutional layers of AlexNet [15].
- The last fully connected layer and the last two residual blocks of ResNet [11]. In each residual block, we independently analyzed the two convolutional operations and the sum after the residual connection.
- The last fully connected layer and the output of the last three dense blocks in DenseNet [12].

For replicability purposes, we adopted publicly available pre-trained models from the Places-365 project[2]. We report selected significant results, but further results and discussions are in the supplementary materials.

4.1 Unit Semantic Alignment

Firstly, we are interested in the semantic alignment of the directions corresponding to distinct neurons in each layer. In doing this, we wish to show how we can replicate Network Dissection [2] within our framework. Furthermore, we illustrate how, in comparison, our proposal increases the number of aligned concepts and enables semantic clustering of units. To recreate the setting introduced by Network Dissection, we consider for each layer l a set of directions

$$D_l = \{(l, [e^{(i)}; \beta_i]) \mid i \in \{0, \ldots, N_l - 1\}\}, \tag{17}$$

where the bias terms β_i are concatenated to each vector $e^{(i)}$ of the canonical basis. For each direction $d = [e^{(i)}; \beta_i]$, we fix the bias term β_i such that

$$\begin{aligned} P(f^l(x)_i > \beta_i) &= P(f^l(x) \cdot e^{(i)} > \beta_i) \\ &= P(f^l(x) \cdot e^{(i)} + \beta_i - \beta_i > \beta_i) \\ &= P(A_d(x) - \beta_i > \beta_i) \\ &= 0.005. \end{aligned} \tag{18}$$

The quantity of alignable concepts is fundamental to producing numerous neural-concept associations. We found that the threshold τ highly affects the number of aligned concepts, which rapidly decays for σ_{IoU}. On the other hand,

[2] https://github.com/CSAILVision/places365.

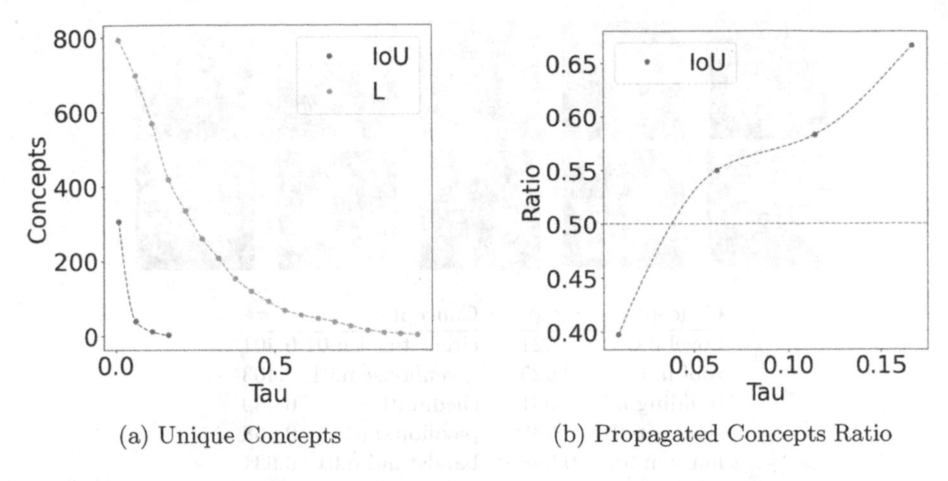

(a) Unique Concepts (b) Propagated Concepts Ratio

Fig. 3. Semantic alignment in AlexNet. For both σ_{IoU} and $\sigma_{\mathcal{L}}$, subfigure (a) plots the number of distinct concepts aligned as the threshold τ varies. Our proposal of adopting $\sigma_{\mathcal{L}}$ results in an higher number of aligned concepts. For σ_{IoU} against increasing τ, Subfigure (b) reports the fraction of aligned concepts obtained by mask propagation over the number of unique concepts. Our propagation strategy produced more than half of the concepts aligned by σ_{IoU} even for small values of τ.

since our proposal of adopting $\sigma_{\mathcal{L}}$ is less restrictive, we found an higher number of aligned concepts (Fig. 3A). Furthermore, we gained empirical confirmation of the advantage in increasing the number of concepts via mask propagation. In all three target networks, we verified how a larger pool of alignable concepts effectively results in a higher number of associations. Remarkably, higher-level concepts account for a significant fraction of the concepts aligned by the IoU measure (Fig. 3B). Therefore, our mask propagation strategy effectively improves the outcome of the Network Dissection approach, by providing concepts that could not have been aligned otherwise.

As expected, we consistently verified how σ_{IoU} and $\sigma_{\mathcal{L}}$ target different aspects of semantic alignment (Fig. 4). The former highlights concepts that activate a unit in an exclusive way, while the latter identifies concepts producing higher than usual activations. Thus, we gained empirical confirmation that the measure $\sigma_{\mathcal{L}}$ is more apt to estimate semantic alignment when a unit responds to multiple visual concepts.

Given the aligned directions, we retrieve neural circuits by linking two concepts if their Jiang-Conrath similarity [13] overcomes a given threshold t_δ. The threshold t_δ also influences the quantity of circuits retrieved: lower values of t_δ cluster Ψ into a fully connected graph, while larger values minimize the number of connections. Furthermore, by aligning more concepts, measure $\sigma_{\mathcal{L}}$ is more apt for the retrieval of neural circuits (Fig. 5).

Zhou et al. [29] tested the importance of hidden neurons in a classifier by ablating them and measuring the most affected classes. We replicate their analysis by considering hidden units clustered by our circuit retrieval strategy.

Concept	σ_{IoU}	Concept	$\sigma_{\mathcal{L}}$
hovel.n.01	0.021	circus'ten t.n.01	0.401
roof.n.03	0.025	greenhouse.n.01	0.403
building.n.01	0.031	shed.n.01	0.469
shelter.n.01	0.035	pavilion.n.01	0.568
house.n.01	0.098	bandstand.n.01	0.631

Fig. 4. Semantic alignment of unit 196 in the last residual block (`layer4.1`) of ResNet-18. We report the ten images from Broden maximally activating the unit and the top-5 aligned concepts according to respectively σ_{IoU} and $\sigma_{\mathcal{L}}$. While both measures identify visual concepts that can be found in these images, $\sigma_{\mathcal{L}}$ produces a list of more specialized concepts within the taxonomy.

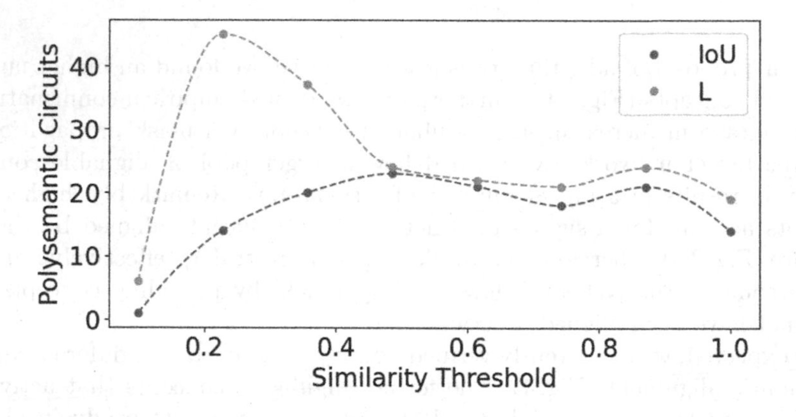

Fig. 5. Number of circuits retrieved in ResNet-18 as t_δ varies. Alignment pairs filtered according to $\tau_{\text{IoU}} = 0.04$ and $\tau_{\mathcal{L}} = 0.2$, resulting in a comparable number of aligned concepts. Our proposed measure $\sigma_{\mathcal{L}}$ produces an higher number of meaningful circuits.

We measure the drop on the Top-5 classification accuracy of the 365 distinct classes from the Places-365 dataset. In general, we found that the ablation of a circuit significantly drops the accuracy of a small number of classes that are, furthermore, related to the aligned concepts (Fig. 6). Targeted accuracy drop highlights how circuits cluster important units for specific tasks, resulting in a valuable instrument to understand which concepts positively affect given outcomes. As control, ablating only the units aligned to the most popular concept in a circuit results in less damaging accuracy drop.

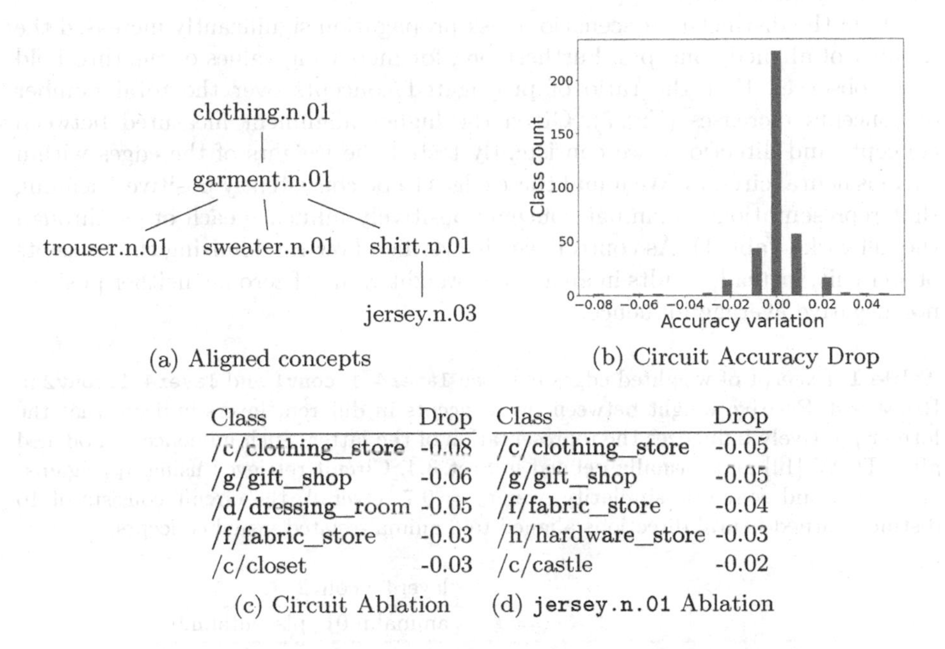

(a) Aligned concepts (b) Circuit Accuracy Drop

Class	Drop
/c/clothing_store	-0.08
/g/gift_shop	-0.06
/d/dressing_room	-0.05
/f/fabric_store	-0.03
/c/closet	-0.03

(c) Circuit Ablation

Class	Drop
/c/clothing_store	-0.05
/g/gift_shop	-0.05
/f/fabric_store	-0.04
/h/hardware_store	-0.03
/c/castle	-0.02

(d) `jersey.n.01` Ablation

Fig. 6. Importance analysis of a circuit found in the hidden layers of AlexNet using our proposed measure $\sigma_{\mathcal{L}}$ against $\tau_{\mathcal{L}} = 0.3$. Similarity between concepts constrained to be over $t_{\delta} = 0.2$. The circuit contains 27 distinct units aligned to 6 clothing-related visual concepts, reported according to the WordNet taxonomy in Subfigure (a). When ablating the circuit, accuracy drop significantly affects only a small number of semantically related classes, as visualized in Subfigure (b). As control, we also ablate the most popular concept in the circuit and verify how the accuracy drop is more sparse and less damaging, as in Subfigure (c). Finally, Subfigure (d) depicts the histogram of categories of the Places-365 dataset as a function of the accuracy drop (on the x-axis).

4.2 Direction Learning

Unlike the TCAV [14] approach, we want to cluster hidden representations of concepts and test their reciprocal influence. To obtain aligned directions, we independently fit a neural direction $d = (v, l)$ for each concept c in each layer l of the network on a sample of the Broden dataset. As discussed in Sect. 3.3, a classifier on the visual concept c should be able to recreate the concept mask $L_c(x)$. We addressed the natural unbalancing of visually segmented datasets by weighting images according to the probability of extracting an example containing the concept. Consequently, we independently split the samples into a training and a validation set, with proportion $4:1$. For each example x in the training set, we applied nearest-neighbor interpolation to each concept mask $L_c(x)$ to match the shape of the activation map $A_d(x)$ and obtain the ground-truth mask.

We trained the classifiers by minimizing the Focal Loss [16] between the concept mask and the activation map. Finally, we estimated their semantic alignment on the validation set using σ_{F1}.

As in the distinct-unit scenario, mask propagation significantly increased the number of aligned concepts. Furthermore, for increasing values of the threshold τ we observed that the ratio of propagated concepts over the total number of concepts increases (Fig. 7). Given the higher alignment measured between concepts and directions, we consistently tested the weights of the edges within various neural circuits. We found these edges to be consistently positive, meaning that representations of similar concepts positively influence each other through the network (Table 1). As control, we also verified how randomizing the concepts of a circuit, instead, results in an average weight value of zero i.e. neither positive nor negative average influence.

Table 1. Excerpt of weighted edges between `layer4.1.conv1` and `layer4.1.conv2` in ResNet-18. Positive weight between two concepts in different layers indicate that the former positively influences the representation of the latter. Such influence is modelled after TCAV [14] and formally defined in Sect. 3.4. Circuit retrieved using σ_{F1} against $\tau_{F1} = 0.5$ and semantic similarity over $t_\delta = 0.7$. Overall, the circuit consists of 16 distinct learned neural directions aligned to 6 animal-related visual concepts.

		layer4.2.conv2	
		animal.n.01	placental.n.01
layer4.1.conv1	animal.n.01	0.114	0.210
	vertebrate.n.01	0.358	0.368
	placental.n.01	0.061	0.259

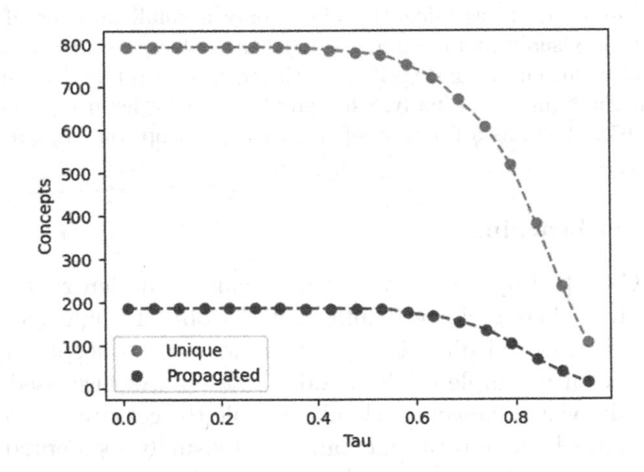

Fig. 7. For different values of the threshold τ, we measure how many of the learned directions are sufficiently semantically aligned according to the measure σ_{F1} on the validation set. In the plot, we also highlight the number of concepts obtained by concept mask propagation (in green) and the overall number of concepts (in red). (Color figure online)

5 Conclusion

We introduced a novel framework for the semantic alignment of CNNs with a complete visual ontology. Overall, we bring three key innovative contributions. Firstly, we defined a propagation strategy to align concepts that lack an explicit annotation in the alignment dataset. Secondly, we generalized previous work on the alignment of single units and neural directions into a unified framework. Finally, we introduced an algorithm to identify connected neural circuits composed of meaningful directions. We experimentally validated our approach by aligning the WordNet ontology with three popular convolutional architectures for image classification. To this end, we considered two datasets: an original extension of the Broden dataset with ontological annotations and a bounding-box annotated subset of ImageNet. We publicly release the extended Broden dataset, the library implementing our approach, and the code used to reproduce our experiments. The experiments highlighted how our methodology can effectively capture semantic alignment. Furthermore, we assessed the emergence of semantically related neural circuits and studied their role in the overall network. This last aspect constitutes the most valuable contribution of our semantic alignment methodology: an innovative instrument to inquire about the nature of neural representations, highlighting semantically related human-interpretable features across the network and their influence towards both network outcomes and other conceptual representations.

Acknowledgments. This research was partially supported by TAILOR, a project funded by EU Horizon 2020 research and innovation programme under GA No 952215.

References

1. Araujo, A., Norris, W., Sim, J.: Computing receptive fields of convolutional neural networks. Distill (2019). https://doi.org/10.23915/distill.00021, https://distill.pub/2019/computing-receptive-fields
2. Bau, D., Zhou, B., Khosla, A., Oliva, A., Torralba, A.: Network dissection: quantifying interpretability of deep visual representations. In: Proceedings of the IEEE Conference on Computer Vision and Pattern Recognition (CVPR), July 2017
3. Deng, J., Dong, W., Socher, R., Li, L., Kai Li, Li Fei-Fei: Imagenet: a large-scale hierarchical image database. In: 2009 IEEE Conference on Computer Vision and Pattern Recognition, pp. 248–255 (2009). https://doi.org/10.1109/CVPR.2009.5206848
4. Euzenat, J., Shvaiko, P.: Classifications of Ontology Matching Techniques. In: In: Ontology Matching, pp. 73–84. Springer, Heidelberg (2013). https://doi.org/10.1007/978-3-642-38721-0_4
5. Euzenat, J., Shvaiko, P.: The Matching Problem. In: In: Ontology Matching, pp. 25–54. Springer, Heidelberg (2013). https://doi.org/10.1007/978-3-642-38721-0_2
6. Fong, R.C., Vedaldi, A.: Interpretable explanations of black boxes by meaningful perturbation. In: 2017 IEEE International Conference on Computer Vision (ICCV), pp. 3449–3457, October 2017. https://doi.org/10.1109/ICCV.2017.371, iSSN: 2380-7504

7. Frege, G.: Function und Begriff. Hermann Pohle, Jena (1891)
8. Goh, G., et al.: Multimodal neurons in artificial neural networks. Distill (2021). https://doi.org/10.23915/distill.00030. https://distill.pub/2021/multimodal-neurons
9. Guarino, N., Oberle, D., Staab, S.: What is an ontology? In: Staab, S., Studer, R. (eds.) Handbook on Ontologies. IHIS, pp. 1–17. Springer, Heidelberg (2009). https://doi.org/10.1007/978-3-540-92673-3_0
10. Guidotti, R., Monreale, A., Ruggieri, S., Turini, F., Giannotti, F., Pedreschi, D.: A survey of methods for explaining black box models. ACM Comput. Surv. (CSUR) **51**(5), 1–42 (2018)
11. He, K., Zhang, X., Ren, S., Sun, J.: Deep residual learning for image recognition. CoRR abs/1512.03385 (2015). http://arxiv.org/abs/1512.03385
12. Huang, G., Liu, Z., Van Der Maaten, L., Weinberger, K.Q.: Densely connected convolutional networks. In: Proceedings of the IEEE Conference on Computer Vision and Pattern Recognition, pp. 4700–4708 (2017)
13. Jiang, J.J., Conrath, D.W.: Semantic similarity based on corpus statistics and lexical taxonomy. In: Proceedings of the 10th Research on Computational Linguistics International Conference, pp. 19–33. The Association for Computational Linguistics and Chinese Language Processing (ACLCLP), Taipei, Taiwan, August 1997. https://aclanthology.org/O97-1002
14. Kim, B., Wattenberg, M., Gilmer, J., Cai, C., Wexler, J., Viegas, F., Sayres, R.: Interpretability Beyond Feature Attribution: Quantitative Testing with Concept Activation Vectors (TCAV). arXiv:1711.11279 [stat], June 2018. http://arxiv.org/abs/1711.11279,arXiv: 1711.11279
15. Krizhevsky, A.: One weird trick for parallelizing convolutional neural networks. CoRR abs/1404.5997 (2014). http://arxiv.org/abs/1404.5997
16. Lin, T.Y., Goyal, P., Girshick, R., He, K., Dollár, P.: Focal loss for dense object detection. In: Proceedings of the IEEE International Conference on Computer Vision, pp. 2980–2988 (2017)
17. Miller, G.A.: Wordnet: a lexical database for English. Commun. ACM **38**(11), 39–41 (1995)
18. Miller, G.A., Hristea, F.: Wordnet nouns: classes and instances. Comput. Linguist. **32**(1), 1–3 (2006). https://doi.org/10.1162/coli.2006.32.1.1
19. Mu, J., Andreas, J.: Compositional explanations of neurons. In: Larochelle, H., Ranzato, M., Hadsell, R., Balcan, M.F., Lin, H. (eds.) Advances in Neural Information Processing Systems, vol. 33, pp. 17153–17163. Curran Associates, Inc. (2020). https://proceedings.neurips.cc/paper/2020/file/c74956ffb38ba48ed6ce977af6727275-Paper.pdf
20. Olah, C., Cammarata, N., Schubert, L., Goh, G., Petrov, M., Carter, S.: Zoom. In: An introduction to circuits. Distill (2020). https://doi.org/10.23915/distill.00024.001. https://distill.pub/2020/circuits/zoom-in
21. Olah, C., Mordvintsev, A., Schubert, L.: Feature visualization. Distill (2017). https://doi.org/10.23915/distill.00007. https://distill.pub/2017/feature-visualization
22. Otero-Cerdeira, L., Rodríguez-Martínez, F.J., Gómez-Rodríguez, A.: Ontology matching: a literature review. Expert Syst. Appl. **42**(2), 949–971 (2015)
23. Page, M.: Connectionist modelling in psychology: a localist manifesto. Behavioral Brain Sci. **23**(4), 443–467 (2000). https://doi.org/10.1017/S0140525X00003356

24. Russakovsky, O., Deng, J., Su, H., Krause, J., Satheesh, S., Ma, S., Huang, Z., Karpathy, A., Khosla, A., Bernstein, M., Berg, A.C., Fei-Fei, L.: ImageNet large scale visual recognition challenge. Int. J. Comput. Vision **115**(3), 211–252 (2015). https://doi.org/10.1007/s11263-015-0816-y
25. Zhou, B., Khosla, A., Lapedriza, À., Oliva, A., Torralba, A.: Object detectors emerge in deep scene cnns. In: Bengio, Y., LeCun, Y. (eds.) 3rd International Conference on Learning Representations, ICLR 2015, San Diego, CA, USA, 7–9, May 2015, Conference Track Proceedings (2015). http://arxiv.org/abs/1412.6856
26. Zhou, B., Lapedriza, A., Khosla, A., Oliva, A., Torralba, A.: Places: a 10 million image database for scene recognition. IEEE Trans. Pattern Anal. Mach. Intell. (2017)
27. Zhou, B., Lapedriza, A., Xiao, J., Torralba, A., Oliva, A.: Learning deep features for scene recognition using places database. In: Ghahramani, Z., Welling, M., Cortes, C., Lawrence, N., Weinberger, K.Q. (eds.) Advances in Neural Information Processing Systems, vol. 27. Curran Associates, Inc. (2014). https://proceedings.neurips.cc/paper/2014/file/3fe94a002317b5f9259f82690aeea4cd-Paper.pdf
28. Zhou, B., Sun, Y., Bau, D., Torralba, A.: Interpretable basis decomposition for visual explanation. In: Ferrari, V., Hebert, M., Sminchisescu, C., Weiss, Y. (eds.) ECCV 2018. LNCS, vol. 11212, pp. 122–138. Springer, Cham (2018). https://doi.org/10.1007/978-3-030-01237-3_8
29. Zhou, B., Sun, Y., Bau, D., Torralba, A.: Revisiting the importance of individual units in cnns via ablation. CoRR abs/1806.02891 (2018). http://arxiv.org/abs/1806.02891

Neural Networks with Feature Attribution and Contrastive Explanations

Housam K. B. Babiker[1,3(✉)], Mi-Young Kim[2,3], and Randy Goebel[1,3]

[1] Department of Computing Science, University of Alberta, Edmonton, Canada
{khalifab,rgoebel}@ualberta.ca

[2] Department of Science, Augustana Faculty, University of Alberta, Edmonton, Canada
miyoung2@ualberta.ca

[3] Alberta Machine Intelligence Institute, Edmonton, Canada

Abstract. Interpretability is becoming an expected and even essential characteristic in GDPR Europe. In the majority of existing work on natural language processing (NLP), interpretability has focused on the problem of explanatory responses to questions like "*Why p?*" (identifying the causal attributes that support the prediction of "*p*.)" This type of local explainability focuses on explaining a single prediction made by a model for a single input, by quantifying the contribution of each feature to the predicted output class. Most of these methods are based on post-hoc approaches. In this paper, we propose a technique to learn centroid vectors concurrently while building the black-box in order to support answers to "*Why p?*" and "*Why p and not q?*," where "*q*" is another class that is contrastive to "*p*." Across multiple datasets, our approach achieves better results than traditional post-hoc methods.

Keywords: Interpretability · NLP · Text classification

1 Introduction

Research on making deep learning models more interpretable and explainable is receiving much attention. One of the main reasons is the application of deep learning models to high-stake domains. In general, interpretability is an essential component for deploying deep learning models. Interpretability in the context of deep learning can be used to tackle a variety of problems: (i) the detection of biased views in a deep learning model, (ii) evaluation of the fairness of a deep learning model, (iii) faithfully explaining the predictions of the classifier, i.e., the construction of accurate explanation that explains the underlying causal phenomena [13] and (iv) the use of explanations as a proxy for model debugging, which allows researchers/engineers to construct models better or debug existing models. Non-linear deep neural networks come at the cost of model interpretability. Most existing related research has focused on identifying feature attribution (e.g., possible causal attributes) to explain the prediction of a black-box neural network. This type of explanation is defined as answers to "*why-questions*."

M.-R. Amini et al. (Eds.): ECML PKDD 2022, LNAI 13713, pp. 372–388, 2023.
https://doi.org/10.1007/978-3-031-26387-3_23

"*Why-questions*," are generally thought of as causal-like explanations [11]. Existing techniques to *why-questions* rely on using a post-hoc approach to identify the causal attributes for a single black-box prediction. Post-hoc methods generally do not always provide accurate explanations [20]. There are many possible reasons for this limitation; for instance, feature attributions typically suffer from noisy gradients in back-propagation techniques [8]. Studies in philosophy and social science show that humans, in general, prefer contrastive explanations, i.e., the explanation of an event is based on explaining the fact (p) in contrast to another event (q) [12,16]. Here "p" represents the model prediction, and "q" represents an alternative class we would use for a contrastive explanation. A contrastive explanation is an essential property of an explanation: 1) humans ask a contrastive question when they are surprised by an event and expect a different outcome, and 2) the contrastive event is what they expect to happen [7,12,16]. Majority of existing post-hoc techniques are only limited to providing answers to "*why p?*" and cannot provide answers to "*why p and not q?*". For instance, gradient-based methods. Contrastive explanations are relatively new in NLP [9]. Our work focuses on building an inherently interpretable model that can support answers to both kinds of questions: "*why p?*," and "*why p, not q?*." In general, a contrastive explanation provides an explanation for why an instance had the current output (fact) rather than a targeted outcome of interest (foil) [25]. An example of our proposed neural network model is shown in Fig. 1.

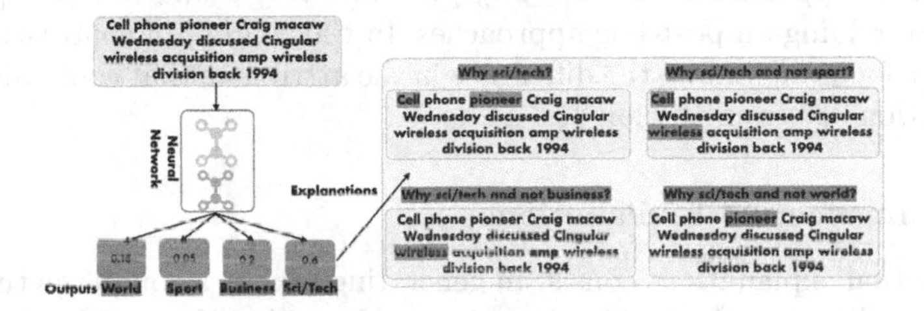

Fig. 1. An example of the proposed neural network model with answers to "*why p?*" and "*why p and not q?*" questions. Here we visualize the top salient attributes.

1.1 Contrastive vs. Counterfactual

The evolving discussions of explainable AI (XAI) have articulated several distinguishing aspects of explanation (e.g., [16]), including a difference between contrastive (e.g., what made the *difference* between the students who failed the exam and those who did not fail?) and counterfactual (e.g., will we reduce climate change *if we reduce fuel consumption?*) explanations. Contrastive explanations are different from counterfactual explanations [15]. In general, contrastive and counterfactual reasoning emphasize different aspects of causation [5]. In counterfactual reasoning, we focus on instances in which the salient causal attributes are absent (missing from the text). In contrast, in a contrastive explanation (our focus here), one considers the difference in attributes between two predictions. The difference between the two approaches is in the knowledge support required for the explanation. For instance, a counterfactual explanation focuses on the question of "*What if?*," while a contrastive explanation focuses on "the difference."

The contributions of this paper can be summarized as follows: (i) we propose an interpretable (intrinsic) neural model that focuses on learning deep discriminative embedding features, (ii) our neural model provides two types of explanations (e.g., non-contrastive explanations and contrastive explanations) using feature attribution, and (iii) we proposed a metric to evaluate the quality of the contrastive explanations. An intrinsic neural model is better than using traditional post-hoc explanations because: (i) we can find faithful explanations, (ii), we do not need an additional complex computation to find an explanation for a single prediction.

2 Related Work

2.1 Contrastive Explanations

With contrastive explanations, we aim to expose an alternative to any given model prediction. In [9], they proposed a post-hoc approach that relies on a projection matrix to devise explanations. Similarly, [18] used SHAP to generate a contrastive explanation. Our approach is different; we propose an intrinsic neural model which supports answers to "*why p?*" and "*why p and not q?*" questions, rather than relying on post-hoc approaches. In the context of contrastive explanations, we focus on finding the difference in the attributes that could distinguish the prediction "*p*" from the foil "*q*."

2.2 Counterfactual Explanations

Counterfactual explanations consist in generating text as a counterfactual example. In general, counterfactual explanations seek to identify a minimal change in model data that "flips" a predictive model's prediction, which is used for explanation. [26] proposed the concept of unconditional counterfactual explanations and introduced a framework for generating counterfactual explanations. For text classification, [29] proposed a method to generate counterfactual text from a pre-trained model for the finance domain. In addition, [6] relied on finding evidence that is discriminative for the target class but not present in the foil class to learn a model to generate counterfactual explanations for why a model predicts class "*p*" instead of "*q*." However, their approach was mainly designed for computer vision.

2.3 Post-hoc Non-contrastive Explanations

One of the most popular techniques for explaining the prediction of a black box is the use of *why p?*. There is much prior work on this topic. For instance, [3] used Shapley approximation and proposed two methods, namely L-Shapley and C-Shapley. Additionally, [22] proposed the integrated gradient method, which relies on using a back-propagation algorithm. Other methods also rely on perturbation techniques such as [19]. Some methods focus on constructing interpretable neural architecture for classification. For instance, [2]'s model learn a rationale as the model's explanation. In general, our approach is different from traditional post-hoc and rationale-based models. We provide two types of explanations using an intrinsic neural model i.e., answers to "*why p?* and *why p and not q?*" questions. Overall, our work is not the first contribution to contrastive explanation nor the first technique for "*why p?*" questions. In [1], authors proposed a knowledge distillation technique which could learn an interpretable vector space model. However our work is different, we focus on building an intrinsic model which can support answers to why p and *why p?*and *why p and not q?* questions.

3 Contrastive Explanation Generation

Our approach is not a post-hoc technique for model's explanation but rather the pursuit of constructing an inherently interpretable neural network. Our intrinsic neural model relies on improving the embedding features (see Fig. 2.) For a given class in the dataset, our network attempts to assign similar texts into a single cluster.

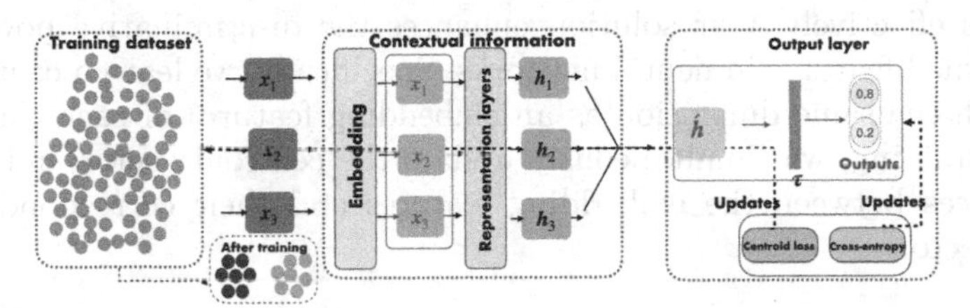

Fig. 2. Our proposed intrinsic neural architecture focuses on clustering texts based on the model predictions. For each class, we define a centroid vector. During training, we used our proposed method to establish a similarity structure between the sentences and the corresponding centroid vector.

3.1 Neural Nets with Feature Attributions and Contrastive Explanations

We propose a multi-task neural network architecture, i.e., a classification task and an explanation task. We jointly optimize the network for both classifications and faithful explanations. For notation, we denote scalars with italic lowercase letters (e.g., x), vectors with bold lowercase letters (e.g., $\boldsymbol{x}$), and matrices with bold uppercase letters (e.g., $\boldsymbol{W}$). In the text classification task, an input sequence $\boldsymbol{x}_1, \boldsymbol{x}_2, ..., \boldsymbol{x}_l \in \mathbb{R}^d$, where l is the length of the text input and d is the vector dimension, is mapped to a distribution over class labels using a parameterized neural network (e.g., a Multi-head attention). In general, the contextual vector $\acute{\boldsymbol{h}} \in \mathbb{R}^d$ is passed to a linear layer with parameters $\boldsymbol{W} \in \mathbb{R}^{d \times n}$ which provides a probability distribution over n classes. The output $\boldsymbol{y}$ is a vector of class probabilities of dimension $\mathbb{R}^n$, where n is the number of classes. The predicted label p of the text input is the index of the maximum element in $\boldsymbol{y}$, i.e., $p = argmax f(\boldsymbol{x})$, $\forall k \in [1, n]$. Here, k iterates over the probabilities and $f(\boldsymbol{x})$ denotes a neural network. During training, an empirical loss (e.g., cross-entropy) $\mathcal{J}(p, y^{'}, \theta)$ is minimized using gradient descent, where $y^{'}$ is the ground truth label and θ represents the network's parameters. We propose to augment the network to provide two types of explanations "*Why p?*" and "*Why p and not q?*." To do so, we first define a randomly initialized centroid vector for each class, and then use the centroid vector as a proxy to explain the black-box prediction.

For instance, if the neural network's prediction is class 1, we use the centroid vector representing that class to calculate the scores for *why p?*. For contrastive explanation, we find the difference between the scores of the centroid vector representing the predicted class and the scores of the centroid vector representing the contrast class (e.g., the centroid vector for class 2). The centroid vector of label p pulls the weighted sentence vector of the text input $\boldsymbol{x}_1, \boldsymbol{x}_2, ..., \boldsymbol{x}_l$ closer. In the following, we discuss the steps for augmenting a neural network with the centroid vectors. Let $\boldsymbol{c}_j (j = 1, 2, ..., n)$ be a collection of randomly initialized centroid vectors, where $\boldsymbol{c}_j \in \mathbb{R}^d$ is a vector representing label $\boldsymbol{y}_j$. We propose a new objective function, namely centroid-loss, to explain the neural network predictions effectively. Our solution enhances the discriminative power of the deeply learned features in neural networks. Specifically, we learn a centroid $\boldsymbol{c}_j$ (a vector with the same dimension as an embedding feature) of each class. In the course of training, we simultaneously update the centroid vector and minimize the distances between the embedding features and their corresponding class' centroid vector.

3.2 Joint Objective

Recall that a supervised learning algorithm input is a set of training instances and the corresponding label. The goal is to learn a function that accurately maps input examples to their desired labels using cross-entropy. Given the prediction p, we learn $\boldsymbol{c}_p \in \mathbb{R}^d$ to pull the sentence vectors representing class p closer. Intuitively, we are minimizing the intra-class variations while keeping the features of different classes separable. In the following, we discuss the optimization objective of our proposed network.

Cross-entropy: term 1 in the optimization objective function is the standard loss function for classification. We denote this loss as $\mathcal{L}^{\text{cls}}$.

Attractive term: term 2 focuses on minimizing the cosine distance between the sentence vector and the corresponding $\boldsymbol{c}_p$. Let $\boldsymbol{X}$ be a matrix consisting of embedding vectors $[\boldsymbol{x}_1, \boldsymbol{x}_2, ..., \boldsymbol{x}_l]$ and the sentence vector of $\boldsymbol{X}$ is $\hat{\boldsymbol{x}} \in \mathbb{R}^d$. Let, $\bar{\boldsymbol{w}} \in \mathbb{R}^l$ be the importance scores, where each.

$$\bar{\boldsymbol{w}}_i = \frac{\boldsymbol{x}_i \cdot \hat{\boldsymbol{x}}}{\|\boldsymbol{x}_i\| \, \|\hat{\boldsymbol{x}}_i\|}, \tag{1}$$

where $\bar{\boldsymbol{w}}_i$ is the importance score of word i, and $\hat{\boldsymbol{x}}$ is the sentence vector of the input $\boldsymbol{X}$. Term 2 minimizes the cosine distance between the weighted sentence vector $\bar{\boldsymbol{x}}$ of each input with the corresponding centroid vector $\boldsymbol{c}_p$. The sentence vector is defined as follows:

$$\bar{\boldsymbol{x}} = \boldsymbol{X}\left(\frac{\exp(\bar{\boldsymbol{w}})}{\sum_{i=1}^{l}\exp(\bar{\boldsymbol{w}}_i)}\right) \tag{2}$$

From Eq. 2, we obtain the weighted sentence vector through multiplying the values in the i-th row of $\boldsymbol{X}$ by $\bar{\boldsymbol{w}}_i$ followed by calculating the sentence vector $\bar{\boldsymbol{x}} \in \mathbb{R}^d$. We define the loss of term 2 as follows:

$$\mathcal{L}^{\text{attr}} = 1 - \frac{\bar{\boldsymbol{x}} \cdot \boldsymbol{c}_p}{\|\bar{\boldsymbol{x}}\| \, \|\boldsymbol{c}_p\|} \tag{3}$$

Term 2 is the second loss of our proposed optimization objective.

Repulsive term: term 3 (the third term in the overall loss function) focuses on maximizing cosine distance of $\bar{\boldsymbol{x}}$ from other centroid vectors, i.e., $\boldsymbol{c}_j$, where $j \neq p$, so that cosine distance between them is maximum. We call this term "repulsive loss" similar to [27] we denote the loss as $\mathcal{L}_{\text{rep}}$.

Pairwise term: term 4 in our objective maximizes the pairwise distance matrix of the centroid vectors. For the distance we proposed to use the squared euclidean distance and we denote the loss as $\mathcal{L}_{\text{pair}}$.

Overall loss: is defined as

$$\mathcal{L} = \mathcal{L}^{\text{cls}} + (\lambda_1 \mathcal{L}^{\text{attr}}) - (\lambda_2 \mathcal{L}^{\text{rep}}) - (\lambda_3 \mathcal{L}^{\text{pair}}) \tag{4}$$

where $(\lambda_1, \lambda_2, \lambda_3)$ are the coefficients. The hyper parameters $(\lambda_1, \lambda_2, \lambda_3)$ are important for minimizing the intra-class variation (to minimize the variance within the same class). More specifically, terms 3 and 4 focus on keeping the features of different classes separable, and term 2 focuses on minimizing the intra-class distances. All of them are essential to our model. We refer to the combination of the new added terms as the centroid loss, i.e., term 2, term 3, and term 4.

3.3 Explanations

We seek to identify a feature with a causal impact on the model prediction decision process. We follow [28]'s definition of intervention: an intervention is an idealized experimental manipulation carried out on some variable $\boldsymbol{x}$ which is hypothesized to be causally related to changes in some other variable p. Any intervention on the text input using attributions on the prediction p is a causal process that changes the model prediction. Therefore, if the intervention changes the model prediction, it is probably due to the adjustment in the causal space of the text input. We will use the idea of "intervention" to understand the effectiveness of our approach for both *why p?* and *why p and not q?*.

Why p? For this type of explanations, we identify potential causal attributes by calculating the cosine similarity between each $\boldsymbol{x}_i$ and the corresponding $\boldsymbol{c}_p$ of class p. A higher score indicates a more informative attribute. The negative scores indicate the features have negatively contributed to the specific class classification and vice-versa. For experiments, we intervene on the text input to remove irrelevant attributes, i.e., replacing each factor with a "<pad>" followed by observing the change in the model's probabilities.

Why p and not q? Given any text instance, a classifier predicts p and a centroid vector $\boldsymbol{c}_p$. A p-contrast question is of the format 'Why [predicted-class (p)] not [desired class (q)]?'. By specifying the desired class, we limit our search space to a single alternative. Given the text input, we estimate attribution scores for "p" using $\boldsymbol{c}_p$. For the desired class q, we calculate the attribution scores of the text input using $\boldsymbol{c}_q$. Please note that, here we also use cosine similarity. We find the attribution scores for contrastive explanations as $\boldsymbol{z}_c = \boldsymbol{z}_p - \boldsymbol{z}_q$, where $\boldsymbol{z}_p$ is the attribution score for the predicted class p obtained using $\boldsymbol{c}_p$ and $\boldsymbol{z}_q$ is the attribution scores for the foil class q obtained using $\boldsymbol{c}_q$. We follow the intervention approach as in "*Why p?,*" to find the candidate attributes for the contrastive explanation.

4 Experiments

To effectively evaluate our approach, we devise a measure to rank the identified causal attributes. Given a prediction "p," for "*why p,*" we rank each attribute by how much it contributes to prediction "p" using $\boldsymbol{c}_p$. As for contrastive explanations, we rank each attribute using $\boldsymbol{z}$ by how contrastively useful it is to the model for choosing "p" against "q." All evaluations follow an interventionist approach defined in Sect. 3.3.

4.1 Setup

Datasets. We adopt the IMDB datasets [14] (train:25000, test:25000 samples) with binary labels, AG news [30](train:102080, test:25520 samples) with four classes, and YELP reviews [23] (train:110400, test:27600 samples) with binary labels. We hold out 10% of the training examples as the development set. We limit the length of the input to 50 for YELP and IMDB and 20 for AG news.

Model. The multi-head model [24] includes an embedding layer and multi-head attention layers. We tokenized sentences and randomly initialized the embedding layer and the centroid vectors. The dimension of the word embedding, centroid vector, and feature vector (at the output layer) is 128. For training the network, we use the Adam optimizer [10] with a batch size of 256 and a learning rate of 0.0001 (We have experimented with different values for the coefficient with interval 5 between 0 and 1000, for the experiments we used (λ_1 : 1000, λ_2 : 10, λ_3 : 1000). The F1-scores for AG news topic classification, IMDB sentiment, and YELP review classification are summarized in Table 1. Performance is in terms of F1-score.

Table 1. Black-box (multi-head) vs. intrinsic multi-head neural network.

Models	Dataset		
	IMDB	YELP	AG news
Black-box (multi-head)	0.81	0.88	0.89
Proposed (multi-head with centroid loss)	0.81	0.88	0.89

4.2 Explainability Metrics

We adopt three metrics from prior work on evaluating word-level attribution (non-contrastive explanation): the area over the perturbation curve (AOPC) from ERASER [4], the log-odds scores [3,21], and the degradation score to the trained model accuracy [17]. We also proposed new evaluation metrics for contrastive explanations. All the metrics measure the local fidelity by deleting or masking top-scored words.

4.3 Evaluating Why p?

We begin first by evaluating the faithfulness of "*Why p?*" questions. Faithfulness means the degree (trust of an explanation) to which an explanation influences the model prediction. ERASER proposes two metrics to measure the quality of the explanations:

Comprehensiveness: Measure whether all required features by the model to make a prediction are selected by the explanation method. To use this metric, we first need to compute a new sentence. For example, given an input text $\boldsymbol{X}$, the new sentence is defined as $\tilde{\boldsymbol{X}} = \boldsymbol{X} - \boldsymbol{R}$, where $\boldsymbol{R}$ is the set of salient features identified by the explanation method. Let $f_\theta(\boldsymbol{X})_p$ be the neural network output for class p. The measure of comprehensiveness is calculated as:

$$\text{Comprehensiveness} = f_\theta(\boldsymbol{X})_p - f_\theta(\tilde{\boldsymbol{X}})_p \tag{5}$$

A higher score implies that the identified tokens included in $\boldsymbol{R}$ were more influential in the model's predictions, compared with other tokens.

Sufficiency: The second metric focuses on evaluating whether the identified features were enough to predict the same label as using the full text or not, and is defined as follows:

$$\text{Sufficiency} = f_\theta(\boldsymbol{X})_p - f_\theta(\boldsymbol{R})_p \tag{6}$$

Under sufficiency metric, lower scores are better. We calculate the AOPC for both comprehensiveness and sufficiency using a variety of token percentages: 5%, 10%, 15%, 20%, and 25%.

Log-Odds: Log-odds score is calculated by averaging the difference of negative logarithmic probabilities on the predicted class over all of the test data before and after masking the top $m\%$ features with zero paddings,

$$\text{Log-odds}(m) = \frac{1}{t}\sum_{i=1}^{t} \log \frac{p(\hat{y}|\boldsymbol{X}_i^{(m)})}{p(\hat{y}|\boldsymbol{X}_i)}, \tag{7}$$

where $\boldsymbol{X}_i^{(m)}$ is the new input based on replacing the top $m\%$ with the special token <pad> in $\boldsymbol{X}_i$ and t is the total number of samples. Lower log-odd scores are better. **Degradation score**: Words are ranked according to "*why p?*" (defined in Subsect. 3.3- *Why p?*). In this way, higher-ranked tokens (features) are recursively eliminated. The degradation score to the trained model accuracy is calculated. We perform this experiment using a variety of token percentages: 5%, 10%, 15%, 20%, and 25%.

Results: We compare our technique with competitive baselines, namely Shapely-based methods (L/C-Shapely) [3], using log-odds, AOPC, and degradation score. The log-odds and degradation scores are shown in Fig. 3. The L/C-Shapley focuses on instance-wise feature importance scores. Shapley values are extremely expensive to compute and L/C-Shapley were proposed to compute approximate Shapley values. We evaluate the explanation on the test set of the datasets.

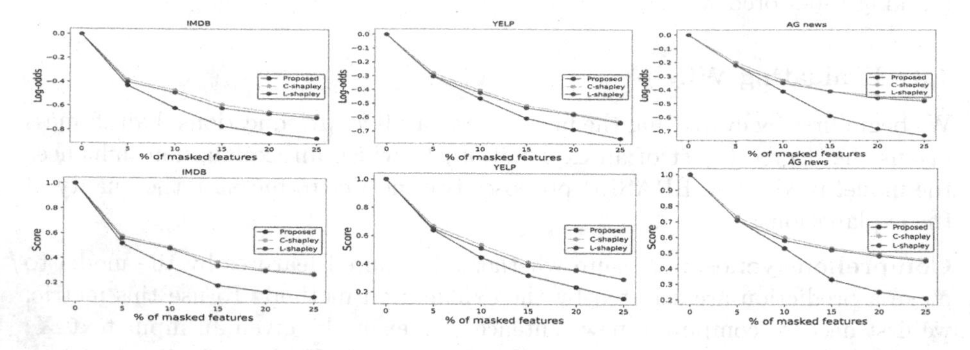

Fig. 3. Log-odds scores as a function of masked features (top). A steeper decline indicates a better performance. Degradation score (y-axis) as a function of removed tokens(bottom). A steeper decline indicates a better performance.

Our approach achieves the best performance on both metrics (log-odds and degradation score). Note that L-Shapley applying approximation Shapley values perform better than C-Shapley. The results also show that the neural network classifier employs less number of features for making predictions. Our method also outperforms Shapley approximation methods on ERASER metrics achieving the best result (Table 2) for both comprehensiveness and sufficiency on the three datasets.

4.4 Evaluating Why p and Not q?

To evaluate the faithfulness of contrastive explanations, we use the following metrics:

Contrastive Overlap Score (COS) (%): we calculate the overlap (%) between the sets of causal attributes of "*Why p?*" and "*Why p and not q?*. Lower % indicates more difference between the explanations of "*Why p?*" and "*Why p and not q?*.

Table 2. Eraser benchmark scores: Comprehensiveness and sufficiency in terms of AOPC

	L-Shapley	C-Shapley	Proposed
IMDB			
Comprehensiveness	0.575	0.554	0.704
Sufficiency	0.1722	0.172	0.112
YELP			
Comprehensiveness	0.494	0.479	0.562
Sufficiency	0.172	0.172	0.112
AG news			
Comprehensiveness	0.384	0.37	0.524
Sufficiency	0.247	0.246	0.086

Contrastive Confidence Score (CCS): For a confidence score, we analyze the change in the probability of the contrastive class "*q*." We remove the attributes that distinguish "*p*" from "*q*" in order of their importance, until the model's prediction is flipped to another class. Please note the scores of the features are obtained using "*why p and not q?*" We calculate the difference in the probability of "*q*" before and after the intervention. An increase in the probability indicates an informative contrastive explanation.

Contrastive Gain (CAG): This metric measures the quality of contrastive explanations compared to the non-contrastive explanations. Here, our explanations for the question "*why p?*" will be called non-contrastive explanations. Given a prediction "*p*" and foil "q," we measure the change in the probability score of "*q*" after removing salient features using attribution-scores obtained from "*why p?*" and also from "*why p and not q?*" explanation. We use our approach as the baseline for "*why p?*", because our method outperformed [3]. For the "*why p and not q?*" explanation, we used the method described in Section (3). A higher contrastive gain indicates that our contrastive explanation is better in answering "*why p and not q?*" questions. In summary, the contrastive gain measures the change in probability of the foil class after removing some features.

Results. We use the AG news dataset to evaluate our contrastive explanation method. For contrastive overlap (COS), the results in Fig. 4 show that most contrastive explanations do have fine-grained differences from "*Why p?*" questions. The result suggests that the model is not using the same reasoning for "*Why p?*" when answering the contrastive questions. We observed that, for multi-class problems, there are fine-grained differences between "*Why p?*" and "*Why p and not q?*" compared to a binary problem such as sentiment classification where there might be a higher similarity between the two explanations.

For (CCS), results shown in Table 3 indicate the effectiveness of our approach in finding contrastive information. Meaning that there is an increase in the score of the foil "*q*" when removing the features that distinguish "*p*" from "*q*".

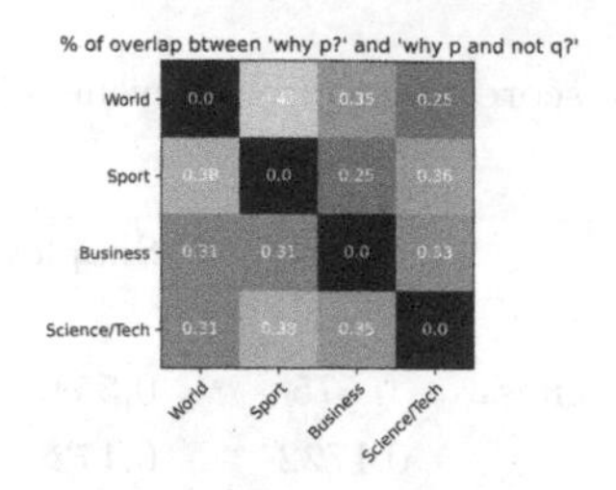

Fig. 4. Overlap score between "*why p?*" and "*why p and not q?*" questions. "0.0" means that we did not consider the contrastive explanations when "*p*" and "*q*" are the same (X-axis: refers to why p? questions and Y-axis: refers to why p and not q? questions.)

Table 3. Contrastive confidence score (CCS). Empty cells mean that we cannot find a contrastive explanation for the same class i.e., the foil should be different from the predicted class. The highlighted cells show the scores of the foil after removing the salient features.

	World(q)		Sport(q)		Business(q)		Sci/Tech(q)	
Class(p)	Before	After	Before	After	Before	After	Before	After
World			0.05	0.22	0.05	0.41	0.01	0.33
Sport	0.07	0.45			0.01	0.35	0.04	0.21
Business	0.06	0.38	0.003	0.16			0.13	0.37
Sci/Tech	0.02	0.32	0.04	0.12	0.13	0.52		

We also show the scores of other classes when using the (CCS) metric. We re-trained the same model again on AG news and re-calculated the CCS. The results are shown in Table 4. We can see that when the foil q is set to "business" and evaluated with different classes (p) such as "world, sport, sci/tech." The probability score for the "business" is higher compared to other classes when using *why p and not business.?* Due to page limit, we only show the results for the "business" class (see Table 4).

Table 4. We compare the scores of other classes when evaluating the CCS for the foil. Here the foil is the business class.

	World	Sport	Business(q)	Sci/Tech
World(p)	0.2	−0.8	0.4	0.1
Sport(p)	0.3	0.1	0.5	−0.8
Sci/Tech(p)	−0.7	0.1	0.4	0.1

Figure 5 compares our non-contrastive explanations and contrastive explanation methods (CAG). We use the AG news data and plot the results for different "*why p and not q?*" questions. The results in Fig. 5 indicate that our contrastive explanations are better capturing the features that contribute prediction of 'not q' than non-contrastive explanations, especially when there are more fine-grained differences. The results show that non-contrastive explanation is not always achieving high contrastive scores when top features are masked.

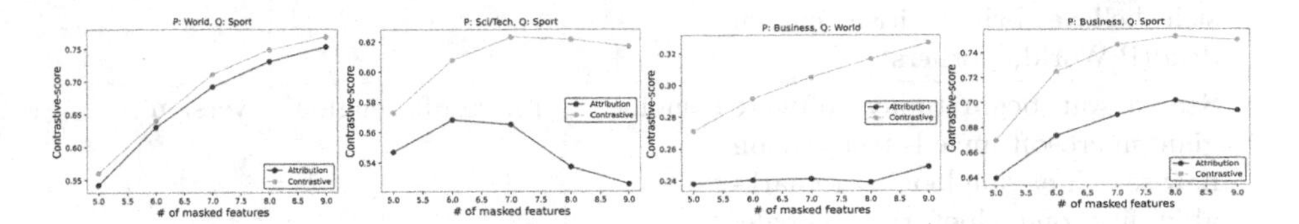

Fig. 5. Contrastive gain as a function of removed tokens. A higher gain indicates that the method was better in capturing contrastive information. Attribution refers to our non-contrastive method.

Instead of tracking the change in probability score of "q" after removing salient as in contrastive gain, we instead calculate the AOPC using different percentages (25%, 30%, 35%, 40%, 45%). The results are summarized in Table 5. Our contrastive-explanation has the highest AOPC compared to our non-contrastive explanation method.

Table 5. Contrastive gain (CAG): Evaluating the effectiveness of using contrastive explanation when there are fined grained differences. We use different percentages (25%, 30%, 35%, 40%, 45%) to calculate the AOPC.

P	Q	AOPC (non-contrastive)	AOPC (contrastive)
World	Business	0.04	0.065
Business	Sci/tech	−0.001	0.002
Sci/tech	World	0.304	0.341
Sci/tech	Business	0.056	0.058
Sport	Business	0.05	0.052
Sport	Sci/tech	0.006	0.009

Highlighting *why p and not q?* Questions: We show qualitative results for interpreting the model predictions using our proposed approach; for example, answers to the "*Why p?*" and "*Why p and not q?*" questions are shown in Table 6. These results show that the model implicitly learns the contrastive information when making the prediction.

Contrastive Explanations Applied to Sentiment Classification. For a contrastive explanation, if there are no fine-grained differences between "p" and "q", then the same reasoning used for "*why p?*" questions will also be used to

Table 6. Contrastive explanations on AG news.

Text	World	Sport	Business	Sci/tech
Record shown mutilated body found iraq kidnapped aid worker margaret hassan british official say still believe british irish citizen dead(P:World,Q:others)	iraq	dead	hassan	kidnapped
Search war begin today software giant microsoft unveils test version new search engine looking remarkably like one chief rival google. (P:Sci/tech,Q:others)	engine	microsoft	engine	version
Version desktop search tool computer run apple computer mac operating system google chief executive eric schmidt said friday(P:Sci/tech,Q:others)	desktop	apple	schmidt	mac
Inflation dozen nation sharing euro slowed initially estimated september company reduced price lure customer store offsetting record energy cost.(P:World,Q:others)	store	inflation	price	lure

answer "*why p and not q?*" questions. We observed this behavior in binary text classification. For instance, we found that the model uses the same reasoning for both questions (see Table 7). We attribute this observation to the fact that "*why p and not q?*" cites the causal difference between p and *not-q*, i.e., consisting of a cause of p and the absence of a corresponding event in the history of q. We also found that explaining "*Why p and not q?*" is not the same as explaining "*Why q and not p?.*" In the case of sentiment classification, we found that these two questions provide different answers, and it is consistent with the work of [12]. To validate our observations in sentiment classification, we focus now on the overlap between "*why p and not q?*" and "*why q and not p?.*" We use the IMDB dataset and calculate the overlap between the attributes (minimum subset of the attributes required to flip the prediction) of "*why p and not q?*" and "*why q and not p?.*" The ratio of similarly was zero, which means the explanations are entirely different.

4.5 Deep Learned Features

In Fig. 6, we apply PCA over the sentence vectors learned via our proposed method. The centroid loss forces the network to learn meaningful representations for the embedding layer. We can see that our current model struggles with negative sentiment reviews according to the number of points (yellow color) appearing in the positive sentiment cluster.

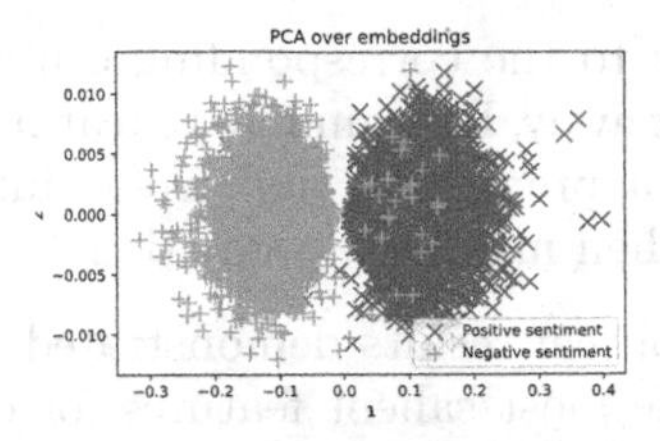

Fig. 6. The distribution of deeply learned features under the centroid loss on IMDB dataset.

Table 7. *Why p and not q?* Contrastive explanation is the same as *why p?* explanations in binary sentiment classification.

Text	Highlight
The story is enjoyable and easy to follow this could have been easily messed up but the actors and director do a great job of keeping it together the actors themselves are fantastic displaying wonderful character and doing a terrific job gotta find a copy somewhere (P:positive,Q:negative)	fantastic
This performance that should elevate the film to a platform where it a place on the best ever lists of courtroom dramas however despite its apparent obscurity sergeant still remains a taut and compelling examination like a book that you just can't put down highly recommended (P:positive,Q:negative)	recommended
Imagined in my mind what i saw on screen was slightly different however it wasn't enough to make me dislike the mini series i recommend this for anyone who has read the novel you will not be disappointed if you have 8 out of 10 stars (P:positive,Q:negative)	8
Provide someone to at well one must do something beside during this film the movie is being sold on vhs now by people on e bay spare yourself the expense and the waste of time a comedy without a laugh a musical without a memorable song or dance (P:negative,Q:positive)	waste

4.6 Discussion

Intrinsic Models. We have introduced an approach for constructing interpretable neural models. We have shown that introducing additional constraints to the learning objective does not sacrifice performance, and it also provides faithful explanations to the black-box predictions. The centroid vectors are used as a proxy to explain the predictions. We found that discriminative features (words) tend to get closer to the corresponding centroid vector and irrelevant features tend to get further away. Discriminative features are the words employed by the network to make a prediction, and irrelevant features are the tokens ignored by the classifier when making a prediction.

Centroid Loss. The empirical results demonstrated the usefulness of the centroid vectors in finding the most salient features for every input. The centroid loss does not require complex recombination of the training samples. Our approach targets the learning objective of the intra-class using term 2, which is very beneficial to discriminative feature learning. We have also shown that our contrastive explanations are helpful when there are fine-grained differences.

5 Conclusion

We have proposed an intrinsic neural model capable of explaining its predictions faithfully. Our network architecture relies on a centroid loss to learn centroid vectors. These centroid vectors are then used to provide two types of explanations: (i) non-contrastive explanations and (ii) contrastive explanations. Our feature attribution method provides a better faithful explanation than Shapley's approximation based on three datasets using three metrics. We have also proposed additional metrics to evaluate contrastive explanations. Our contrastive explanation method can provide additional insights to non-contrastive explanation, resulting in a better understanding of the neural model predictions. We have also shown that interpretability does not affect the predictive accuracy of the neural network. In future work, we would like to study the use of our intrinsic neural model with different tasks in the NLP domain and extend our current solution for producing counterfactual explanations.

Acknowledgements. We would like to acknowledge the support of the Alberta Machine Intelligence Institute (Amii), and the Natural Sciences and Engineering Research Council of Canada (NSERC).

References

1. Bashier, H.K., Kim, M.Y., Goebel, R.: Disk-CSV: distilling interpretable semantic knowledge with a class semantic vector. In: Proceedings of the 16th Conference of the EACL Main Volume, pp. 3021–3030 (2021)
2. Bastings, J., Aziz, W., Titov, I.: Interpretable neural predictions with differentiable binary variables. In: Proceedings of ACL, pp. 2963–2977 (2019)
3. Chen, J., Song, L., Wainwright, M.J., Jordan, M.I.: L-shapley and c-shapley: efficient model interpretation for structured data. In: ICLR 2019 (2018)
4. DeYoung, J., et al.: Eraser: a benchmark to evaluate rationalized NLP models. In: Proceedings of the 58th ACL, pp. 4443–4458 (2020)
5. Einhorn, H.J., Hogarth, R.M.: Judging probable cause. Psychol. Bull. **99**(1), 3 (1986)

6. Hendricks, L.A., Hu, R., Darrell, T., Akata, Z.: Generating counterfactual explanations with natural language. In: ICML Workshop on Human Interpretability in Machine Learning, pp. 95–98 (2018)
7. Hilton, D.J.: Conversational processes and causal explanation. Psychol. Bull. **107**(1), 65 (1990)
8. Ismail, A.A., Corrada Bravo, H., Feizi, S.: Improving deep learning interpretability by saliency guided training. Adv. Neural Inf. Process. Syst. **34**, 26726–26739 (2021)
9. Jacovi, A., Swayamdipta, S., Ravfogel, S., Elazar, Y., Choi, Y., Goldberg, Y.: Contrastive explanations for model interpretability. arXiv preprint arXiv:2103.01378 (2021)
10. Kingma, D.P., Ba, J.L.: Adam: a method for stochastic optimization. Cornell University Library. arXiv preprint arXiv:1412.6980 (2017)
11. Koura, A.: An approach to why-questions. Synthese **74**(2), 191–206 (1988)
12. Lipton, P.: Contrastive explanation. Royal Inst. Philos. Suppl. **27**, 247–266 (1990)
13. Lipton, Z.C.: The mythos of model interpretability: in machine learning, the concept of interpretability is both important and slippery. Queue **16**(3), 31–57 (2018)
14. Maas, A.L., Daly, R.E., Pham, P.T., Huang, D., Ng, A.Y., Potts, C.: Learning word vectors for sentiment analysis. In: Proceedings of ACL, pp. 142–150. ACL (2011)
15. McGill, A.L., Klein, J.G.: Contrastive and counterfactual reasoning in causal judgment. J. Pers. Soc. Psychol. **64**(6), 897 (1993)
16. Miller, T.: Explanation in artificial intelligence: insights from the social sciences. Artif. Intell. **267**, 1–38 (2019)
17. Nguyen, D.: Comparing automatic and human evaluation of local explanations for text classification. In: Proceedings of the 2018 Conference of the North American Chapter of the Association for Computational Linguistics: Human Language Technologies, Volume 1 (Long Papers), pp. 1069–1078 (2018)
18. Rathi, S.: Generating counterfactual and contrastive explanations using shap. arXiv preprint arXiv:1906.09293 (2019)
19. Ribeiro, M.T., Singh, S., Guestrin, C.: Why should i trust you?: explaining the predictions of any classifier. In: Proceedings of the 22nd ACM SIGKDD International Conference on Knowledge Discovery and Data Mining, pp. 1135–1144. ACM (2016)
20. Rudin, C.: Please stop explaining black box models for high stakes decisions. In: 32nd Conference on Neural Information Processing Systems (NIPS 2018), Workshop on Critiquing and Correcting Trends in Machine Learning (2018)
21. Shrikumar, A., Greenside, P., Kundaje, A.: Learning important features through propagating activation differences. In: Proceedings of the 34th International Conference on Machine Learning-Volume 70, pp. 3145–3153. JMLR. org (2017)
22. Sundararajan, M., Taly, A., Yan, Q.: Axiomatic attribution for deep networks. In: Proceedings of International Conference on Machine Learning (ICML), pp. 3319–3328 (2017)
23. Tang, D., Qin, B., Liu, T.: Document modeling with gated recurrent neural network for sentiment classification. In: Proceedings of the 2015 Conference on EMNLP, pp. 1422–1432 (2015)
24. Vaswani, A., et al.: Attention is all you need. In: Advances in Neural Information Processing Systems, pp. 5998–6008 (2017)
25. van der Waa, J., Robeer, M., van Diggelen, J., Brinkhuis, M., Neerincx, M.: Contrastive explanations with local foil trees. arXiv preprint arXiv:1806.07470 (2018)

26. Wachter, S., Mittelstadt, B., Russell, C.: Counterfactual explanations without opening the black box: automated decisions and the GDPR. Harv. JL Tech. **31**, 841 (2017)
27. Wang, Y., Huang, H., Rudin, C., Shaposhnik, Y.: understanding how dimension reduction tools work: an empirical approach to deciphering t-SNE, UMAP, TriMap, and PaCMAP for data visualization. J. Mach. Learn. Res. **22**(201), 1–73 (2021)
28. Woodward, J.: Making Things Happen: A Theory of Causal Explanation. Oxford University Press, Oxford (2005)
29. Yang, L., Kenny, E., Ng, T.L.J., Yang, Y., Smyth, B., Dong, R.: Generating plausible counterfactual explanations for deep transformers in financial text classification. In: Proceedings of the 28th International Conference on Computational Linguistics, pp. 6150–6160 (2020)
30. Zhang, X., Zhao, J., LeCun, Y.: Character-level convolutional networks for text classification. In: Advances in Neural Information Processing Systems, pp. 649–657 (2015)

Explaining Predictions by Characteristic Rules

Amr Alkhatib[(✉)], Henrik Boström, and Michalis Vazirgiannis

KTH Royal Institute of Technology, Electrum 229, 164 40 Kista, Stockholm, Sweden
{amak2,bostromh,mvaz}@kth.se

Abstract. Characteristic rules have been advocated for their ability to improve interpretability over discriminative rules within the area of rule learning. However, the former type of rule has not yet been used by techniques for explaining predictions. A novel explanation technique, called CEGA (Characteristic Explanatory General Association rules), is proposed, which employs association rule mining to aggregate multiple explanations generated by any standard local explanation technique into a set of characteristic rules. An empirical investigation is presented, in which CEGA is compared to two state-of-the-art methods, Anchors and GLocalX, for producing local and aggregated explanations in the form of discriminative rules. The results suggest that the proposed approach provides a better trade-off between fidelity and complexity compared to the two state-of-the-art approaches; CEGA and Anchors significantly outperform GLocalX with respect to fidelity, while CEGA and GLocalX significantly outperform Anchors with respect to the number of generated rules. The effect of changing the format of the explanations of CEGA to discriminative rules and using LIME and SHAP as local explanation techniques instead of Anchors are also investigated. The results show that the characteristic explanatory rules still compete favorably with rules in the standard discriminative format. The results also indicate that using CEGA in combination with either SHAP or Anchors consistently leads to a higher fidelity compared to using LIME as the local explanation technique.

Keywords: Explainable machine learning · Rule mining

1 Introduction

Machine learning algorithms that reach state-of-the-art performance, in domains such as medicine, biology, and chemistry [6], often produce non-transparent (black-box) models. However, understanding the rationale behind predictions is, in many domains, a prerequisite for the users placing trust in the models. This can be achieved by employing algorithms that produce interpretable (white-box) models, such as decision trees and generalized linear models, but in many cases, with a substantial loss of predictive performance [30]. As a consequence, explainable machine learning has gained significant attention as a means to obtain transparency without sacrificing performance [7].

M.-R. Amini et al. (Eds.): ECML PKDD 2022, LNAI 13713, pp. 389–403, 2023.
https://doi.org/10.1007/978-3-031-26387-3_24

Explanation techniques are either model-agnostic, i.e., they allow for explaining any underlying black-box model [16], or model-specific, i.e., they exploit properties of the underlying black-box model to generate the explanations, targeting e.g., random forests [13,14] or deep neural networks [19,20]. Along another dimension, the explanation techniques can be divided into local and global approaches [7]. Local approaches, such as LIME [1] and SHAP [2] aim to explain a single prediction of a black-box model [1,2], while global approaches, such as SP-LIME [1] and MAME [9], aim to provide an understanding of how the model behaves in general [7]. Many explanation techniques produce explanations in the form of (additive) feature importance scores. Such explanations do however not directly lend themselves to verification, due to lack of an established, general and objective way of concluding whether the scores (or rankings imposed by them) are correct or not [29]. In contrast, some techniques, such as Anchors [3], produce explanations in the form of rules. Since each such rule can be used to make predictions, the agreement (fidelity) of the rule to the underlying black-box model can be measured, e.g., using independent test instances. However, in some cases, the produced rules may be very specific [8], which in practice precludes verification due to the limited coverage of the rules.

Setzu et al. proposed GLocalX [10] as a solution to the above problem, by which multiple local explanations (rules) are merged into fewer, more general (global) rules. Similar to all local explanation techniques that output rules, GLocalX produces discriminative rules, which, according to Fürnkranz [17], provide a quick and easy way to distinguish one class from the others using a small number of features. Characteristic rules, on the other hand, capture properties that are common for objects belonging to a specific class, rather than highlighting the differences (only) between objects belonging to different classes. See Fig. 1 for an illustration of discriminative and characteristic rules. Although most rule learning approaches have targeted the former type of rule, also a few approaches for characteristic rule learning have been developed [17,21]. As stated in [17, p.871]:

> "Even though discriminative rules are easier to comprehend in the syntactic sense, we argue that characteristic rules are often more interpretable than discriminative rules."

Characteristic rules could hence be a potentially useful format also for explanations. Until now, however, the use of characteristic rules for explaining predictions have, to the best of our knowledge, not been considered.

The main contributions of this study are:

- a novel technique for explaining predictions by characteristic rules, called **C**haracteristic **E**xplanatory **G**eneral **A**ssociation rules (CEGA)
- an empirical investigation comparing the fidelity and complexity of explanations in the form of characteristic rules, as produced by CEGA, and explanations in the form of discriminative rules, as generated by Anchors and GLocalX

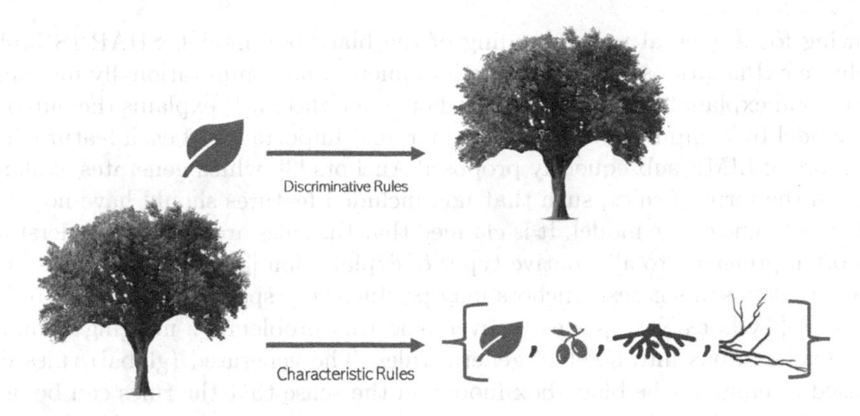

Fig. 1. Discriminative rules distinguish one class from the others using a few features, while characteristic rules learn all the features that characterize the class.

- an ablation study of CEGA in which the format of the rules are changed from characteristic to standard discriminative rules and in which three different options for the local explanation technique are considered; LIME, SHAP and Anchors.

In the next section, we briefly discuss related work. In Sect. 3, we describe the proposed association rule mining approach for extracting general (and hence verifiable) explanations in the form of characteristic rules. In Sect. 4, we present and discuss the results of a large-scale empirical investigation, in which the proposed method is compared to the baselines. Finally, in Sect. 5, we summarize the main findings and outline directions for future work.

2 Related Work

In this section, we start out by discussing model-agnostic explanation techniques that work with any algorithm for generating the black-box model. We then continue with model specific explanation techniques that are explicitly designed for certain underlying models. We also cover some related work on rule learning.

Explainable machine learning is a research area that has gained quite some attention recently, in particular following the introduction of the popular local explanation technique LIME (Local Interpretable Model-agnostic Explanations) [1]. In addition to the original algorithm, a variant called SP (Submodular Pick) LIME was proposed, which produces more general explanations. LIME trains a white-box model using perturbed instances, which are weighted by proximity to the instance of interest. The trained white-box model acts as a faithful explainer locally but not globally. On the other hand, SP-LIME uses a set of representative instances to generate explanations with high coverage, allowing for a general understanding of the black-box model. SHAP (SHapley Additive exPlanations) [2] is another prominent, and computationally more efficient, local explanation technique based on game theory. It explains the outcome of a model by computing the additive marginal importance of each feature. The inventors of LIME subsequently proposed Anchors [3], which generates explanations in the form of rules, such that non-included features should have no effect on the outcome of the model. It is claimed that the rules are easier to understand and often preferred to alternative types of explanation [3]. However, it has been shown that in some cases, Anchors may produce very specific (long) rules of low fidelity [8]. GLocalX [10] tries to overcome this problem by merging (similar) local explanations into a set of general rules. The generated (global) rules can be used to emulate the black-box model, in the sense that the rules can be used for making predictions.

In addition to the model-agnostic approaches, algorithm-specific methods for explaining predictions have also been proposed. An example of such a technique was proposed in [13], which generates explanations of random forest predictions by mining association rules from itemsets corresponding to paths in the trees. Similar approaches were proposed in [25] and [14], however not primarily for explaining predictions, but for extracting (global) rule sets that approximate the underlying tree ensembles. The prominent RuleFit algorithm [15] also uses the prediction paths of the tree ensemble to form a global model, but instead of a rule set, a linear model is fit, using the Lasso to limit the complexity of the model. Again, the goal is not to explain predictions but to obtain a transparent and accurate predictive model.

Highly related to techniques for explaining predictions by rules is the area of (inductive) rule learning [31]. Early work in this area was presented in [32], in which two different types of rule, discriminative and characteristic rules, were described. As described in [18], rules of the former type are on the form `IF <conditions> THEN <class>`, where `<conditions>` is a set of conditions, each referring to a feature, an operator and a value, and `<class>` is a class label. They can differentiate one class from the other classes "quickly" according to Kliegr et al. [18]. On the other hand, the characteristic rules have the opposite direction `IF <class> THEN <conditions>`, which capture properties that are common for instances that belong to a certain class. The area of discriminative rule learning (or propositional rule induction) has been investigated to quite some extent, with contributions in the form of rule learning algorithms, such as CN2 [22], RIPPER [23], and PRIM [24]. In contrast, characteristic rule learning approaches are less common, one example being CaracteriX [21], which generates characteristic rules that can be applied to spatial and multi-relational databases.

3 From Local Explanations to General Characteristic Rules

In this section, we describe the proposed method to explain predictions by characteristic rules. We first outline details of the method and then show how we can easily change the format of the generated rules from characteristic to discriminative, allowing for directly investigating the effect of the rule format without changing any other parts of the algorithm.

3.1 Explanation Mining and Rules Selection

The proposed method (CEGA) takes as input a set of objects (instances), a black-box model, a local explainer, and minimum thresholds for the confidence and support of generated rules. The output of CEGA is a set of characteristic rules. The method is summarized in Algorithm 1. Below we will discuss the different components in detail.

Algorithm 1. CEGA

1: **Input**: A set of objects X, a local explainer L, a black-box model B; confidence threshold c; minimum rule support s; class labels $\{C_1, \ldots, C_k\}$
2:
3: $E \leftarrow$ *Generate-Explanation-Itemsets*(X, L, B)
4:
5: $R \leftarrow$ *Find-Association-Rules*(E, s, c)
6:
7: $R \leftarrow$ *Filter-Rules*$(R, \{C_1, \ldots, C_k\})$
8:
9: **Output**: General characteristic rules (R)

The function *Generate-Explanation-Itemsets* uses the black-box model to obtain predictions (class labels) for the given objects and then employs the local explainer to obtain explanations for those predictions. The explanations are subsequently transformed into a list of itemsets, called *explanation itemsets*, where each itemset represents a local (instance-specific) explanation. Since CEGA is agnostic to the local explanation technique, the user may quite freely select the desired local explanation technique. However, different explanation techniques provide explanations in different formats, e.g., feature scores or rules. Consequently, each explanation format requires to be properly processed to produce the itemsets as required by CEGA. In case the local explanation technique outputs rules, such as Anchors, one processing option, which is the one that we will consider in this study, is to form an itemset from all the conditions of a rule together with the predicted class label. In the case that the local explanation technique outputs feature scores, such as LIME and SHAP, we will consider the option of forming itemsets by including the top-ranked features (in favor of the predicted class) together with the predicted class label. We binarized the categorical features using one-hot encoding, and the binarized feature names that appear in an explanation are added to the itemset (e.g., feature_A_V, where feature_A is the name and V is the value), and the same preprocessing step is applied to the data at prediction time. Continuous features were discretized using equal-width binning, using five bins. As a side note, we extract more than one explanatory itemset per example, one per class, where each feature is added to an itemset with the class label it supports. Therefore, we obtain the same number of itemsets per class, which is particularly useful for highly imbalanced datasets to avoid extracting explanations solely for the dominant class. One final preprocessing step, for binary classification tasks only, is to add the binarized categorical feature to the itemset of the opposite class if the feature value is zero, which is motivated by the explainer associating the absence of one feature with the predicted class. Consequently, the presence of the same feature will favor the other class. This preprocessing step may hence result in that multiple values of a categorical feature may appear in characteristic explanations for the same class. In this study, we will consider binary classification problems only; therefore, the ranking of the features is with respect to the predicted class label. To avoid including features with negligible effect on a prediction, a threshold will be employed to filter out low-ranking features, which also reduces the computational cost when later performing association rule mining on the corresponding itemsets. Now the explanation itemsets can be used as input to the association rule mining algorithm.

An association rule mining algorithm is applied to the explanation itemsets within the *Find-Association-Rules* function, using the specified confidence and support thresholds. It should be noted that CEGA is again agnostic to which algorithm is used for conducting this step.

Using the *Filter-Rules* function, CEGA aims to find a set of rules that characterize each class. The characteristic rules can be obtained from the set of discovered association rules by keeping only the rules for which a single class label appear in the antecedent (condition part) and some set of conditions in the consequent. Moreover, to simplify the resulting set of rules, they are processed in the following way. If there are two rules that have the same class label, while the conditions of one of them is a subset of the conditions of the other rule, and if the former rule has higher confidence, then the latter rule is removed. This last step is motivated by that it is likely to reduce the complexity and the number of conditions of the resulting rules while increasing the coverage of the resulting rule set. As discussed in [17], this may however not necessarily lead to that the resulting rules are indeed more interpretable.

3.2 Discriminative vs. Characteristic Rules

The confidence of a discriminative rule, where the antecedent consists of conditions and the consequent consists of a class label, is the probability that an object belongs to the class given that it satisfies the conditions [4] (Eq. 1).

The confidence of a characteristic rule is conversely the probability that an object satisfies the conditions given that it belongs to the class (Eq. 2). It should be noted that by just a slight modification of the function *Filter-Rules*, concerning whether the conditions and the label should appear in the antecedent and consequent (or vice versa), CEGA will produce discriminative instead of characteristic rules.

$$confidence(Conditions \rightarrow Class) = P(Class|Conditions) = \frac{P(Conditions, Class)}{P(Conditions)} \quad (1)$$

$$confidence(Class \rightarrow Conditions) = P(Conditions|Class) = \frac{P(Conditions, Class)}{P(Class)} \quad (2)$$

The confidence in (Eq. 2) for a characteristic rule measures how characteristic is a set of features given a class. For example, a rule with 100% confidence means that the items (features) are shared between all class objects.

4 Empirical Evaluation

In this section, we compare CEGA to Anchors and GLocalX, two state-of-the-art approaches for explaining predictions with discriminative rules. The methods will be compared with respect to fidelity (agreement with the explained black-box model) and complexity (number of rules). We then conduct an ablation study, in which we change the rule format of CEGA from characteristic to discriminative rules (as explained in the previous section) and also the technique used to generate the local explanations.

4.1 Experimental Setup

The quality of the generated explanations will be estimated by their fidelity to the underlying black-box model. The fidelity can however be measured in different ways. Guidotti et al. [12] define fidelity as the ability of an interpretable model to replicate the output of the black box model, as measured using some predictive performance metric, such as accuracy, F1 score, and recall. We follow this definition and provide a comprehensive set of fidelity measurements, including accuracy, F1-score, and area under the ROC curve (AUC), to report how well an explanation technique approximates the black-box model.

This means that for each object from which we have obtained a black-box prediction, we need to form also a prediction from the rules that are used to explain the prediction. We employ a naive Bayes strategy with Laplace correction [5] to obtain predictions for both discriminative and characteristic rules.

The experiments have been conducted using 20 public datasets[1]. Each dataset is split into training, development, and testing sets, where the black-box model is trained on the first, the explanations (rules) are generated using the black-box predictions on the second set, and finally, the quality of the produced rules is evaluated on the third set. All datasets concern binary classification tasks except for Compas[2], which originally contains three classes (Low, Medium, and High), which were reduced into two by merging Low and Medium into one class. The black-box models are generated by XGBoost [11]. Some of its hyperparameters (learning rate, number of estimators and the regularization parameter lambda) were tuned by grid search using 5-fold cross-validation on the training set. CEGA requires two additional hyperparameters; support and confidence. The former was set to 10 in the case of LIME and SHAP and set to 4 in the case of Anchors, while the confidence of the characteristic rules was tuned based on the fidelity as measured on the development set. In the second experiment, where CEGA was used to produce also discriminative rules, the confidence was set to 100% for this rule type to keep the number of generated rules within a reasonable limit.

In the first experiment, Anchors, GLocalX and CEGA are compared, where the two latter use Anchors at the local explanation technique. Anchors is used with the default hyperparameters, and the confidence threshold has been set to 0.9. GLocalX is tested as well using the default values except for the alpha hyperparameter, in which values between 50 and 95 have been tested with step 5, and the best result is reported. In a second experiment, we consider using also SHAP and LIME as local explanation techniques for CEGA. As described in Sect. 3.1, the output of these techniques require some preprocessing to turn them into itemsets; the threshold to exclude low-ranking features has here been set to 0.01. In all experiments, CEGA will employ the Apriori algorithm [4] for association rule mining[3].

4.2 Baseline Experiments

The results from comparing the characteristic rules of CEGA to GLocalX and Anchors are summarized in Table 1. It can be seen that CEGA performs on par with Anchors with respect to fidelity, while producing much fewer unique rules. At the same time, CEGA generally obtains a higher fidelity than GLocalX.

To test the null hypothesis that there is no difference in fidelity, as measured by AUC, between Anchors, GLocalX, and CEGA, the Friedman test [26] followed by pairwise posthoc Nemenyi tests [27] are employed. The first test rejects the

[1] All the datasets were obtained from https://www.openml.org except Adult, German credit, and Compas.

[2] https://github.com/propublica/compas-analysis.

[3] CEGA is available at: https://github.com/amrmalkhatib/CEGA.

Table 1. Fidelity, number of rules and coverage for Anchors, GLocalX and CEGA.

Dataset	Anchors					GLocalX					CEGA				
	Acc.	AUC	F1	#Rules*	Cov.**	Acc	AUC	F1	#Rules	Cov.	Acc	AUC	F1	#Rules	Cov.
ada	0.90	**0.92**	0.86	120	1.0	0.84	0.86	0.8	11	0.33	0.87	0.91	0.81	5	1.0
Adult***	0.91	**0.93**	0.85	378	1.0	0.85	0.87	0.81	23	0.3	0.88	0.91	0.82	7	1.0
Bank Marketing	0.93	0.67	0.48	88	1.0	0.94	0.61	0.65	57	0.39	0.91	**0.80**	0.65	11	1.0
Blood Transfusion	0.91	**0.97**	0.91	15	1.0	0.95	**0.97**	0.9	3	0.19	0.93	0.92	0.86	4	0.92
BNG breast-w	0.98	0.99	0.98	96	1.0	0.67	0.5	0.4	2	0.0	0.97	**1.00**	0.97	8	0.99
BNG tic-tac-toe	0.84	**0.87**	0.79	842	0.998	0.28	0.33	0.28	158	0.33	0.75	0.77	0.72	4	1.0
Compas	0.89	**0.83**	0.73	168	1.0	0.9	0.76	0.74	78	0.72	0.86	0.82	0.67	10	1.0
Churn	0.89	0.67	0.47	143	1.0	0.88	**0.71**	0.69	80	0.34	0.81	0.66	0.54	20	1.0
German Credit****	0.78	0.75	0.58	39	1.0	0.43	0.59	0.43	25	0.43	0.79	**0.79**	0.72	6	1.0
Internet Advertisements	0.91	0.78	0.78	191	1.0	0.87	0.5	0.46	16	0.57	0.91	**0.80**	0.77	4	1.0
Jungle Chess 2pcs	1.0	**1.0**	1.0	24	1.0	0.45	0.5	0.31	1	0.51	0.89	0.91	0.89	6	1.0
kc1	0.89	**0.86**	0.72	111	1.0	0.87	0.79	0.68	27	1.0	0.85	**0.86**	0.72	4	0.99
mc1	0.96	0.69	0.53	9	0.82	0.97	0.7	0.53	2	0.03	0.97	**0.95**	0.54	19	0.99
mofn-3-7-10	0.89	0.96	0.79	43	1.0	0.64	0.75	0.62	16	0.7	1.0	**1.0**	1.0	7	0.99
Mushroom	0.99	**1.0**	0.99	89	1.0	0.83	0.84	0.83	3	0.63	0.88	0.93	0.88	3	1.0
Phishing Websites	0.93	**0.98**	0.93	107	1.0	0.42	0.5	0.3	3	0.53	0.92	0.96	0.92	5	1.0
socmob	0.95	0.94	0.91	36	1.0	0.18	0.5	0.15	3	0.04	0.96	**0.99**	0.93	7	1.0
Spambase	0.94	**0.98**	0.93	193	0.998	0.6	0.5	0.37	4	0.41	0.90	0.97	0.90	8	0.93
Steel Plates Fault	0.81	**1.00**	0.81	71	1.0	0.72	0.77	0.72	4	0.6	0.77	0.82	0.77	4	1.0
Telco Customer Churn	0.89	0.93	0.84	167	1.0	0.82	0.86	0.79	24	0.37	0.89	**0.95**	0.85	8	1.0
Average rank*****	1.5	1.55	1.55	2.95	1.425	2.525	2.775	2.65	1.6	2.925	1.975	1.675	1.8	1.45	1.65

* The number of rules

** The coverage is the percentage of instances in the dataset that are covered by at least one rule

*** Available at: https://archive.ics.uci.edu/ml/datasets/Adult

**** Available at: https://archive.ics.uci.edu/ml/datasets/statlog+(german+credit+data)

***** The average rank shows which method is better on average, with 1 being the best and 3 the worst result

null hypothesis and the result of the post hoc tests are summarized in Fig. 2. It can be observed that GLocalX is significantly outperformed with respect to fidelity by both Anchors and CEGA. In Fig. 3, the result of the same test is shown when comparing the number of unique rules produced by the three approaches. This time, Anchors is significantly outperformed (produces more rules) than the two other approaches. In summary, it can be concluded that with CEGA, we can keep a level of fidelity that is not significantly different from using Anchors, while reducing the number of rules significantly; from hundreds of rules to just a handful.

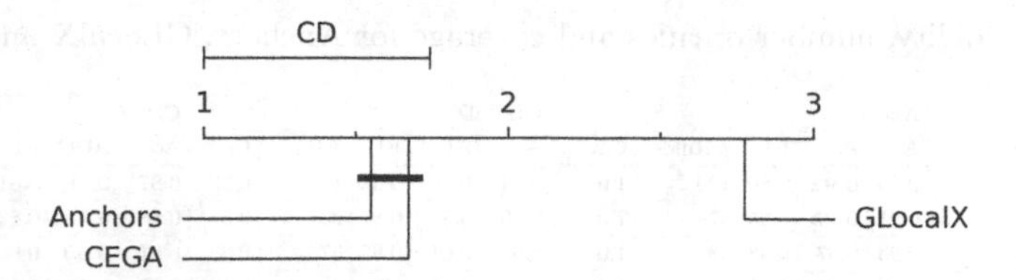

Fig. 2. Comparing average ranks with respect to fidelity measured by AUC (lower rank is better) of Anchors, GLocalX, and CEGA, where the critical difference (CD) represents the largest difference that is not statistically significant.

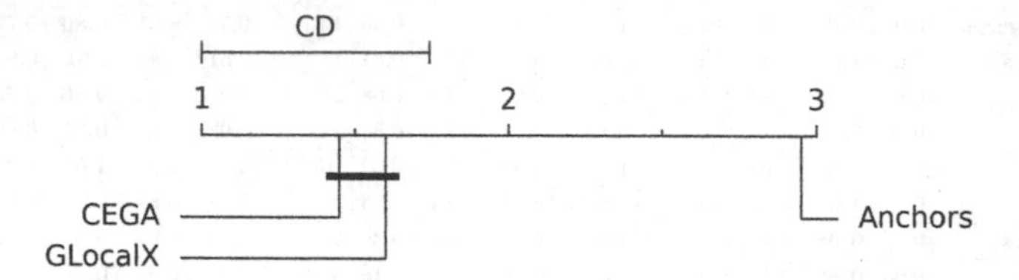

Fig. 3. Comparing average ranks of Anchors, GLocalX, and CEGA with respect to the number of rules

4.3 Comparing Discriminative and Characteristic Rules

Since CEGA allows for changing the rule format with just a minor modification to the algorithm, as described above, we can easily investigate whether CEGA's improved trade-off between fidelity and complexity (measured by the number of rules) compared to Anchors and GLocalX is due to the use of characteristic rules, or potentially comes from the other components, e.g., the use of association rule mining to generalize the rules. We will in this section present results from such a comparison. We will first however illustrate the differences between characteristic and discriminative explanatory rules. To this end, samples of characteristic and discriminative rules are shown in Table 2 and Table 3, respectively. In this study, we are not drawing any conclusions on their relative interpretability, as judged by human users, but will focus mainly on fidelity and complexity. In Table 4, the fidelity, measured by accuracy, AUC and F1 score, of CEGA producing characteristic and discriminative rules is presented, together with the number of rules and the coverage. Due to its computational efficiency, SHAP is here used as a local explanation technique for both versions of CEGA (instead of Anchors in the baseline experiments Sect. 4.2). One can observe that the fidelity of the characteristic rules tends to be higher than for the discriminative rules, while the characteristic rules are clearly fewer; in other words, very similar to what we could observe when comparing (standard) CEGA to Anchors and GLocalX. Since we only compare two methods in this experiment, the Wilcoxon signed-rank test [28] is employed instead of the Friedman test to investigate if the observed difference in fidelity, as measured by AUC, is significant or not. It turns out that the null hypothesis may be rejected at the 0.05 level also here.

Table 2. The top 11 characteristic rules output by CEGA for the German Credit dataset when using SHAP as the local explainer.

Label	Conditions	Confidence
good	credit_history = critical/other existing credit	1.0
bad	credit_history = no credits/all paid	1.0
good	personal_status = male single	0.993
good	purpose = used car	0.993
bad	property_magnitude = no known property	0.993
bad	purpose = education	0.993
good	other_parties = guarantor	0.987
good	other_payment_plans = none	0.987
bad	purpose = new car	0.987
good	checking_status = no checking	0.98
bad	credit_history = all paid	0.953

Table 3. The top 11 discriminative rules output by CEGA for the German Credit dataset when using SHAP as the local explainer.

Conditions	Label	Confidence
other_payment_plans = none & checking_status = no checking	good	1.0
other_payment_plans = none & purpose = radio/tv	good	1.0
checking_status = no checking & purpose = radio/tv	good	1.0
other_payment_plans = none & housing = own	good	1.0
checking_status = no checking & housing = own	good	1.0
property_magnitude = real estate & checking_status = no checking	good	1.0
property_magnitude = real estate & other_payment_plans = none	good	1.0
housing = own & purpose = radio/tv	good	1.0
property_magnitude = real estate & housing = own	good	1.0
other_payment_plans = none & savings status no known savings	good	1.0
property_magnitude = life insurance & credit_history = existing paid	bad	1.0

Table 4. Fidelity, number of rules and coverage of characteristic and discriminative rules output by the standard and the modified versions of CEGA using SHAP.

Dataset	Characteristic rules					Discriminative rules				
	Acc.	AUC	F1	#Rules	Cov.	Acc.	AUC	F1	#Rules	Cov.
ada	0.88	**0.92**	0.83	24	1.00	0.85	0.90	0.79	187	0.99
Adult	0.89	**0.90**	0.81	7	0.90	0.77	0.79	0.65	177	1.00
Bank Marketing	0.90	**0.86**	0.75	2	0.89	0.88	0.73	0.55	54	1.00
Blood Transfusion	0.91	**0.93**	0.83	2	0.86	0.90	0.85	0.78	11	1.00
BNG breast-w	0.96	0.99	0.95	7	0.98	0.95	0.99	0.95	73	1.00
BNG tic-tac-toe	0.75	**0.82**	0.72	10	0.99	0.71	0.78	0.69	13	1.00
Compas	0.88	0.75	0.67	53	0.92	0.75	**0.80**	0.63	511	0.76
Churn	0.89	**0.78**	0.69	4	0.91	0.87	0.77	0.65	224	0.99
German Credit	0.80	**0.82**	0.69	29	1.00	0.78	0.81	0.62	110	1.00
Internet Advertisements	0.96	0.92	0.91	6	1.00	0.95	**0.94**	0.89	24	1.00
Jungle Chess 2pcs	0.70	**0.68**	0.70	3	0.89	0.64	0.67	0.63	73	1.00
kc1	0.79	0.70	0.62	6	0.98	0.82	**0.72**	0.61	184	0.97
mc1	0.99	**0.84**	0.73	6	1.00	0.99	0.49	0.50	74	1.00
mofn-3-7-10	1.0	1.0	1.0	8	1.00	1.0	1.0	1.0	9	1.00
Mushroom	0.97	1.0	0.97	9	1.00	0.95	1.0	0.95	30	1.00
Phishing Websites	0.92	**0.98**	0.92	29	1.00	0.72	0.80	0.71	162	1.00
socmob	0.88	0.89	0.83	5	0.76	0.89	**0.90**	0.83	11	0.84
Spambase	0.91	**0.95**	0.91	15	0.81	0.84	0.91	0.84	88	1.00
Steel Plates Fault	1.0	**1.0**	1.0	9	1.00	0.63	0.66	0.60	136	1.00
Telco Customer Churn	0.85	**0.92**	0.79	4	0.74	0.85	0.91	0.78	51	0.99
Average rank	1.175	1.275	1.075	1	1.675	1.825	1.725	1.925	2	1.325

4.4 Comparing Local Explanation Techniques

Since CEGA is agnostic to the local explanation technique, it is possible to compare the rules obtained using different local explanation techniques, which allows for evaluating the local explainer on a specific task and selecting the one with the highest fidelity. Accordingly, we compare the fidelity, the number of rules and coverage of CEGA when used together with Anchors, SHAP, and LIME, respectively. The result from this comparison is shown in Table 5. The results indicate that the characteristic rules produced by CEGA when using Anchors and SHAP tend to provide higher fidelity than when using LIME for most datasets.

Table 5. Fidelity, number of rules and coverage of characteristic rules output by CEGA using Anchors, SHAP, or LIME

Dataset	Anchors					SHAP					LIME				
	Acc.	AUC	F1	#Rules	Cov.	Acc	AUC	F1	#Rules	Cov.	Acc	AUC	F1	#Rules	Cov.
ada	0.87	0.91	0.81	5	1.00	0.88	**0.92**	0.83	24	1.00	0.80	0.82	0.71	27	1.00
Adult	0.88	**0.91**	0.82	7	1.00	0.89	0.90	0.81	7	0.90	0.84	0.88	0.76	47	1.00
Bank Marketing	0.91	0.80	0.65	11	1.00	0.90	**0.86**	0.75	2	0.89	0.87	0.68	0.55	36	1.00
Blood Transfusion	0.93	0.92	0.86	4	0.92	0.91	**0.93**	0.83	2	0.86	0.88	0.92	0.72	5	0.8
BNG breast-w	0.97	1.00	0.97	8	0.99	0.96	0.99	0.95	7	0.98	0.98	1.00	0.97	10	0.93
BNG tic-tac-toe	0.75	0.77	0.72	4	1.00	0.75	**0.82**	0.72	10	0.99	0.73	0.76	0.69	18	1.00
Compas	0.86	**0.82**	0.67	10	1.00	0.88	0.75	0.67	53	0.92	0.87	0.74	0.64	13	1.00
Churn	0.81	0.66	0.54	20	1.00	0.89	**0.78**	0.69	4	0.91	0.87	0.66	0.58	10	1.00
German Credit	0.79	0.79	0.72	6	1.00	0.80	**0.82**	0.69	29	1.00	0.80	0.80	0.71	35	1.00
Internet Advertisements	0.91	0.80	0.77	4	1.00	0.96	**0.92**	0.91	6	1.00	0.81	0.73	0.67	72	0.81
Jungle Chess 2pcs	0.89	0.91	0.89	6	1.00	0.70	0.68	0.70	3	0.89	0.95	**0.96**	0.95	6	1.00
kc1	0.85	**0.86**	0.72	4	0.99	0.79	0.70	0.62	6	0.98	0.82	0.76	0.59	26	1.00
mc1	0.97	**0.95**	0.54	19	0.99	0.99	0.84	0.73	6	1.00	0.97	0.92	0.50	13	1.00
mofn-3-7-10	1.00	1.00	1.00	7	0.99	1.00	1.00	1.00	8	1.00	1.00	1.00	1.00	21	1.00
Mushroom	0.88	0.93	0.88	3	1.00	0.97	**1.00**	0.97	9	1.00	0.93	0.97	0.93	24	1.00
Phishing Websites	0.92	0.96	0.92	5	1.00	0.92	**0.98**	0.92	29	1.00	0.77	0.86	0.76	28	1.00
socmob	0.96	**0.99**	0.93	7	1.00	0.88	0.89	0.83	5	0.76	0.91	0.97	0.86	55	1.00
Spambase	0.90	**0.97**	0.90	8	0.93	0.91	0.95	0.91	15	0.81	0.89	0.95	0.88	34	1.00
Steel Plates Fault	0.77	0.82	0.77	4	1.00	1.00	**1.00**	1.00	9	1.00	0.92	0.98	0.92	12	1.00
Telco Customer Churn	0.89	**0.95**	0.85	8	1.00	0.85	0.92	0.79	4	0.74	0.83	0.89	0.75	27	1.00
Average rank	1.975	1.825	1.75	1.55	1.775	1.675	1.775	1.725	1.675	2.425	2.35	2.4	2.525	2.775	1.8

5 Concluding Remarks

We have proposed CEGA, a method to aggregate local explanations into general characteristic explanatory rules. CEGA is agnostic to the local explanation technique and can work with local explanations in the form of rules or feature scores, given that they are properly converted to itemsets. We have presented results from a large-scale empirical evaluation, comparing CEGA to Anchors and GLocalX, with respect to three fidelity metrics (accuracy, AUC and F1 score), number of rules and coverage. CEGA was observed to significantly outperform GLocalX with respect to fidelity and Anchors with respect to the number of generated rules. We also investigated changing the rule format of CEGA to discriminative rules and using SHAP, LIME, or Anchors as the local explanation technique. The main conclusion of the former investigation is that indeed the rule format has a significant effect; the characteristic rules result in higher fidelity and fewer rules compared to when using discriminative rules. The results from the second follow-up investigation showed that CEGA combined with either SHAP or Anchors generally result in rules with higher fidelity compared to when using LIME as the local explanation technique.

One direction for future work would be to complement the functionally-grounded (objective) evaluation of the quality of the explanations with user-grounded evaluations, e.g., asking users to indicate whether they actually can follow the logic behind the predictions or solve some tasks using the output rules.

Another direction for future work concerns investigating additional ways of forming itemsets from which general (characteristic or discriminative) rules are formed. This could for example include combining the output of multiple local explanation techniques. Another important direction concerns quantifying the uncertainty of the generated rules, capturing to what extent one can expect a rule to agree with the output of a black-box model. The investigated applications may also include datasets beyond regular tabular data, e.g., text documents and images.

Acknowledgement. This work was partially supported by the Wallenberg AI, Autonomous Systems and Software Program (WASP) funded by the Knut and Alice Wallenberg Foundation. HB was partly funded by the Swedish Foundation for Strategic Research (CDA, grant no. BD15-0006).

References

1. Ribeiro, M., Singh, S., Guestrin, C.: "Why should I trust you?": explaining the predictions of any classifier. In: Proceedings of the 22nd ACM SIGKDD International Conference on Knowledge Discovery and Data Mining, San Francisco, CA, USA, 13–17 August 2016, pp. 1135–1144 (2016)
2. Lundberg, S., Lee, S.: A unified approach to interpreting model predictions. Adv. Neural. Inf. Process. Syst. **30**, 4765–4774 (2017)
3. Ribeiro, M., Singh, S., Guestrin, C.: Anchors: high-precision model-agnostic explanations. In: AAAI Conference on Artificial Intelligence (AAAI) (2018)
4. Agrawal, R., Srikant, R.: Fast algorithms for mining association rules in large databases. In: Proceedings of the 20th International Conference on Very Large Data Bases, pp. 487–499 (1994)
5. Kohavi, R., Becker, B., Sommerfield, D.: Improving simple Bayes. In: European Conference On Machine Learning (1997)
6. Linardatos, P., Papastefanopoulos, V., Kotsiantis, S.: Explainable AI: a review of machine learning interpretability methods. Entropy **23** (2021)
7. Molnar, C.: Interpretable machine learning: a guide for making black box models explainable (2019)
8. Delaunay, J., Galárraga, L., Largouët, C.: Improving anchor-based explanations. In: CIKM 2020–29th ACM International Conference on Information and Knowledge Management, pp. 3269–3272, October 2020
9. Natesan Ramamurthy, K., Vinzamuri, B., Zhang, Y., Dhurandhar, A.: Model agnostic multilevel explanations. Adv. Neural. Inf. Process. Syst. **33**, 5968–5979 (2020)
10. Setzu, M., Guidotti, R., Monreale, A., Turini, F., Pedreschi, D., Giannotti, F.: GLocalX - from local to global explanations of black box AI models. Artif. Intell. **294**, 103457 (2021)
11. Chen, T., Guestrin, C.: XGBoost: a scalable tree boosting system (2016,8)
12. Guidotti, R., Monreale, A., Ruggieri, S., Turini, F., Giannotti, F., Pedreschi, D.: A survey of methods for explaining black box models. ACM Comput. Surv. **51** (2018)
13. Boström, H., Gurung, R., Lindgren, T., Johansson, U.: Explaining random forest predictions with association rules. Arch. Data Sci. Ser. A (Online First). **5**, A05, 20 S. online (2018)

14. Bénard, C., Biau, G., Veiga, S., Scornet, E.: Interpretable random forests via rule extraction. In: Proceedings of the 24th International Conference on Artificial Intelligence and Statistics, vol. 130, pp. 937–945, 13 April 2021
15. Friedman, J., Popescu, B.: Predictive learning via rule ensembles. Ann. Appl. Stat. **2**, 916–954 (2008)
16. Ribeiro, M., Singh, S., Guestrin, C.: Model-agnostic interpretability of machine learning. In: ICML Workshop on Human Interpretability in Machine Learning (WHI) (2016)
17. Fürnkranz, J., Kliegr, T., Paulheim, H.: On cognitive preferences and the plausibility of rule-based models. Mach. Learn. **109**(4), 853–898 (2020). https://doi.org/10.1007/s10994-019-05856-5
18. Kliegr, T., Bahník, Š, Fürnkranz, J.: A review of possible effects of cognitive biases on interpretation of rule-based machine learning models. Artif. Intell. **295**, 103458 (2021)
19. Shrikumar, A., Greenside, P., Kundaje, A.: Learning important features through propagating activation differences. Proceedings of the 34th International Conference on Machine Learning, vol. 70, pp. 3145–3153, 6 August 2017
20. Wang, Z., et al.: CNN explainer: learning convolutional neural networks with interactive visualization. IEEE Trans. Visual. Comput. Graph. (TVCG) (2020)
21. Turmeaux, T., Salleb, A., Vrain, C., Cassard, D.: Learning characteristic rules relying on quantified paths. In: Knowledge Discovery in Databases: PKDD 2003, 7th European Conference On Principles and Practice of Knowledge Discovery in Databases, Cavtat-Dubrovnik, Croatia, 22–26 September 2003, Proceedings, vol. 2838, pp. 471–482 (2003)
22. Clark, P., Boswell, R.: Rule induction with CN2: some recent improvements. In: Kodratoff, Y. (ed.) EWSL 1991. LNCS, vol. 482, pp. 151–163. Springer, Heidelberg (1991). https://doi.org/10.1007/BFb0017011
23. Cohen, W.: Fast effective rule induction. In: Proceedings of the Twelfth International Conference on Machine Learning, pp. 115–123 (1995)
24. Friedman, J., Fisher, N.: Bump hunting in high-dimensional data. Stat. Comput. **9**, 123–143, Apr 1999. https://doi.org/10.1023/A:1008894516817
25. Deng, H.: Interpreting tree ensembles with in Trees. Int. J. Data Sci. Anal. **7**(4), 277–287 (2018). https://doi.org/10.1007/s41060-018-0144-8
26. Friedman, M.: A correction: the use of ranks to avoid the assumption of normality implicit in the analysis of variance. J. Am. Stat. Assoc. **34**, 109–109 (1939)
27. Nemenyi, P.: Distribution-Free Multiple Comparisons. Princeton University (1963)
28. Wilcoxon, F.: Individual comparisons by ranking methods. Biometrics Bull. **1**(6), 80–83 (1945). http://www.Jstor.Org/stable/3001968
29. Slack, D., Hilgard, S., Jia, E., Singh, S., Lakkaraju, H.: Fooling LIME and SHAP: adversarial attacks on post hoc explanation methods. In: AAAI/ACM Conference on AI, Ethics, and Society (AIES) (2020)
30. Loyola-González, O.: Black-box vs. white-box: understanding their advantages and weaknesses from a practical point of view. IEEE Access **7**, 154096–154113 (2019)
31. Fürnkranz, J., Gamberger, D., Lavrac, N.: Foundations of Rule Learning. Springer, Heidelberg (2012). https://doi.org/10.1007/978-3-540-75197-7
32. Michalski, R.: A theory and methodology of inductive learning. Artif. Intell. **20**, 111–161 (1983). https://www.sciencedirect.com/science/article/pii/0004370283900164

Session-Based Recommendation Along with the Session Style of Explanation

Panagiotis Symeonidis[1](✉), Lidija Kirjackaja[2], and Markus Zanker[3,4]

[1] University of the Aegean, Mytilene, Greece
psymeon@aegean.gr
[2] Vilnius Gediminas Technical University, Vilnius, Lithuania
[3] Free University of Bolzano, Bolzano, Italy
mzanker@unibz.it
[4] University of Klagenfurt, Klagenfurt, Austria
Markus.Zanker@aau.at

Abstract. Explainability of recommendation algorithms is becoming an important characteristic in GDPR Europe. There are algorithms that try to provide explanations over graphs along with recommendations, but without focusing in user session information. In this paper, we study the problem of news recommendations using a heterogeneous graph and try to infer similarities between entities (i.e., sessions, articles, etc.) for predicting the next user click inside a user session. Moreover, we exploit meta paths to reveal semantic context about the session-article interactions and provide more accurate article recommendations along with robust explanations. We have experimentally compared our method against state-of-the-art algorithms on three real-life datasets. Our method outperforms its competitors in both accuracy and explainability. Finally, we have run a user study to measure the users' satisfaction over different explanation styles and to find which explanations really help users to make more accurate decisions.

Keywords: Session-based recommendations · Explanations

1 Introduction

Session-based neural network recommendation algorithms [3,5], are like "black boxes", failing to adequately explain their suggestions. Researchers [11,20] tried to provide explanations by extracting them from Knowledge Graphs (KGs). However, since KGs are based on triplets (i.e., entity1, relation, entity2), they require additional effort from the domain expert to capture more sophisticated relationships. In contrast, these semantically-rich relationships are defined easily with meta paths in Heterogeneous Information Networks (HINs). To the best of our knowledge, there is no related work, that provides session-based explanations for recommendations.

In this paper, we provide both session-based and explainable recommendations in the news domain: (i) by exploiting user sessions to infer time-aware similarities among users, articles or sessions and (ii) by combining meta paths

M.-R. Amini et al. (Eds.): ECML PKDD 2022, LNAI 13713, pp. 404–420, 2023.
https://doi.org/10.1007/978-3-031-26387-3_25

extracted from a HIN to better explain our predicted article recommendations using hybrid meta path-based explanations. In particular, we consider a 5-partite HIN (users, sessions, articles, categories, and locations), as shown in Fig. 1, which is able to capture the long-term relations (i.e., user-category, user-article, etc.), and the short-term user preferences (user-session, session-article). Category refers to the kind of a news story (e.g., politics, sports, etc.), whereas location refers to the region that a news article is written about. As shown Fig. 1, we create a new type of node, called *session* (S) node, which is associated with the co-click of two or more articles from a user in a specific short period of time (i.e., user session).

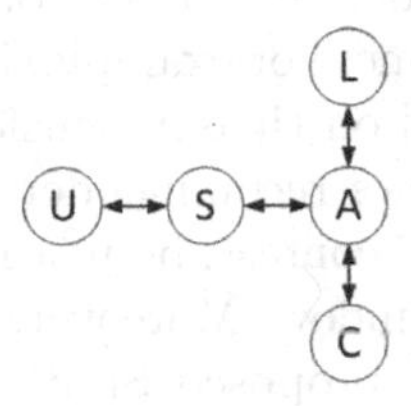

Fig. 1. Network schema for news

By exploiting meta paths in our news network structure, we can better infer similarities between entities. A meta path is a sequence of relations among different entity types, which reveals a different semantic context about the users' interactions. We combine the inferred similarity of different meta paths, to provide better recommendations and more enhanced hybrid meta path-based explanations. For example, let us assume that a user starts his reading session and clicks on article A107. Then, by combining meta path AUA (Article, User, Article) with meta path ACA (Article, Category, Article), we can provide a hybrid meta path-based explanation as follows: *"We recommend you article A250, because: (i) it was viewed by 5 users that also viewed A107 (AUA), and (ii) it belongs to the same category (i.e., politics) like your current article (ACA)"*.

The contributions of this paper are summarized as follows: **(i)** We exploit user sessions to reveal the last moment intentions of users. **(ii)** We first introduce a new style of explaining a recommendation based on user sessions, denoted as the "session style of explanation". **(iii)** We compared our method with 7 other state-of-the-art algorithms on 3 data sets. As will be shown experimentally later, our proposed method achieves an improvement in terms of accuracy and explainability.**(iv)** We performed a user study to identify users' favourite explanation style, and those explanation styles that help users to make more accurate decisions.

The rest of this paper is organized as follows. Section 2 summarizes the related work. Section 3 presents our proposed method. Section 4 describes our three recommendation strategies, whereas Sect. 5 describes our hybrid meta path explanations. Experimental results are presented in Sect. 6. Finally, Sect. 7 concludes the paper.

2 Related Work

There are three generally known fundamental resources used for explaining recommendations [12,16,18] such as *users*, *items* and item *features*, which can be classified into the following explanation styles: (i) *User* Style, which provides explanations based on similar users, (ii) *Item* style, which is based on choices made by users on similar items, and (iii) *Feature* Style, which explains the recommendation based on item features (content). Please notice that any combination of the aforementioned style could result to a multi-dimensional hybrid explanation style.

In news recommendation domain, related work [2] has shown that a way to increase accuracy is to consider the context of the user and the fact that the user's preference evolves over time. For example, Epure et al. [2] considered three levels of reading interests based on time dimension: short-, medium-, and long-term. Moreover, Ludmann's news recommender system [10], denoted as Ody4, won the CLEF NewsREEL 2017 contest, by just recommending the most clicked articles of a 12-h sliding time window. Moreover, in the area of similarity search in graphs, Jeh and Widom [8] proposed SimRank. SimRank is based on the idea that two nodes are similar if they are referenced by similar nodes. Another path-based measure is HeteSim [13], which measures the relatedness of objects in heterogeneous graphs.

Finally, session-based recommendations have been modelled with Recurrent Neural Networks (RNNs). Hidasi et al. [5] presented a recommender system based on Gated Recurrent Unit (GRU), which learns when and how much to update the hidden state of their GRU4REC model. However, a more recent study [7] has shown that a simple k-nearest neighbor (kNN) scheme adapted for session-based recommendations often outperforms the GRU4REC model. Several adjustments were proposed during last years that improve the performances of the initial GRU4REC model [4]. Recently, Xu et al. [21] proposed a graph contextualized self-attention model (GC-SAN), which utilizes both graph neural network and self-attention mechanism, for session-based recommendation.

3 Our Proposed Method

In this Section, we describe our method, which is inspired by the work of Sun et al. [22], who proposed the novel idea of measuring similarities between network objects by analysing meta-paths, through which objects are connected. In a heterogeneous graph, two objects can be connected through different paths as defined in the following:

Definition 1. ***Information Network*** *[22]. An information network is defined as a directed graph $\mathcal{G} = (\mathcal{V}, \mathcal{E})$ with an object type mapping function $\phi : \mathcal{V} \rightarrow \mathcal{Q}$ and a link type mapping function $\psi : \mathcal{E} \rightarrow \mathcal{R}$, where each object $v \in \mathcal{V}$ belongs to one particular object type $\phi(v) \in \mathcal{Q}$, and each link $e \in \mathcal{E}$ belongs to a particular relation $\psi(e) \in \mathcal{R}$.*

For example, in news media, two articles can be connected through the path "article-category-article" (content-based similarity), "article-session-article" (session based similarity), "article-session-user-session-article" (collaborative filtering similarity). Using different paths, different similarities are observed. These paths are called *meta paths* and are formally defined as follows:

Definition 2. *Meta Path* *[22]. A meta path $\mathcal{P}$ is a path defined on the graph of network schema $T_G = (\mathcal{Q}, \mathcal{R})$, and is denoted in the form of $Q_1 \xrightarrow{R_1} Q_2 \xrightarrow{R_2} \ldots \xrightarrow{R_l} Q_{l+1}$, which defines a composite relation $R = R_1 \circ R_2 \circ \ldots \circ R_l$ between type Q_1 and Q_{l+1}, where $\circ$ denotes the composition operator on relations.*

There are various meta paths that can be built on the USACL network, which is shown in Fig. 1. If we start from the article type of the node, we can build the following paths: ACA, ASA, ALA, ASUSA, etc. If we start from sessions, we can build: SAS, SACAS, SALAS to infer similarities, and consequently SASA, SACASA, and SALASA to recommend articles from similar sessions. For example, for the SACASA case, the finally recommended article follows the path: S $\xrightarrow{\text{contains}}$ A $\xrightarrow{\text{belongs to}}$ C $\xrightarrow{\text{is assigned to}}$ A $\xrightarrow{\text{is read within}}$ S $\xrightarrow{\text{contains}}$ A.

A well-known similarity measure that is able to capture the semantics of similarity among network objects by using meta paths is **PathSim** [22].

3.1 Meta Path-Based Similarity

Definition 3. *PathSim: A Single Meta path-based similarity measure* [22]. *Given a symmetric meta path P, PathSim between two objects of the same type x and y is*

$$s(x, y) = \frac{2 * |p_{x \rightsquigarrow y} : p_{x \rightsquigarrow y} \in P|}{|p_{x \rightsquigarrow x} : p_{x \rightsquigarrow x} \in P| + |p_{y \rightsquigarrow y} : p_{y \rightsquigarrow y} \in P|}, \tag{1}$$

where $p_{x \rightsquigarrow y}$ is a path instance between x and y, $p_{x \rightsquigarrow x}$ is that between x and x, and $p_{y \rightsquigarrow y}$ is that between y and y.

PathSim captures the nodes' *visibility* in the network, bringing the nodes that share similar visibility closer, in contrast to SimRank and P-PageRank (RWR), that favour more popular items in the network. However, for the news recommendation domain, we should not penalise popular articles because of the nature of this domain. Thus, to overcome this characteristic of PathSim, we propose its variation, denoted as *xPathSim*, that transforms it into a simple transition probability measure as follows:

Definition 4. *xPathSim: A variation of PathSim similarity measure adapted for the news recommendation domain.* *Given a symmetric meta path P, xPathSim between two objects of the same type x and y is*

$$s(x, y) = \frac{|p_{x \rightsquigarrow y} : p_{x \rightsquigarrow y} \in P, x \neq y|}{\sum\limits_{z \in \mathcal{V}: \phi(z) = \phi(x)} |p_{x \rightsquigarrow z} : p_{x \rightsquigarrow z} \in P|}, \tag{2}$$

where $p_{x \rightsquigarrow y}$ *is a path instance between* x *and* y*, and* $p_{x \rightsquigarrow z}$ *is that between* x *and* z*, where* z *is any object of the same type as* x*. The range of* $s(x, y)$ *is [0,1].*

3.2 Recommendation List Creation by Considering One Meta Path

In this Section, we will describe how we produce a recommendation list and how we rank the articles within the list, by using the similarities inferred from a single meta path. Let us use the following running example.

Consider the SAC (session, article, categories) heterogeneous graph of Fig. 2. We assume that session S5 is the current session of an anonymous user, for whom we have to provide article recommendations. As shown, the anonymous user has already viewed 2 articles (A4 and A5) during his current session and we have to select articles from the remaining set of three (A1, A2, and A3) and decide on their order alongside their meta path explanations.

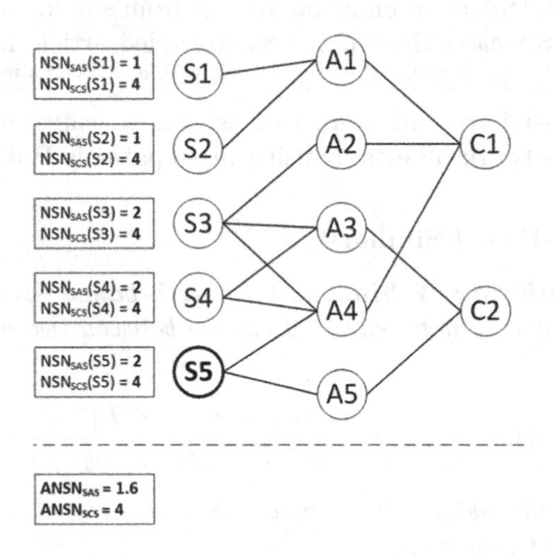

Fig. 2. Similar session nodes using meta-paths SAS and SCS

In our running example, there is no simple way of providing a ranked list of recommendations along with explanations. For example, it is clear that sessions holding exactly the same articles as the user's currently viewed ones (SAS) are more strongly connected to the current session, than the sessions holding just articles from the same category (SCS). Also, there are much less connections of meta-path SAS, than of meta-paths SCS. In order to find the most frequent item inside the connections of a given meta path, we have to compute its occurrences inside the meta path.

Definition 5. ***Number of Votes for an item inside the Neighborhood of Similar Nodes.*** *Given a node* $x_{current}$*, a symmetric meta path* P*, and a user-defined threshold* T*, we want to find the most frequent items inside the*

neighborhood of similar nodes $x \in V$, that belong to the same entity as $x_{current}$ ($\phi(x) = \phi(x_{current})$) and satisfy the constraint $xPathSim_P(x_{current}, x) > T$. This number of appearances of each item, which is connected with a similar node of $x_{current}$ is denoted as number of votes *for article a following the meta path P ($NV_P(a)$).*

In our running example, by using Eq. 2, we can find the nodes that are similar to node S5. Please notice that $\phi(x_{current}) = S$ means that we consider a session node (i.e., S5) as a starting node in our modeling. Then, inside this neighborhood of similar nodes we compute the majority vote of the articles that appear inside the neighborhood of similar nodes and are connected with them inside the meta path. The results for both meta paths for our running example are summarized in Table 1. As shown in Table 1, by using the meta path SAS, when recommending top-3 articles the ranking of the recommendation list is {A3, A2, A1}, whereas by using the meta path SCS the articles' ranking is {A1, A3, A2} or {A3, A1, A2}.

Table 1. Number of votes for an item.

Article	NV_{SAS}	NV_{SCS}
A1	0	2
A2	1	1
A3	2	2

Thus, if we use the meta path SAS to infer similarity between sessions, then we recommend to the anonymous user in session S5 of our running example article A3 alongside with the following explanation: "*We recommend you article A3, because it was viewed in 2 other sessions (S3 and S4) together with article A4, which appears in your current session.*

3.3 Recommendation List Creation by Using Multiple Meta Paths

Next, we will provide hybrid explanations along with recommendations by inferring similarity between two nodes based on multiple meta paths. To measure how strongly connected a starting node is with the destination node based on a specific meta path, we need to compute their average connectivity of this node type over the whole graph. Thus, for a given meta path, we define the *Average Number of Similar Nodes (ANSN)* in the whole graph as follows:

Definition 6. ***Average Number of Similar Nodes in a Graph.*** *Given a target entity $E \in \mathcal{Q}$ of graph nodes $\mathcal{V}$, and meta-path P, the average number of similar nodes (ANSN) for this meta path, is computed based on the number of similar nodes $NSN_P(x)$ of the target node x following the meta path P, as follows:*

$$ANSN_P = \frac{1}{|x : x \in V, \phi(x) = E|} \cdot \sum_{x \in V, \phi(x)=E} NSN_P(x) \quad (3)$$

In our running example, as it is shown in Fig. 2, $NSN_{SAS}(S5)$ is equal to 2, as the session S5 is connected with two sessions S3 and S4 via article A4, and $NSN_{SCS}(S5)$ is equal to 4, as it is possible to reach any other session node of the network from the session S5, following the path SCS (e.g., S5-A4-C1-A1-S1, S5-A4-C1-A1-S2, S5-A4-C1-A2-S3, S5-A5-C2-A3-S4). Consequently, the average numbers of similar nodes in the network for each meta path are calculated by taking all session nodes of the network into consideration, i.e., $ANSN_{SAS} = \frac{1+1+2+2+2}{5} = 1.6$, $ANSN_{SCS} = \frac{4+4+4+4+4}{5}$=4. Thus, each session is *on average* with 1.6 other sessions via articles, and with 4 other sessions via article categories connected.

Next, we propose to measure the *Candidate Item Relevance* in order to rank candidate items based on multiple meta paths. It is the ratio of the number of connections through which a candidate item is connected to the target node of a meta path in relation to the average number of connections of this type of meta path in the whole graph. Therefore, we need to normalize the contribution of each meta path based on its overall presence inside the network. Formally, it is defined as follows:

Definition 7. ***Candidate Item Relevance:*** *Given a node $x_{current}$, a list of meta-paths $\mathcal{P} = [P_1, P_2, ..., P_n]$ that form a combined similarity meta path measure, and the majority vote of the articles that appear in each meta path P_i, which we denote as* number of votes *for article a in meta path P_i ($NV_{P_i}(a)$), the candidate item relevance (CIR) to $x_{current}$ is computed as follows:*

$$CIR(a) = \sum_{i=1}^{|\mathcal{P}|} \frac{1}{ANSN_{P_i}} \cdot NV_{P_i}(a) \quad (4)$$

In our running example, as shown in Table 2, A3 has the biggest relevance score and will be recommended first. To compute its value, we sum the number of votes for each meta path that appears, as shown in Table 2. $CIR(A3) = \frac{2}{1.6} + \frac{2}{4} = 1.75$.

Table 2. Candidate item relevance

Article	CIR
A1	$\frac{0}{1.6} + \frac{2}{4} = 0.5$
A2	$\frac{1}{1.6} + \frac{1}{4} = 0.875$
A3	$\frac{2}{1.6} + \frac{2}{4} = 1.75$

The explanation to support the recommendation would be the following: "*We recommend you article A3, because (i) it was viewed in 2 other sessions (S3 and*

S4) together with article A4, which appears in your current session and (ii) it belongs in category C2 together with article A5, which appears in your current session.

4 Recommendation Strategies and Single Explanations

In this Section, in addition to the 3 resources that can be used in an explanation, as described in the Related Work Section, we provide a fourth type of explanation style, denoted as *Session* Style, which is based on the concept of co-occurrence of items inside a user session. Moreover, we have three recommendation strategies (i.e., **item-based**, **user-based** and **session-based**) depending on the node type used to infer similarities among entities (i.e., user-user, item-item, or session-session similarities).

First, to provide item-based (IB) recommendations, we find articles similar to the one that the user has just clicked, then we rank them based on their similarity to the target article and recommend a top-N list. Alternatively, for predicting the next article, the whole user's last session can be considered to identify his short-term intentions and to recommend those articles that best match user's presumed preferences. This way, IB similarities can be identified by running xPathSim on ASA, AUA, ACA and ALA meta-paths. For example, for meta path ASA, in Fig. 3, we show how articles are ranked inside a top-5 recommendation list.

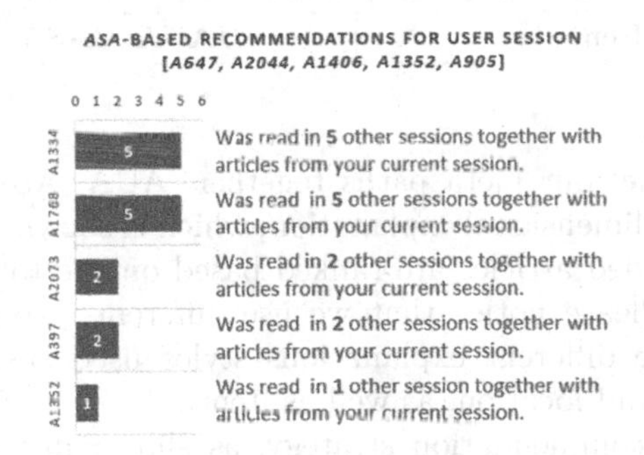

Fig. 3. ASA-based top-5 article recommendations

As it is shown in Fig. 3, we provide recommendations that reflect the overall user interest inside a session along with explanations (i.e., why an article is recommended to a user). This is a very intuitive and user-friendly way of explaining the relation between the recommended article and the article that used for explaining it. e.g., *These two articles are read together in 10 different user sessions.*

Next, to provide user-based (UB) recommendations, we identify users similar to the target user, and recommend the most frequent articles inside the neighborhood of similar users. Finally, in order to provide session-based (SB) recommendations, we firstly identify the sessions that are similar to the current user's session and recommend the most frequent articles inside the neighborhood of the similar sessions.

5 Hybrid Meta Path-Based Explanation

In this Section, we combine several meta paths to provide hybrid (multi-dimensional) explanations alongside with recommendations. For the IB recommendation strategy, as shown in the second column of Table 3, four meta-paths are used to support a recommended item: (AUA - two articles are read by the same user; ASA - two articles are read within the same session; ACA - two articles belong to the same category; ALA - two articles belong to the same location).

Table 3. Meta-paths used in three recommendation strategies.

Explanation styles	IB	UB	SB
User	A**U**A	–	–
Session	A**S**A	–	–
Feature (category)	A**C**A	U**C**UA	S**C**SA
Feature (location)	A**L**A	U**L**UA	S**L**SA
Item	–	U**A**UA	S**A**SA

By combining the four meta paths together [AUA, ASA, ACA, ALA], we result to a hybrid 4-dimensional explanation, which is shown in Fig. 4. As shown, the top-5 recommended articles are ranked based on the total number of meta path connections. Please notice that we use different colours for horizontally stacked bars for the different explanations styles used (user, session, feature signifying category and location, as well as item).

Next, for UB recommendation strategy, as shown in the third column of Table 3, three meta paths are used for inferring similarities between users: (UAU, UCU and ULU). Please note that the relations used when providing recommendations of articles using meta path UAU are known as collaborative filtering: U $\xrightarrow{\text{has read}}$ A $\xrightarrow{\text{was read by}}$ U $\xrightarrow{\text{has \textit{recently} read}}$ A. An example of a hybrid 3-D meta path-based explanation for the UB strategy is as follows: *"Article A1334 is recommended to you because it was recently (i)read by 4 users who read similar articles as you, (ii) read by 3 users who are interested in the same article categories as you, and (iii) read by 3 users who are interested in the same locations as you"*.

Finally, for the SB recommendation strategy, as shown in the last column of Table 3, there are three meta-paths for retrieving similarities of sessions and to

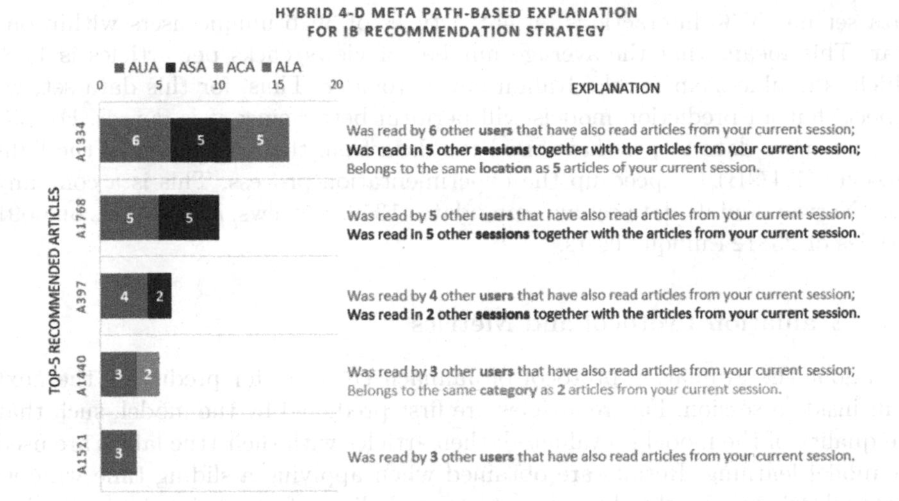

Fig. 4. Hybrid 4-D meta path explanation for IB recommendation strategy.

support a recommended item: (SAS, SCS, and SLS). An example of a hybrid 3-D meta path-based explanation for SB strategy is: *"Article A1522 is recommended to you because it was recently (i) read in 4 sessions together with the articles from your current session, (ii) read in 2 sessions that had articles of the same category as those in your current session, and (iii) read in 1 session that had articles from the same location as those in your current session"*.

6 Experimental Evaluation

In this Section, we compare our xPathSim method with state-of-the art algorithms on three real-life data sets. We also present a user study, which shows how users perceive the meta-path based explanations in comparison with other styles of explanation.

6.1 Real-Life Datasets

In this Section, we will describe the basic characteristics and statistics of three real-life data sets, which are obtained from an Italian, a German, and a Norwegian language news providers. Please notice that the first two are operating in the region of Alto Adige in Italy. For the Italian news provider, the data set accommodates 14367 interactions/events/views on 2081 articles of 10421 unique users in one year (i.e. from 1st April 2016 to 30th March 2017). This means that the average number of views/clicks per article is 6.9, which will affect the prediction of all models due to sparsity. The interactions of each session are logged with the following information: the user session's identifier, the interaction's time stamp and duration, the article's textual content. For the German news provider, the data set has 5536 interactions on 468 articles of 3626 unique users within one year. This means that the average number of views/clicks per articles is 11.8, which is double than for the Italian news provider. Thus, for this data set, we expect that all prediction models will perform better since it is denser. For the Adressa news data set, we have used the data from the first 2 days of the light version[1] (1.4 GB) to speed up the experimentation process. This is a company from Norway and its data set accommodates 1356987 views/interactions on 6091 articles of 238124 unique users.

6.2 Evaluation Protocol and Metrics

We adopt the evaluation protocol of Jannach et al. [6] for predicting the next item inside a session. Future articles are first predicted by the model, such that the quality of the model is evaluated; then articles with their true labels are used for model learning. Results are obtained when applying a sliding time window protocol, where we split the data into several slices of equal size. In particular, for the first two data sets, since we have data over a year, we split the time into 12 months ($N_t = 12$), such that we can aggregate the precision results for each different time period t_p. For the Adressa data set, we split data in 7 time periods. For all data set, we set the sliding time window size $w = 2$, since we got the best results. Finally, we evaluate the precision when we recommend top-2 articles for each next item prediction inside a session. Moreover, we use the *nDCG* metric which is proposed by Song et al. [14] for the prequential evaluation of RSs. nDCG is a fine-grained version of precision, that takes also into account the position of a correct item in the list. Moreover, to measure how explainable is a recommended item i for a target user u, we adapt the proposed metric used in [17] to the session-based recommendation task. Thus, for each different data source d (user, item, feature, session), which is used inside a hybrid explanation, we build a user-item explainability matrix $E_d(u, i)$, which holds the information of how frequent is a data source d based on the previous interactions of the target user u with items, item features, sessions and other users. Then, for a user u that receives a recommendation list L, the *explain coverage* for the justification list J is defined as follows:

$$Explain\ \ coverage(u, J) = \frac{\sum\limits_{\forall (i, f_d^i) \in J} \min\{f_d^i, E_d(u, i)\}}{\sum\limits_{\forall d \in D} E_d(u, i)}, \tag{5}$$

where the pair (i, f_d^i) denotes the overall frequency f of data source d with respect to item i inside the justification list J. Moreover, $E_d(u, i)$ is the frequency of i in the explainability matrix $E_d(u, i)$ of u over data source d. *Explain coverage* takes values in the range [0, 1], whereas values closer to 1 correspond to better coverage. The explain coverage captures how frequent is in the user's profile,

[1] http://reclab.idi.ntnu.no/dataset/.

each of the data source (user, item, feature, and session), which is used to build a hybrid explanation for each recommended item.

6.3 xPathSim Sensitivity Analysis

In this Section, we compare the performance of xPathSim with the three recommendation strategies (IB, UB, and SB) in terms of precision and explain coverage. Due to space limitations, we present only results that concern the Italian news provider data set, but we have verified similar results for both other two data sets.

As shown in Table 4, when we use a combination of multiple meta paths in all three recommendation strategies (UB, IB, SB) the precision is always significantly better than any other single meta path-based recommendation. SB is the most effective recommendation strategy by attaining 34.4% precision, followed by 26.9% and 23.3% of UB and IB strategies, respectively. The reported results are tested for the difference of means between SB with UB and IB strategies, respectively, and found statistically significant based on one-sided t-test at the 0.05 level. The reason is that SB strategy is more appropriate for the news recommendation domain, since user sessions are usually anonymous. Thus, the other two strategies (IB, UB) do not have enough data to build a prediction model, since most of users have only a small number of sessions i.e., 1.23, 1.17 and 1.03 sessions per user in Italian, German and the Norwegian dataset, respectively. Please notice that both meta paths (ACA and ALA) are very ineffective in terms of precision, since all articles that belong to the same category or location get the same prediction score, and thus the selection of top-N recommended items is always random.

Table 4. xPathSim's sensitivity for different rec. strategies and meta paths.

Metric	Expl. type	Item-based rec. (IB)				User-based rec. (UB)			Session-based rec. (SB)		
		AUA	ASA	ACA	ALA	UAUA	UCUA	ULUA	SASA	SCSA	SLSA
Precision	Single metapath	0.199	0.197	0.075	0.083	0.072	0.190	0.232	0.197	0.223	0.272
	Many metapaths	0.233				0.269			**0.344**		
Expl. cov.	Single metapath	0.194	0.182	0.124	0.101	0.197	0.107	0.116	0.182	0.114	0.128
	Many metapaths	0.201				0.230			**0.264**		

As far as explain coverage is concerned, the meta paths AUA, ASA UAUA and SASA, which are related to collaborative filtering, achieve the best results, whereas the meta paths ACA, ALA, UCUA, ULUA, SCSA, and SLSA, which are related to content-based filtering, have much lower effectiveness. The reason is, as also described in Sect. 3.1, that collaborative filtering meta paths can better capture the user preferences and can be found more frequently in the data set. Please also notice that again the SB recommendation strategy with multiple meta paths provides the best results when compared to UB and IB, and their difference of means between SB and the rest strategies is statistically significant based on one-sided t-test at the 0.05 level.

6.4 Comparison with Other Methods

In this Section, we compare xPathSim, against graph-based (i.e., RWR [19], SimRank [8], PathCount [15], PathSim [22]), session-based (i.e., GRU4REC [5], Session-knn [7]), and collaborative with content-based filtering (CAT-TPM) algorithm [9], in terms of precision, nDCG and explain coverage. The parameters we used to evaluate the performance of the comparison partners are similar to those reported in the original papers and for our data sets were hypertuned so as to get the best results for these methods among all three recommendation strategies. As it is shown in Table 5, our method $\text{xPathSim}_{\text{SB}}$ outperforms all comparison partners in both metrics and all the three data sets, because it combines multiple meta paths with the session-based recommendation strategy, which makes it both accurate and explainable. The reported results are tested for the difference of means between the best method and each of the rest comparison partners and found statistically significant based on one-sided t-test at the 0.05 level. Please notice that precision, nDCG and explain coverage of all methods is higher for the German than the Italian news data set. The reason is that news articles on the German platform are viewed twice more than the Italian ones. For the Adressa data set, we have the smallest precision, nDCG, and explain coverage because it is very sparse data set.

Table 5. Methods' comparison.

Model	Italian provider			German provider			Adressa data set		
	Prec.	nDCG	Expl. cov.	Prec.	nDCG	Expl. cov.	Prec.	nDCG	Expl. cov.
$\text{xPathSim}_{\text{SB}}$	**0.344**	**0.218**	**0.264**	**0.431**	**0.309**	**0.342**	**0.141**	**0.113**	**0.197**
Cat – TPM [9]	0.329	0.201	0.232	0.416	0.287	0.294	0.114	0.082	0.179
$\text{RWR}_{\text{USACL}}$ [19]	0.304	0.182	0.217	0.371	0.253	0.281	0.102	0.077	0.159
Session – kNN [7]	0.271	0.179	0.135	0.357	0.239	0.143	0.074	0.052	0.103
GRU4REC [5]	0.245	0.152	0.125	0.333	0.211	0.126	0.053	0.041	0.104
$\text{PathCount}_{\text{AUA}}$ [15]	0.231	0.142	0.176	0.313	0.191	0.225	0.043	0.031	0.189
$\text{PathSim}_{\text{AUA}}$ [22]	0.225	0.132	0.165	0.303	0.182	0.213	0.035	0.021	0.187
$\text{SimRank}_{\text{USACL}}$ [8]	0.141	0.075	0.166	0.213	0.096	0.192	0.024	0.012	0.169

6.5 User Study

In this Section, we present a user study, which measures the users' satisfaction over different (i) meta paths for each recommendation strategy and (ii) explanation styles in helping them to make more accurate decisions.

Research Question 1: Which is Users' Favorite Explanation Style? A group of 34 students (17 males and 17 females) from our university were invited to answer which meta paths they find useful. In particular, we provided to users article recommendations along with their meta-path based explanations for all three recommendation strategies (IB, UB, SB). Next, we asked them to evaluate separately each meta path on a liker scale [1–3]. An example of a survey question to be answered is as follows: *You are recommended article B, because it was read by other similar users who have read the same 30 articles with you.* This stands for a UB recommendation strategy, along with its UAU meta path-based explanation. The average rating values μ and standard deviations σ are summarized in Table 6. We run a one-sided t-test for the difference of means between the most favorite meta path and each of the rest meta paths per strategy and found them statistically significant at 0.05 level. As expected, in Table 6, for IB strategy, the ACA-based explanation is the users' most favourite, because it combines both the feature and the item style of explanation (i.e., it is hybrid). For UB and IB strategies, the UAU-based and the SAS meta paths are the users' favorite, respectively.

Table 6. Users' favorite explanation per rec. strategy.

	μ	σ	μ	σ	μ	σ
IB	AUA		**ACA**		ALA	
	1.69	0.86	**2.44**	**0.8**	1.88	0.61
UB	**UAU**		UCU		ULU	
	2.41	**0.76**	2.19	0.74	1.41	0.61
SB	**SAS**		SCS		SLS	
	2.36	**0.79**	2.22	0.71	1.63	0.79

Research Question 2: Which Explanation Style Helps Users to Make More Accurate Decisions? Previous user studies [1] have not evaluated the session style of explanation. Thus, we followed their methodology, to find which explanation style helps users to make more accurate decisions. First, we measured users' satisfaction on the recommended articles, but without showing to them any explanation (denoted as *Actual rating A*). Our hypothesis is that the hybrid explanation style (i.e., ACA) will help users to more accurately estimate their *actual rating,* because it is more informative, since it combines two other explanation styles (item and feature). The results are illustrated in Table 7.

As shown in Table 7, the best explanation is the one that allows users to best approximate their *actual rating.* That is, the difference of means μ_d between *Explanation ratings* μ_E and *Actual ratings* μ_A should be centered around 0. These values, for each explanation style, are presented in the last two columns of Table 7. The null hypothesis $H_0(\mu_d = 0)$, which states that the difference of means μ_d is equal to zero, is accepted at the 0.05 significance level for the item, feature and hybrid explanation styles, whereas the alternative hypothesis $H_a(\mu_d \neq 0)$, which states that the difference of means is different than

Table 7. Explanation styles: satisfaction vs. promotion.

Explanation style	μ_E	σ_E	μ_A	σ_A	μ_d	σ_d
Item	1.69	0.86	1.88	1.04	**0.19**	0.14
User	2.41	0.76	2.1	1.13	0.30	0.18
Feature	2.1	0.68	2.3	1.06	**0.20**	0.16
Session	2.36	0.79	2.07	1.12	0.29	0.19
Hybrid	2.44	0.8	2.36	0.98	**0.08**	0.12

zero, is verified for the user and session explanation styles. As shown in Table 7, the hybrid explanation style has the smallest μ_d value equal to 0.08. This is as expected, since with the hybrid style users receive explanations with richer information, as also reported in [12]. We also measured the pearson correlation between the *Actual and Explanation ratings* for the item, feature and hybrid styles, and find that they follow similar patterns (i.e., they are positively correlated). Finally, both the user and session explanation styles, make users to overestimate the quality of the articles (i.e. they are more persuasive explanations), as already reported by Bilgic and Mooney [1] for the case of the user style, which makes them more suitable for business marketing and product promotion purposes.

7 Conclusion

In this paper, we combined multiple meta paths to reveal semantically-rich relationships over a graph and provide both accurate and explainable recommendations. To the best of our knowledge, we first introduce the session style of explanation. As future work, we want to apply our meta path-based explanation framework to other network structures such as explainable AI for personalized recommendations in health.

References

1. Bilgic, M., Mooney, Explaining recommendations: satisfaction vs. promotion. In: Proceedings of Recommender Systems Workshop (IUI Conference) (2005)
2. Epure, E.V., Kille, B., Ingvaldsen, J.E., Deneckere, R., Salinesi, C., Albayrak, S.: Recommending personalized news in short user sessions. In: Proceedings of the Eleventh ACM Conference on Recommender Systems, pp. 121–129. ACM (2017)
3. He, X., Liao, L., Zhang, H., Nie, L., Hu, X., Chua, T.-S.: Neural collaborative filtering. In: Proceedings of the 26th International Conference on World Wide Web, pp. 173–182. International World Wide Web Conferences Steering Committee (2017)
4. Hidasi, B., Karatzoglou, A.: Recurrent neural networks with top-k gains for session-based recommendations. arXiv preprint arXiv:1706.03847 (2017)

5. Hidasi, B., Karatzoglou, A., Baltrunas, L., Tikk, D.: Session-based recommendations with recurrent neural networks. CoRR, abs/1511.06939 (2015)
6. Jannach, D., Lerche, L., Jugovac, M.: Adaptation and evaluation of recommendations for short-term shopping goals. In: Proceedings of the Ninth ACM Conference on Recommender Systems, RecSys 2015. ACM, New York (2017)
7. Jannach, D., Ludewig, M.: When recurrent neural networks meet the neighborhood for session-based recommendation. In: Proceedings of the Eleventh ACM Conference on Recommender Systems, RecSys '17, pp. 306–310. ACM, New York (2017)
8. Jeh, G., Widom, J.: SimRank: a measure of structural-context similarity. In: Proceedings of 8th ACM SIGKDD International Conference on Knowledge Discovery and Data Mining (KDD'2002), pp. 538–543, Edmonton, Canada (2002)
9. Liu, J., Dolan, P., Pedersen, E.R.: Personalized news recommendation based on click behavior. In: Proceedings of the 15th International Conference on Intelligent User Interfaces, pp. 31–40. ACM (2010)
10. Ludmann, C.: Recommending news articles in the clef news recommendation evaluation lab with the data stream management system Odysseus. In: Working Notes of the 8th International Conference of the CLEF Initiative, Dublin, Ireland. CEUR Workshop Proceedings (2017)
11. Musto, C., Narducci, F., Lops, P., De Gemmis, M., Semeraro, G.: Explod: a framework for explaining recommendations based on the linked open data cloud. In: Proceedings of the 10th ACM Conference on Recommender Systems, pp. 151–154. ACM (2016)
12. Papadimitrou, A., Symeonidis, P., Manolopoulos, Y.: A generalized explanations styles taxonomy for the traditional and the social recommender systems. Data Min. Knowl. Discov. (2012)
13. Shi, C., Kong, X., Yu, P.S., Xie, S., Wu, B.: Relevance search in heterogeneous networks. In: Proceedings of the 15th International Conference on Extending Database Technology, pp. 180–191. ACM (2012)
14. Song, W., Xiao, Z., Wang, Y., Charlin, L., Zhang, M., Tang, J.: Session-based social recommendation via dynamic graph attention networks. In: Proceedings of the Twelfth ACM Conference on Web Search and Data Mining (2017)
15. Sun, Y., Barber, R., Gupta, M., Aggarwal, C.C., Han, J.: Co-author relationship prediction in heterogeneous bibliographic networks. In: 2011 International Conference on Advances in Social Networks Analysis and Mining (ASONAM), pp. 121–128. IEEE (2011)
16. Symeonidis, P., Nanopoulos, A., Papadopoulos, A.N., Manolopoulos, Y.: Collaborative filtering: fallacies and insights in measuring similarity. In: Berendt, B., Hotho, A., Mladenic, D., Semeraro, G. (Chairs), Proceedings of the 17th European Conference on Machine Learning and 10th European Conference on Principles and the Practice of Knowledge Discovery in Databases Workshop on Web Mining, pp. 56–67 (2006)
17. Symeonidis, P., Nanopoulos, A., Manolopoulos, Y.: Moviexplain: a recommender system with explanations. In: Proceedings of 3nd ACM Conference in Recommender Systems (RecSys'2009), pp. 317–320, New York, NY (2)009
18. Symeonidis, P.: User recommendations based on tensor dimensionality reduction. In: Iliadis, Maglogiann, Tsoumakasis, Vlahavas, Bramer (eds.) AIAI 2009. IFIP International Federation for Information Processing, vol. 296, pp. 331–340. Springer, Boston (2009). https://doi.org/10.1007/978-1-4419-0221-4_39

19. Tong, H., Faloutsos, C., Pan, J.: Fast random walk with restart and its applications. In: Proceedings 6th International Conference on Data Mining (ICDM'2006), pp. 613–622, Hong Kong, China (2006)
20. Wang, X., Wang, D., Xu, C., He, X., Cao, Y., Chua, T.-S.: Explainable reasoning over knowledge graphs for recommendation. In: Proceedings of the AAAI Conference on Artificial Intelligence, vol. 33, pp. 5329–5336 (2019)
21. Xu, C., et al.: Graph contextualized self-attention network for session-based recommendation. In: IJCAI 2019, pp. 3940–3946 (2019)
22. Yizhou, S., Jiawei, H., Xifeng, Y., Philip S.Y., Tianyi, W.: Pathsim: meta path-based top-k similarity search in heterogenuos information networks. : Proceedings of the VLDB Endowment (VLDB'2011), Seattle, Washigton (2011)

ProtoMIL: Multiple Instance Learning with Prototypical Parts for Whole-Slide Image Classification

Dawid Rymarczyk[1,2(✉)], Adam Pardyl[1], Jarosław Kraus[1], Aneta Kaczyńska[1], Marek Skomorowski[1], and Bartosz Zieliński[1,2]

[1] Faculty of Mathematics and Computer Science, Jagiellonian University, 6 Łojasiewicza Street, 30-348 Kraków, Poland
{dawid.rymarczyk,adam.pardyl,jarek.kraus, aneta.kaczynska}@student.uj.edu.pl,
{marek.skomorowski,bartosz.zielinski}@uj.edu.pl

[2] Ardigen SA, 76 Podole Street, 30-394 Kraków, Poland

Abstract. The rapid development of histopathology scanners allowed the digital transformation of pathology. Current devices fastly and accurately digitize histology slides on many magnifications, resulting in whole slide images (WSI). However, direct application of supervised deep learning methods to WSI highest magnification is impossible due to hardware limitations. That is why WSI classification is usually analyzed using standard Multiple Instance Learning (MIL) approaches, that do not explain their predictions, which is crucial for medical applications. In this work, we fill this gap by introducing ProtoMIL, a novel self-explainable MIL method inspired by the case-based reasoning process that operates on visual prototypes. Thanks to incorporating prototypical features into objects description, ProtoMIL unprecedentedly joins the model accuracy and fine-grained interpretability, as confirmed by the experiments conducted on five recognized whole-slide image datasets.

Keywords: Multiple instance learning · Digital pathology · Interpretable deep learning

1 Introduction

A typical supervised learning scenario assumes that each data point has a separate label. However, in Whole Slide Image (WSI) classification, only one label is usually assigned to a gigapixel image due to the laborious and expensive labeling. Because of the hardware limitations, the direct application of supervised

A. Pardyl, J. Kraus and A. Kaczyńska–Equal contribution.

Supplementary Information The online version contains supplementary material available at https://doi.org/10.1007/978-3-031-26387-3_26.

M.-R. Amini et al. (Eds.): ECML PKDD 2022, LNAI 13713, pp. 421–436, 2023.
https://doi.org/10.1007/978-3-031-26387-3_26

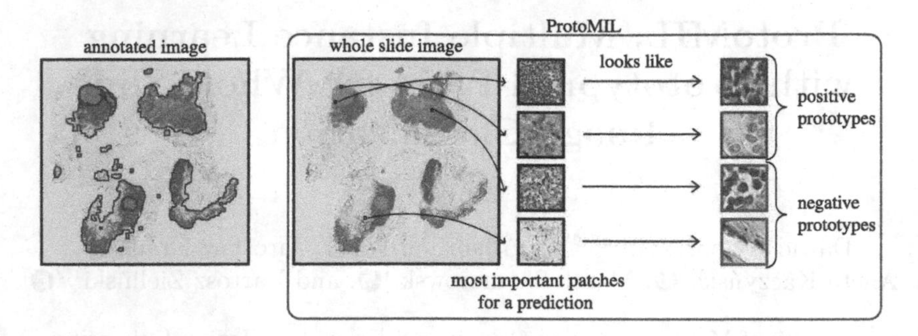

Fig. 1. ProtoMIL divides the whole slide image into patches and analyzes their similarity to the reference prototypical parts that describe the given data class. As a result, it can provide a visual explanation of its prediction. One can observe that ProtoMIL identifies the most important patches with attention weights, that can appear both inside and outside a cancer region (marked as green and blue areas, respectively). Moreover, these patches are described by cancer or healthy tissue prototypes (corresponding to patches in green and red frames, respectively), showing their resemblance to the training examples. (Color figure online)

deep learning methods to WSI two highest magnification is impossible. That is why recent approaches [24] divide the WSI into smaller patches (instances) and process them separately to obtain their representations. Such representations form a bag of instances associated with only one label, and it is unspecified which instances are responsible for this label [15]. This kind of problem, called Multiple Instance Learning (MIL) [12], appears in many medical problems, such as the diabetic retinopathy screening [30,31], bacteria clones identification using microscopy images [7], or identifying conformers responsible for molecule activity in drug design [42,47].

In recent years, with the rapid development of deep learning, MIL is combined with many neural network-based models [14,20,24,27,34,38,39,43–45]. Many of them embed all instances of the bag using a convolutional block of a deep network and then aggregate those embeddings. Moreover, some aggregation methods specify the most important instances that are presented to the user as prediction interpretation [20,24,27,34,39]. However, those methods usually only exhibit instances crucial for the prediction and do not indicate the cause of their importance. Naturally, there were attempts to further explain the MIL models [6,7,25], but overall, they usually introduce additional bias into the explanation [33] or require additional input [25].

To address the above shortcomings of MIL models, we introduce *Prototypical Multiple Instance Learning* (ProtoMIL). It builds on case-based reasoning, a type of explanation naturally used by humans to describe their thinking process [23]. More precisely, we divide each WSI into patches and analyze how similar they are to a trainable set prototypical parts of positive and negative data classes, as defined in [8]. Since, the prototypes are trainable, they are automatically derived by ProtoMIL. Then, we apply an attention pooling operator to accumulate those similarities over instances. As a result, we obtain bag-level representation classified with an additional neural layer. This approach significantly differs from non-MIL approaches because it applies an aggregation layer and introduces a novel regularization technique that encourages the model to derive prototypes from the instances responsible for the positive label of a bag. The latter is a challenging problem because those instances are concealed and underrepresented. Lastly, the prototypical parts are pruned to characterize the data classes compactly. This results in detailed interpretation, where the most important patches according to attention weights are described using prototypes, as shown in Fig. 1.

To show the effectiveness of our model, we conduct experiments on five WSI datasets: Bisque Breast Cancer [16], Colon Cancer [41], Camelyon16 Breast Cancer [13], Lung cancer subtype identification TCGA-NSCLC [5] and Kidney cancer subtype classification [2]. Additionally, in the Supplementary Materials, we show the universal character of our model in different scenarios such as MNIST Bags [20] and Retinopathy Screening (Messidor dataset) [11]. The results we obtain are usually on par with the current state-of-the-art models. However, at the same time, we strongly enhance interpretation capabilities with prototypical parts obtained from the training set. We made our code publicly available at https://github.com/apardyl/ProtoMIL.

The main contributions of this work are as follows:

- Introducing the ProtoMIL method, which substantially improves the interpretability of existing MIL models by introducing case-based reasoning.
- Developing a training paradigm that encourages generating prototypical parts from the underrepresented instances responsible for the positive label of a bag.

The paper is organized as follows. In Sect. 2, we present recent advancements in Multiple Instance Learning and deep interpretable models. In Sect. 3, we define the MIL paradigms and introduce ProtoMIL. Finally, in Sect. 4, we present the results of conducted experiments, and Sect. 5 summarizes the work.

2 Related Works

Our work focuses on classification of whole slide images which is described using Multiple Instance Learning (MIL) framework. Additionally, we develop an interpretable method which relates to eXplainable Artificial Intelligence (XAI). We briefly describe both fields in the following subsections.

2.1 Multiple Instance Learning

Before the deep learning era, models based on SVM, such as MI-SVM [3], were used for MIL problems. However, currently, MIL is addressed with numerous deep models. One of them, Deep MIML [14], introduces a sub-concept layer that is learned and then pooled to obtain a bag representation. Another example is mi-Net [44], which pools predictions from single instances to derive a bag-level prediction. Other architectures adapted to MIL scenarios includes capsule networks [45], transformers [38] and graph neural networks [43]. Moreover, many works focus on the attention-based pooling operators, like AbMILP introduced in [20] that weights the instances embeddings to obtain a bag embedding. This idea was also extended by combining it with mi-Net [24], clustering similar instances [27], self-attention mechanism [34], and sharing classifier weights with pooling operator [39]. However, the above methods either do not contain an XAI component or only present the importance of the instances. Hence, our ProtoMIL is a step towards the explainability of the MIL methods.

2.2 Explainable Artificial Intelligence

There are two types of eXplainable Artificial Intelligence (XAI) approaches, post hoc and self-explaining methods [4]. Among many *post hoc* techniques, one can distinguish saliency maps showing pixel importance [32,36,37,40] or concept activation vectors representing internal network state with human-friendly concepts [9,17,21,46]. They are easy to use since they do not require any changes in the model architecture. However, their explanations may be unfaithful and fragile [1]. Therefore *self-explainable* models were introduced like Prototypical Part Network [8] with a layer of prototypes representing the activation patterns. A similar approach for hierarchically organized prototypes is presented in [18] to classify objects at every level of a predefined taxonomy. Moreover, some works concentrate on transforming prototypes from the latent space to data space [26] or focus on sharing prototypical parts between classes and finding semantic similarities [35]. Other works [28] build a decision tree with prototypical parts in the nodes or learn disease representative features within a dynamic area [22]. Nonetheless, to our best knowledge, no fine-grained self-explainable method, like ProtoMIL, exists for MIL problems.

3 ProtoMIL

Due to the large resolution of whole slide images, which should not be scaled down due to loss of information, we first divide an image into patches. However, we do not know which patches correspond to the given disease state. Therefore, this problem boils down to Multiple Instance Learning (MIL), where there is a bag of instances (in our case patches) and only one label for the whole bag. This bag is passed trough the four modules of ProtoMIL (see Fig. 2): convolutional network f_{conv}, prototype layer f_{proto}, attention pooling a, and fully connected last layer g. Convolutional and prototype layers process single instances, whereas attention pooling and the last layer work on a bag level. More precisely, given a bag of patches $X = \{\mathbf{x}_1, \dots, \mathbf{x}_k\}$, each $\mathbf{x} \in X$ is forwarded through convolutional layers to obtain low-dimensional embeddings $F = \{f_{conv}(\mathbf{x}_1), \dots, f_{conv}(\mathbf{x}_k)\}$. As $f_{conv}(\mathbf{x}) \in H \times W \times D$, for the clarity of description, let $Z_{\mathbf{x}} = \{\mathbf{z}_j \in f_{conv}(\mathbf{x}) : \mathbf{z}_j \in \mathbb{R}^D, j = 1..HW\}$. Then, the prototype layer computes vector $\mathbf{h}$ of similarity scores [8] between each embedding $f_{conv}(\mathbf{x})$ and all prototypes $\mathbf{p} \in P$ as

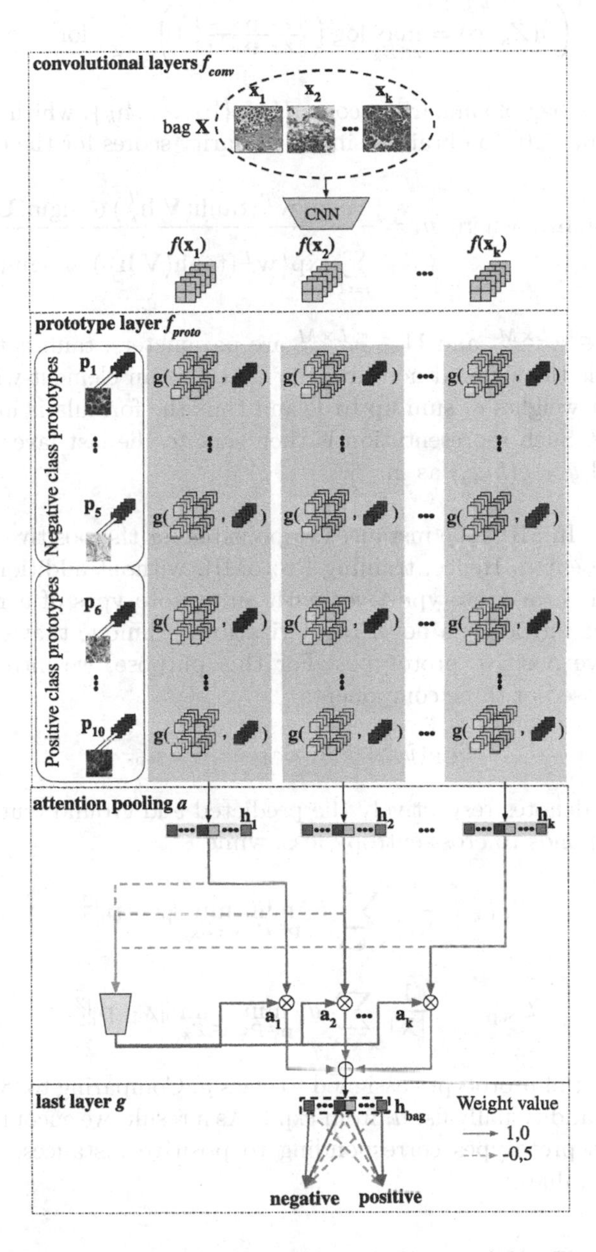

Fig. 2. ProtoMIL passes a bag of patches through four modules. First, convolutional layer f_{conv} generates embeddings for each patch. Then, the prototype layer f_{proto} calculates similarities between patches representations and its prototypes. The similarities are aggregated using the attention pooling a to obtain the bag similarity scores classified using the last layer g. Notice that particular colors in vectors $\mathbf{h_i}$ and $\mathbf{h_{bag}}$ correspond to prototypes similarities.

$$\mathbf{h} = \left(g(Z_{\mathbf{x}}, \mathbf{p}) = \max_{\mathbf{z} \in Z_{\mathbf{x}}} \log \left(\frac{\|\mathbf{z}-\mathbf{p}\|^2+1}{\|\mathbf{z}-\mathbf{p}\|^2+\varepsilon} \right) \right)_{\mathbf{p} \in P} \quad \text{for } \varepsilon > 0.$$

This results in a bag of similarity scores $H = \{\mathbf{h}_1, \ldots, \mathbf{h}_k\}$, which we pass to the attention pooling [20] to obtain a single similarity scores for the entire bag

$$\mathbf{h}_{bag} = \sum_{i=1}^{k} a_i \, \mathbf{h}_i, \quad \text{where } a_i = \frac{\exp\{\mathbf{w}^T(\tanh(\mathbf{V}\,\mathbf{h}_i^T) \odot \text{sigm}(\mathbf{U}\,\mathbf{h}_i^T)\}}{\sum_{j=1}^{k} \exp\{\mathbf{w}^T(\tanh(\mathbf{V}\,\mathbf{h}_j^T) \odot \text{sigm}(\mathbf{U}\,\mathbf{h}_j^T)\}}, \quad (1)$$

$\mathbf{w} \in \mathbb{R}^{L \times 1}$, $\mathbf{V} \in \mathbb{R}^{L \times M}$, and $\mathbf{U} \in \mathbb{R}^{L \times M}$ are parameters, tanh is the hyperbolic tangent, sigm is the sigmoid non-linearity and $\odot$ is an element-wise multiplication. Note that weights a_i sum up to 1, and thus the formula is invariant to the size of the bag. Such representation is then sent to the last layer to obtain the predicted label $\breve{y} = g(h_{bag})$ as in [8].

Regularization. In MIL, the instances responsible for the positive label of a bag are underrepresented. Hence, training ProtoMIL without additional regularizations can result in a prototype layer with only prototypes of a negative class. That is why we introduce a novel regularization technique that encourages the model to derive positive prototypes. For this purpose, we introduce the loss function composed of three components

$$\mathcal{L}_{\text{CE}}(\breve{y}, y) + \lambda_1 \, \mathcal{L}_{\text{Clst}} + \lambda_2 \, \mathcal{L}_{\text{Sep}},$$

where $\breve{y}$ and y denotes respectively the predicted and ground truth label of bag X, $\mathcal{L}_{\text{CE}}$ corresponds to cross-entropy loss, while

$$\mathcal{L}_{\text{Clst}} = \frac{1}{|X|} \sum_{\mathbf{x}_i \in X} a_i \min_{\mathbf{p} \in P^y} \min_{\mathbf{z} \in Z_{\mathbf{x}_i}} \|\mathbf{z} - \mathbf{p}\|_2^2,$$

$$\mathcal{L}_{\text{Sep}} = -\frac{1}{|X|} \sum_{\mathbf{x}_i \in X} a_i \min_{\mathbf{p} \notin P^y} \min_{\mathbf{z} \in Z_{\mathbf{x}_i}} \|\mathbf{z} - \mathbf{p}\|_2^2,$$

where P^y is a set of prototypes assigned to class y. Comparing to [8], components $\mathcal{L}_{\text{Clst}}$ and $\mathcal{L}_{\text{Sep}}$ additionally use a_i from Eq. 1. As a result, we encourage the model to create more prototypes corresponding to positive instances, which usually have higher a_i values.

4 Experiments

We test our ProtoMIL approach on five datasets, for which we train the model from scratch in three steps: (i) *warmup* phase with training all layers except the last one, (ii) prototype projection, (iii) and fine-tuning with fixed f_{conv} and f_{proto}. Phases (ii) and (iii) are repeated several times to find the most optimal set of prototypes. All trainings use Adam optimizer for all layers with $\beta_1 = 0.99$,

$\beta_2 = 0.999$, weight decay 0.001, and batch size 1. Additionally, we use an exponential learning rate scheduler for the *warmup* phase and a step scheduler for prototype training. All results are reported as an average of all runs with a standard error of the mean. In the subsequent subsections, we describe experiment details and results for each dataset.

Across all datasets we use convolutional block from ResNet-18 followed by two additional 1×1 convolutions as the convolutional layer f_{conv}. We use ReLU as the activation function for all convolutional layers except the last layer, for which we use the sigmoid activation function. The prototype layer stores prototypes shared across all bags, while the attention layer implements AbMILP. The last layer is used to classify the entire bag. Weights between similarity scores of prototypes corresponding class logit are initialized with 1, while other connections are set to -0.5 as in [8]. Together with the specific training procedure, such initialization results in a positive reasoning process (we rather say "this looks like that" instead of saying "this does not look like that").

4.1 Bisque Breast Cancer and Colon Cancer Datasets

Experiment Details. We experiment on two histological datasets: Colon Cancer and Bisque Breast Cancer. The former contains 100 H&E images with 22,444 manually annotated nuclei of four different types: epithelial, inflammatory, fibroblast, and miscellaneous. To create bags of instances, we extract 27×27 nucleus-centered patches from each image, and the goal is to detect if the bag contains one or more epithelial cells, as colon cancer originates from them. On the other hand, the Bisque dataset consists of 58 H&E breast histology images of size 896×768, out of which 32 are benign, and 26 are malignant (contain at least one cancer cell). Each image is divided into 32×32 patches, resulting in 672 patches per image. Patches with at least 75% of the white pixels are discarded, resulting in 58 bags of various sizes.

We apply extensive data augmentation for both datasets, including random rotations, horizontal and vertical flipping, random staining augmentation, staining normalization, and instance normalization. We use ResNet-18 convolutional parts with the first layer modified to 3×3 convolution with stride 1 to match the size of smaller instances. We set the number of prototypes per class to 10 with a size of $128 \times 2 \times 2$. Warmup, fine-tuning, and end-to-end training take 60, 20, and 20 epochs, respectively. 10-fold cross-validation with 1 validation fold and 1 test fold is repeated 5 times.

Results. Table 1 presents our results compared to both traditional and attention-based MIL models. On the Bisque dataset, our model significantly outperforms all baseline models. However, due to the small size of the Colon Cancer dataset, ProtoMIL overfits, resulting in poorer AUC than attention-based models. Nevertheless, in both cases, ProtoMIL provides finer explanations than all baseline models (see Fig. 3 and Supplementary Materials).

Table 1. Results for small histological datasets, where ProtoMIL significantly outperforms baseline methods on the Bisque dataset. However, it achieves worse results for the Colon Cancer dataset, probably due to its small size. Additionally, interpretability of the methods is noted and further discussed in Sect. 4.6. Notice that values for comparison indicated with "*" and "**" comes from [20] and [34], respectively.

METHOD	BISQUE		COLON CANCER		
	ACCURACY	AUC	ACCURACY	AUC	INTER.
INSTANCE+MAX*	61.4% ± 2.0%	0.612 ± 0.026	84.2% ± 2.1%	0.914 ± 0.010	+
INSTANCE+MEAN*	67.2% ± 2.6%	0.719 ± 0.019	77.2% ± 1.2%	0.866 ± 0.008	–
EMBEDDING+MAX*	60.7% ± 1.5%	0.650 ± 0.013	82.4% ± 1.5%	0.918 ± 0.010	–
EMBEDDING+MEAN*	74.1% ± 2.3%	0.796 ± 0.012	86.0% ± 1.4%	0.940 ± 0.010	–
ABMILP*	71.7% ± 2.7%	0.856 ± 0.022	88.4% ± 1.4%	0.973 ± 0.007	++
SA-ABMILP**	75.1% ± 2.4%	0.862 ± 0.022	**90.8% ± 1.3%**	**0.981 ± 0.007**	+
PROTOMIL (OUR)	**76.7% ± 2.2%**	**0.886 ± 0.033**	81.3% ± 1.9%	0.932 ± 0.014	+++

4.2 Camelyon16 Dataset

Experiment Details. The Camelyon16 dataset [13] consists of 399 whole-slide images of breast cancer samples, each labeled as *normal* or *tumor*. We create MIL bags by dividing each slide 20x resolution image into 224 × 224 patches, rejecting patches that contain more than 70% of background. This results in 399 bags with a mean of 8, 871 patches and a standard deviation of 6, 175. Moreover, 20 largest bags are truncated to 20, 000 random patches to fit into the memory of a GPU. The positive patches are again highly imbalanced, as only less than 10% of patches contain tumor tissue.

Due to the size of the dataset, we preprocess all samples using a ResNet-18 without two last layers, pre-trained on various histopathological images using self-supervised learning from [10]. The resulting embeddings are fed into our model to replace the feature backbone net. ProtoMIL is trained for 50, 40, and 10 epochs in warmup, fine-tuning, and end-to-end training, respectively. The number of prototypes per class is limited to 5 with no data augmentation. The experiments are repeated 5 times with the original train-test split.

Results. We compare ProtoMIL to other state-of-the-art MIL techniques, including both traditional mean and max MIL pooling, RNN, attention-based MIL pooling, and transformer-based MIL pooling [38]. ProtoMIL performs on par in terms of accuracy and slightly outperforms other models on AUC metric (Table 2) while providing a better understanding of its decision process, as presented in Fig. 4 and Supplementary Materials.

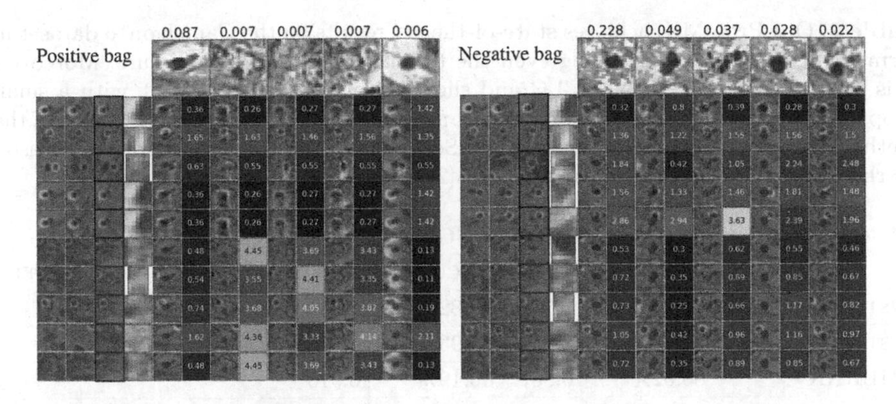

Fig. 3. Similarity scores between five crucial instances of a bag (columns) and ten prototypical parts (rows) for a positive and negative bag (left and right side, respectively) from the Colon Cancer bags. Each prototypical part is represented by a part of the training image and three nearest training patches, and each instance is represented by the patch and the value of its attention weight a_i. Moreover, each cell contains a similarity score and a heatmap corresponding to prototype activation. One can observe that instances of a negative bag usually activate prototypes of a negative class (four upper prototypes in red brackets), while the instances of positive bags mostly activate positive prototypes (four bottom prototypes in green brackets). (Color figure online)

4.3 TCGA-NSCLC Dataset

Experiment details. TCGA-NSCLC includes two subtype projects, i.e., Lung Squamous Cell Carcinoma (TGCA-LUSC) and Lung Adenocarcinoma (TCGA-LUAD), for a total of 956 diagnostic WSIs, including 504 LUAD slides from 478 cases and 512 LUSC slides from 478 cases. We create MIL bags using WSI Segmentation and Patching from [27] with default parameters, except patch-level parameter set to 1. Each slide image is cropped into a series of 224×224 patches. This results in 1,016 bags with a mean of 3,961 patches. We randomly split the data in the ratio of train:valid:test equal 60:15:25 and assure that there is no case overlap between the sets, and use the same ProtoMIL settings as in the Camelyon16 dataset are used. The results are reported for 4-fold cross validation.

Results. Results for the TCGA-NSCLC dataset are presented in Table 2 alongside results of other state-of-the-art approaches from [38]. ProtoMIL performs slightly lower on the Area Under the ROC Curve (AUC) and accuracy metrics than the powerful transformer-based model TransMIL but still is competitive to other CNN-based approaches. However, the advantage of ProtoMIL is its capability to provide a detailed explanation of predictions as presented in Fig. 5 and Supplementary Materials.

Table 2. Our ProtoMIL achieves state-of-the-art results on the Camelyon16 dataset in terms of AUC metric, surpassing even the transformer-based architecture. Moreover, it is competitive on TCGA-NSCLC and slightly worse on TCGA-RCC, with a small drop of accuracy and AUC compared to TransMIL. Additionally, interpretability of the methods is noted and further discussed in Sect. 4.6. Notice that values for comparison marked with "*" and "**" are taken from [24] and [38], respectively.

Method	Camelyon16		TCGA-NSCLC		TCGA-RCC		
	Accuracy	AUC	Accuracy	AUC	Accuracy	AUC	Inter.
instance+mean*	79.84%	0.762	72.82%	0.840	90.54%	0.978	–
instance+max*	82.95%	0.864	85.93%	0.946	93.78%	0.988	+
MILRNN*	80.62%	0.807	86.19%	0.910	–	–	–
ABMILP*	84.50%	0.865	77.19%	0.865	89.34%	0.970	++
DSMIL*	86.82%	0.894	80.58%	0.892	92.94%	0.984	++
CLAM-SB**	87.60%	0.881	81.80%	0.881	88.16%	0.972	+
CLAM-MB**	83.72%	0.868	84.22%	0.937	89.66%	0.980	+
TransMIL**	**88.37%**	0.931	**88.35%**	**0.960**	**94.66%**	**0.988**	+
ProtoMIL (our)	87.29%	**0.935**	83.66%	0.918	92.79%	0.961	+++

4.4 TCGA-RCC Dataset

Experiment details. TCGA-RCC consists of three unbalanced classes: Kidney Chromophobe Renal Cell Carcinoma (TGCA-KICH, 111 slides from 99 cases), Kidney Renal Clear Cell Carcinoma (TCGA-KIRC, 489 slides from 483 cases), and Kidney Renal Papillary Cell Carcinoma (TCGA-KIRP, 284 slides from 264 cases) for a total of 884 WSIs. We create MIL bags using WSI Segmentation and Paching from [27] with default parameters and a patch-level parameter set to 1. Each slide image is cropped into a series of 224×224 patches. This results in 884 bags with a mean of $4,309$ patches. A separate model is trained for each class, and scores are averaged for all classes. Other experiment settings are identical as for TCGA-NSCLC described above.

Results. We compare ProtoMIL to other state-of-the-art MIL techniques, including both traditional mean and max MIL pooling, attention-based MIL pooling, and transformer-based MIL pooling [38]. ProtoMIL performs on par in terms of accuracy and AUC metric (Table 2) while providing a better understanding of its decision process, as presented in Supplementary Materials.

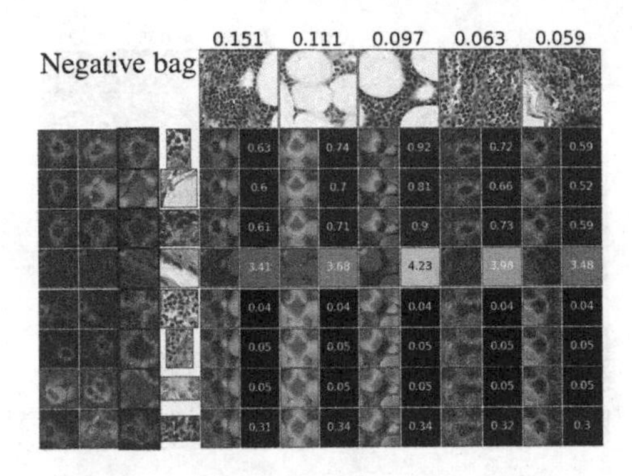

Fig. 4. Similarity scores between five crucial instances of a bag (columns) and eight prototypical parts (rows) for a negative bag from the Camelyon16 dataset. One can observe that ProtoMIL strongly activates only one prototype and focuses mainly on nuclei when analyzing the healthy parts of the tissue. Please refer to Fig. 3 for a detailed description of the visualization.

Table 3. The influence of ProtoMIL pruning on the accuracy and AUC score. One can notice that even though the pruning removes around 30% of the prototypes, it usually does not noticeably decrease the AUC and accuracy of the model.

Dataset	Before pruning			After pruning		
	Proto. #	Accuracy	AUC	Proto. #	Accuracy	AUC
Bisque	20 ± 0	76.7% ± 2.2%	0.886 ± 0.033	13.6 ± 0.25	73.0% ± 2.4%	0.867 ± 0.022
Colon Cancer	20 ± 0	81.3% ± 1.9%	0.932 ± 0.014	15.69 ± 0.34	81.8% ± 2.4%	0.880 ± 0.022
Camelyon16	10 ± 0	87.3% ± 1.2 %	0.935 ± 0.007	6.4 ± 0.24	85.9% ± 1.5%	0.937 ± 0.007
TCGA-NSCLC	10 ± 0	83.66% ± 1.6%	0.918 ± 0.003	7.6 ± 1.2	81.1% ± 1.4%	0.880 ± 0.003
TCGA-RCC	10 ± 0	94.66% ± 1.0%	0.988 ± 0.009	6.2 ± 1.2	91.5% ± 1.2%	0.955 ± 0.006

4.5 Pruning

Experiment Details. We run prototype pruning experiments on all the datasets to remove not class-specific prototypical parts and check their influence on the model performance. For each of them, we use the model trained in the previously described experiments. As pruning parameters, we use $k = 6$ and $l = 40\%$ and fine-tuned for 20 epochs. Details about pruning operation are described in the Supplementary Materials.

Results. The accuracy and AUC in respect to the number of prototypes before and after pruning are presented in Table 3. For all datasets, the number of prototypes after pruning has decreased around 30% on average. However, it does not result in a noticeable decrease in accuracy or AUC, except for Colon Cancer, where we observe a significant drop in AUC. Most probably, it is caused by the high visual resemblance of nuclei patches (especially between *epithelial* and *miscellaneous*) that after prototype projection may be very close to each other in the latent space.

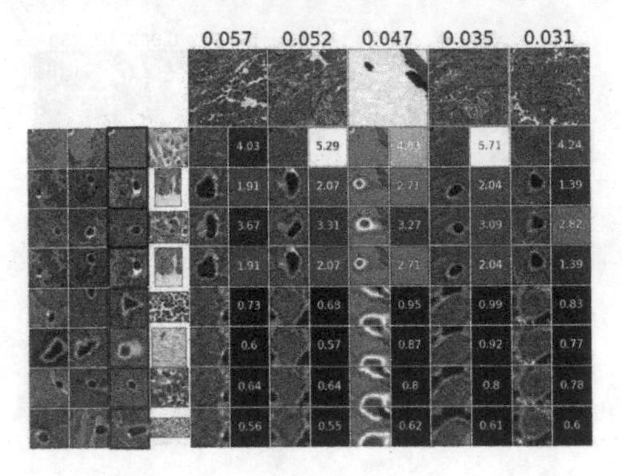

Fig. 5. Similarity scores between five crucial instances of a bag (columns) and eight prototypical parts (rows) for a LUAD type bag from the TCGA-NSCLC dataset.

4.6 Interpretability of MIL Methods

Column *Inter.* in Tables 1, and 2 indicates how interpretable are the considered models. Instances and embeddings-based methods, except instance-max, are not interpretable, similarly to MILRNN, since they lose information about instances crucial for the prediction. On the other hand, the AbMILP [20] identifies crucial instances within a bag and can present the local explanation to the users. However, other attention-based methods, such as SA-AbMILP [34], TransMIL [38] and CLAMs [27] perform additional operations, like self-attention, requiring more effort from the user to analyze the explanation. That is why those methods have been assigned with lower interpretability. Moreover, DS-MIL [24] finds a decision boundary on the bag level and can produce a more detailed explanation than AbMILP, but only for a single prediction (local explanations). In contrast, the ProtoMIL can produce both local (see Fig. 3) and global explanations (see Supplementary Materials).

5 Discussion and Conclusions

In this work, we introduce Prototypical Multiple Instance Learning (ProtoMIL), a method for Whole Slide Image classification that incorporates a case-based reasoning process into the attention-based MIL setup. In contrast to existing MIL methods, ProtoMIL provides a fine-grained interpretation of its predictions. For this purpose, it uses a trainable set of prototypical parts correlated with data classes. The experiments on five datasets confirm that introducing fine-grained interpretability does not reduce the model's effectiveness, which is still on par with the current state-of-the-art methodology. Moreover, the results can be presented to the user with a novel visualization technique.

The experiments show that ProtoMIL can be applied to a challenging problem like Whole-Slide Image classification. Therefore, in future works, we plan to generalize our method to multi-label scenarios and multimodal classification problems since WSI often comes with other medical data like CT and MRI.

5.1 Limitations

ProtoMIL limitations are inherited from the other prototype-based models, such as non-obvious prototype meaning. Ergo, prototype projection might still result in uncertainty on which attributes it represents. However, there are methods mitigating these, e.g. explainer defined in [29].

5.2 Negative Impact

Our solution is based on prototypical parts that are susceptible to different types of adversarial attacks such as [19]. That is why practitioners shall address this risk in a deployed system with ProtoMIL. What is more, it may be used in information war to disinform societies when prototypes are obtained with spoiled data or are shown without appropriate comment, especially in fields like medicine.

Acknowledgments. This work was founded by the POIR.04.04.00-00-14DE/18-00 project carried out within the Team-Net programme of the Foundation for Polish Science cofinanced by the European Union under the European Regional Development Fund. This research was funded by the National Science Centre, Poland (research grant no. 2021/41/B/ST6/01370). For the purpose of Open Access, the authors have applied a CC-BY public copyright licence to any Author Accepted Manuscript (AAM) version arising from this submission.

References

1. Adebayo, J., Gilmer, J., Muelly, M., Goodfellow, I., Hardt, M., Kim, B.: Sanity checks for saliency maps. In: Advances in Neural Information Processing Systems, pp. 9505–9515 (2018)
2. Akin, O., et al.: Radiology data from the cancer genome atlas kidney renal clear cell carcinoma [tcga-kirc] collection. Cancer Imaging Arch. (2016)
3. Andrews, S., Tsochantaridis, I., Hofmann, T.: Support vector machines for multiple-instance learning. In: Advances in neural information processing systems. vol. 2, p. 7 (2002)
4. Arya, V., et al.: One explanation does not fit all: a toolkit and taxonomy of AI explainability techniques. arXiv preprint arXiv:1909.03012 (2019)
5. Bakr, S., et al.: A radiogenomic dataset of non-small cell lung cancer. Sci. Data **5**(1), 1–9 (2018)
6. Barnett, A.J., et al.: Iaia-bl: a case-based interpretable deep learning model for classification of mass lesions in digital mammography. arXiv preprint arXiv:2103.12308 (2021)
7. Borowa, A., Rymarczyk, D., Ochońska, D., Brzychczy-Włoch, M., Zieliński, B.: Classifying bacteria clones using attention-based deep multiple instance learning interpreted by persistence homology. In: International Joint Conference on Neural Networks (2021)
8. Chen, C., Li, O., Tao, C., Barnett, A.J., Su, J., Rudin, C.: This looks like that: deep learning for interpretable image recognition. arXiv preprint arXiv:1806.10574 (2018)

9. Chen, Z., Bei, Y., Rudin, C.: Concept whitening for interpretable image recognition. Nat. Mach. Intell. **2**(12), 772–782 (2020)
10. Ciga, O., Martel, A.L., Xu, T.: Self supervised contrastive learning for digital histopathology. arXiv preprint arXiv:2011.13971 (2020)
11. Decencière, E., et al.: Feedback on a publicly distributed image database: the Messidor database. Image Anal. Stereol. **33**(3), 231–234 (2014)
12. Dietterich, T.G., Lathrop, R.H., Lozano-Pérez, T.: Solving the multiple instance problem with axis-parallel rectangles. Artif. Intell. **89**(1–2), 31–71 (1997)
13. Ehteshami Bejnordi, B., et al.: Diagnostic assessment of deep learning algorithms for detection of lymph node metastases in women with breast cancer. JAMA **318**(22), 2199–2210 (2017). https://doi.org/10.1001/jama.2017.14585
14. Feng, J., Zhou, Z.H.: Deep miml network. In: Proceedings of the AAAI Conference on Artificial Intelligence, vol. 31 (2017)
15. Foulds, J., Frank, E.: A review of multi-instance learning assumptions. Knowl. Eng. Rev. **25**(1), 1–25 (2010)
16. Gelasca, E.D., Byun, J., Obara, B., Manjunath, B.: Evaluation and benchmark for biological image segmentation. In: 2008 15th IEEE International Conference on Image Processing, pp. 1816–1819. IEEE (2008)
17. Ghorbani, A., Wexler, J., Zou, J.Y., Kim, B.: Towards automatic concept-based explanations. In: Advances in Neural Information Processing Systems, pp. 9277–9286 (2019)
18. Hase, P., Chen, C., Li, O., Rudin, C.: Interpretable image recognition with hierarchical prototypes. In: Proceedings of the AAAI Conference on Human Computation and Crowdsourcing, vol. 7, pp. 32–40 (2019)
19. Hoffmann, A., Fanconi, C., Rade, R., Kohler, J.: This looks like that... does it? Shortcomings of latent space prototype interpretability in deep networks. arXiv preprint arXiv:2105.02968 (2021)
20. Ilse, M., Tomczak, J., Welling, M.: Attention-based deep multiple instance learning. In: International Conference on Machine Learning, pp. 2127–2136. PMLR (2018)
21. Kim, B., et al.: Interpretability beyond feature attribution: quantitative testing with concept activation vectors (TCAV). In: International Conference on Machine Learning, pp. 2668–2677. PMLR (2018)
22. Kim, E., Kim, S., Seo, M., Yoon, S.: Xprotonet: diagnosis in chest radiography with global and local explanations. In: Proceedings of the IEEE/CVF Conference on Computer Vision and Pattern Recognition, pp. 15719–15728 (2021)
23. Kolodner, J.: Case-Based Reasoning. Morgan Kaufmann, Burlington (2014)
24. Li, B., Li, Y., Eliceiri, K.W.: Dual-stream multiple instance learning network for whole slide image classification with self-supervised contrastive learning. In: Proceedings of the IEEE/CVF Conference on Computer Vision and Pattern Recognition, pp. 14318–14328 (2021)
25. Li, G., Li, C., Wu, G., Ji, D., Zhang, H.: Multi-view attention-guided multiple instance detection network for interpretable breast cancer histopathological image diagnosis. IEEE Access (2021)
26. Li, O., Liu, H., Chen, C., Rudin, C.: Deep learning for case-based reasoning through prototypes: a neural network that explains its predictions. In: Proceedings of the AAAI Conference on Artificial Intelligence, vol. 32 (2018)
27. Lu, M.Y., Williamson, D.F., Chen, T.Y., Chen, R.J., Barbieri, M., Mahmood, F.: Data-efficient and weakly supervised computational pathology on whole-slide images. Nat. Biomed. Eng. **5**(6), 555–570 (2021)

28. Nauta, M., van Bree, R., Seifert, C.: Neural prototype trees for interpretable fine-grained image recognition. In: Proceedings of the IEEE/CVF Conference on Computer Vision and Pattern Recognition, pp. 14933–14943 (2021)
29. Nauta, M., Jutte, A., Provoost, J., Seifert, C.: This looks like that, because... explaining prototypes for interpretable image recognition. In: Kamp, M., et al. (eds.) ECML PKDD 2021. CCIS, vol. 1524, pp. 441–456. Springer, Cham (2021). https://doi.org/10.1007/978-3-030-93736-2_34
30. Quellec, G., et al.: A multiple-instance learning framework for diabetic retinopathy screening. Med. Image Anal. **16**(6), 1228–1240 (2012)
31. Rani, P., Elagiri Ramalingam, R., Rajamani, K.T., Kandemir, M., Singh, D.: Multiple instance learning: robust validation on retinopathy of prematurity. Int. J. Ctrl. Theory Appl. **9**, 451–459 (2016)
32. Rebuffi, S.A., Fong, R., Ji, X., Vedaldi, A.: There and back again: revisiting backpropagation saliency methods. In: Proceedings of the IEEE/CVF Conference on Computer Vision and Pattern Recognition, pp. 8839–8848 (2020)
33. Rudin, C.: Stop explaining black box machine learning models for high stakes decisions and use interpretable models instead. Nat. Mach. Intell. **1**(5), 206–215 (2019)
34. Rymarczyk, D., Borowa, A., Tabor, J., Zielinski, B.: Kernel self-attention for weakly-supervised image classification using deep multiple instance learning. In: Proceedings of the IEEE/CVF Winter Conference on Applications of Computer Vision, pp. 1721–1730 (2021)
35. Rymarczyk, D., Struski, Ł., Tabor, J., Zieliński, B.: Protopshare: prototype sharing for interpretable image classification and similarity discovery. In: Proceedings of the 27th ACM SIGKDD Conference on Knowledge Discovery and Data Mining (KDD 2021) (2021). https://doi.org/10.1145/3447548.3467245
36. Selvaraju, R.R., Cogswell, M., Das, A., Vedantam, R., Parikh, D., Batra, D.: Grad-cam: visual explanations from deep networks via gradient-based localization. In: Proceedings of the IEEE/CVF International Conference on Computer Vision, pp. 618–626 (2017)
37. Selvaraju, R.R., et al.: Taking a hint: leveraging explanations to make vision and language models more grounded. In: Proceedings of the IEEE/CVF International Conference on Computer Vision, pp. 2591–2600 (2019)
38. Shao, Z., et al.: Transmil: transformer based correlated multiple instance learning for whole slide image classication. arXiv preprint arXiv:2106.00908 (2021)
39. Shi, X., Xing, F., Xie, Y., Zhang, Z., Cui, L., Yang, L.: Loss-based attention for deep multiple instance learning. In: Proceedings of the AAAI Conference on Artificial Intelligence, vol. 34, pp. 5742–5749 (2020)
40. Simonyan, K., Vedaldi, A., Zisserman, A.: Deep inside convolutional networks: visualising image classification models and saliency maps. arXiv:1312.6034 (2013)
41. Sirinukunwattana, K., Raza, S.E.A., Tsang, Y.W., Snead, D.R., Cree, I.A., Rajpoot, N.M.: Locality sensitive deep learning for detection and classification of nuclei in routine colon cancer histology images. IEEE Trans. Med. Imaging **35**(5), 1196–1206 (2016)
42. Straehle, C., Kandemir, M., Koethe, U., Hamprecht, F.A.: Multiple instance learning with response-optimized random forests. In: 2014 22nd International Conference on Pattern Recognition, pp. 3768–3773. IEEE (2014)
43. Tu, M., Huang, J., He, X., Zhou, B.: Multiple instance learning with graph neural networks. arXiv preprint arXiv:1906.04881 (2019)
44. Wang, X., Yan, Y., Tang, P., Bai, X., Liu, W.: Revisiting multiple instance neural networks. Pattern Recogn. **74**, 15–24 (2018)

45. Yan, Y., Wang, X., Guo, X., Fang, J., Liu, W., Huang, J.: Deep multi-instance learning with dynamic pooling. In: Asian Conference on Machine Learning, pp. 662–677. PMLR (2018)
46. Yeh, C.K., Kim, B., Arik, S.O., Li, C.L., Pfister, T., Ravikumar, P.: On completeness-aware concept-based explanations in deep neural networks. In: Advances in Neural Information Processing Systems (2019)
47. Zhao, Z., et al.: Drug activity prediction using multiple-instance learning via joint instance and feature selection. BMC Bioinform. **14**, S16 (2013). Springer

VCNet: A Self-explaining Model for Realistic Counterfactual Generation

Victor Guyomard[1,2(✉)], Françoise Fessant[1], Thomas Guyet[3], Tassadit Bouadi[2], and Alexandre Termier[2]

[1] Orange Labs, Lannion, France
victor.guyomard@orange.com
[2] University of Rennes, Inria, CNRS, IRISA, Rennes, France
[3] Inria, Centre de Lyon, Villeurbanne, France

Abstract. Counterfactual explanation is a common class of methods to make local explanations of machine learning decisions. For a given instance, these methods aim to find the smallest modification of feature values that changes the predicted decision made by a machine learning model. One of the challenges of counterfactual explanation is the efficient generation of realistic counterfactuals. To address this challenge, we propose VCNet – Variational Counter Net – a model architecture that combines a predictor and a counterfactual generator that are jointly trained, for regression or classification tasks. VCNet is able to both generate predictions, and to generate counterfactual explanations without having to solve another minimisation problem. Our contribution is the generation of counterfactuals that are close to the distribution of the predicted class. This is done by learning a variational autoencoder conditionally to the output of the predictor in a join-training fashion. We present an empirical evaluation on tabular datasets and across several interpretability metrics. The results are competitive with the state-of-the-art method.

Keywords: Interpretability · Counterfactual explanation · Realistic counterfactuals · Join training · Conditional VAE · Generative network

1 Introduction

Improvements of machine learning techniques for decision systems has led to the rise of applications in various domains such as healthcare, credit or justice. The eventual sensitivity of such domains, as well as the black-box nature of the algorithms, has motivated the need for methods that explain why some prediction was made. For example, if a person's loan is rejected as a result of a model decision, the bank must be able to explain why. In such a context, it might be interesting to provide an explanation of what that person should change to influence the model's decision. As suggested by Wachter et al. [27], one way to build this type of explanation is through the use of counterfactual explanations. A counterfactual is defined as the smallest modification of feature values that

M.-R. Amini et al. (Eds.): ECML PKDD 2022, LNAI 13713, pp. 437–453, 2023.
https://doi.org/10.1007/978-3-031-26387-3_27

changes the prediction of a model to a given output. In addition, the explanation also provides important feedback to the user. In the context of a denied credit, a counterfactual is a close individual for whom his credit is accepted and the feature modifications suggested by a counterfactual acts as recourse for the user. For privacy reason or simply because there is no similar individual with an opposite decision, we aim to generate synthetic individuals as counterfactuals. In order to provide a meaningful recourse, the counterfactual is expected to be realistic, i.e. close to the existing examples and respecting the observed correlation among features. Furthermore, in order to be representative of its predicted class, it is interesting to obtain a counterfactual close to the existing examples but relative to the counterfactual class. A counterfactual instance is usually found by iteratively perturbing the input characteristics of the original example until the desired prediction is achieved, which is like minimizing a loss function using an optimization algorithm. Many methods proceed in this way, but differ in their definition of the loss function and optimization method [16]. These approaches appear to be computationally expensive. Indeed, for each instance to explain, a new optimisation problem has to be solved. Most of the counterfactual methods apply to already trained decision models and treat them as black boxes in the post-hoc paradigm. However, dissociating the prediction of the model from its explanation can lead to an explanation of poor quality [22].

Self-explaining models which incorporate an explanation generation module into their architecture, such that they provide explanations for their own predictions, can be an alternative to the previous approaches. In general, the predictor and explanation generator are trained jointly, hence the presence of the explanation generator is influencing the training of the predictor [6]. In this spirit, Guo et al. [8] propose CounterNet, a neural network based architecture for the prediction and counterfactual generation along with a novel variant of back propagation to ensure the stabilization of the training process. Compared to a post-hoc approach, they are able to produce counterfactuals with higher validity. A counterfactual is said to be valid if it succeeds in reaching a different prediction. A limitation of the CounterNet approach is that counterfactuals it generates may lack realism w.r.t. the data points of the class where they are positioned. The proposed approach, VCNet, tackles this issue: similarly to CounterNet, it combines a predictor and a counterfactual generator that are jointly trained. The difference lies in the counterfactual generator based on conditional variational autoencoder (cVAE) whose latent space can be controlled and tweaked to generate realistic counterfactuals. Our approach is inspired from John et al. work about learning disentangled latent spaces in the context of text style transfer [9].

Our main contribution is the proposal of a cVAE for counterfactual generation in order to generate realistic counterfactuals. Our second contribution is a self-explainable architecture of a classifier that embeds a cVAE, used as a counterfactual generator. In this architecture, both the model and the cVAE, are jointly trained with an efficient single back-propagation procedure. After recalling the properties of the variational autoencoder models (Sect. 3) that interest us in the context of this paper, we describe our proposal (Sect. 4). Then we present extensive experimental studies on synthetic and real data (Sect. 5). We compare the quality of the counterfactuals produced with those of CounterNet on different datasets through state-of-the-art metrics. The focus is on tabular data but we also show that our architecture is interesting on images.

2 Related Work

Our work is concerned with the search for counterfactual explanation that is usually found by iteratively perturbing the input features of the instance of interest until the desired prediction is reached. In practice, the search of counterfactuals is usually posed as an optimization problem. It consists of minimizing an objective function which encodes desirable properties of the counterfactual instance with respect to the perturbations. Wachter et al. [27] propose the generation of counterfactual instances by minimizing the distance between the instance to be explained and the counterfactual while pushing the new prediction toward the desired class. Other algorithms optimize other aspects with additional terms in the objective function such as actionability [25], diversity [17,23], realism [26].

All aforementioned techniques search for counterfactual example by solving a separate optimization problem for each instance to be explained. This optimization problem is computationally intensive, making it impractical for large numbers of instances. To address this issue, several frameworks based on generative models have been proposed. A generative counterfactual model is trained to predict the counterfactual perturbations or instances directly. Many of these frameworks use the latent space of variational autoencoder models to generate counterfactuals with linear interpolation [2], latent feature selection [5], perturbation [21] or incorporation of the target class in the latent space [18]).

All these counterfactual generation techniques are post-hoc as they assume that the explaining task is done after the decision task with a fixed black-box model. In this post-hoc paradigm, the counterfactual search process is completely uninformed from the decision process. Post-hoc explanations may be the only option for already-trained models.

Another approach is to design models that optimize both the decision and an explanation of that decision during the learning process [1]. In the context of counterfactual explanations, such a strategy has been recently proposed by Guo et al. [8] with CounterNet, a framework in which prediction and explanation are jointly learned. The optimization of the counterfactual example generation only once together with the predictive model allows a better alignment between predictions and counterfactual explanations. This leads to explanations of better quality and significantly reduces the generation process time. Our architecture is inspired by theirs but we have chosen to use a conditional variational autoencoder as a generative model of counterfactuals allowing us to exploit text style-transfer techniques [9,18].

3 Backgrounds

In this section, we introduce some notations, problematic and backgrounds necessary for the presentation of the VCNet architecture in the next section.

Let $\mathcal{X} \subseteq \mathbb{R}^p$ represents the p-dimensional feature space and $l \geq 2$ a number of classes. A training dataset, denoted $D = \{(\boldsymbol{x}_i, y_i)\}_{i=1}^n$, is such that $\boldsymbol{x}_i \in \mathcal{X}$ and $y_i \in \{1, \dots, l\}$ for each $i \in \{1, \dots, n\}$. For a new example $\boldsymbol{x}$, the prediction consists in deciding to which class $\hat{y}$ the example $\boldsymbol{x}$ belongs. For more generality, we consider probabilistic prediction: the prediction is a probability vector, denoted $\hat{\boldsymbol{p}} \in [0,1]^p$. Then, the predicted class is the most likely class according to $\hat{\boldsymbol{p}}$. The counterfactual generation yields an example $\boldsymbol{x}'$ which is close to $\boldsymbol{x}$ and whose prediction $\hat{y}'$ is different from $\hat{y}$ in case the counterfactual is valid.

Our problem is both to learn from D an accurate predictor, denoted f, and a generator of counterfactual, denoted g.

Now that we have presented our problem, we introduce the notion of VAE [11] and cVAE [24] which our proposal relies on.

3.1 Variational Autoencoder (VAE)

A variational autoencoder is a generative model where a latent parameterized distribution is learned. If samples are drawn in the latent space according to this distribution, the decoded samples are expected to be distributed according to the training data distribution (an approximate distribution of the training data distribution is learned) [11]. Formally, let $\boldsymbol{z}$ be a latent vector (drawn from the latent distribution) and $\boldsymbol{x}$ be an example, we denote by $q_\phi(\boldsymbol{z} \mid \boldsymbol{x})$ the encoder distribution and by $p_\theta(\boldsymbol{x} \mid \boldsymbol{z})$ the decoder distribution. Training a VAE is finding the parameters θ and ϕ that minimize the following objective function, i.e. the opposite of the Evidence Lower Bound (ELBO):

$$\mathcal{L}_{\text{VAE}}(\theta, \phi) = -\mathbb{E}_{q_\phi(\boldsymbol{z}|\boldsymbol{x})}\left[\log(p_\theta(\boldsymbol{x} \mid \boldsymbol{z}))\right] + \mathrm{D}_{KL}\left[q_\phi(\boldsymbol{z} \mid \boldsymbol{x}) \middle\| p(\boldsymbol{z})\right] \tag{1}$$

Generally, distributions are chosen to be Gaussian, meaning that $q_\phi(\boldsymbol{z} \mid \boldsymbol{x}) \sim \mathcal{N}(\mu_\phi, \Sigma_\phi)$ and $p_\theta(\boldsymbol{x} \mid \boldsymbol{z}) \sim \mathcal{N}(\boldsymbol{x} \mid \mu_\theta, \Sigma_\theta)$ and distribution parameters are estimated thanks to back-propagation. The first term of Eq. 1 encourages reconstructing $\boldsymbol{x}$ at the output of the decoder ($\hat{\boldsymbol{x}}$). The second term encourages the regularization of the latent space by pushing $q_\phi(\boldsymbol{z} \mid \boldsymbol{x})$ to a Gaussian prior $p \sim \mathcal{N}(0, I)$.

3.2 Conditional Variational Autoencoder (cVAE)

A conditional variational autoencoder is a VAE where distributions are conditioned on a given variable c [24]. The architecture is the same as a standard VAE but c is given as input of the encoder and also as input of the decoder. Then the objective function of Eq. 1 can be rewritten as:

$$\mathcal{L}_{\text{cVAE}}(\theta, \phi) = -\mathbb{E}_{q_\phi(\boldsymbol{z}|\boldsymbol{x},c)}\left[\log(p_\theta(\boldsymbol{x} \mid \boldsymbol{z}, c))\right] + \mathrm{D}_{KL}\left[q_\phi(\boldsymbol{z} \mid \boldsymbol{x}, c) \middle\| p(\boldsymbol{z})\right] \tag{2}$$

The encoder distribution becomes $q_{\phi}(\boldsymbol{z} \mid \boldsymbol{x}, c)$ and is pushed to the Gaussian prior $p \sim \mathcal{N}(\boldsymbol{z} \mid \boldsymbol{0}, \boldsymbol{I})$ by the regularization term regardless of the value of c. The decoder reconstructs $\boldsymbol{x}$ from the concatenation of $\boldsymbol{z}$ with c.

The initial objective of conditional variational autoencoder is to enrich the expressiveness of the model in supervised settings. In this article, we use a property of the cVAE to disentangle the class specification from the content of the data [12].[1] Intuitively, the latent variable $\boldsymbol{z}$ does not need to model the example category, then it can focus on modeling the content of the examples, which is shared by all the categories. To illustrate the effective disentanglement of category and content, Kingma et al. show that the decoder $p_{\theta}(\boldsymbol{x} \mid \boldsymbol{z}, c)$ can be used to generate images of the ten digits with the same shared content (let say the handwriting) by changing the class c and keeping the same random values for $\boldsymbol{z}$. The same property has been applied to text style transfer [9]. In this context, the style is the category, and the content is the wording. For tabular data, the notion of "content" and "style" can be illustrated in the context of the loan decision. The "style" characterizes the category of people loan (*accepted* or *rejected*) and the "content" characterizes the other features. More specifically, the later models correlations between variables that are independent from the loan decisions.

In our proposal, we exploit the modeling properties of a cVAE to generate counterfactuals. Considering that the cVAE disentangles the category and the content, the decoder of a cVAE can be tweaked in a flexible way. For $\boldsymbol{z}$ the encoding of an example $\boldsymbol{x}$ of class c, $p_{\theta}(\boldsymbol{x} \mid \boldsymbol{z}, c')$ generates examples of a category $c' \neq c$. In addition, $(\boldsymbol{z}, c')$ is likely the element of class c that is the closest to $(\boldsymbol{z}, c)$ in the latent space. As this space is regularized, $p_{\theta}(\boldsymbol{x} \mid \boldsymbol{z}, c')$ generates examples that are similar to $\boldsymbol{x}$, but belonging to a different class. This fits exactly the expectations of counterfactuals and will be assessed in Sect. 5.1.

4 A Join Training Model

VCNet is a self-explainable[2] model through counterfactual generation. Inspired by CounterNet [8], the VCNet model is made of a predictor, $f(\cdot)$, and a counterfactual generator, $g(\cdot, \cdot)$, that are jointly trained. In the inference phase, each part can be used on demand: on the one side, to get the prediction $f(\boldsymbol{x})$ for some new example $\boldsymbol{x}$ and, on the other side, to generate its counterfactual $g(\boldsymbol{x}, c)$ for another class c. VCNet can be used as a self-explainable model and generates $(f(\boldsymbol{x}), g(\boldsymbol{x}, c))$, i.e. the couple of the prediction and its counterfactual.

The trick of VCNet is to not train a counterfactual generator, but a supervised autoencoder, i.e. a cVAE. The cVAE is trained as an autoencoder and used as a counterfactual generator in inference.

[1] For Kingma et al. [12], what we call the "content" in this paper is denoted the "style". It refers to the writing style of digits in MNIST-like datasets.

[2] By Self-explainable model here we mean that the predictor is constrained by the counterfactual generator during training but the explanation is not directly used to produce model output as in [1].

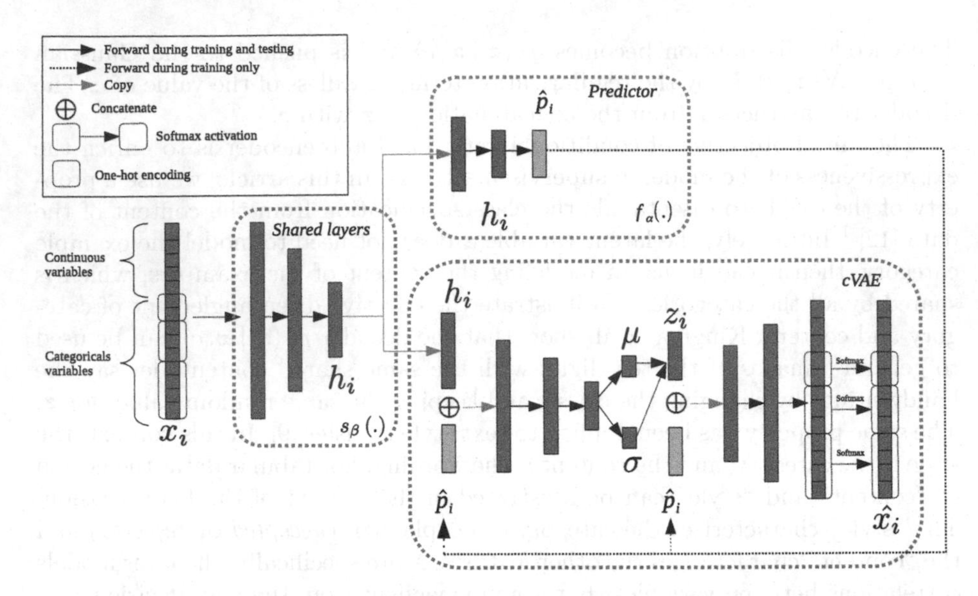

Fig. 1. VCNet architecture is composed of three blocks: Shared layers that transform the input into a latent representation (blue square), a predictor that outputs the prediction (brown square), and a conditional variational autoencoder that acts as a counterfactual generator during testing (red square). (Color figure online)

We start by presenting the architecture of our network, then we detail the training problem by defining the loss which is optimized and finally, we detail how the trained model is used to generate counterfactuals.

4.1 VCNet Architecture

VCNet is a neural network architecture. Figure 1 illustrates this architecture made of three main blocks:

1. Shared layers, s_β, that transform the input into a dense latent representation. We use fully connected layers with ELU activation functions.
2. A predictor network f_α that takes the shared latent representation and returns a probability vector corresponding to the probabilistic prediction. We use fully connected layers with ELU activation functions.
3. A conditional variational autoencoder that takes as input the shared latent representation and also the probability vector given by the predictor. The cVAE reconstructs examples and integrates additional layers to handle categorical variables (see details below).

During training, an example $\boldsymbol{x}_i$ is passed through the shared layers to generate a dense latent representation $\boldsymbol{h}_i = s_\beta(\boldsymbol{x}_i)$. This representation is then shared with both the predictor network and the autoencoder. On the one hand,

$\boldsymbol{h}_i$ is passed through the predictor f_α in order to obtain the probability vector $\hat{\boldsymbol{p}}_i$. Then, the prediction of an example $\boldsymbol{x}_i$ is obtained by the function $f_{\alpha,\beta}(\boldsymbol{x}_i) = f_\alpha \circ s_\beta(\boldsymbol{x}_i)$. On the other hand, $\boldsymbol{h}_i$ is passed through the cVAE. It is first concatenated with $\hat{\boldsymbol{p}}_i$ and then is passed through the encoder of the cVAE and samples a latent vector $\boldsymbol{z}_i$. This latent vector concatenated with $\hat{\boldsymbol{p}}_i$ is passed through the decoder, so as to obtain a reconstructed example $\hat{\boldsymbol{x}}_i$.

It can be noticed that the cVAE slightly differs from the original cVAE [24]. Indeed, formally, the encoder of an end-to-end cVAE includes the shared layer, s_β. In our architecture, the condition is introduced at an intermediary latent representation of the examples. The idea behind this architecture is to enforce the dense latent representation to be as independent of the class as possible.

In addition, we adopt the same preprocessing as Guo et al. [8] to handle categorical variables. At the input of the network, each categorical variable is encoded with a one-hot. At the output of the cVAE, we add a softmax activation function for each one-hot categorical variable in order to obtain a one-hot encoding format by taking the *argmax*. Finally, continuous variables are scaled to have all variables in [0, 1].

4.2 Loss Function and Training Procedure

The objective of the training is to jointly learn the predictor and the cVAE. Then, the loss to minimize is made of two terms.

The first term is derived from the loss of a cVAE defined in Eq. 2. In our case, the cVAE is conditioned by the probability vector obtained at the output of the predictor. Then we can rewrite the objective function as:

$$\begin{aligned}\mathcal{L}_{cVAE}(\theta,\phi,\beta;\boldsymbol{x}_i) = &-\lambda_3\mathbb{E}_{q_\phi(\boldsymbol{z}_i|s_\beta(\boldsymbol{x}_i),\hat{\boldsymbol{p}}_i)}\left[\log(p_\theta(\boldsymbol{x}_i \mid \boldsymbol{z}_i,\hat{\boldsymbol{p}}_i))\right] \\ &+\lambda_1\mathrm{D}_{KL}\left[q_\phi\left(\boldsymbol{z}_i \mid s_\beta(\boldsymbol{x}_i),\hat{\boldsymbol{p}}_i\right) \middle\| p(\boldsymbol{z}_i)\right] \end{aligned} \quad (3)$$

λ_1 and λ_3 are weights to control the impact of each term during the training phase.

The second term enables us to learn the predictor. We use cross-entropy as classification loss between the output of the predictor $\hat{\boldsymbol{p}}_i = f_\alpha \circ s_\beta(\boldsymbol{x}_i)$ and the true label y_i:

$$\mathcal{L}_{pred}(\alpha,\beta;\boldsymbol{x}_i,y_i) = \sum_{k=1}^{l} -\mathbb{1}_{[y_i=k]}\log\left(\left[f_\alpha \circ s_\beta(\boldsymbol{x}_i)\right]_k\right) \quad (4)$$

where $[f_\alpha \circ s_\beta(\boldsymbol{x}_i)]_k$ denotes the predicted probability that $\boldsymbol{x}_i$ belongs to the k-th class.

Then, the loss function on the training set (D) is a weighted sum of the losses from Eq. 4 and Eq. 3 as follows:

$$\mathcal{L}(\theta,\alpha,\beta,\phi;D) = \sum_{i=1}^{n}\mathcal{L}_{cVAE}(\theta,\phi,\beta;\boldsymbol{x}_i) + \lambda_2\frac{1}{n}\sum_{i=1}^{n}\mathcal{L}_{pred}(\alpha,\beta;\boldsymbol{x}_i,y_i)$$

As mentioned at the beginning of this section, it is worth noticing that our problem is not to learn to generate counterfactuals. Then, contrary to CounterNet that has divergent objectives to optimize, the minimization of $\mathcal{L}$ is a simple optimization problem solved by back-propagation.

Note that λ_1,λ_2,λ_3 are hyperparameters to tune for training.

4.3 Counterfactual Generation

Since our model does not directly produce counterfactuals, some modifications are needed for inference. An example $\boldsymbol{x}_i$ is passed through the prediction network to obtain both its predicted probability vector ($\hat{\boldsymbol{p}}_i$) and its dense vector representation ($\boldsymbol{h}_i$). This dense vector representation ($\boldsymbol{h}_i$) and the predicted probability vector ($\hat{\boldsymbol{p}}_i$) are given to the encoder of the cVAE to produce a latent vector $\boldsymbol{z}_i$. Then, the decoder of the cVAE plays the role of a counterfactual generator. Because we want to generate an example with a different predicted class we need a probability vector $\boldsymbol{p}_c$ such that the class with maximum probability is different from the one of the prediction, i.e. $\arg\max(\hat{\boldsymbol{p}}_i) \neq \arg\max(\boldsymbol{p}_c)$. In a binary-classification problem setup, we decided to use a one-hot vector where the probability is 0 for the predicted class of x_i and 1 for the opposite class. The reason for this choice is that we want to generate counterfactuals for which the confidence in the predicted class is the highest for the predictor. In the case of a multi-class classification problem, we propose to select the class with the second-highest probability in $\hat{\boldsymbol{p}}_i$ and to switch the values with the predicted class. For example, if we obtain a probability vector $\hat{\boldsymbol{p}}_i = [0.6, 0.3, 0.1]$ then $\boldsymbol{p}_c = [0.3, 0.6, 0.1]$. An alternative solution would be to let the user select the class for which he/she is interested in having a counterfactual.

This vector $\boldsymbol{p}_c$ and the dense latent representation $\boldsymbol{z}_i$ are passed through the cVAE decoder in order to infer a new predicted class to obtain a counterfactual[3] $\boldsymbol{x}_c$. As explained in Sect. 3.2, the intuition is to benefit from the disentanglement of the latent space of a cVAE: $\boldsymbol{z}_i$ contains non-class-specific information about $\boldsymbol{x}_i$ and $\boldsymbol{p}_c$ encodes information for the desired class. As such, the decoder generates a new example $\boldsymbol{x}_c$ that is similar to $\boldsymbol{x}_i$ and that belongs to a different class.

5 Experiments and Results

Our experiments are organized in four steps. Our main objective is to show that VCNet generates counterfactuals that are both valid (counterfactuals actually belong to another class) and realistic (counterfactuals are close to examples of the same class). In the first set of experiments, we present results of a cVAE on a synthetic dataset to confirm the actual disentanglement of the content and the class. These experiments also aim at providing some intuition about the reason why VCNet works. In the second set of experiments, we compare the results of

[3] Note that the quality of the generated counterfactual depends on the quality of the learned latent space.

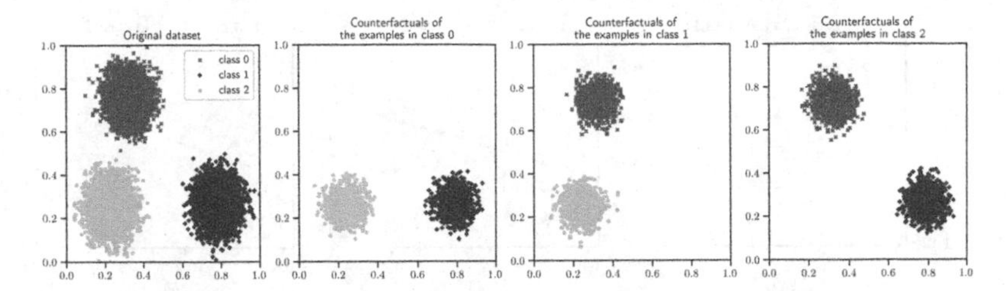

Fig. 2. Comparison of the examples/counterfactuals distributions for synthetic data. All graphics represent the projection of the examples or counterfacturals on the first two dimensions of the examples space ($\mathbb{R}^8$). The graphic on the left represents the original dataset. The three other graphics represent counterfactuals generated from the examples of the test set belonging to each class. For this three rightmost graphics, the same examples have been used to generate counterfactuals with the two other classes. The color/shape of the point represents a class information: the class an example belongs to (graphic on the left) and the class requested for counterfactual generation (3 graphics on the right). (Color figure online)

VCNet with CounterNet, the state-of-the-art algorithm for self-explainable counterfactual generation. The reader interested in the results of more counterfactual generation systems may refer to the original article of Guo et al. [8]. Our experiments use the same datasets and data preprocessing. In the third experiment, we evaluate the impact of join training on the quality of counterfactuals. We propose a post-hoc version of our framework and compare the results obtained with the jointly trained VCNet. Finally, we present some qualitative experiments on the MNIST dataset. We choose to present experiments on MNIST firstly because it has been widely used in the field of counterfactual generation and, secondly, because it illustrates that VCNet may be applied on different types of data (tabular, images, time series, ...).

The hyperparameters and architectures of the models used in these experiments are detailed in Supplementary material. The code to reproduce the results of this section is also provided in supplementary material.

5.1 cVAE for Counterfactual Generation

Our proposal is based on using a cVAE to generate counterfactuals. It relies on the capability of a cVAE to actually disentangle the class and its content such that the decoder can be used to generate counterfactual examples by changing the class conditioning. In this experiment, we generate a synthetic dataset of examples in $\mathbb{R}^8$ with three classes. Each class is distributed according to a multidimensional Gaussian distribution (see Fig. 2 on the left).

A cVAE, i.e. a couple of an encoder $\varphi(\boldsymbol{x}, c)$ and a decoder $\psi(\boldsymbol{z}, c)$, is trained on a set of $10k$ examples. The complete code of these experiments is available in supplementary material for the sake of reproducibility.

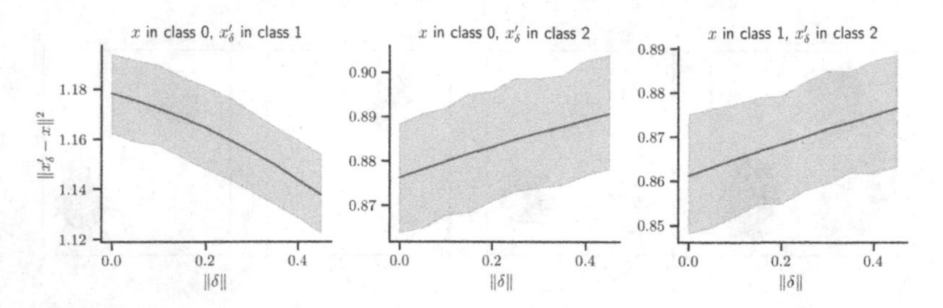

Fig. 3. Distance between $\boldsymbol{x}$, an example of class c to explain, and $\boldsymbol{x}' = \psi(\boldsymbol{z} + \boldsymbol{\delta}, c')$, a counterfactual for class c' perturbed in latent space by $\boldsymbol{\delta}$. Each graphic illustrates this distance with respect to $\|\boldsymbol{\delta}\|$ depending on the class c and c' (on the right: $c = 0$ and $c' = 1$; in the middle: $c = 0$ and $c' = 2$; on the left: $c = 1$ and $c' = 2$). The mean and variance are computed on 10 random δ.

Figure 2 illustrates the capability of a cVAE to generate realistic examples when changing the class that conditions the decoder. The figure on the left illustrates the dataset. Each colored group of point corresponds to a class. The three figures on the right illustrate datasets that have been generated $\boldsymbol{x}' = \psi(\varphi(\boldsymbol{x}, c), c')$ where $\boldsymbol{x}$ is an example from the test that belongs to class c (origin class). Each figure corresponds to one origin class, the colors of the point correspond to the conditioning class (c'). We observe that all figures look similar. This means that taking $\psi(\varphi(\boldsymbol{x}, c), c')$ generates an example that looks similar to an example of class c' whatever the origin class of $\boldsymbol{x}$. Thus, it demonstrates that cVAE can be used to generate realistic counterfactuals. Moreover, $\boldsymbol{x}'$ is a good counterfactual if it is similar to $\boldsymbol{x}$. The question is whether $\boldsymbol{z} = \varphi(\boldsymbol{x}, c)$ is a better choice to generate an example $\psi(\boldsymbol{z}, c')$ than any other example $\psi(\boldsymbol{z}', c')$ (which also likely belongs to c'). To assess this behavior, we randomly generate 10 latent representations $\boldsymbol{z}' = \boldsymbol{z} + \boldsymbol{\delta}$ for each $\boldsymbol{x}$, and compute the Euclidean distance between $\boldsymbol{x}' = \psi(\boldsymbol{z}', c')$ and $\boldsymbol{x}$.

Figure 3 shows three graphs: one per couple of classes (the class of the example to explain and the class requested as counterfactual). Each graph illustrates the mean Euclidean distance between $\boldsymbol{x}'$ and $\boldsymbol{x}$ with respect to $\|\boldsymbol{\delta}\|$. When $\|\boldsymbol{\delta}\| = 0$, it uses the latent representation of $\boldsymbol{x}$ as input of the cVAE decoder. Intuitively, we expect to have $\psi(\varphi(\boldsymbol{x}, c), c')$ closer to $\boldsymbol{x}$ than $\psi(\varphi(\boldsymbol{x}, c) + \boldsymbol{\delta}, c')$ and thus, that the higher $\|\boldsymbol{\delta}\|$, the higher the mean distance to the original example. The two graphics on the right illustrates the expected behavior. In these cases, the latent representation of the example seems to generate a counterfactual that is the most similar among the examples of the opposite class. Nonetheless, we observe that it is not always the case. We can note that the distance decreases when perturbing the latent representation of examples in class 0 and regenerating counterfactual examples in class 1. This may be explained by the proximity between the two classes.

5.2 Comparison Between VCNet and CounterNet

This section compares the quality of counterfactuals of VCNet against CounterNet. We conduct evaluations on six binary-classification datasets with various properties: Adult [13], HELOC [7], OULAD [14], Breast Cancer Wisconsin [3], Student performance [4] and Titanic [10]. Some of these datasets contain only numerical variables but some others, such as Adult, contain both numerical and categorical variables. More details about the datasets can be found in the supplementary material of the article.

To evaluate the counterfactual quality, we used the following four metrics that are classical in the literature.

Prediction Gain: The prediction gain is given by the difference between the predicted probability of the counterfactual $\boldsymbol{x}'$ and the predicted probability of the example $\boldsymbol{x}$, according to the counterfactual class [19].

$$\text{Gain} = [f_{\text{pred}}(\boldsymbol{x}')]_{y_i} - [f_{\text{pred}}(\boldsymbol{x})]_{y_i}$$

where y_i denotes the predicted class for the counterfactual. A higher prediction gain means being more confident in the class change of the counterfactual.

Validity: A counterfactual is valid if it achieves to obtain a different predicted class [17,20]. Then:

$$\text{Validity} = \begin{cases} 0, & \text{if } y_i = y_0 \\ 1, & \text{if } y_i \neq y_0 \end{cases}$$

where y_i denotes the predicted class for the counterfactual and y_0 the predicted class for the example to explain.

Proximity: The proximity is the L_1 distance between an example, $\boldsymbol{x}$ and its counterfactual, $\boldsymbol{x}'$ [17,27].

$$\text{Proximity} = \|\boldsymbol{x}' - \boldsymbol{x}\|_1 = \|\boldsymbol{\delta}\|_1$$

A low value indicates fewer changes of features to apply to the original example to obtain the counterfactual.

Proximity Score: This metric is inspired from Laugel et al. [15] to quantify the distance of a counterfactual to examples of the same predicted class:

$$P_s(\boldsymbol{x}') = \frac{d(\boldsymbol{x}', a_0)}{\frac{1}{\|H\|(\|H\|-1)/2} \sum_{\boldsymbol{x}_i, \boldsymbol{x}_j \in H} d(\boldsymbol{x}_i, \boldsymbol{x}_j)}$$

where $d(\boldsymbol{x}', a_0)$ is the Euclidean distance of the counterfactual to the closest example that has the same predicted class (a_0) and H is the set of existing examples that have the same predicted class as $\boldsymbol{x}'$. The insight behind this metric is that the counterfactual should be close to an existing example of the same predicted class relative to the rest of the data. Note that to be evaluated, this metric requires a set of m examples $X \in \mathbb{R}^{m \times p}$.

Table 1. Comparison of quality metrics of counterfactuals and predictive accuracy for three methods: VCNet, CounterNet and Post-hoc VCNet. Bold items give the best metric values among the three methods.

Datasets	Metrics	Methods		
		VCNet	CounterNet	Post-hoc VCNet
Adult	Validity	**1.0**	0.99	0.84
	Proximity	7.71 ± 2.11	$\mathbf{7.16 \pm 2.13}$	7.28 ± 2.23
	Prediction gain	$\mathbf{0.76 \pm 0.15}$	0.61 ± 0.17	0.47 ± 0.35
	Proximity score	$\mathbf{0.04 \pm 0.11}$	0.31 ± 0.28	0.06 ± 0.14
	Accuracy	**0.83**	**0.83**	**0.83**
OULAD	Validity	**1.0**	0.99	0.74
	Proximity	11.66 ± 2.46	11.96 ± 2.40	$\mathbf{11.22 \pm 2.54}$
	Prediction gain	$\mathbf{0.93 \pm 0.12}$	0.74 ± 0.13	0.66 ± 0.44
	Proximity score	$\mathbf{0.38 \pm 0.18}$	0.46 ± 0.16	$\mathbf{0.38 \pm 0.18}$
	Accuracy	**0.93**	**0.93**	**0.93**
HELOC	Validity	**1.0**	0.99	0.77
	Proximity	5.60 ± 2.11	$\mathbf{4.41 \pm 1.80}$	5.09 ± 1.71
	Prediction gain	$\mathbf{0.64 \pm 0.13}$	0.56 ± 0.15	0.24 ± 0.25
	Proximity score	$\mathbf{0.23 \pm 0.21}$	0.49 ± 0.35	0.40 ± 0.32
	Accuracy	0.71	**0.72**	0.71
Student	Validity	0.96	**1.0**	0.46
	Proximity	19.90 ± 3.21	19.86 ± 2.78	$\mathbf{19.68 \pm 3.03}$
	Prediction gain	$\mathbf{0.86 \pm 0.27}$	0.76 ± 0.05	0.41 ± 0.46
	Proximity score	$\mathbf{0.70 \pm 0.08}$	0.73 ± 0.06	0.75 ± 0.08
	Accuracy	0.90	**0.92**	0.90
Titanic	Validity	0.92	**0.99**	0.38
	Proximity	15.43 ± 3.79	$\mathbf{15.15 \pm 4.05}$	15.56 ± 5.23
	Prediction gain	$\mathbf{0.69 \pm 0.31}$	0.66 ± 0.15	0.26 ± 0.36
	Proximity score	$\mathbf{0.71 \pm 0.21}$	0.80 ± 0.16	1.21 ± 0.26
	Accuracy	0.82	**0.83**	0.82
Breast-cancer	Validity	**1.0**	**1.0**	0.59
	Proximity	5.27 ± 1.47	$\mathbf{1.51 \pm 1.01}$	7.71 ± 1.67
	Prediction gain	$\mathbf{0.95 \pm 0.11}$	0.69 ± 0.15	0.60 ± 0.45
	Proximity score	$\mathbf{0.28 \pm 0.03}$	0.72 ± 0.48	0.94 ± 0.07
	Accuracy	**0.96**	**0.96**	**0.96**

For each dataset, we choose a random sample of size 25% for counterfactual generation. Then, we compute the mean and standard deviation of each metric for every selected random sample.

Table 1 provides results of VCNet and CounterNet [8]. More information on the reproducibility of CounterNet results is available in Supplementary material. It is worth noting that Table 1 contains additional results for post-hoc VCNet that will be discussed in Sect. 5.3.

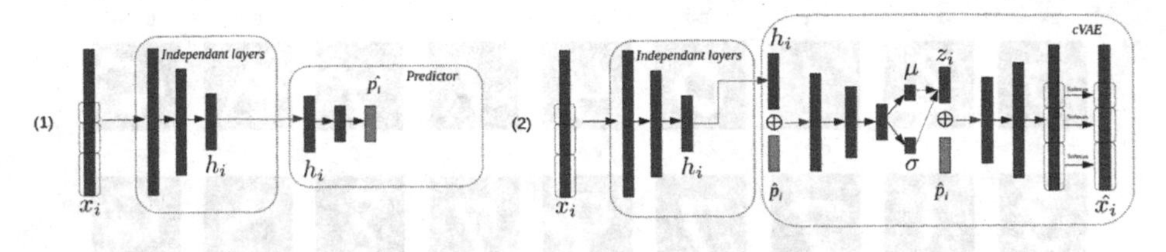

Fig. 4. Post-hoc version of VCNet. (1) is the prediction model and (2) is the cVAE model.

Counterfactual Quality: VCNet achieves perfect validity for 4 of the 6 datasets, and a lower validity of respectively 4.5% and 7.5% for the 2 other datasets (Student and Titanic) compared to CounterNet.

As far as prediction gain and proximity score are concerned, VCNet outperforms CounterNet for all the 6 datasets. The higher the prediction gain, the more confidence one can have in the prediction related to the class change of the counterfactual. At the same time, a low proximity score reflects the achievement of counterfactuals that are close to real examples belonging to the same class as predicted for the counterfactual.

For the last evaluated metric that is proximity, we observe that VCNet achieves higher values than CounterNet on 5 of the 6 datasets. A larger proximity value means that the counterfactuals are obtained at the cost of larger changes in the input space.

Predictive Accuracy: Both CounterNet and VCNet are self-explainable models, and if the previous results show that VCNet generates better counterfactuals, it can not be at the expense of the prediction accuracy. Thus, Table 1 also presents model accuracies. We observe that VCNet achieved similar performances on 3 of the 6 datasets. On the other hand, the accuracies for the other datasets are lower by 0.4% (HELOC) to 2% (Student), which shows that our method still performs very well in terms of prediction.

5.3 Impact of Join-Training on Counterfactual Quality

We derived our architecture to a post-hoc version (see Fig. 4). Its training procedure is the following: 1) we first train a prediction model. For our comparisons here, it is composed of the concatenation of the shared layers block and the predictor block of VCNet, but in practice it can be any machine learning model that outputs a probability score. Once the model is trained, we obtain a probability vector $\hat{\boldsymbol{p}}_i$ by forwarding an example $\boldsymbol{x}_i$ to the model. 2) then we train a cVAE model conditionally to the probability vector ($\hat{\boldsymbol{p}}_i$) output by the predictor learned during Step 1. The cVAE is composed of the same shared layers block than VCNet, but it is not shared with the predictor.

We compare VCNet with its post-hoc version (Post-hoc VCNet) in order to study the impact of join-training on counterfactuals. Table 1 provides the results

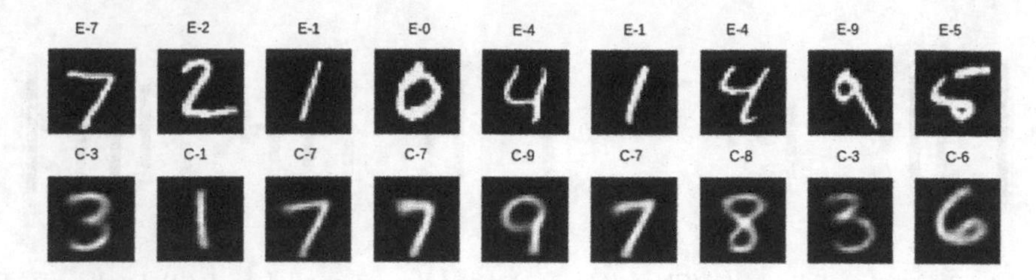

Fig. 5. Counterfactuals obtained with VCNet for the MNIST dataset. The top line corresponds to the examples to explain, the bottom to the corresponding counterfactuals.

of the post-hoc version compared to those obtained previously with the join-training approach. We observe a drop of validity for every dataset, which justifies that the join-training approach allows a better alignment of the explanation task with the prediction task. We also observe a significant decrease in prediction gain for all datasets, which means less changes between an example to explain and its counterfactual. Besides, proximity is higher for 3 datasets (Adult, OULAD, Breast-cancer) and lower by respectively 9%, 1% and 0.8% on the remaining datasets (HELOC, Student, Titanic). Thus, we can argue that this drop of prediction gain does not benefit to closer counterfactuals w.r.t. examples to explained. In terms of model accuracies, training the prediction model alone leads to comparable results, which indicates that the join-training approach is not at the cost of a lower accuracy.

5.4 Qualitative Results on MNIST Dataset

We evaluated VCNet on the MNIST dataset[4] with metrics suggested in Sect. 5.2. This experiment illustrates that VCNet is adaptable to several types of data including image datasets. As CounterNet was not applied to image data in the original paper, we do not offer a comparison with this model here. This avoids a poor adaptation of CounterNet and an unfair comparison.

VCNet gives a mean validity of 0.99, a mean prediction gain of 0.98 and an accuracy of 0.98. The counterfactual quality metrics on MNIST show that VCNet is a good model to generate realistic and valid counterfactuals and, at the same time, to make accurate predictions. These results suggest that VCNet has also good capabilities to generate counterfactuals for images, and not only for tabular data. Figure 5 presents some examples to explain (in the first line) and corresponding counterfactuals generated with VCNet (second line). Each example to explain and its counterfactual is complemented by the predicted class, for instance **E-7** means an example predicted class 7 and **C-3** means a counterfactual predicted class 3. We first notice that each counterfactual "looks like" an example that matches the predicted class. Moreover, we observe that the orientation of the digits is often preserved, for example **E-1** is converted in

[4] http://yann.lecun.com/exdb/mnist/.

C-7 by keeping the orientation of **E-1**. **C-6** is interesting as it shows that VCNet is able to provide a realistic counterfactual even if the class of the example to explain is not trivial.

6 Conclusion

In this article, we propose VCNet, a new architecture for self-explainable classification based on counterfactuals examples. Our architecture generates at the same time the decision and a counterfactual that can be used by the analyst to understand the algorithm decision. The first advantage of VCNet is to generate explanations and decisions in a simple feed forward pass of the examples. Contrary to post-hoc counterfactuals explanation, VCNet does not require expensive optimization to generate counterfactuals.

The main contribution of this article is the use of a cVAE as a counterfactual generator in our model. The choice of a cVAE yields realistic counterfactuals and it is simple to train jointly with the prediction model.

We extensively evaluated the quality of the counterfactuals and compared them to CounterNet. We conclude that VCNet generates valid and realistic counterfactuals. The VCNet counterfactuals are realistic in the sense that they are close to existing examples of the same predicted class (VCNet has better proximity scores than CounterNet) and also the result of a higher confidence in the class change (VCNet has better prediction gains than CounterNet).

Finally, VCNet is simple to train because the training procedure is not based on counterfactuals directly, but on the disentanglement of the class and the content of examples by the cVAE. It allows proposing a simple optimisation procedure which makes our approach easier to put in practice. This is illustrated by its successful application to a dataset of images. In addition, it is also assessed in terms of model accuracy. Our experiments show that our join-training approach keeps its competitive prediction performance against CounterNet. A future work is to compare VCNet performances against state-of-the-art post-hoc counterfactuals methods and also to include actionability constraints.

References

1. Alvarez Melis, D., Jaakkola, T.: Towards robust interpretability with self-explaining neural networks. In: Proceedings of the Conference on Advances in Neural Information Processing Systems (NIPS), pp. 7786–7795 (2018)
2. Barr, B., Harrington, M.R., Sharpe, S., Bruss, C.B.: Counterfactual explanations via latent space projection and interpolation. Preprint arXiv:2112.00890 (2021)
3. Blake, C.: UCI repository of machine learning databases (1998). http://www.ics.uci.edu/mlearn/MLRepository.html
4. Cortez, P., Silva, A.M.G.: Using data mining to predict secondary school student performance. In: Proceedings of Annual Future Business Technology Conference, pp. 5–12 (2008)

5. Downs, M., Chu, J.L., Yacoby, Y., Doshi-Velez, F., Pan, W.: Cruds: counterfactual recourse using disentangled subspaces. In: ICML Workshop on Human Interpretability in Machine Learning (WHI), pp. 1–23 (2020)
6. Elton, D.C.: Self-explaining AI as an alternative to interpretable AI. In: Proceedings of the International Conference on Artificial General Intelligence (AGI), pp. 95–106 (2020)
7. FICO: Explainable machine learning challenge (2018). https://community.fico.com/s/explainable-machine-learning-challenge
8. Guo, H., Nguyen, T., Yadav, A.: CounterNet: end-to-end training of counterfactual aware predictions. In: ICML Workshop on Algorithmic Recourse (2021)
9. John, V., Mou, L., Bahuleyan, H., Vechtomova, O.: Disentangled representation learning for non-parallel text style transfer. In: Proceedings of the Annual Meeting of the Association for Computational Linguistics (ACL), pp. 424–434 (2019)
10. Kaggle: Titanic - machine learning from disaster (2018). https://www.kaggle.com/c/titanic/overview
11. Kingma, D.P., Welling, M.: Auto-encoding variational Bayes. In: Proceedings of the International Conference on Learning Representations (ICLR) (2014)
12. Kingma, D.P., Mohamed, S., Jimenez Rezende, D., Welling, M.: Semi-supervised learning with deep generative models. In: Proceedings of International Conference on Neural Information Processing Systems (NIPS), pp. 3581–3589 (2014)
13. Kohavi, R., Becker, B.: UCI machine learning repository: adult data set (1996)
14. Kuzilek, J., Hlosta, M., Zdrahal, Z.: Open university learning analytics dataset. Sci. Data **4**, 170171 (2017)
15. Laugel, T., Lesot, M.J., Marsala, C., Renard, X., Detyniecki, M.: The dangers of post-hoc interpretability: unjustified counterfactual explanations. In: Proceedings of the International Joint Conference on Artificial Intelligence (IJCAI), pp. 2801–2807 (2019)
16. Molnar, C.: Interpretable Machine Learning. C. Molnar, 2nd edn (2022). https://christophm.github.io/interpretable-ml-book
17. Mothilal, R.K., Sharma, A., Tan, C.: Explaining machine learning classifiers through diverse counterfactual explanations. In: Proceedings of the Conference on Fairness, Accountability, and Transparency (FAT), pp. 607–617 (2020)
18. Nangi, S.R., Chhaya, N., Khosla, S., Kaushik, N., Nyati, H.: Counterfactuals to control latent disentangled text representations for style transfer. In: Proceedings of the Annual Meeting of the Association for Computational Linguistics (ACL), pp. 40–48 (2021)
19. Nemirovsky, D., Thiebaut, N., Xu, Y., Gupta, A.: CounteRGAN: generating realistic counterfactuals with residual generative adversarial nets. preprint arXiv:2009.05199 (2020)
20. de Oliveira, R.M.B., Martens, D.: A framework and benchmarking study for counterfactual generating methods on tabular data. Appl. Sci. **11**(16), 7274 (2021)
21. Pawelczyk, M., Broelemann, K., Kasneci, G.: Learning model-agnostic counterfactual explanations for tabular data. In: Proceedings of the Web Conference (WWW'20), pp. 3126–3132 (2020)
22. Rudin, C.: Stop explaining black box machine learning models for high stakes decisions and use interpretable models instead. Nat. Mach. Intell. **1**, 206–215 (2019)
23. Russell, C.: Efficient search for diverse coherent explanations. In: Proceedings of the Conference on Fairness, Accountability, and Transparency (FAT). Association for Computing Machinery, New York (2019)

24. Sohn, K., Lee, H., Yan, X.: Learning structured output representation using deep conditional generative models. In: Proceedings of the Conference on Advances in Neural Information Processing Systems (NIPS), pp. 3483–3491 (2015)
25. Ustun, B., Spangher, A., Liu, Y.: Actionable recourse in linear classification. In: Proceedings of the Conference on Fairness, Accountability, and Transparency (FAT), pp. 10–19 (2019)
26. Van Looveren, A., Klaise, J.: Interpretable counterfactual explanations guided by prototypes. In: Proceedings of the European Conference on Machine Learning and Knowledge Discovery in Databases (ECML/PKDD), pp. 650–665 (2021)
27. Wachter, S., Mittelstadt, B.D., Russell, C.: Counterfactual explanations without opening the black box: automated decisions and the GDPR. Harv. J. Law Technol. **31**(2), 841–887 (2018)

Ranking and Recommender Systems

A Recommendation System for CAD Assembly Modeling Based on Graph Neural Networks

Carola Gajek[1(✉)], Alexander Schiendorfer[2], and Wolfgang Reif[1]

[1] Institute for Software and Systems Engineering, University of Augsburg, Augsburg, Germany
{gajek,reif}@isse.de

[2] Institute AImotion Bavaria, Technische Hochschule Ingolstadt, Ingolstadt, Germany
Alexander.Schiendorfer@thi.de

Abstract. In computer-aided design (CAD), software tools support design engineers during the modeling of assemblies, i.e., products that consist of multiple components. Selecting the right components is a cumbersome task for design engineers as they have to pick from a large number of possibilities. Therefore, we propose to analyze a data set of past assemblies composed of components from the same component catalog, represented as connected, undirected graphs of components, in order to suggest the next needed component. In terms of graph machine learning, we formulate this as graph classification problem where each class corresponds to a component ID from a catalog and the models are trained to predict the next required component. In addition to pretraining of component embeddings, we recursively decompose the graphs to obtain data instances in a self-supervised fashion without imposing any node insertion order. Our results indicate that models based on graph convolution networks and graph attention networks achieve high predictive performance, reducing the cognitive load of choosing among 2,000 and 3,000 components by recommending the ten most likely components with 82–92% accuracy, depending on the chosen catalog.

Keywords: Graph machine learning · Recommendation · Computer-aided design · AI-aided design

1 Recommending Components in Assembly Modeling

Computer-aided design (CAD) most generally refers to using computers to support the creation, modification, analysis, or optimization of three-dimensional mechanical designs [15]. More specifically, we consider the problem of recommending existing CAD *components* to design engineers during assembly modeling, i.e., the design of new assemblies composed of those components. Consider,

Research leading to this paper was funded by the Bavarian State Ministry for Economic Affairs, Regional Development and Energy (StMWi), Project KOGNIA, in cooperation with Cadenas GmbH.

M.-R. Amini et al. (Eds.): ECML PKDD 2022, LNAI 13713, pp. 457–473, 2023.
https://doi.org/10.1007/978-3-031-26387-3_28

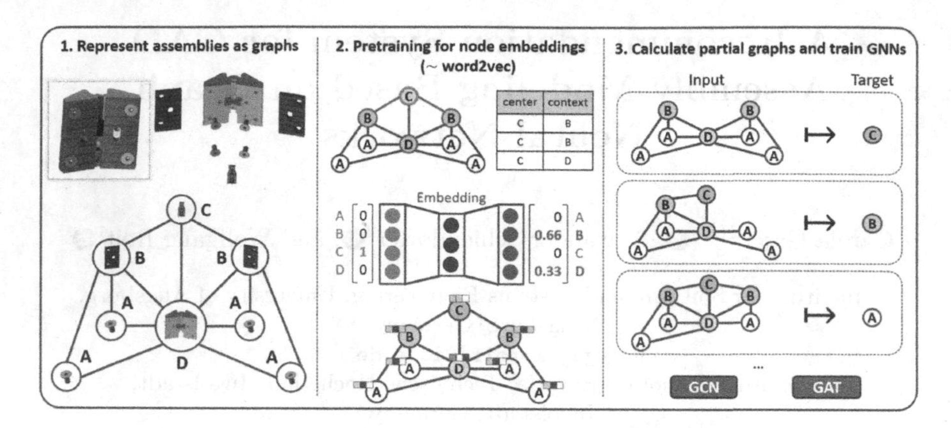

Fig. 1. Our overall approach: CAD assemblies are converted to graphs which are enriched by component representations obtained from a variation of word2vec [14]. Data instances for component recommendation are derived from CAD assemblies to train GNNs to predict the next required components during construction. Best viewed in color on screen.

as a simplified example, a cabinet that consists of five plates, a hinge, a handle, screws, etc. A governing assumption of our approach is that components that are *used together frequently* have a causal relationship that is captured in the data. For example, a heavy hinge might often be combined with a heavy door and similarly for lighter components. Such information could lead to *recommendations* regarding the next component to be inserted. Rather than having design engineers (i.e., CAD system users) manually maintain such logical and domain-specific rules (a tedious task that is likely to be neglected in practice), we strive for a data-driven solution that adapts to the sets of assemblies on which it is trained – even if that entails imperfect recommendations in some cases. That way, experienced designers' knowledge could be extracted from existing assemblies and speed up the design process.

Unfortunately, assumptions made for standard recommendation systems (collaborative or content-based filtering) do not hold in this use case. First, personalized information about the design engineers (e.g., their typical area of work, their recent designs, or even their current intention) that would be necessary for collaborative filtering is typically not available. Second, what constitutes a good recommendation for the next component depends on the intended design as opposed to simply going for a rather static "liking" of elements.

Therefore, we treat mechanical designs as undirected graphs, where nodes correspond to CAD components and edges denote connections between them. Edges can result from so-called "mating" conditions[1] in CAD systems or can be read off by geometric proximity. Even though a graph can be thought of as being generated by a sequence of node and edge insertions (and therefore be

[1] Mating conditions define relative positions of components to each other.

amenable to recurrent models), such sequential information is not stored in CAD designs. Moreover, an assembly can emerge in any order – also depending on the designer's preferences – which is why we prefer a model that is invariant to permutations. Therefore, we propose the following approach, as shown in Fig. 1:

1. Extract only components and connections as connected, undirected graphs from CAD assembly models – ignoring other metadata
2. Pretrain to get low-dimensional component embeddings using a technique based on word2vec [14], choosing an appropriate context size instead of random walks like in node2vec [5]
3. Generate data instances for graph neural networks (GNN) by means of "cutting off" nodes in a self-supervised fashion, resulting in pairs consisting of partial graph and the cut-off node
4. Train GNNs to predict the next component given a partial graph, i.e., learn a discriminative model $p(next\_node \mid partial\_graph)$ that is part of an autoregessive model which, when unrolled, leads to a generative process [9]
5. Evaluate the performance of GNNs by the top-10 rate on an unseen test set

Our approach contributes to the ongoing trend of extending CAx (computer-aided processes like design, engineering and manufacturing) to AIAx (artificial intelligence-aided processes) [3,7,20,24]. In particular, it falls into the realm of AI-aided design that supports design engineers in a data-driven fashion. Whereas most existing AIAD approaches are concerned with adequately modeling the 3D geometry [3,24], our model is only concerned with components and connections found in assemblies and deriving useful recommendations only from their usage patterns in shared component catalogs.

Related Work. The cognitive load imposed onto design engineers working with a complex CAD system has already been recognized [11]. The authors apply collaborative filtering to suggest useful software commands to improve the workflow. Our approach is based on the same motivation but focuses on components that are likely to be inserted into the current assembly instead of commands. Applying principles of search engines and information retrieval to support CAD design engineers is called assembly retrieval [13]. There, different notions of similarity (usage similarity, component overlaps, etc.) are considered to enable queries for similar assembly designs, instead of suggesting how to extend an assembly like in our approach. Recently, transformer-based generative models have been applied to CAD models by generating sequences of CAD-typical geometrical operations such as "sketching", "extruding", "boolean subtracting" [17]. While this method addresses component design tasks, our method focuses on usage similarity of components reoccurring in assemblies, abstracts away from geometrical features, and – most notably – does not require a linear order of operations.

Deep learning on non-Euclidean domains such as graphs or manifolds is often referred to as *geometric deep learning* [2]. This is not limited to the geometry in the sense of a 3D (mesh) model but stems from the fact that certain equivariances or invariances are derived from the symmetry properties prevalent in the data –

such as permutation equivariance for the neighboring nodes in a graph. In that sense, our approach belongs to geometric deep learning: a component graph that we extract from a CAD model should lead to the same hidden representation, regardless of how the components are (arbitrarily) ordered, a property that is found in graph neural networks as well as in deep sets [22].

The applicability of graph neural networks to recommendation systems has also been recognized [18], essentially since the domain of many recommendation tasks (e.g., movies to users, products to customers, etc.) can be formalized as a (hyper)graph consisting of all users and items, connecting users with their liked or purchased items. However, the goal here is to predict new connections, i.e., to recommend items to users, rather than adding new nodes to the graph.

The task of predicting the next component for an assembly can be seen as a generative model for graphs, if rolled out step by step. Existing approaches to generative deep graph models, however, tend to learn the probability distribution of the observed graphs at once, called "one-shot generating" in [6], using variational autoencoders, generative adversarial networks, or normalizing flows. The authors of [12] propose to learn a sequence of node and edge insertions (called structure building actions). Their approach, in particular the neural network architecture that they use to map an intermediate graph to the next insertion action, bears similarities in terms of the architecture we use in our approach. However, the edges can appear in any order which may lead to graphs that are not connected - which we want to explicitly exclude. Our focus on individual component insertion steps also removes the need for recurrent structures that [12] employs. Moreover, training general generative graph models is reported to be more difficult to balance which is why we focus on the discriminative task of predicting $p(next\_node \mid partial\_graph)$ but manage the generation of appropriate data via self-supervision outside the training loop (cf. Fig. 1).

2 Graph Neural Networks

In our approach, we use undirected graphs to represent assemblies. A graph $G = (\mathcal{N}, \mathcal{E})$ consists of a set of nodes $\mathcal{N}$ and a set of edges $\mathcal{E} \subseteq \mathcal{N} \times \mathcal{N}$ where an edge $e = (i, j) \in \mathcal{E}$ denotes that nodes i and j are connected. For undirected graphs, all edges are bidirectional, i.e., $(i, j) \in \mathcal{E}$ implies $(j, i) \in \mathcal{E}$. The neighbors $N(i) \subseteq \mathcal{N}$ of a node i are given by the set of nodes adjacent to it, i.e., $N(i) = \{j \in \mathcal{N} \mid (i, j) \in \mathcal{E}\}$. A node i is called a *leaf node* if it has exactly one neighbor, i.e., if $|N(i)| = 1$. We define the removal $G \setminus \{i\}$ of a node i as the graph that results from removing i from $\mathcal{N}$ and, consequently, removing all edges involving i from $\mathcal{E}$. We call a node *cohesive* if its removal would result in a graph composed of multiple connected components.

Graph neural networks (GNN) operate on graphs to perform tasks such as graph classification, node classification, or link prediction [19]. Each node $i \in \mathcal{N}$ is described by a feature vector $\mathbf{x}_i$ of dimension d (e.g., one-hot encodings of numeric indices, pretrained embeddings, or other features describing i). Stacking the feature vectors of all nodes in a graph gives rise to an input matrix $\mathbf{X}$ with dimensions $|\mathcal{N}| \times d$. The edges $\mathcal{E}$ are represented by an $|\mathcal{N}| \times |\mathcal{N}|$-adjacency matrix $\mathbf{A}$ where $\mathbf{A}_{i,j} = 1 \Leftrightarrow (i,j) \in \mathcal{E}$. In addition, features can be assigned to edges, for example, to describe different relations between users of a social network.

The representation of $\mathcal{N}$ and $\mathcal{E}$ as matrices $\mathbf{X}$ and $\mathbf{A}$ is useful due to the efficient interoperability with existing deep learning frameworks but it enforces the set $\mathcal{N}$ to be ordered. Therefore, graph neural networks require functions f that are applied to any matrix representation of the same graph to be either *permutation equivariant* or *permutation invariant*. Equivariance is used for node classification and states that permuting the output of f is the same as applying f to the permuted input, i.e., $f(\mathbf{PX}) = \mathbf{P}f(\mathbf{X})$ for a permutation matrix $\mathbf{P}$. In contrast, invariance is used for graph classification and describes that permuting the input does not affect the output, i.e., $f(\mathbf{PX}) = f(\mathbf{X})$. In our use case, permutation invariance is crucial as the CAD components have no inherent order and their insertion sequence is not given. In a GNN layer, the node representations $\mathbf{h}_i$ are updated by a function over their neighbors' features. From the perspective of a node i, the update in layer $l+1$ (called message-passing) is given by

$$\mathbf{h}_i^{(l+1)} = \phi\left(\mathbf{h}_i^{(l)}, \bigoplus_{j \in N(i)} \psi\left(\mathbf{h}_j^{(l)}\right)\right) \tag{1}$$

where ϕ and ψ are parameterized (i.e., trainable) mappings, e.g., fully connected layers followed by a nonlinearity [19]. Permutation equivariance is obtained by having ψ depend only on a single node, i.e., it is applied node-wise, and permutation invariance results from using a commutative and associative operation $\bigoplus$ such as the sum, the average, or the maximum. Since the features of each node are processed with the same transformation ψ (parameter sharing) and the aggregation can be performed with any number of elements, GNNs can handle graphs with different structures and even different sizes.

Stacking multiple GNN layers enables propagating node features to more distant nodes. Thus, a single node can contain information of all adjacent nodes within a range of l hops after l layers. In addition to message-passing layers, so-called *readout layers* combine node representations resulting from a message-passing layer into a representation on graph-level. Hence, with a subsequent fully connected layer and Softmax activation, a graph can be classified. The literature provides different forms of GNN layers, which can be divided into three flavors: convolutional, attentional, and generic message-passing [2]. The latter is a generalization of the first two, where edge features are incorporated in addition to node features. Since assemblies are represented by graphs with only one edge type and no additional edge features, we consider only the first two forms:

Graph Convolutional Layer (GCN). The node-wise update rule of a graph convolutional layer [10] is given by

$$\mathbf{h}_i^{(l+1)} = \sigma\left(\sum_{j \in N(i)} \frac{1}{c_{ij}} \mathbf{W}^{(l)} \mathbf{h}_j^{(l)}\right) \tag{2}$$

where $\sigma(\cdot)$ denotes a nonlinear activation function such as ReLU and $\mathbf{W}^{(l)}$ is the weight matrix of the fully connected layer transforming the node representations. The normalization constant $c_{ij} = \sqrt{|N(i)|}\sqrt{|N(j)|}$ is derived from the nodes' degrees only and thus heavily depends on the structure of the graph. Consequently, all neighbor nodes are weighted equally for aggregation. It can be interpreted as the importance of node j to the representation of i.

Graph Attention Layer (GAT). In our use case, some components of the current assembly may be more important for recommending next components than others. For this reason, we also investigate GATs [16] that individually weight neighbor nodes according to an attention mechanism:

$$\mathbf{h}_i^{(l+1)} = \sigma\left(\sum_{j\in N(i)} \alpha_{ij}^{(l)} \mathbf{W}^{(l)} \mathbf{h}_j^{(l)}\right) \tag{3}$$

The attention weights $\alpha_{ij}^{(l)}$ result from a learned function over the node features of i and $N(i)$ and serve as coefficients of an convex combination of transformed node features. Thus, graphs of the same structure but with different node features will typically lead to different attention weights. As with other attention-based architectures, multiple independent attention mechanisms can be used to extend the learning process (multi-head attention).

3 Graph-Based Recommendations for Assemblies Using Pretrained Embeddings

In our proposed approach (depicted in Fig. 1), we represent CAD assemblies as undirected graphs with component embeddings as node features and train GNNs for recommending the next required components during construction. We model this learning problem as a *graph classification problem* where the classes correspond to component IDs – one graph is mapped to the ID of the next needed component. By using Softmax as activation function in a final fully connected layer, we get normalized scores over all components that can be interpreted as a ranking of the components' recommendations.

Since embeddings are used to represent in the input graph's components, one may intuitively be inclined to use the same representation in the output, i.e., to predict component embeddings. However, this modeling has some disadvantages:

a) For component recommendations, the predicted embedding must always be assigned to a component. It may happen that there is no corresponding component in the proximity of the prediction in the embedding space because no component satisfies the desired properties – which component should be taken in this case?
b) The number of desired recommendations determines the architecture of the model ($k \cdot$ *embedding_size* output neurons for k recommendations), so if more

recommendations are desired, the model must be re-built and re-trained. In our modeling, we can simply determine the desired number of most likely elements from the distribution.

The GNNs presented in this paper do not employ edge features due to the fact that the given assemblies only specify which components are connected but do not describe the nature of this connection in more detail. Furthermore, we assume all possible component IDs to be included in the training data - we do not yet consider new or updated components in this paper.

3.1 Pretraining of Component Embeddings (comp2vec)

In some machine learning domains such as natural language processing (NLP), representing discrete objects as continuous vectors, so-called embeddings, instead of one-hot vectors has proven to be advantageous [14]. Since assembly modeling has many parallels to NLP, we investigate whether using component embeddings as node features contributes to the performance of our recommendation models: Just as documents are composed of words or letters, assemblies consist of components. Moreover, neighboring words (predecessor and successor) correspond to the possibly larger set of adjacent nodes in a graph (see Sect. 1 for representing assemblies as graphs). Due to the heterogeneity of available attributes of components in catalogs, our approach uses the only information that is always available: a unique component identifier. However, the approach is sufficiently generic in that the resulting node features can be embedding vectors combined with additional, problem-specific features. This also applies to the inclusion of the 3D geometry of components, which we decided to exclude in our approach.

word2vec [14] is a popular method for word embeddings, since only plain text without annotations is needed for training. The basic idea is that the meaning of a word is defined by its context words. Transferring this to assemblies, connected components define the purpose of a component – this information stored in embeddings could be very helpful for component recommendations. node2vec [5] is a modification of word2vec for network graphs based on the homophily assumption, i.e., that connected components are similar, such as friends in a social network. However, this assumption is clearly violated in our case: Two components serving *different purposes* are most likely connected in mechanical designs. Therefore, we generalized word2vec from sequences to graphs with a small modification in creating instances, and refer to the resulting model as *comp2vec* in the following. Word2vec (in the Skip-gram variant) trains a mapping from words to their context words within a defined context window. Its architecture is similar to an autoencoder with two layers, encoding each word in a small internal vector representation. The last layer is only used in training and truncated for inference as the embeddings are calculated in the hidden layer – whose dimension (a hyperparameter) determines the compression of the representation. The equivalent to n-distant words of a center word are n-hop distant nodes of a center node in a graph. For a graph, this approach may result in more generated instances. However, the training process itself remains the same as in word2vec.

Because representation learning is a subdomain of unsupervised learning, the difficulty is that the model cannot be evaluated directly due to the lack of labels. In this case, a downstream model (extrinsic evaluation) or domain knowledge (intrinsic evaluation) is used instead. In NLP, for example, pairs of synonym terms [1] are used for that purpose. This is, in principle, transferable to CAD assembly modeling, but requires high manual effort from design experts, since such intrinsic evaluations have to be created individually for each component catalog. Fortunately, in word2vec the embeddings are learned by solving a supervised task (predicting context words from a word), which allows to directly evaluate by loss. We assume that the unsupervised task is well solved if the corresponding supervised task is well solved, which is why we optimize for loss. As a consequence, different embedding sizes can be compared, since the output dimension is defined by the task and thus does not depend on hyperparameters.

3.2 Generating Data Instances for Component Recommendation

For the component recommendation task, we need the intermediate states of the assemblies during construction, hereafter referred to as *partial assembly* or *partial graph*, as well as the components that were directly assembled to them. As we do not know the original sequence of node and edge insertions that led to the final design and, even if we did, the order can still depend on the designer's preferences, we create instances for *every* possible creation sequence in a self-supervised fashion by Algorithm 1: Starting from a complete assembly (graph G), we iteratively cut off non-cohesivenodes until the remaining graph contains a minimum number of nodes *min_nodes*. The partial graphs together with the corresponding cut-off components form the data instances stored in the multiset D. Finally, in order to process them with GNNs, component embeddings are used as node features of the graphs and the target component is replaced by its ID.

Algorithm 1. Decomposition of Graphs into Instances

```
1: procedure DecomposeGraph(G = (N, E), D)
2:     if |N| > min_nodes then                    ▷ domain-specific hyperparameter
3:         for every non-cohesivenode n ∈ N do                    ▷ see Section 2
4:             add (G \ {n}, n) to D                              ▷ see Section 2
5:             DecomposeGraph(G \ {n}, D)
6:     remove duplicate instances from D
```

Regarding complexity, the decomposition produces $O(|\mathcal{N}|!)$ graph-node instances which is prohibitive for large assemblies. The designs we considered in our experiments (composed of up to 70 components, cf. Table 1) were highly sequential, i.e. each individual (partial) assembly contained sufficiently few leaf nodes, so that this was not an issue. To scale up the approach, we would expect a sampling-based approach that only performs a subset of the removals or a random walk-based approach like used for large network graphs [8] to work well. This remains to be tested in future work.

3.3 Frequency-Based Baseline Model

The task of component recommendation for assembly modeling bears some similarity to market basket analysis in that we want to recommend additional candidates for a given collection of elements in both cases. Most market basket analysis approaches search for frequent itemsets and corresponding associative rules via counting. However, elements in a market basket are not related per se, whereas components of an assembly are explicitly connected. To the best of our knowledge, no suitable established technique taking into account such relations (and thus, the graph structure) exists in literature.

Therefore, we developed an instance-based model inspired by market basket analysis as a baseline: It stores a relative frequency distribution over the assembled components for each partial assembly of the training set. During inference for a given query graph, it looks up which components were most frequently attached to it. If the graph has not been seen in the training data, it looks for previously seen subgraphs that together form the query graph. These subgraphs should be as large as possible to most widely cover the context of the assembly and consequently provide good component recommendations. To exclude redundant subgraphs, the model determines the *minimal* set of its largest seen subgraphs subsuming the query graph. Thereafter, the relative frequency distributions over components for the subgraphs are aggregated to an overall distribution by determining the component-wise maximum. Finally, the k most frequent components are identified from the distribution.

4 Experiments

To test the applicability of our approach, we performed separate experiments on each of three different catalogs corresponding to different manufacturers. Each catalog contains nearly 12,000 assemblies. The catalogs differ in the number of unique components as well as the designs in the number components per assembly and graph diameters, as shown in Table 1. The assemblies of each catalog were split 60:20:20 into training, validation, and test graphs and afterwards transformed into instances consisting of partial graphs and expected component ID (see Sect. 3.2).

We set out to answer the following research questions:

1. Are GNNs better at predicting the next component needed than the frequency-based baseline model?
2. Does pretraining of embeddings using comp2vec increase accuracy, or are GNNs with one-hot encodings as node features just as accurate?
3. Are GATs or GCNs better suited for the task? By what margin?

Table 1. Key facts of the three used component catalogs. For each metric column, the front part indicates the range of values and the back part the corresponding average.

Catalog	#graphs	#comp.	Node degrees	#nodes	#edges	Graph diameter
A	11,826	1,930	1–9; ∅ 1.7	4–33; ∅ 6.1	3–32; ∅ 5.1	2–32; ∅ 4.45
B	11,895	3,099	1–13; ∅ 1.9	4–69; ∅ 18.2	3–68; ∅ 17.2	2–38; ∅ 10.06
C	11,943	1,924	1–16; ∅ 1.7	4–20; ∅ 6.7	3–19; ∅ 5.7	2–6; ∅ 2.94

4.1 Experimental Setup

In a preliminary study, we investigated suitable embedding sizes for the three component catalogs. First, the assemblies were divided 80:20 into training and validation sets and afterwards transformed into instances (see Sect. 3.1). To rate the models, we computed the sum of training loss and gap between training and validation loss [4, p. 425]. In comparison, different embedding sizes behaved the same for each catalog: Sizes between 20 and 90 as well as from 100 upwards each led to the same error level, consequently, we chose the minimum per range (20 and 100, respectively) as fixed embedding sizes for each catalog. In order to investigate the influence of pretrained embeddings on the performance of the task, we additionally generated an one-hot encoding of the components.

Based on the preliminary study, we examined seven models per catalog: the frequency-based baseline model as well as the two flavors of GNN presented in Sect. 2, each based on the three types of component representation. Both models were implemented using the respective layers in DGL[2], the precise hyperparameters were determined using hyperparameter search and can be found in a publicly available repository[3].

4.2 How Well Can GNNs Learn the Task at Hand?

The main evaluation metric is the top-k rate, referring to the percentage of the true component ID (i.e., the target) being in the top-k predictions of a model. We are especially interested in $k = 10$, as this number of recommendations can be well integrated into a CAD system and offers a wide choice of components for designers. Since we are dealing with a task and data that has been little researched, we would like to be able to better assess the results of the models by estimating their performance from above and below. If we take a closer look at Algorithm 1 for generating the recommendation instances (in particular, lines 3 and 4), it is apparent that there can be several target components for one partial graph (e.g., by removing two nodes in both orders). This implies that a model (without being an oracle) cannot reach 100% at the top-1 rate, and depending on the number of different targets for the same input, this may even affect higher values of k. Therefore, for each catalog we determine the *upper*

[2] Deep Graph Library https://www.dgl.ai/.
[3] https://github.com/isse-augsburg/ecml22-grape.

bound of the top-k rate a perfect, but non-oracle, model could reach for all k. Furthermore, a very simple model called *evergreen* serves as lower bound: This model predicts the k most common labels seen during training (based on a frequency distribution of the labels from the training set), independent of the specific input. The comparison with this model is to show whether the GNNs are capable of processing contextual information from the input graph and to prove that the prediction task is indeed non-trivial.

Table 2. Summary of results for component recommendation per catalog: Top-k rate on the test set for $k = 1, \ldots, 20$ recommendations. The identifier following the GNN model architecture indicates the component representation.

Catalog	Model \k	1	2	3	5	10	15	20
A	Upper bound	94.1%	99.8%	99.9%	99.9%	100%	100%	100%
	GAT-100	46.7%	71.4%	79.7%	85.2%	**90.0%**	**92.2%**	**93.3%**
	GAT-20	47.9%	**72.2%**	**80.4%**	**85.5%**	89.8%	91.5%	92.5%
	GAT-one-hot	46.0%	70.3%	79.0%	84.6%	89.6%	91.5%	92.6%
	GCN-100	47.4%	71.0%	79.3%	84.9%	89.6%	91.5%	92.7%
	GCN-20	46.6%	70.9%	79.5%	84.8%	89.4%	91.3%	92.4%
	GCN-one-hot	45.4%	69.3%	78.1%	83.8%	88.6%	90.7%	92.0%
	Baseline	**57.5%**	64.3%	66.9%	69.0%	70.2%	70.7%	70.8%
	Evergreen	4.5%	6.3%	8.1%	11.3%	15.9%	19.6%	22.2%
B	Upper bound	63.5%	81.5%	92.5%	99.4%	99.99%	100%	100%
	GAT-100	**30.8%**	**50.6%**	**63.1%**	**73.8%**	**82.1%**	**86.1%**	**88.4%**
	GAT-20	30.0%	50.1%	62.6%	73.4%	82.0%	85.9%	88.0%
	GAT-one-hot	30.2%	49.5%	61.8%	72.5%	80.7%	84.6%	86.6%
	GCN-100	30.5%	49.2%	61.2%	72.4%	81.8%	85.9%	88.0%
	GCN-20	28.0%	46.4%	58.8%	71.4%	81.3%	85.6%	88.1%
	GCN-one-hot	27.8%	45.1%	56.8%	69.2%	79.8%	84.3%	86.7%
	Baseline	24.5%	32.9%	39.1%	46.6%	52.8%	53.7%	53.8%
	Evergreen	13.3%	15.5%	18.4%	22.2%	30.0%	36.2%	41.7%
C	Upper bound	74.0%	91.1%	98.0%	99.9%	100%	100%	100%
	GAT-100	28.5%	**49.3%**	**65.0%**	**81.8%**	**92.8%**	**95.8%**	96.9%
	GAT-20	27.9%	48.5%	64.0%	80.7%	91.9%	95.1%	96.3%
	GAT-one-hot	27.5%	48.4%	64.0%	80.6%	91.2%	94.0%	95.2%
	GCN-100	27.9%	48.3%	63.8%	80.5%	92.0%	95.5%	**97.1%**
	GCN-20	27.5%	48.4%	64.2%	81.2%	92.5%	**95.8%**	**97.1%**
	GCN-one-hot	27.0%	47.1%	62.4%	79.2%	90.8%	94.2%	95.7%
	Baseline	**30.1%**	42.1%	50.9%	60.2%	63.9%	64.2%	64.2%
	Evergreen	14.4%	19.9%	23.3%	27.8%	37.0%	44.6%	48.1%

Table 2 presents an overview of the performance of the models as well as the upper and lower bounds for the top-k rate. Additionally, Fig. 2 visualizes the performance of the best models and bounds on catalog A and B. Both the

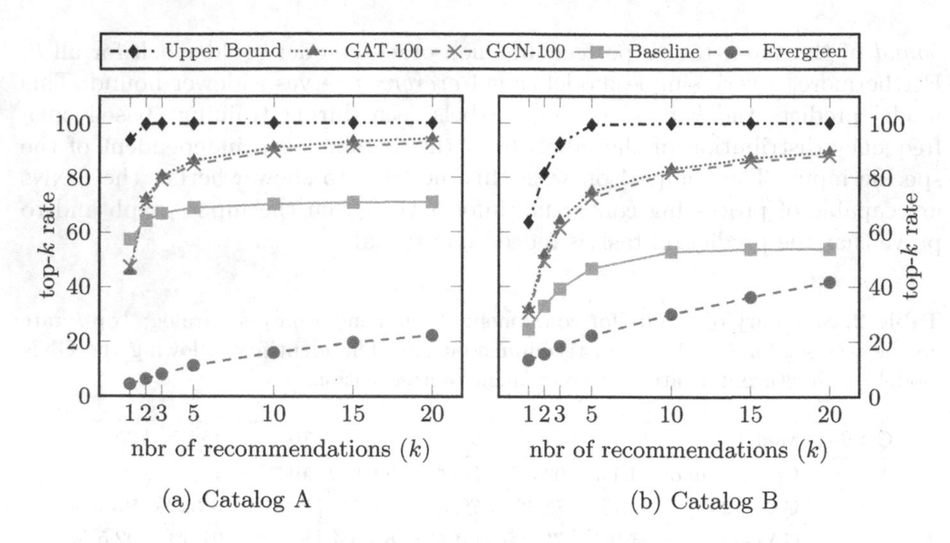

Fig. 2. Visual comparison of component prediction models: for GCN and GAT, the best models (embedding size 100) were used for each catalog.

baseline and the GNN models stand out clearly from the simple ever-green model for each catalog, with the latter performing considerably better. When evaluating the baseline model, a direct lookup of the input graph could only rarely be performed, on catalog B only for 1% of the test instances. Thus, in most cases subgraphs had to be found, which together form the input graph. In contrast to GNNs, the baseline model does not scale in terms of the size of the catalogs or the size of assemblies, making GNN models more suitable for larger data sets.

In all but the case of only one recommendation ($k = 1$), the GNN models outperform the baseline by a large margin, demonstrating that they are capable of generalizing beyond exact subgraph pattern-matching. Since our application scenario focuses on presenting the design engineers with more than one recommendation, the case $k = 1$ is negligible. Furthermore, the performance of the GNN models increases significantly more with growing number of recommendations k. Regarding the top-10 rate, the best performing model (GAT-100) achieved over 90% on catalogs A and C, and still 82.1% for catalog B despite the fact that this catalog is more ambiguous than the other two since it comprises more components – which is also reflected in the lower and upper bound rates. This demonstrates that the GNN models can reliably cut down the candidates for needed components from 1,930 (catalog A), 3,099 (catalog B), and 1,924 (catalog C) to 10, providing a useful preselection for design engineers.

4.3 Are Component Embeddings Better Than One-Hot Node Features?

Regarding the used representation of the components, the results show that pretraining embeddings pays off compared to starting with one-hot node features.

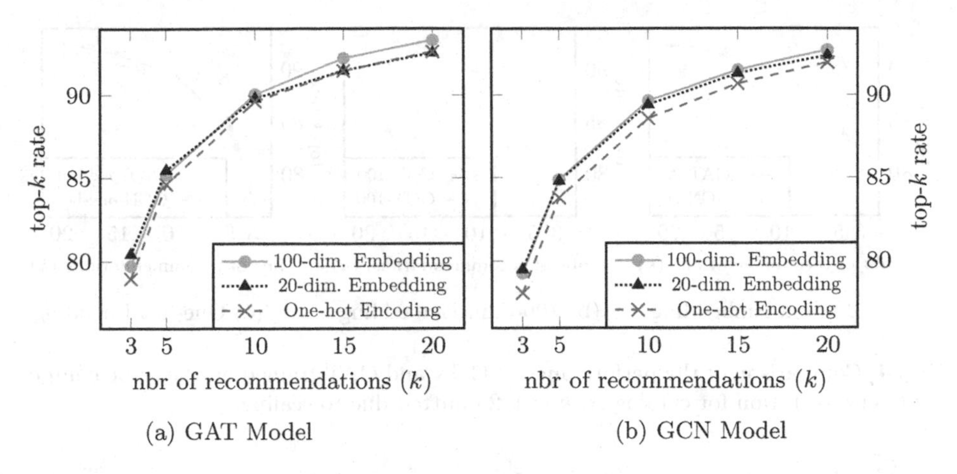

Fig. 3. Effects of the three component representations on the performance (top-k-rate) of both GNN models for catalog A. $k = 1, 2$ omitted due to scaling.

In particular, the GNNs based on 100-dimensional embeddings perform slightly better than those using 20-dimensional ones, respectively. The improvements with increasing embedding dimensions is consistent for both GCNs and GATs, as visualized in Fig. 3. This suggests that an even higher embedding dimension would lead to a further performance improvement, but the preliminary study did not show any improvement in the pretraining task with higher embedding dimensions. Therefore, we decided not to investigate them further in the component prediction task.

Thinking of the recommendation models as extrinsic evaluation for the component embeddings, i.e., using the downstream supervised model to evaluate the embeddings, the representation in 100 dimensions is preferable. Since in the preliminary study sizes of 100 and above resulted in lower loss values than only 20 dimensions, this confirms our assumption from Sect. 3.1 that we can measure the quality of an embedding by the associated supervised task of comp2vec. Concluding, pretraining turned out to be advantageous for the recommendation task, although one-hot-based GNNs also performed well which could be relevant for practical purposes (e.g., lack of time for setting up a pretraining pipeline).

4.4 Comparing GAT and GCN

In a direct comparison of the GNNs, the GAT models yield slightly better results based on the same embedding, as shown in Fig. 4. This makes the GAT with embedding size 100 the overall winner. The difference between the two models lies in the weighting of the neighbors: while in GCN all neighbors are weighted with the same constant factor depending on the graph structure (cf. Eq. (2)), in GAT an individual weight for each neighbor node is determined based on the node features, i.e., the attention scores (cf. Eq. (3)). Whether a GAT indeed weights certain neighbor nodes higher (i.e., makes use of the attention mecha-

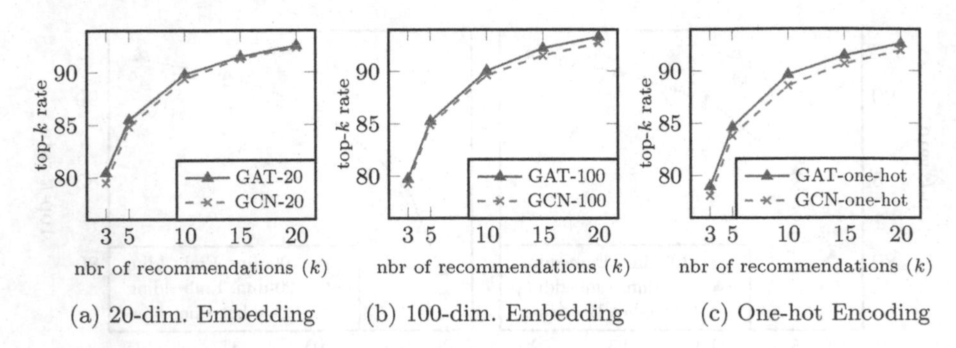

Fig. 4. Comparison of the performance of GCN and GAT trained on the same component representation for catalog A. $k = 1, 2$ omitted due to scaling.

nism) can be investigated by comparing the attention distribution to the uniform distribution, as suggested in [23]. For each graph instance and each of its nodes, the attention distribution over the neighbor nodes as well as the corresponding uniform distribution is determined. Subsequently, the Shannon entropy is calculated for both distributions. Figure 5 shows a histogram of the calculated entropy values (attention distribution and uniform distribution), excluding nodes with exactly one neighbor node (i.e., leaf nodes), since for those the learned attention distribution always corresponds to the uniform distribution. Since the characteristics were similar for all GAT models, the figure visualizes only one particular model (GAT-100 for catalog B). The high prevalence of entropy 0 scores indicates that for many neighborhoods in the assemblies, the attention scores tend to focus all attention on a single (or very few) neighbors – which result in a very different distribution from the uniform one, and explains why GATs are superior to GCNs on these data sets.

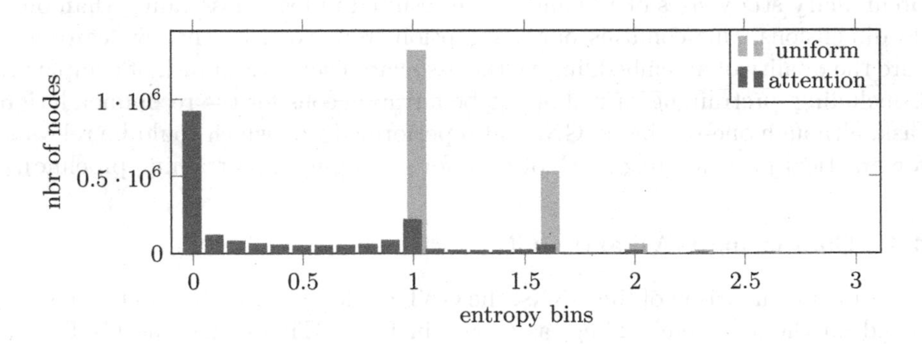

Fig. 5. Histogram of calculated Shannon-Entropy of the attention distribution and the corresponding uniform distribution evaluated on the test set. The attention values originate from an arbitrarily selected head of the last GAT layer from GAT-100 trained on catalog B. Nodes with only one neighbor were excluded, since their distribution equals the uniform distribution.

Finally, we want to stress that the GATs behaved more robustly in our experiments: Slight changes in the hyperparameters led to strong fluctuations in performance for GCNs, while this phenomenon was much less noticeable for GATs. In summary, after analyzing the resulting attention scores and evaluating the performance, we favor GATs over GCNs for our recommendation task.

5 Conclusion and Future Work

We proposed a recommendation system based on Graph Neural Networks for assembly modelling that recommends next required components during construction. For this purpose, we developed an approach to construct instances in a self-supervised fashion by recursively cutting off nodes from assembly graphs which then serve as target for the resulting partial graph instance. Our experiments on three different data sets proved that graph neural networks are well suited for this task, outperforming a self-developed frequency-based baseline model. Further, pretraining low-dimensional representations of components as node features turned out to be beneficial for recommendation, although one-hot encoded node features also led to satisfying results.

In the presented approach, the components are recommended without locating which node of the current assembly they should be attached to. While this is acceptable for small assemblies, for designing larger ones conveniently, it is mandatory to highlight the node where a component should be attached. In future work, we want to integrate these *connection components* into the recommendation. Moreover, as the catalogs get updated over time, the models are likely to be confronted with new, unknown components during inference – here a suitable representation of the new components has to be found. Literature provides several other models, e.g., [21] and [22], to process graph-like data, each of them having a different inductive bias. We plan to further investigated them in terms of their applicability for our use case.

References

1. Bakarov, A.: A survey of word embeddings evaluation methods (2018). https://doi.org/10.48550/ARXIV.1801.09536
2. Bronstein, M.M., Bruna, J., Cohen, T., Veličković, P.: Geometric deep learning: grids, groups, graphs, geodesics, and gauges (2021). https://doi.org/10.48550/ARXIV.2104.13478
3. Cunningham, J.D., Simpson, T.W., Tucker, C.S.: An investigation of surrogate models for efficient performance-based decoding of 3D point clouds. J. Mech. Des. **141**(12) (2019). https://doi.org/10.1115/1.4044597
4. Goodfellow, I., Bengio, Y., Courville, A.: Deep Learning. MIT Press, Cambridge (2016)
5. Grover, A., Leskovec, J.: Node2vec: scalable feature learning for networks. In: Proceedings of the 22nd ACM SIGKDD International Conference on Knowledge Discovery and Data Mining. KDD '16, pp. 855–864. Association for Computing Machinery, New York (2016). https://doi.org/10.1145/2939672.2939754

6. Guo, X., Zhao, L.: A systematic survey on deep generative models for graph generation (2020). https://doi.org/10.48550/ARXIV.2007.06686
7. Guo, X., Li, W., Iorio, F.: Convolutional neural networks for steady flow approximation. In: Proceedings of the 22nd ACM SIGKDD International Conference on Knowledge Discovery and Data Mining. KDD '16, pp. 481–490. Association for Computing Machinery, New York (2016). https://doi.org/10.1145/2939672.2939738
8. Hamilton, W., Ying, Z., Leskovec, J.: Inductive representation learning on large graphs. In: Advances in Neural Information Processing Systems, vol. 30. Curran Associates, Inc. (2017)
9. Hamilton, W.L.: Graph representation learning. Synth. Lect. Artif. Intell. Mach. Learn. **14**(3), 1–159 (2020)
10. Kipf, T.N., Welling, M.: Semi-supervised classification with graph convolutional networks (2017). https://doi.org/10.48550/ARXIV.1609.02907
11. Li, W., Matejka, J., Grossman, T., Konstan, J.A., Fitzmaurice, G.: Design and evaluation of a command recommendation system for software applications. ACM Trans. Comput.-Hum. Interact. (TOCHI) **18**(2), 1–35 (2011). https://doi.org/10.1145/1970378.1970380
12. Li, Y., Vinyals, O., Dyer, C., Pascanu, R., Battaglia, P.: Learning deep generative models of Graphs (2018). https://doi.org/10.48550/ARXIV.1803.03324
13. Lupinetti, K., Pernot, J.P., Monti, M., Giannini, F.: Content-based CAD assembly model retrieval: survey and future challenges. Comput. Aided Des. **113**, 62–81 (2019). https://doi.org/10.1016/j.cad.2019.03.005
14. Mikolov, T., Chen, K., Corrado, G., Dean, J.: Efficient estimation of word representations in vector space (2013). https://doi.org/10.48550/ARXIV.1301.3781
15. Sarcar, M., Rao, K.M., Narayan, K.L.: Computer Aided Design and Manufacturing. PHI Learning Pvt Ltd., New Delhi (2008)
16. Veličković, P., Cucurull, G., Casanova, A., Romero, A., Liò, P., Bengio, Y.: Graph attention networks (2018). https://doi.org/10.48550/ARXIV.1710.10903
17. Wu, R., Xiao, C., Zheng, C.: DeepCAD: a deep generative network for computer-aided design models. In: Proceedings of the IEEE/CVF International Conference on Computer Vision (ICCV), pp. 6772–6782 (2021)
18. Wu, S., Sun, F., Zhang, W., Xie, X., Cui, B.: Graph neural networks in recommender systems: a survey (2020). https://doi.org/10.48550/ARXIV.2011.02260
19. Wu, Z., Pan, S., Chen, F., Long, G., Zhang, C., Yu, P.S.: A comprehensive survey on graph neural networks. IEEE Trans. Neural Netw. Learn. Syst. **32**(1), 4–24 (2021). https://doi.org/10.1109/tnnls.2020.2978386
20. Yoo, S., Lee, S., Kim, S., Hwang, K.H., Park, J.H., Kang, N.: Integrating deep learning into CAD/CAE system: generative design and evaluation of 3D conceptual wheel. Struct. Multidiscip. Optim. **64**(4), 2725–2747 (2021)
21. Yun, S., Jeong, M., Kim, R., Kang, J., Kim, H.J.: Graph transformer networks. In: Advances in Neural Information Processing Systems, vol. 32. Curran Associates, Inc. (2019)
22. Zaheer, M., Kottur, S., Ravanbakhsh, S., Poczos, B., Salakhutdinov, R.R., Smola, A.J.: Deep sets. In: Advances in Neural Information Processing Systems, vol. 30. Curran Associates, Inc. (2017)

23. Zhang, H., Li, M., Wang, M., Zhang, Z.: Understand graph attention network, February 2022. https://www.dgl.ai/blog/2019/02/17/gat.html. Accessed 06 Apr 2022
24. Zhang, Z., Jaiswal, P., Rai, R.: Featurenet: machining feature recognition based on 3d convolution neural network. Comput. Aided Des. **101**, 12–22 (2018). https://doi.org/10.1016/j.cad.2018.03.006

AD-AUG: Adversarial Data Augmentation for Counterfactual Recommendation

Yifan Wang[1], Yifang Qin[1], Yu Han[2], Mingyang Yin[2], Jingren Zhou[2], Hongxia Yang[2(✉)], and Ming Zhang[1(✉)]

[1] Peking University, Beijing, China
{yifanwang,qinyifang,mzhang_cs}@pku.edu.cn
[2] DAMO Academy, Alibaba Group, Hangzhou, China
{hanyu.han,hengyang.ymy,jingren.zhou,yang.yhx}@alibaba-inc.com

Abstract. Collaborative filtering (CF) has become one of the most popular and widely used methods in recommender systems, but its performance degrades sharply in practice due to the sparsity and bias of the real-world user feedback data. In this paper, we propose a novel counterfactual data augmentation framework AD-AUG to mitigate the impact of the imperfect training data and empower CF models. The key idea of AD-AUG is to answer the counterfactual question: "what would be a user's feedback if his previous purchase history had been different?". Our framework is composed of an augmenter model and a recommender model. The augmenter model aims to generate counterfactual user feedback based on the observed ones, while the recommender leverages the original and counterfactual user feedback data to provide the final recommendation. In particular, we design two adversarial learning-based methods from both "bottom-up" data-oriented and "top-down" model-oriented perspectives for counterfactual learning. Extensive experiments on three real-world datasets show that the AD-AUG can greatly enhance a wide range of CF models, demonstrating our framework's effectiveness and generality.

Keywords: Counterfactual augmentation · Collaborative filtering · Recommending systems

1 Introduction

With an unprecedented number of products and services available on online platforms, it becomes challenging and time-consuming for users to discover interested products from overwhelming alternatives. Recommender systems have become essential tools to solve this information overloading problem by generating a personalized recommendation list for different users. Especially CF-based methods

Y. Wang and Y. Qin–Both authors contributed equally to this work.

Supplementary Information The online version contains supplementary material available at https://doi.org/10.1007/978-3-031-26387-3_29.

M.-R. Amini et al. (Eds.): ECML PKDD 2022, LNAI 13713, pp. 474–490, 2023.
https://doi.org/10.1007/978-3-031-26387-3_29

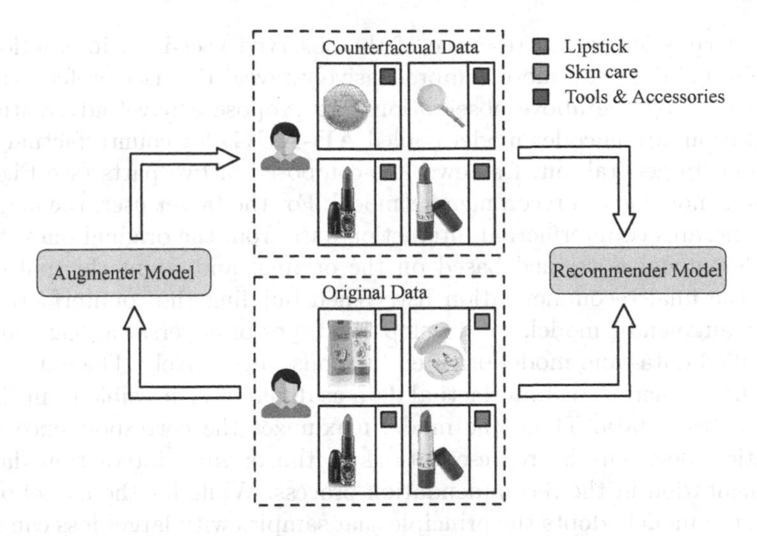

Fig. 1. An illustration of our framework for counterfactual data-augmented recommendation. The picture of each product is downloaded from https://www.amazon.com/, and the categories of the products are also presented for reference.

(e.g., matrix factorization[15]), which have been extensively used for recommendation, assuming users who have made similar choices tend to have similar preferences in the future.

The core of CF methods are how to parameterize users and items with effective feature vectors based on their observed historical interactions. Notably autoencoder architecture, which has served as an effective solution for dimensionality reduction by learning underlying patterns from the interaction data between users and items, has recently applied to CF with strong performance improvements over several competitive approaches. CDAE [29] utilizes a denoising autoencoder by mapping user feedback to embeddings for reconstructing the user's preference. MultVAE [16] subsequently improves CDAE by extending model to variational autoencoders (VAE) and the likelihood to multinomial distributions. MacridVAE [18] further employs VAE to learn disentangled representations representing different user intentions.

These pure CF methods are quite efficient and effective to obtain satisfactory representations relying on the original user-item interaction data. Unfortunately, in real-world applications, most users can only access a limited number of items with large bias, as a result, pure CF can hardly capture these users' preference. Figure 1 illustrates a toy example for the product recommendation scenario. We may observe that a user has clicked Lipsticks and Skin care products influenced by the products' popularity instead of her own interest in original data. Meanwhile, although this user is in need of makeup tools and accessories, these data may not be recorded for various reasons. From the perspective of causal inference, these unrecorded data provide a key counterfactual question: "What would be a user's feedback if his previous purchase history had been different?". As a significant complementary resource of the observed user-item interactions, the counterfactual data can more comprehensively reveal the user preference.

Motivated by the above observations, we propose a novel adversarial data augmentation autoencoder model (called **AD-AUG**) for counterfactual recommendation. In general, our framework is composed of two parts (see Fig. 1), an augmenter model and a recommender model. For the target user, the augmenter model generates counterfactual interaction data from the original ones, the recommender model is trained based on the original and counterfactual data to provide the final recommendation list. When building the counterfactual data from the augmenter model, we develop two types of adversarial learning-based model called data- and model-oriented methods, respectively. The data-oriented method first generates counterfactual data as different as possible from the original interaction data. Then, the model maximizes the correspondence/mutual information between the representations of the original interaction data and its augmentation in the recommendation process. While for the model-oriented method, the model adopts the principle that samples with larger loss can usually provide more knowledge to widen the experience and aims to generate counterfactual data that maximizes the information provided to the recommender model.

To summarize, in this paper we make the following contributions:

- We propose a novel adversarial training framework to empower recommendation models with counterfactual data and our framework can support a wide range of different CF models.
- We implement the above idea in two ways by developing the augmenter from both data and model perspectives to generate the counterfactual data for the recommender.
- We conduct experiments on three real-world datasets to evaluate the proposed approach. Experimental results show the effectiveness and generality of our framework.

2 Related Work

2.1 Autoencoder-Based CF

Autoencoder (AE) has emerged as an important architecture to enable the CF techniques by mapping user-item interactions into latent low-dimensional representations. The goal of AEs are to minimize the reconstruction error for the user's feedback vector [23]. As the variants of AE, denoising autoencoders (DAEs) [27] and VAEs [14,22] are widely used for CF. CDAE [29] utilizes a DAE by corrupting the input feed back vector randomly. MultVAE [16] extends VAE to CF with multinomial distribution in the likelihood. MacridVAE [18] employs VAE to learn disentangled representation representing different interests of the user. RecVAE [24] proposes a new composite prior for training based on alternating updates to enhance performance. Our framework aims to design two leaning-based intervention methods to improve AE-based CF from two inseparable aspects, i.e., "data" and "model".

2.2 Counterfactual Data Augmentation

Counterfactual thinking is a concept describing the human introspection behaviors with typical question: "what would ... if ...?". It has been recently leveraged to alleviate the training data insufficiency problem in the machine learning community, e.g., computer vision [2,3,6] and natural language processing [33]. For recommendation, CASR [28] generates counterfactual user behavior sequences for sequential recommendation. To alleviate the problem of extreme sparse and imbalanced training data, CPR [32] simulates user preference and generates counterfactual samples. Instead of generating counterfactual data and make recommendation in separate two steps, our work focuses on learning counterfactual data augmentation as well as recommendation process in one union step with adversarial training.

2.3 Adversarial Training

The basic idea of adversarial training is to introduce an opponent part into the model optimization process, where two models try to detrimentally influence each other's performance and as a result, both models improve by competing against each other. Adversarial training [5] has demonstrated their abilities and potentials on a number of machine learning applications, such as image generation [12,31], language generation [17], graph representation learning [30] and robust recommender system [9]. Recently, there are some works [25] trying to combine information theory [10,26] to adversarial training. Apart from leading the training target, in this paper, we borrow the idea of adversarial training to data augmentation. The generator serves as a counterfactual data augmenter that samples challenging user interactions to optimize the recommender model.

3 Preliminaries

3.1 Problem Definition

Given a set $\mathcal{U}$ of M users and a set $\mathcal{I}$ of N items, we have a binary rating matrix $X \in \{0,1\}^{M\times N}$, where $x_{u,i} = 1$ indicates that user u explicitly adopts item i, otherwise $x_{u,i} = 0$ and it indicates a missing feedback. Given a user u, $x_u = \{x_{u,i}|i \in \mathcal{I}\}$ represents user u's history feedback vector. The goal is to learn a recommendation model $\mathcal{A}$ with x_u as input to infer user u's preference score and retrieve a ranked list of the top-N items that u prefers the most. However, accurately estimating $\mathcal{A}$ usually suffers from the data sparsity and selection bias problems. Therefore, for user u, we aim to generate sufficient "real" interaction data $\hat{x}_u$ to augment x_u.

3.2 Autoencoder CF Framework

A standard AE is trained to reproduce the input data in an output layer via a compressed latent representation. The encoder part of the framework encodes the input x_u to a d-dimensional latent representation z_u. And the decoder part of the framework takes z_u as input and outputs the reconstructed user feedback.

$$z_u = f(x_u), x'_u = g(z_u), \tag{1}$$

where $f(.)$ and $g(.)$ are encoder and decoder network respectively, which can be multiple layers neural network. And the optimization loss can be defined as,

$$\mathcal{L}_{AE}(\mathcal{A}(x_u)) = d(x'_u, x_u), \tag{2}$$

where $d(.,.)$ is the reconstruction loss. Instead of outputting the latent vectors, the encoder of VAE outputs user representation with a prior distribution. And the optimisation objective is the evidence lower bound to estimate the intractable marginal log-likelihood. For a single user u, we have:

$$\begin{aligned} &\log p(x_u) \geq \mathcal{L}_{VAE}(\mathcal{A}(x_u)) = \\ &\mathbb{E}_{z_u \sim q(z_u|x_u)} \log p(x_u|z_u) - \beta KL(q(z_u|x_u)\|p(z_u)), \end{aligned} \tag{3}$$

where KL is the Kullback-Leibler divergence distance measuring the difference between the prior distribution $p(z_u)$ and posterior distribution $q(z_u|x_u)$ parameterized by encoder function $f(.)$, $p(x_u|z_u)$ is the generated distribution conditioned on z_u parameterized by decoder $g(.)$. β is the regularization hyperparameter that balances the latent channel capacity (i.e., reconstruction accuracy) against independence constraints [10].

4 The Proposed Model

4.1 Model Overview

Inspired by the human introspection behaviors, the basic idea of our proposed model is to remove redundant information from u's positive feedback $\mathcal{I}_u^+ = \{i \in \mathcal{I} | x_{u,i} = 1\}$ and add pseudo information from missing feedback $\mathcal{I}_u^- = \{i \in \mathcal{I} | x_{u,i} = 0\}$ for the recommender learning. As shown in Fig. 2 and Fig. 3, beyond training a recommender model $\mathcal{A}$, our AD-AUG framework introduces an augmenter model $\mathcal{S}$ with the same model structure as $\mathcal{A}$, to answer the counterfactual question by generating counterfactual user feedback data. For the target user u, we use feed back vector x_u as input of $\mathcal{S}$, and the produced counterfactual interaction vector $\hat{x}_u$ are leveraged to optimize $\mathcal{A}$. We implement the model from both model and data perspectives, which will be introduced in the following contents.

4.2 Data-Oriented Counterfactual Learning

As redundant especially incorrect user's feedback in data causes troubles for models to identify user's true preference. We introduce information bottleneck (IB) [10], which is a common practice in contrastive learning [4,30], that requests the model to capture the *minimal sufficient* information for recommendation. For

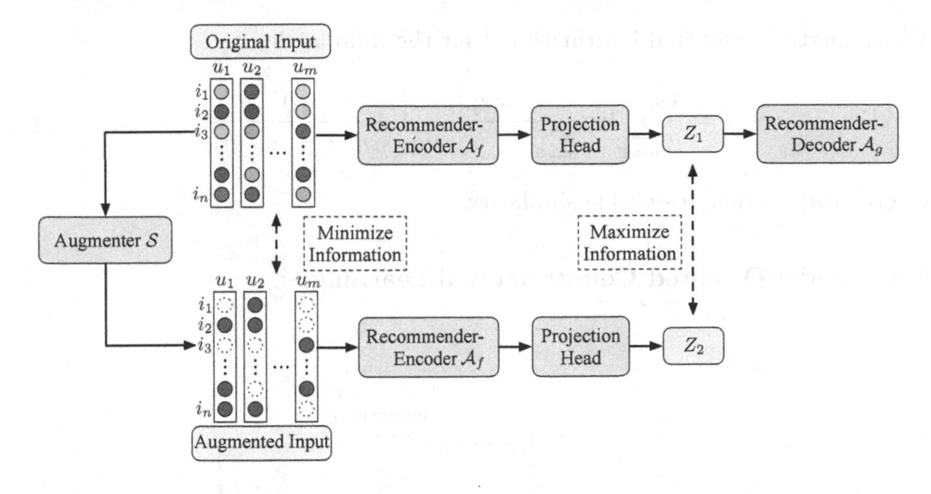

Fig. 2. The schematic view of the Data-oriented method.

the target user u, IB minimizes the information from the original feedback data x_u while maximizing the information for recommendation to remove redundant information as well as noises. As the misleading information gets removed, the model learnt by IB tends to be more robust.

Formally, we learn data-oriented counterfactual model with min-max principle,

$$\min_{\mathcal{A}} \max_{\mathcal{S}} \mathcal{L}(\mathcal{A}(x_u)) - \lambda I(\mathcal{A}_f(x_u), \mathcal{A}_f(\mathcal{S}(x_u))), \tag{4}$$

where $\mathcal{L}(\mathcal{A}(x_u))$ denotes the loss of the recommender model with x_u as input and the loss can be $\mathcal{L}_{AE}(\cdot)$ or $\mathcal{L}_{VAE}(\cdot)$ depending on the recommender model. $\mathcal{A}_f$ denotes the encoder part of the recommender model, $\mathcal{S}(x_u)$ denotes the outputs of the augmenter model, $I(\mathcal{A}_f(x_u), \mathcal{A}_f(\mathcal{S}(x_u)))$ denotes the mutual information between the original user's feedback x_u and the augmented user's feedback $\hat{x}_u$, λ is the hyper-parameter to balance the mutual information and the loss.

Given the optimization target, both models are thus trained in adversarial style. During each iteration, the recommender model $\mathcal{A}$ tends to maximize the mutual information between latent representation and target labels, i.e. to minimize $\mathcal{L}(\mathcal{A}(x_u)$, while minimize the mutual information between the generated and the original data, i.e., to maximize $I(\mathcal{A}_f(x_u), \mathcal{A}_f(\mathcal{S}(x_u)))$. On the other hand, the augmenter model $\mathcal{S}$ is encouraged to generate hard negative samples, i.e., to minimize $I(\mathcal{A}_f(x_u), \mathcal{A}_f(\mathcal{S}(x_u)))$, to be distinguished from original data.

Specifically, we adopt InfoNCE as the estimator [21] for mutual information, which is known to be a lower bound of the mutual information and is frequently used for contrastive learning. Formally, during the training, given a minibatch of b users, let $z_{u,1} = h(\mathcal{A}_f(x_u))$ and $z_{u,2} = h(\mathcal{A}_f(\mathcal{S}(x_u)))$, where $h(.)$ is the projection head implemented by a 2-layer MLP as suggested in previous work [4].

We estimate the mutual information $\hat{I}$ for the minibatch,

$$\hat{I} = \frac{1}{b}\sum_{u=1}^{b} \log \frac{\exp(sim(z_{u,1}, z_{u,2}))}{\sum_{u'=1,u'\neq i}^{b} \exp(sim(z_{u,1}, z_{u',2}))}, \tag{5}$$

where $sim(.,.)$ denotes cosine similarity.

4.3 Model-Oriented Counterfactual Learning

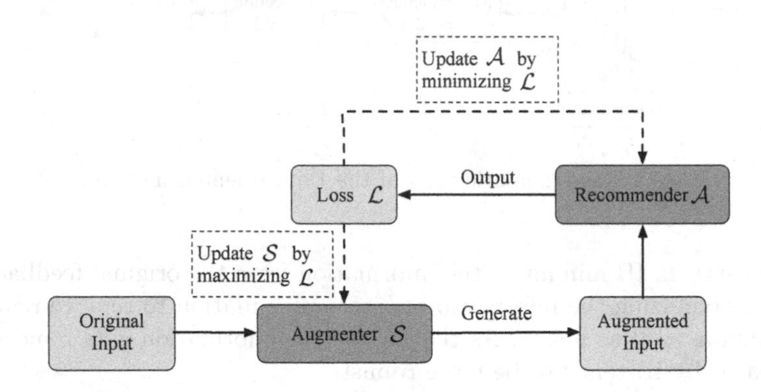

Fig. 3. The schematic view of the Model-oriented method.

Besides augmenting user's feedback from the data perspective, we leverage the loss of the model to augment user's feedback from the model perspective. Motivated by the previous work [1,6,7], which follows the principle that samples with larger loss can usually provide more knowledge to widen the model's experience and improve the performance. In the model-oriented method, we learn the counterfactual feedback data via maximizing the loss $\mathcal{L}$ of the recommender.

Formally, we learn the model with min-max principle defined as follows,

$$\min_{\mathcal{A}} \max_{\mathcal{S}} \mathcal{L}(\mathcal{A}(\mathcal{S}(x_u))), \tag{6}$$

where $\mathcal{L}(\mathcal{A}(\mathcal{S}(x_u)))$ denotes the loss of the recommender with augmented user's feedback $\mathcal{S}(x_u)$, i.e., $\hat{x}_u$, as input, and the loss can be $\mathcal{L}_{AE}(\cdot)$ or $\mathcal{L}_{VAE}(\cdot)$ depending on the recommender model.

4.4 Implementation of Augmenter Model

For the user's feedback data x_u, we introduce a practical instantiation of the augmenter model $\mathcal{S}$. The goal of $\mathcal{S}$ is to remove redundant information from u's positive feedback $\mathcal{I}_u^+ = \{i \in \mathcal{I} | x_{u,i} = 1\}$ and add pseudo information from missing feedback $\mathcal{I}_u^- = \{i \in \mathcal{I} | x_{u,i} = 0\}$. Specifically, each positive/missing feedback i of user u will be associated with a random variable $p_{u,i} \sim \text{Bernoulli}(\omega_{u,i})$,

Algorithm 1: Learning Algorithm of AD-AUG

Input: Training user binary rating matrix $X \in \{0,1\}^{M \times N}$
Initialize: Models $\mathcal{A}$ and $\mathcal{S}$ to different initial conditions, $i \leftarrow 0$
while i <*MaxIteration* **do**
1 Sample one batch user feedbacks from X;
2 Assign α to control changed feedback amount ; // Eq.11
3 Learn the counterfactual user feedback $\hat{x}_u$ by $\mathcal{S}$; // Eq.7
4 **Data-oriented Method:**
5 Encode x_u and $\hat{x}_u$ to learn mutual information I;
if $i\%2 = 0$ **then**
6 Update $\mathcal{S}$ by maximizing $-\lambda\hat{I}+\alpha\mathcal{L}_{\text{reg}}$;
else
7 Update $\mathcal{A}$ by minimizing $\mathcal{L}-\lambda\hat{I}$; // Eq.9
end
8 **Model-oriented Method:**
if $i\%2 = 0$ **then**
9 Update $\mathcal{S}$ by maximizing $\mathcal{L}+\alpha\mathcal{L}_{\text{reg}}$;
else
10 Update $\mathcal{A}$ by minimizing $\mathcal{L}$; // Eq.10
end
11 $i \leftarrow i+1$;
end

where $w_{u,i}$ denotes the probability of the occurrence of the interaction between u and i. For positive feedback, (u,i) is kept if $p_{u,i} = 1$ and dropped otherwise. For missing feedback, (u,i) is added as pseudo feedback in $\hat{x}_u$ if $p_{u,i} = 1$ and kept missing otherwise.

We parameterize the Bernoulli weight $\omega_{u,i}$ by leveraging another AE-based CF framework, i.e., the augmenter $\mathcal{S}$, to take x_u as input to get ω_u. In order to train $\mathcal{S}$ in an end-to-end fashion, we relax the discrete $p_{u,i}$ to be a continuous variable in $[0,1]$ and utilize the Gumbel-Max reparameterization trick [13]. Formally,

$$p_{u,i} = \text{Sigmoid}(\frac{\log\sigma - \log(1-\sigma) + \omega_{u,i}}{\tau}), \tag{7}$$

where $\sigma \sim \text{Uniform}(0,1)$, τ is the temperature hyper-parameter.

Meanwhile, a reasonable augmenter model $\mathcal{S}$ should keep a certain amount of information from original user feedback. Hence, we regularize the ratio of feedback being changed per user by enforcing the constraint defined as,

$$\mathcal{L}_{\text{reg}} = \frac{1}{|\mathcal{U}|}\sum_{u\in\mathcal{U}}\sum_{i}\frac{x_{u,i}\omega_{u,i} + (1-x_{u,i})(1-\omega_{u,i})}{|\mathcal{I}_u^+| + |\mathcal{I}_u^-|}. \tag{8}$$

Table 1. Descriptive statistics of four datasets. *Amazon-Digital Music* and *Amazon-Beauty* are simplified as *A-Music* and *A-Beauty.*

Dataset	#Users	#Items	#Interactions	Sparsity
ML-1M	6,040	3,706	1,000,209	4.47%
A-Music	5,541	3,568	64,706	0.33%
A-Beauty	22,363	121,01	198,502	0.07%

4.5 Curriculum Adversarial Learning

By adding the constraint, the final objectives for the two counterfactual learning models are as follows. For data-oriented counterfactual learning, we have:

$$\min_{\mathcal{A}} \max_{\mathcal{S}} \mathcal{L}(\mathcal{A}(x_u)) - \lambda I(\mathcal{A}_f(x_u), \mathcal{A}_f(\mathcal{S}(x_u))) + \alpha \mathcal{L}_{\text{reg}}, \tag{9}$$

where α is the hyper-parameter to control the amount of user-item interaction changed from the original feedback. And for model-oriented counterfactual learning, we have:

$$\min_{\mathcal{A}} \max_{\mathcal{S}} \mathcal{L}(\mathcal{A}(\mathcal{S}(x_u))) + \alpha \mathcal{L}_{\text{reg}}. \tag{10}$$

In order to learn $\mathcal{A}$ and $\mathcal{S}$, we propose a curriculum learning method on the designed coursed, via an easy-to-difficult process. Specially, an annealing mechanism is applied:

$$\alpha = \rho * \gamma^k, \tag{11}$$

where ρ is the initial weight, γ denotes the decay ratio, and k denotes the current curriculum number. In this way, as the learned courses becomes difficult, i.e., the amount of changed feedback increase, the learned model can be gradually improved. The complete learning algorithm of our framework is shown in Algorithm 1.

5 Experiment

5.1 Experimental Settings

Datasets. We validate the proposed framework on three public available datasets. In specific, *MovieLens*[1] is a widely used benchmark dataset in movie recommendation, we conduct experiments on a widely used subset of this dataset, *MovieLens-1M*. *Amazon*[2] is a widely used benchmark dataset for product recommendation [8]. We select *Digital Music* and *Beauty* subsets from the collection. For each dataset, we treat each review as an interaction between the user and item to transform the it into implicit data. The statistics of the datasets are summarized in Table 1.

[1] MovieLens: https://grouplens.org/datasets/movielens/.
[2] Amazon: http://jmcauley.ucsd.edu/data/amazon/.

Table 2. Results of effectiveness experiments on four different datasets. We use "D-X" and "M-X" to represent the data- and model-oriented counterfactual learning when the backbone model is "X". Statistical significance of pairwise differences of AD-AUG vs. the backbone model is determined by a paired t-test (* for $p \leq 0.01$).

Datasets	MovieLens-1M			A-Music			A-Beauty		
Metrics	R@20	R@50	N@100	R@20	R@50	N@100	R@20	R@50	N@100
WMF	0.1072	0.1977	0.1492	0.1722	0.2534	0.1295	0.0532	0.0867	0.0435
SLIM	0.1153	0.2037	0.1589	0.1578	0.2003	0.1106	0.0166	0.0194	0.0129
CDAE	0.0929	0.1718	0.1410	0.0750	0.1366	0.0671	0.0267	0.0491	0.0232
D-CDAE	0.1044	0.1868	0.1466	0.0885	0.1541	0.0748	0.0319	**0.0571***	**0.0271***
M-CDAE	**0.1052***	**0.1891***	**0.1477***	**0.1061***	**0.1767***	**0.0872***	**0.0323***	0.0551	0.0269
MultDAE	0.1095	0.2060	0.1616	0.2021	0.3208	0.1566	0.0747	0.1224	0.0580
D-MultDAE	**0.1142***	**0.2164***	**0.1657**	**0.2160***	**0.3350***	**0.1628***	0.0784	0.1270	0.0600
M-MultDAE	0.1128	0.2114	0.1636	0.2111	0.3326	0.1603	**0.0791***	**0.1288***	**0.0611***
MultVAE	0.1132	0.2142	0.1659	0.2062	0.3241	0.1579	0.0782	0.1245	0.0588
D-MultVAE	0.1167	0.2169	0.1681	0.2178	**0.3398***	**0.1648***	0.0786	0.1281	0.0605
M-MultVAE	**0.1180**	**0.2196***	**0.1697***	**0.2192***	0.3354	0.1643	**0.0793**	**0.1294***	**0.0609***
MacridVAE	0.1130	0.2167	0.1658	0.2413	0.3626	0.1803	0.1036	0.1559	0.0753
D-MacridVAE	0.1176	**0.2220***	0.1691	**0.2485***	**0.3716***	**0.1861***	0.1082	0.1642	0.0790
M-MacridVAE	**0.1185***	0.2215	**0.1707***	0.2478	0.3699	0.1844	**0.1087***	**0.1652***	**0.0796***

Baselines. To demonstrate the effectiveness, we compare AD-AUG with the following representative models.

- **WMF** [11] is a weighted matrix factorization method, which decomposes the implicit user feedback similar to SVD but with confidence weights defined as number of times user interacted with item.
- **SLIM** [20] learns a sparse matrix of aggregation coefficient that corresponds to the weight of rated items aggregated to produce recommendation scores.
- **CDAE** [29] is an AE-based CF model which uses a demonising autoencoder for recommendation.
- **MultDAE** [16] extends CDAE by using a multinomial likehood for the data distribution.
- **MultVAE** [16] is a VAE-based CF model which uses a multinomial likelihood for VAE to improve the recommendation performance.
- **MacridVAE** [18] employs VAE to learn disentangled representation representing different interests of the user.

Evaluation Metrics. For each user of the dataset, we rank the interactions in chronological order and select the first 80% of historical interactions as the training set with the remaining 10%, 10% as the validation and test set respectively. For testing, we regard all unrated items as candidates and employ three metrics, *Recall (R)@K* with $K \in \{20, 50\}$ and *Normalized Discounted Cumulative Gain (NDCG or N)@K* with $K = 100$, which are computed based on rank of test interactions in top-K ranked list.

Implementation Detail. We implement our AD-AUG in Pytorch[3]. The embedding size of user representation is fixed to 100 for all experiments. The encoder and decoder consist of two layers with [500, 300] and [300, 500] respectively, each with ReLU activation. For our method, the hyper-parameter $\tau = 1.0$, $\beta = 0.2$, $\gamma = 0.99$, and dropout with probability $p = 0.5$ is employed to the input. We set $\lambda = 1.0$ and the initial value $\rho = 10$ for data oriented method while $\rho = 1000$ for model oriented method in curriculum adversarial learning. We optimize AD-AUG with Adam optimizer with the learning rate as 0.001 to both augmenter and recommender and using early stopping with a patience of 50, i.e. we stop training if NDCG@100 on the validation set does not increase for 50 successive epochs. For baseline methods, we split exactly the same training, validation and test set as AD-AUG and apply a gird search for optimal hyper-parameters.

5.2 Experimental Result

Overall Comparison. We summarize the results by comparing the performance of all the methods. As shown in Table 2, MacridVAE performs best among all the AE-based CF methods, which demonstrates the effectiveness of disentangled representation modeling different user interests for recommendation. Meanwhile, compared with these baselines, the data- and model-oriented models have achieved a significant improvement over all datasets, especially model-oriented method. This is actually not surprising, since the counterfactual user feedback generated from the model-oriented method is more targeted, which is tailored for improving the recommender. However, the model-oriented method is achieved by using the information of the recommender, when the loss of the recommender model is unavailable (as black box for augmenter), it is better to use data-oriented method.

The Statistics of Data Augmentation. As there are two proposed methods of data augmentation, we study the augmentation results during whole training process respectively. The ratio of interactions changed by augmenter model is recorded during training process under the sets of hyper-parameters where each model achieves its best performance. As shown in Fig. 4, the ratio of generated counterfactual interactions decrease until they converge during training process. Meanwhile, the ratio differs between data- and model- oriented methods during the training process. The change ratio of data-oriented method is significantly higher than model-oriented method on each epoch. On the one hand, this observation indicates that data-oriented method, which leverages mutual information to augment training data, would prefer more changes on the interaction data to assist the recommender to extract more information from training data. On the other hand, the fewer change ratio of model-oriented method means a few interactions are added or dropped by the augmenter to reach the best performance. This is because that original user feedback is sparse, and this causes the number of adding/deleting user interactions from the original feedback small for the model-oriented method.

[3] The implementations are available at https://github.com/Fang6ang/AD-AUG.

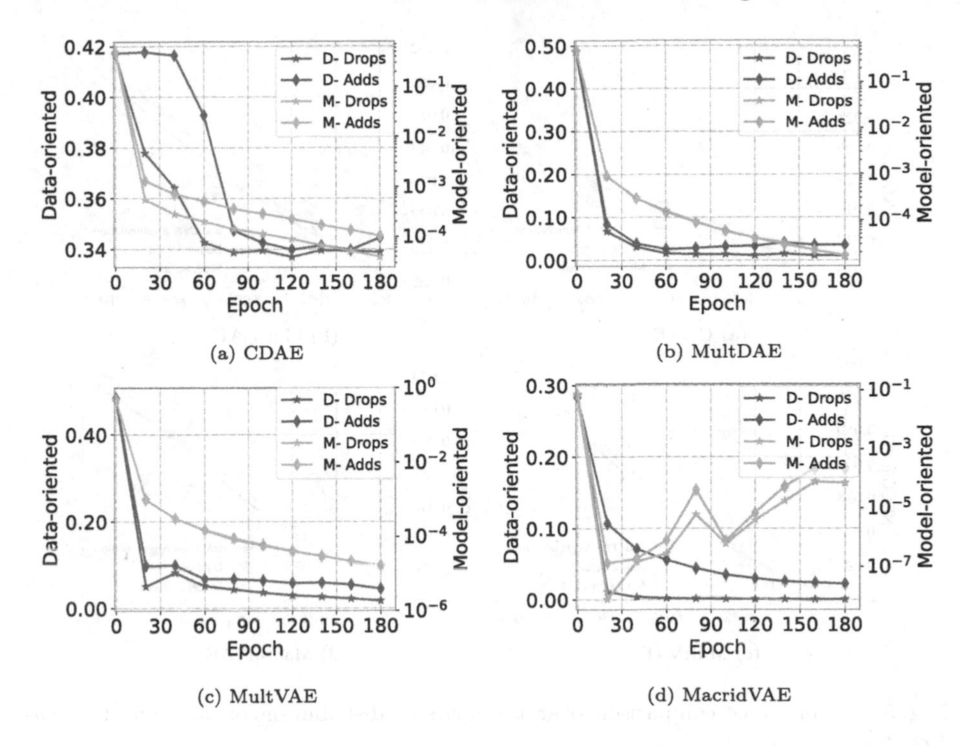

Fig. 4. Training dynamics of drop/add ratio in augmented data on *Amazon-Beauty*.

The Effect of Data Augmentation. The number of user feedback is an important factor that affects the recommendation performance since fewer user-item interactions are insufficient to generate high-quality representations. We study whether our data augmentation methods can alleviate this sparsity issue. Towards this end, we divide the feedback data of user in the training data into five equal folds and vary the amount of training data from one fold to four folds, corresponding to 20%, 40%, 60%, 80% of entire training data as training sets. Figure 5 illustrates the performance w.r.t. different sparsity distribution of data on *Amazon-Beauty*, the performance substantially drops when less training data is used. Meanwhile, we can see our data- and model-oriented counterfactual learning can enhance the performance of each AE-based CF models, and the improvements are particularly significant when the user feedback is relatively sparse (40% to 80% of user feedback). The result indicates that AD-AUG helps improve recommendation for inactive users by generating the counterfactual user feedback.

Influence of Curriculum Learning. To investigate how the curriculum adversarial learning affects the performance, we compare the adversarial learning process under three different annealing configurations: our complete method, our model without annealing that set α with fixed initial value ρ (Fixed), our model that randomly set α between 0 to ρ under each curriculum step (Random). As shown

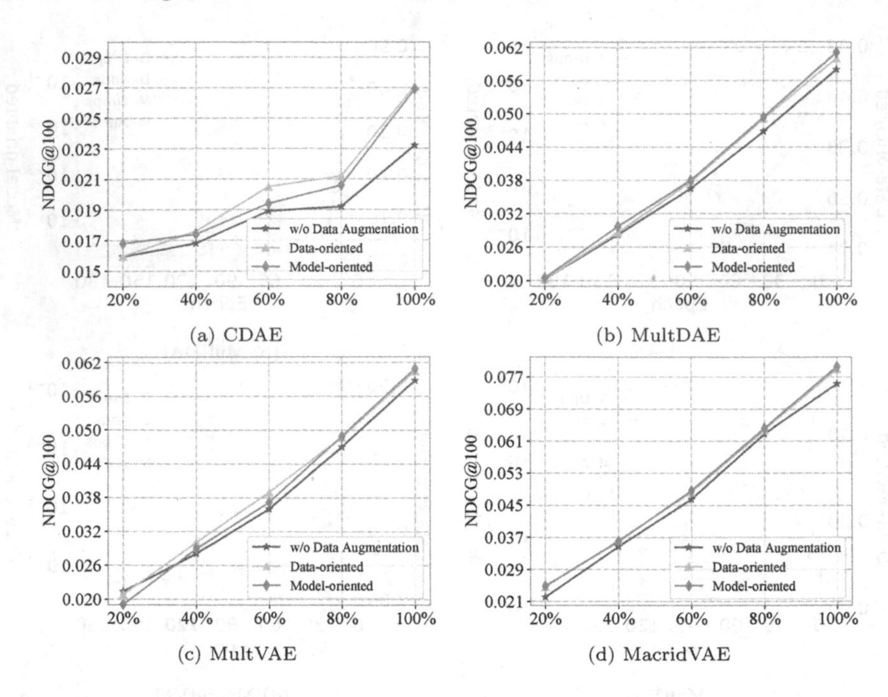

Fig. 5. Performance comparison over the sparsity distribution of data on *Amazon-Beauty*.

in Fig. 6, for the model-oriented counterfactual learning, the performance has slightly been affected without the curriculum learning. In contrast, curriculum learning has a more significant effect on data-oriented counterfactual learning. The best results are attained by considering the easy-to-difficult curriculum pattern, which indicates that curriculum adversarial learning makes the objective smoother, thus more easily reaching the global optimal.

Ablation Study. We conduct an ablation study to verify the effectiveness of the two proposed data augmentation methods. Figure 7 compares the recommender model's performance on *Amazon-Beauty* when the augmenter model applies only one kind of change to the original data, i.e., adding or dropping interactions in user's behavior history. As shown in Fig. 7, model's performance declines when there is only one kind of change is applied to the interaction data. The results indicate that both adding and dropping is required in data augmentation: adding interactions helps the recommender discover implicit feedback from original data, while dropping interactions removes noises from observed interaction history. The model performance suffers more under w/o drops settings. We suspect that it is because data denoising plays a more fundamental role for recommender model in terms of discovering user's true interests, while just adding interactions may lead to more noisy data.

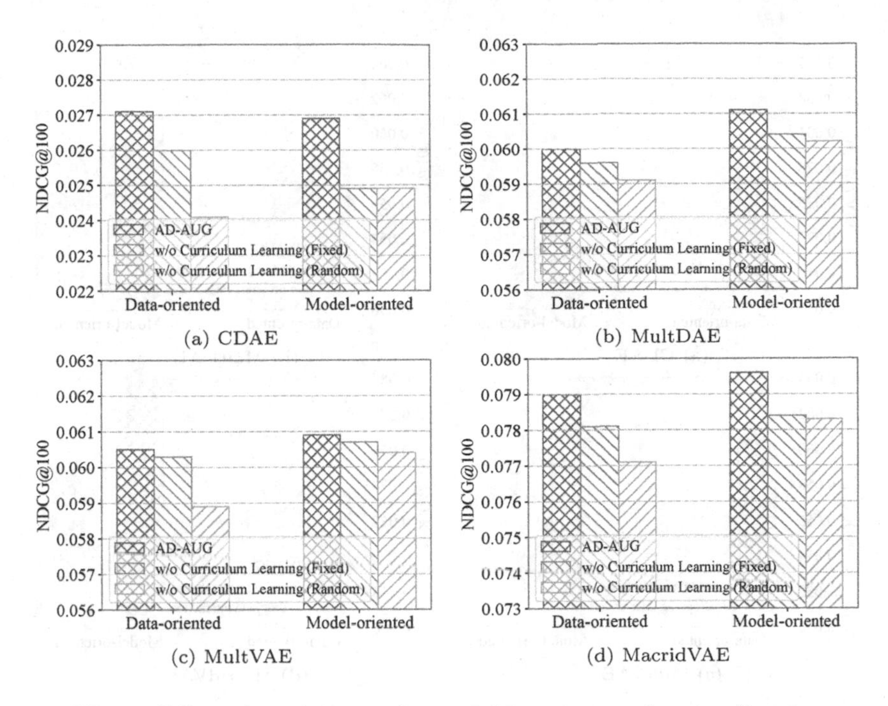

Fig. 6. Effect of curriculum adversarial learning on *Amazon-Beauty*.

Visualization and Case Study. To further investigate how our counterfactual data augmentation framework facilitates the user representation learning, we visualize the learnt hidden representations and conduct case study on *Amazon-Beauty*. We use MacridVAE as backbone and visualize the high-dimensional user representations learned by MacridVAE and our model. Then we randomly select a user (67) and present this user's Top-3 add/delete products for the original feedback. As shown in Fig. 8, we treat each learned disentangle component of a user as an individual point and the k-th component is colored according to k. Compared with backbone model, the data- and model-oriented learning frameworks show different cluster structures, especially data-oriented method, which can form clearer clusters. Additionally, as the random selected user(67) mostly focus on nail makeups and relevant tools, the two counterfactual frameworks remove some of the nail tools and add bath equipment and other makeup products to the feedback. It indicates that the counterfactual data makes the model learned more personalized characteristics.

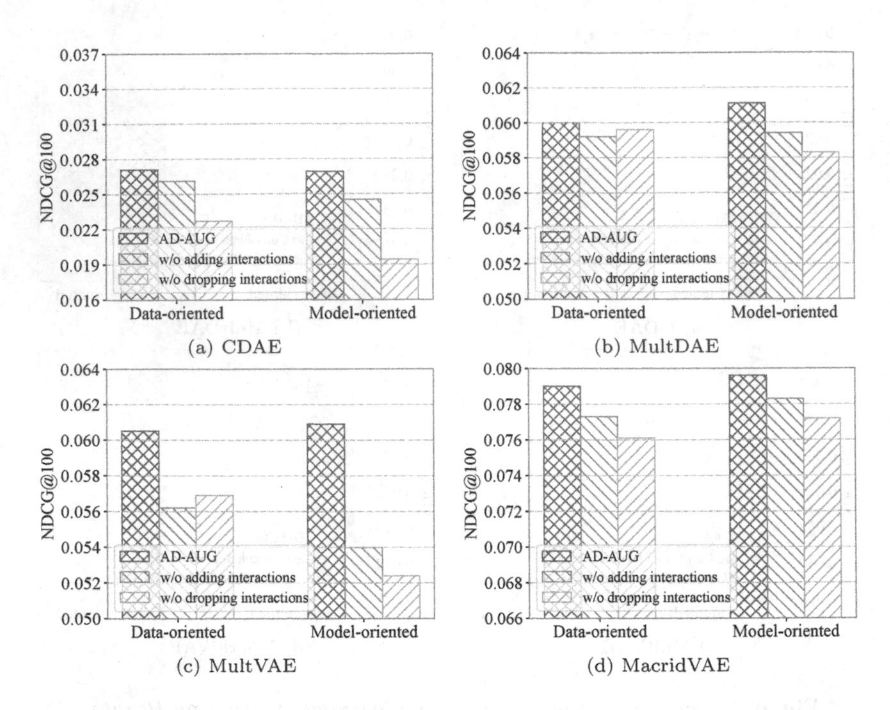

Fig. 7. Ablation study about different types of data augmentation on *Amazon-Beauty.*

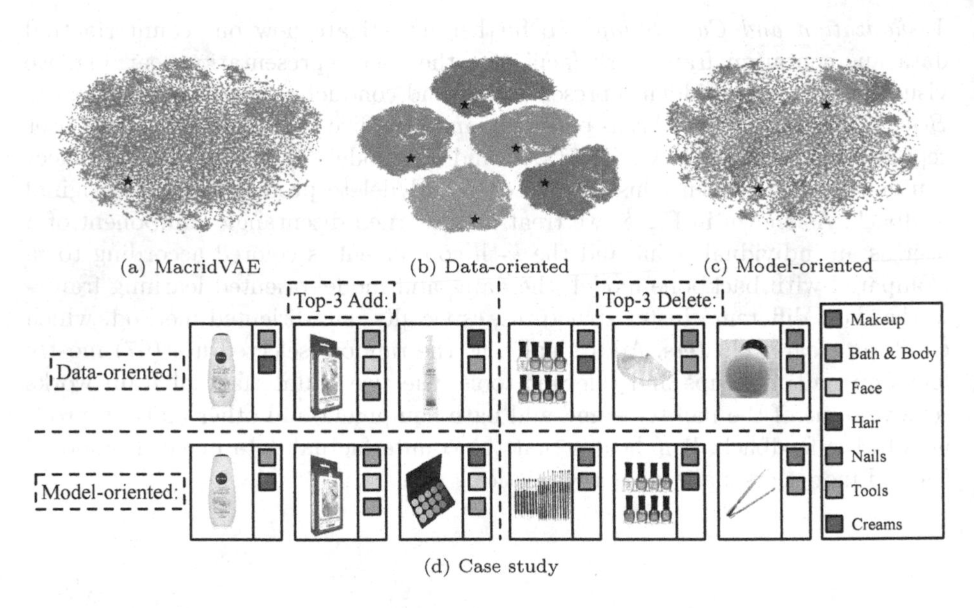

Fig. 8. Visualization of the learned t-SNE [19] transformed user representations on *Amazon-Beauty*, where the marked stars represent a user (67). And Fig. 8d represents the case study of this user.

6 Conclusion

In this paper, we propose to improve CF performance by enriching the user feedback based on the idea of counterfactual thinking. To achieve goal, we design two adversarial learning-based data augmentation methods to generate the counterfactual user feedback data for recommendation. Experiments demonstrate that our proposed AD-AUG model achieves considerable improvement compared with state-of-the-art models.

Acknowledgments. This research is partially supported by National Key Research and Development Program of China with Grant No. 2018AAA0101902, the National Natural Science Foundation of China (NSFC Grant 62106008 & 62006004).

References

1. Abbasnejad, E., Teney, D., Parvaneh, A., Shi, J., van den Hengel, A.: Counterfactual vision and language learning. In: CVPR, pp. 10044–10054 (2020)
2. Ashual, O., Wolf, L.: Specifying object attributes and relations in interactive scene generation. In: ICCV, pp. 4561–4569 (2019)
3. Chen, L., Zhang, H., Xiao, J., He, X., Pu, S., Chang, S.F.: Counterfactual critic multi-agent training for scene graph generation. In: ICCV, pp. 4613–4623 (2019)
4. Chen, T., Kornblith, S., Norouzi, M., Hinton, G.: A simple framework for contrastive learning of visual representations. In: ICML, pp. 1597–1607 (2020)
5. Creswell, A., White, T., Dumoulin, V., Arulkumaran, K., Sengupta, B., Bharath, A.A.: Generative adversarial networks: an overview. IEEE Sig. Process. Mag. **35**(1), 53–65 (2018)
6. Fu, T.-J., Wang, X.E., Peterson, M.F., Grafton, S.T., Eckstein, M.P., Wang, W.Y.: Counterfactual vision-and-language navigation via adversarial path sampler. In: Vedaldi, A., Bischof, H., Brox, T., Frahm, J.-M. (eds.) ECCV 2020. LNCS, vol. 12351, pp. 71–86. Springer, Cham (2020). https://doi.org/10.1007/978-3-030-58539-6_5
7. Goyal, Y., Wu, Z., Ernst, J., Batra, D., Parikh, D., Lee, S.: Counterfactual visual explanations. In: ICML,pp. 2376–2384. PMLR (2019)
8. He, R., McAuley, J.: Ups and downs: modeling the visual evolution of fashion trends with one-class collaborative filtering. In: WWW, pp. 507–517 (2016)
9. He, X., He, Z., Du, X., Chua, T.S.: Adversarial personalized ranking for recommendation. In: SIGIR, pp. 355–364 (2018)
10. Higgins, I., et al.: Beta-VAE: learning basic visual concepts with a constrained variational framework. In: ICLR (2017)
11. Hu, Y., Koren, Y., Volinsky, C.: Collaborative filtering for implicit feedback datasets. In: ICDM, pp. 263–272 (2008)
12. Isola, P., Zhu, J.Y., Zhou, T., Efros, A.A.: Image-to-image translation with conditional adversarial networks. In: CVPR, pp. 1125–1134 (2017)
13. Jang, E., Gu, S., Poole, B.: Categorical reparameterization with gumbel-softmax. In: ICLR (2017)
14. Kingma, D.P., Welling, M.: Auto-encoding variational Bayes. In: ICLR (2014)
15. Koren, Y., Bell, R.M., Volinsky, C.: Matrix factorization techniques for recommender systems. IEEE Comput. **42**(8), 30–37 (2009)

16. Liang, D., Krishnan, R.G., Hoffman, M.D., Jebara, T.: Variational autoencoders for collaborative filtering. In: WWW, pp. 689–698 (2018)
17. Lin, K., Li, D., He, X., Zhang, Z., Sun, M.T.: Adversarial ranking for language generation. In: NeuIPS, pp. 3155–3165 (2017)
18. Ma, J., Zhou, C., Cui, P., Yang, H., Zhu, W.: Learning disentangled representations for recommendation. In: NeuIPS, pp. 5712–5723 (2019)
19. Van der Maaten, L., Hinton, G.: Visualizing data using t-SNE. J. Mach. Learn. Res. **9**(11) (2008)
20. Ning, X., Karypis, G.: Slim: sparse linear methods for top-n recommender systems. In: ICDM, pp. 497–506 (2011)
21. Poole, B., Ozair, S., Van Den Oord, A., Alemi, A., Tucker, G.: On variational bounds of mutual information. In: ICML, pp. 5171–5180 (2019)
22. Rezende, D.J., Mohamed, S., Wierstra, D.: Stochastic backpropagation and approximate inference in deep generative models. In: ICML, pp. 1278–1286. PMLR (2014)
23. Sedhain, S., Menon, A.K., Sanner, S., Xie, L.: Autorec: autoencoders meet collaborative filtering. In: WWW, pp. 111–112 (2015)
24. Shenbin, I., Alekseev, A., Tutubalina, E., Malykh, V., Nikolenko, S.I.: RecVAE: a new variational autoencoder for top-n recommendations with implicit feedback. In: WSDM, pp. 528–536 (2020)
25. Suresh, S., Li, P., Hao, C., Neville, J.: Adversarial graph augmentation to improve graph contrastive learning, pp. 15920–15933 (2021)
26. Tian, Y., Sun, C., Poole, B., Krishnan, D., Schmid, C., Isola, P.: What makes for good views for contrastive learning? In: NeurIPS, pp. 6827–6839 (2020)
27. Vincent, P., Larochelle, H., Bengio, Y., Manzagol, P.A.: Extracting and composing robust features with denoising autoencoders. In: ICML, pp. 1096–1103 (2008)
28. Wang, Z., et al.: Counterfactual data-augmented sequential recommendation. In: SIGIR, pp. 347–356 (2021)
29. Wu, Y., DuBois, C., Zheng, A.X., Ester, M.: Collaborative denoising auto-encoders for top-n recommender systems. In: WSDM, pp. 153–162 (2016)
30. Xu, D., Cheng, W., Luo, D., Chen, H., Zhang, X.: Infogcl: Information-aware graph contrastive learning. In: NeurIPS, pp. 30414–30425 (2021)
31. Xu, T., et al.: AttnGAN: fine-grained text to image generation with attentional generative adversarial networks. In: CVPR, pp. 1316–1324 (2018)
32. Yang, M., Dai, Q., Dong, Z., Chen, X., He, X., Wang, J.: Top-n recommendation with counterfactual user preference simulation. In: CIKM, pp. 2342–2351 (2021)
33. Zmigrod, R., Mielke, S.J., Wallach, H., Cotterell, R.: Counterfactual data augmentation for mitigating gender stereotypes in languages with rich morphology. arXiv preprint arXiv:1906.04571 (2019)

Bi-directional Contrastive Distillation for Multi-behavior Recommendation

Yabo Chu[1], Enneng Yang[1], Qiang Liu[2], Yuting Liu[1], Linying Jiang[1], and Guibing Guo[1(✉)]

[1] Software College, Northeastern University, Shenyang, Liaoning, People's Republic of China
{2071268,20185246}@stu.neu.edu.cn, ennengyang@stumail.neu.edu.cn
{jiangly,guogb}@swc.neu.edu.cn

[2] Center for Research on Intelligent Perception and Computing (CRIPAC), National Laboratory of Pattern Recognition (NLPR), Institute of Automation, Chinese Academy of Sciences, Beijing, People's Republic of China
qiang.liu@nlpr.ia.ac.cn

Abstract. Multi-behavior recommendation leverages auxiliary behaviors (e.g., view, add-to-cart) to improve the prediction for target behaviors (e.g., buy). Most existing works are built upon the assumption that all the auxiliary behaviors are positively correlated with target behaviors. However, we empirically find that such an assumption may not hold in real-world datasets. In fact, some auxiliary feedback is too noisy to be helpful, and it is necessary to restrict its influence for better performance. To this end, in this paper we propose a **B**i-directional **C**ontrastive **D**istillation (BCD) model for multi-behavior recommendation, aiming to distill valuable knowledge (about user preference) from the interplay of multiple user behaviors. Specifically, we design a forward distillation to distill the knowledge from auxiliary behaviors to help model target behaviors, and then a backward distillation to distill the knowledge from target behaviors to enhance the modelling of auxiliary behaviors. Through this circular learning, we can better extract the common knowledge from multiple user behaviors, where noisy auxiliary behaviors will not be involved. The experimental results on two real-world datasets show that our approach outperforms other counterparts in accuracy.

Keywords: Recommender system · Contrastive distillation · Multi-behavior recommender

1 Introduction

Modern applications heavily rely on recommender systems as an essential tool to overcome the issue of information overload and improve user experience and satisfaction. Conventional recommenders aim to learn users preference from their target behaviors on items (e.g., 'buy' in e-commerce, 'watch' in movies). Recently, it has become a hot research topic to involve auxiliary behaviors of

M.-R. Amini et al. (Eds.): ECML PKDD 2022, LNAI 13713, pp. 491–507, 2023.
https://doi.org/10.1007/978-3-031-26387-3_30

users (e.g., 'view' and 'add-to-cart' in e-commerce) for performance enhancement. The basic assumption is that auxiliary behaviors are positively correlated with target ones, and can directly reveal user interest to some extent. Hence, most existing research takes into account all the auxiliary behaviors [1,17,20], and implicitly works on the same conversion paths among user behaviors [3,8]. Take e-commerce as an example, the general conversion paths are 'view → add-to-cart → buy' and 'view → buy'. That is, users generally browse products on the website or apps, and then add the products of interest into the shopping cart (and then purchase) or directly purchase them without add-to-cart. For simplicity, hereafter we use 'cart' to represent the 'add-to-cart' behavior.

In this paper, we revisit the above assumption on two real-world datasets (**Taobao**[1], **Beibei**[2]), and find that the assumption may not hold in real applications. Specifically, we conduct data analysis by applying a funnel model on the two conversion paths. The results show that conversion paths on Beibei are valid while those on Taobao are less helpful: 'cart' is too noisy to be involved in a conversion path. By removing 'cart' from 'view' data, we obtain a refined conversion path that reaches higher conversion rate than the original one. As a conclusion, not all auxiliary behaviors are positively correlated with target behaviors and some noisy behaviors should be removed from existing conversion paths for better conversion rates.

Therefore, we propose a novel **B**i-directional **C**ontrastive **D**istillation (BCD) model for multi-behavior recommendation, aiming to distill valuable knowledge (about user preference) from the refined conversion paths. Specifically, we design a forward distillation to learn the knowledge from auxiliary feedback to help model target behaviors, and a backward distillation to learn the knowledge from target behaviors to enhance auxiliary ones. In this way, we can highlight the common knowledge (about user preference) from both kinds of user behaviors, and thus improve recommendation performance. To sum up, the main contributions of this paper are summarized as follows:

- We conduct a thorough data analysis on two real datasets and find that the previous assumption may not hold, since some auxiliary behaviors are too noisy to be involved in the conversion paths.
- We propose a novel bi-directional contrastive distillation model to distill and transfer the knowledge from one kind of user behaviors to help model the other kind of user behaviors, whereby better representations of users and items can be learned.
- We conduct extensive experiments on two real-world datasets, and the experimental results demonstrate the effectiveness of our proposed approach in comparison with five competing methods.

2 Data Analysis

In this section, we will revisit the underlying assumption of existing multi-behavior recommenders based on two real-world datasets, that is, all auxiliary

[1] https://github.com/chenchongthu/GHCF.
[2] https://tianchi.aliyun.com/dataset/dataDetail?dataId=649.

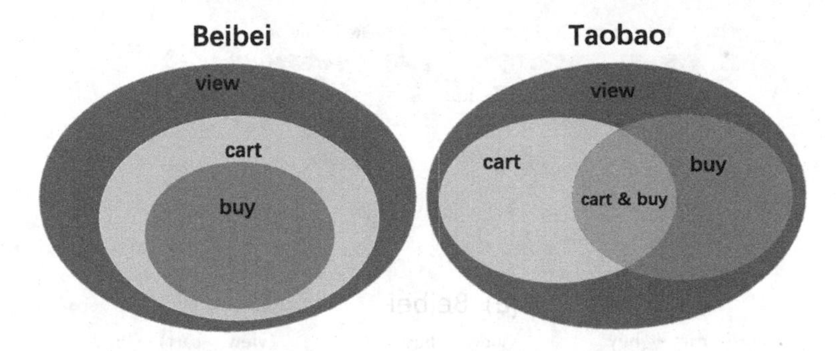

Fig. 1. The distribution of different kinds of user behaviors on the Beibei and Taobao datasets.

user behaviors are positively correlated with target behaviors. The two datasets are Taobao and Beibei. Beibei is a vertical e-commerce platform in China specializing in maternal and infant products, while Taobao is one of China's largest integrated e-commerce platforms. Both datasets consist of three different types of user behaviors: view, cart (i.e., add-to-cart) and buy. The distributions of user behaviors are given in Fig. 1.

For the ease of discussion, we first introduce a number of notations to describe the datasets. Let $U = \{u_1, u_2, \ldots, u_m\}$ and $V = \{v_1, v_2, \ldots, v_n\}$ be the set of users and items, where m and n are the number of users and items, respectively. Let $B = \{b_1, b_2, \ldots, b_k\}$ be the set of behavior types, where k is the number of behavior types and b_k is the k-th type of user behaviors. Each user may have multiple interactions with a same item, resulting in multiple types of user behaviors. Let V_{u,b_k} be the set of items that user u has interacted with by behavior type b_k. In our experiments, we have three different sets of items for user u, namely $V_{u,view}, V_{u,cart}, V_{u,buy}$.

From Fig. 1, we can observe that the relations of user behaviors follow $V_{u,buy} \subset V_{u,cart} \subset V_{u,view}$ in Beibei, indicating that all the purchased items were added to the shopping cart and browsed by the user in the first place. In this respect, all the auxiliary behaviors are positively correlated with target behaviors. In other words, the assumption in question holds in Beibei. However, this assumption does not hold in Taobao. Specifically, although the area of 'cart' and 'buy' has some overlaps, the ratio is less than 40%. That is, more than 60% Taobao users have only two kinds of behaviors (rather than all of them) on items, either 'view/cart' or 'view/buy'. Users in Taobao have only purchased a small portion of products added to their carts. Hence, we may conclude that 'cart' is likely to be a noisy behavior since 'cart' in most cases does not imply 'buy'. As a general e-commerce platform, Taobao covers such a large variety of product categories that users may not have a strong shopping intent when they go window-shopping online (maybe for exploring interesting products). In the contrast, users likely bring a strong intent to meet their needs when visiting a specialized platform like Beibei.

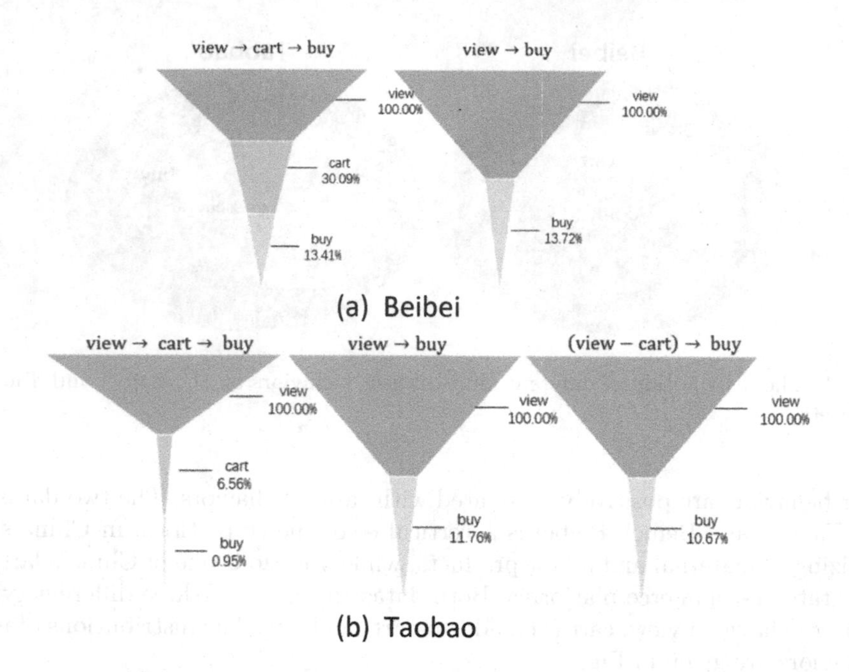

Fig. 2. Funnel model diagram of different conversion paths on the Beibei dataset and Taobao dataset.

To further validate our inference, we apply a funnel model [11] to study the conversion rate of two general conversion paths (i.e., 'view→cart→buy' and 'view→buy'). The purchase funnel is a consumer-focused marketing model that depicts the hypothetical customer journey toward buying a product or service [11]. Since 'cart' is possibly noisy in Taobao, we devise a refined conversion path by considering the products viewed but not added to cart (and finally purchased), denoted by '(view-cart)→buy'. The conversion rate can be calculated as follows:

$$r_{b_1 \to b_2} = \frac{1}{|U|} \sum_u \frac{|V_{u,b_1} \cap V_{u,b_2}|}{|V_{u,b_1}|}$$

$$r_{b_1 \to b_2 \to b_3} = r_{b_1 \to b_2} \times r_{b_2 \to b_3}$$

where $r_{b_1 \to b_2}$ denotes the conversion rate from behavior b_1 to b_2, and $r_{b_1 \to b_2 \to b_3}$ represents the conversion rate from behavior b_1 to behavior b_3 through behavior b_2. $|\cdot|$ is the cardinality of a given set. The experimental results on Beibei and Taobao are illustrated in Fig. 2.

Specifically, in Fig. 2(a) the conversion rates from both paths ('view→cart →buy' and 'view→buy') are very close, indicating that users in Beibei have similar purchase pattern, i.e., either immediately buy products after browsing or firstly add products to cart and then purchase. In fact, there is 40% conversion from 'cart' to 'buy'. In Fig. 2(b), the conversion rate from 'view→cart→buy' is extremely small in comparison with other conversion paths. It shows that users in Taobao only purchase a small portion of products added to cart, and the conversion rate from 'cart' to 'buy' is around 14.5%. By removing 'cart' from 'view', as illustrated in Fig. 3, the conversion rate of path '(view-cart)→buy' is close to that of path 'view→buy', which is much greater than that of path 'view→cart→buy', implying that 'cart' is quite a noisy behavior. This conclusion will be further validated by our experiments in Sect. 4.6.

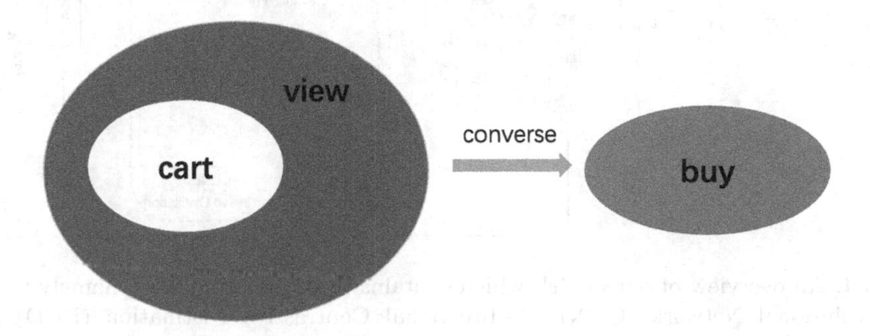

Fig. 3. Illustration of our refined conversion path. The white part of the view indicates that we have removed the items that the user added to the cart from the collection of items she viewed.

To sum up, we find that not all auxiliary behaviors are valuable in providing knowledge about user preference. Their usefulness is domain dependent, that is, a same auxiliary behavior (e.g., 'cart') is helpful in one dataset but maybe noisy in another one. The potential noise involved in auxiliary behaviors motivates us to design a bi-directional contrastive learning for common knowledge distillation from multiple user behaviors.

3 Our Proposed Model

The overall architecture of our proposed BCD model is illustrated in Fig. 4. It contains three main components: graph convolutional network (GCN), Bi-directional Contrastive Distillation (BCD) and prediction modules. Specifically, the GCN module is to learn representations of users, items and behaviors from the graph structure of user-item interactions by each kind of behaviors. The BCD module further refines those representations by applying contrastive learning, i.e., to increase the similarity among the same kind of behaviors and highlight the difference among different kinds of behaviors. Then, a forward distillation process is used to distill valuable knowledge (about user preference) from auxiliary to target behaviors. After that, a backward distillation is designed to distill knowledge from target to auxiliary behaviors, to strength the common knowledge among user behaviors and to enhance the task of auxiliary behavior prediction. Lastly, the prediction module is to predict user's possible behaviors on a given item. We will elaborate each module in next subsections.

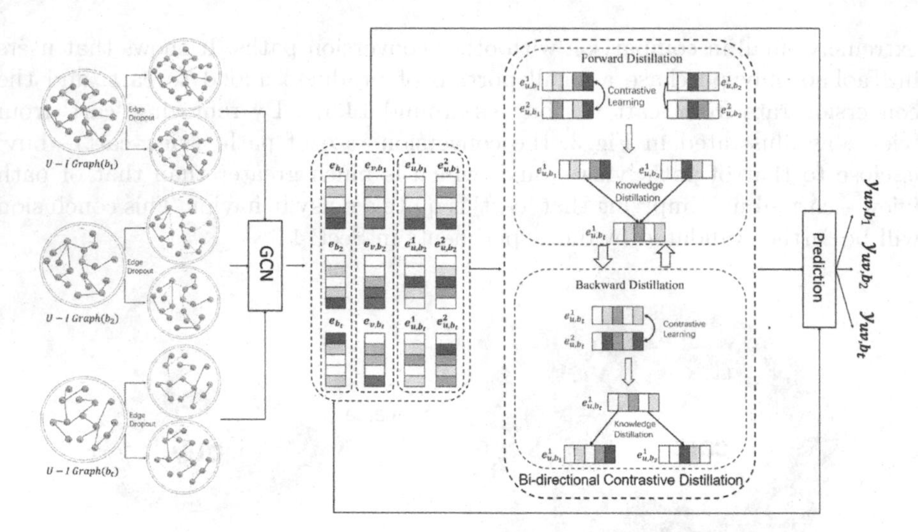

Fig. 4. An overview of our model, which contains three main modules, namely Graph Convolutional Network (GCN), Bi-directional Contrastive Distillation (BCD) and multi-behavior prediction.

3.1 Multi-behavior GCN

Since each user may have multiple behaviors on a same item, we construct an indirected graph to accommodate user-item interactions for each kind of behaviors. The graph nodes are users and items, and the edges are interactions among users and items. We aim to learn user and item representations from the graph structure through graph convolutional networks (GCN) [9,22], which models nodes propagation based on message-passing architecture. Specifically, user u's embedding $\mathbf{e}_{u,b_k}^{(l)}$ in layer l (under behavior b_k) can be learned by aggregating the representations of his/her neighbors (i.e., interacted items in layer $(l-1)$), which is formally defined as follows:

$$\mathbf{e}_{u,b_k}^{(l)} = \sigma\left(\sum_{v\in\mathcal{N}_u} \frac{1}{\sqrt{|\mathcal{N}_u|\,|\mathcal{N}_v|}} \mathbf{W}^{(l)} \phi\left(\mathbf{e}_v^{(l-1)} \odot \mathbf{e}_{b_k}^{(l-1)}\right)\right)$$

where $\mathcal{N}_u$ and $\mathcal{N}_v$ are the set of immediate neighbors of user u and item v, respectively; $\mathbf{W}^{(l)}$ is a weight matrix in the l-th propagation step, ϕ is a composition operator to incorporate behavior embeddings. $\sigma(\cdot)$ is the LeakyReLU [18] activation function. $\odot$ denotes the element-wise product of two vectors. The normalization term $\frac{1}{\sqrt{|\mathcal{N}_u||\mathcal{N}_v|}}$ is used to avoid the scale of embeddings increasing with graph convolution operations. The representation of item v can be learned in the similar way. Besides, we update behavior embedding $\mathbf{e}_{b_k}^{(l)}$ in layer l by applying a linear transformation to its representation in previous layer, defined by:

$$\mathbf{e}_{b_k}^{(l)} = \mathbf{W}_{b_k}^{(l)} \mathbf{e}_{b_k}^{(l-1)}$$

where $\mathbf{W}_{b_k}^{(l)}$ is a layer-specific weight matrix representing the linear transformation. In this way, behaviors are also transformed into the same embedding space as users and items, whereby mathematical operations can be imposed on them. To start with, we adopt an ID embedding layer to initialize the embeddings of users, items and behaviors in the first hop, denoted as $\mathbf{e}_{u,b_k}^{(0)}, \mathbf{e}_{v,b_k}^{(0)}, \mathbf{e}_{b_k}^{(0)}$.

3.2 Bi-directional Contrastive Distillation

The previous module separately learns multiple representations for each user and item by applying GCN technology on a constructed user-item interaction graph (for each kind of behaviors). We now proceed to refine those representations by taking into account two kinds of relations among user behaviors, namely intra-behavior and inter-behavior relations. Specifically, for the intra-behavior relation, we devise a contrastive learning strategy [21,30,32] to strengthen the similarity between two different forms of the same user, and to increase the difference between two different users in the meanwhile. Hence, we design *edge dropout* on the user-item interaction graph by randomly dropping out a certain ratio of edges. By varying different dropout ratios, we obtain two variants of original interaction graph, from which two forms of user presentations (i.e., e_{u,b_k}^1 and e_{u,b_k}^2) can be learnt by the GCN module. For clarity, we use symbols b_a, b_t, b_k to denote auxiliary behavior, target behavior and any kind of user behaviors, respectively. For auxiliary behavior b_a, the objective function of contrastive learning can be formulated as follows.

$$\mathcal{L}_{b_a \to b_a} = -\sum_{u \in U} \mathbb{E}\left[\log \frac{\exp\left(\text{sim}\left(\mathbf{e}_{u,b_a}^1, \mathbf{e}_{u,b_a}^2\right)/\tau\right)}{\sum_{p \in U} \exp\left(\text{sim}\left(\mathbf{e}_{u,b_a}^1, \mathbf{e}_{p,b_a}^1\right)/\tau\right)}\right]$$

where $\text{sim}(\cdot,\cdot)$ is a similarity function to measure the closeness of two input vectors, and inner product is often used for easy computation. τ is a temperature parameter. The numerator term is to maximize the similarity between two embeddings of user u, while the denorminator term is to maximize the difference between user u and any other user p. To avoid mode collapse [14], hereafter we only adopt the first variant of user embeddings during the model learning other than the numerator term.

For the inter-behavior relation, we aim to distill valuable knowledge about user preference from auxiliary behaviors b_a to guide the learning of target behavior b_t. The basic idea is similar with that of intra-behavior relation, i.e., to maximize the similarity between auxiliary and target behaviors of a same user, and meanwhile to maximize the difference of auxiliary and target behaviors between two different users. Formally, the objective loss function of contrastive learning from auxiliary behavior b_a to target behavior b_t can be formulated as follows:

$$\mathcal{L}_{b_a \to b_t} = -\sum_{u \in U} \mathbb{E}\left[\log \frac{\exp\left(\text{sim}\left(\mathbf{e}_{u,b_a}^1, \mathbf{e}_{u,b_t}^1\right)/\tau\right)}{\sum_{p \in U} \exp\left(\text{sim}\left(\mathbf{e}_{u,b_a}^1, \mathbf{e}_{p,b_t}^1\right)/\tau\right)}\right]$$

Through the above two-stage operations which we denote as *forward distillation*, the knowledge from auxiliary behaviors can be well learned and transferred to model target behaviors. Inspired by the bi-directional sequence learning in natural language processing [6,19], we design a *backward distillation* to distill the knowledge from target behavior and help model auxiliary behaviors in return (see Fig. 4). Specifically, we apply contrastive learning on target behaviors to refine its representation and then perform knowledge distillation where (modelling of) target behavior is used as a teacher and auxiliary behaviors as a student. This curriculum learning strategy is to better extract the common knowledge between the two kinds of user behaviors and enhance not only the predictive task of target behavior, but also the prediction of auxiliary behaviors.

Hence, the overall objective of our BCD module is to minimize the following loss function:

$$\mathcal{L}_{\text{BCD}} = \underbrace{\sum_{b_a} (\mathcal{L}_{b_a \to b_a} + \mathcal{L}_{b_a \to b_t})}_{\text{forward distillation}} + \underbrace{\sum_{b_a} (\mathcal{L}_{b_t \to b_t} + \mathcal{L}_{b_t \to b_a})}_{\text{backward distillation}}$$

3.3 Prediction and Learning

We first adopt weighted average across all the behavior-specific embeddings to get the final representations of both users and items, defined by:

$$\mathbf{e}_u = \sum_{b_k} \lambda_{b_k} \mathbf{e}_{u,b_k}, \quad \mathbf{e}_v = \sum_{b_k} \lambda_{b_k} \mathbf{e}_{v,b_k}$$

where λ_{b_k} indicates the importance of behavior b_k relative to target behavior b_t. The setting of $(\lambda_{view}, \lambda_{cart}, \lambda_{buy})$ follows the suggestions given by [1] on two dataset. Then the likelihood that user u will perform the k-th behavior on item v can be estimated by:

$$\hat{y}_{uv,b_k} = \mathbf{e}_u^\top \cdot \text{diag}\left(\mathbf{e}_{b_k}\right) \cdot \mathbf{e}_v = \sum_{i}^{d} e_{u,i} e_{b_k,i} e_{v,i}$$

where diag(·) is a function that converts an input vector into a diagonal matrix, and d is the embedding size.

The main purpose of behavior prediction is to minimize the error of prediction and ground truth, defined by:

$$\mathcal{L}_{b_k} = \sum_{u \in U} \sum_{v \in V} c_{uv,b_k} \left(y_{uv,b_k} - \hat{y}_{uv,b_k}\right)^2$$

where c_{uv,b_k} denotes the weight of entry y_{uv,b_k}. To learn model parameters more effectively and stably, we apply the efficient non-sampling learning technique

[1,7,16] to optimize our model. Specifically, we simplify c_{uv,b_k} to c_{v,b_k} and reformulate the above loss function as follows:

$$\mathcal{L}_{b_k} = \sum_{u \in U} \sum_{v \in V_{u,b_k}} \left(\left(c^{+}_{v,b_k} - c^{-}_{v,b_k} \right) \hat{y}^{2}_{uv,b_k} - 2 c^{+}_{v,b_k} \hat{y}_{uv,b_k} \right)$$

$$+ \sum_{i=1}^{d} \sum_{j=1}^{d} \left(\left(e_{b_k,i} e_{b_k,j} \right) \left(\sum_{u \in U} e_{u,i} e_{u,j} \right) \left(\sum_{v \in V_u} c^{-}_{v,b_k} e_{v,i} e_{v,j} \right) \right)$$

where V_u represents the items set that user u has interacted with; c^{+}_{v,b_k} and c^{-}_{v,b_k} are the weights if item v has been interacted with behavior b_k and other behaviors, respectively.

The final loss function consists of three components: loss value of behavior predictions, loss value of bi-directional contrastive learning and regularisation terms, which is:

$$\mathcal{L}(\Theta) = \sum_{b_k} \lambda_{b_k} \mathcal{L}_{b_k} + \mathcal{L}_{\text{BCD}} + \mu \|\Theta\|_2^2$$

where Θ is the set of model parameters; μ is a regularisation parameter, and $\|\cdot\|_2$ denotes the Frobenius norm.

4 Experiments

4.1 Datasets

As discussed before, Beibei and Taobao are used for our experiments. Their statistics is presented in Table 1. Both datasets are publicly available and the training, validation and test sets are given as well. Specifically, the last purchase records of users are used as test set, the second last records are used as validation set, and the remaining records are used for training. The same two datasets are also used in previous works [1].

Table 1. Statistics of our experimental datasets.

Dataset	#User	#Item	#View	#Add-to-cart	#Purchase
Beibei	21,716	7,977	2,412,586	642,622	304,576
Taobao	48,749	39,493	1,548,126	193,747	259,747

4.2 Comparison Methods

To demonstrate the effectiveness of our BCD model, we compare it with several state-of-the-art methods. The baselines are classified into two categories based on whether they utilize single-behavior or heterogeneous data. The compared single-behavior methods include:

- **ENMF** [2]: This is a state-of-the-art nonsampling recommendation method for Top-N recommendation.
- **LightGCN** [10]: This is a state-of-the-art graph neural network model that simplifies the design of GNNs and thus makes them more suitable for single-behavior recommendation.

The second category that leverages heterogeneous data are as follows:

- **NMTR** [8]: This is a state-of-the-art method which combines the recent advances of NCF modeling and the efficacy of multi-task learning.
- **EHCF** [3]: This is a multi-behavior recommendation algorithm which correlates the prediction of each behavior in a transfer way and adopts non-sampling learning for multi-behavior recommendation.
- **CML** [27]: This is a state-of-the-art multi-behavior recommendation algorithm which proposes a multi-behavior contrastive learning framework to distill transferable knowledge across different types of behaviors via the constructed contrastive loss.
- **GHCF** [1]: This is a state-of-the-art graph neural network model that simulates high-order heterogeneous connectivities beneath each behavior in the user-item integration graph.

4.3 Parameter Settings

We empirically search for optimal parameter settings on the validation set. To speed up searching, model-specific parameters are initialized by the suggested values in the original papers. Other parameters are set as follows. Batch size is 256, the dimension of embeddings is 64, and learning rate is set to 0.0001. For sampling-based methods (LightGCN, NMTR), we set the negative sampling ratio to 1:1. For non-sampling methods (EHCF, GHCF, BCD), we set the negative weight to 0.01 on Beibei and 0.1 on Taobao. The number of graph layers is set to 4 on Beibei and 2 on Taobao to avoid over-fitting, the dropout ratio is set to 0.8 on both datasets. For Beibei, the temperature parameters of forward distillation are set to (2, 1.5) and those for backward distillation is set to (0.5, 3). For Taobao, we set the temperature parameters of forward distillation to (1.5, 4) and those for backward distillation to (1, 1.5).

4.4 Evaluation Metrics

We adopt three popular evaluation metrics, namely HR (Hit Ratio) [12], NDCG (Normalized Discounted Cumulative Gain) [26], and MRR (Mean Reciprocal Rank) [23]. Specifically, HR measures to what extent a recommendation list contains items that users actually like. NDCG gives more weights to the relevant items if being ranked top in the recommendation list. MRR scores high if the first relevant item appears early in the recommendation list (Table 2).

Table 2. Performance of all the comparison methods on the Taobao and Beibei datasets. Row '*p*-value' indicates the significance score of BCD relative to the second best approach (i.e., GHCF) on each evaluation metric.

Taobao	HR@K				NDCG@K				MRR@K			
	K=3	K=5	K=10	K=15	K=3	K=5	K=10	K=15	K=3	K=5	K=10	K=15
ENMF	.0116	.0152	.0227	.0283	.0091	.0106	.0129	.0144	.0082	.0090	.0100	.0104
LightGCN	.0195	.0258	.0371	.0443	.0155	.0181	.0217	.0236	.0141	.0155	.0170	.0176
NMTR	.0263	.0317	.0417	.0549	.0189	.0232	.0268	.0284	.0184	.0197	.0219	.0231
EHCF	.0286	.0358	.0482	.0572	.0234	.0264	.0304	.0327	.0216	.0233	.0249	.0256
CML	.0346	.0469	.0693	.0885	.0284	.0294	.0413	.0432	.0236	.0268	.0299	.0328
GHCF	.0380	.0521	.0777	.0977	.0291	.0349	.0432	.0485	.0261	.0293	.0327	.0343
BCD	**.0385**	**.0533**	**.0799**	**.0983**	**.0297**	**.0360**	**.0442**	**.0492**	**.0266**	**.0299**	**.0333**	**.0348**
p-value	0.125	2.4e-4	7.6e-5	0.006	0.018	7.0e-4	2.5e-4	0.002	0.078	0.002	6.8e-4	0.002
Beibei	HR@K				NDCG@K				MRR@K			
	K=3	K=5	K=10	K=15	K=3	K=5	K=10	K=15	K=3	K=5	K=10	K=15
ENMF	.0129	.0196	.0356	.0493	.0099	.0126	.0177	.0213	.0089	.0104	.0124	.0135
LightGCN	.0227	.0295	.0325	.0390	.0177	.0206	.0216	.0232	.0159	.0176	.0180	.0185
NMTR	.0221	.0356	.0679	.1005	.0157	.0212	.0316	.0401	.0135	.0166	.0207	.0233
EHCF	.0688	.0966	.1483	.1862	.0531	.0645	.0811	.0911	.0477	.0540	.0608	.0638
CML	.0783	.1069	.1785	.2163	.0574	.0706	.0911	.1018	.0496	.0595	.0673	.0704
GHCF	.0814	.1214	.1900	.2343	.0600	.0765	.0983	.1103	.0527	.0618	.0708	.0744
BCD	**.0836**	**.1233**	**.1906**	**.2380**	**.0614**	**.0777**	**.0996**	**.1120**	**.0538**	**.0628**	**.0718**	**.0755**
p-value	0.033	0.007	2.1e-5	0.001	3e-4	0.015	0.012	0.013	0.078	0.040	0.048	0.043

4.5 Performance Comparison

Table 1 summarizes the performance of all comparison methods on the two datasets. The results show that multi-behavior recommendation methods are superior to single-behavior ones in terms of all evaluation metrics, implying the usefulness of multiple auxiliary behaviors. Among multi-behavior recommenders, our approach BCD consistently outperforms the other methods. Since other methods may also adopt the non-sampling technique for model learning other than multiple user behaviors, we believe it is our bi-directional contrastive distillation module that leverages both intra- and inter-behaviors relations of user behaviors and thus improves the recommendation performance. Specifically, the average improvements relative to the second best approach (i.e., GHCF) are around 1.61% and 1.89% on Beibei and Taobao, respectively. The larger improvements on Taobao can be explained by the fact that we refine the original conversion path by removing the noisy auxiliary behavior 'cart'.

We conduct statistical significance test (paired t-tests, confidence 0.95) between our approach and GHCF on both datasets, and the results are presented in the last row of Table 1. The results (all *p*-values much smaller than 0.05) demonstrate that our approach is statistically significant in comparison with the second best comparison method.

Table 3. Effects of two important components, where 'BCD-ib' and 'BCD-bd' indicate the variants of BCD without the consideration of intra-behavior relations and backward distillation, respectively.

Methods	Beibei				Taobao			
	HR@5	@10	NDCG@5	@10	HR@5	@10	NDCG@5	@10
BCD-ib	0.1196	0.1880	0.0755	0.0976	0.0530	0.0789	0.0356	0.0439
BCD-bd	0.1191	0.1882	0.0753	0.0976	0.0531	0.0783	0.0356	0.0438
BCD	**0.1233**	**0.1906**	**0.0777**	**0.0996**	**0.0533**	**0.0799**	**0.0360**	**0.0442**

4.6 Ablation Study

Intra-behavior Learning and Backward Distillation. In this section, we will study two important components of our approach, namely intra-behavior contrastive learning and backward distillation. We denote 'BCD-ib' and 'BCD-bd' as the variants of our BCD model without the component of intra-behavior contrastive learning and that of backward distillation, respectively. The results on two datasets are presented in Table 3. It can be observed that BCD outperforms both BCD-ib and BCD-bd variants. We may conclude that (1) it is beneficial to take into account intra-behavior relations for recommendation performance. We can obtain better representations of users/items by applying contrastive learning on the same behaviors. (2) backward distillation is indeed useful to distill the knowledge from target behaviors to enhance the modelling of auxiliary behaviors.

Auxiliary Behaviors for Inter-behavior Learning. The inter-behavior distillation will learn better user preference from the knowledge distillation of auxiliary behaviors on target behaviors. As discussed in Sect. 2, not all auxiliary behaviors are helpful. We further validate this finding by conducting a series of experiments based on different auxiliary behaviors. Specifically, we select different behaviors for the inter-behavior distillation to investigate their usefulness. The results are given in Table 4. On Beibei, we can find that both inter-behaviors ('view→buy' and 'cart→buy') have similar results and are slightly smaller than our BCD method, which considers both kinds of inter-behavior relations. On Taobao, the effect of inter-behavior relation ('cart→buy') is worse than the other ones, indicating the noisy information brought by 'cart' behaviors. By removing 'cart' information from 'view' (i.e., the relation 'view-cart→buy'), we can improve the recommendation performance.

Table 4. Effect of Inter-behavior Learning, where 'v-c→buy' is short for 'view-cart→buy'.

Inter-behavior	Beibei				Taobao			
	HR@5	@10	NDCG@5	@10	HR@5	@10	NDCG@5	@10
view→buy	0.1206	0.1900	0.0758	0.0982	0.0528	0.0792	0.0358	0.0441
cart→buy	0.1206	0.1897	0.0759	0.0985	0.0520	0.0784	0.0354	0.0436
v-c→buy	-	-	-	-	0.0525	0.0791	0.0356	0.0439
BCD	**0.1233**	**0.1906**	**0.0777**	**0.0996**	**0.0533**	**0.0799**	**0.0360**	**0.0442**

4.7 Parameter Analysis

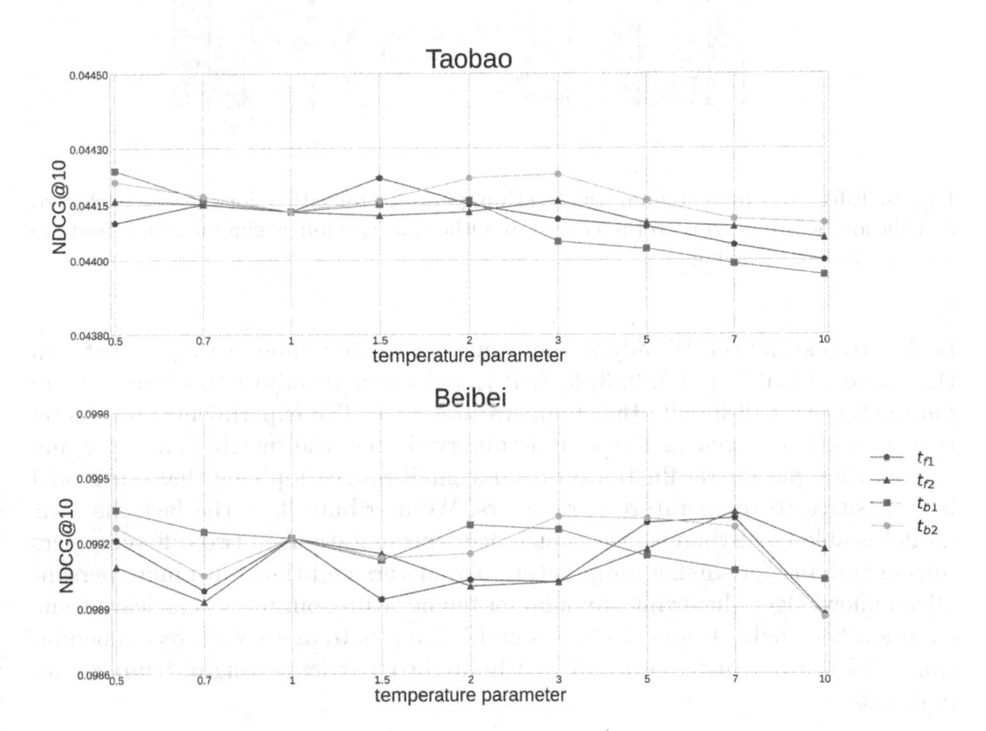

Fig. 5. Effect of temperature parameters on two datasets. (t_{f1}, t_{f2}) are the temperature parameters of forward distillation and (t_{b1}, t_{b2}) are for backward distillation.

Analyses on Temperature Parameter. The temperature parameter has been a key parameter for most existing knowledge distillation-based recommendation algorithms. In this section, we intend to analyze the effect of temperature parameters on our recommendation performance. Specifically, we denote (t_{f1}, t_{f2}) as the temperature parameters of forward distillation and (t_{b1}, t_{b2}) of

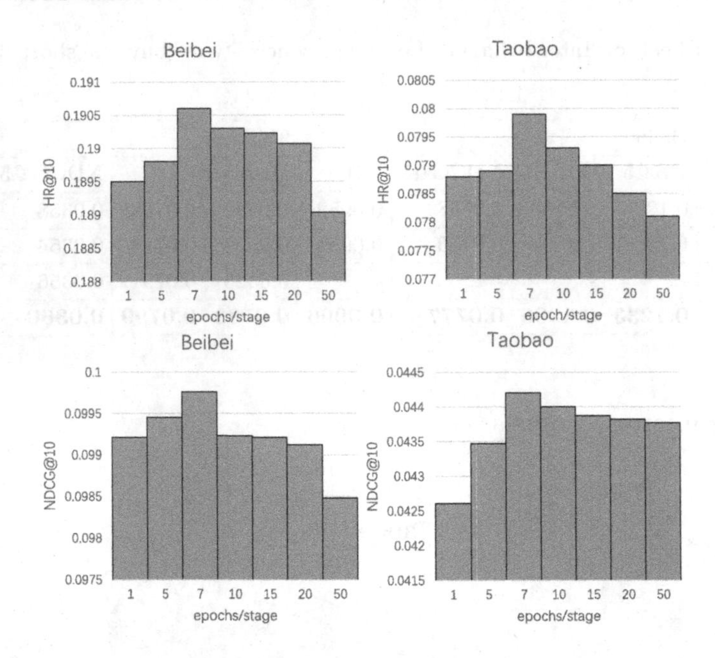

Fig. 6. Effect of different number of training epochs for either forward or backward distillation before moving to the next step. Other metrics follow similar trends and are omitted for space saving.

backward distillation. We adjust the value of a single temperature parameter in the range {0.5, 0.7, 1, 1.5, 2, 3, 5, 7, 10}, and when we adjust this temperature parameter, we will fix all other temperatures to 1. The experimental results on two datasets as given in Fig. 5. It is observed that the metric values for any temperature parameter fluctuate within a small range, implying that our model is insensitive to temperature parameters. We attribute it to the fact that our model is able to distinguish the preference representations of two different users during training for smaller temperature parameters and thus learn more personalized knowledge. Meanwhile, for a larger temperature, our model can learn some common knowledge from other users and behaviors to assist with recommendations. As a result, our approach is not limited to a strict setting of temperature parameters.

Parameter Analysis of Stage Epochs. An important parameter for our model training is the number of training epochs for either forward or backward distillation. The results are illustrated in Fig. 6. We can find that the best performance can be reached when each stage of distillation is trained 7 epochs before moving to the next step. Other parameter settings will decrease the performance to a certain extent.

5 Related Work

5.1 Multi-behavior Recommendation

We will briefly review a number of representative multi-behavior recommenders that are most relevant to our work. These works are mainly built upon conversion paths among user behaviors. For example, ChainRec [24] explores the monotonic behavior chains based on conversion paths to model the effect of auxiliary behaviors on target behaviors. NMTR [8] construct a cascade prediction structure from conversion paths, which is then used to predict subsequent behaviors according to the prediction of previous behaviors. EHCF [3] propose to share parameters based on cascading relationships of conversion paths for evolutionary knowledge learning. MRIG [25] opt to construct a graph structure (rather than a cascading chain) from auxiliary behaviors to enhance the prediction of target behaviors. MBGCN [13] learns discriminative behavior representations using graph convolutional network. MB-GMN [29] uses graph meta network for learning the heterogeneity and diversity among different behaviors. GHCF [1] further adopt the operation of graph convolution to capture higher-order representations of users under each behavior. CML [28] proposes a multi-behavior contrastive learning framework to distill transferable knowledge across different types of behaviors via the constructed contrastive loss.

However, all the above works implicitly assume that all the auxiliary behaviors are positively correlated with target ones, which may not hold in real datasets as discussed in Sect. 2. Our work argues that it is better not to involve noisy auxiliary behaviors for performance improvement.

5.2 Contrastive Distillation in Recommendations

Contrastive learning, which aims to learn high-quality representations via self-supervised pretext tasks, recently achieves remarkable success in the field of computer vision [4,5]. Till now, only few works have been proposed to leverage contrastive distillation for recommendation. DE-RDD [15] designs a contrastive distillation loss function to make better use of the knowledge from a teacher model to guide the learning of a student recommendation model. MICRO [31] adopts contrastive distillation to maximize the agreement between item representations under each modality and the fused multi-modal representation, whereby a more precision item representation can be obtained. Hence, our work focuses on different problem settings (i.e., multi-behavior recommendation) from the above two relevant works (one for model compression and the other for multi-modal recommendation).

6 Conclusions and Future Work

In this paper, we conducted thorough data analysis on two real datasets, and found that not all auxiliary behaviors were positively correlated with target behaviors. We proposed a Bi-directional Contrastive Distillation (BCD) model to distill the common knowledge from multiple user behaviors via forward and backward distillation. We conducted a series of experiments and verified that our approach beat other methods in terms of ranking accuracy. For future work, we plan to explore the manners of data augmentation to enhance recommendation performance under each auxiliary behavior. Besides, we will try to reduce the computational cost due to the additional graph convolution operations on users' interaction graph and further speed up our algorithm.

Acknowledgement. This work is partially supported by the National Natural Science Foundation of China under Grant (No. 62032013, 61972078), and Hebei Natural Science Foundation No. G2021203010.

References

1. Chen, C., et al.: Graph heterogeneous multi-relational recommendation. In: Proceedings of AAAI (2021)
2. Chen, C., et al.: An efficient adaptive transfer neural network for social-aware recommendation. In: Proceedings of SIGIR (2019)
3. Chen, C., Zhang, M., Zhang, Y., Ma, W., Liu, Y., Ma, S.: Efficient heterogeneous collaborative filtering without negative sampling for recommendation. In: Proceedings of AAAI (2020)
4. Chen, T., Kornblith, S., Norouzi, M., Hinton, G.: A simple framework for contrastive learning of visual representations. In: Proceedings of ICML (2020)
5. Chen, X., Fan, H., Girshick, R., He, K.: Improved baselines with momentum contrastive learning. arXiv preprint arXiv:2003.04297 (2020)
6. Devlin, J., Chang, M.W., Lee, K., Toutanova, K.: BERT: pre-training of deep bidirectional transformers for language understanding. arXiv preprint arXiv:1810.04805 (2018)
7. Devooght, R., Kourtellis, N., Mantrach, A.: Dynamic matrix factorization with priors on unknown values. In: Proceedings of KDD (2015)
8. Gao, C., et al.: Neural multi-task recommendation from multi-behavior data. In: 2019 IEEE 35th International Conference on Data Engineering (ICDE) (2019)
9. Hamilton, W., Ying, Z., Leskovec, J.: Inductive representation learning on large graphs. In: Proceedings of NeurIPS (2017)
10. He, X., Deng, K., Wang, X., Li, Y., Zhang, Y., Wang, M.: LightGCN: simplifying and powering graph convolution network for recommendation. In: Proceedings of SIGIR (2020)
11. Jansen, B.J., Schuster, S.: Bidding on the buying funnel for sponsored search and keyword advertising. J. Electron. Commerce Res. (2011)
12. Järvelin, K., Kekäläinen, J.: Cumulated gain-based evaluation of IR techniques. ACM Trans. Inf. Syst. (TOIS) (2002)
13. Jin, B., Gao, C., He, X., Jin, D., Li, Y.: Multi-behavior recommendation with graph convolutional networks. In: Proceedings of SIGIR (2020)
14. Jing, L., Vincent, P., LeCun, Y., Tian, Y.: Understanding dimensional collapse in contrastive self-supervised learning. arXiv preprint arXiv:2110.09348 (2021)
15. Kang, S., Hwang, J., Kweon, W., Yu, H.: De-RRD: a knowledge distillation framework for recommender system. In: Proceedings of CIKM (2020)

16. Liang, D., Charlin, L., McInerney, J., Blei, D.M.: Modeling user exposure in recommendation. In: Proceedings of WWW (2016)
17. Liu, Q., Wu, S., Wang, L.: Multi-behavioral sequential prediction with recurrent log-bilinear model. IEEE Trans. Knowl. Data Eng. (2017)
18. Maas, A.L., Hannun, A.Y., Ng, A.Y.: Rectifier nonlinearities improve neural network acoustic models. In: Proceedings of ICML (2013)
19. Schuster, M., Paliwal, K.K.: Bidirectional recurrent neural networks. IEEE Trans. Sig. Process. (1997)
20. Tang, L., Long, B., Chen, B.C., Agarwal, D.: An empirical study on recommendation with multiple types of feedback. In: Proceedings of KDD (2016)
21. Tian, Y., Krishnan, D., Isola, P.: Contrastive multiview coding. In: Vedaldi, A., Bischof, H., Brox, T., Frahm, J.-M. (eds.) ECCV 2020. LNCS, vol. 12356, pp. 776–794. Springer, Cham (2020). https://doi.org/10.1007/978-3-030-58621-8_45
22. Veličković, P., Cucurull, G., Casanova, A., Romero, A., Liò, P., Bengio, Y.: Graph attention networks. In: Proceedings of ICLR (2018)
23. Voorhees, E.M., et al.: The trec-8 question answering track report. In: TREC (1999)
24. Wan, M., McAuley, J.: Item recommendation on monotonic behavior chains. In: Proceedings of the 12th ACM Conference on Recommender Systems (2018)
25. Wang, W., et al.: Beyond clicks: modeling multi-relational item graph for session-based target behavior prediction. In: Proceedings of WWW (2020)
26. Wang, Y., Wang, L., Li, Y., He, D., Liu, T.Y.: A theoretical analysis of NDCG type ranking measures. In: Conference on Learning Theory (2013)
27. Wei, W., Huang, C., Xia, L., Xu, Y., Zhao, J., Yin, D.: Contrastive meta learning with behavior multiplicity for recommendation. In: Proceedings of WSDM (2022)
28. Wei, Y., et al.: Contrastive learning for cold-start recommendation. In: Proceedings of ACM MM (2021)
29. Xia, L., Xu, Y., Huang, C., Dai, P., Bo, L.: Graph meta network for multi-behavior recommendation. In: Proceedings of SIGIR (2021)
30. Xu, G., Liu, Z., Li, X., Loy, C.C.: Knowledge distillation meets self-supervision. In: Vedaldi, A., Bischof, H., Brox, T., Frahm, J.-M. (eds.) ECCV 2020. LNCS, vol. 12354, pp. 588–604. Springer, Cham (2020). https://doi.org/10.1007/978-3-030-58545-7_34
31. Zhang, J., Zhu, Y., Liu, Q., Zhang, M., Wu, S., Wang, L.: Latent structures mining with contrastive modality fusion for multimedia recommendation. arXiv preprint arXiv:2111.00678 (2021)
32. Zhang, J., Xu, X., Shen, F., Lu, H., Liu, X., Shen, H.T.: Enhancing audio-visual association with self-supervised curriculum learning. In: Proceedings of AAAI (2021)

Improving Micro-video Recommendation by Controlling Position Bias

Yisong Yu[1,2], Beihong Jin[1,2(✉)], Jiageng Song[1,2], Beibei Li[1,2], Yiyuan Zheng[1,2], and Wei Zhuo[3]

[1] State Key Laboratory of Computer Science, Institute of Software, Chinese Academy of Sciences, Beijing, China
Beihong@iscas.ac.cn
[2] University of Chinese Academy of Sciences, Beijing, China
[3] MX Media Co., Ltd., Singapore, Singapore

Abstract. As the micro-video apps become popular, the numbers of micro-videos and users increase rapidly, which highlights the importance of micro-video recommendation. Although the micro-video recommendation can be naturally treated as the sequential recommendation, the previous sequential recommendation models do not fully consider the characteristics of micro-video apps, and in their inductive biases, the role of positions is not in accord with the reality in the micro-video scenario. Therefore, in the paper, we present a model named PDMRec (Position Decoupled Micro-video Recommendation). PDMRec applies separate self-attention modules to model micro-video information and the positional information and then aggregate them together, avoid the noisy correlations between micro-video semantics and positional information being encoded into the sequence embeddings. Moreover, PDMRec proposes contrastive learning strategies which closely match with the characteristics of the micro-video scenario, thus reducing the interference from micro-video positions in sequences. We conduct the extensive experiments on two real-world datasets. The experimental results shows that PDMRec outperforms existing multiple state-of-the-art models and achieves significant performance improvements.

Keywords: Recommender systems · Micro-video recommendation · Contrastive learning

1 Introduction

Micro-video streaming platforms are a kind of newly-emerging applications to provide entertainment services, where users can not only produce micro-videos and upload them to the platforms but also watch micro-videos in their spare time or fragmentary time by the apps on their smart phones. With micro-video apps, such as TikTok, Kwai, etc., increasingly popular, the numbers of micro-videos and users show a rapid growing trend.

M.-R. Amini et al. (Eds.): ECML PKDD 2022, LNAI 13713, pp. 508–523, 2023.
https://doi.org/10.1007/978-3-031-26387-3_31

In general, a micro-video app displays a single video in full-screen mode at a time and automatically plays it in a repetitive way. A user slides his finger on the screen to bring the next micro-video which is what the app recommends. If the micro-videos exposed to a user do not fall within the scope of his/her interests, then the user might give up watching and leave the app. Therefore, the recommendation, whose results will greatly affect the user experiences, becomes the key part in a micro-video streaming platform.

We note that micro-videos are large in quantity and short in time, usually in tens of seconds. Most of micro-videos are lack of the side information (such as genre, director, actors). By a micro-video app, a user can interact with a micro-video in several different ways, such as playing or replaying a micro-video, sharing or liking a micro-video, etc. These interactions can be divided into explicit feedback (e.g. share or like) and implicit feedback (e.g. play or replay), where the former is few in number and the latter is large. In particular, while examining how the interaction sequences are generated, two phenomena need to be taken notice of. Firstly, the playing order of micro-videos is designated by the platform, instead of a result that a user proactively chooses. If the platform recommends another micro-video sequence, then a user has to browse or watch the micro-videos in that sequence in turn, leaving the totally different watching records. Secondly, the behaviors that a user browses micro-videos are driven by his/her interests, usually having no specific purposes, in comparison with the shopping in e-commerce scenarios. As a result, adjacent micro-videos in an interaction sequence have no strong inherent connections.

Existing micro-video recommendation models [1–5] exploit multi-modal information including visual, acoustic, and textual features for recommending micro-videos. However, the multi-modal information of a micro-video is not always available, which renders these models impossible to effectively work. Even if having multi-modal information of each micro-video, we hold the view that these models are more appropriate for the micro-video ranking rather than the matching, where the matching and ranking are two standard successive phases in large-scale recommender systems. The reason behind our view is that the acquisition of multi-modal features requires a large amount of calculation. During the matching, the millions or even billions of micro-videos need to be processed, thus it is not practicable to recall micro-videos using these models in terms of time cost. Alternatively, sequential recommendation models [6–9,11,13,14] can be used for recalling micro-videos. However, so far, the characteristics inherent in interactions between a user and micro-videos have not been taken fully advantage of. Moreover, in most existing models [6,9,11], items in a user-item interaction sequence are believed to have unidirectional associations, that is, an item is dependent on one or more previous items in an interaction. Unfortunately, such an inductive bias is not matched exactly with what happens in the micro-video scenario.

In this paper, we propose a recommendation model named PDMRec (Position Decoupled Micro-video Recommendation) to recall micro-videos. We improve micro-video recommendations from two aspects. Firstly, we enhance the representation of the sequence embeddings by postponing the fusion of micro-video embedding and positional embedding so as to capture the nature of the order between micro-videos. Secondly, we argue that in the micro-video scenario, an original interaction sequence and its reordered sequence are semantically equivalent in the space of user interests, and further two reordered sequences from a same sequence are semantically equivalent. Thus, we can adopt the contrastive learning to reduce the interference from micro-video positions.

To summarize, the main contributions are as follows.

- We take a divide-and-merge policy to generate sequence embeddings, i.e., employ different multi-head attention to model micro-videos and their positional information in the sequences, respectively and then aggregate them into sequence embeddings, which can reflect the actual role of micro-video positions.
- We construct semantically equivalent sequences by reordering operations for a given interaction sequence and present the reordering sequence loss for two newly-generated sequences, which can eliminate the implicit bias that the positional information brings out.
- We conduct extensive experiments on two real-world datasets. The experimental results show PDMRec outperforms the other six models, i.e., GRU4Rec, STAMP, SASRec, BERT4Rec, CL4SRec and DuoRec, in terms of Recall and NDCG.

The rest of the paper is organized as follows. Section 2 introduces the related work. Section 3 describes the PDMRec model in detail. Section 4 gives the experimental evaluation. Finally, the paper is concluded in Sect. 5.

2 Related Work

Our work is related to the research under three non-orthogonal topics: micro-video recommendation, sequential recommendation and contrastive learning.

Micro-video Recommendation. Micro-video recommendation has attracted much attention of many researchers [1–5]. For example, Chen et al. [1] characterizes both short-term and long-term correlations implied in user behaviors, and profiles user interests at both coarse and fine granularities. Li et al. [2] present a temporal graph-based LSTM model to route micro-videos. Wei et al. [3] design a Multi-Modal Graph Convolution Network (MMGCN) which can yield modal-specific representations of users and micro-videos. Liu et al. [4] propose the User-Video Co-Attention Network (UVCAN) which learns multi-modal information from both users and micro-videos using an attention mechanism. Jiang et al. [5] propose a multi-scale time-aware user interest modeling framework, which learns user interests from fine-grained interest groups. These models, as we point out, utilize the multi-modal information of micro-videos but ignore the characteristics of interactions in micro-video apps.

Sequential Recommendation. Sequential recommendation is to model user interaction sequences to predict the next item the user will interact with, where an item can be a micro-video. A natural idea of modeling sequences is to employ RNNs. For example, GRU4Rec [6] and its improved version [7] adopt multiple GRUs to predict the next item the user is most likely to interact with. Subsequently, various Graph Neural Networks (GNNs), such as SR-GNN [8], are applied to the sequential recommendation. However, comparing to the goods in e-commerce scenarios, micro-videos are akin to disposable products and might be forgotten by users quickly, which mismatches with the advantages of GNNs. The alternative way to achieve the sequential recommendation is to design attention mechanisms. For example, STAMP [9] provides an attention network to calculate the coefficient for each item in a sequence, and then generate general user interest and current user interest. Kang et al. apply Transformer [10] in NLP to the sequential recommendation and present SASRec [11]. SASRec adopts a self-attention mechanism to calculate the coefficients of items and presents positional embeddings to indicate the positions of items. Along with the advent of BERT [12], Sun et al. present BERT4Rec [13], which trains the bidirectional model to model sequential data using the cloze task. Fan et al. [14] improve the self-attention module, which scales linearly w.r.t. the user's historical sequence length in terms of time and space, and make the model more resilient to over-parameterization.

Contrastive Learning. More recently, contrastive learning has attracted a great deal of attention. It augments data to discover the implicit supervision signals in the input data and maximize the agreement between differently augmented views of the same data in a certain latent space. After the contrastive learning achieves first success in computer vision, it has been introduced to recommender systems [15–19]. For example, CL4SRec [15] proposes three methods to generate new sequences form raw data, and then utilize them to improve the base model. DuoRec [16] utilizes contrastive learning to resolve the representation degeneration, improving the recommendation accuracy. Moreover, the contrastive learning also has been applied to reduce bias [17] and decrease noise [19], and alleviate the cold start problem [18].

Compared to existing work, our model is tightly bound to the micro-video scenario. It generates embeddings of sequences in which positional information among micro-videos are encoded separately. It utilizes the scenario characteristic i.e., position independence to augment interaction sequences and narrow their semantic gap.

3 Our Model

3.1 Overview

For a micro-video scenario, let $\mathcal{U}$ and $\mathcal{V}$ denote a user set and a micro-video set, respectively. For a user $u \in \mathcal{U}$, his/her positive interactions refer to the

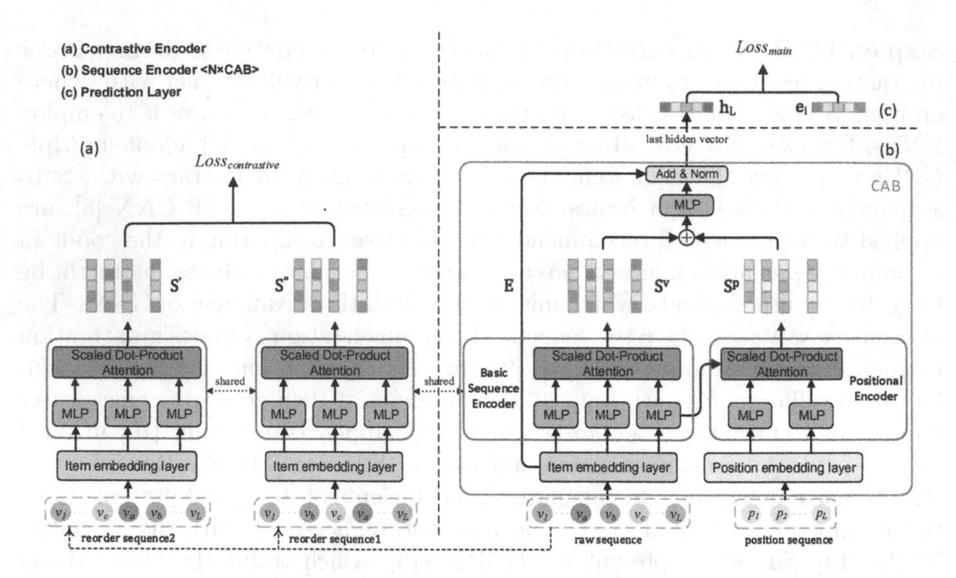

Fig. 1. Architecture of PDMRec.

explicit/implicit feedback that meets some constraint conditions. The examples of positive interactions are given in Sect. 4.1. We choose positive interactions from his/her watching records and order them by the interaction time, thus form a positive interaction sequence, denoted by $s_u = [v_1, v_2, ..., v_{|s_u|}]$.

Our goal is to predict the next micro-video that the user u is most likely to be satisfied with, given the positive interaction sequence s_u of the user u, where the criterion of user satisfaction is that his/her interaction with the micro-video belongs to a positive interaction.

Our model PDMRec exploits positive interaction sequences of users for recalling micro-videos, putting aside multi-modal information of micro-videos. For convenience of processing, we reconstruct each positive interaction sequence to be a sequence with fixed length L. Specifically, if a sequence has the length greater than or equal to L, then we choose latest L interactions, otherwise we pad 0 from the end of sequence to the length L. Hereinafter, we also denote this fixed-length sequence as s_u.

Given the set $\mathcal{S}$ of fixed-length interaction sequences of users in $\mathcal{U}$, PDMRec learns a d-dimensional real-valued embedding $\mathbf{e}_i$ for each of the item i in $\mathcal{V}$ and generates the sequence embedding matrix $\mathbf{S} \in \mathbb{R}^{d \times |\mathcal{S}|}$ by the sequence encoder, and enhances item embeddings by the contrastive encoder. Then in the prediction layer, PDMRec generates $\hat{y} = \{\hat{y}_1, \hat{y}_2, ..., \hat{y}_{|\mathcal{V}|}\}$ for each user, where $\hat{y}_i$ denotes the score for the item v_i in $\mathcal{V}$. Finally, PDMRec is trained as a classifier, taking the micro-videos with top-k scores as recommendations for each user. Figure 1 gives the architecture of the model.

3.2 Sequence Encoder

For generating sequence embeddings, the existing self-attention based recommendation models [11,13] adopt the method similar to that in Transformer [10] to gather the information of items in a sequence and positional information among these items. However, we find that these models take the addition of item embeddings and positional embeddings as the input of self-attention, which leads to same projection matrices to be applied to different relationships, i.e., item-item and position-position relationship, and will bring the mixed and noisy correlations. This defect is inherited from Transformer [20]. We think it should be avoided in the scenarios of applying Transformer.

In the sequence encoder, we introduce two slightly different attention modules, i.e., a basic sequence encoder and a positional encoder, to generate the basic sequence embedding and the positional embedding, respectively.

In a basic sequence encoder, we use a fixed-length sequence $[v_1, v_2, ..., v_L]$ as the input of item embedding layer and obtain an item embedding matrix $\mathbf{E} = [\mathbf{e}_1, \mathbf{e}_2, ..., \mathbf{e}_L], \mathbf{E} \in \mathbb{R}^{d \times L}$, where d is the dimension of item embeddings.

To capture the semantic dependence information among micro-videos in the sequence, we borrow a typical multi-head self-attention from Transformer and apply it with a different way from existing self-attention based recommendation models [11,12].

For the basic sequence embedding, we adopt the following multi-head attention module.

$$\mathbf{S}_h^v = \text{softmax}\left(\frac{\mathbf{E}^T\mathbf{W}_{Qh}\left(\mathbf{E}^T\mathbf{W}_{Kh}\right)^T}{\sqrt{d/hd}}\right)\mathbf{E}^T\mathbf{W}_{Vh} \tag{1}$$

$$\mathbf{S}^v = \text{concat}(\mathbf{S}_1^v, ..., \mathbf{S}_h^v) \tag{2}$$

where $h \in [1, hd]$, hd is the number of attention heads, $\mathbf{W}_{Qh}$, $\mathbf{W}_{Kh}$, $\mathbf{W}_{Vh} \in \mathbb{R}^{d \times dh}$ are learnable parameters, and $dh = d/hd$.

Further, in the positional encoder, we introduce another self-attention module to exclusively handle positional embeddings for extracting the relationship among the positions of micro-videos.

We take the position sequence as input of the position embedding layer and obtain the positional embedding matrix $\mathbf{P} \in \mathbb{R}^{d \times L}$, and then perform the following calculation.

$$\mathbf{S}_h^p = \text{softmax}\left(\frac{\mathbf{P}^T\mathbf{W}_{Qh}^p\left(\mathbf{P}^T\mathbf{W}_{Kh}^p\right)^T}{\sqrt{d/hd}}\right)\mathbf{E}^T\mathbf{W}_{Vh} \tag{3}$$

$$\mathbf{S}^p = \text{concat}(\mathbf{S}_1^p, ..., \mathbf{S}_h^p) \tag{4}$$

where $h \in [1, hd]$, hd is the number of attention heads, $\mathbf{W}_{Qh}^p$, $\mathbf{W}_{Kh}^p \in \mathbb{R}^{d \times dh}$ are learnable parameters, and $dh = d/hd$. Note that the number of head in the positional encoder is set to the same as the one in the sequence encoder. $\mathbf{S}^p$ is the resulting positional embeddings.

Having $\mathbf{S}^v$ and $\mathbf{S}^p$ in hand, we perform the following aggregation strategy.

$$\mathbf{S} = \text{LayerNorm}(\text{dropout}(\text{MLP}(\mathbf{S}^v + \mathbf{S}^p)) + \mathbf{E})) \tag{5}$$

As shown in Eq. 5, we apply MLP, dropout, skip connection and layer normalization to aggregate $\mathbf{S}^v$ and $\mathbf{S}^p$, thus obtaining the sequence embedding matrix $\mathbf{S}$.

The basic sequence encoder, the positional encoder and the aggregator compose a context-aware block (*CAB*).

In the sequence encoder, we stack N context-aware blocks (*CABs*). Let $\mathbf{S}^1$ be the output of the first *CAB* (i.e., $CAB^{(1)}$), the output of n-th *CAB* i.e., $CAB^{(n)}$ will be $\mathbf{S}^n = CAB^{(n)}(\mathbf{S}^{(n-1)}, \mathbf{P})$, where $n \in \{1, 2, \ldots, N\}$ and $\mathbf{S}_0 = \mathbf{E}$. $\mathbf{S}^N$ can be regarded as a set of L hidden vectors, that is, $\mathbf{S}^N = [\mathbf{h}_1^N, \mathbf{h}_2^N, ..., \mathbf{h}_L^N]$, the hidden vector $\mathbf{h}_L^N$ is taken as the representation of the user sequence.

3.3 Contrastive Encoder

In the micro-video scenario, we think the relative locations among a group of micro-videos that a user interacts with are not of great importance. Following the calculation in the sequence encoder, unidirectional associations between items are encoded into item embeddings, which is not amenable to the micro-video scenario. To eliminate impact of unidirectional association between items on item embeddings and reflect the essence of the order between micro-videos, we build a contrastive encoder using contrastive learning, which consists of two basic sequence encoders, as illustrated on the left side of Fig. 1.

We generate new sequences which are semantically equivalent to the real interaction sequences by a reordering operation. More concretely, for a given positive interaction sequence s_u of the user u, i.e., $s_u = [v_1, v_2, ..., v_{|s_u|}]$, we randomly shuffle a continuous sub-sequence $[v_r, v_{r+1}, ..., v_{r+L_r-1}]$, which starts at r with length $L_r = \lfloor \alpha * |s_u| \rfloor$, to $[\hat{v}_r, \hat{v}_{r+1}, ..., \hat{v}_{r+L_r-1}]$, where α is the proportion of reordering. As a result, we get the reordered sequence $s_r = [v_1, v_2, \ldots, \hat{v_r}, ..., \hat{v}_{r+L_r-1}, ..., v_{|s_u|}]$.

After applying the reordering operation twice, we generate two new sequences s_{r1} and s_{r2} for s_u. The reordering operation does not increase or decrease the number of items in the sequence, i.e., $|s_{r1}| = |s_{r2}| = |s_u|$. Next, we feed these two reordered sequences to the basic sequence encoders, respectively and obtain corresponding sequence representations $\mathbf{S}^{'} = [\mathbf{h}_1^{N'}, \mathbf{h}_2^{N'}, ..., \mathbf{h}_{|s_u|}^{N'}]$ and $\mathbf{S}^{''} = [\mathbf{h}_1^{N''}, \mathbf{h}_2^{N''}, ..., \mathbf{h}_{|s_u|}^{N''}]$. Here, we only send micro-video sequence information into the basic sequence encoder, deliberately ignoring the positional information of the sequence. This is for contrasting two augmented sequences without any disturbance from the positional information. This enables us to disentangle the contrastive loss and positional information modeling, as a result, the parameter updating inducing by the contrastive loss and the updating of positional embeddings do not affect each other.

Then, by the concatenation operation, we obtain $\hat{\mathbf{h}}^{'} = \text{concat}(\mathbf{h}_1^{N'}, \mathbf{h}_2^{N'}, ..., \mathbf{h}_{|s_u|}^{N'})$ and $\hat{\mathbf{h}}^{''} = \text{concat}(\mathbf{h}_1^{N''}, \mathbf{h}_2^{N''}, ..., \mathbf{h}_{|s_u|}^{N''})$. Finally, in order to minimize the gap between the representations of two sequences derived from the same original interaction sequence, we define the contrastive loss function in Eq. 6 as the reordering sequence loss, where we adopt the dot-product operation to calculate the similarity of two embeddings, i.e., for two embeddings $\mathbf{a}$, and $\mathbf{b}$, $\text{sim}(\mathbf{a}, \mathbf{b}) = \mathbf{a}^T\mathbf{b}$.

$$\mathcal{L}_{\text{cl}}\left(\hat{\mathbf{h}}', \hat{\mathbf{h}}''\right) = -\log \frac{\exp\left(\text{sim}\left(\hat{\mathbf{h}}', \hat{\mathbf{h}}''\right)\right)}{\exp\left(\text{sim}\left(\hat{\mathbf{h}}', \hat{\mathbf{h}}''\right)\right) + \sum_{s_* \in S^-} \exp\left(\text{sim}\left(\hat{\mathbf{h}}', \hat{\mathbf{h}}_*\right)\right)} \tag{6}$$

While defining the reordering sequence loss, we think that it is more reasonable to take $\hat{\mathbf{h}}^{'}$, $\hat{\mathbf{h}}^{''}$, the results of concatenating all hidden vectors of the corresponding sequence representation, as parameters, because what we want the reordering sequence loss to do is to measure the semantic difference between two sequences. We have tried to replace $\hat{\mathbf{h}}^{'}$, $\hat{\mathbf{h}}^{''}$ with $\mathbf{h}_L^{N'}$, $\mathbf{h}_L^{N''}$ i.e., the last hidden vectors, which is proved ineffective by experimental results.

Moreover, we adopt the following negative sampling method. Assume that there are M sequences in a batch. Since we apply the reordering operation to the original sequence (e.g., s_u) to generate two new sequences (e.g., s_{r1}, s_{r2}) , we have a list of sequences with the length of $2M$. We treat (s_{r1}, s_{r2}) as a positive pair and the other sequences in the same batch as negatives, the latter forms the set of negatives, denoted by S^-. $\hat{\mathbf{h}}_*$ is the concatenation of all the hidden vectors in the representation of s_*.

3.4 Prediction and Loss Function

We calculate the score for each candidate micro-video v_i by conducting the dot product of sequence embedding $\mathbf{h}_L^n$ and embedding $\mathbf{e}_i$ of v_i, as shown in Eq. 7.

$$\text{Score}(v_i|\{v_1, \cdots, v_L\}) = \text{sim}(\mathbf{h}_L^N, \mathbf{e}_i) \tag{7}$$

We adopt the negative log likelihood with full softmax as the main loss function. It can be written as follows.

$$\mathcal{L}_{\text{main}} = -\log \frac{\exp\left(\text{sim}\left(\mathbf{h}_L^N, \mathbf{e}_g\right)\right)}{\sum_{i=1}^{|\mathcal{V}|} \exp\left(\text{sim}\left(\mathbf{h}_L^N, \mathbf{e}_i\right)\right)} \tag{8}$$

where $\mathbf{e}_g$ is the embedding of the ground-truth micro-video. Besides, we have the reordering sequence loss as an auxiliary loss. Finally, the total loss function is shown in Eq. 9, where coefficient λ is a hyperparameter.

$$\mathcal{L}_{total} = \mathcal{L}_{main} + \lambda \mathcal{L}_{cl} \tag{9}$$

3.5 Complexity Analysis

Space Complexity. In PDMRec, the learnable parameters are mainly from the embeddings of micro-videos, i.e., $\{\mathbf{e}_i, i \in [1, |\mathcal{V}|]\}$, the positional embedding matrix $\mathbf{P} \in \mathbb{R}^{d \times L}$, the parameters in multi-head self-attention, i.e., $\mathbf{W}_{Qh}$,

$\mathbf{W}_{Kh}, \mathbf{W}_{Vh}, \mathbf{W}^p_{Qh}, \mathbf{W}^p_{Kh} \in \mathbb{R}^{d\times dh}$, and the parameters in multiple MLPs, whose number are $|\mathcal{V}| \times d$, Ld, d^2/hd and d^2, respectively. Therefore, the total space complexity of PDMRec is $O\left(|\mathcal{V}|d + Ld + d^2\right)$.

For example, if training the model using WeChat dataset, the number of the learnable parameters is 2264144, where the number of item embedding is 2129472, which illustrates that item embeddings account for the bulk of all the parameters.

Time Complexity. The computation amount of PDMRec is mainly concentrated on the self-attention module and the MLPs, whose time complexity is $O\left(L^2 d\right)$ and $O\left(Ld^2\right)$, respectively. Thus, the total time complexity is $O\left(L^2 d + Ld^2\right)$.

A favorite property of PDMRec is that the computation in each self-attention module is fully parallelizable, which is suitable to GPU acceleration. Finally, it should be noted that our model does not increase the computational cost, compared to the state-of-the-art model such as DuoRec.

3.6 Discussion

Avoiding Noisy Correlations. Our model can avoid noisy correlations being involved into sequence embeddings. For Eq. 5, we know sequence embeddings comes from the aggregation of $\mathbf{S}^v + \mathbf{S}^p$. Now, we expand $\mathbf{S}^v + \mathbf{S}^p$ in Eq. 10.

$$\begin{aligned} \mathbf{S}^v + \mathbf{S}^p &= \text{concat}(\mathbf{S}^v_1, ..., \mathbf{S}^v_h) + \text{concat}(\mathbf{S}^p_1, ..., \mathbf{S}^p_h) \\ &= \text{concat}(\mathbf{S}^v_1 + \mathbf{S}^p_1, ..., \mathbf{S}^v_h + \mathbf{S}^p_h) \end{aligned} \tag{10}$$

We continue to expand Eq. 10 and obtain Eq. 11 as follows. For $j \in [1, h]$,

$$\begin{aligned} \mathbf{S}^v_j + \mathbf{S}^p_j = &\text{softmax}\left(\frac{\mathbf{E}^T\mathbf{W}_{Qj} \cdot (\mathbf{E}^T\mathbf{W}_{Kj})^T}{\sqrt{d/hd}}\right)\mathbf{E}^T\mathbf{W}_{Vj} + \\ &\text{softmax}\left(\frac{\mathbf{P}^T\mathbf{W}^p_{Qj} \cdot (\mathbf{P}^T\mathbf{W}^p_{Kj})^T}{\sqrt{d/hd}}\right)\mathbf{E}^T\mathbf{W}_{Vj} \end{aligned} \tag{11}$$

As shown in Eq. 11 , we apply linear projection matrices $\mathbf{W}_{Qj}$,$\mathbf{W}_{Kj}$ on item embedding matrix $\mathbf{E}$, apply linear projection matrices $\mathbf{W}^p_{Qj}$ and $\mathbf{W}^p_{Kj}$ on positional embedding matrix $\mathbf{P}$, and calculate the attention scores of $\mathbf{E}$ and $\mathbf{P}$ in different latent spaces. These projection matrices are different from each other and trained independently. Such the method adds no noisy correlations into sequence embeddings, reflecting the actual role of micro-video positions.

Augmenting Equivalent Sequences. For data augmentation method, besides the reordering operation, masking and cropping operations can also be applied on the sequences. Given a sequence s_u, the masking operation refers to randomly masking a proportion γ of items and the cropping operation is to randomly delete a continuous sub-sequence of length $\eta * |s_u|$. However, in the micro-video scenario, applying masking or cropping operations cannot guarantee generating semantically-equivalent sequences. Therefore, we do not choose these two operations. Experimental results in Sect. 4.4 also illustrate that augmenting sequences by masking or cropping operations cannot achieve good recommendations.

4 Experimental Evaluation

4.1 Experimental Setup

Datasets. We adopt the following two datasets for experiments.

- **WeChat-Channels**: a dataset released by WeChat Big Data Challenge 2021[1]. It contains 14-day interactions from anonymous users on WeChat Channels, a popular micro-video streaming platform in China.
- **TikTok**: a dataset released by Short Video Understanding Challenge 2019[2] hosted by ByteDance, one of the largest companies in the world engaged in the micro-video streaming platform. It contains more than 275 million user interactions with TikTok app.

Both datasets include engagement interactions such as watching and satisfaction interactions such as like and favorite. However, not every interaction between a user and a micro-video in the dataset reflect that the user likes the micro-video, so we set some constraints on interactions and choose those satisfying these constraints as positive interactions. More concretely, according to the characteristics of different micro-video streaming platforms (e.g. the average micro-video duration in WeChat and TikTok is 34.4 s and 10.5 s, respectively), we use different criteria to obtain positive interactions on TikTok and WeChat. From WeChat-Channels, we choose all the satisfaction interactions (such as likes and comments) and interactions whose loop times is greater than 1.1 or watching time is greater than 45s to form WeChat dataset. From TikTok, we randomly sample 10% of users and choose their interactions whose loop times is greater than 1.0, forming TikTok1. Also from TikTok, we choose all the interactions that users like micro-videos to form TikTok2 dataset. Next, we remove users and micro-videos with fewer than five interactions, respectively so as to guarantee that each user/micro-video has enough interactions. Then, for each user, we sort his/her historical micro-videos by the interacted timestamp to obtain his/her interaction sequence. These sequences compose the datasets used in experiments. The statistics of processed datasets are shown in Table 1.

We adopt the leave-one-out strategy to divide the dataset into train/ validation/ test sets. That is, for each user u, we split his/her historical sequence $s = [v_1^u, v_2^u, ..., v_{|s|}^u]$ into three parts: (1) the most recent interaction $v_{|s|}^u$ for testing, (2) the second most recent interaction $v_{|s|-1}^u$ for validation, and (3) all remaining interactions for training.

[1] https://algo.weixin.qq.com/problem-description.

[2] https://www.biendata.xyz/competition/icmechallenge2019/.

Table 1. Statistics of datasets.

Dataset	#Users	#Micro-videos	#Interactions	Density	Avg. length of sequence
WeChat	19998	77985	2464798	0.16%	123.3
TikTok1	19960	55954	2434289	0.22%	121.9
TikTok2	30473	35716	684216	0.06%	25.5

Metrics. We employ two common top-k metrics, Recall@K and NDCG@K, to evaluate recommendation performance. Recall@K calculates the proportion of test items in top-k items of prediction scoring, while NDCG@K is a position-aware metric which assigns larger weights on higher positions. Since we only have one test item for each user, Recall@K is equivalent to Hit@K. In this paper, the value K is set to 20, 50, 100.

In order to reduce the time cost of metric calculation, lots of previous work [6, 11, 12] adopts sampled negative items to calculate metrics. However, this method may lead to inconsistent with exact metrics [21]. Therefore, we compute metrics on the whole item set to evaluate the model performance. That is, for each user, we rank all the micro-videos he/she has not interacted with by their scores rather than only rank the sampled negative items.

Implementation Details. We implement our model with PyTorch, initializing all parameters by the normal distribution with mean 0 and standard deviation 0.02. Embedding size is set to 64. The number of heads (i.e., hd) is set to 2., the number of CAB (i.e., N) is set to 2. The proportion rate α of reordering is set to 0.2. Coefficient λ in the loss function is set to 0.1. Dropout rate is set to 0.5. We use Adam as the optimizer with the learning rate of 0.001. Batch size (i.e., M) is set to 512. The sequence length L is set to 100 for WeChat and TikTok1 datasets, and 50 for TikTok2 dataset. Our code is available publicly on GitHub[3] for reproducibility.

We train the model by an early stopping technique, that is, when Recall@50 on the validation set has not been improved in 15 consecutive epochs, we stop training the model.

4.2 Performance Comparison

We conduct comparative experiments, comparing our model with the following six models.

- GRU4Rec$^+$ [7]. It uses GRU modules to model user preferences by interaction sequences and is improved with prefix sub-sequences as data augmentation and a method to account for shifts in the input data distribution.
- STAMP [9]. It is a short-term attention/memory priority model which aims to capture long-term user preferences from previous interactions and short-term user preferences from the last interaction in a sequence.

[3] https://github.com/Ethan-Yys/PDMRec.

- SASRec [11]. It firstly introduces the Transformer encoder to learn user representations. It models use preferences through a self-attention mechanism and achieves state-of-the-art performance at that time.
- BERT4Rec [13]. It learns via the BERT structure, that is, employs the deep bidirectional self-attention to model user behavior sequences.
- CL4SRec [15]. It firstly uses contrastive learning to enhance user representations in sequential recommendation. Specifically, item cropping, masking and reordering are applied to an original sequence to generate sequences for calculating the contrastive loss.
- DuoRec [16]. It is also a method based on contrastive learning. It uses both supervised and unsupervised methods to generate sequences for calculating contrastive loss.

Among these competitors, GRU4Rec$^+$, STAMP, SASRec and BERT4Rec are from a popular open-source recommendation framework RecBole[4]. CL4SRec and DuoRec are from the implementation of Zhao et al.[5] To be fair, in all the models, we set the dimension of item embedding to 64 and batch size to 512. Particularly, we adopt the 2-layer encoder for the models which apply self-attention.

Experimental results on WeChat, TikTok1 and TikTok2 are listed in Table 2, where the number in a bold type is the best performance in each row and the underlined number is the second best in each row.

From the results, we have the following observations.

- The attention-based sequential recommendation models are inferior to ones which integrate additional contrastive learning module in terms of both Recall and NDCG. It makes us believe that augmenting data more targeted for the specific scenario and optimizing in different latent spaces do improve performance.
- The classic baseline i.e., GRU4Rec$^+$ performs not bad, which is beyond our expectation. As shown in Table 2, the worse one in performance on three datasets is STAMP instead of GRU4Rec$^+$. In particular, on datasets WeChat and TikTok1 which contain not only explicit feedback but also implicit feedback, GRU4Rec$^+$ is even superior to several attention-based models such as SASRec. It is supposed that GRU4Rec$^+$ is more appropriate to modeling long sequences.
- Most importantly, PDMRec surpasses all the competitors in all metrics on the three datasets. For example, on the TikTok1 dataset, PDMRec outperforms the second best model, i.e., DuoRec, about 6.23% on Recall@50 and 6.02% on NDCG@50. We think the good performance of PDMRec stems from two steps of position decoupling: learning positional embedding independently and optimizing representations of reordered sequences for position-independence semantic conformity.

[4] https://github.com/RUCAIBox/RecBole.

[5] https://github.com/RuihongQiu/DuoRec.

Table 2. Recommendation performance on three datasets.

Metrics		GRU4Rec$^+$	STAMP	SASRec	BERT4Rec	CL4SRec	DuoRec	PDMRec	Improv. (%)
WeChat									
Recall	@20	0.1093	0.0888	0.1069	0.0892	0.1035	0.1108	**0.1157**	4.42%
	@50	0.2125	0.1727	0.2095	0.1762	0.2057	0.2169	**0.2224**	2.54%
	@100	0.3270	0.2647	0.3224	0.2744	0.3161	0.3263	**0.3423**	4.90%
NDCG	@20	0.0407	0.0339	0.0388	0.0322	0.0386	0.0410	**0.0422**	2.93%
	@50	0.0610	0.0504	0.0591	0.0494	0.0587	0.0619	**0.0632**	2.10%
	@100	0.0795	0.0652	0.0774	0.0654	0.0766	0.0796	**0.0826**	3.77%
TikTok1									
Recall	@20	0.1003	0.0815	0.1120	0.0842	0.1066	0.1139	**0.1190**	4.48%
	@50	0.1915	0.1503	0.2028	0.1609	0.2005	0.2071	**0.2200**	6.23%
	@100	0.2991	0.2308	0.3087	0.2525	0.3062	0.3114	**0.3281**	5.36%
NDCG	@20	0.0395	0.0342	0.0462	0.0319	0.0424	0.0465	**0.0481**	3.44%
	@50	0.0574	0.0477	0.0640	0.0470	0.0609	0.0648	**0.0687**	6.02%
	@100	0.0748	0.0607	0.0812	0.0618	0.0780	0.0817	**0.0861**	5.39%
TikTok2									
Recall	@20	0.0514	0.0422	0.0777	0.0623	0.0727	0.0808	**0.0830**	2.70%
	@50	0.0979	0.0795	0.1240	0.1045	0.1177	0.1266	**0.1332**	5.20%
	@100	0.1547	0.1213	0.1752	0.1548	0.1698	0.1781	**0.1894**	6.34%
NDCG	@20	0.0205	0.0175	0.0475	0.0314	0.0432	0.0479	**0.0491**	3.55%
	@50	0.0297	0.0248	0.0566	0.0398	0.0520	0.0569	**0.0590**	3.69%
	@100	0.0388	0.0316	0.0645	0.0479	0.0604	0.0653	**0.0681**	4.29%

4.3 Ablation Study

We conduct the ablation study on our model to observe the effectiveness of different components. We compare our model with three variants, i.e., PDMRec1, PDMRec2 and PDMRec3. PDMRec1 and PDMRec2 are PDMRec models which remove the contrastive encoder and the positional encoder, respectively. PDMRec3 is the PDMRec model which removes the positional encoder but adopts the addition of item embedding and positional embedding as the input of the model. The results are shown in Fig. 2.

From Fig. 2, we find the variant PDMRec1 suffers severe declines in performance, which illustrates that optimizing the reordering sequence loss has a favorable effect on improving the performance. Moreover, PDMRec2 shows the worst performance while comparing with PDMRec1 and other variants. It proves that lacking position information will lead to sharply declining in performance. Further, we compare the performance of PDMRec and PDMRec3, where they provide the different ways of fusing position information into sequence embeddings. Obviously, PDMRec has better performance, which shows that encoding positional embeddings separately is necessary.

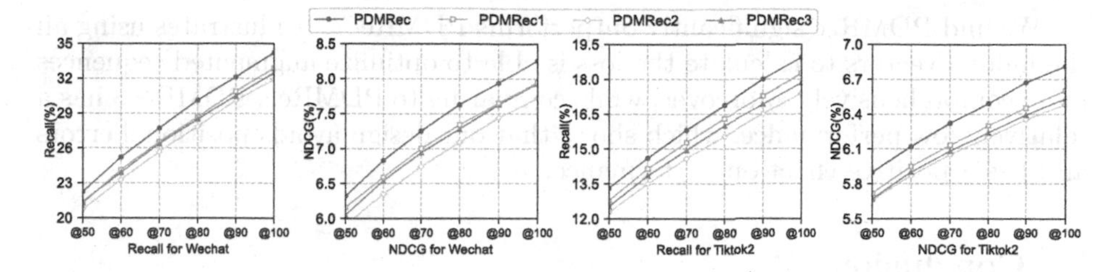

Fig. 2. Ablation study on two datasets

4.4 Impact of Contrastive Learning Strategies

We conduct experiments to evaluate the rationality of contrastive learning strategies adopted in this paper. The contrastive learning strategies involve data augmentation and contrastive learning loss calculation.

By replacing the reordering operation in PDMRec with other operations in CL4SRec, we can obtain different variants: PDMRec4 using the masking operation, PDMRec5 using the cropping operation, and PDMRec6 using the masking, cropping and reordering operations. In experiments, we set γ to 0.5 and η to 0.5, the values which lead to the best performance of experiments in [15]. The performance of these variants is listed in the columns 3–5 of Table 3. From the results, we find that three variants fall far behind PDMRec in terms of Recall and NDCG. These results illustrate that using the reordering operation to generate new sequences is effective.

For the contrastive learning loss calculation, in our model, we use $\hat{\mathbf{h}}^{'}$ and $\hat{\mathbf{h}}^{''}$, which are the output of the Scaled Dot-Product Attention in the basic sequence encoder, to calculate the loss. An alternative is to take last vector $\hat{\mathbf{h}}_L^{N'}$, $\hat{\mathbf{h}}_L^{N''}$ in $\mathbf{S}^{'}$ and $\mathbf{S}^{''}$ to calculate the loss, which is the way CL4SRec adopts. Another alternative is to take the output of Add&Norm layer to calculate the loss, which is the way DuoRec adopts. The corresponding variants are denoted by PDMRec7 and PDMRec8. The performance of these variants is listed in the columns 6–7 of Table 3.

Table 3. Recommendation performance of different contrastive learning strategies.

Metrics		PDMRec	PDMRec4	PDMRec5	PDMRec6	PDMRec7	PDMRec8
Recall	**@20**	**0.1157**	0.1118	0.1068	0.1103	0.1081	0.1099
	@50	**0.2224**	0.2157	0.2144	0.2138	0.2134	0.2183
	@100	**0.3423**	0.3277	0.3311	0.3309	0.3297	0.3282
NDCG	**@20**	**0.0422**	0.0416	0.0393	0.0400	0.0393	0.0405
	@50	**0.0632**	0.0620	0.0606	0.0604	0.0600	0.0619
	@100	**0.0826**	0.8010	0.0794	0.0794	0.0788	0.0796

We find PDMRec significantly outperforms PDMRec7. It illustrates using all the hidden vectors to calculate the loss is able to optimize augmented sequences more comprehensively. Moreover, while comparing to PDMRec, PDMRec8 has a relatively low performance, which shows that our design avoids positional errors and has a positive effect on performance.

5 Conclusion

In this paper, we examine the characteristics of the micro-video scenario, rethink the role of positions in interaction sequences, and then propose a micro-video recommendation model PDMRec. For improving the representations of sequences and avoiding the noise being added into, PDMRec encodes micro-videos and positions in sequences, respectively, computing the micro-video contextual correlations and positional correlations with different parameterizations. Further, PDMRec adopts the reordering operation to augment interaction sequences and presents a reordering sequence loss to remedy the negative impact brought by micro-video positions in sequences. Results of experiments on real-world datasets show that our PDMRec model is effective in terms of Recall and NDCG, and outperforms the state-of-the-art baselines.

Acknowledgements. This work was supported by the National Natural Science Foundation of China under Grant No. 62072450 and the 2021 joint project with MX Media.

References

1. Chen, X., Liu, D., Zha, Z.J., Zhou, W., Xiong, Z., Li, Y.: Temporal hierarchical attention at category- and item-level for micro-video click-through prediction. In: Proceedings of the 26th ACM International Conference on Multimedia. pp. 1146–1153. MM '18 (2018)
2. Li, Y., Liu, M., Yin, J., Cui, C., Xu, X.S., Nie, L.: Routing micro-videos via a temporal graph-guided recommendation system. In: Proceedings of the 27th ACM International Conference on Multimedia. pp. 1464–1472. MM '19 (2019)
3. Wei, Y., Wang, X., et al.: MMGCN: Multi-modal graph convolution network for personalized recommendation of micro-video. In: Proceedings of the 27th ACM International Conference on Multimedia. pp. 1437–1445. MM '19 (2019)
4. Liu, S., Chen, Z., Liu, H., Hu, X.: User-video co-attention network for personalized micro-video recommendation. In: The World Wide Web Conference. pp. 3020–3026. WWW '19 (2019)
5. Jiang, H., Wang, W., Wei, Y., Gao, Z., Wang, Y., Nie, L.: What aspect do you like: Multi-scale time-aware user interest modeling for micro-video recommendation. In: Proceedings of the 28th ACM International Conference on Multimedia. pp. 3487–3495 (2020)
6. Hidasi, B., Karatzoglou, A., Baltrunas, L., Tikk, D.: Session-based recommendations with recurrent neural networks. In: International Conference on Learning Representations, ICLR 2016 (2016)

7. Tan, Y.K., Xu, X., Liu, Y.: Improved recurrent neural networks for session-based recommendations. In: Proceedings of the 1st Workshop on Deep Learning for Recommender Systems. pp. 17–22. DLRS 2016 (2016)
8. Wu, S., Tang, Y., Zhu, Y., Wang, L., Xie, X., Tan, T.: Session-based recommendation with graph neural networks. Proceedings of the AAAI Conference on Artificial Intelligence **33**(01), 346–353 (2019)
9. Liu, Q., Zeng, Y., Mokhosi, R., Zhang, H.: STAMP: short-term attention/memory priority model for session-based recommendation. In: Proceedings of the 24th ACM SIGKDD International Conference on Knowledge Discovery & Data Mining. pp. 1831–1839 (2018)
10. Vaswani, A., Shazeer, N., Parmar, N., Uszkoreit, J., Jones, L., Gomez, A.N., Kaiser, L., Polosukhin, I.: Attention is all you need. In: Proceedings of the 31st International Conference on Neural Information Processing Systems. pp. 6000–6010. NIPS'17 (2017)
11. Kang, W.C., McAuley, J.: Self-attentive sequential recommendation. In: 2018 IEEE International Conference on Data Mining (ICDM). pp. 197–206 (2018)
12. Devlin, J., Chang, M.W., Lee, K., Toutanova, K.: BERT: Pre-training of deep bidirectional transformers for language understanding. In: Proceedings of the 2019 Conference of the North American Chapter of the Association for Computational Linguistics: Human Language Technologies, Volume 1. pp. 4171–4186 (2019)
13. Sun, F., Liu, J., Wu, J., Pei, C., Lin, X., Ou, W., Jiang, P.: Bert4rec: Sequential recommendation with bidirectional encoder representations from transformer. In: Proceedings of the 28th ACM International Conference on Information and Knowledge Management. pp. 1441–1450. CIKM '19 (2019)
14. Fan, X., Liu, Z., Lian, J., Zhao, W.X., Xie, X., Wen, J.: Lighter and better: Low-rank decomposed self-attention networks for next-item recommendation. In: SIGIR '21: The 44th International ACM SIGIR Conference on Research and Development in Information Retrieval, Virtual Event, Canada, July 11–15, 2021. pp. 1733–1737. ACM (2021)
15. Xie, X., Sun, F., Liu, Z., Gao, J., Ding, B., Cui, B.: Contrastive pre-training for sequential recommendation. CoRR abs/2010.14395 (2020)
16. Qiu, R., Huang, Z., Yin, H., Wang, Z.: Contrastive learning for representation degeneration problem in sequential recommendation. In: Proceedings of the Fifteenth ACM International Conference on Web Search and Data Mining. pp. 813–823. WSDM '22 (2022)
17. Zhou, C., Ma, J., Zhang, J., Zhou, J., Yang, H.: Contrastive learning for debiased candidate generation in large-scale recommender systems. In: KDD '21: The 27th ACM SIGKDD Conference on Knowledge Discovery and Data Mining, Virtual Event. pp. 3985–3995 (2021)
18. Wei, Y., Wang, X., Li, Q., Nie, L., Li, Y., Li, X., Chua, T.S.: Contrastive learning for cold-start recommendation. In: Proceedings of the 29th ACM International Conference on Multimedia. pp. 5382–5390 (2021)
19. Qin, Y., Wang, P., Li, C.: The world is binary: Contrastive learning for denoising next basket recommendation. In: SIGIR '21: The 44th International ACM SIGIR Conference on Research and Development in Information Retrieval, Virtual Event, Canada, July 11–15, 2021. pp. 859–868. ACM (2021)
20. Ke, G., He, D., Liu, T.Y.: Rethinking positional encoding in language pretraining. In: International Conference on Learning Representations, ICLR 2021 (2021)
21. Krichene, W., Rendle, S.: On sampled metrics for item recommendation. In: Proceedings of the 24th ACM SIGKDD International Conference on Knowledge Discovery & Data Mining. pp. 1748–1757. KDD'20 (2020)

Mitigating Confounding Bias for Recommendation via Counterfactual Inference

Ming He(✉), Xinlei Hu, Changshu Li, Xin Chen, and Jiwen Wang

Beijing University of Technology, Beijing, China
heming@bjut.edu.cn,
{huxl,lichangshu,chenxin,wangjiwen}@emails.bjut.edu.cn

Abstract. Recommender systems usually face the bias problem, which creates a gap between recommendation results and the actual user preference. Existing works track this problem by assuming a specific bias and then develop a method to mitigate it, which lack universality. In this paper, we attribute the root reason of the bias problem to a causality concept: confounders, which are the variables that influence both which items the user will interact with and how they rate them. Meanwhile, the theory around causality says that some confounders may remain unobserved and are hard to calculate. Accordingly, we propose a novel ***C****ounterfactual* ***I****nference for* ***D****econfounded* ***R****ecommendation* (CIDR) framework that enables the analysis of causes of biases from a causal perspective. We firstly analyze the causal-effect of confounders, and then utilize the biased observational data to capture a substitute of confounders on both user side and item side. Finally, we boost counterfactual inference to eliminate the causal-effect of such confounders in order to achieve a satisfactory recommendation with the help of user and item side information (e.g., user post-click feedback data, item multi-model data). For evaluation, we compare our method with several state-of-the-art debias methods on three real-world datasets, in addition to new causal-based approaches. Extensive experiments demonstrate the effectiveness of our proposed method.

Keywords: Recommender systems · Causal inference · Unobserved confounders

1 Introduction

Recommender systems (RS) aim to provide personalized suggestions to users in a wide spectrum of online applications, such as E-commerce (Amazon, Alibaba), social networks (Twitter, Facebook), and search engines (Yahoo, Bing), by mining user preferences from the user-item interactions (e.g., clicks, views). However, interaction data are observational rather than experimental, which makes various biases widely exist in the data [5] such as popularity bias [28], position bias [11], exposure bias [24], etc. Worse still, the recommendation models trained on such biased data may not only inherit the bias but also amplify it.

M.-R. Amini et al. (Eds.): ECML PKDD 2022, LNAI 13713, pp. 524–540, 2023.
https://doi.org/10.1007/978-3-031-26387-3_32

Recent years have witnessed the success of incorporating causality into recommender systems. [2] used uniform data to guide the model for learning unbiased causal embedding to eliminate the bias. [1] proposed a method to combine dual learning and Inverse Propensity Weighting (IPW) and constructed an unbiased propensity model to remedy the position bias. [24] considered the exposure bias from the perspective of item exposure features, and leveraged causal inference to eliminate the exposure bias. However, these methods do not consider of the root reason of the bias. Inspired by [18], we argue that the existence of most data biases is attributed to a causal inference concept-*confounders*. Confounders are variables that impact both the treatment assignments (which items the users view) and the outcomes (how they rate them). The reason for considering confounding bias as the root reason for the data bias problem is that biases in RS, such as popularity bias, position bias, and exposure bias, all have the same mechanism of effect, i.e., they all affect what items the user were recommended (treatments) and the user's decision (outcomes). Therefore, we treat confounding bias as a collection of various data biases from the causal-effect perspective. Because RS data are observational rather than experimental, it is difficult to conduct a fine-grained analysis of the differences between covariates and confounders. To simplify this issue, we considered confounders as a set of covariates [26]. Hence, eliminating the confounding bias is crucial to enhancing the recommendation effectiveness.

Confounders are problems that exist in real-world applications, while many confounders cannot be directly observed [21], which leads to bias learning issues. From a causal perspective, the conventional rating prediction can be treated as the result of two causal-effect paths: user-rate path and item-rate path (Fig. 1(a)). Unfortunately, embedding-based recommendation models, such as matrix factorization (MF) [13], does not provide an unbiased inference of rating prediction by these two paths due to the hidden confounder.

To overcome these obstacles, we first analyze the causality in RS and abstract a causal diagram (Fig. 1(b)), where we introduced the cause of confounding bias, confounders, as a node. In our view, there are three paths affecting the probability of an interaction: user-rate path, item-rate path and confounder-rate path. However, existing works mainly concentrate on finding a better calculation for user/item-rate paths [12], thus ignore the effects of confounders. As mentioned in [26], the confounders are hidden in the exposure data, which is a form of implicit data (e.g., clicks, views) in recommender systems. So we argue that confounders exist in the implicit data, which can be represented by fitting an embedding using implicit data. In this regard, we propose the *Counterfactual Inference for Deconfounded Recommendation* (CIDR), which uses implicit data to estimate a substitute for the confounders, unitizes side data (i.e., user post-click feedback data and item multi-model data) to extract the true user/item features, and obtains the negative impact of confounders on the generation of recommendations by causal-effect analysis. In the end, CIDR applies counterfactual inference to reduce the bad effect of confounders and emphasizes the true user/item feature. In this way, we attain a robust and unbiased rating prediction model. We instantiate CIDR on MMGCN [29], which is a state-of-the-art multi-model recommendation model that takes full advantage of user and item extra features.

Extensive experiments over three benchmarks demonstrate that our CIDR not only alleviates confounders effectively but also improves the recommendation accuracy over the backbone models.

The main contributions of this work are as follows:

- We construct a causal graph and then analyze how confounders affect the RS models, which reveals the relationship between confounders and the bias problem from a causal-effect perspective.
- We propose a novel CIDR framework with counterfactual learning to capture a substitute representation of confounders in order to mitigate the bad effect of it and emphasize the true user preference via side information.
- We instantiate the proposed framework on MMGCN and validate it on three widely used real-world datasets. Extensive experiments are conducted to validate the effectiveness of our proposal.

2 Methodology

2.1 Preliminary

Basic Notations. In this paper, we use capital letters to represent random variables (e.g., X, Y, M). We use lowercase letters to represent the specific realizations of such random variables (e.g., x, y, m). We use uppercase calligraphy to denote a set (e.g., $\mathcal{U}, \mathcal{V}$). We utilize bold-font lowercase letters to represent the latent vector embedding of users, items, and other elements, such as $\boldsymbol{u}$, $\boldsymbol{v} \in \mathbb{R}^K$, where $\mathbb{R}^K$ are the dimension of embedding vectors.

Basic Concept of Causal Inference

Definition 1 (Causal Graph). *A causal graph is a directed acyclic graph (DAG) G=($\mathcal{N}, \mathcal{E}$), which consists of a set of nodes ($\mathcal{N}$) representing the variables in $\mathcal{U}$ and $\mathcal{V}$, and a set of edges ($\mathcal{E}$) between the nodes representing the functions in $\mathcal{F}$, which is a collection of function that assign a value to each variable V based on the values of other variables in the model.*

Definition 2 (Counterfactual Inference). *Counterfactual inference is a method for estimating what the descendant variables would be if the value of one treatment variable was different from its real value in the factual world.*

We simply use $Y_{x^*} = y$ to denote the counterfactual situation, "Y" would be y had X been x^*, though the observed value of X in the real world is not x^*.

Definition 3 (Causal Effect[1]). *The causal effect of a binary random variable X on another random variable Y is defined as,*

$$P(y|do(X=1)) - P(y|do(X=0)), \tag{1}$$

which reflects comparisons between two potential outcomes of the same variable given two different treatments

[1] In this work, we follow [28] and [17] to define the causal effect on individual rather than a population.

In the next section, we will further analysis the recommendation from the view of such causal effect and explore the causal effect of confounders on recommendation results.

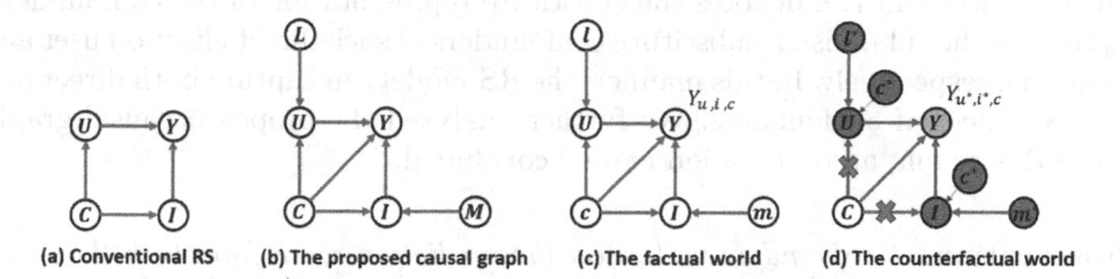

Fig. 1. The causal graph of conventional recommendation models and our proposed CIDR, where node U, I, C, L, M, and Y denote the user features, item features, user preference, item content features, substitute confounders, and prediction score, respectively. And * represents the reference value of random variables. (a) Example of a conventional recommendation models. (b) Our proposed causal graph. (c) The factual world. (d) The counterfactual world.

2.2 Causal Look in Recommendation

Link to Conventional Recommendation. In Fig. 1(a), we first abstract a causal graph to describe the process of data generation in conventional RS models, where node U, I, C, and Y denote the user features, item features, confounders, and prediction score, respectively. We use edges to denote causal effects, where $C \rightarrow U(I)$ means the direct effect comes from confounders on user (item) features, and $U(I) \rightarrow Y$ means user (item) features will directly affect the prediction score. $C \rightarrow U(I) \rightarrow Y$ denotes the indirect effect comes from confounders to the predictions via the mediators U and I.

However, the lack of consideration of C makes these predictions made by conventional RS models mismatch the historical interactions and misrepresent user preference and item attribution. For instance, considering an item at the top of the recommended list, users may click on it simply because they trust the recommendation systems rather than like this item. From a causal perspective, there is a direct effect of position (i.e., confounder) on clicking behavior.

Link to Deconfounded Recommendation. To bridge this gap, we scrutinize a new causal graph (Fig. 1(b)), where we add a new direct path from confounders to prediction score ($C \rightarrow Y$). Meanwhile, recall that the presence of confounders will cause user/item representations to be skewed. So we also add two new nodes, M and L, two new direct edges, $L \rightarrow U$ and $M \rightarrow I$, to obtain more accurate user/item features. M and L denote content features extracted from item multi-model data and user preferences learned from positive post-click feedback, respectively. Formally, we formulate the prediction score as:

$$\hat{y}_{uic} = Y_{u,i,c} = Y(U = u, I = i, C = c), where \\ u = U(C = c_u, L = l), i = I(C = c_i, M = m), \quad (2)$$

where $Y()$, $U()$, and $I()$ denote as the score function of Y, and the aggregation function of U and I. $\boldsymbol{c}$ denotes the substitute representation of the confounders, $\boldsymbol{c_u}, \boldsymbol{c_i}$, are the subdivision substitute confounders of such direct effect on user and item side, respectively. In this manner, the RS model can capture both direct and indirect effect of confounders. We further analyzed the proposed causal graph. From this graph, an observation can be concluded.

Observation. *Confounders will affect the prediction score from both the direct and indirect path.*

The indirect effect $C{\to}(U,I){\to}Y$ denotes that confounders alter the exposure likelihood of items and users' activities, causing the user (item)'s representation to be disrupted. Finally, the indirect effect is included in the predictions. Such an indirect effect, however, may be reasonable. For example, a user may particularly enjoy romantic films, occasionally like action films, and dislike other genres (i.e., genres as a confounder). As a result, it's natural to recommend more romantic movies to him or her. In other words, making decisions with such an indirect effect in mind is appropriate.

In contrast, the direct path $C{\to}Y$ indicates that confounders blend the prediction score with confounding biases, which means that the recommendation model gives recommendations only with the consideration of confounders. In the prior example, such a direct effect may lead to the model only recommending romantic movies regardless of action or other movies-something that is illogical and should be avoided. As the conventional RS model usually ignores such effects, such model will inevitably emphasize the bad effect of confounders while missing the actual user preference and item attribution. Based on the above analyses, we make the following hypothesis:

Hypothesis. *There is a lack of consideration of such a direct effect, resulting in the recommendation model capturing skewed user preference and amplifying the bad effect of confounders. Thus, such an effect has to be eliminated when formulating the predictive model.*

2.3 Deconfounded Analysis

Following the definitions in [18–20], we formulate the *total effect* (TE) from confounders C to prediction score Y as:

$$TE = Y_{u,i,c} - Y_{u^*,i^*,c^*}, \quad (3)$$

where u^* and i^* are the reference value of U and I, which indicates that the U and I does not have their own characteristics, and the corresponding attributes are all the same for all users and items. Note that we treat the average value as the reference value.

Recall that RS models may suffer from the spurious correlation between confounders and prediction scores, resulting in unsatisfactory recommendations. Therefore, we expect RS models to eliminate the direct effect of confounders. Based on the causal graph in Fig. 1(b) and the *intervention* [17], such an effect can be denoted by the *natural direct effect* (NDE) of confounders on predictions:

$$NDE = Y_{u^*,i^*,c} - Y_{u^*,i^*,c^*}, \tag{4}$$

$Y_{u^*,i^*,c}$ stands for the prediction score under no-treatment intervention $do(U{=}u^*)$ and $do(I{=}i^*)$, and corresponds to a counterfactual thinking: *What the prediction score would be if the recommendation result can only be affected by the confounding bias?* Y_{u^*,i^*,c^*} is the reference value (i.e., constant for any users and items).

Since the effect of C on the mediator $U(I)$ is blocked, NDE explicitly captures the confounding bias. Furthermore, the reduction of confounding bias can be realized by subtracting NDE from TE and then get the *total indirect effect* (TIE). The implication is that RS models should not make predictions based on confounders alone, but formulate predictions taking a combination of confounders and user (item) features. We formulate the TIE as:

$$TIE = TE - NDE = Y_{u,i,c} - Y_{u^*,i^*,c}. \tag{5}$$

2.4 Deconfounded Recommendation Model

The causal effect is estimated by comparing two states of Y, the prediction score, for example, $Y_{u,i,c}$ and $Y_{u^*,i^*,c}$ in Eq. 5 correspond to the prediction score of normal strategies, and the Y affected by the effect of confounders alone (Fig. 1(d)), respectively. In words, we formulated the causal effect based on comparing the value of Y in different states. Now, we use the TIE from Eq. 5 as the deconfounded prediction score.

$$\hat{y}_{uic} = TIE = Y_{u,i,c} - Y_{u^*,i^*,c}. \tag{6}$$

By using TIE for recommendation prediction, we can mitigate spurious correlation between confounders and prediction scores. So the next key question is how we can obtain such alternative representations of confounders to estimate the causal effect of it. In this work, we treat the exposure data as the evidence of confounders exists in the data (refer [9,26] for details and proof which we omit for brevity). We use MMGCN to train on both such exposure (confounded) data and user post-click feedback data to obtain confounded , preference (content) representations of user (item). Formally, we obtain the user preference representation $\boldsymbol{l}$, user confounded representation $\boldsymbol{c_u}$, item content feature $\boldsymbol{m}$, and item confounded representation $\boldsymbol{c_i}$.

Once we obtain the above representations, we can calculate the prediction score now. Specifically, we split $Y_{u,i,e}$ and $Y_{u^*,i^*,c}$ into six different scores, $Y_{u,i}$, Y_{u,c_i},$Y_{c_u,i}$, Y_{u^*,i^*},Y_{u^*,c_i}, and Y_{c_u,i^*}. Then, we calculate the values of these components by: $Y_{u,i}{=}\boldsymbol{l}^\top \boldsymbol{m}$, $Y_{u,c_i}{=}\boldsymbol{l}^\top \boldsymbol{c_i}$, $Y_{c_u,i}{=}\boldsymbol{c_u}^\top \boldsymbol{m}$, $Y_{u^*,i^*}{=}\mathbb{E}_{Y_{u^*,i^*}}$, $Y_{u^*,c_i}{=}\mathbb{E}_{Y_{u^*,c_i}}$, $Y_{c_u,i^*}{=}\mathbb{E}_{Y_{c_u,i^*}}$.

Fusion Strategy. Next, with the motivation of integrating the newly added confounded score $Y_{u,c_i}, Y_{c_u,i}, Y_{u^*,c_i}$, and Y_{c_u,i^*} with conventional score $Y_{u,i}$ and Y_{u^*,i^*}. Inspired by [17], we calculate the values of these two predictions via a classical fusion strategy, Multiplication, which can be formulated as:

$$\begin{aligned} Y_{u,i,c} &= f(Y_{u,i}, Y_{u,c_i}, Y_{c_u,i}) = Y_{u,i} * (\sigma(Y_{u,c_i}) + \sigma(Y_{c_u,i})), \\ Y_{u^*,i^*,c} &= f(Y_{u^*,i^*}, Y_{u^*,c_i}, Y_{c_u,i^*}) = Y_{u^*,i^*} * (\sigma(Y_{u^*,c_i}) + \sigma(Y_{c_u,i^*})), \end{aligned} \tag{7}$$

where σ represent the non-linear sigmoid function.

Algorithm 1: CIDR Algorithm

Input: Dataset, including observed interactions as user-item pair (p, q), the user preference representation $\boldsymbol{l}$ and user confounded representation $\boldsymbol{c_u}$ for all users, and the item content feature $\boldsymbol{m}$ and item confounded representation $\boldsymbol{c_i}$ for all items; regularization parameters α, β for the two joint tasks; L2-norm regularization weight: λ_θ.

Output: Prediction Y

1 **while** *stop condition is not reached* **do**
2   Read batch of training samples ;
3   **for** *each batch in training set* **do**
4         $Loss \leftarrow 0$;
5   **for** *each training sample* (p, q) *in the batch* **do**
6             $Y_{u,i} \leftarrow \boldsymbol{l_p}^\top \boldsymbol{m_q}$;
7             $Y_{u,c_i} \leftarrow \boldsymbol{l_p}^\top \boldsymbol{c_i^q}$;
8             $Y_{c_u,i} \leftarrow \boldsymbol{c_u^p}^\top \boldsymbol{m_q}$;
9             $k \leftarrow Y_{u^*,i^*} \leftarrow \mathbb{E}_{Y_{u^*,i^*}}$;
10             $Y_{u^*,c_i} \leftarrow \mathbb{E}_{Y_{u^*,c_i}}$;
11             $Y_{c_u,i^*} \leftarrow \mathbb{E}_{Y_{c_u,i^*}}$;
12             $Loss \leftarrow Loss + L_{u,i} + \alpha * L_{c_u} + \beta * L_{c_i} + \lambda_\theta ||\Theta||_2^2$
13   **end**
14         $Y \leftarrow Y_{u,i,c} - k * \sigma(Y_{u^*,c_i}) - k * \sigma(Y_{c_u i^*})$
15   **end**
16 **end**
17 **return** Y

Deconfounded Training. Based on the analysis before, we need to optimize the prediction score $Y_{u,i,c}$ from the factual world and $Y_{u^*,i^*,c}$ from the counterfactual world. We adopt multi-task learning to train the CIDR model with the following learning objective function:

$$\begin{aligned} \mathcal{L} &= L_{ui} + \alpha * L_{c_u} + \beta * L_{c_i} + \lambda_\theta ||\Theta||_2^2 \\ &= l(Y_{u,i}, \overline{Y}) + \alpha * l(Y_{c_u,i}, \overline{Y}) + \beta * l(Y_{u,c_i}, \overline{Y}) + \lambda_\theta ||\Theta||_2^2 \end{aligned} \tag{8}$$

where α and β are the hyper-parameters to tune the relative weight of tasks. $L_{u,i}$ is the loss function of task of capturing user preference and item content features from side data. L_{c_u} and L_{c_i} are the loss function of user side task and item side task for capturing the confounded user and item representation. l is the BPR loss and $\overline{Y}$ is the label from the training epochs. Y_{u,c_i} and $Y_{c_u,i}$ can be estimated by forcing the model to accept the c_i and c_u as the item and user representation, respectively. Θ and λ_θ denote parameters of the model and the L2-normalization weight.

Counterfactual Inference. To eliminate the direct effect of causal path $C \rightarrow Y$ from the rating prediction. We calculate predictions and perform the following reduction to eliminate such bad effects:

$$Y = Y_{u,i,c} - Y_{u^*,i^*,c} = Y_{u,i,c} - k * \sigma(Y_{u^*,c_i}) - k * \sigma(Y_{c_u,i^*}) \tag{9}$$

where k is the expectation of Y_{u^*,i^*}, formulated as $k{=}E(Y_{u^*,i^*})$, which means that the model does not accept user/item feature as input. The overall algorithm is provided in Algorithm 1.

3 Experiments

In this section, we conduct experiments to evaluate the effectiveness of the proposed CIDR framework by answering the following three research questions:

- **RQ1:** Does CIDR outperform state-of-the-art methods?
- **RQ2:** How do the different components (i.e., user bran-ch, item branch) and hyper-parameters affect the performance of CIDR?
- **RQ3:** Does CIDR handle the confounder problems universally?

Table 1. Statistics of three different datasets.

Dataset	#Users	#Items	#Interactions	#Sparsity	#Likes
Adressa	31,123	4,895	1,437,540	0.00944	998,612
Tiktok	18,855	34,756	1,493,532	0.00228	589,008
Coat	290	300	6960	0.08	1905

3.1 Experimental Settings

Datasets. We use three real-world datasets for evaluation, with their statistics shown in Table 1. **Tiktok** [29] is a micro-video dataset released by ByteDance. Notably, actions of favorite and finish are used as the positive post-click feedback. **Adressa** [10] is a news reading dataset from Adressavisen. We use the pre-trained Bert [6] to derive the textual features. It should be noted that the dwell

Table 2. Performance comparison between CIDR and the baselines for the two real-world datasets (* indicate $p < 0.05$ in t-test). The best results are highlighted in bold, and underlined values denote the best results of baseline methods.

Method	Adressa						Tiktok					
	P@10	P@20	R@10	R@20	N@10	N@20	P@10	P@20	R@10	R@20	N@10	N@20
BPRMF	0.0480	0.0401	0.0941	0.1503	0.0781	0.0992	0.0206	0.0203	0.0314	0.0606	0.0311	0.0399
MMGCN	0.0501	0.0415	0.0975	0.1612	0.0817	0.1059	0.0256	0.0231	0.0357	0.0635	0.0333	0.0430
CT	0.0493	0.0428	0.0951	0.1611	0.0799	0.1051	0.0217	0.0194	0.0295	0.0520	0.0294	0.0372
NR	0.0499	0.0415	0.0970	0.1610	0.0814	0.1058	0.0239	0.0216	0.0346	0.0605	0.0329	0.0424
RR	0.0521	0.0415	0.1007	0.1612	0.0831	0.1059	0.0264	0.0231	0.0383	0.0635	0.0367	0.0430
IPS	0.0419	0.0361	0.0804	0.1378	0.0663	0.0883	0.0230	0.0210	0.0334	0.0582	0.0314	0.0406
MACR	0.0524	0.0432	0.1023	0.1677	0.0863	0.1111	0.0313	0.0277	0.0472	0.0812	0.0443	0.0563
CR	0.0532	0.0439	0.1045	0.1712	0.0878	0.1133	0.0269	0.0242	0.0393	0.0683	0.0376	0.0476
PDA	0.0578	0.0478	0.1057	0.1729	0.0914	0.1167	0.0340	0.0301	0.0466	0.0789	0.0461	0.0573
CIDR*	**0.0631**	**0.0509**	**0.1280**	**0.2041**	**0.1072**	**0.1362**	**0.0361**	**0.0322**	**0.0526**	**0.0841**	**0.0514**	**0.0628**
%Improv.	9.17%	6.49%	21.1%	18.1%	17.3%	16.7%	6.18%	6.98%	11.4%	3.57%	11.5%	9.60%

time>30 s reflects the like of users [33]. **Coat** [25] follows the MNAR and the MAR assumption for training and test datasets (i.e., uniform test dataset).

For Adressa and Tiktok, we randomly split the interactions into the training, validation, and testing subsets with a ratio of 80%, 10%, and 10%, respectively. For each click, we randomly choose an item the user has never interacted with as the negative sample to create the triples for training.

Since the original observational data are biased by the confounders [5], in this work, we treat the post-click feedback data (i.e., like) as the evaluation tool. For Coat, we use the original uniform test data as the test set (i.e., uniform data for evaluation). The motivation for using different test set settings is to check the performance of CIDR on multiple tasks.

Baselines. Due to the need for item multi-model data, we implemented our CIDR with the SOTA multi-modal recommender model MMGCN [29], and compared CIDR with various SOTA methods. In consideration of fairness, all methods are instantiated on MMGCN or add pre-trained side embeddings. **BPRMF** [13] is the classical MF model with BPR loss. **MMGCN** [29] is the normal train of MMGCN. **CT** We train the MMGCN with a clean train (CT) setting [24], where only items of user like are treated as positive samples. **NR** [30] adopt the Negative feedback Re-weighting (NR) to reweight the samples. **RR** [16] re-rank the top 20 items recommended by MMGCN. Besides, we also compared four SOTA causal inference models. **IPS** [21] adds the Inverse Propensity Score to reweight samples. **CR** [24] leverages counterfactual reasoning to alleviate the clickbait issue. **PDA** [34] performs deconfounded training with causally intervenes the popularity bias. **MACR** [28] utilizes counterfactual reasoning to eliminate the popularity bias.

Evaluation Metrics and Hyper-parameters. We adopt three popular metrics including Precision(P@K), Recall(R@K), and NDCG@K (N@K) to evaluate the performance of the recommendation models.

Table 3. Performance comparison over coat dataset (* indicate $p < 0.05$ in t-test). The best results are highlighted in bold, and underlined values denote the best results of baseline methods.

Method	Coat					
	P@10	P@20	R@10	R@20	N@10	N@20
BPRMF	0.0497	0.0515	0.0320	0.0604	0.0507	0.0591
MMGCN	0.0505	0.0520	0.0327	0.0620	0.0531	0.0599
CT	0.0488	0.0499	0.0317	0.0600	0.0493	0.0570
NR	0.0511	0.0533	0.0330	0.0622	0.0528	0.0601
RR	0.0523	0.0558	0.0329	0.0611	0.0541	0.0619
IPS	0.0494	0.0506	0.0301	0.0609	0.0502	0.0588
MACR	<u>0.0596</u>	<u>0.0580</u>	<u>0.0372</u>	0.0650	<u>0.0611</u>	<u>0.0658</u>
CR	0.0535	0.0561	0.0344	0.0630	0.0547	0.0621
PDA	0.0589	0.0577	0.0369	<u>0.0661</u>	0.0604	0.0631
CIDR*	**0.0629**	**0.0617**	**0.0415**	**0.0685**	**0.0691**	**0.0707**
%Improv.	5.54%	6.38%	11.6%	3.63%	13.1%	7.45%

For a fair comparison, all methods are optimized by BPR loss. We optimize all models with the Adam optimizer with a batch size of 1,024 and default choice of learning rate (i.e., 0.001). For our CIDR framework, the multi-task trade-off parameters α and β in Eq. (8) are tuned in $\{0, 1, 2, 3, 4, 5, 6\}$ and set to 2 and 1 through extensive experiments. Other hyper-parameters for our method and baselines are tuned by grid search. Furthermore, we use the early stopping strategy that stops training if Recall@10 on the validation data does not increase for 10 epochs. The L2 regularization coefficient is set to 0.001 by default.

3.2 Performance Comparison (RQ1)

Tables 2 and 3 presents the overall performance comparison of the baseline models on three datasets. From the table we can conclude the following:

- RR can achieve a better result than MF and other conventional methods, which validates the effectiveness of leveraging post-click feedback data to alleviate the effect of confounders. Although post-click feedback data are introduced into both NR and CT, they perform worse than RR, e.g., the P@10 and N@10 of NR decrease by 10.5% and 6.94% on Adressa dataset, which can be explained by the sparsity of post-click feedback. Due to the lack of training interactions, it is hard to generalize better user and item representations through such models.
- In most cases, there is a gap between the performance of conventional methods and causal inference approaches. For example, CR, PDA, and MACR outperform conventional methods. The performance gain indicates that the theory of causality has huge improvements on debias recommendation. However, IPS

produces the worst result compared to all baseline methods on the Adressa dataset and attains a second-to-last result on the Tiktok dataset. We ascribe such inferior performance to the inaccurate estimation and high variance of propensity scores, and the lack of stability on different datasets. PDA achieves the best results over other baseline models due to the leveraging of popularity bias, which improves the probability of recommend an item that has the potential to be popular (i.e., desired popularity bias).

- Without any doubt, in all cases, our CIDR framework consistently outperforms all baselines across all datasets and metrics. Specifically, CIDR makes over 10% relative improvements with respect to R@10 and N@10 over three datasets. This verifies that our CIDR can learn better debias configurations and user preferences than the others, which is attributed to the effectiveness of our multi-task learning, causal effect analysis of confounders, and counterfactual inference. In particular, CIDR outperforms SOTA causal inference method CR, PDA, and MACR. This can be explained by the idea of capturing the causal effect of substitute confounders on rating predictions and the use of side information, which can better capture the users and items representations, thereby improving the effectiveness of counterfactual inference to eliminate the effect of confounders.

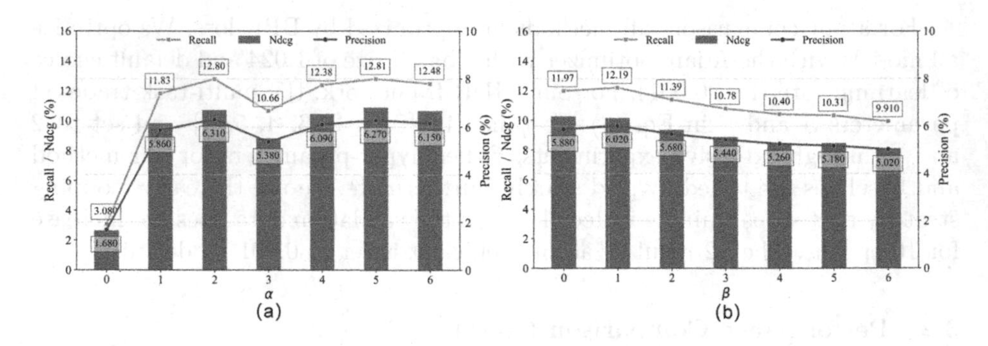

Fig. 2. Effect of Hyper-parameters α and β.

3.3 Case Study (RQ2)

We further evaluate the performance of CIDR on Adressa dataset as an example.

Effect of Hyper-parameters. To explore the influence of these two side losses, we conduct experiments with the value of α, β within $\{0, 1, 2, 3, 4, 5, 6\}$. It should be noted that while varying one parameter, the other one is set as constant 1. The effect of α and β on the performance is visualized in Fig. 2. Analyzing this figure from top to bottom, we can observe that:

- The performance of CIDR enhances with the increase of α from 0 to 2, which demonstrates the importance of capturing the effect of confounders over user side. Similarly, while varying β from 0 to 1, the performance improves, which means that the benefit of capturing the item side substitute confounders.

- However, when α and β surpasses a threshold (2 or 1), the performance deteriorates with the increase of these parameters. We ascribe such inferior performance to the overthinking of confounders affecting the user and item representation. Therefore, we need to control α and β within a certain threshold.

Table 4. Comparison for different model variants on Adressa.

Method	P@20	R@20	NDCG@20
CIDR	**0.0509**	**0.2041**	**0.1362**
CIDR w/o inference	0.0435	0.1704	0.1197
CIDR w/o user branch	0.0459	0.1792	0.1202
CIDR w/o item branch	0.0479	0.1856	0.1268
CIDR w/o side information	0.0477	0.1866	0.1247

Effect of Inference Strategy and User/Item Branch. We first attempt to answer the two following questions: *Is it necessary to conduct the inference strategy? How do user branch and item branch affect the performance of our CIDR?* To this end, five variants are given: "CIDR w/o inference", where we remove the inference of counterfactual reasoning and predict the ratings via TE of L=l, C=c, M=m; "CIDR w/o user branch" and "CIDR w/o item branch", where we remove the process of capturing user (or item) confounded representations for evaluation (i.e., remove the confounded causal effect for inference); "CIDR w/o side information", where we simply remove the task of $L_{u,i}$ on training but retain its effect on prediction. Table 4 shows the performance comparison of these variants. We can observe that CIDR w/o inference outperforms MMGCN, which can be explained by the effectiveness of calculating the prediction from a causality view (i.e., TE). Meanwhile, both user and item inference improve the performance, which indicates that eliminating the direct effect of confounders indeed helps the RS models to perform better. Moreover, incorporating side information does provide a more accurate representation of user preference and item content features, but its generalization ability is obviously insufficient, resulting in poor results (same for CT). This may result in a lack of diversity in the recommendation results.

3.4 Deconfounding Capability (RQ3)

Here, we thoroughly investigate the capability of CIDR for eliminating the confounding bias issue. Specifically, we discuss the capability of CIDR with respect to the following two tasks:

- **Does CIDR achieve a satisfying recommendation?**
  Following [24], we use likes/clicks ratios to express the user satisfying. In Fig. 3, we present the comparison on Adressa dataset, where we abstract top-20 items in the recommended lists and measure the frequency of

recommended. We observed that CIDR achieves the best performance, where items with ratios lower than 0.4 are seldom recommended to the user, and CIDR more likely to give recommendations with ratio higher than 0.6. These enhancements will be more significant in real-world massive data scenarios.

- **If confounding bias is the collection of various biases, does CIDR eliminate these biases such as popularity bias?** We report the relative recommendation rate (RR) of different groups in Fig. 4 (a), and we conduct in Fig. 4 (b) an experiment to show the average recall over groups. We divided items into five groups according to their popularity, and recommend 20 items for each user to calculate the average item recall over the groups.
  As shown in Fig. 4 (b),
  our CIDR has higher recall on each group than MMGCN and CR, which indicates the superiority of our method. However, the relative recommendation rate of our CIDR model on the last group is reduced, which can be explained by the fact that our model mitigates the problem of over-exposure of popular items and makes more recommendations based on user preferences and item attributions rather than the popularity.

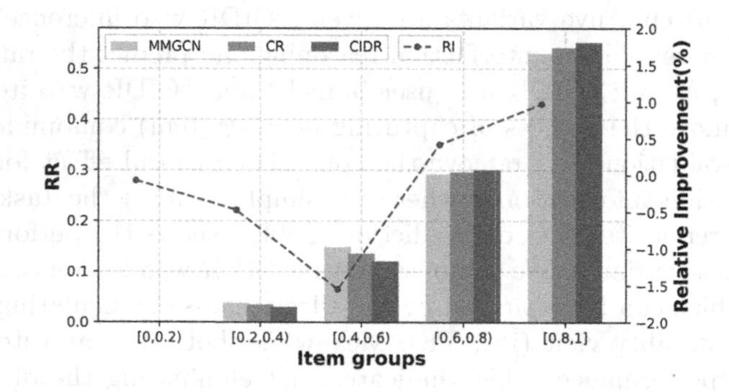

Fig. 3. Visualization of the averaged recommendation rate (RR) and the relative improvement (RI) of CIDR over CR for item groups with different likes/clicks ratios.

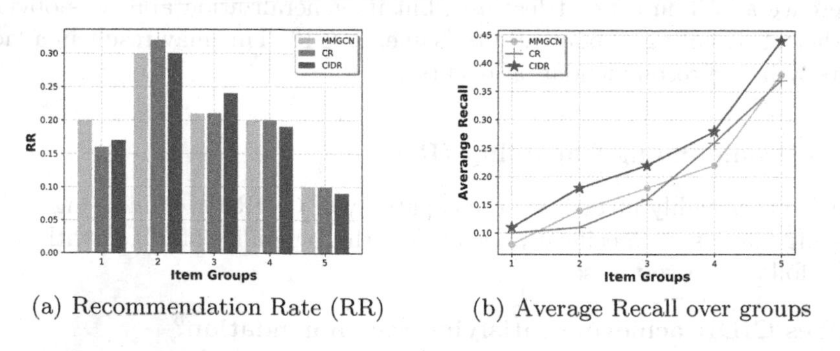

(a) Recommendation Rate (RR)

(b) Average Recall over groups

Fig. 4. Average item Recall and RR over groups on Adressa.

4 Related Work

In this section, we explore existing works that are highly relevant to our paper, which can be summarized to three topics: confounders and bias problem in the recommender system, and causal recommendation.

4.1 Debiasing in Recommender Systems

In order to bridge the gap between observation data and actual user preferences, previous researchers focused on eliminating the effect of various biases. A line of existing works mainly study the effects of bias from the aspect of data. For example, the causal embedding-based methods [2] utilize a small unbiased data by intervention to eliminate popularity bias. Data distribution aspect method [22] focused on solving the data miss-not-at-random problem to eliminating selection bias and exposure bias. Besides, meta-learning [4] and knowledge distillation [15] also achieve state-of-the-art debias performance. Another line of research pays attention to removing the influence of such bias on rating prediction, where two type of methods are mostly used: IPS based methods [35] aim to re-weight each instance as the inverse of corresponding item bias (e.g., popularity, position, exposure) score, and re-rank methods utilize regularization [30].

4.2 Deconfounded in Recommender Systems

In recent years, many researches have explored the confounding bias problem [27] from various perspectives; some attributed the problem of confounders to the recommendation model and algorithm [3]. Other studies [27,32] tried to learn a substitute of confounders by a factor model. Different from these approaches, we explored the confounders issues from a novel causal view and leveraged counterfactual inference to eliminate the negative causal-effect of such confounders.

4.3 Causal Recommendation

The theory of causality has been used in various domains, as for its application for RS [8,14,31,32], most studies are paid attention to debias such as position bias [11], conformity bias [35], and popularity bias [28,34]. The most popular methods can be divided into three types: The first comprises the aforementioned IPS based methods. The second includes causal intervention methods that cut the influence of prior biases from prediction score. However, because the sample space is too large, their approximation of scores over the intervention terms is subject to large variance and lacks stability. The last method is counterfactual inference that adjusts the rating prediction by reducing the effect of confounder or bias [7]. Another line of works are dedicated to using causality to improve the interpretability of the recommender system [23]. However, the current causal recommendation seldom leverages the causal effect of both user side and item side information, and simply treats the observational data as the true representation of user preference and item attribution. Moreover, they ignore how confounded observations influence the recommendation effectiveness.

5 Conclusion and Future Work

In this work, we attribute the root reason of bias problem to a causal concept, namely, confounders. We further present the CIDR framework that first captures substitute confounders represented both on user side and item side from click data, and then leverages the theory about causality to measure the effect of confounders in the recommendation model. Finally, we boost counterfactual inference to eliminate such effects from the prediction score.

References

1. Ai, Q., Bi, K., Luo, C., Guo, J., Croft, W.B.: Unbiased learning to rank with unbiased propensity estimation. In: The 41st International ACM SIGIR Conference on Research & Development in Information Retrieval, pp. 385–394 (2018)
2. Bonner, S., Vasile, F.: Causal embeddings for recommendation. In: Proceedings of the 12th ACM Conference on Recommender Systems, pp. 104–112 (2018)
3. Chaney, A.J., Stewart, B.M., Engelhardt, B.E.: How algorithmic confounding in recommendation systems increases homogeneity and decreases utility. In: Proceedings of the 12th ACM Conference on Recommender Systems, pp. 224–232 (2018)
4. Chen, J., et al.: AutoDebias: learning to debias for recommendation. arXiv preprint arXiv:2105.04170 (2021)
5. Chen, J., Dong, H., Wang, X., Feng, F., Wang, M., He, X.: Bias and debias in recommender system: a survey and future directions. arXiv preprint arXiv:2010.03240 (2020)
6. Devlin, J., Chang, M.W., Lee, K., Toutanova, K.: BERT: pre-training of deep bidirectional transformers for language understanding. arXiv preprint arXiv:1810.04805 (2018)
7. Ding, S., Feng, F., He, X., Liao, Y., Shi, J., Zhang, Y.: Causal incremental graph convolution for recommender system retraining. arXiv preprint arXiv:2108.06889 (2021)
8. Dong, Z., et al.: Counterfactual learning for recommender system. In: Fourteenth ACM Conference on Recommender Systems, pp. 568–569 (2020)
9. Greenland, S., Pearl, J., Robins, J.M.: Confounding and collapsibility in causal inference. Stat. Sci. **14**(1), 29–46 (1999)
10. Gulla, J.A., Zhang, L., Liu, P., Özgöbek, Ö., Su, X.: The adressa dataset for news recommendation. In: Proceedings of the International Conference on Web Intelligence, pp. 1042–1048 (2017)
11. Guo, H., Yu, J., Liu, Q., Tang, R., Zhang, Y.: Pal: a position-bias aware learning framework for CTR prediction in live recommender systems. In: Proceedings of the 13th ACM Conference on Recommender Systems, pp. 452–456 (2019)
12. He, X., Liao, L., Zhang, H., Nie, L., Hu, X., Chua, T.S.: Neural collaborative filtering. In: Proceedings of the 26th International Conference on World Wide Web, pp. 173–182 (2017)
13. Koren, Y., Bell, R., Volinsky, C.: Matrix factorization techniques for recommender systems. Computer **42**(8), 30–37 (2009)
14. Li, Q., Wang, X., Xu, G.: Be causal: de-biasing social network confounding in recommendation. arXiv preprint arXiv:2105.07775 (2021)
15. Liu, D., Cheng, P., Dong, Z., He, X., Pan, W., Ming, Z.: A general knowledge distillation framework for counterfactual recommendation via uniform data (2020)

16. Liu, W., Guo, J., Sonboli, N., Burke, R., Zhang, S.: Personalized fairness-aware re-ranking for microlending. In: Proceedings of the 13th ACM Conference on Recommender Systems, pp. 467–471 (2019)
17. Niu, Y., Tang, K., Zhang, H., Lu, Z., Hua, X.S., Wen, J.R.: Counterfactual VGA: a cause-effect look at language bias. In: Proceedings of the IEEE/CVF Conference on Computer Vision and Pattern Recognition, pp. 12700–12710 (2021)
18. Pearl, J.: Causality. Cambridge University Press (2009)
19. Pearl, J., Glymour, M., Jewell, N.P.: Causal inference in statistics: a primer. John Wiley & Sons (2016)
20. Pearl, J., Mackenzie, D.: The book of why: the new science of cause and effect. Basic books (2018)
21. Rosenbaum, P.R., Rubin, D.B.: The central role of the propensity score in observational studies for causal effects. Biometrika **70**(1), 41–55 (1983)
22. Saito, Y., Yaginuma, S., Nishino, Y., Sakata, H., Nakata, K.: Unbiased recommender learning from missing-not-at-random implicit feedback. In: Proceedings of the 13th International Conference on Web Search and Data Mining, pp. 501–509 (2020)
23. Tan, J., Xu, S., Ge, Y., Li, Y., Chen, X., Zhang, Y.: Counterfactual explainable recommendation. In: Proceedings of the 30th ACM International Conference on Information & Knowledge Management, pp. 1784–1793 (2021)
24. Wang, W., Feng, F., He, X., Zhang, H., Chua, T.S.: Clicks can be cheating: counterfactual recommendation for mitigating clickbait issue. In: Proceedings of the 44th International ACM SIGIR Conference on Research and Development in Information Retrieval, pp. 1288–1297 (2021)
25. Wang, X., Zhang, R., Sun, Y., Qi, J.: Doubly robust joint learning for recommendation on data missing not at random. In: International Conference on Machine Learning, pp. 6638–6647. PMLR (2019)
26. Wang, Y., Blei, D.M.: The blessings of multiple causes. J. Am. Stat. Assoc. **114**(528), 1574–1596 (2019)
27. Wang, Y., Liang, D., Charlin, L., Blei, D.M.: Causal inference for recommender systems. In: Fourteenth ACM Conference on Recommender Systems, pp. 426–431 (2020)
28. Wei, T., Feng, F., Chen, J., Wu, Z., Yi, J., He, X.: Model-agnostic counterfactual reasoning for eliminating popularity bias in recommender system. In: Proceedings of the 27th ACM SIGKDD Conference on Knowledge Discovery & Data Mining, pp. 1791–1800 (2021)
29. Wei, Y., Wang, X., Nie, L., He, X., Hong, R., Chua, T.S.: MMGCN: multi-modal graph convolution network for personalized recommendation of micro-video. In: Proceedings of the 27th ACM International Conference on Multimedia, pp. 1437–1445 (2019)
30. Wen, H., Yang, L., Estrin, D.: Leveraging post-click feedback for content recommendations. In: Proceedings of the 13th ACM Conference on Recommender Systems, pp. 278–286 (2019)
31. Xu, S., Ge, Y., Li, Y., Fu, Z., Chen, X., Zhang, Y.: Causal collaborative filtering. arXiv preprint arXiv:2102.01868 (2021)
32. Xu, S., Tan, J., Heinecke, S., Li, J., Zhang, Y.: Deconfounded causal collaborative filtering, pp. 40–43 (2021). http://arxiv.org/abs/2110.07122
33. Yi, X., Hong, L., Zhong, E., Liu, N.N., Rajan, S.: Beyond clicks: dwell time for personalization. In: Proceedings of the 8th ACM Conference on Recommender systems, pp. 113–120 (2014)

34. Zhang, Y., et al.: Causal intervention for leveraging popularity bias in recommendation. arXiv preprint arXiv:2105.06067 (2021)
35. Zheng, Y., Gao, C., Li, X., He, X., Li, Y., Jin, D.: Disentangling user interest and conformity for recommendation with causal embedding. In: Proceedings of the Web Conference 2021, pp. 2980–2991 (2021)

Recommending Related Products Using Graph Neural Networks in Directed Graphs

Srinivas Virinchi(✉), Anoop Saladi, and Abhirup Mondal

International Machine Learning, Amazon, Bengaluru, India
{virins,saladias,mabhirup}@amazon.com

Abstract. Related product recommendation (RPR) is pivotal to the success of any e-commerce service. In this paper, we deal with the problem of recommending related products i.e., given a query product, we would like to suggest top-k products that have high likelihood to be bought together with it. Our problem implicitly assumes asymmetry i.e., for a phone, we would like to recommend a suitable phone case, but for a phone case, it may not be apt to recommend a phone because customers typically would purchase a phone case only while owning a phone. We also do not limit ourselves to complementary or substitute product recommendation. For example, for a specific night wear t-shirt, we can suggest similar t-shirts as well as track pants. So, the notion of relatedness is subjective to the query product and dependent on customer preferences. Further, various factors such as product price, availability lead to presence of selection bias in the historical purchase data, that needs to be controlled for while training related product recommendations model. These challenges are orthogonal to each other deeming our problem nontrivial. To address these, we propose DAEMON, a novel Graph Neural Network (GNN) based framework for related product recommendation, wherein the problem is formulated as a node recommendation task on a directed product graph. In order to capture product asymmetry, we employ an asymmetric loss function and learn dual embeddings for each product, by appropriately aggregating features from its neighborhood. DAEMON leverages multi modal data sources such as catalog metadata, browse behavioral logs to mitigate selection bias and generate recommendations for cold-start products. Extensive offline experiments show that DAEMON outperforms state-of-the-art baselines by 30–160% in terms of HitRate and MRR for the node recommendation task. In the case of link prediction task, DAEMON presents 4–16% AUC gains over state-of-the-art baselines. DAEMON delivers significant improvement in revenue and sales as measured through an A/B experiment.

Keywords: Related product recommendation · Graph neural networks · Directed graphs · Selection bias

M.-R. Amini et al. (Eds.): ECML PKDD 2022, LNAI 13713, pp. 541–557, 2023.
https://doi.org/10.1007/978-3-031-26387-3_33

1 Introduction

Related product[1] recommendation (RPR) plays a vital role in helping customers easily find right products on e-commerce websites and hence, critical to their success. It not only helps customers discover new related products, but also simplifies their shopping effort, thereby delivering a great shopping experience. In this paper, we are interested in related product recommendation problem: *given a query product, the goal is to recommend top-k products that have a high likelihood to be bought together with it.* For notation purpose, let a, b and c be any three products. Let R_{cp}, R_{cv} and R_{like} be binary relationships where, $aR_{cp}b$ represents that a is co-purchased with b, $aR_{cv}b$ represents that a is co-viewed with b, and $aR_{like}b$ represents that a is similar to b based on its product features. Consider a sample product graph in Fig. 1, where a node corresponds to a product and an edge corresponds to a co-purchase or co-view relationship between the products (refer to Sect. 4.1 for graph construction). RPR problem entails few challenges which we illustrate using Fig. 1 as follows: 1) *product relevance:* Given a query product P, we would like to suggest related products such as an adapter AC, phone case PC, AirPods AP, screen guard SC. 2) *product asymmetry:* For a phone P, we would like to recommend a suitable phone case PC, but for a phone case PC, it may not be apt to recommend a phone P because customers typically would purchase a phone case only while owning a phone. Formally, product asymmetry can be represented as $aR_{cp}b \nRightarrow bR_{cp}a$. 3) *selection bias* is inherent to historical purchase data due to several factors like product availability, price etc. For example, while shopping for a phone case for phone P, customer c_1 might browse phone cases PC and $PC1$ ($PC2$ and $PC3$ are not shown to him owing to some of the factors mentioned above). Another customer c_2 might browse phone

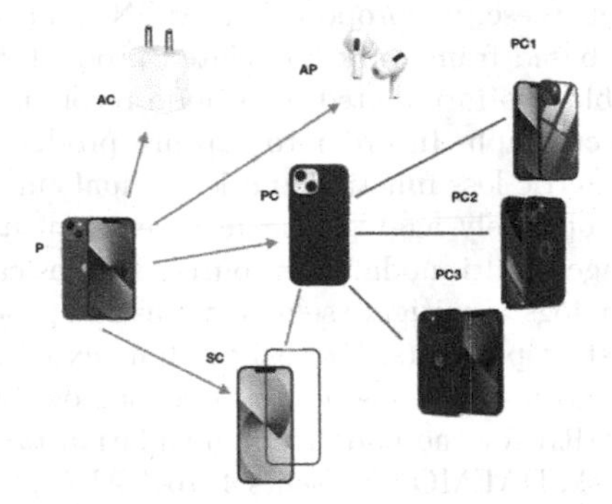

Fig. 1. Sample product graph. Green and blue edges represent co-purchase and co-view edges respectively. P refers to an iPhone 13, AC is an adapter, AP refers to AirPods and SC refers to a screen guard. Phone case PC is co-viewed with other phone cases $PC1$, $PC2$ and $PC3$; $PC1$, $PC2$ and $PC3$ are similar to PC. Undirected edges are bidirectional. (Color figure online)

[1] We use product, item and node interchangeably in this paper.

cases PC and $PC2$ ($PC1$ and $PC3$ are not presented to him). A third customer c_3 might browse phone cases PC and $PC3$ ($PC1$ and $PC2$ are not presented to him). Assume that all three customers eventually purchase PC; this could also be due to a bias in the list of products shown to each customer. In our example, across different customers, $PC, PC1, PC2$ and $PC3$ are co-viewed (similar). In order to correct for selection bias, we would like to recommend not only phone case PC, but also $PC1$, $PC2$ and $PC3$ as related products given a query product P. Formally, we want to uncover relationships of the form: a) $aR_{cp}b \wedge bR_{cv}c \implies aR_{cp}c$, b) $aR_{cv}b \wedge bR_{cp}c \implies aR_{cp}c$ in order to mitigate selection bias. 4) *cold-start product recommendations:* We not only need to suggest recommendations for existing products, but also suggest recommendations for newly launched i.e. cold-start products. Formally, given two existing products a and b, and a cold-start product c, we want to uncover relationships of the form: a) $cR_{like}a \wedge aR_{cp}b \implies cR_{cp}b$, b) $aR_{cp}b \wedge bR_{like}c \implies aR_{cp}c$. These two rules correspond to the case when c is the query product and the recommended product respectively. Observe that mitigating selection bias and tackling cold-start products deal with modelling transitive relationships, while preserving edge asymmetry. 5) Dealing with millions of products in our catalog demands a *scalable* solution for the recommendation task. These challenges are uncorrelated with each other making our problem non-trivial.

Given multi-modal data sources such as catalog metadata, product co-purchase data, anonymized browse behavioral logs, etc. product graphs serve as an excellent abstraction to seamlessly capture relationships between products. RPR boils down to the node recommendation problem [9,21] in directed product graphs. Recent work on node representation learning in directed graphs such as HOPE [9], APP [21], NERD [5], DGGAN [22] and Gravity Graph VAE [11] learn two embeddings for each node and utilize them for downstream tasks. Further, GNN models like DGCN [16] and DiGraphIB [15] generate real valued embeddings for each node, while MagNet [20] generates complex valued embeddings to preserve edge strength and direction. Selection bias in recommendation systems has been well studied [1] for different applications. Prior work has limitations across different dimensions as follows: a) there is no straightforward approach to extend popular GNN models like GCN [7], GraphSage [3], GAT [18], RGCN [12] to model directed graphs as they do not model product asymmetry. b) GNN models for directed graphs like DGCN [16], DGGAN [22], DiGraphIB [15] and MagNet [20] do not address the problem of recommendation for cold-start products. c) prior work does not address selection bias inherent to historical purchase data, and make specific assumption regarding the availability of unbiased data [19].

In this paper, we formulate RPR as a node recommendation task on a directed product graph. We propose DAEMON, **D**irection **A**war**E** Graph Neural Network **MO**del for **N**ode recommendation. In order to capture product asymmetry, our model generates dual embeddings for each product, by appropriately aggregating features from its neighborhood. We leverage customer co-view pairs from anonymized browse behavioral logs to recommend relevant products that customers clicked but didn't purchase to tackle selection bias. Specifically, during model training, we employ a novel asymmetric loss function which explicitly considers product co-purchases to capture product asymmetry, and product co-views to estimate co-purchase likelihood for products that were previously clicked but not purchased. We exploit product catalog metadata to generate product embeddings for cold-start products, and consequently generate recommendations for them. Further, as a byproduct, the proposed model is also able to suggest substitute product recommendations.

We perform exhaustive experiments offline on real-world datasets to evaluate the performance of our model. Results show that DAEMON outperforms state-of-the-art baselines by 30–160% in terms of HitRate and MRR for the node recommendation task which is the primary task of interest. For link prediction tasks, DAEMON outperforms state-of-the-art baselines by 4–16% and 3–6.5% in terms of AUC for the existence [20] and direction [20] link prediction task respectively. DAEMON delivers significant improvement in revenue and sales as measured through an A/B experiment. To summarize, we make the following contributions:

1. We formulate RPR as a node recommendation task in a directed product graph and propose a Graph Neural Network (GNN) based framework for related product recommendation. To this end, we present DAEMON, a novel GNN model, that leverages dual embeddings to capture node asymmetry in directed graphs.
2. In order to train the model, we employ a novel asymmetric loss function that explicitly deals with modelling co-purchase likelihood, product asymmetry and selection bias. We exploit product catalog metadata to deal with the issue of cold-start products. This is the first work that jointly addresses product asymmetry, generating cold start recommendations and mitigating selection bias in historical purchase data.
3. Offline evaluation shows that DAEMON outperforms state-of-the-art models for node recommendation and link prediction tasks. DAEMON derives significant improvement in product sales and revenue as measured through an A/B experiment.

2 Related Work

Our problem conceptually relates to the node recommendation problem [9,21] in directed graphs. Prior work uses random walk based techniques to model node relationships in directed graphs. VERSE [17], HOPE [9] propose learning two embeddings for each node to preserve higher order proximity and consequently, node asymmetry in directed graphs. APP [21] captures asymmetry by preserving Rooted PageRank between nodes by relying on random walk with restart strategy. ATP [13] addresses the problem of question answering by embeddings nodes of directed graph by preserving asymmetry. However, their approach is restricted to only directed acyclic graphs (DAGs), while an e-commerce product graph can be cyclic. NERD [5] learns a pair of role specific embeddings for each node using a alternating random walk strategy to model edge strength and direction in directed graphs.

Recent work has seen GNN's being designed for directed graphs. DGCN [16] extends the spectral-based GCN model to directed graphs by using first and second-order proximity to expand the receptive field of the convolution operation. APPNP [8] uses a GCN model to approximate personalized PageRank. DGCN [16] uses one first-order matrix and two second order proximity matrices to model asymmetry. DiGraphIB [15] builds upon the ideas of [16] and constructs a directed Laplacian of a PageRank matrix. It uses an inception module to share information between receptive fields. Gravity GAE [11] is inspired from Graph Auto Encoders [6] by applying the idea of gravity to address link prediction in directed graphs. DGGAN [22] is based on Generative Adversarial Network by using a discriminator and two generators to jointly learn each node's source and target embedding. MagNet [20] proposes a GNN for directed graphs based on a complex Hermitian matrix. The magnitude of entries in the complex matrix encodes the graph structure while the directional aspect is captured in the phase parameter. Prior work in this space does not deal with the the issue of cold-start products. Selection bias in recommendation systems has been well studied [1] for different applications. However, prior work does not address the bias inherent to historical purchase data. Further, we make no assumption regarding the availability of unbiased data in order to combat selection bias unlike [19]. Ours is the first work that jointly: a) learns node representations in directed graphs to capture edge strength and direction, b) generates recommendations for cold-start products by leveraging catalog metadata, and c) mitigates selection bias in product co-purchase directed graphs.

3 Related Product Recommendation Problem

We present relevant notation in Table 1. Let G be a directed product graph with products P as the nodes and directed edges corresponding to a co-purchase or co-view relationship between the products (refer Sect. 4.1). Further, every product i ($i \in P, \forall i$) has an input feature X_i from the product catalog metadata. Given a query node[2] q, the goal is to recommend R_k^q, top-k related products that have a high likelihood to be bought together with q.

This corresponds to the node recommendation problem [9,21] in directed graphs. Note that while generating related product recommendations, we must jointly capture product co-purchase likelihood and preserve product asymmetry.

4 Proposed Framework

We show the proposed framework in Fig. 2. We leverage the co-purchase data CP and co-view data CV to create the product graph G. We represent each product by an input feature based on its metadata from catalog (product name, product

[2] We use product and node interchangeably in this paper based on the context.

Table 1. Notation

Notation	Description
P	set of products in catalog
$i \in P$	product i in catalog
X_i	input feature of product i
θ_i^s	source embedding of product i
θ_i^t	target embedding of product i
G	directed product graph
$q \in P$	query product
R_k^q	top-k related products for query product q
CP	product co-purchase pairs
CV	product co-view pairs
E_{cp}	product co-purchase edges
E_{cv}	product co-view edges

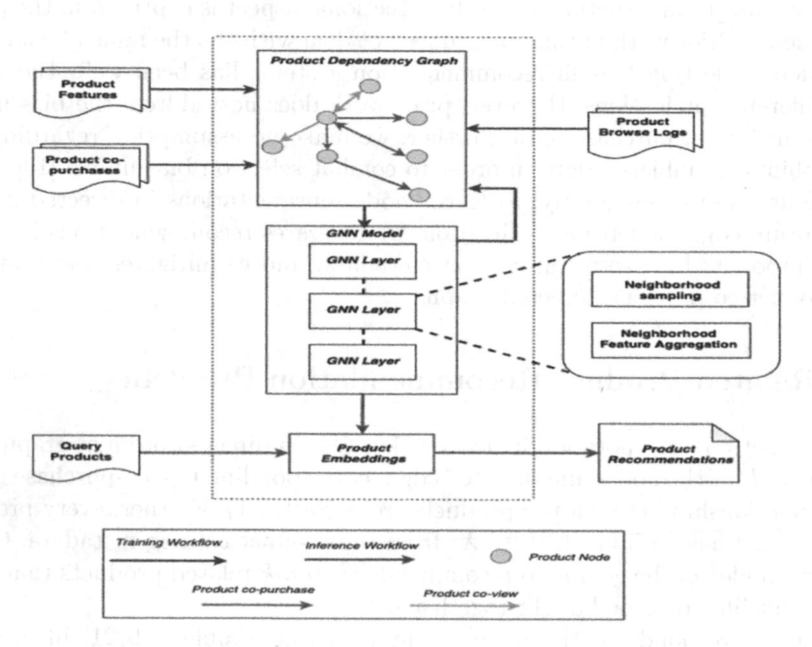

Fig. 2. Proposed Framework for Related Product Recommendation. The computational graphs corresponding to the source (s) and target (t) embedding of node A in the product graph is shown.

description, product type etc.). In order to jointly model product co-purchase likelihood and product asymmetry, we represent each product using dual i.e. source and target embeddings. We train DAEMON using an asymmetric loss function. The trained model generates product embeddings by appropriately aggregating features from its neighborhood in G. We show the computational graphs corresponding to the source and target embedding of node A in Fig. 2.

For every node u, we generate a pair of embeddings (denoted by u-s and u-t in Fig. 2). In order to address node asymmetry, we use the following rationale: a) while generating the source embedding of a node, we aggregate information from the target embedding of its out-neighbors, b) while generating the target embedding of a node, we aggregate information from the source embedding of its in-neighbors. Compared to undirected graphs, the computational graph corresponding to the source and target embedding of each node is different in directed graphs. Consequently, this trick enables us to model edge strength and edge direction efficiently in directed graphs. Given a query product q, we generate its source embedding and perform a nearest neighbor lookup in the target embedding space to recommend top-k related products for q. We explain each component in detail next.

4.1 Product Graph Construction

Given a set of product co-purchase pairs CP, we create the co-purchase edges $E_{cp} = \{(u, v) | \forall\ u, v \in P \wedge uR_{cp}v\}$. We use E_{cp} to model the product co-purchase likelihood. However, E_{cp} is prone to selection bias.

In order to combat this issue, we exploit anonymized browse behavioral logs, to create product co-view pairs CV, which we use to create the set of co-view edges $E_{cv} = \{(u, v) | \forall\ u, v \in P \wedge uR_{cv}v\}$. This lets us construct a directed product graph $G = (P, \{E_{cp} \cup E_{cv}\})$, which contains both co-purchase and co-view relationships between products.

For any three products $a, b, c \in P$, G contains product relationships of the form: 1) $aR_{cp}b \nRightarrow bR_{cp}a$ 2) $aR_{cp}b \wedge bR_{cv}c \implies aR_{cp}c$ 3) $aR_{cv}b \wedge bR_{cp}c \implies aR_{cp}c$. The goal is to design a GNN model that learns these kind of relationships from G, while preserving co-purchase likelihood; rule 1 corresponds to product asymmetry, and rules 2-3 are specific to the case of selection bias. *Observe that E_{cp} and E_{cv} are not correlated with each other, and we need to provide differential treatment to E_{cp} and E_{cv} during neighborhood feature aggregation.*

4.2 DAEMON: Proposed GNN Model

We first describe the product embedding generation procedure i.e. forward pass of the model assuming that the model is already trained (Sect. 4.2.1). We then describe how the model parameters can be learned using stochastic gradient descent and backpropagation techniques (Sect. 4.2.2).

4.2.1 Product Embedding Generation: Forward Pass

Algorithm 1 describes the forward pass to generate the source and target embedding of each product. It expects the product graph G and product features X as input. We set the initial source and target embedding of each product to its input feature. Let $(h_u^s)^l$ and $(h_u^t)^l$ denote a product's source and target representation in the lth step of Algorithm 1. First, for each product u, extract its co-purchase and co-view neighbors. The source hidden node representation of u

Algorithm 1. DAEMON product embedding generation (i.e. forward pass)

Input: product graph $G = (P, \{E_{cp} \cup E_{cv}\})$; input product features $\{X_u, \forall \in P\}$; Number of GNN layers L; weight matrices $W^l, \forall l \in \{1, 2, .., L\}$
Output: source embedding θ_u^s and target embedding θ_u^t, $\forall u \in P$
1: $(h_u^s)^0 \leftarrow X_u$; $(h_u^t)^0 \leftarrow X_u \forall u \in P$
2: **for** l=1,...L **do**
3: **for** $u \in P$ **do**
4: $(h_u^s)^l \leftarrow \sigma\left(\sum_{(u,v)\in E_{cp}} (h_v^t)^{l-1} W^l\right) + \sigma\left(\sum_{(u,v)\in E_{cv}} (h_v^s)^{l-1} W^l\right)$
5: $(h_u^t)^l \leftarrow \sigma\left(\sum_{(v,u)\in E_{cp}} (h_v^s)^{l-1} W^l\right) + \sigma\left(\sum_{(v,u)\in E_{cv}} (h_v^t)^{l-1} W^l\right)$
6: **end for**
7: $(h_u^s)^l \leftarrow (h_u^s)^l / ||(h_u^s)^l||_2, \forall u \in P$
8: $(h_u^t)^l \leftarrow (h_u^t)^l / ||(h_u^t)^l||_2, \forall u \in P$
9: **end for**
10: $\theta_u^s \leftarrow (h_u^s)^L, \forall u \in P$
11: $\theta_u^t \leftarrow (h_u^t)^L, \forall u \in P$

in the lth step is aggregated as a linear combination of two non-linear terms i.e. non-linear aggregate of target representation of its co-purchased out-neighbors and non-linear aggregate of source representation of its co-viewed out-neighbors from the $(l-1)^{th}$ step. W^l corresponds to fully connected layer at step l with non-linear ReLU activation function σ (line 4). We perform a similar neighborhood aggregation to estimate a node's target representation during the lth step (line 5). The source and target node representations of step l is used in the next step. We normalize the embeddings to a unit norm (lines 7, 8). We repeat this process for L steps to generate the final source and target product representation of products $\{\theta_u^s, \theta_u^t\}$ $\forall u \in P$ (line 10, 11). We leverage the generated product embeddings to recommend related products. We also show relevant properties of the embeddings in Lemmas 1 and 2.

Related Product Recommendation: Given a query product q, we use θ_q^s, the source embedding of q, to perform a nearest neighbor lookup in the target embedding space of all the products to recommend top-k set of related products denoted by R_k^q. Specifically, for a query product $q \in P$, we compute a relevance score with respect to a candidate product $v \in P$ as shown in Eq. 1.

$$rel(q, v) = (\theta_q^s)^\intercal (\theta_v^t) \tag{1}$$

Observe that $rel(q, v) \neq rel(v, q)$ which helps capturing product asymmetry.

Cold-start Related Product Recommendation: Our model can also be used for *cold-start* product recommendation. Given a cold-start product c, we perform a neighborhood lookup to find similar existing (warm-start) products i.e. $\{c_1, c_2, ..c_k\}$ based on its input product features (X). We augment the edges of the form $\{(c, c_1), (c, c_2), ..(c, c_k)\}$ to the product graph G. Further, c has an input feature X_c from catalog metadata. We pass the subgraph corresponding to the cold-start product c to the GNN model and generate the source and target embedding of c using Algorithm 1. We use θ_c^s to probe the target embeddings to recommend related products for c using Eq. 1.

4.2.2 Learning DAEMON Parameters: Backward Pass

In order to train the model in an unsupervised manner, we employ a novel asymmetric loss function shown in Eq. 2. This helps us capture different aspects as follows:

1. When $(u, v) \in E_{cp}$, the model forces the source embedding of u to be similar to the target embedding of v, and distant from the target embedding of a disparate product z (terms 1 and 2). Negative samples are generated using a uniform distribution P_r. This models product co-purchase likelihood and asymmetry.
2. In order to force product asymmetry, for every one-way directed co-purchase edge i.e. $(u, v) \in E_{cp} \wedge (v, u) \notin E_{cp}$, our model assigns a high score to the edge (u, v) and a low score to the edge (v, u) (terms 3 and 4).
3. For similar products i.e. $(u, v) \in E_{cv}$, our model forces both the source and target embeddings of u and v to be similar (terms 5 and 6). This property is useful in cold-start recommendation and mitigating selection bias.

$$loss = -\left\{\sum_{(u,v)\in E_{cp}} log(\sigma(\theta_u^s \cdot \theta_v^t)) + \sum_{\substack{n_s=1 \\ z\sim P_r(P), u\neq z,}}^{n_k} log(\sigma(1-\theta_u^s \cdot \theta_z^t))\right\}$$
$$-\left\{\sum_{\substack{(u,v)\in E_{cp}\wedge \\ (v,u)\notin E_{cp}}} log(\sigma(\theta_u^s \cdot \theta_v^t)) + \sum_{\substack{(u,v)\in E_{cp}\wedge \\ (v,u)\notin E_{cp}}} log(\sigma(1-\theta_v^s \cdot \theta_u^t))\right\}$$
$$-\left\{\sum_{(u,v)\in E_{cv}} log(\sigma(\theta_u^s \cdot \theta_v^s)) + \sum_{(u,v)\in E_{cv}} log(\sigma(\theta_u^t \cdot \theta_v^t))\right\} \quad (2)$$

Lemma 1. *The embeddings generated by DAEMON capture product co-purchase likelihood and product asymmetry.*

Proof. When $(a, b) \in E_{cp} \leftrightarrow aR_{cp}b$, our model assigns a high score to $\theta_a^s.\theta_b^t$ compared to $\theta_a^s.\theta_z^t$ i.e. $\theta_a^s.\theta_b^t >> \theta_a^s.\theta_z^t$ (for a random product z). This implies that the co-purchase likelihood between related products a and b is preserved. Further, when $(a, b) \in E_{cp} \wedge (b, a) \notin E_{cp} \leftrightarrow aR_{cp}b \wedge \sim bR_{cp}a$, our model will assign a higher score to $\theta_a^s.\theta_b^t$ compared to $\theta_b^s.\theta_a^t$ i.e. $\theta_a^s.\theta_b^t >> \theta_b^s.\theta_a^t$, thereby capturing product asymmetry.

Lemma 2. *The embeddings generated by DAEMON helps mitigating selection bias.*

Proof. For $a, b, c \in P$, G consists paths of the form $\{a, b, c\}$ where, $(a, b) \in E_{cp} \leftrightarrow aR_{cp}b$ and $(b, c) \in E_{cv} \leftrightarrow bR_{cv}c$. Our model assigns a high score to $\theta_a^s.\theta_b^t$ as $(a, b) \in E_{cp}$. As $(b, c) \in E_{cv}$, our model assigns a high score to $\theta_b^s.\theta_c^s$ and $\theta_b^t.\theta_c^t$. Hence, our model assigns a high score to $\theta_a^s.\theta_c^t = \theta_a^s.\theta_b^t \times \theta_b^t.\theta_c^t$, implying $aR_{cp}c$. We can present a similar argument to show that the embeddings capture $aR_{cv}b \wedge bR_{cp}c \implies aR_{cp}c$. In this manner, our model mitigates selection bias.

In terms of scalability, we employ minibatch sampling both during training and inference procedure to generate product embeddings. We use FAISS [4] to perform efficient nearest neighbor lookup. Our model uses $O(|P|)$ space to store embeddings corresponding to $|P|$ products. We discuss the results in the next section.

5 Experiments

As previously discussed, RPR problem is equivalent to the node recommendation task in directed graphs. In this section, we evaluate the performance of DAEMON against state-of-the-art models in directed graphs. Specifically, we aim to answer the following evaluation questions:

- **EQ1:** Is DAEMON able to improve the node recommendation performance on real-world e-commerce datasets?
- **EQ2:** How effective is DAEMON in capturing the co-purchase likelihood between products?
- **EQ3:** Is DAEMON able to capture product asymmetry?
- **EQ4:** How effective is DAEMON for cold-start product recommendation?
- **EQ5:** Is DAEMON able to combat selection bias?
- **EQ6:** Semantically, what kind of relationships between products can be extracted using the source and target embeddings generated by DAEMON?

To this end, we introduce the datasets, baselines and follow it up with experiments to answer these questions.

5.1 Experimental Setting

5.1.1 Datasets

We extract two real-world datasets sampled from different emerging marketplaces in Amazon. For each dataset, we construct a graph consisting of products as nodes and edges corresponding to either a co-purchase or co-view relationship as explained in Sect. 4.1. Note that we use anonymized browse logs while creating product co-view pairs. The statistics of the graph datasets is shown in Table 2. These graphs consist 2–5.5M nodes and 14–32M edges and are directed ($\sim 75\%$ of the edges are directed). In both the datasets, we represent each product using input features of size 384 and 512 respectively. These datasets are sampled, and not reflective of production traffic in terms of scale.

Table 2. Summary of graph datasets employed

Dataset	# Nodes	# Edges	Average Degree	%Directed Edges	# Co-purchase pairs	# Co-view pairs	Input Product Feature Dimension
G1	1.98 M	14.1 M	7.3	76.33	7 M	7.1 M	384
G2	5.5 M	31.7 M	5.76	79.97	13.2 M	18.5 M	512

5.1.2 Implementation Details

We implemented DAEMON using DGL and PyTorch. We vary the learning rate in $\{10^{-1}, 10^{-2}, 10^{-3}, 10^{-4}\}$ and observed 10^{-4} using Adam's optimizer to work the best. We use $L = 3$ layer GNN model for DAEMON. We use minibatches of size 1024 during model training (Sect. 4.2.2) and inference (Sect. 4.2.1). We also use layer-wise neighborhood sampling, i.e. 20 neighbors for the first layer and 10 random neighbors for the subsequent layers. After generating the product embeddings, we perform nearest neighbor lookup to suggest top-k related products for k values in $\{5, 10, 20\}$. All the experiments were conducted on a 64-core machine with a 488 GB RAM running Linux. For all models, we learn 64 dimensional product embeddings trained for a maximum of 30 epochs. We repeat all the experiments 10 times and report the average value across the runs.

5.1.3 Baselines

We previously discussed state-of-the-art models in directed graphs (Sect. 2). In order to evaluate our proposed model, DAEMON, we choose the most competitive baselines as follows:

1. We choose APP [21] and NERD [5] as they deliver superior performance compared to deepwalk [10], node2vec [2], LINE [14], HOPE [9] and VERSE [17].
2. Gravity GAE [11] and DGGAN [22] outperform deepwalk [10], node2vec [2], LINE [14], APP [21], HOPE [9] and VGAE [6].
3. MagNet [20] is the latest state-of-the-art GNN model which delivers best results compared to GraphSage [3], GAT [18], DGCN [16] and APPNP [8].

In summary, we choose APP [21], NERD [5], Gravity GAE [11], DGGAN [22] and MagNet [20] as the most competitive baselines. We use publicly released code repositories for all the employed baselines. Further, we employ parameter tuning to choose the best parameters for each baseline.

However, all the baselines are suitable only for homogeneous directed graphs. For fair comparison, we restrict the evaluation to co-purchase directed graphs (edges corresponding to E_{cp}), and evaluate DAEMON against the baselines (Sects. 5.2, 5.3). For the case of complete data ($E_{cp} \cup E_{cv}$), we compare DAEMON against state-of-the-art R-GCN [12] model in Sect. 5.4.

5.1.4 Experiment Setup

In order to answer the evaluation questions, we setup the experiments (similar to [3,20,21]) as follows:

- [EQ1]. Node Recommendation Task (Sect. 5.2): For each graph dataset, we use 75%, 5% and 20% non-overlapping edges for training, validation and testing respectively. For a ground-truth recommendation (u, v) (from the test data), where u is the query node, we retrieve top-k node recommendations (R_k^q) suggested by each model. In order to evaluate the quality of recommendations, we use HitRate@k and MRR@k for k in $\{5, 10, 20\}$. This helps answering EQ1.
- [EQ2]. Existential link prediction Task (Sect. 5.3): For each graph dataset, we use 75%, 5% and 20% non-overlapping edges for training, validation and testing respectively. In this task, we want to capture the edge score predicted by each model; a good model assigns higher scores to existing links compared to non-existing links. We evaluate the performance of various models using AUC for this task. This helps answering EQ2.
- [EQ3]. Directed link prediction Task (Sect. 5.3): For each graph dataset, we use 75%, 5% and 20% non-overlapping edges for training, validation and testing respectively. In this task, we consider only one-way directed edges ($(u, v) \in G \wedge (v, u) \notin G$) as the test edges. We reverse the direction of the test edges to create negative links for testing. We want to evaluate how different models capture the edge direction; a good model assigns a high score to the correct edge (positive links in the test set) and a low score to the reverse edge (negative links in the test set). We evaluate the model performance using AUC for this task. This helps answering EQ3.
- [EQ4]. Cold-start Recommendation Task (Sect. 5.4): For each graph dataset, we use subgraphs corresponding 75%, 5% and 20% non-overlapping nodes for training, validation and testing respectively. For evaluation, we take the same approach as employed in the node recommendation task. This helps answering EQ4.
- [EQ5]. Selection-bias Recommendation Task (Sect. 5.4): For each graph dataset, we use edges corresponding 75%, 5% and 20% non-overlapping nodes for training, validation and testing respectively. *Further, we add edges corresponding to transitive relationships $aR_{cp}b \wedge bR_{cv}c \implies aR_{cp}c$ to the test set.* In order to mitigate selection bias, a model needs to recommend c as a related product for a as these transitive relationships are present in the training graph. For evaluation, we take the same approach as employed in the node recommendation task. This helps answering EQ5.

5.2 EQ1. Node Recommendation Task on Co-purchase Data

The results[3] for the node recommendation task is shown in Table 3. We see that DAEMON performs the best for this task in terms of both HitRate and MRR. DAEMON yields more gains on the bigger $G2$ dataset compared to the $G1$ datset. Further, DGGAN fails to complete model training in 48 h. It does not scale to real-world datasets; the paper show their model efficacy on smaller datasets (the biggest graph has 15K nodes). Further, Gravity GAE runs out of

[3] Results are relative to the co-purchase baseline and absolute numbers are not presented due to confidentiality.

memory (488 GB RAM) during model training. This is due to one-hot feature encoding employed by the model to represent node input features i.e. each node is represented as a million valued embedding initially. APP delivers the second best performance. MagNet is not applicable for node recommendation task, and we compare against it for the link prediction tasks.

Table 3. Node recommendation. Best results are in **bold** and second are underlined

Dataset	Model	HitRate@k			MRR@k		
		5	10	20	5	10	20
G1	APP	2.14x	4.01x	6.95x	1.02x	1.26x	1.46x
	NERD	0.62x	1.94x	0.0314	0.61x	0.71x	0.79x
	DAEMON	**3.07x**	**5.54x**	**8.98x**	**1.45x**	**1.77x**	**2.01x**
	%Gain	43.4	38.15	29.2	42.15	40.47	37.6
G2	APP	1.23x	1.99x	3.31x	0.67x	0.77x	0.85x
	NERD	1.28x	1.85x	2.63x	0.71x	0.78x	0.84x
	DAEMON	**3.51x**	**5.87x**	**8.6x**	**1.71x**	**2.02x**	**2.21x**
	%Gain	174	194	159.8	155.2	162.3	160

5.3 [EQ2, EQ3.] Link Prediction Tasks on Co-purchase Data

Table 4. Link prediction (%). Best results are in **bold** and second are underlined

Dataset	Model	Existence Prediction (AUC %)	Direction Prediction (AUC %)
G1	APP	34.9x	0.1x
	NERD	23.3x	7.1x
	MagNet	36.1x	12.65x
	DAEMON	**40.31x**	**14.72x**
	Δ Gain	4.2	2.25
G2	APP	5.27x	0.15x
	NERD	14.45x	2.48x
	MagNet	18.26x	6.62x
	DAEMON	**34.28x**	**12.53x**
	Δ Gain	16	6.6

We present the performance[4] of different models in Table 4 for two link prediction tasks. We see that DAEMON outperforms all baselines on both these tasks.

[4] Results are relative to the co-purchase baseline and absolute numbers are not presented due to confidentiality.

MagNet displays the second best performance. We make a note that predicting existence of links is a simpler task when compared to predicting the correct edge direction; this is also observed in the results. As explained previously, DGGAN fails to complete model training in 48 h and Gravity GAE runs out of memory (488 GB RAM) during model training. These results indicate that the proposed model is able to jointly model co-purchase likelihood and product asymmetry compared to the baselines which answers **EQ2** and **EQ3**.

5.4 [EQ4, EQ5, EQ6.] Ablation Study on G1 Graph

We evaluate[5] DAEMON on the complete graph in this section to analyze its performance for the case of cold-start product recommendation task (Table 5) and the case of mitigating selection bias (Table 6). Observe from Table 5 that DAEMON delivers significant performance gains over R-GCN for cold-start recommendation evaluation. Although, both models were fed with the same cold-start graphs, observe that R-GCN performs poorly. Results indicate how leveraging catalog product information to create subgraphs pertaining to cold-start products aid in boosting the performance of our model. This answers **EQ4**.

Table 5. Cold-start evaluation on G1 graph

Model	HitRate@k			MRR@k		
	5	10	20	5	10	20
R-GCN	x	x	x	x	x	x
DAEMON	**34.14x**	**22.02x**	**15.52x**	**98.05x**	**86.75x**	**51.4x**

Table 6 shows the results pertaining to selection bias. Observe that when the model is trained only the co-purchase edges, it results in a performance dip. This is because the transitive relationships crafted in the test set cannot be captured when trained only on the co-purchase edges. However, R-GCN trained on both co-purchase and co-view edges delivers a poor performance which confirms our claim that we need to provide differential treatments to co-purchase and co-view edges during feature aggregation. DAEMON delivers the best performance compared to R-GCN model. These results demonstrate how DAEMON is able to mitigate selection and this answers **EQ5**.

In order to answer **EQ6**, we assume a query product q. We generate a related product v, when $\theta_q^s.\theta_v^t$ is the highest $\forall v \in P$. Similarly, given a query product q, we generate a similar product v, when $\theta_q^s.\theta_v^s$ is the highest $\forall v \in P$. Observe from Table 7 the kind of recommendations generated. This shows that we can leverage both the source and target embeddings generated by DAEMON to recommend not only related products, but also similar items as a byproduct.

[5] Results are relative to the R-GCN baseline and absolute numbers are not presented due to confidentiality.

Table 6. Selection bias evaluation on G1 graph

Model	HitRate@k			MRR@k		
	5	10	20	5	10	20
R-GCN (co-purchase+co-view)	x	x	x	x	x	x
DAEMON(co-purchase)	3.1x	3.26x	3.5x	4.26x	4.02x	3.94x
DAEMON (co-purchase+co-view)	**3.43x**	**3.5x**	**3.59x**	**4.58x**	**4.32x**	**4.21x**

Table 7. Sample product recommendations generated using DAEMON

Sl. No	Query product	Related product Recommendation	Similar product Recommendation
1	amway attitude sunscreen cream 100 g	amway attitude face wash for dry skin - 100 ml	amway attitudes sunscreen cream 100 g x 2 = 200 gm
2	oppo reno phone 6 pro 5 g stellar black 12 gb ram 256 gb storage	lustree oppo reno 6 pro 5 g bumper back cover case	oppo reno phone 6 5 g stellar black 8 gb ram 128 gb storage
3	paseo tissues pocket hanky 6 packs 3 ply	origami wet wipes wet tissue wet facial tissue	inkulture pocket hanky tissue paper white pack of 10
4	yiwoo 5 pieces false eyelashes curler	ardell duo individual lash adhesive white 7 g	vega premium eye lash curler
5	skinn by titan steele fragrance for men 100 ml	blue nectar natural vitamin c face cream for glowing skin	skinn by titan fragrance for men 20 ml

5.5 Online Platform Performance

We further evaluate the performance of DAEMON in production environment by conducting an A/B test in two different marketplaces. For the control group, we use an incumbent approach based on product co-purchases, while for the treatment group, we show the recommendations generated from DAEMON. We run the experiments for 4 weeks and observe +170% improvement on product sales and +190% improvement on profit gain. All the results are statistically significant with $p\text{-value} < 0.05$. These results show the product recommendations generated from DAEMON can significantly improve customer shopping experience in discovering potentially related products of interest.

6 Conclusion and Future Work

In this paper, we propose DAEMON, a novel Graph Neural Network based framework for related product recommendation, wherein the problem is formulated as a node recommendation task on a directed product graph. In order to train the model, we employ an asymmetric loss function by modelling product co-purchase likelihood, capturing product asymmetry and mitigating selection bias. We leverage the trained GNN model to learn dual embeddings for each product, by appropriately aggregating features from its neighborhood. Extensive offline experiments show that DAEMON outperforms state-of-the-art baselines for the node recommendation and link prediction tasks against state-of-the-art baselines. In the future, we will explore the possibility of representing products using complex valued embeddings and extend DAEMON to the case of complex embeddings to improve the performance.

References

1. Chen, J., Dong, H., Wang, X., Feng, F., Wang, M., He, X.: Bias and debias in recommender system: a survey and future directions. arXiv preprint (2020)
2. Grover, A., Leskovec, J.: node2vec: Scalable feature learning for networks. In: SIGKDD, pp. 855–864 (2016)
3. Hamilton, W., Ying, Z., Leskovec, J.: Inductive representation learning on large graphs. In: NIPS 30 (2017)
4. Johnson, J., Douze, M., Jégou, H.: Billion-scale similarity search with GPUs. IEEE Trans. Big Data **7**(3), 535–547 (2021)
5. Khosla, M., Leonhardt, J., Nejdl, W., Anand, A.: Node representation learning for directed graphs. In: ECML PKDD, pp. 395–411 (2019)
6. Kipf, T.N., Welling, M.: Variational graph auto-encoders. In: NIPS Workshop on Bayesian Deep Learning (2016)
7. Kipf, T.N., Welling, M.: Semi-supervised classification with graph convolutional networks. In: ICLR (2017)
8. Klicpera, J., Bojchevski, A., Günnemann, S.: Predict then propagate: graph neural networks meet personalized pageRank. In: ICLR (2019)
9. Ou, M., Cui, P., Pei, J., Zhang, Z., Zhu, W.: Asymmetric transitivity preserving graph embedding. In: Proceedings of the 22nd ACM SIGKDD (2016)
10. Perozzi, B., Al-Rfou, R., Skiena, S.: Deepwalk: online learning of social representations. In: Proceedings of the 20th ACM SIGKDD, pp. 701–710 (2014)
11. Salha, G., Limnios, S., Hennequin, R., Tran, V.A., Vazirgiannis, M.: Gravity-inspired graph autoencoders for directed link prediction. In: CIKM, pp. 589–598 (2019)
12. Schlichtkrull, M., Kipf, T.N., Bloem, P., van den Berg, R., Titov, I., Welling, M.: Modeling relational data with graph convolutional networks. In: Gangemi, A., et al. (eds.) ESWC 2018. LNCS, vol. 10843, pp. 593–607. Springer, Cham (2018). https://doi.org/10.1007/978-3-319-93417-4_38
13. Sun, J., Bandyopadhyay, B., Bashizade, A., Liang, J., Sadayappan, P., Parthasarathy, S.: ATP: directed graph embedding with asymmetric transitivity preservation. In: AAAI (2019)
14. Tang, J., Qu, M., Wang, M., Zhang, M., Yan, J., Mei, Q.: Line: Large-scale information network embedding. In: WWW, pp. 1067–1077 (2015)
15. Tong, Z., Liang, Y., Sun, C., Li, X., Rosenblum, D., Lim, A.: Digraph inception convolutional networks. NIPS **33**, 17907–17918 (2020)
16. Tong, Z., Liang, Y., Sun, C., Rosenblum, D.S., Lim, A.: Directed graph convolutional network. arXiv preprint arXiv:2004.13970 (2020)

17. Tsitsulin, A., Mottin, D., Karras, P., Müller, E.: Verse: versatile graph embeddings from similarity measures. In: WWW 2018, pp. 539–548 (2018)
18. Veličković, P., Cucurull, G., Casanova, A., Romero, A., Lio, P., Bengio, Y.: Graph attention networks. In: ICLR (2018)
19. Wang, X., Zhang, R., Sun, Y., Qi, J.: Combating selection biases in recommender systems with a few unbiased ratings. In: WSDM, pp. 427–435 (2021)
20. Zhang, X., He, Y., Brugnone, N., Perlmutter, M., Hirn, M.: MagNet: a neural network for directed graphs. In: NIPS 34 (2021)
21. Zhou, C., Liu, Y., Liu, X., Liu, Z., Gao, J.: Scalable graph embedding for asymmetric proximity. In: AAAI (2017)
22. Zhu, S., Li, J., Peng, H., Wang, S., Yu, P.S., He, L.: Adversarial directed graph embedding. In: AAAI (2021)

A U-Shaped Hierarchical Recommender by Multi-resolution Collaborative Signal Modeling

Peng Yi[1], Xiongcai Cai[1,2](✉), and Ziteng Li[1]

[1] UNSW Sydney, Sydney, NSW 2052, Australia
x.cai@unsw.edu.au
[2] Techcul Research, Sydney, Australia

Abstract. Items (users) in a recommender system inherently exhibit hierarchical structures with respect to interactions. Although explicit hierarchical structures are often missing in real-world recommendation scenarios, recent research shows that exploring implicit hierarchical structures for items (users) would largely benefit recommender systems. In this paper, we model user (item) implicit hierarchical structures to capture user-item relationships at various resolution scales resulting in better preferences customization. Specifically, we propose a U-shaped Graph Convolutional Network-based recommender system, namely UGCN, that adopts a hierarchical encoding-decoding process with a message-passing mechanism to construct user (item) implicit hierarchical structures and capture multi-resolution relationships simultaneously. To verify the effectiveness of the UGCN recommender, we conduct experiments on three public datasets. Results have confirmed that the UGCN recommender achieves overall prediction improvements over state-of-the-art models, simultaneously demonstrating a higher recommendation coverage ratio and better-personalized results.

Keywords: Recommender systems · Hierarchical model · Embedding learning

1 Introduction

Recommender systems implemented by a wide range of online businesses are critical in alleviating the information overload problem [30]. The basic point of building a recommender system is to model the user-item relationship based on previous interactions. As one of the most widely used recommendation techniques, model-based collaborative filtering (CF) algorithms extract useful information about user-item connectivity by projecting users and items into a shared latent space and representing them with corresponding low-dimensional embeddings [9,15]. In other words, model-based CF describes user preferences for items through the inner product of the projected embeddings in the shared space [22]. Therefore, the algorithm for projecting users/items into certain embeddings plays a key role in the success of these model-based CF recommenders.

M.-R. Amini et al. (Eds.): ECML PKDD 2022, LNAI 13713, pp. 558–573, 2023.
https://doi.org/10.1007/978-3-031-26387-3_34

There are many approaches to learn user/item embeddings, including classical matrix factorization (MF) [14], neural network-based models, and graph convolutional networks (GCNs)-based models [8,26]. Although these models present competitive performance, they also have the obvious drawback of neglecting the inherent hierarchy of items/users. More specifically, most existing embedding learning models only focus on modeling individual items or users independently, which cannot fully capture the items' (or users') hierarchies information and hence fail to precisely model personalized preference. Take the movie Frozen as an example, it belongs to the subgenre "Family Animation" and can be further categorized into the genre "Animation", demonstrating a hierarchical structure of "individual → sub-genre → genre". Similarly, users may present a similar hierarchy of "individual → occupation → age". Since items of the same subgenre (or genre) are likely to have similar attributes, they are likely to acquire similar preferences [25]. Thus, hierarchical information about items or users can very likely improve the preference modeling process of recommender systems.

It is worth pointing out that in real-world recommendation scenarios, explicit hierarchies are often not applicable [24]. For this reason, some researchers have used MF-based models to learn implicit item/user hierarchies [25]. Specifically, the implicit item/user hierarchy can be easily obtained by decomposing the original item/user embedding matrix into several smaller and more compressed item/user matrices, respectively. The success of these models confirms the validity of exploring implicit hierarchies [2]. However, there are still two limitations that limit the performance of these models, which can be summarized as the limited representation based on a simple MF model and the neglect of diverse collaboration signals in different hierarchical levels. More specifically, recent studies have argued that the basic MF-based models simply adopt the interaction between users and items as the ultimate objective function, but ignore the potential similarity signals stored in the interaction [1]. From this perspective, models based on graphical convolutional networks (GCNs), which can naturally model the user-item relationship in the interaction graph structure, propose a more reasonable and efficient way to build recommender systems. In addition to the limitations imposed by simple models, existing models that focus solely on building two separate multilayer architectures are problematic. More directly, exploring parallel user/item hierarchies only shows the multi-level relationships within a user or item. However, the core information contained in the hierarchy is the multi-resolution collaborative signals stored in different levels. For example, given the interaction that a user "Alice" like a movie "Frozen", exploring the structure of "Alice → Second Grade Nursery Student → Kids" and "Frozen → Family Animation → Animation" merely displays the inner association for a user or an item. Indeed, this interaction not only reveals "Alice's" unique preference for "Frozen", but also represents a shared interest that "Second Grade Nursery Students" may prefer "Family Animation", or announces a more general signal that "Kids" prefer "Animation". Clearly, these multi-resolution collaborative signals will help generate better recommendation predictions.

Based on the above analysis, we propose an implicit user/item hierarchical exploration model, i.e., UGCN, which utilizes a u-shaped hierarchical graph convolution model and is able to capture user-item collaboration signals in different resolution scales. Specifically, the u-shaped structure uses a pooling operation to adaptively compress the original user-item interaction graph into smaller compressed graphs. In other words, by merging similar nodes in the fine-grained graph into a new group node in the coarse-grained graph, collaboration signals at different resolution scales can be better captured by graph convolution operations on different compressed graphs. Then, the captured collaborative signals are gradually projected back to the original nodes through symmetric unpooling operations. With the multi-resolution collaborative signal captured in final item/user embeddings, better personalized recommendations can be ultimately generated.

In summary, this work has the following main contributions.

- We propose the UGCN recommender, a U-shaped hierarchical recommender based on graph convolutional networks, which captures diverse collaborative signal from a stacked multilayer graph architecture.
- We conducted an empirical study on three million-level datasets of recommender systems. The experimental results show that the proposed model achieves the best recommendation performance, obtains significant improvements in recommendation coverage, and at the same time obtains more personalized recommendation results.

2 Related Work

There are many state-of-the-art model-based recommender systems. The most representative one is Matrix Factorization (MF). MF-based models such as PMF [19] and SVD [15] utilize straightforward procedures to generate low-dimensional user (item) embeddings directly from the user (item) one-hot vector. Although these models could achieve reasonable results on some experimental datasets, they are still insufficient for real-world recommender systems due to their simple structure and limited model expressions. With the superior power of representation learning, utilizing neural architecture to learn embeddings has become popular for model-based recommender systems. Zhuang et al. [31] proposed a model named REAP, which applies an autoencoder to generate the latent factors for users and items from the user-item matrix. Xue et al. [28] proposed a Deep Matrix Factorization Model (DMF) by utilizing two parallel neural networks to map the user and item into low-dimensional space from both explicit rating and implicit behavior. Han et al. [7] indicated that using one aspect representations is insufficient and proposed modeling the complicated relationship in recommender systems through Heterogeneous Information Network (HIN). Ebesu et al. [5] believed the user/item neighborhood information can help generate better results and utilized a memory network to model the neighborhood information and generate neighbor-based embeddings. Moreover, Jiang et al. [11] proposed a convolutional neural network-based Gaussian model to represent the users (items) with uncertainty and create better recommendations.

Compared with the traditional collaborative filtering [22] and neural collaborative filtering recommender systems, the GCN-based recommendation methods demonstrated competitive performances [8] and have attracted much research attention. Wang et al. [26] proposed Neural Graph Collaborative Filtering (NGCF), which utilizes a bipartite graph structure to exploit the high-order user-item connectivity and achieve promising recommendation performance. He et al. [8] further simplified the design of GCN to make it more concise and appropriate for recommendation. Moreover, Sun et al. [23] pointed out that directly applying GCNs to process the bipartite graph is sub-optimal since this method neglects the intrinsic differences between user nodes and item nodes. To this end, they proposed NIA-GCN, a new framework that explicitly models relationships between neighboring nodes and exploits the heterogeneous nature of the user-item bipartite graph. Liu et al. [17] mentioned that existing GCN-based recommenders often suffer from the over-smoothing problem. To alleviate that problem, they presented a recommendation model, namely IMP-GCN, which performs high-order graph convolution inside subgraphs and hence limits the neighbor exploration process to similar users (items). Furthermore, Wu et al. [27] explored self-supervised learning on GCN-based recommenders and targeted improving their recommendation accuracy and robustness. Although these models present promising recommendation results, the neglect of item (user) hierarchical structures still prevents them from generating superior predictions.

In fact, item (user) hierarchical structures have been widely explored and utilized in recommendation scenarios [20,29]. For example, Lu et al. [18] utilized item hierarchies stored in side-information and demonstrated the effectiveness of incorporating explicit hierarchical structure in recommender systems. However, real-world recommendation data may not contain detailed external side information or can not directly provide explicit item (user) hierarchy structures. Therefore, exploring and using the implicit hierarchical structures of user-item interaction information has become a common method to solve this problem. Wang et al. [25] proposed a framework, namely IHSR, which can construct implicit item (user) hierarchies by gradually decomposing the item (user) characteristic matrix. Analogously, Li [16] proposed a Hidden Hierarchical Matrix Factorization (HHMF), which learns the hidden hierarchical structures from the user-item scoring record without prior knowledge of the hierarchy. Even though these models achieve competitive results compared to basic model-based recommenders, the straightforward MF-based strategies still limit the model expression and result in sub-optimal predictions.

In summary, most existing embedding learning models neglected the inherent item (user) hierarchical structure and did not fully capture the valuable information stored in user-item interactions. Meanwhile, the explicit hierarchical structures are not often provided in real-world recommendation scenarios, and existing implicit hierarchical models are merely based on simple matrix factorization, which is insufficient for handling increasingly sparse and complex recommendation scenarios. Hence, a better hierarchical model for the recommendation problem is needed.

3 Methodology

3.1 The Basic Collaborative Filtering Model

The basic idea of traditional model-based collaborative filtering algorithms can be summarized as utilizing a model to project an individual user/item into a shared latent space and modeling a user's preference given an item by the inner product of the corresponding user/item embeddings. To make it more clear, we adopt e_u and e_i to represent the learned personalized embedding for user u and item i respectively. Moreover, as classic models merely rely on interactions and neglects high-level hierarchical context information, we formulate e_u and e_i as e_{u-c0} and e_{i-c0}, where $c0$ indicates the context extracted in the $0th$ hierarchy. Then, the preference of user u given item i can be computed as follows:

$$y_{ui} = e_u \cdot (e_i)^T = e_{u-c0} \cdot (e_{i-c0})^T \tag{1}$$

3.2 Modeling Implicit Hierarchies

In the basic model, the interaction is often considered as the unique preference of a single user given a specific item. However, as a member of our modern society, a user's behavior is inevitably influenced by implicit contextual information and would demonstrate some implicit context-based preferences. When it comes to interaction modeling recommender systems, we interpret the implicit contextual information as the group-level collaborative signal. More specifically, according to the example in Sect. 1, a user's interaction is not only demonstrating an individual's personalized interest but also indicates several informative group-level tendencies. As these group-level tendencies (multi-resolution collaborative signal) are useful for understanding the user's behavior and would be helpful for generating customized recommendations, a hierarchical model becomes a more reasonable way to construct general recommender systems. To be more specific, we decompose user embedding e_u and item embedding e_i into the summation of different representations, which corresponds to the preference information learned among different levels of hierarchical architectures. Take a n-levels hierarchical model as an example; the final user embedding e_u and item embedding e_i can be calculated through:

$$e_u = e_{u-c0} + e_{u-c1} + e_{u-c2} + ... + e_{u-cn} \tag{2}$$
$$e_i = e_{i-c0} + e_{i-c1} + e_{i-c2} + ... + e_{i-cn}, \tag{3}$$

where e_{u-c0} and e_{i-c0} represent the personalized embedding for individual user u and item i respectively. While, $e_{u-cx}, x \in (1, n)$ and $e_{i-cy}, y \in (1, n)$ demonstrate the diverse group-level collaborative signals. It is worth noticing that $e_{u-cx}, x \in (1, n)$ and $e_{i-cy}, y \in (1, n)$ would be shared among the group members within the same group at the same hierarchical level. Moreover, the greater x value in $e_{u-cx}, x \in (1, n)$ corresponds to a more compressed group representation, which indicates e_{u-cx} would be shared by a larger number of group members. With the defined e_u and e_i in hierarchical architecture, the user u's preference given item i can be estimated through:

$$y_{ui} = e_u \cdot (e_i)^T = (e_{u-c0} + e_{u-c1} + ... + e_{u-cn}) \cdot (e_{i-c0} + e_{i-c1} + ... + e_{i-cn})^T \quad (4)$$

3.3 UGCN Recommender

As discussed above, the essential point for constructing a hierarchical architecture is to extract multi-resolution collaborative signals and encode these informative signals into corresponding embeddings. To tackle this issue, we adopt the idea of utilizing a hierarchical graph structure to enable generating diverse group-level embeddings on different graph scales. In other words, we aim to gradually compress the original user-item interaction graph into several group-level interaction graphs. Hence, a node in the compressed graph could represent a group of similar users/items, and an edge in the compressed graph would indicate the shared behavior of group members. Taking a step further, the node embedding learned on diverse graph architectures of different scales can naturally capture the corresponding diverse group-level collaborative signals shared among certain group members. Following the analysis, we display the whole hierarchical recommender construction in three parts: item/user hierarchies extraction (i.e., graph architecture compressing), graph node embedding learning (i.e., multi-resolution collaborative signal modeling), and diverse embedding integrating.

Item/User Hierarchies Extraction. To reasonably explore item/user hierarchies and properly compress the original user-item interaction graph, we follow the core point of collaborative filtering that similar users/items might exhibit similar preferences. In other words, by merging the similar nodes in the finer scaled graph into a group node in the coarser scaled graph, the generated diverse graph architectures could successfully reveal the user/item hierarchies relationships. More importantly, the group node in a coarser scaled graph can automatically exhibit the shared interest of those similar nodes in a finer scaled graph. To simplify the discussion, we adopt the common pooling operation to represent the process of compressing a finer graph into a coarser one, which can be defined as:

$$G'(U', I', E') = Pooling(G(U, I, E)), \quad (5)$$

where $G(U, I, E)$ and $G'(U', I', E')$ is the finer scaled graph and coarser scaled graph respectively. Expressly, to implement the pooling operation, we utilize the following two steps: 1) group formation: assembling similar neighbors based on the similarity calculation on 1-hop connections; 2) node merging: generating a new group node based on the assembled members, where the edge of a new group node is mainly determined by the common connections of similar nodes. Meanwhile, a predefined ratio α is utilized to control the percentage of the nodes selected for grouping, enabling the construction of more flexible hierarchical architectures.

Graph Node Embedding Learning. After obtaining the graphs of diverse scales, capturing the diverse collaborative signal in different graphs and learning informative embeddings for each node in each structure become the essential issue. To this end, we apply a graph convolution operation on graph architecture to encode informative collaborative signals into node embeddings. Specifically, the neighborhood aggregation proposed by LightGCN [8] is adopted to simplify the whole convolutional calculation process, which can be represented as:

$$e_u^{(k+1)} = \sum_{i \in \mathcal{N}_u} \frac{1}{\sqrt{|\mathcal{N}_u|}\sqrt{|\mathcal{N}_i|}} e_i^{(k)} \tag{6}$$

$$e_i^{(k+1)} = \sum_{u \in \mathcal{N}_i} \frac{1}{\sqrt{|\mathcal{N}_i|}\sqrt{|\mathcal{N}_u|}} e_u^{(k)}, \tag{7}$$

where $e_u^{(k)}$ and $e_i^{(k)}$ respectively denote the embedding of user u and item i generated after k convolutional operations. Meanwhile, $\mathcal{N}_u$ and $\mathcal{N}_i$ denote the set of items that are interacted by the user u and the set of users that interact with the item i, respectively. The symmetric normalization term $1/(\sqrt{|\mathcal{N}_i|}\sqrt{|\mathcal{N}_u|})$ follows the design of standard GCN [13] to avoid the scale of embeddings increasing with graph convolution operations. The final embeddings after operating m convolutional operations can be computed as:

$$e_u = \frac{1}{m}\sum_{k=0}^{m} e_u^{(k)} \quad ; \quad e_i = \frac{1}{m}\sum_{k=0}^{m} e_i^{(k)} \tag{8}$$

It is worth noticing that the graph convolution operations will be independently implemented on each graph structure. The initialization for graph nodes will only be applied to the original user-item interaction graph. Moreover, the initial node embeddings in the coarser graph can be easily generated by the mean embeddings of similar nodes in the finer graph that are selected and merged into the corresponding group nodes.

Diverse Embedding Aggregation. Following the multi-resolution collaborative signal extraction process, properly propagating the extracted information back to final embeddings is also crucial. Symmetric to pooling operation, we utilize un-pooling operation to gradually restore the coarser scaled architecture back to the finer ones:

$$G(U, I, E) = UnPooling(G'(U', I', E')) \tag{9}$$

As a group-level node in a coarser scaled graph represents all of its group members' collaborative interaction patterns, the collaborative information contained in the group node is directly copied to its group members in the finer scaled graph. In this case, the group-level collaborative signal obtained in a coarser scaled graph can successfully be propagated back to the corresponding group members.

UGCN Architecture and Model Training. To better exhibit the proposed UGCN architecture, we develop a 3-levels UGCN recommender depicted in Fig. 1.

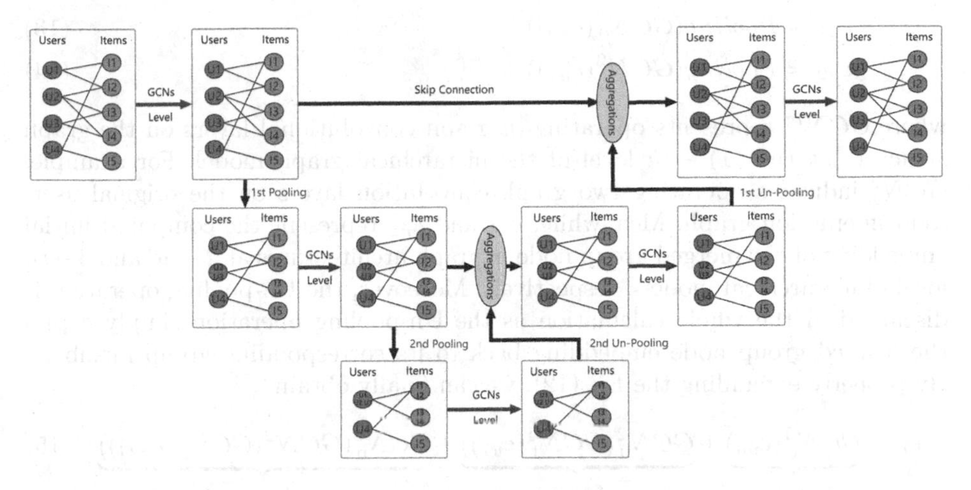

Fig. 1. An illustration of a 3-levels UGCN Recommender. In this example, we hierarchically stack the original user-item interaction graph with two compact graphs generated by Pooling operations. Then, simplified GCN layers are implemented to extract the multi-resolution collaborative signals at three graphs of different scales. The Unpooling operations would gradually restore the finer graph architecture and integrate the informative signals into final user/item embeddings.

To properly train the UGCN recommender, we employ the BPR loss, a pairwise loss that aims to enlarge the prediction differences among positive and negative samples. The final loss function for UGCN is presented below:

$$\hat{x}_{uij} = y_{ui} - y_{uj} \tag{10}$$

$$L_{BPR} = -\sum_{u=1}^{M} \sum_{i \in \mathcal{N}_u} \sum_{j \notin \mathcal{N}_u} \ln \sigma(\hat{x}_{uij}) + \lambda ||\Theta||, \tag{11}$$

where σ denotes the activation function, λ controls the L2 regularization strength, and Θ represents the parameter vector of the UGCN model. As the only trainable parameter is the initial embedding of the user item interaction graph, the Θ in Eq. (11) equals the embedding matrix $\mathbf{E}$.

3.4 Theoretical Analysis for UGCN Recommender

In this section, we offer in-depth analyses of the UGCN Recommender, aiming to answer the question: how does this hierarchical recommender benefit real-world recommendation problems? To answer this question, we utilize a 3-levels UGCN recommender depicted in Fig. 1 as an example.

Given a user u with his/her initial embedding e_{u0}, the final embedding e_u produced by a 3-levels UGCN model can be represented as:

$$e_u = GCN_0^2(GCN_0^2(e_{u0}) + GCN_1^2(GCN_1^2(e_{u1}) + GCN_2^2(e_{u2}))) \quad (12)$$

$$e_{u1} = Pooling(GCN_0^2(e_{u0})) \quad (13)$$

$$e_{u2} = Pooling(GCN_1^2(e_{u1})), \quad (14)$$

where GCN_n^m represents operating m graph convolutional layers on the graph structure at $(n+1)-th$ level of the hierarchical graph model. For example, GCN_0^2 indicates operating two graph convolution layers on the original user-item interaction graph. Meanwhile, e_{u1} and e_{u2} represent the computed initial embedding of the merged group node in graph architectures at $2-nd$ and $3-rd$ level of hierarchical models, respectively. Moreover, the Un-pooling operation is discarded in the whole calculation as the Un-pooling operation simply copies the learned group node embedding back to its corresponding group members. By properly expanding the Eq. (12), we can finally obtain:

$$e_u = \underbrace{GCN_0^4(e_{u0})}_{e_{u-c0}} + \underbrace{GCN_0^2(GCN_1^4(e_{u1}))}_{e_{u-c1}} + \underbrace{GCN_0^2(GCN_1^2(GCN_2^2(e_{u2})))}_{e_{u-c2}} \quad (15)$$

As the final embedding generated by the UGCN recommender can be interpreted as the hierarchical structure discussed in Sect. 3.2, we hence adopt the simplified expressions in our following discussion.

Formally, for user u, the gradient of the BPR loss $w.r.t.$ the trainable embedding e_{u0} is as follows:

$$\frac{\partial BPR}{\partial e_{u0}} \propto \sum_{i \in \mathcal{N}_u} \sum_{j \notin \mathcal{N}_u} \frac{-e^{-\hat{x}_{uij}}}{1+e^{-\hat{x}_{uij}}} \cdot \frac{\partial}{\partial e_{u0}} \hat{x}_{uij} - \lambda \quad (16)$$

According to the Eq. (10), $\hat{x}_{uij}$ can be further expanded as follows:

$$\hat{x}_{uij} = (e_{u-c0} + e_{u-c1} + e_{u-c2}) \cdot [(e_{i-c0} + e_{i-c1} + e_{i-c2}) - (e_{j-c0} + e_{j-c1} + e_{j-c2})] \quad (17)$$

As presented above, the initial group embeddings e_{u1} and e_{u2} in Eq. (15) can be calculated through a linear model contains e_{u0}, we can finally obtain:

$$\frac{\partial}{\partial e_{u0}} \hat{x}_{uij} = \alpha \cdot [(e_{i-c0} - e_{j-c0}) + (e_{i-c1} - e_{j-c1}) + (e_{i-c2} - e_{j-c2})], \quad (18)$$

where α is the constant number computed through the partial derivatives of $(e_{u-c0} + e_{u-c1} + e_{u-c2})$ with respect to e_{u0}. Based on the Eqs. (15) and (18), it is apparent that the U-shape hierarchical model can contain valuable group-level differences information during the model training process. Hence, we believe the UGCN model can produce more informative embeddings and generate better predictions.

4 Experiments

We conduct experiments on three public million-size datasets to answer the following research questions.

- **Q1:** How does the UGCN recommender perform compared to state-of-the-art models? Does UGCN achieve higher accuracy?
- **Q2:** Are multi-resolution collaborative signals helpful for generating more personalized embedding? Can our model produce better recommendations?
- **Q3:** How do different hyper-parameter settings (e.g., the number of model's hierarchical levels) affect the recommendation performance?

4.1 Experimental Settings

Datasets. We conduct experiments on three public million-size datasets, namely Movielens 1M (Ml1m), Gowalla, and Yelp2018. The detailed dataset statistics are shown in Table 1. For each dataset, we randomly split 80% of each user's interactions to construct the training set, while the remainder forms the test set. For each training set, we randomly select 10% of interactions as a validation set to tune hyperparameters. Moreover, according to our pair-wise BPR training strategy, each interaction of an individual user in the training set is treated as a positive sample, and the corresponding negative one will be randomly sampled from the non-interacted items of the same user.

Table 1. Statistics of experimental datasets.

Statistics	Users	Items	Interactions	Density
Ml1m	6022	3043	995,154	0.05431
Gowalla	29,858	40,981	1,027,370	0.00084
Yelp2018	31,668	38,048	1,561,406	0.00130

Evaluation Metrics. To evaluate different aspects of Top@K recommendation results for the proposed model and existing ones, we first adopt **Recall@K** and **NDCG@K** [10] to compare the overall prediction accuracy. Then, we apply an additional publicly accepted method (i.e., Coverage [3]) to assess the personalization performance. As coverage demonstrates the percentage of items that have been recommended to users, we adopt the following equations to compute both recommendation coverage $R-Coverage$ and hit coverage $H-Coverage$:

$$\begin{aligned} R-Coverage(K) &= N_{recommended}/N \\ H-Coverage(K) &= N_{Hit}/N, \end{aligned} \tag{19}$$

where $N_{recommended}$ and N_{Hit} imply the number of different items that have been recommended and hit in one or more users' top@K list, respectively; N is the number of total items. According to common sense, users are always pleased with mixed recommendation results rather than merely recommending the same items [3]. Therefore, we aim to pursue a higher Coverage value.

Comparison Methods. We compare UGCN with several state-of-the-art recommendation models.

- **MF-BPR** [21]: This model presented a generic optimization criterion BPR-Opt for personalized ranking.
- **NeuMF** [9]: This is a state-of-the-art neural Collaborative Filtering model that used implicit user-item interaction.
- **HHMF** [16]: This is a recent proposed hidden hierarchical recommender which learned from the user-item scoring record and does not need prior knowledge.
- **NGCF** [26]: This is a state-of-the-art GCN-based model which integrated the user-item interaction into the embedding process through the graph convolution operation.
- **NIA-GCN** [23]: This is a newly proposed GCN-based model which explicitly exploits the user-user and item-item relationships through a pairwise neighborhood aggregator.
- **LR-GCCF** [4]: This is a state-of-art GCN-based model which utilizes a residual network structure and linear graph convolution operations to alleviate the over-smoothing problem.
- **LightGCN** [8]: This is a newly proposed GCN-based model that simplifies the design of graph convolution operation and only includes neighbor aggregation in convolution operations.
- **SGL** [27]: This is a newly proposed Graph Training Strategy which implements on the LightGCN model and achieves competitive performance.

Parameter Settings. The embedding size is fixed to 64 for all models, and the embedding parameters are initialized with the Xavier method [6]. The default learning rate is 0.002, and the default mini-batch size is 2048. The depth of hierarchical architecture is tested in the range of $[2, 3, 4]$. The number of graph convolution operations is tested from 1 to 3, and the predefined ratio α is validated in the range of [0.01 0.99]. The early stopping and validation strategies are the same as LightGCN. The Adam [12] optimizer is also employed and used in a mini-batch manner. Moreover, we also adopt dropout mechanisms on every GCN layer to mitigate over-fittingm and the default dropout rate is 0.6.

Implementation Details. In the group formation step, we adopt the cosine function to derive the similarity among a pair of users (or items). For node merging in pooling operation, we mainly retain the common interactions of similar users in the same group and randomly delete some individual behaviors. Meanwhile, it is worth pointing out that the pooling operation would run separately and steply on user and item nodes. Moreover, the only trainable parameters of the UGCN recommender are the initial user/item embeddings of the original user-item interaction graph. For the implementation of the baseline models, the default training strategies and hyperparameters settings from the corresponding referenced papers are followed.

4.2 Prediction Accuracy Comparison (QR1)

To evaluate the overall prediction accuracy, we test Recall@20 and NDCG@20 among all different models. The final results are presented in Table 2.

Table 2. The comparison of overall performance among UGCN and baseline methods.

Data-sets	Ml1m		Gowalla		Yelp2018	
	Recall@20	NDCG@20	Recall@20	NDCG@20	Recall@20	NDCG@20
MF-BPR	0.2101	0.1787	0.1291	0.1109	0.0433	0.0354
NeuMF	0.2297	0.1886	0.1399	0.1212	0.0451	0.0363
HHMF	0.2311	0.2025	0.1477	0.1283	0.0498	0.0384
NGCF	0.2513	0.2511	0.1570	0.1327	0.0579	0.0477
NIA-GCN	0.2359	0.2243	0.1359	0.1358	0.0599	0.0491
LR-GCCF	0.2231	0.2124	0.1519	0.1358	0.0561	0.0343
LightGCN	0.2576	0.2427	<u>0.1830</u>	<u>0.1554</u>	0.0649	0.0530
SGL	<u>0.2700</u>	<u>0.2547</u>	0.1781	0.1501	<u>0.0674</u>	<u>0.0553</u>
UGCN	**0.2774**	**0.2633**	**0.1876**	**0.1587**	**0.0689**	**0.0561**
%Improv.	*+2.74%*	*+3.38%*	*+2.51%*	*+2.12%*	*+2.23%*	*+1.45%*

Our UGCN recommender consistently outperforms the baseline methods on all datasets. In particular, UGCN achieves 2.74%, 2.51%, and 2.23% improvement of recall@20 over the best baseline on Ml1m, Gowalla, and Yelp2018, respectively. Meanwhile, the performance of UGCN on NDCG@20 is also outstanding, presenting 3.38%, 2.51%, and 1.45% enhancements. Moreover, compared to MF-based and neural-based models, the significant improvements among GCN-based models reveal the superiority of GCNs in handling embedding learning tasks, which is consistent with the discussion in Sect. 2. The results also demonstrate that the HHMF model and UGCN recommender perform better than basic-MF and existing GCN-based models, respectively, confirming the effectiveness of exploring users/items hierarchies in recommender systems. In contrast to the HHMF model, the better performance of UGCN demonstrates that modeling the diverse resolution collaborative signal would be a better strategy than merely focusing on exploring user/item hierarchies separately. Moreover, since a denser dataset would result in more accurate user/item grouping and lead to more precise group-level signal extraction, this would explain why the most significant improvement is achieved in the ML1m dataset.

All the analyses above exhibit that our proposed UGCN recommender outperforms all the state-of-the-art models and achieves the highest recommendation accuracy.

4.3 Personalization Comparison (QR2)

As discussed above, we adopt Coverage [3] to demonstrate whether integrating diverse group-level collaborative signals can lead to better recommendations. Specifically, we adopt the state-of-the-art models LightGCN [8] and SGL [27] as our compared baseline owing to their competitive results displayed in Table 2. Moreover, we test three top@K recommendation cases, and the comparison results of Coverage among three models are presented in Fig. 2.

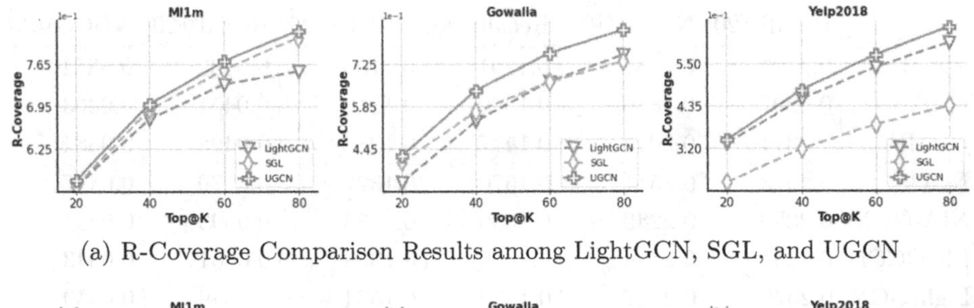

(a) R-Coverage Comparison Results among LightGCN, SGL, and UGCN

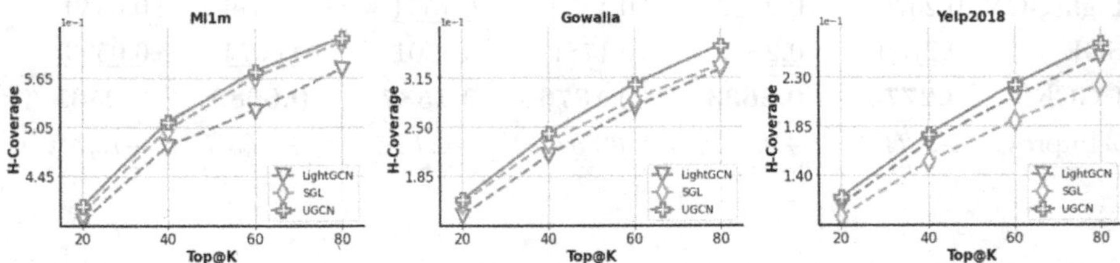

(b) H-Coverage Comparison Results among LightGCN, SGL, and UGCN

Fig. 2. Coverage comparison results

Based on the results in Fig. 2, UGCN achieves higher $R - Coverage$ and $H - Coverage$ in all datasets. Taking the worst-case (top@20) as an example, $R - Coverage$ for UGCN are 57.34%, 42.20%, and 34.39%, which indicate the corresponding 1.60%, 4.95% and 2.81% improvement on three datasets compared to the best baseline. Similarly, $H - Coverage$ results also achieve relatively significant increases. Meanwhile, it is worth noticing that the $Coverage$ result of SGL on the Yelp2018 dataset is under-performed by those of LightGCN, even though SGL obtains a higher recommendation accuracy according to Table 2. These results demonstrate that higher recommendation accuracy might not reveal better recommendations. In comparison, the proposed UGCN recommender obtains higher recommendation accuracy and higher coverage simultaneously, indicating the UGCN's superiority. Considering the difference between

UGCN and those baseline models, we could answer Q2 by concluding that integrating multi-resolution collaborative allows recommender systems to precisely capture user/item personalized preferences and leads to better recommendations.

4.4 Hyper Parameter Analysis (QR3)

The above discussion indicates integrating diverse collaborative signals will benefit the recommender system. Nevertheless, it is still unclear how the number of model's hierarchical levels impact the recommendation performance? To this end, we conducted the comparison experiments on Ml1m and Gowalla. Owing to the limited space, we discard the results on Yelp2018, but the discussion would be applicable for other datasets. The results are presented in Fig. 3.

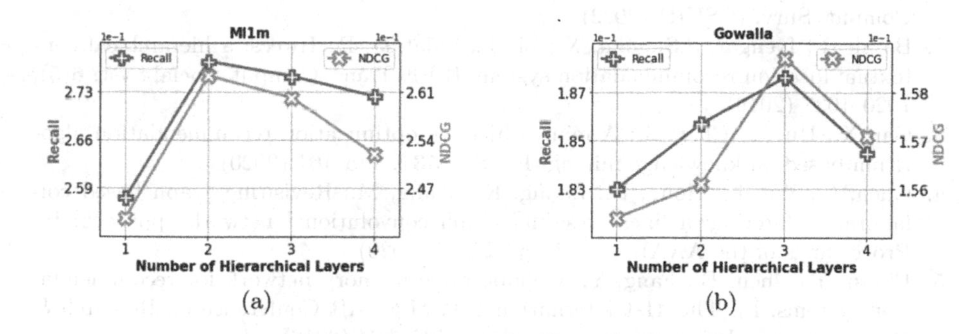

Fig. 3. Hyper-parameters Comparison Results

As shown in Fig. 3, the best performance is achieved when the number of the model's hierarchical levels are set to two and three on Ml1m and Gowalla, respectively. Compared to the one-level UGCN model (i.e., original LightGCN, which does not contain hierarchical architecture), the huge improvement confirms the necessity of modeling diverse collaborative signals. Meanwhile, the different optimal results on different datasets reveal that a fair number of UGCN's hierarchical levels are always needed to be optimized on specific datasets during training.

5 Conclusion

This paper proposes a graph convolutional network-based hierarchical recommender, namely UGCN, on a bipartite graph to capture diverse group-level collaborative signals and produce better recommendation predictions. Precisely, the proposed model utilizes a U-shaped architecture to gradually compress the original user-item interaction graph into smaller graphs by merging similar nodes together. Then, simplified graph convolutional operations can easily extract diverse group-level collaborative signals from compressed graphs. After that, those learned informative signals would be projected back to the final user/item embeddings adaptively. Extensive and comprehensive experiments on three public datasets demonstrate that the UGCN recommender achieves significant overall prediction improvements over state-of-the-art models with better recommendations simultaneously.

Acknowledgment. The first author is funded by the China Scholarship Council (CSC) from the Ministry of Education of P.R. China.

References

1. Alhijawi, B., Awajan, A., Fraihat, S.: Survey on the objectives of recommender system: measures, solutions, evaluation methodology, and new perspectives. ACM Comput. Surv. (CSUR) (2022)
2. Baral, R., Iyengar, S.S., Zhu, X., Li, T., Sniatala, P.: Hirecs: a hierarchical contextual location recommendation system. IEEE Trans. Comput. Social Syst. **6**(5), 1020–1037 (2019)
3. Cai, X., Hu, Z., Chen, J.: A many-objective optimization recommendation algorithm based on knowledge mining. Inf. Sci. **537**, 148–161 (2020)
4. Chen, L., Wu, L., Hong, R., Zhang, K., Wang, M.: Revisiting graph based collaborative filtering: a linear residual graph convolutional network approach. In: Proceedings of the AAAI, vol. 34, pp. 27–34 (2020)
5. Ebesu, T., Shen, B., Fang, Y.: Collaborative memory network for recommendation systems. In: The 41st international ACM SIGIR Conference on Research & Development in Information Retrieval, pp. 515–524 (2018)
6. Glorot, X., Bengio, Y.: Understanding the difficulty of training deep feedforward neural networks. In: Proceedings of the Thirteenth International Conference on Artificial Intelligence and Statistics, pp. 249–256. JMLR Workshop and Conference Proceedings (2010)
7. Han, Z., et al.: Genetic meta-structure search for recommendation on heterogeneous information network. In: Proceedings of the 29th ACM International Conference on Information and Knowledge Management, pp. 455–464 (2020)
8. He, X., Deng, K., Wang, X., Li, Y., Zhang, Y., Wang, M.: LightGCN: simplifying and powering graph convolution network for recommendation. In: Proceedings of the 43rd ACM SIGIR, pp. 639–648 (2020)
9. He, X., Liao, L., Zhang, H., Nie, L., Hu, X., Chua, T.S.: Neural collaborative filtering. In: Proceedings of the 26th International Conference on World Wide Web, pp. 173–182. International World Wide Web Conferences Steering Committee (2017)
10. Isinkaye, F.O., Folajimi, Y.O., Ojokoh, B.A.: Recommendation systems: principles, methods and evaluation. Egypt. Inf. J. **16**(3), 261–273 (2015)
11. Jiang, J., Yang, D., Xiao, Y., Shen, C.: Convolutional gaussian embeddings for personalized recommendation with uncertainty. arXiv preprint arXiv:2006.10932 (2020)
12. Kingma, D.P., Ba, J.: Adam: a method for stochastic optimization. arXiv preprint arXiv:1412.6980 (2014)
13. Kipf, T.N., Welling, M.: Semi-supervised classification with graph convolutional networks. arXiv preprint arXiv:1609.02907 (2016)

14. Koren, Y.: Factorization meets the neighborhood: a multifaceted collaborative filtering model. In: Proceedings of the 14th ACM SIGKDD International Conference on Knowledge Discovery and Data Mining, pp. 426–434. ACM (2008)
15. Koren, Y., Bell, R., Volinsky, C.: Matrix factorization techniques for recommender systems. Computer **42**(8), 30–37 (2009)
16. Li, H., Liu, Y., Qian, Y., Mamoulis, N., Tu, W., Cheung, D.W.: HHMF: hidden hierarchical matrix factorization for recommender systems. Data Min. Knowl. Disc. **33**(6), 1548–1582 (2019)
17. Liu, F., Cheng, Z., Zhu, L., Gao, Z., Nie, L.: Interest-aware message-passing GCN for recommendation. In: Proceedings of the Web Conference 2021, pp. 1296–1305 (2021)
18. Lu, K., Zhang, G., Li, R., Zhang, S., Wang, B.: Exploiting and exploring hierarchical structure in music recommendation. In: Hou, Y., Nie, J.-Y., Sun, L., Wang, B., Zhang, P. (eds.) AIRS 2012. LNCS, vol. 7675, pp. 211–225. Springer, Heidelberg (2012). https://doi.org/10.1007/978-3-642-35341-3_18
19. Mnih, A., Salakhutdinov, R.R.: Probabilistic matrix factorization. In: Advances in Neural Information Processing Systems, vol. 20 (2007)
20. Nikolakopoulos, A.N., Kouneli, M.A., Garofalakis, J.D.: Hierarchical itemspace rank: exploiting hierarchy to alleviate sparsity in ranking-based recommendation. Neurocomputing **163**, 126–136 (2015)
21. Rendle, S., Freudenthaler, C., Gantner, Z., Schmidt-Thieme, L.: BPR: Bayesian personalized ranking from implicit feedback. arXiv preprint arXiv:1205.2618 (2012)
22. Schafer, J.B., Frankowski, D., Herlocker, J., Sen, S.: Collaborative filtering recommender systems. In: Brusilovsky, P., Kobsa, A., Nejdl, W. (eds.) The Adaptive Web. LNCS, vol. 4321, pp. 291–324. Springer, Heidelberg (2007). https://doi.org/10.1007/978-3-540-72079-9_9
23. Sun, J., et al.: Neighbor interaction aware graph convolution networks for recommendation. In: Proceedings of the 43rd ACM SIGIR, pp. 1289–1298 (2020)
24. Wang, S., Tang, J., Wang, Y., Liu, H.: Exploring implicit hierarchical structures for recommender systems. In: Twenty-Fourth International Joint Conference on Artificial Intelligence (2015)
25. Wang, S., Tang, J., Wang, Y., Liu, H.: Exploring hierarchical structures for recommender systems. IEEE Trans. Knowl. Data Eng. **30**(6), 1022–1035 (2018)
26. Wang, X., He, X., Wang, M., Feng, F., Chua, T.S.: Neural graph collaborative filtering. In: Proceedings of the 42nd ACM SIGIR, pp. 165–174 (2019)
27. Wu, J., Wang, X., Feng, F., He, X., Chen, L., Lian, J., Xie, X.: Self-supervised graph learning for recommendation. In: Proceedings of the 44th ACM SIGIR, pp. 726–735 (2021)
28. Xue, H.J., Dai, X., Zhang, J., Huang, S., Chen, J.: Deep matrix factorization models for recommender systems. In: IJCAI, pp. 3203–3209 (2017)
29. Yang, J., Sun, Z., Bozzon, A., Zhang, J.: Learning hierarchical feature influence for recommendation by recursive regularization. In: Proceedings of the 10th ACM Conference on Recommender Systems, pp. 51–58 (2016)
30. Zhang, S., Yao, L., Sun, A., Tay, Y.: Deep learning based recommender system: a survey and new perspectives. ACM Comput. Surv. (CSUR) **52**(1), 1–38 (2019)
31. Zhuang, F., Luo, D., Yuan, N.J., Xie, X., He, Q.: Representation learning with pair-wise constraints for collaborative ranking. In: Proceedings of the Tenth ACM International Conference on Web Search and Data Mining, pp. 567–575 (2017)

Basket Booster for Prototype-based Contrastive Learning in Next Basket Recommendation

Ting-Ting Su[1,2,3], Zhen-Yu He[1,2,3], Man-Sheng Chen[1,2,3], and Chang-Dong Wang[1,2,3](✉)

[1] School of Computer Science and Engineering, Sun Yat-sen University, Guangzhou, China
{sutt8,hezhy65,chenmsh27}@mail2.sysu.edu.cn
[2] Guangdong Province Key Laboratory of Computational Science, Guangzhou, China
[3] Key Laboratory of Machine Intelligence and Advanced Computing, Ministry of Education, Guangzhou, China
changdongwang@hotmail.com

Abstract. Next basket recommendation seeks to model the correlation of items and mine users' interests hidden in basket sequences, and tries to infer a set of items that tend to be adopted in the next session with the mined information. However, the feedback provided by users often involves only a small fraction of millions to billions of items. Sparse data makes it hard for model to infer high-quality representations for basket sequences, which further leads to poor recommendation. Inspired by the recent success of representation learning in some fields, e.g., computer vision and clustering, we propose a basket booster for prototype-based contrastive learning (**BPCL**) in next basket recommendation. A correlative basket booster is designed to mine self-supervised signals just from raw data and make augmentation for baskets. To our best knowledge, this is the first work to promote learning of prototype representation through basket augmentation, which helps overcome the difficulties caused by data sparsity and leads to a better next basket recommendation performance. Extensive experiments on three public real-world datasets demonstrate that the proposed BPCL method achieves better performance than the existing state-of-the-art methods.

Keywords: Contrastive learning · Next basket recommendation · Data augmentation

1 Introduction

With the rapid growth of the number of entities involved in online platforms, it is difficult for users to find the items that meet their demands. Therefore, recommendation systems are widely used to provide users with more proper items by mining information contained in historical data, e.g., user preferences. There are many previous works trying to model user and item portraits by making use of historical interactions as a set [3,5,23,29], regardless of chronological order.

M.-R. Amini et al. (Eds.): ECML PKDD 2022, LNAI 13713, pp. 574–589, 2023.
https://doi.org/10.1007/978-3-031-26387-3_35

However, chronological order often contributes significantly to recommendation, as the interests of users change over time [10]. To capture evolving interests of users, some attempts have been made in sequential recommendation [1,4,10,12] that tries to learn representation of strictly sequential sequence. Note that user interactions are not strictly sequential over a short period in many cases, e.g., multiple items may be purchased with different intentions in the same shopping session. Next basket recommendation [8,16,24,28] breaks through the bottleneck of sequential recommendation by recommending a set of items simultaneously. Figure 1 shows an example for next basket recommendation, in which the last basket is expected to be inferred with the preceding basket sequence with length of 2. Note that both of basket and basket sequence have no fixed size, so the size of the predicted next basket is predefined in next basket recommendation.

Fig. 1. An example for next basket recommendation.

Data sparsity is a severe challenge for recommendation tasks. The feedback provided by users often involves only a small fraction of millions to billions of items, which makes it hard for model to infer high-quality representations for basket sequences. In the recent years, contrastive learning has been shown to perform well in representation learning in the data-sparse tasks, such as computer vision [14,17,18] and clustering [13,20]. The main idea lies in capturing self-supervised signals from raw data with various data augmentations. In computer vision, it is easy to augment data by rotation, changing color or adding noise, but these methods are unsuitable for recommendation tasks because of the different data types. In order to integrate the advantages of contrastive learning into recommendation tasks, some attempts have been made in developing augmentation methods for item sequence in sequential recommendation [22,25]. These methods mainly rely on executing random reordering, clipping or insertion of items, which are not applicable for basket sequence since it would destroy the correlation of items within a basket. The augmentation for basket sequences in next basket recommendation remains an unaddressed issue.

To address the issues mentioned above, we introduce contrastive learning to next basket recommendation by proposing a basket booster for prototype-based contrastive learning (**BPCL**). Basket sequences are modeled via an item correlation graph for constructing correlation matrix. With the correlation matrix, we develop a correlation basket booster for basket augmentation, which maintains the correlation between intra-basket items while introducing randomness. The augmented basket sequences as well as the corresponding prototype basket sequences will be encoded to basket sequence representations for prototype-based contrastive learning to improve the performance of next basket recommendation.

To summarize, our contributions are listed as follows:

- We propose a **BPCL** method that introduces contrastive learning to next basket recommendation to mine self-supervised signals from primitive basket sequences, i.e., prototypes. To our best knowledge, this is the first work to promote learning of prototype representation through basket augmentation.
- We propose a basket booster for prototype-based contrastive learning to maintain the correlation between intra-basket items while introducing randomness. The effectiveness of the booster is demonstrated by ablation study.
- Comprehensive experiments on three public real-world datasets demonstrate that the proposed BPCL method achieves better basket recommendation performance than the state-of-the-art methods in terms of the four metrics.

2 Related Work

In this section, we will briefly review previous works related to our work, namely next basket recommendation and contrastive learning.

2.1 Next Basket Recommendation

Temporal recommendation focusing on modeling interactions with timeline has shown competitive performance in many time-sensitive scenarios [2]. According to the target of recommendation, temporal recommendation can be divided to sequential recommendation and basket recommendation. Specially, next basket recommendation is a type of basket recommendation that predicts next basket without any information from next basket. Sequential recommendation [7,27,30, 32] seeks to predict next item with the representation of item sequence, while next basket recommendation [16,21,24,26] tries to predict next basket with the representation of basket sequence. In many real-world scenarios, interactions do not in strict chronological order over a short period, e.g., a shopping session. Hence, next basket recommendation that fits the situation has gained increasing attention in recent years.

There are some previous works [8,16,31] focusing on modeling qualified representation of basket sequence to guide the prediction of next basket. Hu et al. [8] design a KNN-based method to model basket sequence representation with the information from similar users' interactions. Le et al. [16] model basket sequence representation by developing a hierarchical network considering both inter-basket association and intra-basket association. Recently, advanced representation learning methods such as graph embedding [10,19], attention mechanisms [7,12] are widely applied in sequential recommendation for their outstanding performance in learning representation. However, the performance of these methods depends on data in a high degree, while the feedback provided by users often involves only a small fraction of millions to billions of items leading to data sparsity. It is necessary to develop a method to alleviate the issues caused by sparse data in next basket recommendation.

2.2 Contrastive Learning

Contrastive learning aims to mine useful signals from unlabeled data to alleviate sparse data problems. This method has been widely used in some fields, e.g., computer vision [14,17,18] and clustering [13,20]. For computer vision, contrastive learning can be adopted to promote performance of domain adaptation [11]. For clustering, contrastive learning can help with learning representation of different clusters [20]. When it comes to basket recommendation, only a few attempts has been made in leveraging contrastive learning to alleviate the problem of data sparsity. To our best knowledge, the only attempt for integrating contrastive learning to next basket recommendation is [24]. It tries to split a target basket sequence into two sub-basket sequences according to the correlation between the item in a basket and the item in candidate set, and aims to learn a qualified representation for the filtered pos-basket sequence. However, in next basket recommendation, augmentation method for learning better representation of prototype is less well-studied.

3 The Proposed Method

In this section, we present the proposed BPCL method for next basket recommendation. We start with formulating next basket recommendation problem in Sect. 3.1. Then, we introduce the proposed BPCL method in detail in Sect. 3.2 with six parts, i.e., correlative basket booster, basket encoder, dynamic context encoder, basket predictor, prototype-based contrastive learning and multi-task training. The overall architecture of BPCL is illustrated in Fig. 2.

3.1 Problem Statement

Let $\mathcal{I} = \{i_1, \ldots, i_N\}$ denote the set of items and $\mathcal{B}_t = \{i_1, \ldots, i_{|\mathcal{B}_t|}\}$ denote the basket at time step t, where N is the number of items and $\mathcal{B}_t \subset \mathcal{I}$. Given a basket sequence $S = [\mathcal{B}_1, \ldots, \mathcal{B}_{|S|}]$ consisting of several baskets, where $|S|$ is the length of S, next basket recommendation aims to predict several items to be adopted at time step $(|S| + 1)$ as:

$$\hat{\mathcal{B}}_{|S|+1} = \mathcal{K}_{i \in \mathcal{I}}(\mathrm{P}(i \in \mathcal{B}_{|S|+1}|S)) \tag{1}$$

where $\mathcal{K}$ denotes k items picked with the highest probability. Note that k is a predefined number indicating the size of the next basket, and next basket recommendation predicts all items in the next basket simultaneously.

3.2 BPCL

Correlative Basket Booster. Given a basket $\mathcal{B}_t$, it will be converted to a binary vector as $\mathbf{b}_t \in \{0, 1\}^N$ in which the n-th entry is set to 1 if $i_n \in \mathcal{B}_t$. For basket augmentation, one of the most direct ways is random masking, i.e., randomly set some 1 to 0 in the vector. We argue that using random masking alone

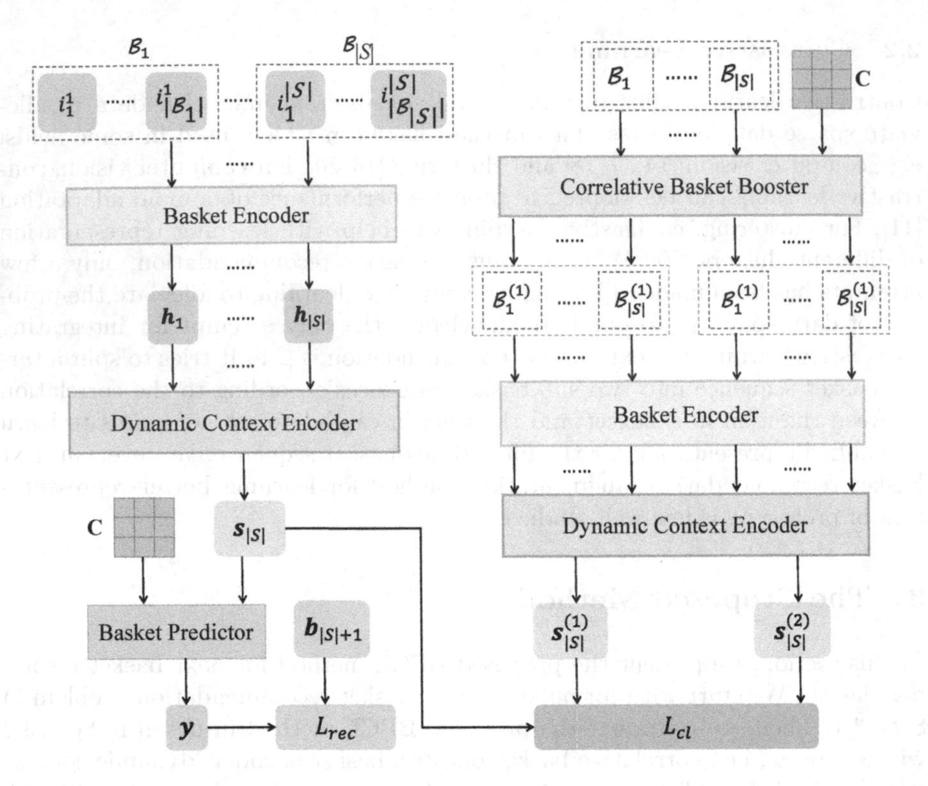

Fig. 2. The architecture of the proposed BPCL method.

can destroy the correlation between items within the same basket. Instead of augmenting basket by random masking, we propose a correlative basket booster to maintain the correlation and introduce randomness for data augmentation.

When modeling correlation between items by a graph, each item can be regarded as a node in the graph. For item p and q, the weight of their connection edge can be set to the number of times that they are in the same basket. And then we can obtain a weighted adjacency matrix $\mathbf{A} \in \mathbb{R}^{|\mathcal{I}|\times|\mathcal{I}|}$ according to the graph. Further, the adjacency matrix can be normalized to correlation matrix $\mathbf{C} \in \mathbb{R}^{|\mathcal{I}|\times|\mathcal{I}|}$, with each $\mathbf{C}_{pq}$ defined as:

$$\mathbf{C}_{pq} = \begin{cases} 0, & p = q; \\ \frac{\mathbf{A}_{pq}}{\sqrt{\sum_p \mathbf{A}_{pq}}\sqrt{\sum_q \mathbf{A}_{pq}}}, & p \neq q. \end{cases} \tag{2}$$

With the correlation matrix, we can distinguish the items associated with a basket in a higher degree for basket augmentation as:

$$\mathbf{b} = \max\{0, \mathbf{b}\cdot\mathbf{C}\}\,. \tag{3}$$

However, it is unreasonable for that $\mathbf{b}$ and $\mathbf{C}$ are in different representation spaces. And we should map them to the same latent space before matching. Finally, the formulation of the proposed correlative basket booster is:

$$\mathbf{b}^{aug} = \mathbf{b}^{'} \circ \mathcal{B}(\mathbf{m}) \tag{4}$$

$$\mathbf{b}^{'} = \max(\mathbf{0}, (\mathbf{b}\mathbf{\Pi} + \boldsymbol{\delta}) \cdot (\mathbf{C}\mathbf{\Upsilon} + \boldsymbol{\xi})^T) \tag{5}$$

where $\circ$ means element-wise product, $\mathbf{m} \in \mathbb{R}^N$ conforms to the uniform distribution, i.e., $m_n \sim U(0,1)$, $\mathbf{\Pi} \in \mathbb{R}^{N \times L}$, $\boldsymbol{\delta} \in \mathbb{R}^L$, $\mathbf{\Upsilon} \in \mathbb{R}^{N \times L}$ and $\boldsymbol{\xi} \in \mathbb{R}^{N \times L}$ are learnable parameters, L is the correlation latent dimension, and $\mathcal{B}$ is defined as:

$$\mathcal{B}(\mathbf{x}_p) = \begin{cases} 0, & \text{if } \mathbf{x}_p \leq \mu; \\ 1, & \text{if } \mathbf{x}_p > \mu \end{cases} \tag{6}$$

where $\mu \in \mathbb{R}^+$ is a hyper-parameter of mask threshold. The correlation matrix contributes to maintain the correlation between items within the same basket, while masking introduces some randomness.

The correlative basket booster will work for the data augmentation in contrastive learning task.

Basket Encoder. Although the vectorized basket representation $\mathbf{b}_t$ of $\mathcal{B}_t$ includes the direct relationships between the basket and all items, the information contained in the high-dimensional vector is far less than the maximum information it can store. In order to reduce data redundancy, we compress it to a low-dimensional space and obtain the hidden representation $\mathbf{h}_t$ of basket $\mathcal{B}_t$ as:

$$\mathbf{h}_t = \mathcal{F}(\mathbf{b}_t \cdot \mathbf{\Phi} + \boldsymbol{\theta}) \tag{7}$$

where $\mathcal{F}$ is an activation function, e.g., Relu, $\mathbf{\Phi} \in \mathbb{R}^{N \times H}$, $\boldsymbol{\theta} \in \mathbb{R}^H$ are learnable parameters, and H is the basket latent dimension.

Dynamic Context Encoder. Since basket sequence consists of a string of chronological sets of interactions, it is convenient to mine dynamic user interest hidden in the sequence to guide next basket recommendation. We utilize a Recurrent Neural Network (RNN) [1,6,15] to model basket sequence as:

$$\mathbf{s}_t = \mathcal{G}(\mathbf{h}_t\mathbf{\Psi} + \mathbf{s}_{t-1}\mathbf{\Omega} + \boldsymbol{\varphi}) \tag{8}$$

where $\mathbf{s}_t$ is the sequence hidden representation at time step t, $\mathcal{G}$ is an activation function, $\mathbf{\Psi} \in \mathbb{R}^{H \times K}$, $\mathbf{\Omega} \in \mathbb{R}^{K \times K}$ and $\boldsymbol{\varphi} \in \mathbb{R}^K$ are learnable parameters, K is the sequence latent dimension. The sequence hidden state at time step $|S|$, i.e., $\mathbf{s}_{|S|}$, encodes the interaction context of basket sequence S.

Basket Predictor. In intuition, we can get the probability that each item belongs to the next basket by linearly mapping the sequence representation $\mathbf{s}_{|S|}$ to probability space. However, the method ignores the importance of correlation between items within the same basket for next basket recommendation. So we utilize the correlation matrix $\mathbf{C}$ defined in Eq. (2) to provide information about the correlation between items:

$$\mathbf{y} = \mathbf{s}_{|S|} \cdot \frac{1}{1 + e^{-\mathbf{\Lambda C}}} \tag{9}$$

where $\mathbf{\Lambda} \in \mathbb{R}^{K \times N}$ is a learnable weight matrix.

We choose the recommendation loss $\mathcal{L}_{rec}$ with the expectation that items belonging to the next basket are assigned with a higher probability, and encourage them to maintain the greatest possible advantage over items not in the next basket. Then the objective function is formulated as:

$$\mathcal{L}_{rec} = -\frac{N - |\mathcal{B}_{|S|+1}|}{|\mathcal{B}_{|S|+1}|} \sum_{p \in \mathcal{B}_{|S|+1}} \log(\frac{1}{1 + e^{-\mathbf{y}_p}}) - \sum_{q \notin \mathcal{B}_{|S|+1}} \log(\frac{e^{r - \mathbf{y}_q}}{1 + e^{r - \mathbf{y}_q}}) \tag{10}$$

where r is the minimum probability of items belonging to the next basket.

To help the recommendation task encode basket sequence, and learn a better representation for final prediction, we try to integrate contrastive learning into recommendation task in the next section.

Prototype-based Contrastive Learning. With the correlative basket booster defined by Eq. (4), two augmentations can be obtained for a basket sequence $S = [\mathcal{B}_1, \ldots, \mathcal{B}_{|S|}]$ by executing augmentation twice, namely $S^{(1)} = [\mathcal{B}_1^{(1)}, \ldots, \mathcal{B}_{|S|}^{(1)}]$ and $S^{(2)} = [\mathcal{B}_1^{(2)}, \ldots, \mathcal{B}_{|S|}^{(2)}]$. There are $(2 \times M)$ augmentations corresponding to a batch of basket sequences with batch size of M. To construct contrastive signals, the two augmentations from the same basket sequence will be treated as a positive pair, and the remaining $2(M-1)$ augmentations from different basket sequences will be treated as negative pairs.

All the augmentations in the batch can be encoded by the context encoder defined in Eq. (8), denoted as $\mathcal{S}_{aug} = \left\{\mathbf{s}_1^{(1)}, \ldots, \mathbf{s}_M^{(1)}, \mathbf{s}_1^{(2)}, \ldots, \mathbf{s}_M^{(2)}\right\}$. The contrastive loss $\mathcal{L}_{cl}$ is designed to maximize the similarity of representations from the same basket, and minimize the similarity of representations from different baskets. To achieve the goal, we can define contrastive loss for the augmented basket sequence $S_p^{(1)}$ in the batch as:

$$\mathcal{L}_{cl}(\mathbf{s}_p^{(1)}, \mathbf{s}_p^{(2)}) = -\log \frac{\exp(\mathbf{s}_p^{(1)} \cdot \mathbf{s}_p^{(2)^T})}{\exp(\mathbf{s}_p^{(1)} \cdot \mathbf{s}_p^{(2)^T}) + \sum_{\mathbf{s}^- \in \mathcal{S}_{aug}^-} \exp(\mathbf{s}_p^{(1)} \cdot \mathbf{s}^{-T})} \tag{11}$$

where $\mathcal{S}_{aug}^-$ is the set of sequence representations of augmentations from different basket sequences with $S_p^{(1)}$ in the batch.

Note that introducing contrastive learning into the recommendation task would help the recommendation model learn the representation of basket sequence that is more conducive to predict recommendation probability. However, augmented basket sequences inevitably lose some information of their primitive basket sequences. In order to preserve as much information as possible of prototype, we further define a prototype-based contrastive learning loss as:

$$\mathcal{L}_{cl}(\mathbf{s}_p, \mathbf{s}_p^{(1)}, \mathbf{s}_p^{(2)}) = -\log \frac{\exp(\mathbf{s}_p^{(1)} \cdot \mathbf{s}_p{}^T)}{\exp(\mathbf{s}_p^{(1)} \cdot \mathbf{s}_p{}^T)) + \sum_{\mathbf{s}^- \in \mathcal{S}^-} \exp(\mathbf{s}_p^{(1)} \cdot \mathbf{s}^{-T})} - \log \frac{\exp(\mathbf{s}_p^{(1)} \cdot \mathbf{s}_p^{(2)}{}^T)}{\exp(\mathbf{s}_p^{(1)} \cdot \mathbf{s}_p^{(2)}{}^T) + \sum_{\mathbf{s}^- \in \mathcal{S}_{aug}^-} \exp(\mathbf{s}_p^{(1)} \cdot \mathbf{s}^{-T})} \quad (12)$$

where $\mathcal{S}^-$ is the set of sequence representations of prototypes and $\mathcal{S}_{aug}^{(1)} = \left\{\mathbf{s}_1^{(1)}, \ldots, \mathbf{s}_M^{(1)}\right\}$ in the batch except $\mathbf{s}_p$ and $\mathbf{s}_p^{(1)}$.

Multi-Task Training. In the previous sections, we introduced the correlative basket booster for augmentation, the part of model for recommendation task, and the part of model for contrastive learning task, respectively. In this section, we adopt a multi-task strategy to combine them by optimizing them jointly as:

$$\mathcal{L} = \mathcal{L}_{rec} + \lambda \mathcal{L}_{cl} \quad (13)$$

where $\lambda \in \mathbb{R}$ is a hyper-parameter to adjust the intensity of contrastive learning.

4 Experiments

In this section, we design extensive experiments to evaluate the performance of the proposed BPCL method against six baseline methods on three real-world datasets. In particular, we aim to answer the following three research questions:

- **RQ1**: How does the proposed BPCL method perform on next basket recommendation compared with existing methods?
- **RQ2**: Whether the prototype-based contrastive learning with correlative basket booster promotes the model to recommend the next basket?
- **RQ3**: How do the key hyper-parameters, i.e., the correlation latent dimension L, the weight of contrastive learning λ and the mask threshold μ, affect model quality?

4.1 Experiments Settings

Datasets. We conduct experiments on three public real-world datasets: Delicious[1], Beauty[2] and TaFeng[3]. Delicious contains tagging information from Nov 2003 to Nov 2010 of a social bookmarking system, in which the tags assigned to the same bookmark is regarded as a basket. Beauty consists of interactions from May 1996 to Jul 2014 of subcategory "Beauty" on Amazon, and TaFeng contains the transaction data of a Chinese grocery store from Nov 2000 to Feb 2001. We define the set of items that are interacted with the same user within the same day as a basket for the two datasets. Each basket sequence consists of the baskets from the same user in chronological order. The statistics of the three datasets are described in Table 1.

[1] http://www.delicious.com.

[2] https://www.kaggle.com/skillsmuggler/amazon-ratings.

[3] https://www.kaggle.com/chiranjivdas09/ta-feng-grocery-dataset.

Table 1. Statistics for the Delicious, Beauty and TaFeng datasets

Dataset	Num of items	Average basket size	Average basket sequence length
Delicious	8920	3.78	31.66
Beauty	19340	1.50	6.33
TaFeng	14313	5.76	4.90

Following [16], we divide the basket sequences into three non-overlapping time periods as training set, validation set and testing set. The items and users with less than 5 interactions, as well as the basket sequences of less than 3 in length are filtered out. For Delicious, the part of interactions before Sep 2010 is treated as training set, the part from Sep to Oct of 2010 is validation set and the part after Oct 2010 is testing set. For Beauty, the part of interactions before Jun 2013 is treated as training set, the part after Jul of 2013 is validation set and the part of 2014 is testing set. For TaFeng, the part of interactions of 2000 is treated as training set, the part before Jan of 2001 is validation set and the part after Feb 2001 is testing set.

Evaluation. Given a basket sequence $S = [\mathcal{B}_1, \ldots, \mathcal{B}_{|S|}]$, the preceding $(|S|-1)$ baskets are used to predict the last basket, and any sequence with more than 30 baskets will be truncated with the prefix cut off. To evaluate the performance of the proposed method and baselines, we adopt the widely used Hit Ratio ($HR@K$) [4,9,22] and F-measure ($F1@K$) [15,24] as evaluation metrics with $K = 5$ and $K = 20$, which can be formulated as:

$$HR@K = \frac{\sum Hit(\mathcal{B}_{pred}^{K})}{\sum_{\mathcal{B}_{target} \in \mathcal{T}} |\mathcal{B}_{target}|} \tag{14}$$

$$F1@K = \frac{2 \times Recall@K \times Precision@K}{Recall@K + Precision@K} \tag{15}$$

where $\mathcal{B}_{target}$ and $\mathcal{B}_{pred}^{K}$ denote a target basket and the corresponding predicted basket with the size of K respectively, Hit counts the number of items that appear in both $\mathcal{B}_{pred}^{K}$ and $\mathcal{B}_{target}$, and $\mathcal{T}$ is the set of the last basket in the testing set. The $Recall@K$ and $Precision@K$ are defined as:

$$Recall@K = \frac{1}{|\mathcal{T}|} \sum_{\mathcal{B}_{target} \in \mathcal{T}} \frac{Hit(\mathcal{B}_{pred}^{K})}{|\mathcal{B}_{target}|} \tag{16}$$

$$Precision@K = \frac{1}{|\mathcal{T}|} \sum_{\mathcal{B}_{target} \in \mathcal{T}} \frac{Hit(\mathcal{B}_{pred}^{K})}{|\mathcal{B}_{pred}^{K}|}. \tag{17}$$

Baselines. We adopt the following six recommendation models for comparison.

- **POP-K** It is a non-personalized recommendation model that recommends K items with the highest popularity in terms of basket for users.

- **MCNet** It learns transition probability between the latest basket and candidate items based on the Markov-chain implemented by a neural network.
- **BSEQ** [15] It recommends next basket based on the corresponding basket sequence representation learned by making use of recurrent neural network.
- **SASRec** [12] It makes an adaption for transformer layer to learn the correlation of items in sequences, and recommends items based on the correlation.
- **CoSeRec** [22] It introduces two informative augmentations for item sequence to construct self-supervised signals, and applies transformer encoder to promote recommendation performance.
- **Beacon** [16] It is a state-of-the-art model for next basket recommendation that utilizes correlation between items to encode basket sequences and conduct next basket recommendation.

Implementation Details. We choose LSTM with 0.3 dropout probability as the type of recurrent layer units, and set the latent dimension L, H and K as 32. The RMSProp optimizer with a learning rate of 0.01 is adopted for optimizing the model. For hyper-parameters, we tune mask threshold μ within the range of $\{0.5, 0.6, 0.7, 0.8, 0.9\}$, and contrastive learning weight λ within the range of $\{0.05, 0.1, 0.2, 0.3, 0.4\}$. To be fair, the hyper-parameters of baseline methods share the same experimental settings as mentioned above. As for the others, we tune them on the validation set applying early stopping with patience of 5 epochs, and report results on the testing set.

4.2 Performance Comparison

Table 2 presents the experimental results of the all methods on the three real-world datasets. We can observe that the proposed BPCL method obtains the best performance on the three datasets in terms of the four metrics. For all datasets, our model achieves at least 7% improvement in every metric, which helps answer **RQ1.** In particular, it achieves at least 44% improvement on Beauty. We can find that Beauty contains the largest number of items and the least average interactions among the three datasets from Table 1, so the experimental results suggest that BPCL can effectively overcome the difficulties caused by data sparsity in next basket recommendation.

Note that CoSeRec is a sequential recommendation model that introduces informative augmentation for contrastive learning, but achieves poor performance comparing to the others on Delicious which contains the most average interactions among the three datasets. This is perhaps due to its intent to predict the next item by modeling item sequence with the assumption that interactions are in strict chronological order which is at odds with the reality, and improper augmentation for item sequences destruct the intra-basket correlation. BSEQ performs well on Delicious and Tafeng, but fails on Beauty, a dataset with large data sparsity. This indicates that although next basket recommendation can improve the recommendation effectiveness by fitting actual situation, it is prone

Table 2. Performance comparison of different methods on next basket recommendation. The best scores on all datasets are highlighted in bold and the second best is labeled with *. The last column is the improvement compared with the best baseline results.

Model	Metrics		POP	BSEQ	MCNet	SASRec	CoSeRec	Beacon	BPCL	**Improve**
Delicious	HR@K (%)	5	1.84	3.28	3.66*	2.89	0.16	3.09	**4.00**	9.29%
		20	7.73	10.5*	10.4	8.86	0.59	9.25	**11.4**	8.57%
	F1@K ($\times 10^2$)	5	1.83	2.24	2.57*	2.19	0.11	2.22	**2.75**	7.00%
		20	2.24	2.93*	2.91	2.50	0.17	2.61	**3.16**	7.85%
Beauty	HR@K (%)	5	0.04	0.00	0.23	0.11	0.15	0.41*	**0.68**	65.9%
		20	0.83	0.19	0.26	0.34	0.38	1.02*	**1.47**	44.1%
	F1@K ($\times 10^2$)	5	0.02	0.00	0.09	0.04	0.06	0.16*	**0.28**	75.0%
		20	0.12	0.03	0.03	0.05	0.05	0.13*	**0.19**	46.2%
Tafeng	HR@K (%)	5	4.65	4.66	4.73	4.32	3.67	5.11*	**5.52**	8.02%
		20	6.03	6.41	6.65	6.24	5.98	7.29*	**7.88**	8.09%
	F1@K ($\times 10^2$)	5	4.79	4.80	4.84	4.40	3.72	5.20*	**5.63**	8.27%
		20	2.75*	2.45	2.52	2.37	2.27	2.75*	**2.98**	8.36%

to be affected by data density and is in great instability. All of the above suggest that the proposed correlative basket booster could promote the model to recommend the next basket. In order to further verify this conjecture, ablation study is conducted.

4.3 Ablation Study

The performance of BPCL and its variants on all three datasets are shown in Table 3. BPCL-CL indicates BPCL without contrastive learning, and BPCL-prototype indicates BPCL adopting contrastive loss without the introduction of prototype, i.e., Eq. (11). It is clear that the performances of BPCL-CL and BPCL-prototype are no better than that of BPCL in terms of the four metrics on all datasets. More specially, sparse data magnifies the advantage, i.e., BPCL achieves the greatest improvement of performance on Beauty. Although BPCL-prototype attains better performance than BPCL-CL because of the integration of contrastive signal, it is still hard to surpass BPCL due to the inevitable loss of prototype information. We can conclude that the proposed prototype-based contrastive learning with correlative basket booster promotes the model to recommend the next basket (**RQ2**).

4.4 Hyper-Parameter Study

In this section, we analyze three key hyper-parameters of the proposed BPCL method to answer **RQ3**, including the correlation latent dimension L, the weight of contrastive learning λ and the mask threshold μ.

Table 3. Performance of the proposed BPCL and its variants for ablation studies.

Dataset	Model	HR@K (%)		F1@K ($\times 10^2$)	
		5	20	5	20
Delicious	BPCL-CL	3.732	10.02	2.608	2.810
	BPCL-prototype	3.237	10.93	2.202	3.036
	BPCL	**3.998**	**11.35**	**2.751**	**3.159**
Beauty	BPCL-CL	0.000	0.000	0.000	0.000
	BPCL-prototype	0.451	1.278	0.192	0.170
	BPCL	**0.677**	**1.466**	**0.280**	**0.194**
Tafeng	BPCL-CL	2.775	4.323	2.785	1.620
	BPCL-prototype	5.517	7.676	5.626	2.886
	BPCL	**5.517**	**7.879**	**5.631**	**2.976**

Firstly, we explore how the correlation latent dimension L influences the performance of BPCL and show the results in Fig. 3. It is obvious that BPCL holds steady performance on the three datasets with different L and keeps a slow lift on the whole as the increase of L until it reaches 32. The similar performance of BPCL with $L = 32$ and BPCL with $L = 8$ implies that a small correlation latent dimension is sufficient for embedding the correlation information, which is consistent with the fact of data sparsity. The performance tends to decrease when L is larger than 32, which indicates that too large latent dimension makes it difficult for the model to capture the most valuable information for augmentation.

Next, we investigate how the different combinations of the weight of contrastive learning λ and the mask threshold μ in correlative basket booster affect recommendation performance. The λ is tuned among $\{0.05, 0.1, 0.2, 0.3, 0.4\}$, while the μ is tuned among $\{0.5, 0.6, 0.7, 0.8, 0.9\}$. We adopt heatmaps to show the results of the proposed BPCL method with different combinations on Beauty visually, and the results in terms of HR are shown in upper Fig. 4 and the results in terms of F1 are shown in lower Fig. 4. In general, better performance is achieved by λ larger than 0.2 with $\mu = 0.5$, suggesting that proper combination of the weight of contrastive learning and randomness helps the model capture useful self-supervised signal from prototype to learn better basket sequence representation indeed, thus can contribute to overcome the difficulties caused by sparsity data and get a better next basket recommendation performance.

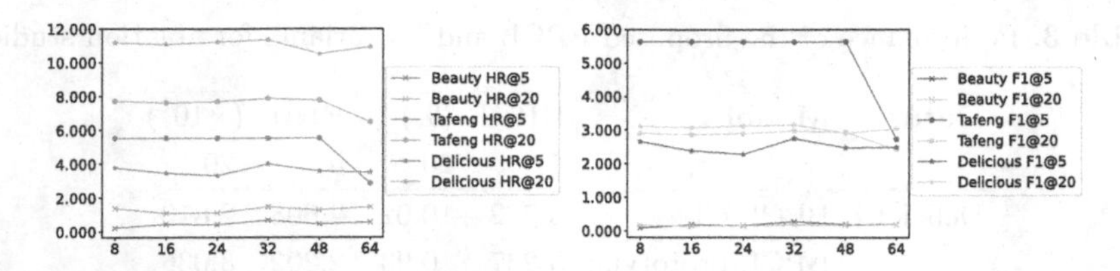

Fig. 3. The performance of BPCL with varying correlation latent dimension L. The left corresponds to $HR@K$ (%) and the right corresponds to $F1@K$ ($\times 10^2$). The x-axis denotes L varies within the range of $\{8, 16, 24, 32, 48, 64\}$, and the y-axis denotes $HR@K$ (%) and $F1@K$ ($\times 10^2$) respectively.

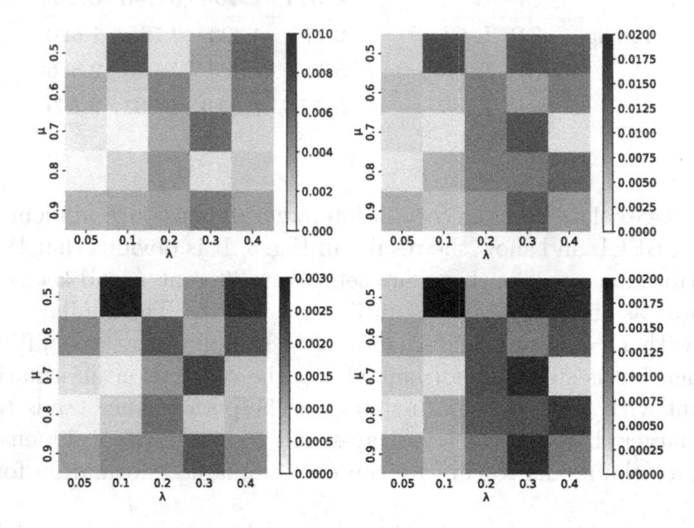

Fig. 4. Heatmap of Hit Ratio (HR) and F1 on Beauty. The upper left corresponds to $HR@5$ (%), the upper right corresponds to $HR@20$ (%), the lower left corresponds to $F1@5$ ($\times 10^2$), the lower right corresponds to $F1@20$ ($\times 10^2$). The x-axis denotes λ varies within the range of $\{0.05, 0.1, 0.2, 0.3, 0.4\}$, and the y-axis denotes μ varies within the range of $\{0.5, 0.6, 0.7, 0.8, 0.9\}$. The darker color represents the better performance.

5 Conclusion

In this paper, we propose a new BPCL method that introduces contrastive learning to next basket recommendation. A correlative basket booster is designed to make augmentation for baskets, which can mine self-supervised signals from primitive basket sequences. The augmentations are utilized by prototype-based contrastive learning for promoting next basket recommendation task. To our best knowledge, this is the first work to promote learning of prototype representation through basket augmentation, which helps overcome the difficulties caused by data sparsity and leads to a better next basket recommendation performance. The proposed method is verified on three public real-world datasets, and show the best performance compared with the baseline methods.

Acknowledgments. This work was supported by NSFC (61876193) and Guangdong Province Key Laboratory of Computational Science at Sun Yat-sen University (2020B1212060032).

References

1. Cui, Q., Wu, S., Liu, Q., Zhong, W., Wang, L.: MV-RNN: a multi-view recurrent neural network for sequential recommendation. IEEE Trans. Knowl. Data Eng. **32**(2), 317–331 (2020)
2. Dai, S., Yu, Y., Fan, H., Dong, J.: Spatio-temporal representation learning with social tie for personalized POI recommendation. Data Sci. Eng. **7**(1), 44–56 (2022)
3. Deng, Z., Huang, L., Wang, C., Lai, J., Yu, P.S.: DeepCF: a unified framework of representation learning and matching function learning in recommender system. In: The Thirty-Third AAAI Conference on Artificial Intelligence, Honolulu, pp. 61–68. AAAI Press (2019)
4. Du, Y., Liu, H., Wu, Z.: Modeling multi-factor and multi-faceted preferences over sequential networks for next item recommendation. In: Oliver, N., Pérez-Cruz, F., Kramer, S., Read, J., Lozano, J.A. (eds.) ECML PKDD 2021. LNCS (LNAI), vol. 12976, pp. 516–531. Springer, Cham (2021). https://doi.org/10.1007/978-3-030-86520-7_32
5. Flanagan, A., Oyomno, W., Grigorievskiy, A., Tan, K.E., Khan, S.A., Ammad-Ud-Din, M.: Federated multi-view matrix factorization for personalized recommendations. In: Hutter, F., Kersting, K., Lijffijt, J., Valera, I. (eds.) ECML PKDD 2020. LNCS (LNAI), vol. 12458, pp. 324–347. Springer, Cham (2021). https://doi.org/10.1007/978-3-030-67661-2_20
6. Gama, R., Fernandes, H.L.: An attentive RNN model for session-based and context-aware recommendations: a solution to the recsys challenge 2019. In: Proceedings of the Workshop on ACM Recommender Systems Challenge, Copenhagen, pp. 6:1–6:5. ACM (2019)
7. He, Z., Zhao, H., Lin, Z., Wang, Z., Kale, A., McAuley, J.J.: Locker: locally constrained self-attentive sequential recommendation. In: The 30th ACM International Conference on Information and Knowledge Management, Virtual Event, pp. 3088–3092. ACM (2021)
8. Hu, H., He, X., Gao, J., Zhang, Z.: Modeling personalized item frequency information for next-basket recommendation. In: Proceedings of the 43rd International ACM SIGIR conference on research and development in Information Retrieval, Virtual Event, pp. 1071–1080. ACM (2020)
9. Islek, I., Oguducu, S.G.: A hybrid recommendation system based on bidirectional encoder representations. In: Koprinska, I., et al. (eds.) ECML PKDD 2020. CCIS, vol. 1323, pp. 225–236. Springer, Cham (2020). https://doi.org/10.1007/978-3-030-65965-3_14
10. Ji, Y., et al.: Temporal heterogeneous interaction graph embedding for next-item recommendation. In: Hutter, F., Kersting, K., Lijffijt, J., Valera, I. (eds.) ECML PKDD 2020. LNCS (LNAI), vol. 12459, pp. 314–329. Springer, Cham (2021). https://doi.org/10.1007/978-3-030-67664-3_19
11. Kang, G., Jiang, L., Wei, Y., Yang, Y., Hauptmann, A.: Contrastive adaptation network for single- and multi-source domain adaptation. IEEE Trans. Pattern Anal. Mach. Intell. **44**(4), 1793–1804 (2022)

12. Kang, W., McAuley, J.J.: Self-attentive sequential recommendation. In: IEEE International Conference on Data Mining, Singapore, pp. 197–206. IEEE Computer Society (2018)
13. Ke, G., Hong, Z., Zeng, Z., Liu, Z., Sun, Y., Xie, Y.: CONAN: contrastive fusion networks for multi-view clustering. In: 2021 IEEE International Conference on Big Data (Big Data), Orlando, pp. 653–660. IEEE (2021)
14. Kim, S., Jeong, S., Kim, E., Kang, I., Kwak, N.: Self-supervised pre-training and contrastive representation learning for multiple-choice video QA. In: Thirty-Fifth AAAI Conference on Artificial Intelligence, 2021, pp. 13171–13179. AAAI Press (2021)
15. Le, D., Lauw, H.W., Fang, Y.: Modeling contemporaneous basket sequences with twin networks for next-item recommendation. In: Proceedings of the Twenty-Seventh International Joint Conference on Artificial Intelligence, Stockholmn, pp. 3414–3420. ijcai.org (2018)
16. Le, D., Lauw, H.W., Fang, Y.: Correlation-sensitive next-basket recommendation. In: Proceedings of the Twenty-Eighth International Joint Conference on Artificial Intelligence, Macao, pp. 2808–2814. ijcai.org (2019)
17. Lee, S., Lee, Y., Lee, G., Hwang, S.: Supervised contrastive embedding for medical image segmentation. IEEE Access **9**, 138403–138414 (2021)
18. Lee, T., Yoo, S.: Augmenting few-shot learning with supervised contrastive learning. IEEE Access **9**, 61466–61474 (2021)
19. Li, C., Hsu, C., Zhang, Y.: FairSR: fairness-aware sequential recommendation through multi-task learning with preference graph embeddings. ACM Trans. Intell. Syst. Technol. **13**(1), 16:1-16:21 (2022)
20. Li, Y., Hu, P., Liu, J.Z., Peng, D., Zhou, J.T., Peng, X.: Contrastive clustering. In: Thirty-Fifth AAAI Conference on Artificial Intelligence, Virtual Event, pp. 8547–8555. AAAI Press (2021)
21. Liu, T., Yin, X., Ni, W.: Next basket recommendation model based on attribute-aware multi-level attention. IEEE Access **8**, 153872–153880 (2020)
22. Liu, Z., Chen, Y., Li, J., Yu, P.S., McAuley, J.J., Xiong, C.: Contrastive self-supervised sequential recommendation with robust augmentation. CoRR abs/2108.06479 (2021)
23. Lu, Y., et al.: Social influence attentive neural network for friend-enhanced recommendation. In: Dong, Y., Mladenić, D., Saunders, C. (eds.) ECML PKDD 2020. LNCS (LNAI), vol. 12460, pp. 3–18. Springer, Cham (2021). https://doi.org/10.1007/978-3-030-67667-4_1
24. Qin, Y., Wang, P., Li, C.: The world is binary: contrastive learning for denoising next basket recommendation. In: The 44th International ACM SIGIR Conference on Research and Development in Information Retrieval, Virtual Event, pp. 859–868. ACM (2021)
25. Qiu, R., Huang, Z., Yin, H., Wang, Z.: Contrastive learning for representation degeneration problem in sequential recommendation. In: WSDM 2022: The Fifteenth ACM International Conference on Web Search and Data Mining, Virtual Event/Tempe, 2022, pp. 813–823. ACM (2022)
26. Rendle, S., Freudenthaler, C., Schmidt-Thieme, L.: Factorizing personalized markov chains for next-basket recommendation. In: Proceedings of the 19th International Conference on World Wide Web, Raleigh, pp. 811–820. ACM (2010)
27. Tong, X., Wang, P., Li, C., Xia, L., Niu, S.: Pattern-enhanced contrastive policy learning network for sequential recommendation. In: Zhou, Z. (ed.) Proceedings of the Thirtieth International Joint Conference on Artificial Intelligence, Virtual Event/Montreal, pp. 1593–1599. ijcai.org (2021)

28. Wan, S., Lan, Y., Wang, P., Guo, J., Xu, J., Cheng, X.: Next basket recommendation with neural networks. In: Poster Proceedings of the 9th ACM Conference on Recommender Systems, Vienna. CEUR Workshop Proceedings, vol. 1441. CEUR-WS.org (2015)
29. Xi, W., Huang, L., Wang, C., Zheng, Y., Lai, J.: BPAM: recommendation based on BP neural network with attention mechanism. In: Proceedings of the Twenty-Eighth International Joint Conference on Artificial Intelligence, Macao, pp. 3905–3911. ijcai.org (2019)
30. Xie, Z., Liu, C., Zhang, Y., Lu, H., Wang, D., Ding, Y.: Adversarial and contrastive variational autoencoder for sequential recommendation. In: The Web Conference 2021, Virtual Event/Ljubljana, pp. 449–459. ACM/IW3C2 (2021)
31. Yu, F., Liu, Q., Wu, S., Wang, L., Tan, T.: A dynamic recurrent model for next basket recommendation. In: Proceedings of the 39th International ACM SIGIR conference on Research and Development in Information Retrieval, Pisa, pp. 729–732. ACM (2016)
32. Yuan, X., Chen, H., Song, Y., Zhao, X., Ding, Z.: Improving sequential recommendation consistency with self-supervised imitation. In: Proceedings of the Thirtieth International Joint Conference on Artificial Intelligence, Virtual Event/Montreal, pp. 3321–3327. ijcai.org (2021)

Graph Contrastive Learning with Adaptive Augmentation for Recommendation

Mengyuan Jing, Yanmin Zhu(✉), Tianzi Zang, Jiadi Yu, and Feilong Tang

Shanghai Jiao Tong University, Shanghai, China
{jingmy,yzhu,zangtianzi,jiadiyu,tang-fl}@sjtu.edu.cn

Abstract. Graph Convolutional Network (GCN) has been one of the most popular technologies in recommender systems, as it can effectively model high-order relationships. However, these methods usually suffer from two problems: sparse supervision signal and noisy interactions. To address these problems, graph contrastive learning is applied for GCN-based recommendation. The general framework of graph contrastive learning is first to perform data augmentation on the input graph to get two graph views and then maximize the agreement of representations in these views. Despite the effectiveness, existing methods ignore the differences in the impact of nodes and edges when performing data augmentation, which will degrade the quality of the learned representations. Meanwhile, they usually adopt manual data augmentation schemes, limiting the generalization of models. We argue that the data augmentation scheme should be learnable and adaptive to the inherent patterns in the graph structure. Thus, the model can learn representations that remain invariant to perturbations of unimportant structures while demanding fewer resources. In this work, we propose a novel **G**raph **C**ontrastive learning framework with **A**daptive data augmentation for **Rec**ommendation (GCARec). Specifically, for adaptive augmentation, we first calculate the retaining probability of each edge based on the attention mechanism and then sample edges according to the probability with a Gumbel Softmax. In addition, the adaptive data augmentation scheme is based on the neural network and requires no domain knowledge, making it learnable and generalizable. Extensive experiments on three real-world datasets show that GCARec outperforms state-of-the-art baselines.

Keywords: Recommender systems · Graph neural network · Contrastive learning · Self-supervised learning

1 Introduction

Recommender systems have been an indispensable component in many online services, such as E-commerce platforms and online entertainment applications, for their effectiveness in alleviating information overloading. Recently, graph convolution network (GCN) has become a state-of-the-art method in recommender systems [2,10,23,24], as it can integrate high-order neighbors in the graphs.

M.-R. Amini et al. (Eds.): ECML PKDD 2022, LNAI 13713, pp. 590–605, 2023.
https://doi.org/10.1007/978-3-031-26387-3_36

Despite the effectiveness, GCN-based recommendation models still suffer from two problems [25]. (1) sparse supervision signal: most GCN-based methods focus on supervised settings [10,23], where the supervision signal is derived from the observed interactions. However, the observed interactions are very sparse in comparison to the entire interaction space [1,34], limiting the performance of these methods. (2) noisy interactions. The interaction data usually contain noises, since users often provide implicit feedback like clicks rather than explicit feedback like ratings. There may be cases where users click or purchase items and find they do not like them [28]. Since GCN-based models learn representations by aggregating information from neighbors, noisy interactions will make GCNs fail to learn reliable representations.

To address these problems, we apply graph contrastive learning [32] in GCN-based recommendation. A general framework of graph contrastive learning is to perform data augmentation on the input graph by uniform node/edge dropout to generate two graph views and then maximize the agreement of representations in these views. It extracts additional supervised signals from the input data and learns representations that remain invariant to the perturbations introduced by the data augmentation, making it possible to solve both problems mentioned above. Although several works [15,25,33] leverage graph constative learning in GCN-based recommendation, data augmentation schemes, a key component of contrastive learning [27] are still rarely explored in graph contrastive learning-based recommendation methods. Specifically, there are currently two main types of data augmentation schemes. (1)stochastic augmentation [25], such as uniformly dropping a portion of edges or nodes from the input graph; (2) artificially designed augmentation [15,33], which usually constructs views from additional domain knowledge, such as social networks and knowledge graphs. We argue that these augmentation schemes have two limitations:

First, they ignore the differences in the impact of nodes and edges when performing data augmentation. Taking uniform edge dropout as an example, dropping important edges (e.g., edges that connect items of great interest to a user) for a user will degrade the quality of its representation. Besides, keeping unimportant edges (e.g., the noisy interactions) can be harmful to representation learning. Second, they choose augmentation schemes either based on domain knowledge or performance evaluation on the validation set. However, domain knowledge is not always available and performance evaluation is usually expensive, limiting the generalization of the model to other datasets. Hence, we propose that the data augmentation scheme should be learnable and adaptive to the intrinsic patterns in the graph structure. Such that, it can guide the model to keep the important connective structures while perturbing the unimportant edges in the graph, as well as demanding fewer resources compared with manual selection.

To this end, we propose a novel **G**raph **C**ontrastive learning framework with **A**daptive data augmentation for **Rec**ommendation (GCARec). Specifically, we design a learnable data augmentation scheme based on the neural network, whose parameters are learned from data during the optimization of the contrastive objective. Meanwhile, as introduced above, the data augmentation scheme should preserve important edges while perturbing unimportant ones. To achieve this goal, we make the preserving probability of important edges higher and the probability of unimportant ones lower. Specifically, the probability is calculated based on the attention mechanism to reflect the importance of each edge. After getting the probability, we sample edges according to them. However, the random sampling strategy is not differentiable and will make the model cannot be trained. Hence, we propose a view generation method based on a Gumbel Softmax [11], which introduces a continuous distribution to approximate categorical samples, to address this issue. After that, we apply contrastive learning based on node self-discrimination to maximize the agreement between the representations in the generated views. Finally, we leverage the contrastive learning task (i.e., node self-discrimination) as the auxiliary task to the recommendation task and jointly train them using the multi-task training strategy.

In summary, the contributions of our work are as follows:

1. We propose a novel graph contrastive learning framework with adaptive data augmentation, which encourages the model to learn important structural information in the graph.
2. We designed a learnable data augmentation scheme that not only requires less human effort but also easily generalizes to different graph-based recommendation scenarios.
3. Extensive experiments are conducted on three real-world datasets. The results indicate that our model outperforms the state-of-art methods and demonstrates the effectiveness of the adaptive data augmentation scheme.

The remainder of this paper is organized as follows. We first introduce the notations and GCN-based recommendation in Sect. 2. We then present details of our model in Sect. 3. Settings and results of conducted experiments are introduced in Sect. 4. We summarize the related work in Sect. 5. Finally, we conclude this work and future work in Sect. 6.

2 Preliminaries

Notations. Let $\mathcal{U} = \{u_1, u_2, \cdots, u_m\}(|\mathcal{U}| = m)$ and $\mathcal{I} = \{i_1, i_2, \cdots, i_n\}(|\mathcal{I}| = n)$ denote the set of users and items, respectively. Let $\mathcal{O}^+ = \{y_{u,i}|u \in \mathcal{U}, i \in \mathcal{I}\}$ denote the observed interactions, where $y_{u,i}$ indicates that user u has interacted with item i. Moreover, a user-item interaction graph is constructed, denoted as $\mathcal{G} = (\mathcal{V}, \mathcal{E})$, where $\mathcal{V} = \mathcal{U} \cup \mathcal{I}$ is the set of nodes and $\mathcal{E} = \{(u, i)|y_{u,i} \in \mathcal{O}^+, u \in \mathcal{U}, i \in \mathcal{I}\}$ is the edge set.

Recap GCN. Generally, at each layer of GCN-based recommendation models, two key computations, i.e., aggregate and update, are involved to generate the node representations, which can be formulated as follows:

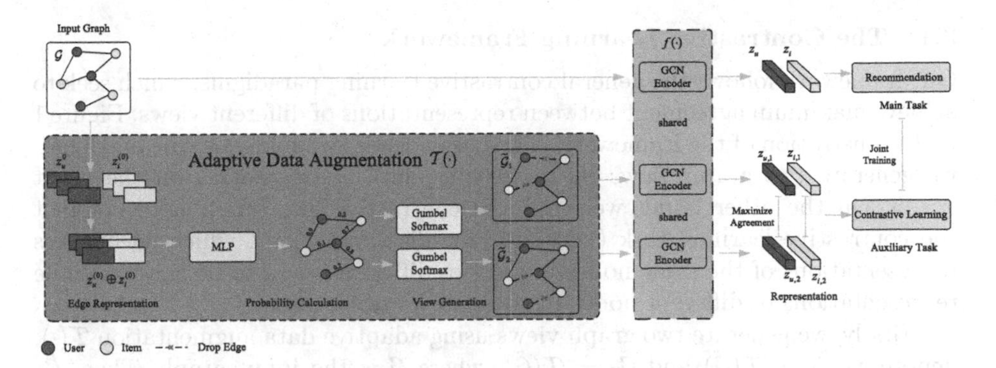

Fig. 1. Overall framework illustration of our proposed GCARec. The upper is the recommendation task, which predicts the preference score of users on items. The bottom is the contrastive learning task, which aims to maximize the agreement of representations in different views.

$$\begin{aligned} a_u^{(l)} &= f_{\text{aggregate}}(\{z_i^{(l-1)} | i \in \mathcal{N}_u\}), \\ z_u^{(l)} &= f_{\text{update}}(z_u^{(l-1)}, a_u^{(l)}), \end{aligned} \tag{1}$$

where $\mathcal{N}_u$ denotes the neighbors of node u and $a_u^{(l)}$ is the aggregated representation of $\mathcal{N}_u$. $z_u^{(l)}$ is the representation of node u at the l-th layer. $z_u^{(0)}$ is initialized by the learnable embedding. To generate the representation of node u, it first aggregates the representations of $\mathcal{N}_u$ and then updates the representation of u from its representation at $(l-1)$-th layer and the aggregated representations. It can be seen that $z_u^{(l)}$ encodes the l-hop neighbors of u. To produce the final representation for the recommendation, a readout function may be used:

$$z_u = f_{\text{readout}}\left(\left\{z_u^{(l)} \mid l = [0, \cdots, L]\right\}\right), \tag{2}$$

where L is the number of layers in the GCN-based model.

After getting the final representations, a prediction layer is built to calculate the preference score, which indicates how likely user u would adopt item i. For fast retrieval, the inner product usually is adopted:

$$\hat{y}_{u,i} = z_u^T z_i, \tag{3}$$

where z_u and z_i are final representations of u and i, respectively.

3 Methodology

In this section, we introduce our method in detail. Firstly, in Sect. 3.1, we describe the overall framework of GCARec. Then in Sect. 3.2 and Sect. 3.3, we present the adaptive data augmentation method to generate different views and the contrastive learning based on node self-discrimination. Finally, in Sect. 3.4, the multi-task training strategy is introduced.

3.1 The Contrastive Learning Framework

Our framework follows the general contrastive learning paradigms, which seek to achieve maximum agreement between representations of different views. Figure 1 is the illustration of the framework. The basic idea is two folds. On the one hand, we generate two graph views by performing data augmentation on the input graph. On the other hand, we employ a contrastive loss function to conduct the contrastive learning task (i.e., node self-discrimination), which encourages representations of the same nodes in the two different views to be similar, while representations of different nodes in those views to be distinct.

Firstly, we generate two graph views using adaptive data augmentation $\mathcal{T}(\cdot)$, denote as $\tilde{\mathcal{G}}_1 = \mathcal{T}(\tilde{\mathcal{G}})$ and $\tilde{\mathcal{G}}_2 = \mathcal{T}(\tilde{\mathcal{G}})$, where $\mathcal{G}$ is the input graph. Then, $\tilde{\mathcal{G}}_1$ and $\tilde{\mathcal{G}}_2$ are fed to the GCN encoder $f(\cdot)$, after which the encoder outputs representations of nodes in the two generated views. $Z_1 = f(\tilde{\mathcal{G}}_1)$ and $Z_2 = f(\tilde{\mathcal{G}}_2)$ denote the representations in the two views, respectively. Specifically, we adopt LightGCN [10] as the GCN encoder in this work.

Next, we perform the contrastive learning task on the representation of generated views. Since we focus on the data augmentation, we simply treat the same nodes in different views as positive pairs and different nodes in different views as negative pairs. To be specific, for a node $u \in \mathcal{U}$, $\{(z_{u,1}, z_{u,2})|u \in \mathcal{U}, z_{u,1} \in Z_1, z_{u,2} \in Z_2\}$ is the positive pair and $\{(z_{u,1}, z_{u',2})|u, u' \in \mathcal{U}, u' \neq u, z_{u,1} \in Z_1, z_{u',2} \in Z_2\}$ are the negative pairs. Technically, we adopt the InfoNCE [7] as the contrastive loss.

Finally, we adopt a multi-task training strategy to improve the recommendation performance. In specific, the recommendation task is the main task and the contrastive learning task is the auxiliary task.

3.2 Adaptive Augmentation

The user-item interaction graph contains several collaborative filtering signals, as it is constructed based on observed interaction. For example, the first-order neighbors reflect user interest and the second-order neighbors exhibit behavior similarity. Hence, it is useful to mine the inherent patterns in the user-item interaction graph structure. In addition, contrastive learning that maximizes the agreement between views aims to learn representations that remain invariant to the perturbations introduced by data augmentation [30]. Therefore, the data augmentation scheme should preserve important connective structures in the user-item interaction graph while perturbing unimportant ones.

To achieve this goal, we propose an adaptive data augmentation scheme that tends to retain the important edges and perturb possibly unimportant edges in the user-item interaction graph. To be specific, we sample edges in the graph with lower preserving probability for unimportant edges and higher preserving probability for important ones. Compared with methods that randomly corrupt views, our method puts more emphasis on important structures, which can guide the model to mine the inherent patterns in the graph structure. Two main processes are included in our proposed adaptive data augmentation: probability calculation and view generation.

Formally, we first calculate the preserving probability of each edge (u, i) as $p_{u,i}$. Then, we sample two subsets $\tilde{\mathcal{E}}_1$ and $\tilde{\mathcal{E}}_2$ from the original edge set $\mathcal{E}$ with the probability of each edge to generate two views. We will introduce the details as follows.

Probability Calculation. As we aim to corrupt unimportant edges and keep important structures in the user-item interaction graph, $p_{u,i}$ is required to reflect the importance of edge (u, i). In GCARec, we calculate the preserving probability of each edge based on the attention mechanism, which has been shown to be effective in modeling the importance of the edge. Following GAT [20], we perform self-attention on the nodes using a Multi-Layer Perception (MLP), and then calculate $p_{u,i}$ according to the following equations:

$$\begin{aligned} \alpha_{u,i} &= \mathbf{W}(z_u^{(0)} \oplus z_i^{(0)}) + \mathbf{b}, \\ p_{u,i} &= \frac{\exp(\alpha_{u,i})}{1 + \exp(\alpha_{u,i})} \end{aligned} \tag{4}$$

where $\alpha_{u,i}$ is the attention coefficient, which indicates the importance of i to u. $\oplus$ is the concatenation operation. $\mathbf{W}$ and $\mathbf{b}$ are trainable parameters. $z_u^{(0)}$ and $z_i^{(0)}$ are representations initialized by the learnable embeddings.

View Generation. After getting the preserving probabilities, we can randomly sample edges according to their preserving probabilities. However, random sampling is not differentiable and makes model cannot be trained well via back-propagation. Inspired by [17], we adopt Gumbel Softmax to be the differential surrogate to address this issue. It can be described as follows:

$$g(u, i) = \frac{\exp\left((\log(p_{u,i}) + g_1)/\tau_1\right)}{\sum_{y=0}^{1} \exp\left((\log(p_{u,i}^y (1 - p_{u,i})^{1-y}) + g_y)/\tau_1\right)}, \tag{5}$$

where g_y is the noise, which is i.i.d sampled from a Gumbel distribution: $g = -\log(-\log(x))$ and $x \sim \text{Uniform}(0, 1)$. τ_1 is the temperature parameter. When $\tau_1 \rightarrow 0$, $\mathcal{G}$ approximates to a one-hot vector. The edge subset of the generated view is $\tilde{\mathcal{E}} = \{(u, i) | g(u, i) > p, (u, i) \in \mathcal{E}\}$, where the threshold p is a hyperparameter to control the removal of unimportant edges.

3.3 Contrastive Learning

As we introduced in Sect. 3.1, we treat the same nodes in the different views as positive pairs and different nodes in the different views as negative pairs. The loss function of the contrastive learning task can be described as follows:

$$\mathcal{L}_{gcl} = \mathcal{L}_{gcl}^{user} + \mathcal{L}_{gcl}^{item}, \tag{6}$$

$$\mathcal{L}_{gcl}^{user} = \sum_{u \in \mathcal{U}} -\log \frac{\exp(sim(z_{u,1}, z_{u,2})/\tau)}{\exp(sim(z_{u,1}, z_{u,2})/\tau) + \sum_{u' \neq u, u' \in \mathcal{U}} sim(z_{u,1}, z_{u',2})/\tau)}, \tag{7}$$

$$\mathcal{L}_{gcl}^{item} = \sum_{i \in \mathcal{I}} -\log \frac{\exp(sim(z_{i,1}, z_{i,2})/\tau)}{\exp(sim(z_{i,1}, z_{i,2})/\tau) + \sum_{i' \neq i, i' \in \mathcal{I}} sim(z_{i,1}, z_{i',2})/\tau)}, \tag{8}$$

where $\mathcal{L}_{gcl}^{user}$ and $\mathcal{L}_{gcl}^{item}$ are the contrastive losses of the user side and item side, respectively. $sim(\cdot, \cdot)$ is the discriminator function, which takes two vectors as the input and then scores the similarity between them. In this work, we set it as cosine similarity function, i.e., $sim(z_1, z_2) = <z_1, z_2> / (||z_1|| \cdot ||z_2||)$. τ is the temperature to amplify the effect of discrimination.

3.4 Multi-task Training

We leverage a multi-task training strategy to optimize the main recommendation task and the auxiliary contrastive learning task jointly.

$$\mathcal{L} = \mathcal{L}_{main} + \lambda_1 \mathcal{L}_{gcl} + \lambda_2 ||\Theta||_2^2, \tag{9}$$

where Θ is the set of model parameters. λ_1 and λ_2 are hyperparameters to control the strengths of contrastive loss and L_2 regularization, respectively. $\mathcal{L}_{main}$ is the loss function of the main recommendation task. In this work, we adopt Bayesian Personalized Ranking (BPR) loss [19]:

$$\mathcal{L}_{main} = \sum_{(u,i,j) \in O} -\log \sigma \left(\hat{y}_{u,i} - \hat{y}_{u,j} \right), \tag{10}$$

where $\hat{y}_{u,i} = z_u^T z_i$ is the preference score. $\sigma(\cdot)$ is the sigmoid function. $\mathcal{O} = \{(u, i, j) \mid (u, i) \in \mathcal{O}^+, (u, j) \in \mathcal{O}^-\}$ denotes the training data, and $\mathcal{O}^-$ is the unobserved interactions.

4 Experiments

To evaluate the effectiveness of our proposed GCARec, we conduct extensive experiments by answering the following questions:

- **RQ1**: Does GCARec outperform state-of-the-art methods for the top-K recommendation?
- **RQ2**: What are the benefits of our proposed adaptive data augmentation scheme for recommendation performance?
- **RQ3**: Is our model sensitive to hyperparameters? How do different hyperparameters influence the recommendation performance?

4.1 Experimental Setup

Datasets. We adopt three real-world datasets, including MovieLens-1M[1], Retailrocket[2] and Yelp2018 [23]. Detailed statistics of them are summarized in Table 1.

Table 1. Statistics of the datasets.

Dataset	#Users	#Items	#interactions	Density
MovieLens-1M	5,949	2,810	571,531	0.03419
Retailrocket	259,531	86,053	931,549	0.00004
Yelp2018	31,668	38,048	1,561,406	0.00130

- **MovieLens-1M**. This dataset contains the ratings (1–5 stars) of users for movies, which were collected through the MovieLens website.
- **Retailrocket**. This dataset contains the interactions of users on a real-world e-commerce website. The interactions include clicks, adding to carts and transactions with items.
- **Yelp2018**. This dataset is the 2018 edition of the Yelp challenge. It contains businesses, reviews and user data. We view the businesses such as restaurants as items.

Evaluation Metrics. To evaluate all methods, we randomly split the interactions into training, validation, and testing sets with a ratio of 7:1:2. Items that the user has not interacted with are treated as negative items. For each user in the test set, each method produces users' preference scores for all items, excluding the positive items used in the training set. We employ the widely-used *Recall@K* and *NDCG@K* as evaluation metrics for the top-*K* recommendation, where *K* is set to 2, 6 and 10.

Compared Methods. We compare GCARec with the following methods:

- **POP**. This method recommends items according to item popularity. The popularity of an item is the number of its interactions. This is a non-personalized method but is still adopted in some scenarios.
- **BPR** [19]. This is a matrix factorization method. It is optimized by Bayesian personalized ranking (BPR) loss to make the preference score of positive items higher than negative items.
- **NGCF** [23]. This is a graph-based recommendation method, which incorporates the second-order neighbors.

[1] https://grouplens.org/datasets/movielens/ .
[2] https://www.kaggle.com/retailrocket/ecommerce-dataset .

Table 2. Performance comparison of all compared methods on three datasets. The best results are bolded and the best results of baselines are underlined.* indicates the significance level p-value<0.01 compared with the best baseline.

Dataset	Metric	POP	BPRMF	NGCF	LightGCN	SGL	GCARec	Improve
MovieLens-1M	Recall@2	0.0212	0.0404	0.0420	0.0452	<u>0.0487</u>	**0.0517***	+6.16%
	NDCG@2	0.1540	0.2585	0.2663	0.2796	<u>0.2938</u>	**0.3080***	+4.83%
	Recall@6	0.0518	0.0989	0.1004	0.1104	<u>0.1170</u>	**0.1238***	+5.81%
	NDCG@6	0.1418	0.2358	0.2422	0.2565	<u>0.2675</u>	**0.2799***	+4.64%
	Recall@10	0.0781	0.1446	0.1464	0.1587	<u>0.1689</u>	**0.1772***	+6.43%
	NDCG@10	0.1388	0.2310	0.2342	0.2520	<u>0.2600</u>	**0.2711***	+3.17%
Retailrocket	Recall@2	0.0018	0.0210	0.0471	0.0650	<u>0.0782</u>	**0.0861***	+10.10%
	NDCG@2	0.0017	0.0393	0.0442	0.0600	<u>0.0732</u>	**0.0800***	+9.29%
	Recall@6	0.0050	0.0631	0.0635	0.1352	<u>0.1616</u>	**0.1742***	+7.78%
	NDCG@6	0.0031	0.0615	0.0635	0.0908	<u>0.1103</u>	**0.1182***	+7.16%
	Recall@10	0.0071	0.1253	0.1277	0.1830	<u>0.2131</u>	**0.2272***	+6.63%
	NDCG@10	0.0038	0.0730	0.0754	0.1056	<u>0.1288</u>	**0.1358***	+5.43%
Yelp2018	Recall@2	0.0026	0.0099	0.0105	0.0130	<u>0.0150</u>	**0.0158***	+5.33%
	NDCG@2	0.0112	0.0463	0.0512	0.0620	<u>0.0707</u>	**0.0742***	+4.95%
	Recall@6	0.0063	0.0245	0.0256	0.0318	<u>0.0366</u>	**0.0377***	+3.01%
	NDCG@6	0.0104	0.0413	0.0444	0.0545	<u>0.0622</u>	**0.0645***	+3.70%
	Recall@10	0.0095	0.0374	0.0383	0.4660	<u>0.0535</u>	**0.0548***	+2.43%
	NDCG@10	0.0109	0.0439	0.0465	0.0564	<u>0.0647</u>	**0.0665***	+2.78%

- **LightGCN** [10]. This is a graph-based recommendation method, which simplifies the design of GCN. It discards the feature transformation and nonlinear activation in GCN and only uses the neighbor aggregation.
- **SGL** [25].This is a state-of-the-art graph-based recommendation method using graph contrastive learning. It designs three data augmentation methods, including node dropout, edge dropout and random walk. In our experiments, we adopt the edge dropout.

Parameter Settings. All models are trained from scratch and optimized by the Adam optimizer. The learning rate is fixed to 0.001 and the batch size is 1024. Parameters are initialized by Xavier initializer [6]. The early stopping strategy is adopted, i.e., models are stopped if the *Recall*@10 on the validation data does not increase for 50 consecutive epochs. The embedding size is fixed to 64 for all models. For NGCF[3], LightGCN[4] and SGL[5], we use the implementation provided by their authors on Github. For all baselines, we tune the parameters and report the best performance. We tune λ_1, λ_2, τ and p within the

[3] https://github.com/xiangwang1223/neural_graph_collaborative_filtering .
[4] https://github.com/kuandeng/LightGCN .
[5] https://github.com/wujcan/SGL .

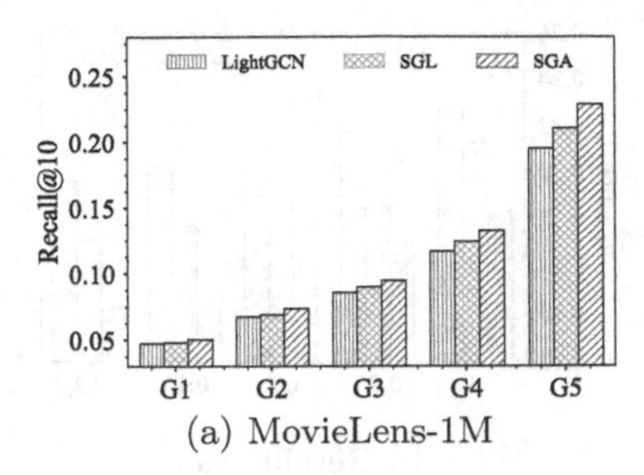

(a) MovieLens-1M

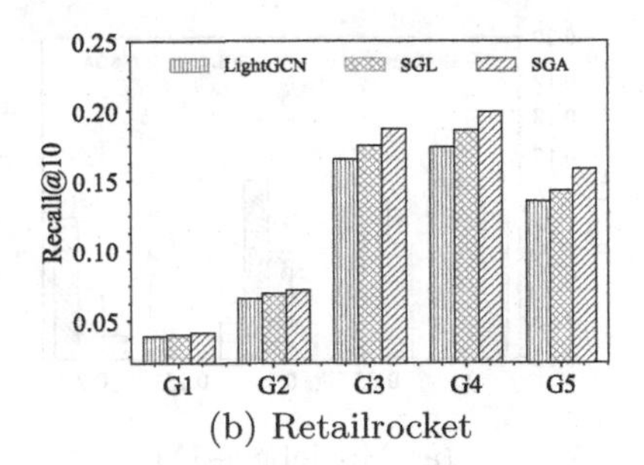

(b) Retailrocket

Fig. 2. Performance comparison over different user groups.

ranges of $\{0, 0.1, 0.2, \cdots, 0.5\}$, $\{0.005, 0.01, 0.05, 0.1, 0.5, 1.0\}$, $\{0.1, 0.2, 0.5, 1.0\}$ and $\{0, 0.1, 0.2, \cdots, 0.5\}$, respectively. The temperature τ_1 in Gumbel Softmax is initialized to 10. Following [11], we adopt the following annealing schedule:

$$\tau_1 \leftarrow \max(0.3, \exp(-rt)), \tag{11}$$

where t is the global training step. r is the decay rate and set to 10^{-4}. In addition, τ_1 is updated every 500 batches. The code is released at https://github.com/my-jing/GCARec.

4.2 Performance Comparison (RQ1)

The performance comparison of our proposed GCARec and compared methods on three datasets is shown in Table 2. From this, we can find that: (1) Graph-based methods achieve better performance compared with conventional methods, i.e., POP and BPR. This shows that incorporating the high-order neighbors helps to improve recommendation performance. LightGCN outperforms NGCF on all datasets, demonstrating the effectiveness of removing the feature transformation and nonlinear activation. (2) SGL and GCARec consistently perform better than other baselines, which shows the effectiveness of graph contrastive learning. (3) Overall, GCARec consistently outperforms other baselines on all datasets. Compared with SGL, it shows the effectiveness of the adaptive data augmentation than the uniform dropout.

4.3 Further Study of GCARec

In this section, we first study the benefits of GCARec from two aspects: (1) data sparsity and (2) robustness to noises. Then, we investigate the effect of hyperparameters including τ, λ_1 and p. Due to the space limitation, we only report the results on MovieLens-1M and Retailrocket, while having similar observations in Yelp2018.

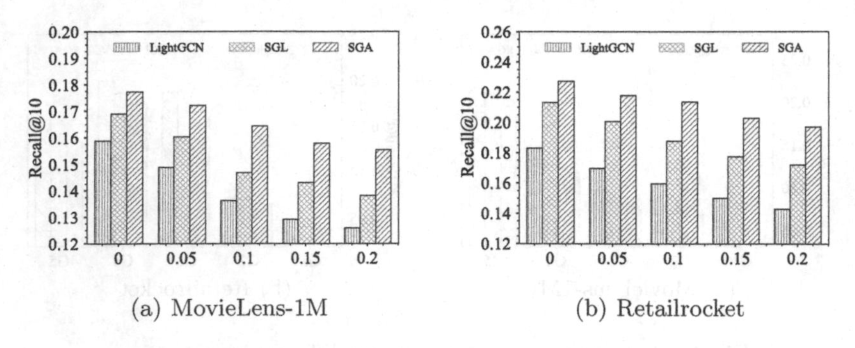

(a) MovieLens-1M (b) Retailrocket

Fig. 3. Performance *w.r.t.* noise ratio.

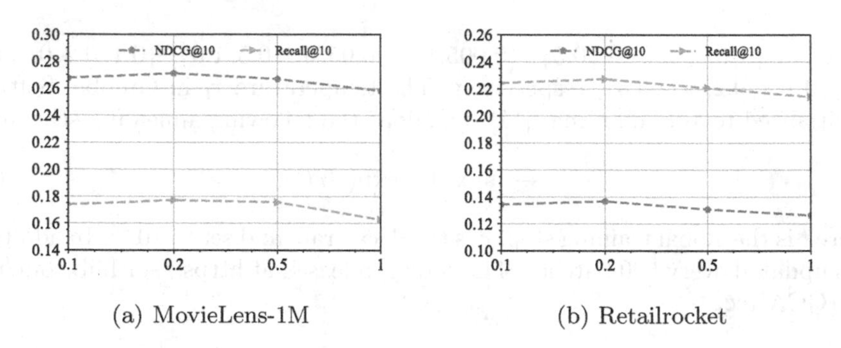

(a) MovieLens-1M (b) Retailrocket

Fig. 4. Performance *w.r.t.* different τ.

Effect of Data Sparsity (RQ2). To verify the effectiveness of GCARec to solve the data sparsity problem, we divide users into five groups according to their interaction numbers and make the total number of interactions in each group the same. The larger the GroupID, the larger the average number of user interactions, i.e., the lower the sparsity level. Figure 2 shows the results of *Recall*@10 on these five groups. From it, we can see that our GCARec consistently outperforms LightGCN and SGL, showing its effectiveness in solving the problem of data sparsity. In addition, as the sparse lever increases, the performance improvement from GCARec increases. This shows that the adaptive data augmentation scheme facilitates GCARec to make recommendations on sparse data.

Robustness to Noisy Interactions(RQ2). To further verify the robustness of GCARec against noisy interactions, we add 5%, 10%, 15% and 20% adversarial examples (negative interactions) to the training set. The testing set is kept unchanged. The results are shown in Fig. 3. We can observe that adding noise data degrades the performance of all models, but the degradation of the performance of GCARec is smaller than that of GCARec and LightGCN. Moreover, the larger the ratio of noise data, the larger the performance degradation gap

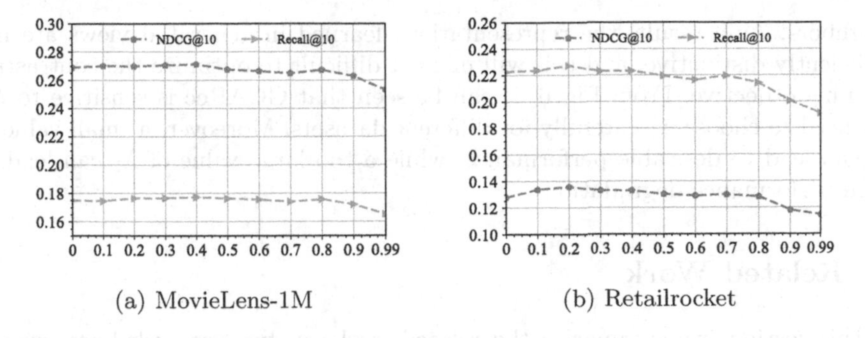

(a) MovieLens-1M (b) Retailrocket

Fig. 5. Performance *w.r.t.* different p.

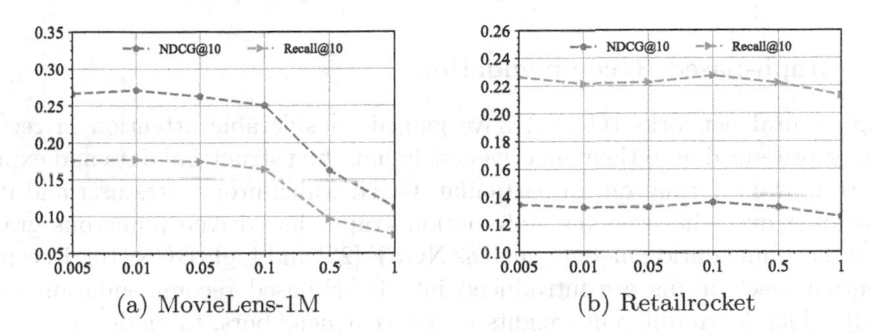

(a) MovieLens-1M (b) Retailrocket

Fig. 6. Performance *w.r.t.* λ_1.

between GCARec and LightGCN. This shows that adaptive data augmentation schemes can identify inherent patterns in the graph structure more effectively.

Parameter Sensitivity Analysis (RQ3). There are three important hyper-parameters used in GCARec: (1) τ defined in Eq. 7 and Eq. 8; (2) p which determines the generation of graph views; (3) λ_1 which controls the strength of contrastive learning loss. To analyze the parameter sensitivity of GCARec, we select representative values for them. When investigating the effect of τ, we fix $p = 0.4, \lambda = 0.01$ on MovieLens-1M and $p = 0.2, \lambda = 0.1$ on Retailrocket. When investigating the effect of p, we fix $\lambda = 0.01$ on MovieLens-1M, $\lambda = 0.1$ on Retailrocket, $\tau = 0.2$ on both datasets. When investigating the effect of λ_1, the settings of p and τ are as the same as those in the previous cases. The results are shown in Figs. 4, 5 and 6.

From Fig. 4, we can observe that GCARec is sensitive to τ. Too large (e.g., 1.0) or too small a value (e.g., 0.1) of τ will reduce the performance of the model. This is consistent with the experimental results in [25]. From Fig. 5, it can be found that the performance of GCARec is relatively stable when p is not too large. Therefore, overall, our model is not sensitive to p. When $p > 0.9$, almost all edges of the graph will be dropped, resulting in isolated nodes in the augmented graph views. In this case, it is difficult to aggregate useful information from neighbors. As a result, the representations learned in generated views are not sufficiently distinctive, and this will make it difficult to optimize the contrastive learning objective. From Fig. 6, it can be seen that GCARec is sensitive to λ_1. We need to choose λ_1 carefully for different datasets. Moreover, a small value of λ_1 can lead to desirable performance, while a too large value of λ_1 can lead to huge performance degradation.

5 Related Work

In this section, we summarize the related works in two research lines: graph-based recommendation and self-supervised learning in recommender systems.

5.1 Graph-based Recommendation

Graph neural networks (GNNs) have gained considerable attention in recommender systems due to their effectiveness in handling structural data and exploring structural information. In particular, GCN, which propagates user and item embedding over the user-item interaction graph, has driven numerous graph-based recommendation models, such as NGCF [23] and LightGCN [10]. Recently, attention mechanisms are introduced into GCN-based recommendation models [4], which learn different weights to different neighbors, to model the importance of the different neighbors to represent the node. In addition to models that only utilize the user-item interaction graph, some graph-based recommendation models use other graphs like DHCN [29] over the session-based graph and DiffNet [26] over the social network. Knowledge graph has also attracted a surge of attention recently [22].

All of these works focus on supervised settings for model training. However, supervised learning relies heavily on expensive labeled data, making them usually suffer from the problem of sparse supervision signal and noisy interactions. Our work explores graph contrastive learning for solving these problems.

5.2 Self-supervised Learning in Recommender Systems

Self-supervised learning [14] is a new paradigm to learn prior knowledge from unlabeled data. It was firstly used in the field of computer vision [3,9,35] and natural language processing [5,12] for tasks like image classification and text classification. Inspired by the success of these works, SSL has been applied in graph representation learning recently [8,16,18,21]. In this line of research, graph contrastive learning is the dominant method, which maximizes agreement between multiple views that are generated from the raw graph through data augmentation.

Inspired by the success of graph contrastive learning, several works apply self-supervised learning in recommendation. S^3-Rec [36] proposes four optimization objectives to learn attributes, item, subsequence and subsequence correlations.

DNN-SSL [31] adopts a two-tower DNN architecture using uniform feature masking and dropout. SGL [25] utilizes uniform node/edge dropout and random walk on user-item interaction graph and applies contrastive learning to the graph-based recommendation. NCL [13] incorporates structural neighbors and semantic neighbors into contrastive pairs. It should be noted that our research focus is different from that of NCL. NCL focuses on sampling strategies, while we focus on data augmentation schemes. They are separate components of contrastive learning. In addition, some studies apply self-supervised learning in other recommendation scenarios such as social recommendation [33].

In this work, we focus on applying graph contrastive learning in the graph-based recommendation method. Compared with SGL, we propose an adaptive and flexible data augmentation scheme that tends to keep the important structures and perturb possibly unimportant edges in the graph. This adaptive augmentation is helpful for the model to preserve the fundamental structure pattern in the graph.

6 Conclusion and Future Work

In this work, we propose a graph contrastive learning for recommendation with adaptive data augmentation. In particular, we propose a learnable and adaptive data augmentation scheme, which tends to retain the important structures and perturb possibly unimportant edges in the user-item interaction graph while generaling easily to other graph-based recommendation. To make the data augmentation scheme learnable, we design it based on the neural network. In specific, we first identify the importance of each edge based on the attention mechanism and calculate the preserving probability based on the importance. Then, edges are sampled according to the importance with a Gumbel Softmax. Extensive experiments on three public datasets demonstrate the effectiveness of our proposed GCARec.

In future work, we will explore other scenarios such as sequential recommendation to apply our work. Besides, we will consider the other part of contrastive learning, i.e., neighbor sampling, to improve the recommendation performance.

Acknowledgements. This research is supported in part by the 2030 National Key AI Program of China 2018AAA0100503, National Science Foundation of China (No. 62072304, No. 61772341, No. 61832013), Shanghai Municipal Science and Technology Commission (No. 19510760500, No. 21511104700, No. 19511120300), the Oceanic Interdisciplinary Program of Shanghai Jiao Tong University (No. SL2020MS032), Scientific Research Fund of Second Institute of Oceanography, the open fund of State Key Laboratory of Satellite Ocean Environment Dynamics, Second Institute of Oceanography, MNR, GE China, and Zhejiang Aoxin Co. Ltd.

References

1. Bayer, I., He, X., Kanagal, B., Rendle, S.: A generic coordinate descent framework for learning from implicit feedback. In: WWW, pp. 1341–1350 (2017)
2. Berg, R.V.D., Kipf, T.N., Welling, M.: Graph convolutional matrix completion. arXiv preprint arXiv:1706.02263 (2017)
3. Chen, T., Kornblith, S., Norouzi, M., Hinton, G.: A simple framework for contrastive learning of visual representations. In: ICML, pp. 1597–1607 (2020)
4. Chen, W., et al.: Semi-supervised user profiling with heterogeneous graph attention networks. In: IJCAI, vol. 19, pp. 2116–2122 (2019)
5. Devlin, J., Chang, M.W., Lee, K., Toutanova, K.: BERT: pre-training of deep bidirectional transformers for language understanding. arXiv preprint arXiv:1810.04805 (2018)
6. Glorot, X., Bengio, Y.: Understanding the difficulty of training deep feedforward neural networks. In: AISTATS, pp. 249–256 (2010)
7. Gutmann, M., Hyvärinen, A.: Noise-contrastive estimation: a new estimation principle for unnormalized statistical models. In: AISTATS, pp. 297–304 (2010)
8. Hassani, K., Khasahmadi, A.H.: Contrastive multi-view representation learning on graphs. In: ICML, pp. 4116–4126. PMLR (2020)
9. He, K., Fan, H., Wu, Y., Xie, S., Girshick, R.: Momentum contrast for unsupervised visual representation learning. In: CVPR, pp. 9729–9738 (2020)
10. He, X., Deng, K., Wang, X., Li, Y., Zhang, Y., Wang, M.: LightGCN: simplifying and powering graph convolution network for recommendation. In: SIGIR, pp. 639–648 (2020)
11. Jang, E., Gu, S., Poole, B.: Categorical reparameterization with gumbel-softmax. arXiv preprint arXiv:1611.01144 (2016)
12. Lan, Z., Chen, M., Goodman, S., Gimpel, K., Sharma, P., Soricut, R.: ALBERT: a lite BERT for self-supervised learning of language representations. arXiv preprint arXiv:1909.11942 (2019)
13. Lin, Z., Tian, C., Hou, Y., Zhao, W.X.: Improving graph collaborative filtering with neighborhood-enriched contrastive learning. arXiv preprint arXiv:2202.06200 (2022)
14. Liu, X., et al.: Self-supervised learning: generative or contrastive. TKDE (2021)
15. Long, X., et al.: Social recommendation with self-supervised metagraph informax network. In: CIKM, pp. 1160–1169 (2021)
16. Peng, Z., et al.: Graph representation learning via graphical mutual information maximization. In: Proceedings of The Web Conference 2020, pp. 259–270 (2020)
17. Qin, Y., Wang, P., Li, C.: The world is binary: contrastive learning for denoising next basket recommendation. In: SIGIR, pp. 859–868 (2021)
18. Qiu, J., et al.: GCC: graph contrastive coding for graph neural network pre-training. In: SIGKDD, pp. 1150–1160 (2020)
19. Rendle, S., Freudenthaler, C., Gantner, Z., Schmidt-Thieme, L.: BPR: Bayesian personalized ranking from implicit feedback. arXiv preprint arXiv:1205.2618 (2012)
20. Velickovic, P., Cucurull, G., Casanova, A., Romero, A., Lio, P., Bengio, Y.: Graph attention networks. Stat **1050**, 20 (2017)
21. Velickovic, P., Fedus, W., Hamilton, W.L., Liò, P., Bengio, Y., Hjelm, R.D.: Deep graph infomax. ICLR (Poster) **2**(3), 4 (2019)
22. Wang, X., He, X., Cao, Y., Liu, M., Chua, T.S.: KGAT: knowledge graph attention network for recommendation. In: SIGKDD, pp. 950–958 (2019)

23. Wang, X., He, X., Wang, M., Feng, F., Chua, T.S.: Neural graph collaborative filtering. In: SIGIR, pp. 165–174 (2019)
24. Wang, X., Jin, H., Zhang, A., He, X., Xu, T., Chua, T.S.: Disentangled graph collaborative filtering. In: SIGIR, pp. 1001–1010 (2020)
25. Wu, J., et al.: Self-supervised graph learning for recommendation. In: SIGIR, pp. 726–735 (2021)
26. Wu, L., et al.: A neural influence diffusion model for social recommendation. In: SIGIR, pp. 235–244 (2019)
27. Wu, M., Zhuang, C., Mosse, M., Yamins, D., Goodman, N.: On mutual information in contrastive learning for visual representations. arXiv preprint arXiv:2005.13149 (2020)
28. Wu, Z., Xiong, Y., Yu, S.X., Lin, D.: Unsupervised feature learning via non-parametric instance discrimination. In: Proceedings of the IEEE Conference on Computer Vision and Pattern Recognition, pp. 3733–3742 (2018)
29. Xia, X., Yin, H., Yu, J., Wang, Q., Cui, L., Zhang, X.: Self-supervised hypergraph convolutional networks for session-based recommendation. In: AAAI, vol. 35, pp. 4503–4511 (2021)
30. Xiao, T., Wang, X., Efros, A.A., Darrell, T.: What should not be contrastive in contrastive learning. arXiv preprint arXiv:2008.05659 (2020)
31. Yao, T., et al.: Self-supervised learning for large-scale item recommendations. In: CIKM, pp. 4321–4330 (2021)
32. You, Y., et al.: Graph contrastive learning with augmentations. In: NIPS, vol. 33, pp. 5812–5823 (2020)
33. Yu, J., Yin, H., Gao, M., Xia, X., Zhang, X., Viet Hung, N.Q.: Socially-aware self-supervised tri-training for recommendation. In: SIGKDD, pp. 2084–2092 (2021)
34. Zang, T., Zhu, Y., Liu, H., Zhang, R., Yu, J.: A survey on cross-domain recommendation: taxonomies, methods, and future directions. arXiv preprint arXiv:2108.03357 (2021)
35. Zbontar, J., Jing, L., Misra, I., LeCun, Y., Deny, S.: Barlow twins: self-supervised learning via redundancy reduction. In: ICML, pp. 12310–12320. PMLR (2021)
36. Zhou, K., et al.: S3-Rec: self-supervised learning for sequential recommendation with mutual information maximization. In: CIKM, pp. 1893–1902 (2020)

Multi-interest Extraction Joint with Contrastive Learning for News Recommendation

Shicheng Wang[1,2], Shu Guo[3(✉)], Lihong Wang[3], Tingwen Liu[1,2], and Hongbo Xu[1,2]

[1] Institute of Information Engineering, Chinese Academy of Sciences, Beijing, China
{wangshicheng,liutingwen,hbxu}@iie.ac.cn
[2] School of Cyber Security, University of Chinese Academy of Sciences, Beijing, China
[3] National Computer Network Emergency Response Technical Team/Coordination Center of China, Beijing, China
guoshu@cert.org.cn, wlh@isc.org.cn

Abstract. News recommendation techniques aim to recommend candidate news to target user that he may be interested in, according to his browsed historical news. At present, existing works usually tend to represent user reading interest using a single vector during the modeling procedure. Actually, it is obviously that users usually have multiple and diverse interest in reality, such as sports, entertainment and so on. Thus it is irrational to represent user sophisticated semantic interest simply utilizing a single vector, which may conceal fine-grained information. In this work, we propose a novel method combining multi-interest extraction with contrastive learning, named MIECL, to tackle the above problem. Specifically, first, we construct several interest prototypes and design a multi-interest user encoder to learn multiple user representations under different interest conditions simultaneously. Then we adopt a graph-enhanced user encoder to enrich user corresponding semantic representation under each interest background through aggregating relevant information from neighbors. Finally, we contrast user multi-interest representations and interest prototype vectors to optimize the user representations themselves, in order to promote dissimilar semantic interest away from each other. We conduct experiments on two real-world news recommendation datasets MIND-Large, MIND-Small and empirical results demonstrate the effectiveness of our approach from multiple perspectives.

Keywords: News recommendation · Multi-interest extraction · Contrastive learning

1 Introduction

Recently, with the rapid development of Internet and online news services [3,8], massive news are released on various news platforms all the time and can make

M.-R. Amini et al. (Eds.): ECML PKDD 2022, LNAI 13713, pp. 606–621, 2023.
https://doi.org/10.1007/978-3-031-26387-3_37

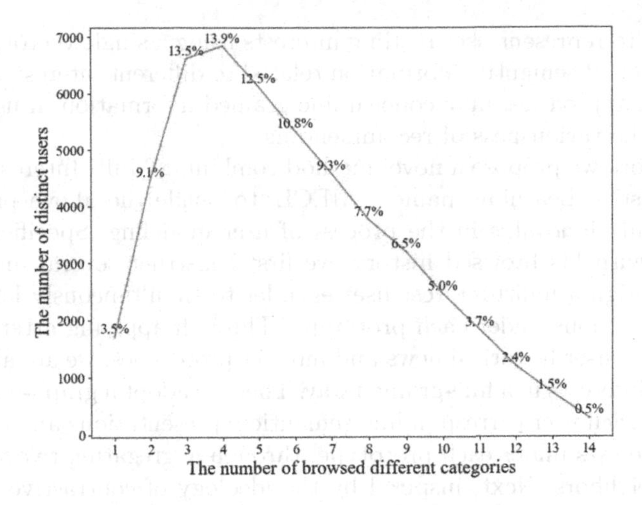

Fig. 1. The number of distinct users with specific number of browsed categories.

users overwhelmed. Thus, personalized news recommendation is necessary for news platforms to help users alleviate information overload as well as improve their reading experience. Briefly, news recommendation aims to recommend candidate news to target user that he may be interested in.

Traditional collaborative filtering methods [3,17] reconstruct interactive matrix to learn user and news representations. However, they suffer from severe cold-start problem due to the short life cycles characteristic of news. Then content-based methods are designed to represent news semantic information and mine user reading interests accurately. Therefore, they share a general framework, including news encoder, user encoder and click predictor. In the light of this framework, existing methods usually adopt BERT [4] or Transformer [14] to learn news representations based on text content such as news titles [1,21]. User interest modeling is another significant procedure in such framework. As deep learning methods are widely concerned, Recurrent Neural Network and Transformer, regarded as effective methods for dealing with sequence modeling, are adopted to model user interests [12,19,20]. In addition, with the huge influence of graph neural networks techniques, some works introduce GAT [15] into user modeling process to enhance user representations by leveraging neighborhood information [5,7]. However, above methods usually represent overall user interests by a single vector based on his/her browsing news history.

Actually, users usually have multiple and diverse interests. For ease of explanation, we utilize category information to represent different interests. We conduct statistic analysis on a real-world dataset MIND-Small [23]. We count the number of distinct users respectively grouped by the number of browsed different categories. As shown in Fig. 1, it is obviously that massive users have browsed multiple categories of news. For example, users may be interested in sports, movies, and finance according to historical news. Nevertheless, previous works usually tend to represent user reading interests using a single vector, resulting in the integration of semantic information related to different interests. As a consequence, such approaches may conceal fine-grained information in user modeling and reduce the variousness of recommending.

In this work we propose a novel method combining Multi-Interest Extraction with Contrastive Learning, named MIECL, to tackle the above problem. Our method mainly innovates in the process of user modeling. Specifically, given a target user with his browsed history, we first construct several interest prototypes and design a multi-interest user encoder to simultaneously learn multiple user representations under each prototype. Through applying attention mechanisms between user historical news and interest prototypes, we are able to model diverse user interests in a fine-grained way. Then we adopt a graph-enhanced user encoder to enrich user corresponding semantic representations and capture their potential interests under each prototype, through aggregating relevant information from neighbors. Next, inspired by the ideology of contrastive learning [2], we optimize the user multi-interest representations through contrasting representations themselves and interest prototypes, in order to promote the multiple representations to be differentiated. Finally we aggregate user multi-interest representations adaptively and calculate click probability between target user and candidate news.

The major contributions of this paper include:

- We design a multi-interest user encoder to explore diverse and multiple user interests in a fine-grained way for more accurate user interest modeling.
- We utilize the superiority of contrastive learning to optimize the above user multi-interest representations, making them more differentiated.
- We conduct experiments on real-world datasets MIND-Large and MIND-Small. The empirical results demonstrate the effectiveness of our approach.

2 Related Work

2.1 Personalized News Recommendation

News recommendation has attracted more and more attention recently with the growth of individual and social needs. Therefore, a variety of methods have been proposed, including collaborative filtering based methods and content based methods. Most traditional methods achieved news recommendation based on collaborative filtering framework [3,17]. They parameterized users and items in a latent space and aimed at reconstructing historical interactions. However, due to the short life cycles characteristic of news articles, CF methods based on IDs always suffered from severe cold start problem, which required us to understand news contents and user interests.

To address this issue, content-based or hybrid methods have been proposed. For example, Okura [10] utilized an auto-encoder to learn news representations from news bodies. Then they applied a GRU network to model user interests from clicked historical news sequences. Finally, they calculated click probability between user interests and news representations based on dot product. NAML [19] leveraged a CNN network to model news semantic representations from news titles and categories. Then they learnt user representations through attentively aggregating clicked news. Similarly, LSTUR [1] also learnt news representations based on CNN network. However, unlike NAML method, they applied GRU neural network to model user short-term interests from clicked history and further model user long-term interests via user ID embeddings. Then they adopted attention mechanisms to integrate the above user representations. NRMS [21] utilized multi-head self-attention networks with similar structures to learn news representations and user representations separately, in order to capture interaction information in word sequences and news sequences respectively. With the development of graph neural network technology, some work proposed to model news contents and user interests based on GCN or GAT. KRED [9] first introduced news titles and entities to construct a news recommendation knowledge graph. Then they applied graph attention network to learn news representations. GNewsRec [7] learnt user short-term interests by applying attentive GRU neural network to clicked history and user long-term interests via graph neural networks. However, there is few work recognizing the importance of modeling diverse user interests explicitly. The above-mentioned methods usually tend to represent user interests using a single vector. Different from these methods, in this paper, we propose to model diverse user interests explicitly in a fine-grained way.

2.2 Contrastive Learning

Contrastive learning, as a branch of self-supervised learning, is devised to learn by comparing among different input samples. The objective of contrastive learning is to map the representations of similar samples close together, while that of dissimilar samples should be further away in the embedding space. According to the scale of the samples involved in the comparison, recent contrastive learning methods can be formulated as global-local contrast and local-local contrast. The global-local contrastive learning focuses on modeling the belonging relationship between the local feature of a sample and its global context representation. For example, Deep Infomax [6] proposed to maximize the mutual information between a local patch and its global context, which provided us with a new paradigm. Deep Graph Infomax [16] introduced the ideology of DIM into graph representation learning, which regarded target node representation as local feature and its high-level summary of graph as global feature.

However, the global-local contrastive learning may generate ill-conditioned representations. Recently local-local contrastive learning discards mutual information and directly studies the relationships between different samples instance-level local representations. For example, CMC [13] proposed to adopt multiple different views of an image as positive samples and sampled another irrelevant image as the negative. SimCLR [2] illustrated the importance of introducing data augmentation operations for contrastive representation learning based on a simple framework. Inspired by the recent prominent advances in local-local

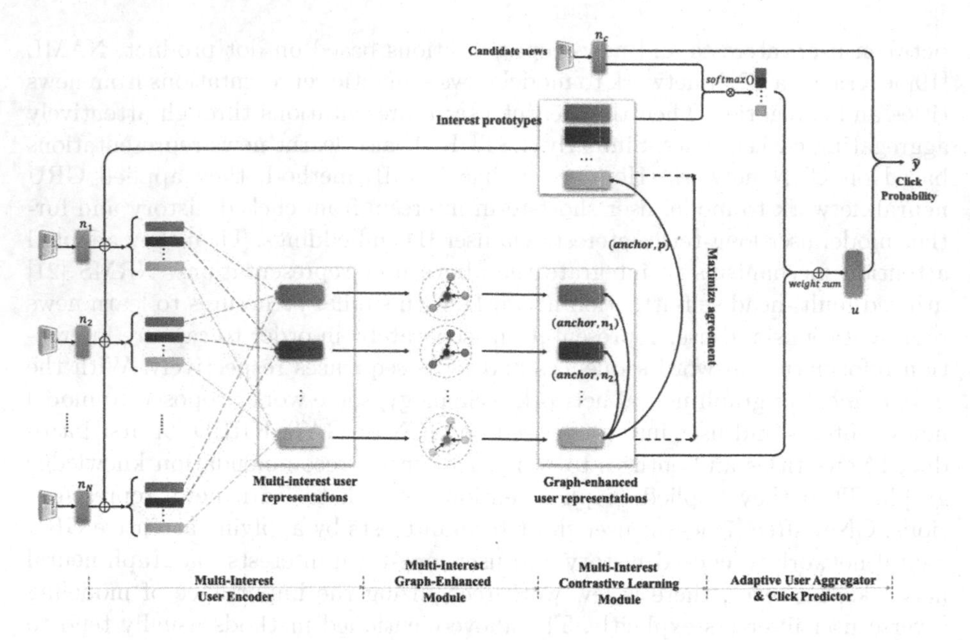

Fig. 2. Overall framework of our MIECL method.

contrast methods, in this work we explore contrastive learning to assist us in differentiating diverse user interests.

3 Methodology

In this section, we first present the problem formulation of personalized news recommendation. Specifically, given a candidate news n_c and a target user u with his clicked news history $[n_1, \cdots, n_N]$, we aim to learn candidate news representation as well as user representation respectively, followed by calculating the relevance score between their representations. Finally we decide whether to recommend n_c or not according to the score. Then we propose a method combining multi-interest extraction with contrastive learning, named MIECL, to model diverse user interests in a fine-gained way. As shown in Fig. 2, our model MIECL is mainly innovative in user modeling, including three modules: multi-interest user encoder, multi-interest graph-enhanced module and multi-interest contrastive learning module. We will elaborate our method in the subsequent sections in detail[1]

3.1 News Encoder

In this section, we introduce how to learn news semantic representations from news titles. We indicate the title word sequence as $[w_1, \cdots, w_M]$, where w_i is

[1] Our source code is available at https://github.com/wangsc2113/MIECL..

denoted as the i-th word in title. Here we encode titles based on the traditional Transformer framework.

First, at the bottom of news encoder module, it applies word embedding layer to convert each word w_i into corresponding vector e_i. Then, it adopts a multi-head self-attention network to capture semantic interactions between title words $[e_1, \cdots, e_M]$. The representation of the word w_i learned by the s-th attention head h_i^s is calculated as:

$$h_i^s = V_s^w \sum_{j=1}^{M} \alpha_{i,j}^s \cdot e_j, \quad \alpha_{i,j}^s = \frac{\exp(e_i^T Q_s^w e_j)}{\sum_{t=1}^{M} \exp(e_i^T Q_s^w e_t)} \tag{1}$$

where $V_s^w \in \mathbb{R}^{d/S \times d}$ and $Q_s^w \in \mathbb{R}^{d \times d}$ are the projection parameters in the s-th self-attention head, and $\alpha_{i,j}$ indicates the interaction score between the word w_i and w_j. Then the multi-head representation of word w_i is concatenated as $h_i \in \mathbb{R}^d$, i.e., $h_i = [h_i^1; h_i^2; \cdots; h_i^S]$, where S denotes the number of separate self-attention heads.

Finally, it applies an additive attention network to aggregate contextual word representations into a news representation n, formulated as:

$$n = \sum_{i=1}^{M} \alpha_i^w \cdot h_i, \quad \alpha_i^w = \frac{\exp(q_\alpha^T \cdot h_i)}{\sum_{j=1}^{M} \exp(q_\alpha^T \cdot h_j)} \tag{2}$$

where $q_\alpha \in \mathbb{R}^d$ is a projection parameter vector.

3.2 Multi-interest User Encoder

We argue that one representation vector has a low potential to reflect user diverse and complex reading interests. Obviously, an intuitive solution is to learn multiple representations to model diverse user interests.

At the beginning, since there is semantic relatedness between news articles browsed by the same user, it applies a multi-head self-attention to enhance the news representations by capturing their relatedness. Given clicked news history $[n_1, \cdots, n_N]$, the news representation learned by the s-th attention head n_i^k is calculated as:

$$n_i^s = V_s^n \sum_{j=1}^{N} \beta_{i,j}^s \cdot n_j, \quad \beta_{i,j}^s = \frac{\exp(n_i^T Q_s^n n_j)}{\sum_{t=1}^{N} \exp(n_i^T Q_s^n n_t)} \tag{3}$$

where $V_s^n \in \mathbb{R}^{d/S \times d}$ and $Q_s^n \in \mathbb{R}^{d \times d}$ are the news-level projection parameters in the s-th self-attention head. Then the multi-head representation of news n_i is defined as n_i, i.e., $n_i = [n_i^1; n_i^2; \cdots; n_i^S]$.

We assume that each user possesses K different interests and each interest has a corresponding prototype vector $I_k \in \mathbb{R}^d$. Interest prototypes are introduced instead of the "hard" category information, since we deem that news is semantic related to the category, even if there is no subordinate relationship between them.

Though similar to the previous content-based user modeling methods but in a fine-grained way, we design to aggregate specific interest-relevant information from historical news to acquire corresponding interest-level user representation.

First, it extracts semantic information n_i^k from history news n_i oriented to specific interest I_k. n_i^k is expected to contain specific interest-relevant information. It simply adopts concatenation here, which is formulated as:

$$n_i^k = W(n_i||I_k), \tag{4}$$

where $W \in \mathbb{R}^{d \times 2d}$ is a learnable parametric matrix for semantic transforming. In this way, for each news in clicked history, we can obtain relevant semantic information under given interest condition.

Next, under different interest conditions, it adopts additive attention networks to generate interest-level user representations $u_k \in \mathbb{R}^d, k \in [1, K]$ through aggregating interest-relevant information respectively, according to user historical news. The formula is defined as:

$$u_k = \sum_{i=1}^{N} \beta_{i,k}^n \cdot n_i^k, \quad \beta_{i,k}^n = \frac{\exp(q_\beta^T \cdot n_i^k)}{\sum_{j=1}^{N} \exp(q_\beta^T \cdot n_j^k)} \tag{5}$$

where $q_\beta \in \mathbb{R}^d$ is a projection vector and $\beta_{i,k}^n$ denotes the attention weight of the i-th clicked news contributing to user representation in condition of k-th interest prototype. Now we obtain K interest-level user representations $[u_1, \cdots, u_K]$ to reveal his/her diverse interests.

3.3 Multi-interest Graph-Enhanced Module

In this section, we introduce graph attention neural networks to enrich user interest-level semantic representations through utilizing neighbor users information. For the given target user u, we search his/her second-order neighbor users based on user-news-user co-occurrence relation in advance. For the convenience of calculation, we randomly choose T neighbors for each user. For each neighbor user u^i, we utilize the Equation (7) to obtain their semantic representations under different prototypes, noted as $[u_1^1, \cdots, u_K^1]$, $\cdots$, $[u_1^T, \cdots, u_K^T]$ respectively. Then under each interest condition, it adopts a separate graph attention network to aggregate interest-specific neighbor information:

$$u_k = \sum_{i=1}^{T \cup \{u\}} \gamma_k^i \cdot u_k^i, \quad \gamma_k^i = \frac{\exp(q_k^\gamma \cdot [u_k||u_k^i])}{\sum_{j=1}^{T \cup \{u\}} \exp(q_k^\gamma \cdot [u_k||u_k^j])} \tag{6}$$

where $q_k^\gamma \in \mathbb{R}^{1 \times 2d}$ is weight matrix and γ_k^i denotes the attention weight between the target user and his i-th neighbor user in condition of k-th interest prototype. Now we obtain the graph-enhanced user multi-interest representations.

3.4 Multi-interest Contrastive Learning Module

In the previous sections we already obtain user multi-interest representations. However, there might be information redundancy among these vectors since we do not impose any constraints on them. This may have a harmful effect on model performance since similar interest-level user representations are difficult to reflect diverse interests.

Inspired by the local-local contrast methods recently [2,13], we design a joint learning paradigm and construct contrastive learning objective to optimize above-mentioned user multi-interest representations, in order to promote dissimilar semantic interests away from each other. According to the universal methods of contrastive learning, we have to construct positive pairs as well as corresponding negative pairs first. Considering the purpose of optimizing target user u multi-interest representations, we randomly choose an interest prototype vector I_k, then we treat (u_k, I_k) as positive pair, and (u_k, u_j) as corresponding negative pairs for all $j \neq k$. Afterwards, we define a interest-level contrastive learning objective function as follows:

$$\mathcal{L}_{ssl} = -\sum_{i \in TS} \log \frac{\exp(f(u_k, I_k))}{\sum_{j \neq k} \exp(f(u_k, u_j))} \quad (7)$$

where TS denotes the training set and $f(\cdot)$ is a scoring function for sample pairs, such as cosine similarity. Through optimizing this objective function, we are able to make interest-specific representation close to the corresponding interest prototype. In the meantime, the interest-level user representations under different conditions become more differentiated. In this way, our method is able to model diverse user interests more accurately.

3.5 Adaptive User Aggregator and Click Predictor

After obtaining the desired fine-grained user multi-interest representations, we learn target user representation u in an adaptive way considering given candidate news n_c. First we calculate δ_k, the normalization of similarity score between candidate news and interest prototype vector I_k, for $k \in [1, K]$. This indicates the probability that the candidate news related to the specific interest.

$$\delta_k = \frac{\exp(f(n_c, I_k))}{\sum_{i=1}^{K} \exp(f(n_c, I_i))} \quad (8)$$

The probability δ_k can also be regarded as the weight of corresponding interest-level user representation when aggregated into final summary representation. Next it adopts a weight summation operation to generate adaptive user representation considering candidate news, i.e., $u = \sum_{k=1}^{K} \delta_k \cdot u_k$.

It can help fine-grained matching between candidate news and target user reading interests. Finally, the click probability score y is computed by the dot product between the target user representation and the candidate news representation, i.e., $y = u^T \cdot n_c$.

Table 1. Statistic information of MIND-Large dataset.

#News	161,013	#Users	1,000,000
#News category	20	#Impression	15,777,377
#Entity	3,299,687	#Click behavior	24,155,470
Avg. title len	11.52	Avg. abstract len	43.00
Avg. body len	585.05		

3.6 Model Training

Following [21], we use negative sampling techniques for model training. Given a positive sample n_i (clicked news, labeled as 1) in the training dataset, we then randomly select P negative samples $[n_{i,1}, \cdots, n_{i,P}]$ (non-clicked news, labeled as 0) from the same impression displayed to target user. Denote the click probability score of the positive and the P negative news as y_i^+ and $[y_{i,1}^-, y_{i,2}^-, \cdots, y_{i,P}^-]$ respectively. The supervised classification loss is formulated as follows:

$$\mathcal{L}_{ce} = -\sum_{i \in TS} \log \frac{\exp(y_i^+)}{\exp(y_i^+) + \sum_{j=1}^{P} \exp(y_{i,j}^-)} \tag{9}$$

Since we have already obtained the main classification loss function and auxiliary contrastive loss function, we define the final loss function in a joint paradigm as:

$$\mathcal{L} = \mathcal{L}_{ce} + \alpha \cdot \mathcal{L}_{ssl} \tag{10}$$

where α is a hyper-parameter that makes a trade-off between classification loss and contrastive loss. Minimizing the joint loss $\mathcal{L}$ helps to obtain fine-grained and differentiated user multi-interest representations.

4 Experiment

4.1 Dataset and Experimental Settings

We conduct extensive experiments on two large-scale real-world datasets, MIND-Large[2] and MIND-Small[3], to evaluate the effectiveness of our method. MIND-Large dataset collected from Microsoft News platform contains two record documents. One document describes text content of news, including titles and abstracts. The other document describes interaction behaviors that each user clicked news and these click behaviors are gathered from October 12 to November 22, 2019 (six weeks). The click behaviors in the first four weeks are regarded as user reading history, the behaviors in the penultimate week is applied for training, and the data in last week is used for performance evaluation. Detailed statistic information about MIND-Large dataset is summarized in Table 1.

[2] https://msnews.github.io/.

[3] A small version of the MIND-Large dataset by randomly sampling 50,000 users and their behavior logs.

Next, we introduce experimental and hyper-parameters settings of our method. For news text content modeling, we utilize the first 30 words of news titles to learn news representations. In addition, a special character [PAD] is used for filling when the length of word sequence does not meet the condition. Besides, we adopt pre-trained Glove embeddings [11] for word initialization. For user interest modeling, we treat the recent 50 clicked news as user reading history. Moreover, news representations and user representations, including user multi-interest representations and adaptive user representations are both 400-dimensional vectors. For hyper-parameters, the number of interest prototypes K is set to 5. The joint learning weight α is set to 1.0. And the number of user neighbors T is merely set to 2 to promote the time efficiency. In addition, we utilize dropout technique and Adam optimizer for training. The dropout rate and learning rate are 0.1 and 0.001 respectively. Following [21], we use four metrics, i.e., AUC, MRR, nDCG@5, and nDCG@10, for performance evaluation. Notably, AUC is the most important one among them.

4.2 Performance Evaluation

We first introduce the baseline methods we compared in experiments, including six sequence-based and three GNN-based methods[4]: (1) EBMR [10] learns user representations from clicked news history via a GRU network. (2) DKN [18] utilizes an adaptive attention network to learn user representations considering relatedness between candidate news and historical news. (3) NPA [20] employs personalized attention networks to learn individual representation for each user. (4) NAML [19] leverages CNN networks to model news semantic representations and learns user representations through attentively aggregating clicked news. (5) LSTUR [1] models short-term user interests via a GRU network and long-term user interests via user ID embeddings. (6) NRMS [21] learns news representations and user representations through utilizing multi-head self-attention networks respectively. (7) GNewsRec [7] models user short-term interests by applying attentive GRU neural network and user long-term interests via graph neural networks based on user-news-topic heterogeneous graph. (8) GERL [5] uses the neighbors of news and users on the user-news graph to enhance their representations. (9) User-as-Graph [22] proposes a heterogeneous graph pooling method to learn user interest representations from the personalized heterogeneous graph.

The purpose of this section is to verify the effectiveness of our method. Thus we first conduct experiments to compare our model with several baseline models on MIND-Large dataset and then apply them to MIND-Small dataset for supplement. The overall performance results are displayed in Table 2 and Table 3 respectively, from which we have several observations: First, in terms of AUC, our proposed MIECL outperforms all baselines on both two datasets. We achieve 1.55% and 2.28% improvement comparing to state-of-the-art result respectively

[4] Due to the limitation of computer resources, we did not use the pretrained language models to encode the news titles and compare with baselines based on pretrained models.

Table 2. Performance of different methods on MIND-Large Dataset (%). *The improvement is significant at the level $p < 0.001$.

Method	AUC	MRR	nDCG@5	nDCG@10
EBNR	65.42	31.24	33.76	39.47
DKN	64.60	31.32	33.84	39.48
NPA	66.69	32.24	34.98	40.68
NAML	66.86	32.49	35.24	40.91
LSTUR	67.73	32.77	35.59	41.34
NRMS	67.76	33.05	35.94	41.63
GNewsRec	67.53	32.68	35.46	41.17
GERL	68.24	33.46	36.38	42.11
User-as-Graph	<u>69.23</u>	**34.14**	<u>37.21</u>	<u>43.04</u>
MIECL*	**70.30**	<u>34.13</u>	**37.87**	**44.31**

Table 3. Performance of different methods on MIND-Small Dataset (%). *The improvement is significant at the level $p < 0.001$.

Method	AUC	MRR	nDCG@5	nDCG@10
EBNR	61.62	28.07	30.55	37.07
DKN	63.99	28.95	31.73	37.07
NPA	64.28	29.64	32.28	38.93
NAML	64.30	29.81	32.64	39.11
LSTUR	65.68	30.44	33.49	39.95
NRMS	65.43	30.74	33.13	39.66
GNewsRec	65.91	30.50	33.56	40.13
GERL	66.22	30.89	34.28	40.50
User-as-Graph	<u>66.71</u>	<u>31.13</u>	<u>34.51</u>	<u>40.95</u>
MIECL*	**68.23**	**32.46**	**36.17**	**42.32**

on each dataset. Besides, our model achieves excellent performance in terms of other evaluation metrics, which shows the effectiveness and adaptability to data scale of our model. Since neither sequence-based methods nor GNN-based methods recognize the significance of modeling diverse user interests, they merely use one vector to represent user reading interests. This result illustrates the necessity of modeling diverse user interests explicitly. Second, GNN-based methods usually perform better than sequence-based methods, since they utilize GNN to obtain high-order neighbor information. However, as we elaborate in the ablation study below, although GNN module does not produce particularly significant improvement to our model, we still acquire even better performance than them. This result further demonstrates the effectiveness and importance of modeling diverse user interests, which deserves our further exploration.

Table 4. Effect of each module perform in our model.

	AUC	MRR	nDCG@5	nDCG@10
MIECL	68.23	32.46	36.17	42.32
w/o multi-interest user encoder	66.56	31.51	34.55	41.00
w/o multi-interest graph-enhanced module	67.93	32.83	36.27	42.43
w/o multi-interest contrastive learning module	67.36	31.83	35.13	41.42

4.3 Ablation Study

In order to further evaluate the effectiveness of each module in our method, we conduct the ablation experiments on MIND-Small dataset and report the results in Table 4. According to the performance of different variants, we then discuss the effect of each module in our method.

w/o Multi-interest User Encoder. The module is the concrete implementation portion of motivation. This variant indicates each user only has a single representation vector, in addition contrastive learning here is not applicable. Compared with the results of the complete model, this variant showed a significant decline of 2.45% in terms of AUC. Therefore we verify the significance of mining fine-grained and diverse user interests and conclude that such modeling method is beneficial to personalized news recommendation.

w/o Multi-interest Graph-Enhanced Module. In this variant, we learn user representations from their clicked historical news without aggregating neighbors information. Excluding the module decreases the AUC by 0.44%. Obviously, utilizing neighbors information can enrich user semantic representations to a certain extent.

w/o Multi-interest Contrastive Learning Module. This variant describes the situation that the original joint learning framework is turned into a separate recommendation task. It is noticed that the performance of this variant drops by 1.28% in terms of AUC. Without the contrastive learning module, the discrimination between user multiple representations is reduced, which leads to user multi-interest representations become similar. To some extent, this is equivalent to model user interests using a single vector. As the first variant discussed, the performance of model will degrade. The result demonstrates the importance of the contrastive learning module.

4.4 Hyper-Parameters Analysis

In order to further evaluate the sensitivity of our method to hyper-parameters, we conduct experiments on MIND-Small dataset with different values of hyper-parameters, including the number of interest prototypes K and the joint learning weight α.

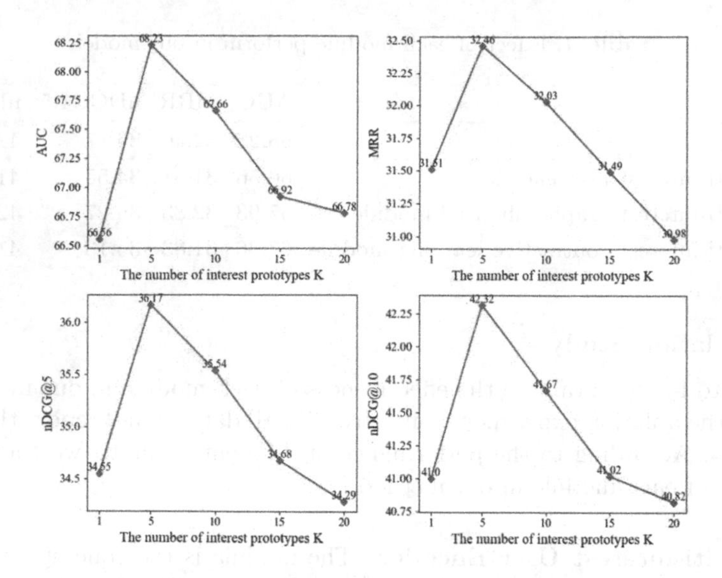

Fig. 3. Performance of MIECL with different values of hypermeter K.

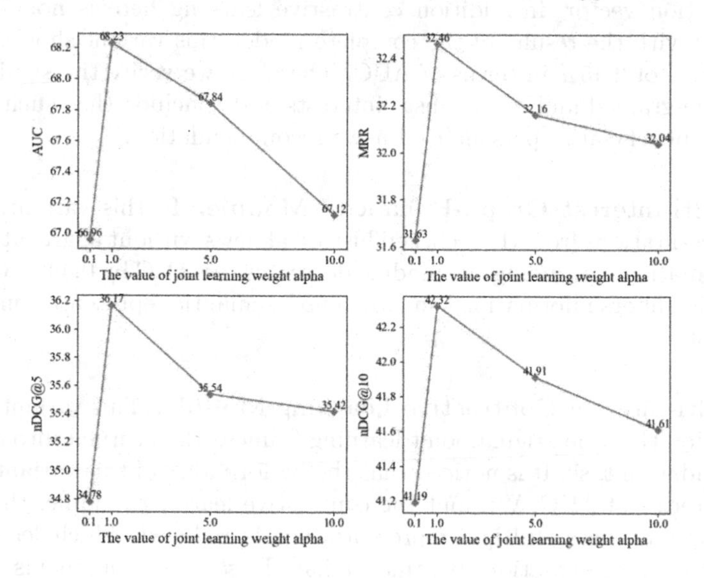

Fig. 4. Performance of MIECL with different values of hypermeter α.

Figure 3 shows the trend variation of our method with various number of interest prototypes. It can be seen that with the increases of K, the performance of MIECL will also improve first in terms of all metrics. This is because user interests is usually diverse and rarely relatively single. Hence, utilizing the fine-grained modeling method can better explore the diversity of user interests. However, when the value of K becomes larger, the performance of our model gradually decreases. We speculate that although user interests is diverse, the number of them can not be particularly numerous. Once the value of K is too large, a lot of interest-level noise information will be introduced while aggregated into the final user representations. This is likely to be detrimental to the recommendation performance. The deduction is also conform to the reality and intuition. Furthermore, we discover that the result echoes with Fig. 1, because most users browse about 5 categories of news.

Figure 4 shows the trend variation of our method with different value of joint learning weight. The abscissa represents the value of α and the ordinate represents the results of evaluation metrics. We can discover that the trend variation is similar to that in Fig. 3. When α is relatively small, the impact of contrastive learning decreases as well. This is because we are hardly to model fine-grained interests once the learnt multi-interest representations are not sufficiently differentiated. Then when α becomes larger, the influence of contrastive learning is extremely exaggerated at this time. The consequence is that the affect of recommendation classification loss will be correspondingly reduced, which is not conducive to the main recommendation task.

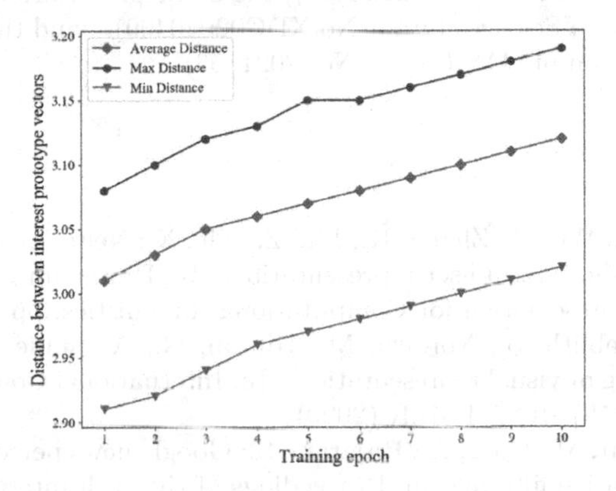

Fig. 5. Statistics of the distances between interest prototype vectors, at different training epoch.

4.5 Statistic Analysis

In this section, we conduct statistical experiments on distances between interest prototypes on MIND-Small dataset. Specifically, we record the representations of interest prototypes after each epoch of training procedure. Then we calculate the euclidean distances between each pair of interest prototype vectors and conduct statistical analysis, at different training epoch. The statistical results are displayed in the Fig. 5.

Quite evidently, with the increases of training epoch, the average (max, min) distance between interest prototype vectors is also gradually increasing. This phenomenon further implies that contrastive learning can distinguish the interest prototypes effectively. In addition, this will be beneficial to learning more differentiated user multi-interest representations.

5 Conclusion

In this paper, we propose a novel method combining multi-interest extraction with contrastive learning, named MIECL, to model diverse user interests effectively. Specifically, first, we construct several interest prototypes and design a multi-interest user encoder to simultaneously learn multiple user representations under each prototype. Then we adopt a graph-enhanced user encoder to enrich user corresponding semantic representation under each interest background. Finally, we contrast user multi-interest representations and interest prototypes to optimize the user representations themselves, in order to promote dissimilar semantic interest away from each other. Extensive experiments on real-world datasets validate the effectiveness of our approach.

Acknowledgements. We sincerely thank all the anonymous reviewers for their comments and suggestions. This work is supported by the National Key Research and Development Program of China (grant No.2021YFB3100600), the National Natural Science Foundation of China (No.62106059), the Strategic Priority Research Program of Chinese Academy of Sciences (grant No.XDC02040400), , and the Youth Innovation Promotion Association of CAS (Grant No. 2021153).

References

1. An, M., Wu, F., Wu, C., Zhang, K., Liu, Z., Xie, X.: Neural news recommendation with long-and short-term user representations. In: Proceedings of the 57th Annual Meeting of the Association for Computational Linguistics, pp. 336–345 (2019)
2. Chen, T., Kornblith, S., Norouzi, M., Hinton, G.: A simple framework for contrastive learning of visual representations. In: International Conference on Machine Learning, pp. 1597–1607. PMLR (2020)
3. Das, A.S., Datar, M., Garg, A., Rajaram, S.: Google news personalization: scalable online collaborative filtering. In: Proceedings of the 16th International Conference on World Wide Web, pp. 271–280 (2007)
4. Devlin, J., Chang, M.W., Lee, K., Toutanova, K.: Bert: pre-training of deep bidirectional transformers for language understanding. arXiv preprint arXiv:1810.04805 (2018)
5. Ge, S., Wu, C., Wu, F., Qi, T., Huang, Y.: Graph enhanced representation learning for news recommendation. In: Proceedings of The Web Conference 2020, pp. 2863–2869 (2020)
6. Hjelm, R.D., et al.: Learning deep representations by mutual information estimation and maximization. In: International Conference on Learning Representations (2018)
7. Hu, L., Li, C., Shi, C., Yang, C., Shao, C.: Graph neural news recommendation with long-term and short-term interest modeling. Inf. Process. Manag. **57**(2), 102142 (2020)
8. Khattar, D., Kumar, V., Varma, V., Gupta, M.: Weave&rec: a word embedding based 3-d convolutional network for news recommendation. In: Proceedings of the 27th ACM International Conference on Information and Knowledge Management, pp. 1855–1858 (2018)

9. Liu, D., et al.: Kred: knowledge-aware document representation for news recommendations. In: Fourteenth ACM Conference on Recommender Systems, pp. 200–209 (2020)
10. Okura, S., Tagami, Y., Ono, S., Tajima, A.: Embedding-based news recommendation for millions of users. In: Proceedings of the 23rd ACM SIGKDD International Conference on Knowledge Discovery and Data Mining, pp. 1933–1942 (2017)
11. Pennington, J., Socher, R., Manning, C.D.: Glove: global vectors for word representation. In: Proceedings of the 2014 Conference on Empirical Methods in Natural Language Processing (EMNLP), pp. 1532–1543 (2014)
12. Qi, T., et al.: Hierec: hierarchical user interest modeling for personalized news recommendation. In: Proceedings of the 59th Annual Meeting of the Association for Computational Linguistics and the 11th International Joint Conference on Natural Language Processing, vol. 1: Long Papers, pp. 5446–5456 (2021)
13. Tian, Y., Krishnan, D., Isola, P.: Contrastive multiview coding. In: Vedaldi, A., Bischof, H., Brox, T., Frahm, J.-M. (eds.) ECCV 2020. LNCS, vol. 12356, pp. 776–794. Springer, Cham (2020). https://doi.org/10.1007/978-3-030-58621-8_45
14. Vaswani, A., et al.: Attention is all you need. Adv. Neural Inf. Process. Syst. **30**, 1–11 (2017)
15. Veličković, P., Cucurull, G., Casanova, A., Romero, A., Lio, P., Bengio, Y.: Graph attention networks. arXiv preprint arXiv:1710.10903 (2017)
16. Veličković, P., Fedus, W., Hamilton, W.L., Liò, P., Bengio, Y., Hjelm, R.D.: Deep graph infomax. In: International Conference on Learning Representations (2018)
17. Wang, C., Blei, D.M.: Collaborative topic modeling for recommending scientific articles. In: Proceedings of the 17th ACM SIGKDD International Conference on Knowledge Discovery and Data Mining, pp. 448–456 (2011)
18. Wang, H., Zhang, F., Xie, X., Guo, M.: DKN: deep knowledge-aware network for news recommendation. In: Proceedings of the 2018 World Wide Web Conference, pp. 1835–1844 (2018)
19. Wu, C., Wu, F., An, M., Huang, J., Huang, Y., Xie, X.: Neural news recommendation with attentive multi-view learning. In: Proceedings of the 28th International Joint Conference on Artificial Intelligence, pp. 3863–3869 (2019)
20. Wu, C., Wu, F., An, M., Huang, J., Huang, Y., Xie, X.: NPA: neural news recommendation with personalized attention. In: Proceedings of the 25th ACM SIGKDD International Conference on Knowledge Discovery & Data Mining, pp. 2576–2584 (2019)
21. Wu, C., Wu, F., Ge, S., Qi, T., Huang, Y., Xie, X.: Neural news recommendation with multi-head self-attention. In: Proceedings of the 2019 Conference on Empirical Methods in Natural Language Processing and the 9th International Joint Conference on Natural Language Processing (EMNLP-IJCNLP), pp. 6389–6394 (2019)
22. Wu, C., Wu, F., Huang, Y., Xie, X.: User-as-graph: user modeling with heterogeneous graph pooling for news recommendation. In: Proceedings of the Thirtieth International Joint Conference on Artificial Intelligence, pp. 1624–1630 (2021)
23. Wu, F., et al.: Mind: a large-scale dataset for news recommendation. In: Proceedings of the 58th Annual Meeting of the Association for Computational Linguistics, pp. 3597–3606 (2020)

Transfer and Multitask Learning

On the Relationship Between Disentanglement and Multi-task Learning

Łukasz Maziarka(✉), Aleksandra Nowak, Maciej Wołczyk, and Andrzej Bedychaj

Jagiellonian University, Kraków, Poland
{lukasz.maziarka,aleksandra.nowak,maciej.wolczyk, andrzej.bedychaj}@ii.uj.edu.pl

Abstract. One of the main arguments behind studying disentangled representations is the assumption that they can be easily reused in different tasks. At the same time finding a joint, adaptable representation of data is one of the key challenges in the multi-task learning setting. In this paper, we take a closer look at the relationship between disentanglement and multi task learning based on hard parameter sharing. We perform a thorough empirical study of the representations obtained by neural networks trained on automatically generated supervised tasks. Using a set of standard metrics we show that disentanglement appears naturally during the process of multi-task neural network training.

Keywords: Multitask learning · Disentangled representation

1 Introduction

Disentangled representations have recently become an important topic in the deep learning community [10,12,26,29,35]. The main assumption in this problem is that the data encountered in the real world is generated by few independent and explanatory factors of variation. It is commonly accepted that such representations are not only more interpretable and robust but also perform better in tasks related to transfer learning and one-shot learning [3,23,28,37].

Intuitively, a disentangled representation encompasses all the factors of variation and as such can be used for various tasks based on the same input space. On the other hand, non-disentangled representations, such as those learned by vanilla neural networks, might focus only on one or a few factors of variations that are relevant for the current task, while discarding the rest. Such a representation may fail when encountering different tasks that rely on distant aspects of variation which have not been captured.

Ł. Maziarka, A. Nowak, M. Worłczyk and A. Bedychaj—All authors contributed equally.

Supplementary Information The online version contains supplementary material available at https://doi.org/10.1007/978-3-031-26387-3_38.

M.-R. Amini et al. (Eds.): ECML PKDD 2022, LNAI 13713, pp. 625–641, 2023.
https://doi.org/10.1007/978-3-031-26387-3_38

Exploiting prevalent features and differences across tasks is also the paradigm of multi-task learning. In a standard formulation of a multi-task setting, a model is given one input and has to return predictions for multiple tasks at once. The neural network might be therefore implicitly regularized to capture more factors of variation than a network that learns only a single task. Based on this intuition, we hypothesize that disentanglement is likely to occur in the latent representations in this type of problem.

This paper aims to test this hypothesis empirically. We investigate whether the use of disentangled representations improves the performance of a multi-task neural network and whether disentanglement itself is achieved naturally during the training process in such a setting.

Our key contributions are:

- Construction of synthetic datasets that allow studying the relationship between multi-task and disentanglement learning.
- Study of the effect of multi-task learning with hard parameter sharing on the level of disentanglement obtained in the latent representation of the model.
- Analysis of the informativeness of the latent representation obtained in the single- and multi-task training.
- Inspection of the effect of disentangled representations on the performance of a multi-task model.

We verify our hypotheses by training multiple models in single- and multi-task settings and investigating the level of disentanglement achieved in their latent representations. In our experiments, we find that in a hard-parameter sharing scenario multi-task learning indeed seems to encourage disentanglement. However, it is inconclusive whether disentangled representations have a clear positive impact on the models performance, as the obtained by us results in this matter vary for different datasets.

Code for our experiments is available at:
https://github.com/gmum/disentanglement-multitask.

2 Related Work

2.1 Disentanglement

Over the recent years, many methods that directly encourage disentanglement have been proposed. This includes algorithms based on variational and Wasserstein auto-encoders [4,13,19,21,39], flow networks [9,38] or generative adversarial networks [8]. The main interest behind disentanglement learning lays in the assumption that such transformation unravels the semantically meaningful factors of variation present in the observations and thus it is desired in training deep learning models. In particular, disentanglement is believed to allow for informative compression of the data that results in a structural, interpretable representation, which is easily adaptable for new tasks [3,23,24,36].

Several of these properties have been experimentally proven in applications in many domains, including video processing tasks [15], recommendation systems [29] or abstract reasoning [40,41]. Moreover, recent research in reinforcement learning concludes that disentangling embeddings of skills allows for faster retraining and better generalization [33]. Finally, disentanglement seems also to be positively correlated with fairness when sensitive variables are not observed [26]. On the other hand, some empirical studies suggest that one should be cautious while interpreting the properties of disentangled representations. For instance, the latest studies in the unsupervised learning domain point that increased disentanglement does not lead to a decreased sample complexity in downstream tasks [27].

Another key challenge in studying disentangled representations is the fact that measuring the quality of the disentanglement is a nontrivial task [10,12,19], especially in a unsupervised setting [27]. This motivates the research on practical advantages of disentanglement representations and their impact on the studied problem in possible future applications, which is the main focus of our work in the case of multi-task learning.

2.2 Multi-task Learning

Multi-task learning aims at simultaneously solving multiple tasks by exploiting common information [34]. The approaches used predominantly to this problem are soft [11] and hard [6] parameter sharing. In hard parameter sharing the weights of the model are divided into those shared by all tasks, and task-specific. In deep learning, this idea is typically implemented by sharing consecutive layers of the network, which are responsible for learning a joint data representation. In soft parameter sharing each task is given a set of separate parameters. The limitations are then imposed by information-sharing or regularizing the distance between the parameters by adding an applicable loss to the optimization objective.

Multi-task learning is widely used in the Deep Learning community, for instance in applications related to natural language processing [25,30], computer vision [32] or molecular property prediction modeled by graph neural networks [5]. One may observe that the premises of multi-task and disentanglement learning are related to each other and thus it is interesting to investigate whether the joint data representation obtained in a multi-task problem exhibits some disentanglement-related properties.

3 Methods

In this section, we describe the methods and datasets used for conducting the experiments.

3.1 Dataset Creation

In order to investigate the relationship between multi-task learning and disentanglement, we require a dataset that fulfills two conditions:

1. It provides access to the true (disentangled) generative factors z from which the observations x are created.

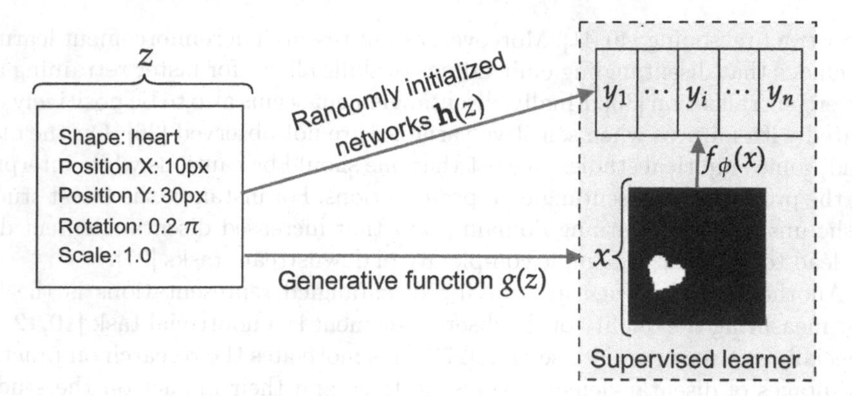

Fig. 1. The setting of our experiments. Given a dataset of pairs (x, z) of observations and their true generative factors, we generate a set of functions $\mathbf{h}(z)_i$ which are aimed to approximate real-world supervised tasks. Then, we train a neural network $f_\phi(x)$ in a multi-task regression setting on pairs $(x, \mathbf{h}(z))$. After the training, we investigate the hidden representations learned by f_ϕ and explore their relation to true factors z.

2. It proposes multiple tasks for a supervised learner by providing labels y_i which non-linearly depend on the true factors z.

The first condition is required in order to measure how well the learned representations approximate the true latent factors z. Access to the true factors allows for full control over the experimental settings and permits a fair comparison through the use of supervised disentanglement metrics. Note that even though unsupervised metrics have been proposed in the literature as well, they typically yield less reliable results, as we further discuss in Sect. 3.3.

The second condition is needed to train a network on multiple nontrivial tasks to approximate the real-world setting of multi-task learning.

To our best knowledge, no nontrivial datasets exist that would abide by both those requirements. Most of the available disentanglement datasets, such as dSprites, Shapes3D, and MPI3D do fulfill the first condition, as they provide pairs (x, z) of observations and their true generative factors. However, those datasets do not offer any type of challenging task on which our model could be trained. On the other hand, many datasets used for supervised multi-task learning fulfill the second condition by providing pairs (x, y), but do not equip the researcher with the latent factors z (ground truth), failing the first condition.

Thus, we aim to create our own datasets which fulfill both conditions by incorporating nontrivial tasks into standard disentanglement datasets. Since in multi-task approaches one often tries to solve tens of tasks at once, designing them by hand is infeasible and as such we decide to generate them automatically in a principled way. In particular, since supervised learning tasks might be formalized as finding a good approximation to an unknown function $h(x)$ given a set of points $(x, h(x))$, we generate random functions $h(z)$ which are then used to obtain targets for our dataset (see Fig. 1).

We require $h(z)$ to be both nontrivial (i.e. non-linear and non-convex) and sufficiently smooth to approximate the nature of real-life tasks. In order to find a family of functions that fulfills those conditions, we take inspiration from the field of extreme learning, which finds that features obtained from randomly initialized neural networks are useful for training linear models on various real-world problems [16]. As such, randomly initialized networks should be able to approximate these tasks up to a linear operation.

In particular, in order to generate the dataset, we define a neural network architecture $h(z, \theta)$. For this purpose, we used an MLP with four hidden layers with 300 units, tanh activations, and an output layer which returns a single number. Then we sample n weight initializations of this network from the Gaussian distribution $\theta_i \sim \mathcal{N}(0, 1)$, where $i \in \{1, \ldots, n\}$. Each of the networks $h(z, \theta_i)$ obtained by random initialization defines a single task in our approach. Thus, for a given dataset $\mathcal{D} = (x, z)$ containing observations and their true generative factors, we obtain a dataset for multi-task supervised learning by applying:

$$\tilde{\mathcal{D}} = \{(x, \mathbf{h}(z)) \mid (x, z) \in \mathcal{D}\} = \{(x, y)\},$$

where $\mathbf{h}(z)$ is a vector of stacked target values for each task, whose element i is given by $\mathbf{h}(z)_i = h(z, \theta_i)$.

We use this data as a regression task, i.e. for a given neural network f_ϕ parameterized by ϕ the goal is to find:

$$\arg\min_{\phi} \sum_{(x,y)\in\tilde{\mathcal{D}}} \|f_\phi(x) - y\|_2^2.$$

We use this process to create multi-task supervised versions of dSprites, Shapes3D, and MPI3D, with 10 tasks for each dataset.

3.2 Models

Multi-task Model. We investigate the relation between disentanglement and multi-task learning based on a hard parameter sharing approach. In this setting, several consecutive hidden layers of the model are shared across all tasks in order to produce a joint data representation. This representation is then propagated to separate task-specific layers which are responsible for computing the final predictions.

In particular, we use a network consisting of a shared convolutional encoder and separate fully-connected heads for each of the tasks. The encoder learns the joint representation by transforming the inputs into a d-dimensional latent space.[1] The heads are implemented by 4-layer MLPs with ReLU activations, in order to match the capacity of the networks used for task generating functions $h_i(x)$. This overview of the model is illustrated in Fig. 2.

[1] We provide the full model summary in **Appendix A**. The architecture of the encoder follows the one from [1], which adopts the work of [27] for the `pytorch` package. We use the implementations from https://github.com/amir-abdi/disentanglement-pytorch.

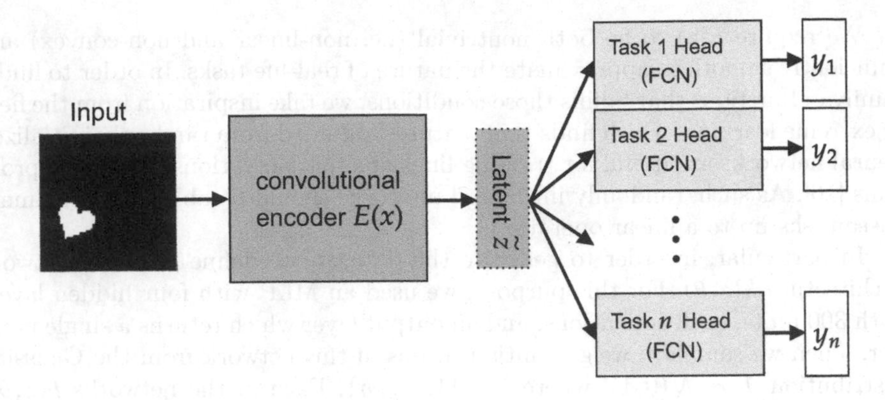

Fig. 2. The model used for multi-task training. The convolutional encoder $E(x)$ transforms the input data x to a latent representation $\tilde{z}$. The parameters of the encoder are shared across all tasks. Next, the produced representation is passed to the task-specific heads, which are implemented by fully-connected networks (FCN).

Auto-encoder Model. In the second part of our experiments we want to understand if disentangled representation provides some benefits for the multi-task problem. In order to produce disentangled representations, we decided to use three different representation-learning algorithms: a vanilla auto-encoder, the (beta)-variational auto-encoder [13,20] and FactorVAE [19].

All these variants of the auto-encoder architecture encompass a similar framework. An auto-encoder imposes a bottleneck in the network which forces a compressed knowledge representation of the original input. In some variants of those models, we additionally try to constrain the latent variables to be highly informative and independent which further correlates to disentanglement, e.g. in models like β-VAE and FactorVAE. We use latent representations from these models to train task-specific heads and evaluate if disentanglement helped to decrease an error for that task.

The vanilla auto-encoder is also used in Sect. 4.2, where we add a decoder with transposed convolutions to pre-trained encoders from Sect. 4.1. This treatment is aimed to decode information for particular encoders in the most efficient way. As such, we find auto-encoders to be a useful tool for investigating disentanglement.

3.3 Disentanglement Metrics

Measuring the qualitative and quantitative properties of the disentanglement representation discovered by the model is a nontrivial task. Due to the fact that the true generating factors of a given dataset are usually unknown, one may assume that decomposition can be obtained only to some extent.

Commonly used unsupervised metrics are based on correlation coefficients which measure the intrinsic dependencies between the latent components. Such measures are widely used in the independent component analysis [2,4,14,17,18,39]. However, uncorrelatedness does not imply stochastical independence.

Furthermore, metrics based on linear correlations may not be able to capture higher-order dependencies and are often ineffective in large dimensional or in over-determined spaces. All this makes the use of such unsupervised metrics questionable.

An alternative solution would be to use supervised metrics, which usually are more reliable [27]. This is of course only possible after assuming access to the true generative factors. Such an assumption is rarely valid for real-world datasets, however, it is satisfied for synthetic datasets. Synthetic datasets present therefore a reasonable baseline for benchmarking disentanglement algorithms.

Frequently used metrics which use supervision are mutual information gap (MIG) [7], the FactorVAE metric [19], Separated Attribute Predictability (SAP) score [22] and disenanglement-completness-informativeness (DCI) [12]. In order to comprehensively assess the level of disentanglement in our experiments, we have decided to use all of the above-mentioned metrics to validate our results. A more detailed description of those metrics is available in **Appendix B**.

4 Results and Discussion

In this section, we describe the performed experiments and discuss the obtained results. For more details on the training regime and experimental setup please refer to **Appendix C**.

4.1 Does Hard Parameter Sharing Encourage Disentanglement?

One of the most common approaches to multi-task learning is hard parameter sharing. The key challenge in this method is to learn a joint representation of the data which is at the same time informative about the input and can be easily processed in more than one task. It is therefore tempting to verify whether disentanglement arises in those representations implicitly, as a consequence of hard parameter sharing.

In order to investigate this problem we build a simple multi-task model described in Sect. 3.2 and evaluate it on the three datasets discussed in Sect. 3.1: dSprites, Shapes3D, and MPI3D, each with 10 artificial tasks. After the training is complete, we calculate each of the disentanglement metrics described in Sect. 3.3 on the latent representation of the input data[2]. We compare the obtained results with the same metrics computed for an untrained (randomly initialized) network and for single task models. In all the cases we use the same architecture and training regime. Note that in the single-model scenario we train a separate model for each of the 10 tasks, which is implemented by utilizing only one, dedicated head in the optimization process. We train all models three times, using a different random seed in the parameters initialization procedure. We report the mean results and standard deviations in Fig. 3.

[2] We use the implementations of [27], which are available at https://github.com/google-research/disentanglement_lib.

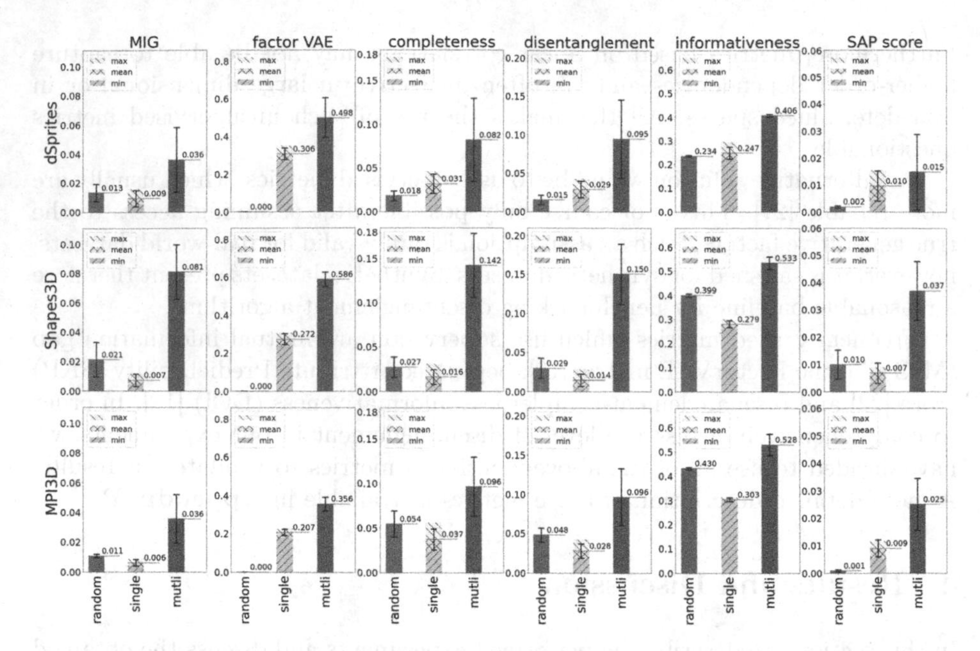

Fig. 3. Different disentanglement metrics computed for random (untrained), single-task and multi-task models evaluated on the three datasets described in Sect. 3.1. The higher the value the better. For the single-task scenario, we report the mean over all task-specific models. Note that in almost every case the multi-task representations (red bars) outperform the random or single-task representations (dark-gray bars and light gray bars, respectively). Additionally, for single-task models, we report the maximal and minimal values over all tasks to show that the performance on multi-task does not rely on any single 'lucky' task. For tabulated results please refer to **Appendix E**.

We observe that disentanglement metrics computed for the representations obtained in the multi-task setting are typically significantly better than the values obtained for single-task or random representations. Note that even the maximum mean result over all ten single-task models is in almost every case further than one standard deviation from the multitask mean. Moreover, this is true for all the tested datasets.

Let us also point out that instead of using separate heads for each of the tasks in the multi-task model one could simply use one head with the output dimension equal to the number of tasks and perform standard multivariate regression (with no parameter sharing). As presented in Fig. 4, the latent representations emerging in such a scenario are less disentangled (in terms of the considered metrics) than the representations obtained when utilizing hard parameter sharing. However, the achieved values are still better than in single-task models. This suggests that even though the increase in the metrics may be partially caused by simply training the network on higher-dimensional targets, the positive influence of hard parameter sharing cannot be ignored. This advocates in favor of

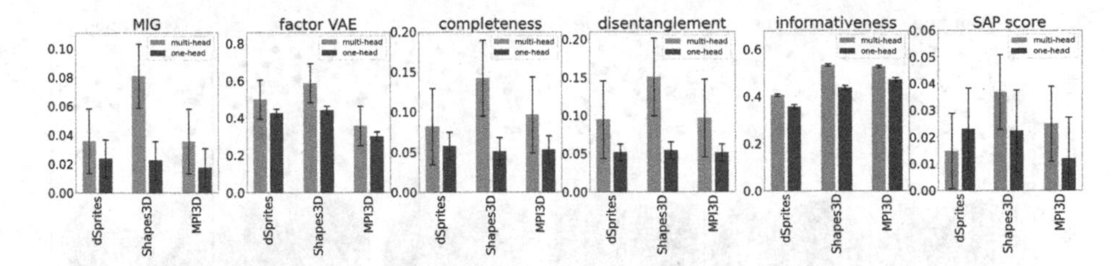

Fig. 4. Different disentanglement metrics computed for the multi-task setting with one head shared between all tasks (one-head) and separate head for every task (multi-head), evaluated on the three datasets described in Sect. 3.1. The higher the value the better. One may observe that multi-head representations perform better than the ones obtained in the standard, one-head multivariate regression task. For tabulated results please refer to **Appendix E**.

the hypothesis that multi-task representations are indeed more disentangled than the ones arising in single-task learning.

4.2 What Are the Properties of the Learned Representations?

The previous section discussed the obtained representations by analyzing quantitative disentanglement metrics. Here, we provide more insights into the characteristics of latent encodings.

UMAP Embeddings. In order to gain intuition behind the differences between the representations obtained in the previous experiment we compute a 2D-embedding of the latent encodings using the UMAP algorithm [31]. The results are presented in Fig. 5.

The embeddings obtained for the multi-task representations are much more semantically meaningful, with easily distinguishable separate clusters. Moreover, the position and internal structure of the clusters correspond to different values of the true factors. This cannot be observed for the untrained or single-task representations, suggesting that the multi-task representations are indeed more successful in encompassing the information about the real values of the generative sources of the data.

Latent Space Traversal. Providing qualitative results of the retrieved factors is a common practice in disentanglement learning [21,27,28,35,38]. In particular, visual presentation of the interpolations over the latent space allows assessing — from a human perspective — the informativeness and decomposition of the obtained representations. Note that such analysis is possible only after adding and training a suitable decoder network, which maps the retrieved factors back to the image space.

In our setting, the decoder mirrors the architecture of the encoder (the convolutions are replaced by transposed convolutions of the same size — see

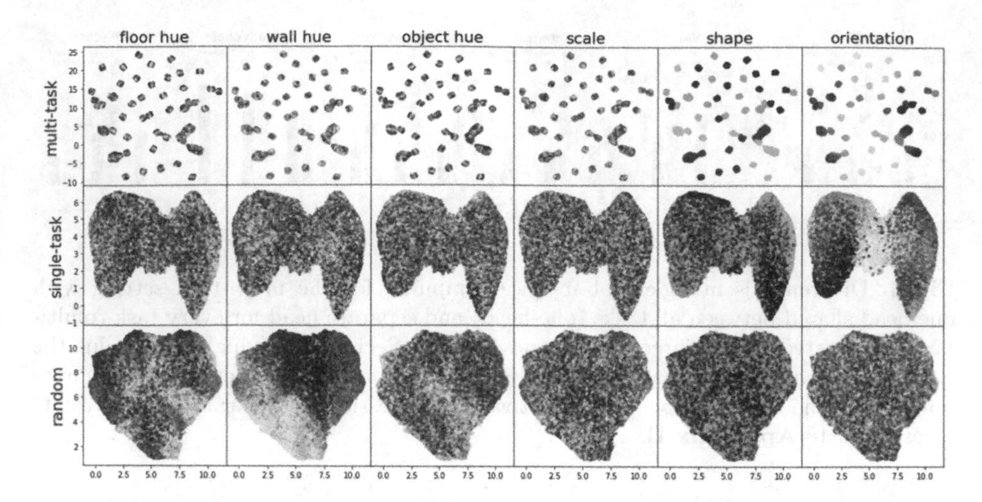

Fig. 5. UMAP embeddings of the latent representations of the Shapes3D test dataset obtained for different models. Change of the color within one subplot presents the change in one particular ground truth component. The embeddings obtained by the multi-task model seem to be most semantically meaningful. See **Appendix D** for plots for other datasets.

Appendix A). Given the latent representations as an input, the decoder optimizes the reconstruction error (as measured by MSE) between its outputs and the original images. We train three separate decoders corresponding to the different encoders from the previous section — a randomly initialized encoder, an encoder produced by one of the single-task models, and a multi-task encoder.

First, let us discuss the reconstruction quality achieved by each of the tested decoders. Results of this experiment are presented in Fig. 6[3]. Reconstructions produced for the multi-task encodings are clearly superior to the ones obtained for the single-task encodings. In the first case, the resulting images are sharp and contain almost no noise. In contrast, the single task reconstructions are blurry and similar to the ones produced for the randomly initialized encoder. We would like to emphasise that all the decoders used the same architecture and that during their optimization the parameters of the corresponding encoders were kept fixed. Therefore the quality of the reconstruction is an important property of a latent representation, as it allows us to assess the compression capacity of the representation. From this perspective, the compression obtained in the multi-task scenario is much more informative about the input than in the single-task scenario.

Another approach to the visualisation of the latent variables is to perform interpolations (traversals) in the latent space. We start by selecting a random sample from the dataset and compute its encoding $\tilde{z} \in \mathbb{R}^d$. By modifying one of the components of vector $\tilde{z}$ from -1 to 1 with 0.1 step and leaving the $d-1$

[3] Numerical values for reconstruction errors are presented in **Appendix D.2** .

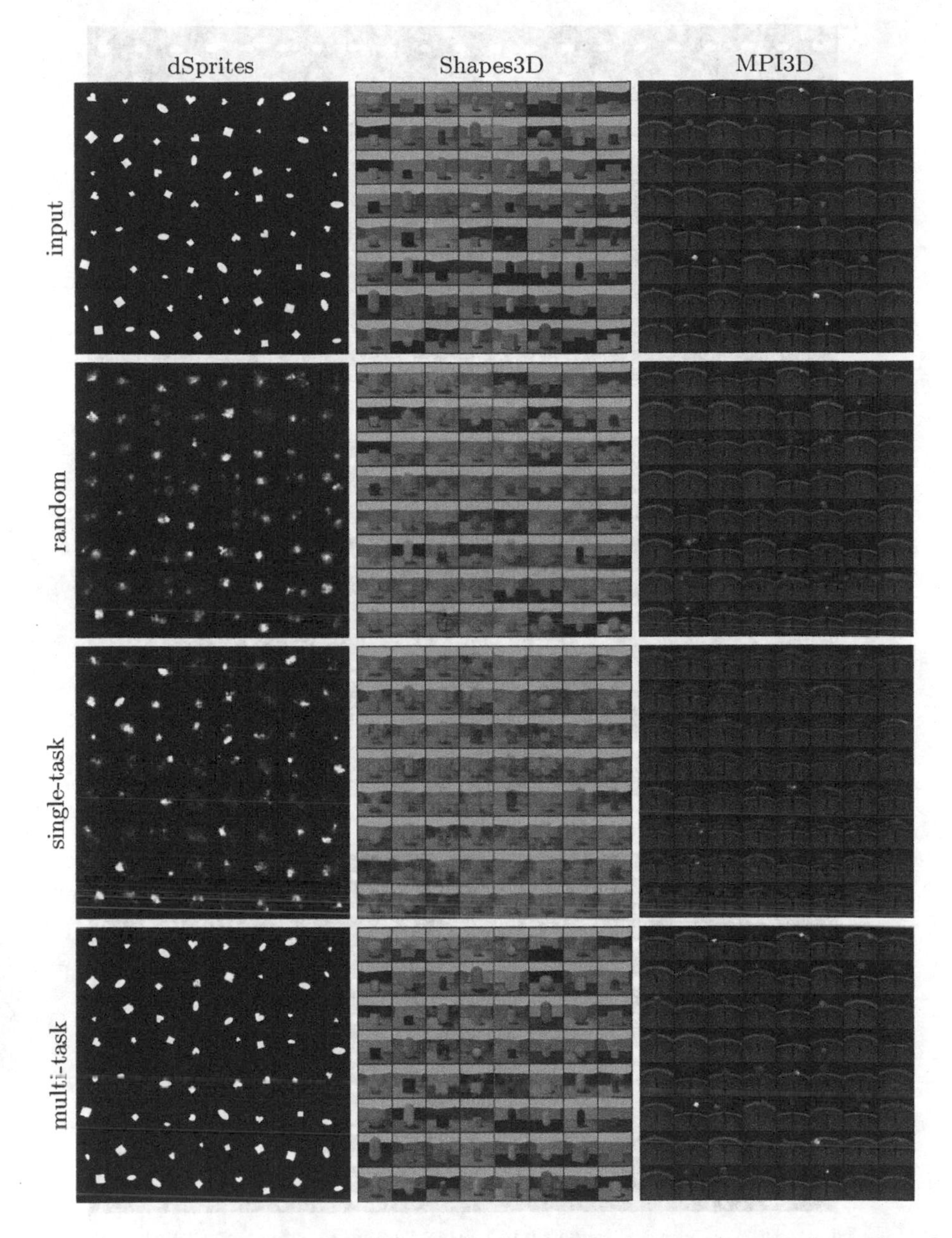

Fig. 6. Reconstructions obtained by the decoders trained on random, single-task, and multi-task encoding. For reference, we provide the original input images in the first row. The quality of the reconstruction for the random and single-task representation is very poor. Contrary, the multi-task encoder provided a latent space that can be successfully decoded into images that closely resemble the corresponding examples from the input. Thus, we conclude that the multi-task representations are more informative about the data and provide better compression.

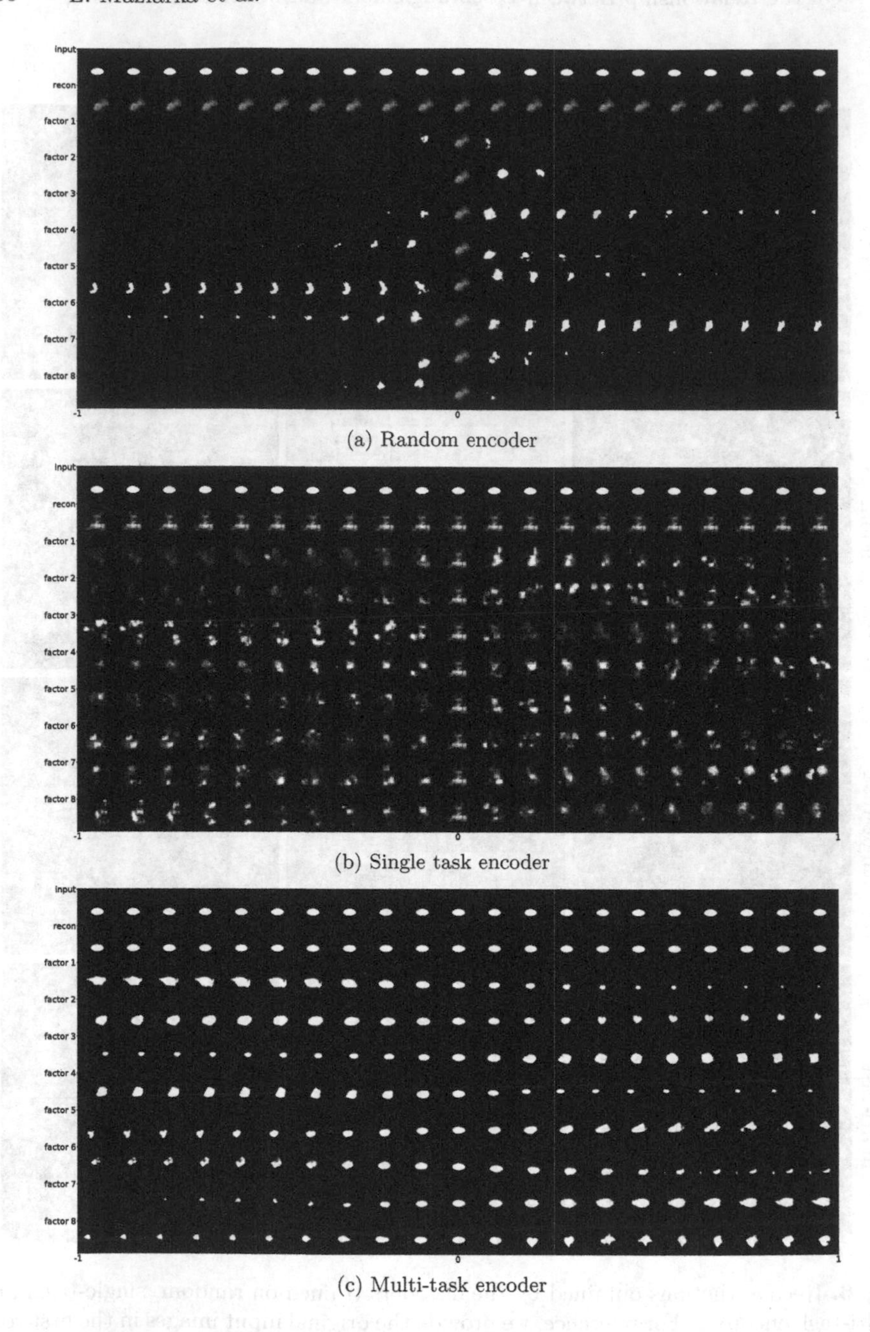

(a) Random encoder

(b) Single task encoder

(c) Multi-task encoder

Fig. 7. Traverses over latent variable produced for a given architecture. The same example was used in all three traverses. The second row of each image shows how the decoder reconstructed this example in a particular setting. The rest of the factors come from latent space generated from each encoder. Visualization of components from the multi-task encoder are sharp and distinguish the generating factors distinctly. The same cannot be said about the latent factors in single-task and random encoders, which are blurry and disconnected from any interpretable ground truth factors. Please refer to **Appendix D** for the results of the traversals over other datasets.

unchanged, we produce a traversal along that particular factor. We repeat this procedure for all the factors in order to capture their impact on the decoded example. Results of such traverses for the dSprites dataset are shown in Fig. 7.

Note that since the models were not trained directly for disentanglement but only to solve a supervision task, it is not surprising that the representations are not as clearly factorized as in specialized methods such as FactorVAE. However, for the multi-task model, certain latent dimensions still appear to be disentangled and one can easily spot the difference in quality between the single and multi-task representations. In the multi-task traversals, we can notice components that are responsible for the position and scale of a given figure (in Fig. 7c, consider the 5th and 7th factors, respectively). In contrast, the results for single task representations demonstrate that even a slight change in any of the single latent dimensions leads to a degradation of the reconstructed examples. As expected, this effect is even more evident for the random (untrained) representations, where the corruption over latent factor is even more prevalent than in the case of a single-task traversal.

4.3 Does Disentanglement Help in Training Multi-task Models?

In the previous sections, we studied whether multi-task learning encourages disentanglement. Here we consider an inverse problem by asking whether using disentangled representation helps in multi-task learning. To investigate this issue, we train an auto-encoder-based model devised specifically to produce disentangled latent representations without access to the true latent factors. Next, we freeze its parameters and use the encoder function to transform the inputs. The obtained latent encodings are then passed directly to the heads of a multi-task network which minimizes the average regression loss given the target values of the artificial tasks.

We consider three different auto-encoder-based algorithms described in Sect. 3.2: a vanilla auto-encoder (AE), a variational auto-encoder (VAE), and the FactorVAE. The vanilla auto-encoder does not directly enforce latent disentanglement during the training. In the VAE model, the prior normal distribution with identity covariance matrix implies some disentanglement. Finally, FactorVAE introduces a new module to the VAE architecture that explicitly induces informative decomposition. Therefore, the representations obtained for each subsequent model should be also naturally ordered by the level of the achieved disentanglement. For the exact values of the calculated metrics please refer to **Appendix F**. In addition, we also study a scenario in which we explicitly provide the true source factors. We trained all regression models three times, using a different random seed in the parameters initialization procedure.

Table 1 summarizes the performance of the multi-task model trained on the representations obtained for the above-discussed methods. Although the representations obtained from FactorVAE are better (see, for instance, MIG or DCI measures in **Appendix F**) than those from VAE and AE, the encodings produced by the vanilla AE are the best among the tested, exceeding the others on Shapes3D and MPI3D and being second on dSprites. Note that these results

Table 1. RMSE of multi-task networks trained on latent representations obtained by different auto-encoder-based methods. For comparison, we added the model trained on ground truth factors. The best results are bolded, and best out of auto-encoder architectures underlined.

Dataset	dSprites	Shapes3D	MPI3D
Ground Truth	150.235 ± 3.754	**72.979 ± 0.193**	**108.568 ± 0.285**
AE	80.062 ± 0.341	114.939 ± 0.160	150.190 ± 0.097
VAE	**63.260 ± 0.260**	132.072 ± 0.169	194.865 ± 15.61
FactorVAE	91.937 ± 0.199	118.396 ± 0.423	151.646 ± 0.336

coincide with observations presented in the literature. For example, [27] compared different models that enforce disentanglement during the training and showed that even a high value of that property within the factors do not constitute a better model performance. However, in two out of three datasets, the use of the ground true factors seems to significantly improve the obtained results. This may suggest that the representations produced by the considered disentanglement methods are not fully factorized. It is therefore inconclusive whether the discrepancy between the obtained results is due to the shortcomings of the used methods or a manifestation of the impracticality of disentanglement.

5 Conclusions

In this paper, we studied the relationship between multi-task and disentanglement representation learning. A fair evaluation of our hypothesis is impossible on real-world datasets, without provided ground truth factors. To evaluate our results we had to introduce synthetic datasets that contain all necessary properties to be seen as a benchmark in this field. Next, we studied the effects of multi-task learning with hard parameter sharing on representation learning. We found that nontrivial disentanglement appears in the representations learned in a multi-task setting. Obtained factors have intuitive interpretations and correspond to the actual ground truth components. Finally, we inverted the question and investigated the hypothesis that disentangled representation is needed for multi-task learning, the results however are not conclusive. We found out that multi-task models benefit from disentanglement only on specific datasets. However, we cannot name an indicator of when this unambiguously applies.

Acknowledgements. The work of Ł. Maziarka was supported by the National Science Centre (Poland) grant no. 2019/35/N/ST6/02125. The work of A. Nowak and M. Wołczyk was supported by Foundation for Polish Science (grant no POIR 04.04.00-00-14DE/18-00) carried out within the Team-Net program co-financed by the European Union under the European Regional Development Fund.

References

1. Abdi, A.H., Abolmaesumi, P., Fels, S.: Variational learning with disentanglement-pytorch. arXiv preprint arXiv:1912.05184 (2019)
2. Bedychaj, A., Spurek, P., Nowak, A., Tabor, J.: WICA: nonlinear weighted ICA (2020)
3. Bengio, Y.: Deep learning of representations: looking forward (2013)
4. Brakel, P., Bengio, Y.: Learning independent features with adversarial nets for non-linear ICA (2017)
5. Capela, F., Nouchi, V., Van Deursen, R., Tetko, I.V., Godin, G.: Multitask learning on graph neural networks applied to molecular property predictions. arXiv preprint arXiv:1910.13124 (2019)
6. Caruana, R.: Multitask learning: a knowledge-based source of inductive bias. In: Proceedings of the Tenth International Conference on Machine Learning, pp. 41–48. Morgan Kaufmann (1993)
7. Chen, R.T., Li, X., Grosse, R.B., Duvenaud, D.K.: Isolating sources of disentanglement in variational autoencoders. In: Advances in Neural Information Processing Systems, pp. 2610–2620 (2018)
8. Chen, X., Duan, Y., Houthooft, R., Schulman, J., Sutskever, I., Abbeel, P.: InfoGAN: interpretable representation learning by information maximizing generative adversarial nets. Adv. Neural. Inf. Process. Syst. **29**, 2172–2180 (2016)
9. Dinh, L., Krueger, D., Bengio, Y.: Nice: non-linear independent components estimation. arXiv preprint arXiv:1410.8516 (2014)
10. Do, K., Tran, T.: Theory and evaluation metrics for learning disentangled representations. In: International Conference on Learning Representations (2020). https://openreview.net/forum?id=HJgK0h4Ywr
11. Duong, L., Cohn, T., Bird, S., Cook, P.: Low resource dependency parsing: Cross-lingual parameter sharing in a neural network parser. In: Proceedings of the 53rd Annual Meeting of the Association for Computational Linguistics and the 7th International Joint Conference on Natural Language Processing (Volume 2: Short Papers), pp. 845–850 (2015)
12. Eastwood, C., Williams, C.K.I.: A framework for the quantitative evaluation of disentangled representations. In: International Conference on Learning Representations (2018). https://openreview.net/forum?id=By-7dz-AZ
13. Higgins, I., et al.: beta-VAE: Learning basic visual concepts with a constrained variational framework. In: ICLR (2017)
14. Hirayama, J., Hyvarinen, A., Kawanabe, M.: Splice: fully tractable hierarchical extension of ICA with pooling. In: Proceedings of the International Conference on Machine Learning, vol. 70, pp. 1491–1500. Machine Learning Research (2017)
15. Hsieh, J.T., Liu, B., Huang, D.A., Fei-Fei, L.F., Niebles, J.C.: Learning to decompose and disentangle representations for video prediction. In: Advances in Neural Information Processing Systems, pp. 517–526 (2018)
16. Huang, G.B., Wang, D.H., Lan, Y.: Extreme learning machines: a survey. Int. J. Mach. Learn. Cybernet. **2**(2), 107–122 (2011)
17. Hyvarinen, A., Morioka, H.: Unsupervised feature extraction by time-contrastive learning and nonlinear ICA. In: Advances in Neural Information Processing Systems, pp. 3765–3773 (2016)
18. Hyvarinen, A., Morioka, H.: Nonlinear ICA of temporally dependent stationary sources. Proc. Mach. Learn. Res. (2017)

19. Kim, H., Mnih, A.: Disentangling by factorising. arXiv preprint arXiv:1802.05983 (2018)
20. Kingma, D.P., Welling, M.: Auto-encoding variational bayes. arXiv preprint arXiv:1312.6114 (2013)
21. Kumar, A., Sattigeri, P., Balakrishnan, A.: Variational inference of disentangled latent concepts from unlabeled observations. arXiv preprint arXiv:1711.00848 (2017)
22. Kumar, A., Sattigeri, P., Balakrishnan, A.: Variational inference of disentangled latent concepts from unlabeled observations (2018)
23. Lake, B.M., Ullman, T.D., Tenenbaum, J.B., Gershman, S.J.: Building machines that learn and think like people. Behav. Brain Sci. 40 (2017)
24. Lipton, Z.C.: The mythos of model interpretability. Queue **16**(3), 31–57 (2018)
25. Liu, X., He, P., Chen, W., Gao, J.: Multi-task deep neural networks for natural language understanding. arXiv preprint arXiv:1901.11504 (2019)
26. Locatello, F., Abbati, G., Rainforth, T., Bauer, S., Schölkopf, B., Bachem, O.: On the fairness of disentangled representations. In: Advances in Neural Information Processing Systems, pp. 14611–14624 (2019)
27. Locatello, F., et al.: Challenging common assumptions in the unsupervised learning of disentangled representations (2019)
28. Locatello, F., Tschannen, M., Bauer, S., Rätsch, G., Schölkopf, B., Bachem, O.: Disentangling factors of variation using few labels. arXiv preprint arXiv:1905.01258 (2019)
29. Ma, J., Zhou, C., Cui, P., Yang, H., Zhu, W.: Learning disentangled representations for recommendation. In: Advances in Neural Information Processing Systems, pp. 5711–5722 (2019)
30. Maziarka, Ł., Danel, T.: Multitask learning using BERT with task-embedded attention. In: 2021 International Joint Conference on Neural Networks (IJCNN), pp. 1–6. IEEE (2021)
31. McInnes, L., Healy, J., Melville, J.: UMAP: uniform manifold approximation and projection for dimension reduction. arXiv preprint arXiv:1802.03426 (2018)
32. Misra, I., Shrivastava, A., Gupta, A., Hebert, M.: Cross-stitch networks for multi-task learning. In: Proceedings of the IEEE Conference on Computer Vision and Pattern Recognition, pp. 3994–4003 (2016)
33. Petangoda, J.C., Pascual-Diaz, S., Adam, V., Vrancx, P., Grau-Moya, J.: Disentangled skill embeddings for reinforcement learning (2019)
34. Ruder, S.: An overview of multi-task learning in deep neural networks. arXiv preprint arXiv:1706.05098 (2017)
35. Sanchez, E.H., Serrurier, M., Ortner, M.: Learning disentangled representations via mutual information estimation (2019)
36. Schmidhuber, J.: Learning factorial codes by predictability minimization. Neural Comput. **4**(6), 863–879 (1992)
37. Schölkopf, B., Janzing, D., Peters, J., Sgouritsa, E., Zhang, K., Mooij, J.: On causal and anticausal learning. arXiv preprint arXiv:1206.6471 (2012)
38. Sorrenson, P., Rother, C., Köthe, U.: Disentanglement by nonlinear ICA with general incompressible-flow networks (GIN) (2020)
39. Spurek, P., Nowak, A., Tabor, J., Maziarka, Ł, Jastrzębski, S.: Non-linear ICA based on cramer-wold metric. In: Yang, H., Pasupa, K., Leung, A.C.-S., Kwok, J.T., Chan, J.H., King, I. (eds.) ICONIP 2020. LNCS, vol. 12534, pp. 294–305. Springer, Cham (2020). https://doi.org/10.1007/978-3-030-63836-8_25

40. Steenbrugge, X., Leroux, S., Verbelen, T., Dhoedt, B.: Improving generalization for abstract reasoning tasks using disentangled feature representations. arXiv preprint arXiv:1811.04784 (2018)
41. Van Steenkiste, S., Locatello, F., Schmidhuber, J., Bachem, O.: Are disentangled representations helpful for abstract visual reasoning? In: Advances in Neural Information Processing Systems, pp. 14245–14258 (2019)

InCo: Intermediate Prototype Contrast for Unsupervised Domain Adaptation

Yuntao Du, Hongtao Luo, Haiyang Yang, Juan Jiang, and Chongjun Wang(✉)

State Key Laboratory for Novel Software Technology at Nanjing University, 210023 Nanjing, China
{duyuntao,mf20330054,hyyang,mf20330037}@smail.nju.edu.cn, chjwang@nju.edu.cn

Abstract. Unsupervised domain adaptation aims to transfer knowledge from the labeled source domain to the unlabeled target domain. Recently, self-supervised learning (e.g. contrastive learning) has been extended to cross-domain scenarios for reducing domain discrepancy in either instance-to-instance or instance-to-prototype manner. Although achieving remarkable progress, when the domain discrepancy is large, these methods would not perform well as a large shift leads to incorrect initial pseudo labels. To mitigate the performance degradation caused by large domain shifts, we propose to construct multiple intermediate prototypes for each class and perform cross-domain instance-to-prototype based contrastive learning with these constructed intermediate prototypes. Compared with direct cross-domain self-supervised learning, the intermediate prototypes could contain more accurate label information and achieve better performance. Besides, to learn discriminative features and perform domain-level distribution alignment, we perform intra-domain contrastive learning and domain adversarial training. Thus, the model could learn both discriminative and invariant features. Extensive experiments are conducted on three public benchmarks (ImageCLEF, Office-31, and Office-Home), and the results show that the proposed method outperforms baseline methods.

Keywords: Unsupervised domain adaptation · Transfer learning · Contrastive learning · Intermediate prototypes

1 Introduction

Deep learning has achieved remarkable performance in various computer vision tasks, such as image classification [13,18], semantic segmentation [21], and object detection [12]. Despite achieving remarkable progress, deep neural networks trained on a specific domain often fail to generalize to new domains because of the domain shift problem [28]. Unsupervised domain adaptation (UDA) could

Y. Du and H. Luo—The first two authors contributed equally.

M.-R. Amini et al. (Eds.): ECML PKDD 2022, LNAI 13713, pp. 642–658, 2023.
https://doi.org/10.1007/978-3-031-26387-3_39

overcome this challenge by transferring knowledge from a fully-labeled source domain to an unlabeled target domain.

Most existing deep domain adaptations fall into two strategies: moment matching and adversarial domain adaptation. The former aims to reduce the domain discrepancy by optimizing the statistical distribution discrepancy, such as Maximum Mean Discrepancy (MMD) distance [22], Joint Maximum Mean Discrepancy (JMMD) distance [24], and Wasserstein distance [32]. The latter reduces the domain discrepancy by the adversarial training across domains where a domain discriminator is introduced to distinguish the source domain from the target domain [10]. As the domain adversarial loss only achieves domain-level alignment and may lead to class mismatch, the following methods focus on class-level alignment [3,23] to achieve better adaptation.

Recently, some works have attempted to bridge the domain gap by extending traditional self-supervised learning (SSL, e.g., contrastive learning (CL)), which is learned from a single domain, to performing SSL across domains [2,17,36,40]. Early methods focus on instance-to-instance contrastive learning [2,17,36]. These methods differ in how to construct positive pairs and negative pairs. For example, CDS [17] proposes a two-stage pipeline (i.e., SSL followed by domain adaptation) and cross-domain instance-based contrastive loss is adopted for learning domain-invariant features across domains. It selects a sample in the other domain as a positive pair and other samples as negative pairs. However, such a method regards each instance as a class, the semantic structure of the data (class information) is not encoded by the learned representations. To overcome this challenge, the following methods focus on semantic aware contrastive learning. TCL [2] and CDCL [36] perform instance-to-instance contrastive learning and they select the samples from the same class of the other domain as positive pairs and the samples from other classes as negative pairs. Besides, different from instance-to-instance manner, PCS [40] proposes prototypical cross-domain self-supervised learning, where cross-domain instance-to-prototype matching is designed to transfer knowledge from source to the target in a more robust manner. The semantic information is encoded in the class prototypes and they regard the corresponding class prototype in the other domain as positive pair and other prototypes as negative pairs. Although achieving remarkable progress, when there is a large shift across domains, these methods would not perform well. As these methods rely on pseudo labels for training, the initial pseudo labels are largely affected by the distribution discrepancy. The larger the shift is, the worse pseudo labels we will get. In such cases, the learned model would be negatively affected by mislabeled positive and negative pairs, leading to poor performance.

In this paper, we would explore how to better perform contrastive learning to bridge the domain gap under a large domain shift. Firstly, we follow the line of instance-to-prototype contrastive learning, as cross-domain instance-to-instance matching is very sensitive to abnormal samples [40]. Secondly, recent progress in UDA [7,26] reveals that intermediate domains across domains could effectively deal with large domain shifts and achieve better performance. These intermediate domain based methods focus on the sample-level intermediate domain. In

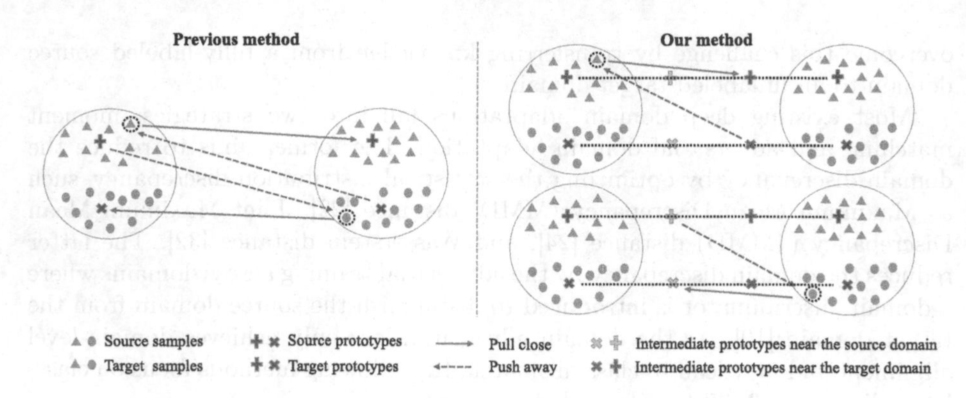

Fig. 1. Illustration of our method. **Left:** previous method performs direct cross-domain instance-to-prototype contrastive learning to bridge domain shift. But it could not perform well when domain shift is large. **Right:** To overcome this problem, we construct multiple intermediate prototypes and perform bidirectional cross-domain instance-to-prototype contrastive learning based on these intermediate prototypes.

this paper, we turn to the prototype-level intermediate domain and construct multiple intermediate prototypes for each class to perform contrastive learning. To achieve this, multiple intermediate class prototypes are constructed by a fixed ratio mixup [41] between the source prototypes and the target prototypes [26]. Compare with the sample-level intermediate domain, the intermediate prototypes would be more robust to outliers in the source domain and could contain the sample relations in each domain. Moreover, as shown in Fig. 1 and similar to the sample-level intermediate domain, an augmented class prototype close to the source domain has more reliable label information but is less similar to the target domain. By contrast, The class prototype close to the target domain has more relevant information about the target domain, but the label information is less accurate.

To this end, we propose **In**termediate prototype **Co**ntrast (**InCo**), a novel UDA method that constructs multiple intermediate prototypes for performing cross-domain instance-to-prototype contrastive learning. InCo contains three major components to learn semantic, domain-invariant, and discriminative features. As the core component of our method, InCo performs the inter-domain contrastive learning based on intermediate prototypes to mitigate the key challenge of large domain shift. Specially, we construct multiple intermediate class prototypes to bidirectionally apply instance-to-prototype contrastive learning. Before constructing intermediate prototypes, we firstly construct the prototypes in both domains. The source prototypes are computed as the mean representations of each class with true labels, while the target prototypes are computed with pseudo labels. The pseudo labels are obtained by clustering where the class center is initiated by source prototypes. Then, we apply a fixed ratio mixup between the source prototypes and target prototypes to construct intermediate prototypes. In inter-domain contrastive learning, for a given sample (either source domain or target domain), the corresponding class prototype near the other domain is selected as the positive pair and other prototypes are selected as negative pairs. Compared with direct cross-domain matching, the label information of intermediate class prototypes is more accurate as the source domain contains true labels. Thus, it could better deal with the large shift. Besides, similar to previous methods, we also adopt contrastive learning in each domain to learn discriminative features. As the instance-to-prototype contrastive learning, to some content, could be regarded as class-level alignment, we also introduce the domain adversarial loss [10] to further decreases the domain-level distribution discrepancy. Combining these losses together, we could learn both invariant and discriminative features.

We conduct extensive experiments to validate the effectiveness of the proposed method on standard DA benchmarks such as ImageCLEF, Office-31, and Office-Home. We also conduct lots of ablation studies to analyze the proposed method. To sum up, the main contributions of this paper are summarized as follows,

- We propose to construct intermediate prototypes by fixed-ratio mixup to perform contrastive learning for adaptation, which could deal with the large domain shift.
- We follow the instance-to-prototype manner and design bidirectional inter-domain contrastive learning to learn invariant features. Besides, with the intra-domain contrastive learning loss and domain adversarial loss, the model could learn both invariant and discriminative semantic features.
- We conduct extensive experiments on three real-world datasets, the results show the effectiveness of the proposed method.

2 Related Work

2.1 Unsupervised Domain Adaptation

A classical domain adaptation theory [2] indicates that it is crucial to reduce the distribution discrepancy across domains to achieve better adaptation. Based on this theory, many domain adaptation methods have been proposed and they are divided into moment matching and adversarial domain adaptation. The goal of the former is to reduce the statistical distribution discrepancy across domains. The widely used statistical measurements include the first-order moment [22], the second-order moment [33], and other statistical measurements [32]. Adversarial domain adaptation reduces the distribution discrepancy in an adversarial manner [4,10]. DANN [10] introduces a domain discriminator which plays a min-max game with the feature extractor by the domain adversarial loss. MCD [31] introduces two classifiers as a discriminator to play a min-max game with the feature extractor. Considering the practical multi-class problem, MDD [45] proposes a margin-based theory, and a new method based on this theory is proposed. As these methods focus on domain-level alignment, following methods [3] adopt the multi-class discriminator and considers both the domain and class information to achieve class-level alignment.

2.2 Contrastive Learning

Contrastive learning is a promising part in unsupervised learning [1,11,27]. The standard manner of contrastive learning is to learn discriminative representations by pulling the query together with positive pairs and pushing apart from negative pairs. Most methods focus on instance-based methods where each sample is regarded as a class. In these methods, the positive pairs are generated by creating different augmentations of each sample and the negative pairs are randomly chosen from different samples. However, the standard contrastive learning [16] methods have not considered task-specific semantic information. To overcome this problem, supervised contrastive learning has been proposed to leverage category labels to select positive and negative pairs. Furthermore, prototype contrastive learning [19] considered the semantic information in an unsupervised setting by clustering the samples to leverage the semantic information.

2.3 Contrastive Learning for Domain Adaptation

Although achieving remarkable progress, existing contrastive learning approaches can not be directly used in the standard UDA setting as they are performed in a single domain. While some methods have attempted to attend standard contrastive learning to cross-domain scenarios and have achieved satisfactory results. CDS [17] proposes a two-stage pipeline (i.e., SSL followed by domain adaptation), and cross-domain instance-based supervised loss is adopted for learning domain-invariant features across domains. However, the semantic structure of the data (class information) is not encoded by the learned representations. To overcome this challenge, the following methods focus on semantic aware contrastive learning. TCL [2] and CDCL [36] perform instance-to-instance contrastive learning where the positive pairs are selected from the same class in the other domain and the negative pairs are from other classes. Besides, different from instance-to-instance based manner, PCS [40] proposes prototypical cross-domain self-supervised learning, where cross-domain instance-to-prototype matching is designed to transfer knowledge from source to the target in a more robust manner. The semantic information is encoded in the class prototypes and they regard the corresponding class prototype in the other domain as positive pair and other prototypes as negative pairs. Although achieving remarkable progress, these methods would not perform well under large shifts.

3 Method

3.1 Problem Definition and Overall Idea

In UDA, we are given a labeled source domain $\mathcal{D}_s = \{(\mathbf{x}_i^s, y_i^s)\}_{i=1}^{N_s}$ and an unlabeled target domain $\mathcal{D}_t = \{(\mathbf{x}_j^t)\}_{j=1}^{N_t}$. The source samples and target samples are from different distributions $P_s(\mathbf{x}, y)$ and $P_t(\mathbf{x}, y)$. $\mathcal{D}_s$ and $\mathcal{D}_t$ contain the shared K categories, i.e., $\mathcal{Y}_s = \mathcal{Y}_t = \{1, ...K\}$. The goal of UDA is to learn a generalized model with $\mathcal{D}_s$ and $\mathcal{D}_t$ that could classify the target samples correctly.

In this section, we describe **InCo** in detail. As shown in Fig. 2, our model consists of four basic modules, a feature extractor g that maps the samples into feature embeddings, a project head h where the contrastive learning is performed, a domain discriminator D that performs domain adversarial training, and a classifier f that classifies the features into K categories. InCo follows the line of instance-to-prototype manner and uses intermediate prototypes to perform cross-domain contrastive learning so that it could effectively bridge the large discrepancy domains. Specially, we construct intermediate prototypes by fixed ratio mixup and design the bidirectional inter-domain contrastive learning based on these intermediate prototypes. Besides, we also perform contrastive learning within each domain to learn discriminative features. Moreover, inter-domain contrastive learning could achieve class-level alignment, we further introduce domain adversarial training to achieve domain-level alignment. Combing these losses together, the model could learn both invariant and discriminative features. In the next subsections, we introduce each loss in detail.

3.2 Revisit of Contrastive Learning

Contrastive learning [1,11,27] aims to learn discriminative features from unlabeled data in the form of positive/negative pairs by a contrastive loss. We denote the query and key vector as q, k, and k^+ and k^- as the positive and negative key for the query q. The goal of contrastive learning is to learn representations such that the query and positive key vector is as close as possible, meanwhile, the query and the negative key vector is far away from each other. A popular framework to achieve this goal is to formulate the contrastive learning as a 'two-class' classification problem, and the loss is formulated as,

$$\mathcal{L}(q, k^+, k^-) = -log \frac{\exp(q \cdot k^+/T)}{\exp(q \cdot k^+/T) + \sum_{k^-} \exp(q \cdot k^-/T)} \tag{1}$$

Here T is the temperature parameter, and $q \cdot k^+$ denotes the inner product between q and k^+.

3.3 Intra-Domain Contrastive Learning

As shown in the above subsection, contrastive learning could learn discriminative features for downstream visual tasks by a contrastive loss. Following that, we perform contrastive learning within each domain to learn discriminative representation for every single domain. As instance-to-instance based CL methods treat each sample as a single class, regardless of the semantic information, we follow the previous method [40] and adopt instance-to-prototype based CL, where the prototypes for all classes are set as the key vectors. By the intra-domain CL loss, representations with intra-class compactness and inter-class discrimination could be learned.

To start with, we define every prototype as the mean representation of each class to convey high-level class information. And we maintain two memory banks $\mathcal{Q}^s$ and $\mathcal{Q}^t$ for source and target prototypes respectively:

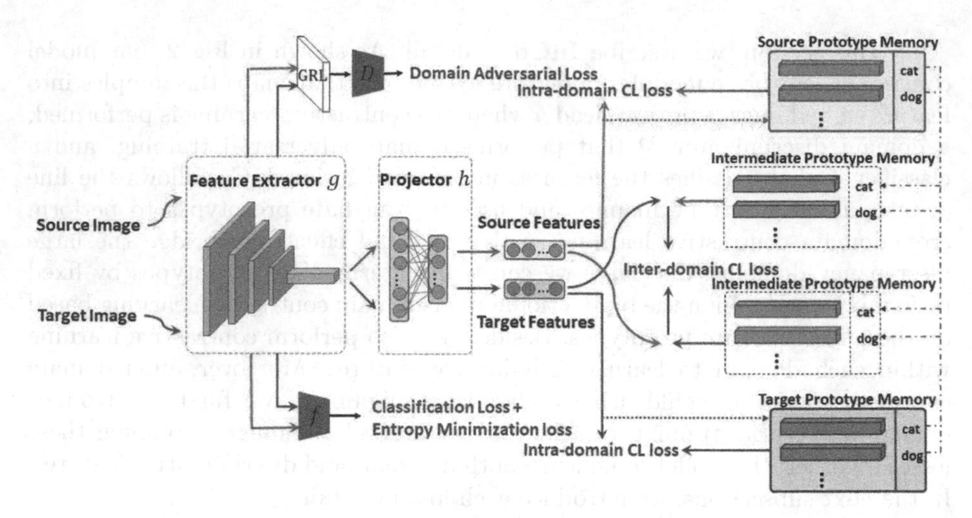

Fig. 2. An overview of InCo. In addition to conventional classification loss on labeled source samples and entropy minimization loss on unlabeled target samples, inter-domain contrastive learning is proposed to bridge the domain gap, where the corresponding prototype near the other domain is selected as the positive pair and other prototypes are selected as negative pairs. We also perform intra-domain contrastive learning within each domain and domain adversarial training, such that the model could learn both invariant and discriminative features.

$$\mathcal{Q}^s = [\mu_1^s, \ldots, \mu_K^s], \quad \mathcal{Q}^t = [\mu_1^t, \ldots, \mu_K^t], \tag{2}$$

where μ_k stores the prototype of class k for each domain. After initialization, the memory banks are updated with a momentum m in every batch during training:

$$\mu_k^s \leftarrow m\mu_k^s + (1-m)\frac{1}{|\mathcal{D}_s^k|}\sum_{\mathbf{x}_i^s \in \mathcal{D}_s^k} \mathbf{v}_i^s, \quad \mu_k^t \leftarrow m\mu_k^t + (1-m)\frac{1}{|\mathcal{D}_t^k|}\sum_{\mathbf{x}_i^t \in \mathcal{D}_t^k} \mathbf{v}_i^t \tag{3}$$

where $\mathbf{v}_i = h(g(\mathbf{x}_i))$ is the L_2-normalized feature embedding of $\mathbf{x}_i$ extracted by the feature extractor g and project head h, and $\mathcal{D}_s^k/\mathcal{D}_t^k$ denote the set of source/target samples whose labels/pseudo labels (described in the later subsection) are k in the current mini-batch.

Given a query sample $\mathbf{x}_i$, intra-domain contrastive learning computes the similarity score distribution $\mathbf{P}_i$ over K classes based on the distances to the prototypes, where the k-th element denotes the probability of the sample $\mathbf{x}_i$ belonging to the class k. For the source domain, we have

$$\mathbf{P}_{i,k}^s = \frac{\exp(\mu_k^s \cdot \mathbf{v}_i^s/T)}{\sum_{j=1}^K \exp(\mu_j^s \cdot \mathbf{v}_i^s/T)}, \tag{4}$$

where T is a temperature parameter. Similar operations are performed on target samples and we will get $\mathbf{P}_i^t$. Then we can write intra-domain contrastive loss as:

$$\mathcal{L}_{intra} = \sum_{i=1}^{N_s} \mathcal{L}_{CE}\left(\mathbf{P}_i^s, y_i^s\right) + \sum_{i=1}^{N_t} \mathcal{L}_{CE}\left(\mathbf{P}_i^t, \hat{y}_i^t\right), \tag{5}$$

where $\hat{y}_i^t$ is the pseudo label for target sample $\mathbf{x}_i^t$. As we can see, the intra-domain contrastive loss can push the query feature $\mathbf{v}_i$ close to the prototype indicated by the ground truth label (or pseudo label for target), and keep it away from other prototypes. Thus, we can learn discriminative feature representations for classification in each domain.

3.4 Inter-Domain Contrastive Learning

The domain discrepancy across domains posits a unique obstacle for learning effective representations that perform well in the target domain. Recently, some methods have attempted to attend the standard contrastive learning to the cross-domain scenario for bridging domain discrepancy. The main challenge of applying contrasting learning to UDA lies in how to construct positive and negative pairs in a cross-domain scenario. To retain semantic information, some methods select the sample from the same class of the other domain as positive pairs and the samples from other classes as negative pairs according to the true labels or pseudo labels. Besides, some methods adopt cross-domain instance-to-prototype based contrastive learning where the prototype (instead of the samples) of the same class in the other domain is selected as the positive pair and the prototypes of other classes are as the negative pairs. However, both strategies have drawbacks. For the former, the instance-to-instance matching is very sensitive to abnormal samples, especially under domain shift. For the latter, under a large domain shift, direct cross-domain contrastive learning would not perform well as the large domain shift would lead to incorrect initial pseudo labels.

To mitigate these problems, we propose to perform an intermediate domain prototypical contrastive learning. We firstly construct multiple intermediate prototypes by fixed ratio mixup between the source class prototypes and the target prototypes. Then, for the source domain, the intermediate prototypes near the target domain are used to perform cross-domain contrastive learning. The positive pair is the intermediate prototype of the same class close to the target domain and the negative pairs are the intermediate prototypes of the other classes close to the target domain. And the similar strategy is used for the target samples. In this manner, we could not only reduce the domain discrepancy but also prevent the semantic structure of the data. Compared with direct cross-domain matching, the label information of the intermediate prototype is more accurate as the source domain contains true labels and could be less affected by initial pseudo labels.

Specially, we construct a pair of intermediate prototypes, $\{\mu_k^{st}\}_{k=1}^K$ and $\{\mu_k^{ts}\}_{k=1}^K$ using source prototypes and target prototypes with a fixed ratio mixup:

$$\mu_k^{st} = \lambda_{st}\mu_k^s + \left(1 - \lambda_{st}\right)\mu_k^t, \quad \mu_k^{ts} = \lambda_{ts}\mu_k^s + \left(1 - \lambda_{ts}\right)\mu_k^t \tag{6}$$

where $\lambda_{st} \in (0.5, 1)$ and $\lambda_{ts} \in (0, 0.5)$ are two fixed mixup ratios. We always set $\lambda_{st} + \lambda_{ts} = 1$ to get a pair of domain-symmetric intermediate prototypes. As we can see, the prototypes μ^{st} is close to the source domain and the prototypes μ^{ts} is close to the target domain.

Taking advantage of the intermediate prototypes, we could alleviate the domain shift with instance-to-prototype contrastive learning. The prototypes μ^{st} close to the source domain have more reliable label information because the source prototypes μ^{s} computed with ground truth source label are account for a large proportion. By contrast, the prototypes μ^{ts} close to the target domain have strong target domain relevance but weak label confidence. Thus, we proposed the inter-domain contrastive loss for bidirectional transfer.

Given a query feature $\mathbf{v}_i^s$ in the source domain, and the intermediate prototypes $\{\mu_k^{ts}\}_{k=1}^K$ close to the target domain, inter-domain contrastive loss first computes the similarity distribution $\mathbf{P}_i^{ts}$, which is,

$$\mathbf{P}_{i,k}^{ts} = \frac{\exp(\mu_k^{ts} \cdot \mathbf{v}_i^s / T)}{\sum_{j=1}^K \exp(\mu_j^{ts} \cdot \mathbf{v}_i^s / T)} \tag{7}$$

As there are true labels in the source domain, we perform cross-entropy loss on the pair of source instances and target intermediate prototypes to fully use ground truth label information,

$$\mathcal{L}_{ts} = \sum_{i=1}^{N_s} \mathcal{L}_{CE}\left(\mathbf{P}_i^{ts}, y_i^s\right) \tag{8}$$

Similarly, we compute $\mathbf{P}_i^{st}$ using the target feature $\mathbf{v}_i^t$ and intermediate prototypes $\{\mu_k^{st}\}_{k=1}^K$ close to the source domain. Then, we perform entropy minimization on the similarity distribution $\mathbf{P}_i^{st}$, which could find the match between the target feature and source intermediate prototypes but rely less on the label information:

$$\mathcal{L}_{st} = -\sum_{i=1}^{N_t} \sum_{k=1}^{K} \mathbf{P}_{i,k}^{st} \log \mathbf{P}_{i,k}^{st} \tag{9}$$

The final inter-domain contrastive loss is:

$$\mathcal{L}_{inter} = \mathcal{L}_{st} + \mathcal{L}_{ts} \tag{10}$$

3.5 Other Losses

Domain Adversrial Loss. The inter-domain contrastive learning could reduce domain discrepancy and achieve domain alignment. But it only focuses on class-level alignment and does not explicitly reduce the domain-level distribution shift across domains. To deal with this problem, InCo follows the adversarial manner [10], and introduces a domain discriminator D to distinguish the source feature and target feature. While the feature extractor g is trained to confuse the domain discriminator. By this adversarial loss, the feature extractor could learn domain-invariant features. The adversarial object between feature extractor g and domain discriminator D can be written as:

$$\mathcal{L}_{adv} = \mathbb{E}_{\mathbf{x}_i^s \sim \mathcal{D}_s}[\log D(g(\mathbf{x}_i^s))] + \mathbb{E}_{\mathbf{x}_i^t \sim \mathcal{D}_t}[\log(1 - D(g(\mathbf{x}_i^t)))] \tag{11}$$

Classification Loss and Entropy Minimization Loss. To capture the source supervised information, the model is trained to minimize the empirical risk on labeled samples as conventional supervised methods. The feature extractor g maps a source sample $\mathbf{x}_i^s$ into the feature. Then, the classifier f would classify the feature into K categories, i.e., $p(y|\mathbf{x}_i^s) = f(g(\mathbf{x}_i^s))$. Then, the cross-entropy loss $\mathcal{L}_{CE}(\cdot, \cdot)$ is adopted to minimize the empirical risk:

$$\mathcal{L}_{cls} = \mathbb{E}_{(\mathbf{x}_i^s, y_i^s) \sim \mathcal{D}_s} \mathcal{L}_{CE}(y_i^s, p(y|\mathbf{x}_i^s)) \tag{12}$$

As there are no labeled samples in the target domain, we adopt the entropy minimization loss to pass through the low-density regions of the target feature space, which is,

$$\mathcal{L}_{ent} = \mathbb{E}_{\mathbf{x}_i^t \sim \mathcal{D}_t} - \sum_{k=1}^{K} p_k(y|\mathbf{x}_i^t) \log p_k(y|\mathbf{x}_i^t) \tag{13}$$

where $p_k(y|\mathbf{x}_i^t)$ is the k-th dimension of $p(y|\mathbf{x}_i^t)$ and $p(y|\mathbf{x}_i^t) = f(g(\mathbf{x}_i^t))$ is the prediction of sample $\mathbf{x}_i^t$ by the model. The combined loss is,

$$\mathcal{L}_{cls-ent} = \mathcal{L}_{cls} + \mathcal{L}_{ent} \tag{14}$$

3.6 Overall

Generation of Pseudo Labels. Since the ground truth labels are not available in the target domain during training, we perform k-means clustering to generate pseudo labels for the target samples. Due to the randomness in clustering, we use class prototypes from the source domain as the initial clustering centers and set the number of clusters as K. In this case, the clustering algorithm can be seen as the distance matching between target features and source prototypes which could better maintain the target data structure and could easily use the cluster label as the target pseudo label $\hat{y}_i^t$.

Training. The InCo learning framework performs intra-domain contrastive loss, inter-domain contrastive loss, domain adversarial loss, and classification loss. Together with the momentum update in the memory bank, the overall learning objective is:

$$\min_{g,h,f} \mathcal{L}_{cls-ent} + \lambda_{intra} \cdot \mathcal{L}_{intra} + \lambda_{inter} \cdot \mathcal{L}_{inter} + \lambda_{adv} \cdot \mathcal{L}_{adv} \tag{15}$$

$$\max_{D} \mathcal{L}_{adv} \tag{16}$$

where λ_{intra}, λ_{inter} and λ_{adv} are hyper-parameters. Following previous method [10], the min-max training procedure in Eq. 15 and 16 is accomplished by applying a Gradient Reversal Layer (GRL). GRL behaves as the identity function during the forward propagation and inverts the gradient sign during the backward propagation, hence driving the parameters to maximize the output loss.

Table 1. Accuracy (%) on the **Office-31** dataset (ResNet-50).

Method	A→W	D→W	W→D	A→D	D→A	W→A	Avg
ResNet-50	68.4	96.7	99.3	68.9	62.5	60.7	76.1
DANN	82.0	96.9	99.1	79.7	68.2	67.4	82.2
MSTN	91.3	98.9	**100.0**	90.4	72.7	65.6	86.5
CDAN+E	94.1	98.6	**100.0**	92.9	71.0	69.3	87.7
DMRL	90.8	99.0	**100.0**	93.4	73.0	71.2	87.9
SymNets	90.8	98.8	**100.0**	93.9	74.6	72.5	88.4
PCS	92.6	96.6	99.4	95.8	76.6	75.8	89.5
GSDA	**95.7**	99.1	**100.0**	94.8	73.5	74.9	89.7
PCT	94.6	98.7	99.9	93.8	77.2	76.0	90.0
InCo	94.0	**99.1**	**100.0**	**95.8**	**77.3**	**77.0**	**90.5**

4 Experiments

4.1 Datasets

We evaluate InCo on three common benchmarks based on previous works [22,23]. **Office-31**[1] is a classical real-world dataset for UDA. It has 4110 images with 31 classes shared with three domains: Amazon (**A**), Webcam (**W**), and DSLR (**D**). In this dataset, six adaptation tasks are constructed. **ImageCLEF**[2] is composed of three domain with 12 classes: Caltech-256 (**C**), ImageNet ILSVRC 2012 (**I**), and Pascal VOC 2012 (**P**). **Office-Home**[3] is a more difficult dataset, which consists of four domains: Artistic (**Ar**), Clipart (**Cl**), Product (**Pr**), and Real-World (**Rw**), containing 15500 images with 65 classes.

4.2 Setup

We use PyTorch to implement the proposed method. We use ResNet-50 [13] pre-trained on ImageNet [30] as the backbones for all datasets. To enable a fair comparison with the existing method [40], we remove the last FC layer in ResNet and implement a projection head h with the default nonlinear projection and an additional hidden layer activated by ReLU as same as SimCLR [1]. The output dimension of h is 512 and L2-normalizing is performed on the output features. Following DANN [10], we use the same architecture for the domain discriminator D and the classifier f. We use SGD with a momentum of 0.9 and weight decay $5e^{-4}$ to train the InCo for all the experiments. The initial learning rate is 0.001 for the pre-trained feature extractor and 0.01 for other modules. Besides, we split large batch size into small parts, and use gradient accumulation in Pytorch which

[1] https://www.hemanthdv.org/officeHomeDataset.html.
[2] https://www.imageclef.org/2014/adaptation.
[3] https://www.hemanthdv.org/officeHomeDataset.html.

Table 2. Accuracies (%) on the **ImageCLEF** dataset (ResNet-50).

Method	I → P	P → I	I → C	C → I	C → P	P → C	Avg
ResNet-50	74.8	83.9	91.5	78.0	65.5	91.2	80.7
DAN	74.5	82.2	92.8	86.3	69.2	89.8	82.5
DANN	75.0	86.0	96.2	87.0	74.3	91.5	85.0
MADA	75.0	87.9	96.0	88.8	75.2	92.2	85.8
iCAN	79.5	89.7	94.7	89.9	78.5	92.0	87.4
CDAN	77.7	90.7	97.7	91.3	74.2	94.3	87.7
A^{2}LP	79.6	92.7	96.7	92.5	78.9	96.0	89.4
CGDM	78.7	93.3	97.5	92.7	79.2	95.7	89.5
ETD	**81.0**	91.7	**97.9**	93.3	79.5	95.0	89.7
SymNets	80.2	93.6	97.0	93.4	78.7	**96.4**	89.9
InCo	79.5	**94.5**	96.5	**94.8**	**80.3**	96.2	**90.3**

could backward gradient after multiple forward iterations to achieve the same effect as large batch size but obtain smoother prototypes with multiple update operations. Specially, we use a batch size of 16 for Office-31 and ImageCLEF and backward loss after four forward iterations. For Office-Home, we use a batch size of 32 and backward after two forward iterations. The temperature parameter T is fixed to 0.3, 0.5, and 0.1 for Office-31, ImageCLEF, and Office-Home. The momentum m is 0.9 for all datasets. The hyper-parameters λ_{intra}, λ_{inter}, and λ_{adv} are all set to 1.0 which is selected from $\{0.5, 1.0, 2.0\}$. We set mixup ration $\lambda_{st} = 0.8$ and $\lambda_{ts} = 0.2$ for all datasets with λ_{st} selected from $\{0.7, 0.8, 0.9\}$.

4.3 Baselines

We compare with InCo with four kinds of baselines:

- **ResNet-50**. This baseline refers to the source-only method, where only the source samples are used for training.
- **Moment matching and adversarial-based methods**, including DAN [22], DANN [10], MADA [29], MCD [31], CDAN [23], MSTN [38], iCAN [42], MDD [45], SymNets [44], GSDA [14], DMRL [37], ETD [20], A^{2}LP [43], MDD+IA [15], BNM [5], BDG [39], GVB [6], SRDC [34], and CGDM [9].
- **Prototype-based methods**, including PCT [35].
- **Contrastive learning based methods**, including PCS [40].

4.4 Results

Table 1 displays the performances of various models on Office-31. Generally, InCo outperforms the baseline method in most transfer tasks (5/6). It is noticed that InCo is especially effective on harder transfer tasks, e.g. W→A and A→D,

Table 3. Classification accuracies (%) on the **Office-Home** dataset (ResNet-50).

Method	Ar→Cl	Ar→Pr	Ar→Rw	Cl→Ar	Cl→Pr	Cl→Rw	Pr→Ar	Pr→Cl	Pr→Rw	Rw→Ar	Rw→Cl	Rw→Pr	Avg
ResNet-50	34.9	50.0	58.0	37.4	41.9	46.2	38.5	31.2	60.4	53.9	41.2	59.9	46.1
MCD	48.9	68.3	74.6	61.3	67.6	68.8	57.0	47.1	75.1	69.1	52.2	79.6	64.1
CDAN	50.7	70.6	76.0	57.6	70.0	70.0	57.4	50.9	77.3	70.9	56.7	81.6	65.8
BNM	52.3	73.9	80.0	63.3	72.9	74.9	61.7	49.5	79.7	70.5	53.6	82.2	67.9
MDD	54.9	73.7	77.8	60.0	71.4	71.8	61.2	53.6	78.1	72.5	60.2	82.3	68.1
BDG	51.5	73.4	78.7	65.3	71.5	73.7	65.1	49.7	81.1	74.6	55.1	84.8	68.7
MDD+IA	56.2	77.9	79.2	64.4	73.1	74.4	64.2	54.2	79.9	71.2	58.1	83.1	69.5
GVB	57.0	74.7	79.8	64.6	74.1	74.6	65.2	55.1	81.0	74.6	59.7	84.3	70.4
SRDC	52.3	76.3	81.0	**69.5**	76.2	78.0	**68.7**	53.8	81.7	**76.3**	57.1	85.0	71.3
PCT	57.1	78.3	81.4	67.6	77.0	76.5	68.0	55.0	81.3	74.7	60.0	85.3	71.8
InCo	**59.2**	**78.6**	**82.5**	67.1	**79.8**	**79.8**	67.3	55.4	**82.7**	74.6	59.3	84.8	**72.6**

where the two domains are substantially different. Moreover, compared with PCS, which adopts direct cross-domain instance-to-prototype contrastive learning, InCo gets an improvement of 1%. This verifies that the intermediate prototype based contrastive learning method is a legitimate solution in the context of domain adaptation under large domain shifts.

Table 2 illustrates the performance comparisons on the six adaption directions of ImageCLEF. InCo again demonstrates strong superiority over its competitors. Particularly, InCo offers a significant performance boost on tasks C→P, P→I, and C→I. Compared with other moment matching and adversarial domain adaptation methods, InCo achieves better performance and the results show that contrastive learning based methods could learn invariant features and achieve domain alignment. Besides, by intra-domain contrastive learning, the model could learn more discriminative features, leading to better performance.

Table 3 reports the classification accuracy of twelve transfer tasks on the Office-Home dataset. We can see that InCo gets the best accuracy in the six categories and obtains comparable results in others. Compared with PCT which is a prototype based method, InCo obtains better results combined with an intermediate prototype based contrastive learning and the improvement is 0.8%. Moreover, Office-Home is a more challenging dataset than the other two datasets, and we get better improvement in this dataset, which shows that InCo could deal with large domain shifts.

4.5 Insight Analysis

Analysis of Intermediate Prototypes. To better understand the role of intermediate prototypes, we conduct lots of ablation studies to analyze them. We compare InCo with the following variants: 1) **No intermediate prototypes**, where we do not construct any intermediate prototypes and perform direct cross-domain instance-to-prototype contrastive learning. 2) **Only one intermediate prototype for each class**, where only one intermediate prototype is constructed for each class and it is used to perform cross-domain instance-to-prototype contrastive learning for samples from both domains. 3) **Two intermediate prototypes for each class**, where two intermediate prototypes are

Table 4. Analysis of intermediate prototypes on **Office-31** dataset.

Settings	Average
No intermediate prototypes	89.41
One intermediate prototype (0.2)	89.31
One intermediate prototype (0.5)	89.59
One intermediate prototype (0.8)	89.76
Two intermediate prototypes (0.1+0.9)	89.94
Two intermediate prototypes (0.3+0.7)	89.75
Two intermediate prototypes (0.2+0.8, ours)	**90.53**

constructed for each class but with a different mixup ratio ($\lambda_{st} = 0.9$,$\lambda_{st} = 0.8$, and $\lambda_{st} = 0.7$). The results are shown in Table 4. As we can see, in most cases, intermediate prototype based contrastive learning methods outperform direct cross-domain contrastive learning methods as the intermediate prototype are more accurate. Besides, two intermediate prototypes could achieve better performance than that of one intermediate prototype for each class ($\lambda_{st} = 0.8$), as two prototypes contain complementary information and could better bridge two domains. Moreover, we obverse that InCo works well with different mixup ratio, and we experimentally find that $\lambda_{st} = 0.8$ and $\lambda_{ts} = 0.2$ is the best value.

Table 5. Ablation study of losses on **Office-31** dataset.

$\mathcal{L}_{cls-ent}$	$\mathcal{L}_{adv}$	$\mathcal{L}_{intra}$	$\mathcal{L}_{inter}$	Average
√				78.53
√	√			87.32
√	√	√		89.27
√	√	√	√	**90.53**

Ablation Study of Losses. In this subsection, we investigate the influence of each component on the overall objective defined in Eq. 15. The results are shown in Table 5. Only adopting the classification loss $\mathcal{L}_{cls}$ and the entropy loss $\mathcal{L}_{ent}$ gets the worst accuracy. After adding domain adversarial loss $\mathcal{L}_{adv}$ to achieve domain-level alignment, the performance is improved to 87.32%. Then, the intra-domain contrastive learning loss $\mathcal{L}_{intra}$ is added to learn discriminative features, the accuracy is improved by 1.9%. Lastly, combining the cross-domain contrastive learning loss $\mathcal{L}_{inter}$, InCo achieves the best performance.

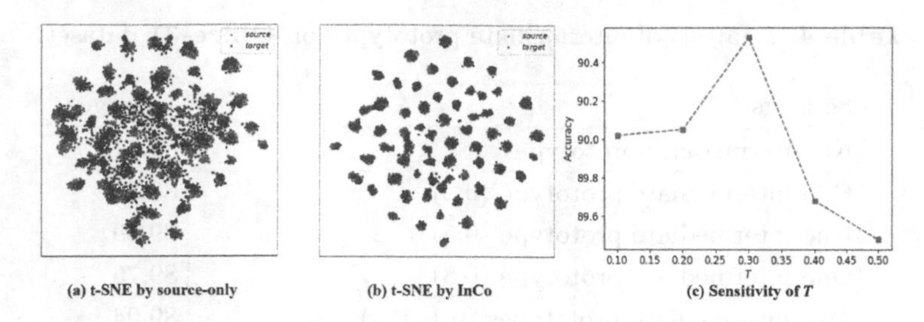

Fig. 3. Visualization of representations learned by source-only model and InCo as well as the parameter sensitivity of T.

Feature Visualization. Fig. 3(a) and 3(b) show the t-SNE [25] visualization of the features from both domains for task Cl→Rw (65 classes) before (source-only) and after alignment, respectively. Before alignment, there exists a large distribution shift between the source domain and the target domain. While after alignment the domain shift is reduced and the features of target samples have become discriminative. Thus, the samples can be easily classified by the classifier.

Parameter Sensitivity of T. We perform parameter sensitivity of the temperature parameter T on Office-31. When $T \leq 1$, the model would sharpen the similarity score in contrastive learning to avoid ambiguous predictions, thus, we set $T \leq 1$, and the results under different values are shown in Fig. 3(c). As we can see, the performance raises firstly and then drops, as a smaller value would be overconfident in the predictions and a larger value would be less confident. We experimentally find that $T = 0.3$ is the best value.

5 Conclusion

In this paper, we propose a novel UDA method InCo, which performs instance-to-prototype contrastive learning based on intermediate prototypes to deal with large domain shifts. The intermediate prototypes are constructed with a fixed ratio mixup between the source prototypes and target prototypes. Compared with direct cross-domain instance-to-prototype contrastive learning, the intermediate prototypes are more accurate and could mitigate the problem of incorrect initial pseudo labels. Together with intra-domain contrastive learning and domain adversarial training, the model could learn both invariant and discriminative semantic features. The results of three real-world datasets show the effectiveness of the proposed method. In the future, we would like to explore more difficult scenarios such as source-free domain adaptation [8].

Acknowledgement. This paper is supported by the National Key Research and Development Program of China (Grant No. 2018YFB1403400), the National Natural Science Foundation of China (Grant No. 61876080), the Key Research and Development Program of Jiangsu (Grant No. BE2019105), the Collaborative Innovation Center of Novel Software Technology and Industrialization at Nanjing University.

References

1. Chen, T., Kornblith, S., Norouzi, M., Hinton, G.E.: A simple framework for contrastive learning of visual representations. In: ICML (2020)
2. Chen, Y., Pan, Y., Wang, Y., Yao, T., Tian, X., Mei, T.: Transferrable contrastive learning for visual domain adaptation. In: Proceedings of the 29th ACM International Conference on Multimedia (2021)
3. Cicek, S., Soatto, S.: Unsupervised domain adaptation via regularized conditional alignment. In: ICCV, pp. 1416–1425 (2019)
4. Cui, F., Chen, Y., Du, Y., Cao, Y., Wang, C.: Joint feature and labeling function adaptation for unsupervised domain adaptation. In: PAKDD (2022)
5. Cui, S., Wang, S., Zhuo, J., Li, L., Huang, Q., Tian, Q.: Towards discriminability and diversity: Batch nuclear-norm maximization under label insufficient situations. In: CVPR, pp. 3940–3949 (2020)
6. Cui, S., Wang, S., Zhuo, J., Su, C., Huang, Q., Tian, Q.: Gradually vanishing bridge for adversarial domain adaptation. In: CVPR, pp. 12452–12461 (2020)
7. Dai, Y., Liu, J., Sun, Y., Tong, Z., Zhang, C., yu Duan, L.: Idm: An intermediate domain module for domain adaptive person re-id. In: ICCV, pp. 11844–11854 (2021)
8. Du, Y., Yang, H., Chen, M., Jiang, J., Luo, H., Wang, C.: Generation, augmentation, and alignment: A pseudo-source domain based method for source-free domain adaptation. ArXiv abs/ arXiv: 2109.04015 (2021)
9. Du, Z., Li, J., Su, H., Zhu, L., Lu, K.: Cross-domain gradient discrepancy minimization for unsupervised domain adaptation. In: CVPR, pp. 3936–3945 (2021)
10. Ganin, Y., Ustinova, E., et al.: Domain-adversarial training of neural networks. In: JMLR (2016)
11. He, K., Fan, H., Wu, Y., Xie, S., Girshick, R.B.: Momentum contrast for unsupervised visual representation learning. In: CVPR, pp. 9726–9735 (2020)
12. He, K., Gkioxari, G., Dollár, P., Girshick, R.B.: Mask r-cnn. IEEE Trans. Pattern Anal. Mach. Intell. **42**, 386–397 (2020)
13. He, K., Zhang, X., et al.: Deep residual learning for image recognition. In: CVPR (2016)
14. Hu, L., Kan, M., Shan, S., Chen, X.: Unsupervised domain adaptation with hierarchical gradient synchronization. In: CVPR, pp. 4042–4051 (2020)
15. Jiang, X., Lao, Q., Matwin, S., Havaei, M.: Implicit class-conditioned domain alignment for unsupervised domain adaptation. In: ICML (2020)
16. Khosla, P., et al.: Supervised contrastive learning. In: NeurIPS (2020)
17. Kim, D., Saito, K., Oh, T.H., Plummer, B.A., Sclaroff, S., Saenko, K.: Cds: Cross-domain self-supervised pre-training. In: ICCV, pp. 9103–9112 (2021)
18. Krizhevsky, A., Sutskever, I., Hinton, G.E.: Imagenet classification with deep convolutional neural networks. Commun. ACM **60**, 84–90 (2012)
19. Li, J., Zhou, P., Xiong, C., Socher, R., Hoi, S.C.H.: Prototypical contrastive learning of unsupervised representations. In: ICLR (2021)
20. Li, M., Zhai, Y., Luo, Y.W., Ge, P., Ren, C.X.: Enhanced transport distance for unsupervised domain adaptation. In: CVPR, pp. 13933–13941 (2020)
21. Long, J., Shelhamer, E., Darrell, T.: Fully convolutional networks for semantic segmentation. In: CVPR (2015)
22. Long, M., Cao, Y., Wang, J., Jordan, M.I.: Learning transferable features with deep adaptation networks. In: ICML (2015)

23. Long, M., Cao, Z., et al.: Conditional adversarial domain adaptation. In: NeruIPS (2018)
24. Long, M., Zhu, H., Wang, J., Jordan, M.I.: Deep transfer learning with joint adaptation networks. In: ICML (2017)
25. van der Maaten, L., Hinton, G.E.: Visualizing data using t-sne. JMLR **9**, 2579–2605 (2008)
26. Na, J., Jung, H., Chang, H., Hwang, W.: Fixbi: Bridging domain spaces for unsupervised domain adaptation. In: CVPR, pp. 1094–1103 (2021)
27. van den Oord, A., Li, Y., Vinyals, O.: Representation learning with contrastive predictive coding. ArXiv abs/ arXiv: 1807.03748 (2018)
28. Pan, S.J., Yang, Q.: A survey on transfer learning. TKDE **22**, 1345–1359 (2010)
29. Pei, Z., Cao, Z., Long, M., Wang, J.: Multi-adversarial domain adaptation. In: AAAI (2018)
30. Russakovsky, O., et al.: Imagenet large scale visual recognition challenge. IJCV **115**, 211–252 (2015)
31. Saito, K., Watanabe, K., Ushiku, Y., Harada, T.: Maximum classifier discrepancy for unsupervised domain adaptation. In: CVPR, pp. 3723–3732 (2018)
32. Shen, J., Qu, Y., Zhang, W., Yu, Y.: Wasserstein distance guided representation learning for domain adaptation. In: AAAI (2018)
33. Sun, B., Feng, J., Saenko, K.: Return of frustratingly easy domain adaptation. In: AAAI (2016)
34. Tang, H., Chen, K., Jia, K.: Unsupervised domain adaptation via structurally regularized deep clustering. In: CVPR, pp. 8722–8732 (2020)
35. Tanwisuth, K., et al.: A prototype-oriented framework for unsupervised domain adaptation. In: NeurIPS (2021)
36. Wang, R., Wu, Z., Weng, Z., Chen, J., Qi, G.J., Jiang, Y.G.: Cross-domain contrastive learning for unsupervised domain adaptation. ArXiv abs/ arXiv: 2106.05528 (2022)
37. Wu, Y., Inkpen, D., El-Roby, A.: Dual mixup regularized learning for adversarial domain adaptation. In: Vedaldi, A., Bischof, H., Brox, T., Frahm, J.-M. (eds.) ECCV 2020. LNCS, vol. 12374, pp. 540–555. Springer, Cham (2020). https://doi.org/10.1007/978-3-030-58526-6_32
38. Xie, S., Zheng, Z., Chen, L., Chen, C.: Learning semantic representations for unsupervised domain adaptation. In: ICML (2018)
39. Yang, G., Xia, H., Ding, M., Ding, Z.: Bi-directional generation for unsupervised domain adaptation. In: AAAI (2020)
40. Yue, X., et al.: Prototypical cross-domain self-supervised learning for few-shot unsupervised domain adaptation. In: CVPR, pp. 13829–13839 (2021)
41. Zhang, H., Cissé, M., Dauphin, Y., Lopez-Paz, D.: mixup: Beyond empirical risk minimization. In: ICLR (2018)
42. Zhang, W., Ouyang, W., Li, W., Xu, D.: Collaborative and adversarial network for unsupervised domain adaptation. In: CVPR, pp. 3801–3809 (2018)
43. Zhang, Y., Deng, B., Jia, K., Zhang, L.: Label propagation with augmented anchors: a simple semi-supervised learning baseline for unsupervised domain adaptation. In: Vedaldi, A., Bischof, H., Brox, T., Frahm, J.-M. (eds.) ECCV 2020. LNCS, vol. 12349, pp. 781–797. Springer, Cham (2020). https://doi.org/10.1007/978-3-030-58548-8_45
44. Zhang, Y., Tang, H., Jia, K., Tan, M.: Domain-symmetric networks for adversarial domain adaptation. In: CVPR, pp. 5026–5035 (2019)
45. Zhang, Y., Liu, T., Long, M., Jordan, M.I.: Bridging theory and algorithm for domain adaptation. In: ICML (2019)

Fast and Accurate Importance Weighting for Correcting Sample Bias

Antoine de Mathelin[1,2](✉), Francois Deheeger[1], Mathilde Mougeot[2], and Nicolas Vayatis[2]

[1] Manufacture Française des Pneumatiques Michelin, Clermont-Ferrand, France
{antoine.de-mathelin-de-papigny,francois.deheeger}@michelin.com
[2] Université Paris-Saclay, CNRS, ENS Paris-Saclay, Centre Borelli, Gif-sur-Yvette, France
{mathilde.mougeot,nicolas.vayatis}@ens-paris-saclay.fr

Abstract. Bias in datasets can be very detrimental for appropriate statistical estimation. In response to this problem, importance weighting methods have been developed to match any biased distribution to its corresponding target unbiased distribution. The seminal Kernel Mean Matching (KMM) method is, nowadays, still considered as state of the art in this research field. However, one of the main drawbacks of this method is the computational burden for large datasets. Building on previous works by Huang et al. (2007) and de Mathelin et al. (2021), we derive a novel importance weighting algorithm which scales to large datasets by using a neural network to predict the instance weights. We show, on multiple public datasets, under various sample biases, that our proposed approach drastically reduces the computational time on large dataset while maintaining similar sample bias correction performance compared to other importance weighting methods. The proposed approach appears to be the only one able to give relevant reweighting in a reasonable time for large dataset with up to two million data.

1 Introduction

The most common assumption in a traditional learning scenario is that training data are independently and identically distributed (iid) and drawn from the same distribution as the target data. However, in real cases, the training dataset often appears to be biased with respect to the target dataset. This happens in particular in medical applications, when, for example, the age distribution of the patients does not match the distribution of the overall population. In product design, predictive models of product performances may be biased by the large amount of data corresponding to outdated products. For both previous cases, it often happens that the learner has access to the unbiased distribution, either because it is known from an external source (the age distribution in the whole population is known) or because he has access to an unbiased dataset (a sample of data of the recent products).

In this paper we assume that the learner owns a sample drawn from a source biased distribution $p_s(x, y)$ as well as a sample coming from the target marginal

M.-R. Amini et al. (Eds.): ECML PKDD 2022, LNAI 13713, pp. 659–674, 2023.
https://doi.org/10.1007/978-3-031-26387-3_40

distribution $p_t(x)$ such that $p_s(x) \neq p_t(x)$. Our goal is to estimate $p_t(y|x)$ or $p_t(y)$ where y is the variable of interest (e.g. patient survival expectancy for a clinical model or product performances for product design). Given the bias on the marginals, the estimation of $p_t(y|x)$ on the target domain will be biased as well.

To correct this type of sample bias, importance weighting methods can be used. These methods seek to reweight the source data to debias the marginals by looking for the weights corresponding to $w(x) = p_t(x)/p_s(x)$. A successful non-parametric method in this field is the Kernel Mean matching (KMM) method [15] which reweighs the sources in order to minimize the MMD distance between the reweighed sources and the targets [13]. Although KMM is one of the first non-parametric method developed to handle sample bias, it is still used nowadays in modern sample bias correction methods for deep learning [10] or for deriving two-sample hypothesis testing under sample bias [1]. KMM solves a quadratic problem for the minimization of the MMD with as many parameters as the number of source instances. Thus, when the number of source data is large, one faces a computational burden because of the large kernel matrix to compute. Some methods have proposed to reduce the problem in batch and to perform a KMM on each of them [6,27]. This lightens the memory, but the computational time remains important as the number of KMM sub-problems to compute increases with the number of data. Other importance weighting methods reduce the number of parameters to be optimized by linking the weights of the source instances by a parametric function as done for KLIEP [33] and ULSIF [16]. These two methods propose to write each of the weights as a linear combination of kernels centered on target points. Thus, the number of parameters is fixed (in general by selecting a hundred centers in the target domain), however, the computational cost of the pairwise distance calculations between the centers and all the source data still remains. Moreover, this large matrix of pairwise distances is used in the resolution of the gradient descent algorithm which slows the optimization. A last method, NearestNeighborsWeighting (NNW) consists in computing the weights of the source instances according to their number of target nearest neighbors [20]. This heuristic, not relying on the minimization of a distance between distributions, is quite efficient, and its computation time is generally less than its KMM counterpart. However, the search for the nearest neighbors requires the computation of pairwise distances between source and target data and despite the optimization algorithms of type KDTree [11] and BallTree [29], the method encounters computational burden for datasets with many instances and features.

Finally, all these algorithms rely on hyper-parameters to be tuned. The choice of the kernel and its bandwidth for the KLIEP and KMM methods are very important, as well as the number of nearest neighbors to consider for NNW. To choose these parameters, a cross-validation procedure using an unsupervised metric (which does not require the y data on the target domain) is mainly used such as the J-score for KLIEP [33], the normalized mean squared error (NMSE) between the actual and estimated density ratios for KMM [26], an information criterion for ULSIF [16], or any divergence metric between distributions such as the linear discrepancy or the domain classifier divergence [2,21,23]. This hyper-parameter selection procedure, necessary to use these methods in practice, adds to the computational time.

Considering the drawbacks of the previous mentioned methods, we propose, in this paper, a new importance weighting algorithm that scales to large datasets. Our goal is to obtain the same level of performance than KMM but with less computational time. To do so, we propose to minimize the objective of KMM, i.e. the MMD, by a batch gradient descent to avoid the memory burden of the huge kernel matrix. However, it should be underlined that optimizing the weights of the source instances individually brings no complexity gain since each source weight is only updated in its corresponding batch. Assuming $w(x) = p_t(x)/p_s(x)$ continuous and regular, for two close source points $x_1 \simeq x_2$, the weights will be similar $w(x_1) \simeq w(x_2)$. Consequently, we propose to optimize at each batch the parameters θ of a parametric and continuous function $W_\theta(x)$ in order to minimize the empirical MMD on the batch. Inspired from recent works of weighting adversarial neural network (WANN) [25], this function W_θ is chosen as a neural network. The advantage of the networks is the fast update of the θ parameters by backpropagation of the gradient through the layers which is highly parallelizable [18]. This avoids working with huge matrices of pairwise kernel as done in the KLIEP algorithm. We show on several datasets that this approach allows to obtain importance weighting at least as efficient as KMM in a drastically reduced time. The source code of the experiments is publicly available on GitHub[1].

Our contributions can be listed as follows:

- We derive a fast and scalable importance weighting algorithm. This is achieved by using batch gradient descent optimizing the MMD and by parameterizing the weights by a neural network.
- The developed algorithm optimizes the kernel parameters of the MMD in the gradient-descent optimization and thus avoid a time consuming CV process to select it.

2 Problem Setting and Proposed Approach

2.1 Learning Scenario

Given an input space $\mathcal{X} \in \mathbb{R}^p$ of dimension $p > 0$ and an output space $\mathcal{Y} \in \mathbb{R}^q$ with $q > 0$, we consider the sample bias scenario, in which the learner has access to a source sample $\mathcal{S} = \{(x_1, y_1), ..., (x_m, y_m)\} \subset \mathcal{X} \times \mathcal{Y}$ drawn iid from a source distribution $p_s(x, y)$ on $\mathcal{X} \times \mathcal{Y}$ and a target sample $\mathcal{T} = \{x_1, ..., x_n)\} \subset \mathcal{X}$ drawn iid according to a target distribution $p_t(x)$ on $\mathcal{X}$. We suppose that $p_s(x) \neq p_t(x)$ and that $p_t(x)$ is absolutely continuous with respect to $p_s(x)$. Finally, we make the covariate-shift assumption [3] which states that the conditional probabilities of $y|x$ remain unchanged for the two distributions: $p_s(y|x) = p_t(y|x)$.

[1] https://github.com/antoinedemathelin/Importance-Weighting-Network.

2.2 MMD

Let's consider $\phi_\sigma : \mathcal{X} \rightarrow \mathcal{F}_\sigma$ with $\mathcal{F}_\sigma$ the RKHS of Gaussian kernel k_σ such that $\forall x, x' \in \mathcal{X}$, $k_\sigma(x, x') = \langle \phi_\sigma(x), \phi_\sigma(x') \rangle = \exp(-\sigma ||x - x'||^2)$ with $\sigma >$. 0. The Maximum Mean Discrepancy (MMD) between the source and target distributions is defined as follows:

$$\text{MMD}_\sigma(p_s(x), p_t(x)) = \left|\left| \mathop{\mathbb{E}}_{x \sim p_s(x)} [\phi_\sigma(x)] - \mathop{\mathbb{E}}_{x \sim p_t(x)} [\phi_\sigma(x)] \right|\right| \tag{1}$$

The MMD is a distance characterizing how close are the two marginal distributions $p_s(x), p_t(x)$. As we consider a Gaussian kernel, $\text{MMD}_\sigma = 0$ if and only if $p_s(x) = p_t(x)$ [13].

As our goal is to correct the sample bias between the source and target distributions with importance weighting, we aim at finding the weights $w(x) \in \mathbb{R}_+$ that solve the following optimization problem:

$$\begin{aligned} &\min_{w : \mathcal{X} \rightarrow \mathbb{R}_+} \left|\left| \mathop{\mathbb{E}}_{x \sim p_s(x)} [w(x)\phi_\sigma(x)] - \mathop{\mathbb{E}}_{x \sim p_t(x)} [\phi_\sigma(x)] \right|\right|^2 \\ &\text{subject to } w(x) > 0 \, \forall x \in \mathcal{X} \text{ and } \mathop{\mathbb{E}}_{x \sim p_s(x)} [w(x)] = 1 \end{aligned} \tag{2}$$

As we consider $p_t(x)$ absolutely continuous with respect to $p_s(x)$, the solution of the optimization problem (2) is the density ratio $w(x) = p_t(x)/p_s(x)$ [15].

In practice, we only have access to samples $\{x_i, .., x_m\}$ and $\{x'_i, .., x'_n\}$ respectively drawn according to both distributions $p_s(x)$ and $p_t(x)$, we then consider the empirical formulation of the previous optimization problem (2) which is written:

$$\begin{aligned} &\min_{w \in \mathbb{R}^m} \frac{1}{m^2} \sum_{i,j}^{m} w_i w_j k_\sigma(x_i, x_j) + \frac{1}{n^2} \sum_{i,j}^{n} k_\sigma(x'_i, x'_j) - \frac{2}{nm} \sum_{i}^{m} \sum_{j}^{n} w_i k_\sigma(x_i, x'_j) \\ &\text{subject to } w_i > 0 \, \forall i \in [|1, m|] \text{ and } \frac{1}{m} \sum_{i}^{m} w_i = 1 \end{aligned} \tag{3}$$

2.3 Importance Weighting Network

The optimization problem (3) is a quadratic optimization problem which can be solved by gradient descent. However computing the MMD requires to compute a kernel matrix of size $\mathcal{O}((n+m)^2)$ which can cause memory burden. We propose, in this paper, to compute the MMD on small batches of size B. At each batch, we impose the constraints on the weights by taking their absolute values and dividing them by their sum. It has been shown that, although self-normalizing the weights creates a biased estimation of the MMD, the estimator is asymptotically unbiased, with the bias decreasing at a rate of $\mathcal{O}(1/B)$ [8,22].

To obtain a fast update of all weights at each batch, we parameterized the weights through a neural network $W_\theta : \mathcal{X} \to \mathbb{R}$ such that $w_i = W_\theta(x_i)$ for each $i \in [|1, m|]$. In this way, at each batch, the parameters θ are updated and then all the parameters w_i are updated with them. Notice that the MMD estimation produced by the batch of size B is approaching the true MMD at a strong rate of $\mathcal{O}(1/\sqrt{B})$ [13] which comforts the idea that the update of θ at each batch will be in favor of finding the optimal weights for all x_i.

It should be stressed that the MMD quantity depends on σ which corresponds to the kernel bandwidth. In the seminal paper of KMM [15] the choice of σ is not clearly motivated, but a method proposed by KLIEP [33] consists in choosing between several predefined σ and compute the optimal weights for each σ value. The value which provides the best matching of the target distribution is finally selected. This type of selection is time consuming and requires fixing a pre-selection of σ values.

We propose, instead, to optimize the σ parameter at the same time as the weights. Inspired by the works on MMD-GAN [19], the kernel parameter σ is modified at each batch in order to maximize the MMD. The idea behind this choice of implementation is to increase the discriminative power of the MMD and thus reduce the risk of estimating, from finite samples, that the source and target distributions are the same when this is not the case. By maximizing over σ, we end with an alternate gradient descent-ascent algorithm, where we aim at finding a saddle point. The final optimization formulation can be written as follows:

$$\max_{\sigma} \min_{\theta} \frac{\sum_{i,j}^{B} |W_\theta(x_i) W_\theta(x_j)| k_\sigma(x_i, x_j)}{\sum_{i,j}^{B} |W_\theta(x_i) W_\theta(x_j)|} + \frac{1}{B^2} \sum_{i,j}^{B} k_\sigma(x'_i, x'_j) - \frac{\sum_{i}^{B} \sum_{j}^{n} |W_\theta(x_i)| k_\sigma(x_i, x'_j)}{B \sum_{i}^{B} |W_\theta(x_i)|} \tag{4}$$

We therefore introduce the Importance Weighting Network (IWN) which searches for the saddle points that solve the above optimization problem (cf Algorithm 1).

3 Related Work

Instance-Based Domain Adaptation. Our work is in line with instance-based unsupervised transfer learning or domain adaptation [30]. Most of the instance-based methods have already been introduced previously as KMM, KLIEP, ULSIF and RULSIF [15,16,33,36]. All of these methods aim to compute the source weights which minimize a distance between the input distributions like the MMD or the Kullback-Leibler. Other methods have also proposed to

Algorithm 1. Importance Weighting Network

Inputs: Source and target datasets $\mathcal{S}_{\mathcal{X}}, \mathcal{T}$, initial bandwidth σ, batch size B, neural network W_θ, learning rate ν

Initialization: Fit W_θ with loss $\mathcal{L} = \sum_i ||W_\theta(x_i) - 1||^2$

while stopping criterion is not reached **do**

Take batches $\{x_1, ..., x_B\} \subset \mathcal{S}_{\mathcal{X}}$ and $\{x'_1, ..., x'_B\} \subset \mathcal{T}$

Forward propagation

$w_i \leftarrow |W_\theta(x_i)| / \sum_i^B |W_\theta(x_i)| \; \forall x_i$

$\text{MMD}_{\sigma,\theta} = \sum_{i,j}^B w_i w_j k_\sigma(x_i, x_j) + \frac{1}{B^2} \sum_{i,j}^n k_\sigma(x'_i, x'_j) - \frac{2}{B^2} \sum_i^B \sum_j^B w_i k_\sigma(x_i, x'_j)$

Backward propagation

$\theta \leftarrow \theta - \nu \nabla_\theta \text{MMD}_{\sigma,\theta}$

$\sigma \leftarrow \sigma + \nu \nabla_\sigma \text{MMD}_{\sigma,\theta}$

take into account the model used to estimate y using appropriate metrics such as the discrepancy [7,21]. Most of the unsupervised instance-based approaches make the assumption of covariate-shift [3].

Importance Weighting and Deep Learning. This work is related to existing works in importance weighting using deep learning. Recently, Fang et al. [10] have developed a task-oriented sample-bias correction method where a KMM is performed at each batch in different depths of the neural network. In a different context from ours, Diesendruck et al. [8] have developed a sample bias correction method for deep generative models. In this approach, the MMD is also minimized by batch, however the weights are not parameterized but assumed to be known (e.g. a uniform proportion of classes is desired). Importance weighting methods have also been used along with deep feature transformation in partial domain adaptation [4], i.e. when the number or the proportion of classes differ between targets and sources. In this category of methods, the output of a domain classifier network is often used to reweight the source instances. The domain classifier is either trained in parallel to the feature transformation [5,37,38] or after it [31,34]. Other methods in this field consider the uncertainty of a task classifier to reweight both source and target instances during the feature transformation [14,35]. These works are interesting from a computational point of view and may be seen as an alternative to MMD minimization. Finally, the weighting adversarial neural network (WANN) [25], explicitly proposes to use a neural network to learn the source weights minimizing a distance between distribution called the $\mathcal{Y}$-discrepancy [28]. Their approach, however, is developed in the supervised context and involves a task network fitted at the same time as the weights. Their approach is then deep learning specific. The present work generalizes this last approach as any estimator can be used once the weights are computed.

4 Experiments

We conduct the experiments on a synthetic dataset and 15 UCI datasets[2] [9] of various size and number of features. The experiments are conducted on a 3.3 Ghz computer with 64G RAM and 24 Cores. The source code of the experiments is available on GitHub[3]..

4.1 IWN Settings

The purpose of IWN is to provide a simple and fast tool to perform importance weighting. We observe that the choice of network has little incidence on the learned weights (see Sect. 4.5), we then arbitrarily choose a three layers neural network with 100 neurons each and a ReLU activation. This architecture is used in all experiments without fine-tuning. We choose the Adam optimizer [17]. The optimization parameters are also fixed for all experiments to a learning rate of 0.001, a batch size of 256 and a maximal number of iterations set to $5 \cdot 10^4$. Early stopping on the objective function is used, if the objective has not improved after $2 \cdot 10^4$ iterations, the learning is stopped.

We remind that the kernel bandwidth σ used to compute the MMD is learned in the gradient descent (cf Algorithm 1) and does not require a cross-validation process. The initial value of σ is set to 0.1 for all experiments.

4.2 Competitors Settings

We consider the following competitors which have already been introduced previously in this paper:

- KMM [15]. We use a Gaussian kernel and the default optimization parameters $B = 1000$ and $\epsilon = \sqrt{m-1}/\sqrt{m}$. The bandwidth σ of the kernel is selected in the set $\{10^{(i-4)}\}_{i \in [|0,8|]}$ with unsupervised cross-validation using the linear discrepancy [21].
- KLIEP [33]. We use a Gaussian kernel and a learning rate of 0.01 with a maximum number of iterations of 1000 as parameters for the gradient descent. These parameters have been selected to obtain an important decrease of the objective function with a fast convergence for most of the datasets. The bandwidth σ of the kernel is selected in the set $\{10^{(i-4)}\}_{i \in [|0,8|]}$ with the native Likelihood Cross-Validation (LCV) procedure of KLIEP.
- NNW [20]. The nearest neighbors are computed with the optimized Nearest-Neighbors algorithm of scikit-learn [32] which optimizes the computation approach between brute force and KD-Ball-Tree in function of the number of features and samples in the dataset. The Euclidean distance is used and the number of nearest neighbors for averaging is chosen in the set $\{1, 5, 10, 20, 50, 100\}$ with unsupervised cross-validation using the linear discrepancy.

[2] https://archive.ics.uci.edu/ml/datasets.php.
[3] https://github.com/antoinedemathelin/Importance-Weighting-Network.

The implementation of the competitors are provided by the ADAPT library [24]. The library also provides the metric for the cross-validation processes.

To offer the best chance to the competitors, the parameters selection with cross-validation are performed with parallel computing for KMM and NNW. For KLIEP, the parallel computing is not available in ADAPT and thus not used which explain why its computational time is higher than the others.

4.3 Synthetic Dataset

We consider the synthetic experiment, inspired from [7], where $p_s(x)$ is a mixture of $M = 10$ Gaussians, i.e. $p_s(x) = \sum_{k=1}^{M} \pi_k \mathcal{N}(\mu_k, 0.2)(x)$ where the centers $\mu_k \in \mathbb{R}^N$ are drawn according to the distribution $\mathcal{N}(0, 1)$ in $\mathbb{R}^N$, the ratios π_k are set such that $\pi_k = 0.8/(M-1)$ $\forall i \neq M$ and $\pi_M = 0.2$. The output variable is written $y = \beta_k^T x$ for any x drawn according to the kth Gaussian. The coefficients $\beta_k \in \mathbb{R}^N$ are drawn according to the distribution $\mathcal{N}(0, 1)$ in $\mathbb{R}^N$. The target distribution $p_t(x)$ is drawn according to the same mixture of Gaussians but with ratios $\pi'_k = 0.1/(M-2)$ $\forall k < M-1, \pi'_{M-1} = 0.1$ and $\pi'_M = 0.8$. We suppose that the learner has access to a sample of size n of source labeled instance $\{(x_i, y_i)\}_{1 \leq i \leq n}$ drawn according to the source distribution $p_s(x, y)$ on $\mathbb{R}^N$ and an unlabeled set of size n, $\{x'_j\}_{1 \leq j \leq n}$ drawn according to the target distribution $p_t(x)$ on $\mathbb{R}^N$. An illustration of the problem for dimension $N = 2$ is given in Fig. 1A.

We conduct several experiments on this synthetic dataset. First, we fit IWN for the setting $N = 32, n = 10000$. We make 4000 batch updates with batch size 256, at each of them, we use the weights w_i returned by the weighting network to fit a weighted Ridge regression model of parameters β on the set $\{(x_i, y_i, w_i)\}_{1 \leq i \leq n}$, we then record the mean absolute error (MAE) of this model on the target dataset: $\frac{1}{n} \sum_i |\beta^T x'_i - y'_i|$. We also record, at each batch, the computed MMD on the batch and the "true" MMD computed with the whole samples. We also record the current value of the parameter σ which is also updated during the optimization (see Sect. 2.3). We report the results of this experiment on Fig. 1B, 1C, 1D. We first represent the final importance weights returned by the weighting network at the end of the 4000 iterations in Fig. 1B. As we can observe, the learned weights are very close to the true sampling probability. On Fig. 1C, we report the evolution of the recorded MAE in plain orange, the batch and true MMD in blue and the value of σ in green. We also report, for comparison with "MAE (IWN)", the MAE of a Ridge model fitted with uniform weights: "MAE (Unif)" and the MAE of the model fitted with the weights obtained using KMM: "MAE (KMM)". We observe that the importance weighting produced by IWN helps to learn the task on the target domain as the error decreases of 20% compared to the error of the model fitted with uniform weights. Concerning the recorded MMD, we observe that both the batch and "true" MMD decrease very fast, but an offset remains between the two due to the estimation error made with finite samples. We then see on the zoom of Fig. 1D that the error decreases very fast as well as the MMD. After 100 iterations, the MMD is minimized and the target error of IWN is on the same level than the error produced by KMM. We notice that the MAE increases a little after some iterations which may indicates some overfitting effect of the weighting network. This observation argues for the use of early stopping based on the evolution of the MMD.

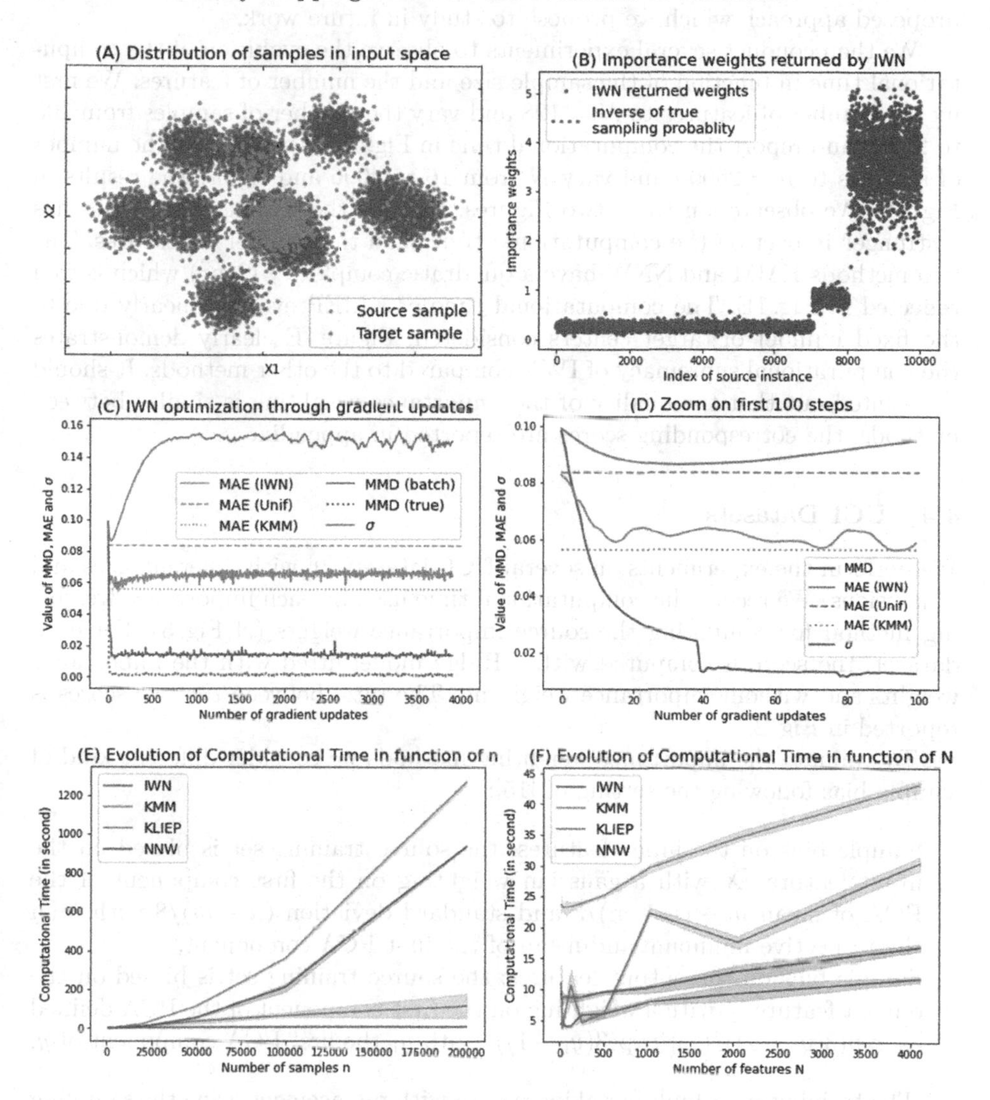

Fig. 1. Visualization of the synthetic experiments. The corresponding experimental settings are: (A) $n = 10000, N = 2$, (B, C, D) $n = 10000, N = 32$, (E) $n \in [100, 2 \cdot 10^5], N = 128$, (F) $n = 2 \cdot 10^4, N \in [16, 4096]$

We finally remark, in Fig. 1C, that the σ parameter becomes relatively stable around 0.15 after some iterations which comforts the idea that the parameter can be efficiently set during the optimization thanks to the adversarial learning (see Sect. 2.3). We could however argue, that the increase of the target error after the 100th iteration is correlated to the increase of σ from 0.1 to 0.15. This is a plausible explanation as σ is updated in order to make the MMD more discriminative which may provoke overfitting. This is a limitation of the proposed approach which we propose to study in future work.

We then conduct several experiments to observe the evolution of the computational time in function of the sample size and the number of features. We first fix the number of features to $N = 128$ and vary the number of samples from 100 to $2 \cdot 10^5$ and report the computational time in Fig. 1E. Then, we fix the number of samples to $n = 20000$ and vary N from 16 to 4096 and report the results in Fig. 1F. We observe, on these two Figures, that for the number of samples has a stronger impact on the computational time than the number of features. The two methods KMM and NNW have a quadratic complexity $\mathcal{O}(n^2)$ which is well reflected in Fig. 1E. The computational time of KLIEP evolves linearly due to the fixed number of target centers considered. Figure 1E clearly demonstrates the computational supremacy of IWN compared to the other methods. It should be pointed out that the quality of the importance weighting is similar between methods, the corresponding scores are reported in appendix.

4.4 UCI Datasets

We perform the experiments on several UCI datasets [9] with different sizes and dimensions. We record the computational time used by each importance weighting method for computing the source importance weights (cf Fig. 3). For each dataset, the score is computed with a Ridge model fitted with the importance weights and without importance weighting. The ratio between the two scores is reported in Fig. 2.

To evaluate the importance weighting methods we consider different kind of sample bias following the setting of [15]:

- Sample bias on the input features: the source training set is biased on the input features X with a gaussian weighting on the first component of the PCA of mean $m + (\mu - m)/3$ and standard deviation $(\mu - m)/8$ with m, μ the respective minimum and mean of the first PCA component.
- Sample bias on the output features: the source training set is biased on the output features y with a weighting on the first component of the PCA defined as $\exp(3(y_1 - 1))/(1 + \exp(3(y_1 - 1)))$ with y_1 the first PCA component of y.

The training set is built by taking n data with replacement using the sampling bias as probabilities of selection. The target set is the original dataset without selection bias. For each experiment, we apply the following preprocessing: standard scaling of the numerical inputs (using the mean and standard deviation of the unbiased inputs) and one-hot-encoding of the categorical inputs. The dimension p of the input space corresponds to the dimension after preprocessing.

To evaluate the weighting scheme of each method, we fit a Ridge model with trade-off parameters α selected by leave-one-out process between values

$\{10^{(i-4)}\}_{i \in [|0,8|]}$. For regression datasets, we compute the mean absolute error (MAE) on the target testing set (the original dataset without bias). We then fit another Ridge model with the uniformly weighted biased source data and compute the MAE on the target set. We can the compute a score ratio for each method as follows:

$$\text{Score Ratio} = \frac{\sum_i |\beta_{IW}^T x_i' - y_i'|}{\sum_i |\beta_{Unif}^T x_i' - y_i'|} \tag{5}$$

where β_{IW}, β_{Unif} are respectively the coefficients of the Ridge models fitted with the importance weights and the uniform weights.

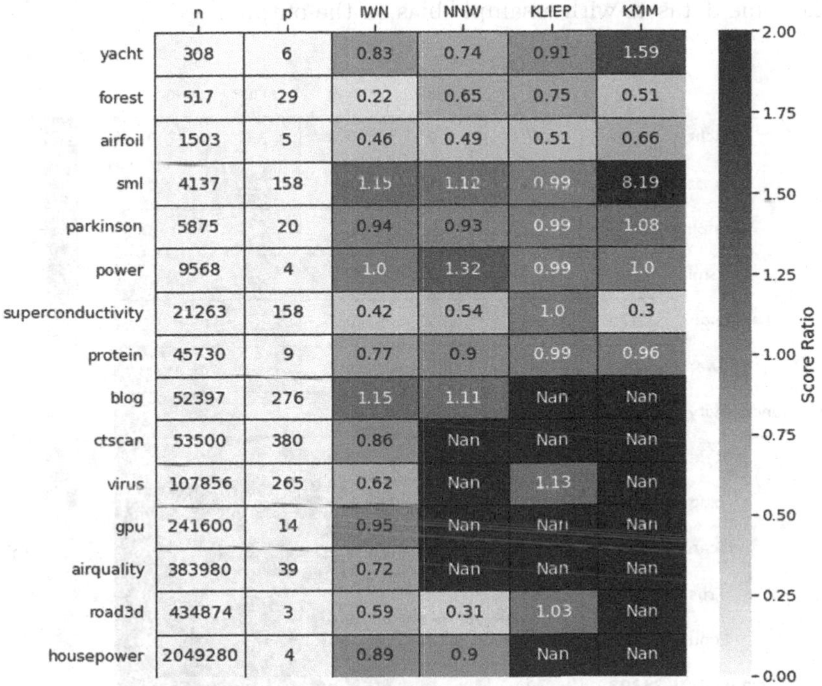

	n	p	IWN	NNW	KLIEP	KMM
yacht	308	6	0.83	0.74	0.91	1.59
forest	517	29	0.22	0.65	0.75	0.51
airfoil	1503	5	0.46	0.49	0.51	0.66
sml	4137	158	1.15	1.12	0.99	8.19
parkinson	5875	20	0.94	0.93	0.99	1.08
power	9568	4	1.0	1.32	0.99	1.0
superconductivity	21263	158	0.42	0.54	1.0	0.3
protein	45730	9	0.77	0.9	0.99	0.96
blog	52397	276	1.15	1.11	Nan	Nan
ctscan	53500	380	0.86	Nan	Nan	Nan
virus	107856	265	0.62	Nan	1.13	Nan
gpu	241600	14	0.95	Nan	Nan	Nan
airquality	383980	39	0.72	Nan	Nan	Nan
road3d	434874	3	0.59	0.31	1.03	Nan
housepower	2049280	4	0.89	0.9	Nan	Nan

Fig. 2. Score ratio for the experiments with the sample bias applied on input features. In each column, the values correspond to the ratios between the mean absolute error (MAE) of the corresponding methods (IWN, NNW, KLIEP and KMM) and the MAE of the Uniform Weighting approach where no reweighting is performed. The MAEs are computed on the target dataset with a Ridge model fitted on the reweighted source data. The two first columns n and p are respectively the number of sample and number of features of the dataset. The lightest colors correspond to the best ratios. The experiment stopped after 500 s are marked with Nan.

We repeat each experiment 10 times and report the results of the experiments with the sample bias on the input features in Figs. 2 and 3, the standard deviation over the 10 repetitions are given in appendix. The first Figure presents the score ratios computed with Eq. (5). We observe that the quality of the weighting scheme provided by IWN is competitive with other methods. IWN is in the top 2 best ratios for all experiments except one. Even more impressive are the computational time of IWN compared to other methods (Fig. 3), IWN provides a fast computation of no more than 25 s for sample of size $< 5 \times 10^5$. Whereas the other methods failed to provide importance weights in a reasonable time for sample above 5×10^4 samples and > 100 features. The same observations are made on Table 1 which reports the summary results of the experiments conducted on the same datasets with a sample bias on the outputs.

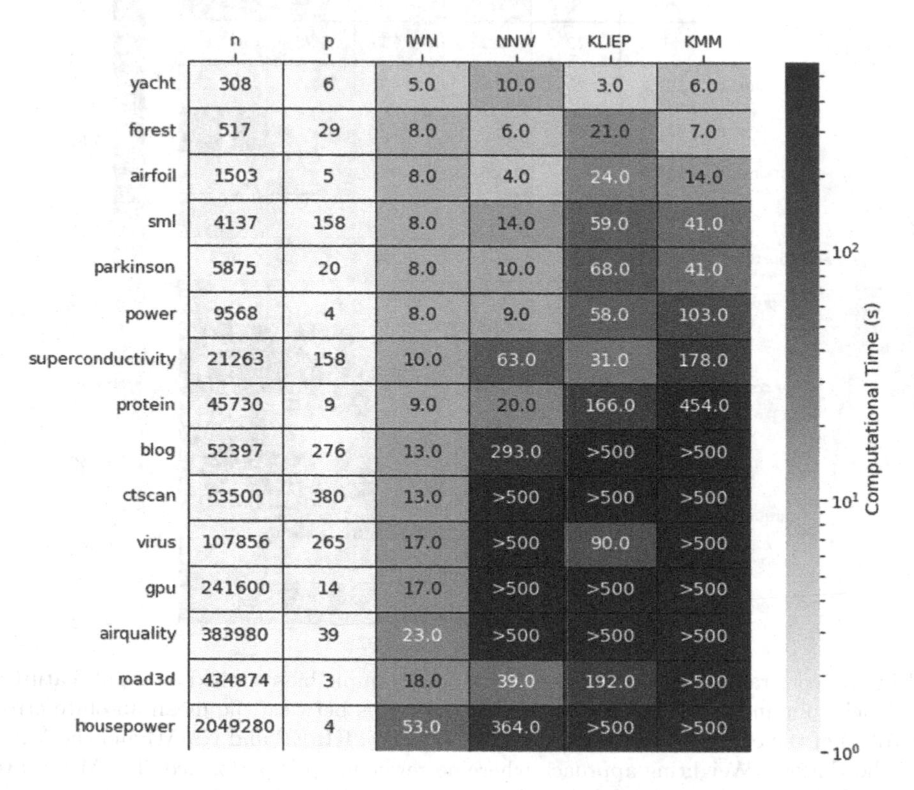

	n	p	IWN	NNW	KLIEP	KMM
yacht	308	6	5.0	10.0	3.0	6.0
forest	517	29	8.0	6.0	21.0	7.0
airfoil	1503	5	8.0	4.0	24.0	14.0
sml	4137	158	8.0	14.0	59.0	41.0
parkinson	5875	20	8.0	10.0	68.0	41.0
power	9568	4	8.0	9.0	58.0	103.0
superconductivity	21263	158	10.0	63.0	31.0	178.0
protein	45730	9	9.0	20.0	166.0	454.0
blog	52397	276	13.0	293.0	>500	>500
ctscan	53500	380	13.0	>500	>500	>500
virus	107856	265	17.0	>500	90.0	>500
gpu	241600	14	17.0	>500	>500	>500
airquality	383980	39	23.0	>500	>500	>500
road3d	434874	3	18.0	39.0	192.0	>500
housepower	2049280	4	53.0	364.0	>500	>500

Fig. 3. Computational times (in second) for the experiments with the sample bias applied on input features. The two first columns n and p are respectively the number of sample and number of features of the dataset. The computational time are given in second. The lightest colors correspond to the lowest computational time. The experiment stopped after 500 s are marked (>500).

Table 1. Summary of the results of the output sample bias experiments (extensive results are reported in appendix).

Method	Avg Score Ratio	Avg Rank	Avg Comp. Time (in sec.)
IWN	0.9	**2.0**	**8.06**
NNW	0.92	2.37	182.81
KLIEP	1.03	2.93	223.16
KMM	**0.87**	2.7	290.01

4.5 Impact of Network Architecture and Batch Size

Finally, we study the impact of the network architecture and the batch size on the solution of IWN. We conduct the experiments on the CTscan dataset[4] [12] biased through the sample bias on the input features described previously. First, we fix the batch size to 256 and vary the number of hidden layers of the weighting network from 0 to 4 and the number of neurons per layer between [10, 100, 300]. Then, we fix the number of hidden layers to 3 and the number of neurons to 100 and vary the batch size on a geometric scale of ratio 4 from 16 to 4096. We repeat each experiment 10 times and report the means and standard deviations of the scores in Table 2. We observe that the architecture of the network has little impact on the score, however, we observe a slight improvement of the score between the simplest architectures and the more complex ones (0.88 for 0 hidden layer and 0.85 for 4 layers and 300 neurons). The impact of the batch size is more significant on the performance of IWN, enlarging the batches produce better corrections of sample bias. This is due to the more accurate estimation of the MMD made with larger batch. However, increasing the batch size comes with an increase of the computational time as shown in the last column of Table 2.

Table 2. Summary of the study on the impact of weighting network architecture and batch size. Standard deviation over the 10 repetitions are given in brackets.

Neural network architecture				Batch size		
Layers	Neurons					
	10	100	300	Size	Score	Time (s)
0	0.88 (0.02)	0.88 (0.02)	0.88 (0.02)	16	0.96 (0.01)	9.7 (8.2)
1	0.86 (0.02)	0.86 (0.02)	0.85 (0.02)	64	0.90 (0.01)	6.6 (0.2)
2	0.86 (0.02)	0.86 (0.02)	0.85 (0.02)	256	0.85 (0.02)	7.4 (0.3)
3	0.86 (0.02)	0.85 (0.02)	0.85 (0.02)	1024	0.82 (0.03)	12.1 (0.1)
4	0.86 (0.02)	0.85 (0.02)	0.85 (0.02)	4096	0.83 (0.04)	117.9 (1.3)

[4] https://archive.ics.uci.edu/ml/datasets/Relative+location+of+CT+slices+on+axial+axis.

5 Conclusion

This work introduces a novel algorithm for importance weighting called Importance Weighting Network and shows that sample biases can be efficiently corrected by fitting a weighting neural network with the MMD as loss function. This approach appears to provide very competitive results with state-of-the-art instance-based domain adaptation methods for a minimal cost in term of computational time.

References

1. Bellot, A., van der Schaar, M.: A kernel two-sample test with selection bias. In: Uncertainty in Artificial Intelligence, pp. 205–214. PMLR (2021)
2. Ben-David, S., Blitzer, J., Crammer, K., Pereira, F.: Analysis of representations for domain adaptation. In: Schölkopf, B., Platt, J.C., Hoffman, T. (eds.) Advances in Neural Information Processing Systems, vol. 19, pp. 137–144. MIT Press (2007)
3. Bickel, S., Brückner, M., Scheffer, T.: Discriminative learning under covariate shift. J. Mach. Learn. Res. **10**(9), 2137–2155 (2009)
4. Cao, Z., Long, M., Wang, J., Jordan, M.I.: Partial transfer learning with selective adversarial networks. In: Proceedings of the IEEE Conference on Computer Vision and Pattern Recognition, pp. 2724–2732 (2018)
5. Cao, Z., You, K., Long, M., Wang, J., Yang, Q.: Learning to transfer examples for partial domain adaptation. In: Proceedings of the IEEE Conference on Computer Vision and Pattern Recognition, pp. 2985–2994 (2019)
6. Chandra, S., Haque, A., Khan, L., Aggarwal, C.: Efficient sampling-based kernel mean matching. In: 2016 IEEE 16th International Conference on Data Mining (ICDM), pp. 811–816. IEEE (2016)
7. Cortes, C., Mohri, M.: Domain adaptation and sample bias correction theory and algorithm for regression. Theoretical Computer Science 519 (2014)
8. Diesendruck, M., Elenberg, E.R., Sen, R., Cole, G.W., Shakkottai, S., Williamson, S.A.: Importance weighted generative networks. In: Brefeld, U., Fromont, E., Hotho, A., Knobbe, A., Maathuis, M., Robardet, C. (eds.) ECML PKDD 2019. LNCS (LNAI), vol. 11907, pp. 249–265. Springer, Cham (2020). https://doi.org/10.1007/978-3-030-46147-8_15
9. Dua, D., Graff, C.: UCI machine learning repository (2017). http://archive.ics.uci.edu/ml
10. Fang, T., Lu, N., Niu, G., Sugiyama, M.: Rethinking importance weighting for deep learning under distribution shift. Adv. Neural. Inf. Process. Syst. **33**, 11996–12007 (2020)
11. Friedman, J.H., Bentley, J.L., Finkel, R.A.: An algorithm for finding best matches in logarithmic expected time. ACM Trans. Math. Software (TOMS) **3**(3), 209–226 (1977)
12. Graf, F., Kriegel, H.-P., Schubert, M., Pölsterl, S., Cavallaro, A.: 2D image registration in CT images using radial image descriptors. In: Fichtinger, G., Martel, A., Peters, T. (eds.) MICCAI 2011. LNCS, vol. 6892, pp. 607–614. Springer, Heidelberg (2011). https://doi.org/10.1007/978-3-642-23629-7_74
13. Gretton, A., Borgwardt, K.M., Rasch, M.J., Schölkopf, B., Smola, A.: A kernel two-sample test. J. Mach. Learn. Res. **13**(1), 723–773 (2012)

14. Guan, D., Huang, J., Xiao, A., Lu, S., Cao, Y.: Uncertainty-aware unsupervised domain adaptation in object detection. IEEE Trans. Multimedia **24**, 2502–2514 (2021)
15. Huang, J., Gretton, A., Borgwardt, K., Schölkopf, B., Smola, A.J.: Correcting sample selection bias by unlabeled data. In: Schölkopf, B., Platt, J.C., Hoffman, T. (eds.) Advances in Neural Information Processing Systems, vol. 19, pp. 601–608. MIT Press (2007)
16. Kanamori, T., Hido, S., Sugiyama, M.: A least-squares approach to direct importance estimation. J. Mach. Learn. Res. **10**, 1391–1445 (2009)
17. Kingma, D.P., Ba, J.: Adam: A method for stochastic optimization. In: Bengio, Y., LeCun, Y. (eds.) 3rd International Conference on Learning Representations, ICLR 2015, San Diego, CA, USA, May 7–9, 2015, Conference Track Proceedings (2015)
18. LeCun, Y., Bengio, Y., Hinton, G.: Deep learning. Nature **521**(7553), 436–444 (2015)
19. Li, C.L., Chang, W.C., Cheng, Y., Yang, Y., Póczos, B.: MMD GAN: towards deeper understanding of moment matching network. In: Advances in Neural Information Processing Systems, vol. 30 (2017)
20. Loog, M.: Nearest neighbor-based importance weighting. In: 2012 IEEE International Workshop on Machine Learning for Signal Processing, pp. 1–6. IEEE (2012)
21. Mansour, Y., Mohri, M., Rostamizadeh, A.: Domain adaptation: learning bounds and algorithms. In: COLT (2009)
22. Martino, L., Elvira, V., Louzada, F.: Effective sample size for importance sampling based on discrepancy measures. Signal Process. **131**, 386–401 (2017)
23. de Mathelin, A., Deheeger, F., Mougeot, M., Vayatis, N.: Handling distribution shift in tire design. In: NeurIPS 2021 Workshop on Distribution Shifts: Connecting Methods and Applications (2021)
24. de Mathelin, A., Deheeger, F., Richard, G., Mougeot, M., Vayatis, N.: Adapt: Awesome domain adaptation python toolbox. arXiv preprint arXiv:2107.03049 (2021)
25. de Mathelin, A., Richard, G., Deheeger, F., Mougeot, M., Vayatis, N.: Adversarial weighting for domain adaptation in regression. arXiv preprint arXiv:2006.08251 (2020)
26. Miao, Y.Q., Farahat, A.K., Kamel, M.S.: Auto-tuning kernel mean matching. In: 2013 IEEE 13th International Conference on Data Mining Workshops, pp. 560–567. IEEE (2013)
27. Miao, Y.Q., Farahat, A.K., Kamel, M.S.: Ensemble kernel mean matching. In: 2015 IEEE International Conference on Data Mining, pp. 330–338. IEEE (2015)
28. Mohri, M., Muñoz Medina, A.: New analysis and algorithm for learning with drifting distributions. In: Bshouty, N.H., Stoltz, G., Vayatis, N., Zeugmann, T. (eds.) ALT 2012. LNCS (LNAI), vol. 7568, pp. 124–138. Springer, Heidelberg (2012). https://doi.org/10.1007/978-3-642-34106-9_13
29. Omohundro, S.M.: Five balltree construction algorithms. International Computer Science Institute Berkeley (1989)
30. Pan, S.J., Yang, Q.: A survey on transfer learning. IEEE Trans. Knowl. Data Eng. **22**(10), 1345–1359 (2010). https://doi.org/10.1109/TKDE.2009.191
31. Park, S., Bastani, O., Weimer, J., Lee, I.: Calibrated prediction with covariate shift via unsupervised domain adaptation. In: International Conference on Artificial Intelligence and Statistics, pp. 3219–3229. PMLR (2020)
32. Pedregosa, F., et al.: Scikit-learn: machine learning in Python. J. Mach. Learn. Res. **12**, 2825–2830 (2011)

33. Sugiyama, M., Nakajima, S., Kashima, H., Bünau, P.V., Kawanabe, M.: Direct importance estimation with model selection and its application to covariate shift adaptation. In: Proceedings of the 20th International Conference on Neural Information Processing Systems. NIPS 2007, pp. 1433–1440, Red Hook, NY, USA. Curran Associates Inc. (2007)
34. Wang, X., Long, M., Wang, J., Jordan, M.: Transferable calibration with lower bias and variance in domain adaptation. Adv. Neural. Inf. Process. Syst. **33**, 19212–19223 (2020)
35. Wen, J., Zheng, N., Yuan, J., Gong, Z., Chen, C.: Bayesian uncertainty matching for unsupervised domain adaptation. arXiv preprint arXiv:1906.09693 (2019)
36. Yamada, M., Suzuki, T., Kanamori, T., Hachiya, H., Sugiyama, M.: Relative density-ratio estimation for robust distribution comparison. In: Advances in Neural Information Processing Systems, vol. 24 (2011)
37. You, K., Long, M., Cao, Z., Wang, J., Jordan, M.I.: Universal domain adaptation. In: Proceedings of the IEEE Conference on Computer Vision and Pattern Recognition, pp. 2720–2729 (2019)
38. Zhang, J., Ding, Z., Li, W., Ogunbona, P.: Importance weighted adversarial nets for partial domain adaptation. In: Proceedings of the IEEE Conference on Computer Vision and Pattern Recognition, pp. 8156–8164 (2018)

Overcoming Catastrophic Forgetting via Direction-Constrained Optimization

Yunfei Teng[1(✉)], Anna Choromanska[1], Murray Campbell[2], Songtao Lu[2], Parikshit Ram[2], and Lior Horesh[2]

[1] New York University, New York, USA
yt1208@nyu.edu
[2] IBM Research, New York, USA

Abstract. This paper studies a new design of the optimization algorithm for training deep learning models with a *fixed* architecture of the classification network in a continual learning framework. The training data is non-stationary and the non-stationarity is imposed by a sequence of distinct tasks. We first analyze a deep model trained on only one learning task in isolation and identify a region in network parameter space, where the model performance is close to the recovered optimum. We provide empirical evidence that this region resembles a cone that expands along the convergence direction. We study the principal directions of the trajectory of the optimizer after convergence and show that traveling along a few top principal directions can quickly bring the parameters outside the cone but this is not the case for the remaining directions. We argue that catastrophic forgetting in a continual learning setting can be alleviated when the parameters are constrained to stay within the intersection of the plausible cones of individual tasks that were so far encountered during training. Based on this observation we present our direction-constrained optimization (DCO) method, where for each task we introduce a linear autoencoder to approximate its corresponding top forbidden principal directions. They are then incorporated into the loss function in the form of a regularization term for the purpose of learning the coming tasks without forgetting. Furthermore, in order to control the memory growth as the number of tasks increases, we propose a memory-efficient version of our algorithm called compressed DCO (DCO-COMP) that allocates a memory of fixed size for storing all autoencoders. We empirically demonstrate that our algorithm performs favorably compared to other state-of-art regularization-based continual learning methods. The codes are publicly available at https://github.com/yunfei-teng/DCO.

Keywords: Continual/Lifelong learning · Deep learning · Optimization

A. Choromanska—Senior lead.

Supplementary Information The online version contains supplementary material available at https://doi.org/10.1007/978-3-031-26387-3_41.

M.-R. Amini et al. (Eds.): ECML PKDD 2022, LNAI 13713, pp. 675–692, 2023.
https://doi.org/10.1007/978-3-031-26387-3_41

1 Introduction

A key characteristic feature of intelligence is the ability to continually learn over time by accommodating new knowledge and transferring knowledge between correlated tasks while retaining previously learned experiences. This ability is often referred to as *continual or lifelong learning.* In a continual learning setting one needs to deal with a continual acquisition of incrementally available information from non-stationary data distributions (online learning) and avoid *catastrophic forgetting* [28], i.e., a phenomenon that occurs when training a model on currently observed task leads to a rapid deterioration of the model's performance on previously learned tasks. In the commonly considered scenario of continual learning the tasks come sequentially and the model is not allowed to inspect again the samples from the tasks seen in the past [29]. Within this setting, there exist two types of approaches that are complementary and equally important in the context of solving the continual learning problem: i) methods that assume fixed architecture of deep model and focus on designing the training strategy that allows the model to learn many tasks and ii) methods that rely on existing training strategies (mostly SGD [5] and its variants, which themselves suffer catastrophic forgetting [12]) and focus on expanding the architecture of the network to accommodate new tasks. In this paper we focus on the first framework.

Training a network in a continual learning setting, when the tasks arrive sequentially, requires solving many optimization problems, one per task. A space of solutions (i.e., network parameters) that correspond to good performance of the network on all encountered tasks determine a common manifold of plausible solutions for all these optimization problems. In this paper we seek to understand the geometric properties of this manifold. In particular we analyze how this manifold is changed by each new coming task and propose an optimization algorithm that efficiently searches through it to recover solutions that well-represent all previously-encountered tasks. The contributions of our work could be summarized as follows:

- We empirically analyse the deep learning loss landscape and show that there is a cone in the network parameter space where the model performance is close to the recovered optimum.
- We propose a new regularization-based continual learning algorithm that explicitly encourages the model parameters to stay inside the plausible cone by identifying a few top forbidden principal directions for each task.
- We propose an autoencoder architecture that significantly reduces the memory complexity to save the top forbidden principal directions for a given task.
- We design a compression method to control the memory growth and avoid introducing a new autoencoder per task, thereby requiring only a constant size memory overhead irrespective of the number of tasks.

The paper is organized as follows: Sect. 2 reviews recent progress in the research area of continual learning, Sect. 3 provides empirical analysis of the geometric properties of the deep learning loss landscape and builds their relation to the continual learning problem, Sect. 4 introduces our algorithm that we call DCO since it is based on the idea of direction-constrained optimization, Sect. 5 contains empirical evaluations, and finally Sect. 6 concludes the paper. Additional results are contained in the Supplement.

2 Related Work

Continual learning and the catastrophic forgetting problem has been addressed in a variety of papers. A convenient literature survey dedicated to this research theme was recently published [29]. The existing approaches can be divided into three categories [10, 29] listed below.

Regularization-based methods modify the objective function by adding a penalty term that controls the change of model parameters when a new task is observed. In particular these methods ensure that when the model is being trained on a new task, the parameters stay close to the ones learned on the tasks seen so far. EWC [18] approximates the posterior of the model parameters after each task with a Gaussian distribution and uses tasks' Fisher information matrices to measure the overlap of tasks. The idea is extended in [33] where the authors introduce Kronecker factored Laplace approximation. SI [42] introduces the notion of synaptic importance, enabling the assessment of the importance of network parameters when learning sequences of classification tasks, and penalizes performing changes to the parameters with high importance when training on a new task in order to avoid overwriting old memories. Relying on the importance of the parameters of a neural network when learning a new task is also a characteristic feature of another continual learning technique called MAS [1]. The RWALK method [6] is a combination of an efficient variant of EWC and a modified SI technique that computes a parameter importance score based on the sensitivity of the loss over the movement on the Riemannian manifold. Additionally, RWALK stores a small subset of representative samples from the previous tasks and uses them while training the current task, which is essentially a form of a replay strategy described later in this section. The recently proposed OGD algorithm [9] and its variant GPM [36] rely on constraining the parameters of the network to move within the orthogonal space to the gradients of previous tasks. [9] is memory-consuming and not scalable as it requires saving the gradient directions of the neural network predictions on previous tasks. Finally, the recursive method of [24] modifies the gradient direction of each step to minimize the expected forgetting by introducing an additional projection matrix which requires a per-step update with linear memory complexity in the number of model parameters. All methods discussed so far constitute a family of techniques that keep the architecture of the network fixed. The algorithm we propose in this paper also belongs to this family.

Another regularization method called LwF [23] optimizes the network both for high accuracy on the next task and for preservation of responses on the network outputs corresponding to the past tasks. This is done using only examples for the next task. The encoder-based lifelong learning technique [30] uses per-task under-complete autonecoders to constraint the features from changing when the new task arrives, which has the effect of preserving the information on which the previous tasks are mainly relying. Both these methods fundamentally differ from the aforementioned techniques and the approach we propose in this paper in that they require a separate network output for each task. Finally, P&C [37] builds upon EWC and takes advantage of the knowledge distillation mechanism to preserve and compress the knowledge obtained from the previous tasks. Such a mechanism could as well be incorporated on the top of SI, MAS, or our technique.

The next two families of continual learning methods are not directly related to the setting considered in this paper and are therefore reviewed only briefly.

Dynamic architecture methods either expand the model architecture [2,22,35,41] to allocate additional resources to accommodate new tasks (they are typically memory expensive) or exploit the network structure by parameter pruning or masking [26,27]. Some techniques [16] interleaves the periods of network expansion with network compression, network pruning, and/or masking phases to better control the growth of the model.

Replay methods are designed to train the model on a mixture of samples from a new task and samples from the previously seen tasks. The purpose of replaying old examples is to counter-act the forgetting process. Many replay methods rely on the design of sampling strategies [3,17]. Other techniques, such as GEM [25], A-GEM [7] and MER [32], use replay specifically to encourage positive transfer between the tasks (increasing the performance on preceding tasks when learning a new task). ORTHOG-SUBSPACE [8] reduces the interference between tasks by learning the tasks in different subspaces. Replay methods typically require large memory. Deep generative replay technique [34,38] addresses this problem and employs a generative model to learn a mixed data distribution of samples from both current and past tasks. Samples generated this way are used to support the training of a classifier. Finally, note that the setting considered in our paper does not rely on the replay mechanism.

In addition to the above discussed research directions, very recently authors started to look at task agnostic and multi-task continual learning where no information about task boundaries or task identity is given to the learner [14, 15,31,40,43]. These approaches lie beyond the scope of this work.

Remark: Regularization-based methods and replay methods are usually implemented with a fixed architecture of the classification network, but they require additional memory to save regularization terms or data samples. Conversely, dynamic architecture methods do not explicitly keep extra information in the memory, but they rely on the expansion and modification of the network architecture itself. Our approach falls into the family of regularization-based methods since we do not allow the architecture of the classification network to dynamically change and we also do not allow replay.

3 Loss Landscape Properties

The experimental observations provided in this section extend and complement the behavior characterization of SGD [11] connecting its dynamics with random landscape theory that stems from physical systems. The results that will be presented here were obtained on MNIST and CIFAR-10 data sets (CIFAR-10 results are deferred to the Supplement). The details of the experimental setup of this section can be found in the Supplement (Sect. 9). Consider learning only one task. We analyze the top principal components of the trajectory of SGD *after convergence*, i.e., after the optimizer reached a saturation level[1]. Let x^* denotes the value of the parameters in the beginning of the saturation phase. The convergence trajectory will be represented as a sequence of optimizer steps, where each step is represented by the change of model parameters that the optimizer induced (gradient). We consider n steps after model convergence and compute the gradient of the loss function at these steps that we refer to as $\nabla L(x_1;\zeta_1), \nabla L(x_2;\zeta_2), \ldots, \nabla L(x_n;\zeta_n)$ (x_i denotes the model parameters at the ith step and ζ_i denotes the data mini-batch for which the gradient was computed at that step). We use them to form a matrix $G \in \mathbb{R}^{d\times n}$ (i-th column of the matrix is $\nabla L(x_i;\zeta_i)$) and obtain the eigenvectors $\{v_i\}$ of GG^T.[2] We furthermore define the averaged gradient direction $\bar{g} = \frac{1}{n}\sum_{i=1}^{n} \nabla L(x_i;\zeta_i)$. We first study the landscape of the deep learning loss function along directions v_i and $\bar{g}$, i.e., we analyze the function

$$f(\alpha,\beta,v_i) = L(x^* - \alpha\bar{g} + \beta v_i;\zeta), \tag{1}$$

where α and β are the step sizes along $-\bar{g}$ and v_i respectively and ζ denotes the entire training data set. We will show how this function is connected to our algorithm later in Sect. 4.4.

Remark: Below, the eigenvector with the lower-index corresponds to a larger eigenvalue.

Observation 1: Behavior of the loss for $\alpha = 0$ and changing β For each eigenvector v_i, we first fix α to 0 and change β in order to study the behavior of $f(0,\beta,v_i)$. Figure 1a captures the result. It can be observed that as the model parameters move away from the optimal point x^* the loss gradually increases. At the same time, the rate of this increase depends on the eigendirection that is followed and grows faster while moving along eigenvectors with the lower-index. Thus we have empirically shown that *the loss changes more slowly along the eigenvectors with the high index, i.e., the landscape is flatter along these directions.*

Observation 2: Behavior of the loss in the subspaces spanned by groups of eigenvectors Here we generalize Observation 1 to the subspaces spanned by a set of eigenvectors. For the purpose of this observation only we consider the following metric instead of the one given in Eq. 1:

$$h(\sigma,V_s) = \mathbb{E}_{\delta\sim\mathcal{N}(0,\frac{\sigma^2}{d}I)} L(x^* + V_s V_s^T\delta;\zeta), \tag{2}$$

[1] The optimization process is typically terminated when the loss starts saturating but we argue that running the optimizer further gives benefits in the continual learning setting.

[2] The explanation of the difference between GG^T and the Fisher information matrix underlying the EWC method is deferred to the Supplement (Sect. 8).

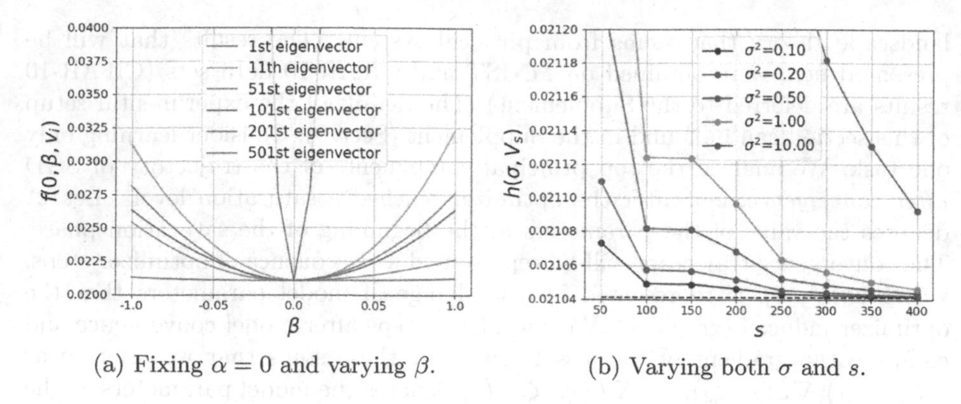

(a) Fixing $\alpha = 0$ and varying β.

(b) Varying both σ and s.

Fig. 1. left (a): The behavior of the loss function for $\alpha = 0$ and varying β when moving along different eigenvectors on MNIST (the complementary plot obtained on CIFAR-10 can be found in the Supplement, Fig. 9); **right (b)**: The behavior of the loss function when varying σ and s on MNIST (the complementary plot obtained on CIFAR-10 can be found in the Supplement, Fig. 10).

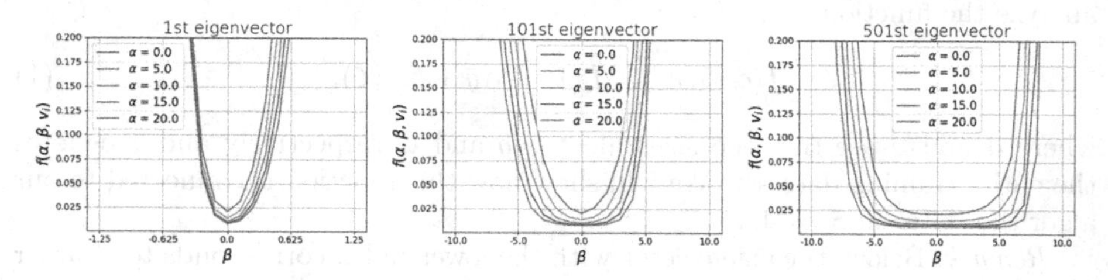

Fig. 2. The behavior of the loss function when both α and β are changing for eigenvectors with different index on MNIST (the complementary plot obtained on CIFAR-10 can be found in the Supplement, Fig. 11).

where δ is the random perturbation, σ is the standard deviation, and $V_s = [v_{s-49}, v_{s-48}, \cdots, v_s]$ is the matrix of eigenvectors of 50 consecutive indexes. To be more concrete, we locally (in the ball of radius σ around x^*) sample the space spanned by the eigenvectors in V_s. The expectation is computed over 3000 random draws of δ. In Fig. 1b we examine the behavior of $h(\sigma, V_s)$ for various values of σ and s. The plot confirms what was shown in Observation 1 that the loss landscape becomes flatter in the subspace spanned by the eigenvectors with high index.

Observation 3: Behavior of the loss for changing α and β We generalize Observation 1 and examine what happens with $f(\alpha, \beta, v_i)$ when both α and β change. Figure 2 captures the result. We can see that as α increases, or in other words *as we go further along the averaged gradient direction, the loss landscape becomes flatter. This property holds for an eigenvector with an arbitrary index.* Thus for larger values of α we can go further along eigenvector directions without significantly changing the loss. This can be seen as a *cone* that expands along $-\bar{g}$. Furthermore, the findings of Observation 1 are also confirmed in Fig. 2. For the eigenvectors with higher index the loss changes less rapidly (the cone is wider along these directions). These properties underpin the design of new continual learning algorithm proposed in this work. When adding the second task, the algorithm constrains the optimizer to stay within the cone of the first task. Intuitively this can be done by first pushing the optimizer further into the cone along $-\bar{g}$ and then constraining the optimizer from moving along eigenvectors with low indexes in order to prevent forgetting the first task. This procedure can be generalized to an arbitrary number of tasks as will be shown in the next section.

4 Algorithm

In Sect. 3 we analyzed the loss landscape for a single task and discovered the existence of the cone in the model's parameter space where the model sustains good performance. We then discussed the consequence of this observation in the continual learning setting. In this section we propose a *tractable* continual learning algorithm that for each task finds its cone and uses it to constrain the optimization problem of learning the following tasks. We refer to the model that is trained in the continual learning setting as $\mathcal{M}$. The proposed algorithm relies on identifying the top directions along which the loss function for a given task increases rapidly and then constraining the optimization from moving along these directions (we will refer to these directions as "prohibited") when learning subsequent tasks. Note that each new task adds prohibited directions. In order to efficiently identify and constrain the prohibited directions we use *reduced linear autoencoders* whose design was tailored for the purpose of the proposed algorithm. We train separate autoencoders for each learned task. The jth autoencoder admits on its input gradients of the loss function that are obtained when training the model $\mathcal{M}$ on the jth task. The intuitive idea behind this approach is that autoencoder with small feature vector will capture the top directions of the gradients it is trained on. We refer to our method as *direction-constrained optimization* (DCO) method. Furthermore, we will show that we can relax the need to allocate new memory for each task and propose a memory-efficient version of our algorithm called *compressed* DCO (DCO-COMP) which requires a memory of fixed size for storing all autoencoders.

4.1 Loss Function

We next explain the loss function that is used to train the model $\mathcal{M}$ in a continual learning setting. From Sect. 3, we recognize that the matrix G formed by the gradients obtained from the current task can be used to describe the properties of the loss landscape. Furthermore, we can incorporate it as a regularization term into the loss function to prevent the increment of loss for the current task when training for a future task. Therefore, the loss function that is used to train the model on the ith task takes the form:

$$L_i(x;\xi) = L_{ce}(x;\xi) + \lambda \sum_{j=1}^{i-1} \left\| (G^j)^T (x - x_j^*) \right\|_2^2, \tag{3}$$

where ξ is a training example, L_{ce} is a cross-entropy loss, λ is a hyperparameter controlling the strength of the regularization, G^j is the regularization matrix whose columns are the sampled gradients, and x_j^* are the parameters of model $\mathcal{M}$ obtained at the end of training the model on the jth task. However, directly saving matrix G^j will make the algorithm become *intractable*. Thus, we instead introduce an autoencoder ENC^j to approximate Eq. 3 by:

$$L_i(x;\xi) = L_{ce}(x;\xi) + \lambda \sum_{j=1}^{i-1} \left\| ENC^j (x - x_j^*) \right\|_2^2, \tag{4}$$

where $ENC^j(\cdot)$ is the operation of the encoder of the autoencoder trained on task j. The linear autoencoders with appropriate regularization are able to recover the principal components of gradients [4]. The top principal components correspond to the directions where the loss changes the quickest and we consider these directions as the prohibited directions. This is well-aligned with our observations from Sect. 3.

4.2 Reduced Linear Autoencoders

In our algorithm, the role of autoencoder is to identify the top k directions of the optimizer's trajectory after convergence, where this trajectory is defined by gradient steps, obtained during training the model $\mathcal{M}$. A traditional linear autoencoder, consisting of two linear layers, would require $2 \times d \times k$ number of parameters, where d denotes the number of parameters of the model $\mathcal{M}$. Commonly used deep learning models however contain millions of parameters [13,20,39], which makes a traditional autoencoder not tractable for this application. In order to reduce the memory footprint of the autoencoder we propose an architecture that is inspired by the singular value decomposition. The proposed autoencoder admits a matrix on its input and is formulated as

$$AE(M) = U diag(U^\top M V) V^\top, \tag{5}$$

where $diag(U^\top MV)$ is a matrix formed by zeroing out the non-diagonal elements of $U^\top MV$, M is an autoencoder input matrix of size $m \times n$, and U and V are autoencoder parameters of size $m \times k$ and $n \times k$ respectively. Thus, the total number of parameters of the proposed autoencoder is $k(n + m)$, which is significantly lower than in case of traditional autoencoder (knm), especially when n and m are large. We call this architecture a *reduced linear autoencoder*.

We use a separate encoder ENC_l and decoder DEC_l for each layer l of the model $\mathcal{M}$. We couple them between layers using a common "feature vector" which is created by summing outputs of all encoders. This way the feature vector

Algorithm 1. DCO/DCO-COMP Algorithm

Require: η and η_a: learning rates of the model and autoencoders respectively. $\gamma_1, \gamma_2 \in (0, 1]$: pulling strengths that controls the searching scope of the model parameters. N: number of additional epochs used to train the model after saturation. C: number of points to average. θ: step size for pushing the optimizer inside the cone ($\theta \leq 1$ corresponds to parameter interpolation; $\theta > 1$ corresponds to parameter extrapolation). m: the size of the batch of gradients fed into autoencoders. τ: the period of updates of the model parameters in step 3. n: number of tasks. $\mathcal{T} = \{\mathcal{T}_1, \ldots, \mathcal{T}_n\}$: training data from task $1, 2, \ldots, n$. $|\mathcal{T}_i|$: number of iterations (mini-batches) required to process all data samples from task i.

Procedure:
for $i = 1$ **to** n **do**
  # step 1: train model until convergence
  $x_0 \leftarrow x$
  repeat
    $\xi \leftarrow$ randomly sample from $\mathcal{T}_i$
    $x \leftarrow x - \eta \nabla_x L_i(x; \xi) - \gamma_1 (x - x_0)$
  until convergence

  # step 2: push the model parameters into the cone
  $x_1 \leftarrow 0, x_2 \leftarrow 0$
  for $j = 1$ **to** $N \times |\mathcal{T}_i|$ **do**
    $\xi \leftarrow$ randomly sample from $\mathcal{T}_i$
    $x \leftarrow x - \eta \nabla_x L_{ce}(x; \xi)$
    if $j \leq C$ **then** $x_1 \leftarrow x_1 + x$
    else if $j > N \times |\mathcal{T}_i| - C$ **then** $x_2 \leftarrow x_2 + x$
  end for
  $x_i^* \leftarrow x_1 - \theta \frac{(x_1 - x_2)}{\|x_1 - x_2\|}$ {push into the cone}

  # step 3: train autoencoder until convergence
  repeat
    $g \leftarrow 0, G \leftarrow \{\}$
    for $j = 1$ **to** m **do**
      $\xi \leftarrow$ randomly sample from $\mathcal{T}_i$
      $g \leftarrow g + \nabla_x L_{ce}(x; \xi)$; $G \leftarrow G \cup \nabla_x L_{ce}(x; \xi)$
      if τ **divides** j **then** $x \leftarrow x - \eta g$; $g \leftarrow 0$
    end for
    $G \leftarrow \frac{G}{\sqrt{\|G\|_2^2 / m}}$ {Normalize batch of gradients}
    $W \leftarrow W - \frac{\eta_a}{m} \nabla_W L_{mse}(W; G)$; $x \leftarrow x - \gamma_2 (x - x_i^*)$
  until convergence

  # step 4: store autoencoder parameters
  $W^i \leftarrow W$
  if use DCO-COMP **then**
    compress and update $\{W^1, \cdots, W^i\}$ by Equation 15 and Equation 16.
  end if
end for

Output: x_n^*

will contain the information from all layers. The proposed autoencoder is then formulated as

$$AE(G) = \{DEC_1(ENC(G)), \ldots, DEC_L(ENC(G))\}, \tag{6}$$

where

$$ENC(G) = \sum_l ENC_l(G_l), \tag{7}$$

$$ENC_l(G_l) = diag(U_l^\top G_l V_l), \tag{8}$$

$$DEC_l(ENC(G)) = U_l ENC(G) V_l^\top, \tag{9}$$

$G = \{G_1, G_2, \ldots, G_L\}$ is a set of matrices such that each matrix contains gradients of the model for a given layer, and L is number of layers in the model. Finally, in order to enable processing the gradients of the convolutional layers we reshape them from their original size $o \times i \times w \times h$ to $o \times iwh$, where o is number of output channels, i is number of input channels, and w and h are width and height of the kernel of the convolutional layer. We train the autoencoder with standard mean square error loss

$$L_{mse}(W; G) = \|AE(G) - G\|_2^2, \tag{10}$$

where $W = \{U_1, V_1, \ldots, U_L, V_L\}$ is set of autoencoder's parameters. In the next section we propose a memory efficient variant DCO-COMP and show a compression scheme which allows us to avoid scaling the memory size as the number of tasks increases and results in a solution with a fixed memory size.

Remark: Using one autoencoder per task in a continual learning context has been explored in the literature before. For example, [30] uses an autoencoder to capture the features that are crucial for its corresponding task. Authors show experiments for only two tasks. Another method [2] embeds autoencoders into the classification network to identify the tasks and make predictions. How do we differ from these approaches? First, we utilize autoencoders to encode optimizer directions. Second, as opposed to [2] the autoencoders are not used within the classification network, thus they are not utilized at testing, but only at training. Third, as opposed to [30] we demonstrate experiments on multiple tasks. Finally, note that autoencoders have not been used before to support parameter-wise regularization-based continual learning frameworks.

4.3 Compression of Autoencoders

To avoid scaling the memory size as the number of tasks increases, we compress the autoencoders recursively so that only a *constant* memory of size $k \times (m+n)$ is required to store all autoencoders during training. More specifically, after training on the ith task, all autoencoders are compressed together such that each autoencoder keeps $\frac{1}{2i} \times k \times (m+n)$ parameters separately. On top of that,

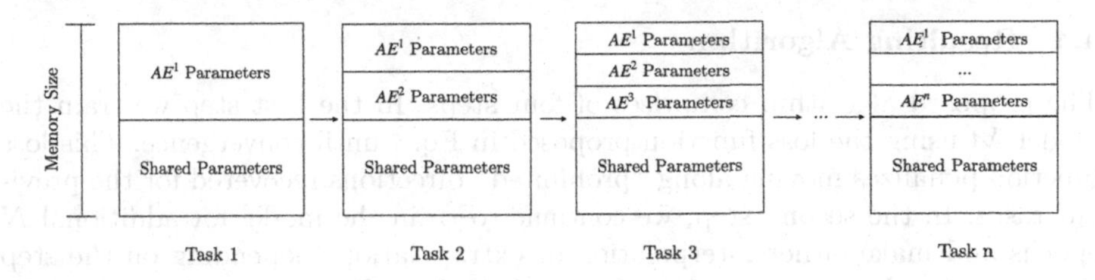

Fig. 3. The memory size of the autoencoders remains unchanged across the tasks.

by introducing shared parameters across autoencoders, whose size is fixed to $\frac{1}{2} \times k \times (m+n)$, we ensure that the information that would be lost due to compression is instead partially absorbed by the shared parameters. The memory allocation of autoencoders on each task is illustrated in Fig 3.

Denote the shared parameters and the jth compressed autoencoder's parameters as $\bar{W} = \{\bar{U}_1, \bar{V}_1, \ldots, \bar{U}_L, \bar{V}_L\}$ and $\tilde{W}^j = \{\tilde{U}_1^j, \tilde{V}_1^j, \ldots, \tilde{U}_L^j, \tilde{V}_L^j\}$, respectively. Similar as before, we use a separate encoder $\widehat{ENC}_l$ and decoder $\widehat{DEC}_l$ for each layer l. The formulation of the jth compressed autoencoder is given as

$$\widehat{AE}^j(W^j) = \{\widehat{DEC}_1^j(\widehat{ENC}^j(W^j)), \ldots, \widehat{DEC}_L^j(\widehat{ENC}^j(W^j))\}, \tag{11}$$

where

$$\widehat{ENC}^j(W_l^j) = \sum_l \widehat{ENC}_l^j(W_l^j), \tag{12}$$

$$\widehat{ENC}_l^j(W_l^j) = diag\left(\bar{U}_l^\top U_l^j (V_l^j)^\top \bar{V}_l + (\tilde{U}_l^j)^\top U_l^j (V_l^j)^\top \tilde{V}_l^j\right), \tag{13}$$

$$\widehat{DEC}_l^j(\widehat{ENC}^j(W^j)) = \tilde{U}_l^j \widehat{ENC}^j(W^j)(\tilde{V}_l^j)^\top \tag{14}$$

where $W^j = \{U_1^j, V_1^j, \ldots, U_L^j, V_L^j\}$ is set of jth autoencoder's uncompressed parameters. We train the compressed autoencoders with standard mean square error loss:

$$L_{mse}(\tilde{W}^1, \cdots, \tilde{W}^i, \bar{W}; W^1, \cdots, W^i) = \sum_{j=1}^{i} \left\| \widehat{AE}^j(W^j) - W^j \right\|_2^2 \tag{15}$$

Then we assign the jth autoencoder with a new set of parameters:

$$W^j = \{(\bar{U}_1, \tilde{U}_1^j), (\bar{V}_1, \tilde{V}_1^j), \cdots, (\bar{U}_L, \tilde{U}_L^j), (\bar{V}_L, \tilde{V}_L^j)\} \tag{16}$$

Remark : We are not able to re-access the model gradients from previous tasks anymore, but the learned prohibited directions for each task could be recovered from the corresponding autoencoder parameters. Thus we make compression directly on the autoencoder parameters.

4.4 Resulting Algorithm

The proposed algorithm comprises of four steps. In the first step we train the model $\mathcal{M}$ using the loss function proposed in Eq. 4 until convergence. This loss function penalizes moving along "prohibited" directions recovered for the previous tasks. In the second step, we continue to train the model for additional N epochs and make either interpolation or extrapolation (depending on the step size) between the averages of the first and the last C points on the optimizer's trajectory. This step is equivalent to pushing the model parameters deeper into the cone and aligns well with Sect. 3 (see conclusions resulting from Fig. 2). In the third step we train the autoencoder to recover "prohibited" directions for the current task. This again aligns well with Sect. 3 (see conclusions resulting from Fig. 1a and 1b). Finally, we store the autoencoder parameters with or without compression. The algorithm's pseudo code is captured in Algorithm 1.

5 Experiments

In this section we compare the performance of DCO and DCO-COMP with state-of-the-art regularization-based continual learning methods such as EWC [18], SI [42], RWALK [6] and GPM [36], as well as the vanilla SGD [5]. We use open source codes[3][4] for the experiments.

5.1 Data Sets and Architectures

In our experiments we consider commonly used continual learning data sets: (1) **Permuted MNIST**. For each task we used a different permutation of the pixels of images from the original MNIST data set [21]. We generated 5 data sets this way corresponding to 5 tasks. (2) **Split MNIST**. We divide original MNIST data set into 5 disjoint subsets corresponding to labels $\{\{0,1\},\{2,3\},\{4,5\},\{6,7\},\{8,9\}\}$. (3) **Split CIFAR-100**. We divide original CIFAR-100 data set [19] into 10 disjoint subsets corresponding to labels $\{\{0-9\},\cdots,\{90-99\}\}$. Additionally, we consider a cross-domain learning scenario, where tasks come from different domains. For a cross-domain learning experiment (**MNIST/Fashion MNIST**) we combine MNIST and Fashion MNIST data sets together.

For Permuted MNIST, Split MNIST, and MNIST/Fashion MNIST we use the original image of size $1\times28\times28$. We then normalize each image by mean (0.1307) and standard deviation (0.3081). For Split CIFAR-100, we use the original image of size $3\times32\times32$. We then normalize each image by mean $(0.5071, 0.4867, 0.4408)$ and standard deviation $(0.2675, 0.2565, 0.2761)$. Also, in the experiments with Split MNIST and Split CIFAR-100 we use a multi-head setup [6,42] and we provide task descriptors [7,25] to the model at both training and testing.

[3] https://github.com/facebookresearch/agem.
[4] https://github.com/sahagobinda/GPM.

Finally, for Permuted MNIST and MNIST/Fashion MNIST experiments we use a Multi-Layer Perceptron (MLP) with two hidden layers, each having 256 units with ReLU activation functions (we refer to this architecture as MLP-256). For Split MNIST, we use a MLP with two hidden layers each having 100 units with ReLU activation functions (we refer to this architecture as MLP-100). For Split CIFAR-100, we use the same convolutional neural network as in [6] (we refer to this architecture as ConvNet). In all architectures we turn off the biases.

5.2 Training Details

To train the model, we use SGD optimizer [5] with momentum of 0.9. The batch sizes are set to 128, 128, 128 and 64 respectively for MNIST/Fashion MNIST, Permuted MNIST, Split MNIST and Split CIFAR-100.

For MNIST/Fashion MNIST, we use a constant learning rate of 1×10^{-3}. For Permuted MNIST and Split MNIST, we use a constant learning rate of 1×10^{-3} and add weight decay penalty of 0.001. For Split CIFAR-100, we use a learning rate of 1×10^{-2} for the first task and then drop it by a factor of 0.1 for the remaining tasks.

For DCO on Split CIFAR-100, we also clip the l_2 norm of the gradients induced by regularization terms with a threshold of 1 to avoid exploding gradient problem.

5.3 Hyperparameters

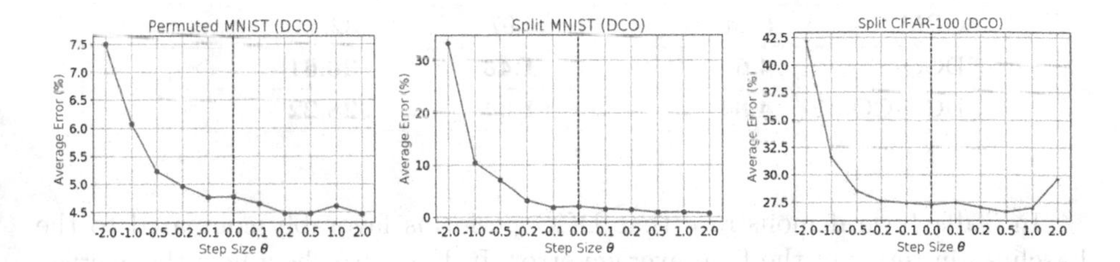

Fig. 4. Average error versus the step size θ (**left:** Permuted MNIST **middle:** Split MNIST **right:** Split CIFAR-100).

The values of the hyperparameters explored in the experiments are reported in the Supplement (Sect. 10.1). Here we illustrate how the final average error of DCO varies as the step size θ increases on Permuted MNIST, Split MNIST and Split CIFAR-100 (see Fig. 4). The figure further confirms our findings in Sect. 3 that when moving deeper inside the cone, the performance improves. After some point the performance however eventually starts dropping, as can been seen on the right plot. This is most likely because of falling outside the cone, due to either inaccurate estimation of the direction pointing towards the center of the cone or the fact that the cone is bounded.

5.4 Metric and Results

We denote $e_{i,j}$ as *test* classification error of the model on j^{th} task after getting trained on i tasks and evaluate the performance of each method based on the following metrics:

1. **Average error**, which represents the average performance on all tasks learned so far. The average error E_i^A on the ith task ($j \leq i$) is defined as $E_i^A = \frac{1}{i}\sum_{j=1}^{i} e_{i,j}$.
2. **Forward interference (FWI) error**, which shows how preserving the knowledge of the previous tasks impairs the model's learning ability for a new task. The final FWI error is defined as $E_n^{FWI} = \frac{1}{n}\sum_{j=1}^{n} e_{j,j}$.
3. **Backward transfer (BWT) error** [25], which directly reflects how much the model has forgotten the previously learned tasks at the end of training. The final BWT error is defined as $E_n^{BWT} = \frac{1}{n}\sum_{j=1}^{n} e_{n,j} - e_{j,j}$.

Table 1. Average Error E_n^A (%) for Permuted MNIST, Split MNIST, and Split CIFAR-100.

Method	Permuted MNIST	Split MNIST	Split CIFAR-100
SGD	17.41	9.68	33.13
EWC	7.29	2.7	30.23
SI	6.38	2.29	29.91
RWALK	6.7	5.67	29.08
GPM	5.84	4.97	32.93
DCO	**4.68**	**1.46**	**26.61**
DCO-COMP	**4.84**	**1.34**	**28.22**

In Table 1 we demonstrate that DCO performs favorably compared to the baselines in terms of the final average error. In Fig. 5 we show how the average error behaves when adding new tasks. The figure reveals that DCO consistently outperforms other methods. In most cases DCO-COMP performs similarly to DCO and across all experiments, just as DCO, it is superior to other techniques.

In Fig. 6 we report both FWI error and BWT error for each method. In most cases DCO(-COMP) obtains the lowest FWI and BWT error among the regularization-based methods. Since the average error is the sum of FWI error and BWT error, we can conclude that DCO(-COMP) shows strong forward-learning ability while being the most efficient in alleviating the effect of catastrophic forgetting among all considered techniques.

In Fig. 7 we report the memory-performance trade-off of DCO-COMP on permuted MNIST. As the number of prohibited directions k increases, the FWI error slightly increases but the BWT error drops dramatically. Consequently, the DCO-COMP with largest k shows the lowest final average error.

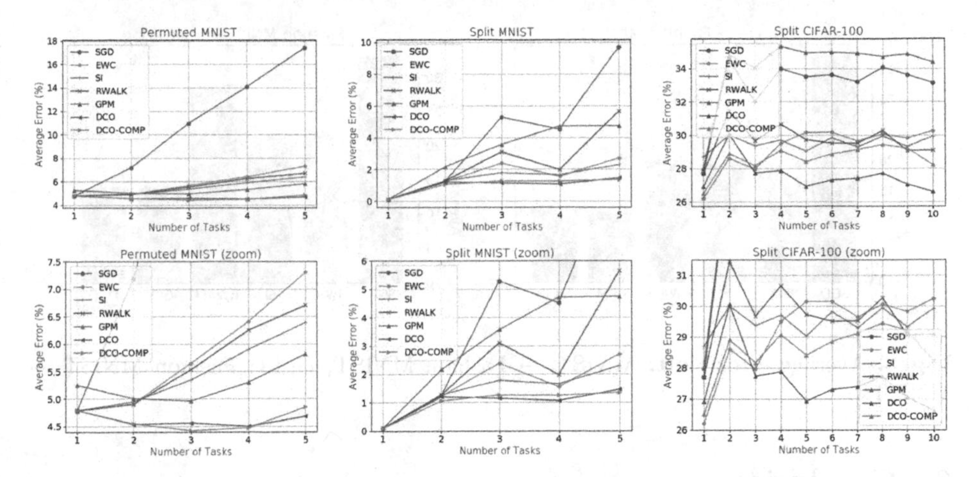

Fig. 5. Average error versus the number of tasks (original plots are on the top and zoomed are on the bottom; **left:** Permuted MNIST, **middle:** Split MNIST, **right:** Split CIFAR-100).

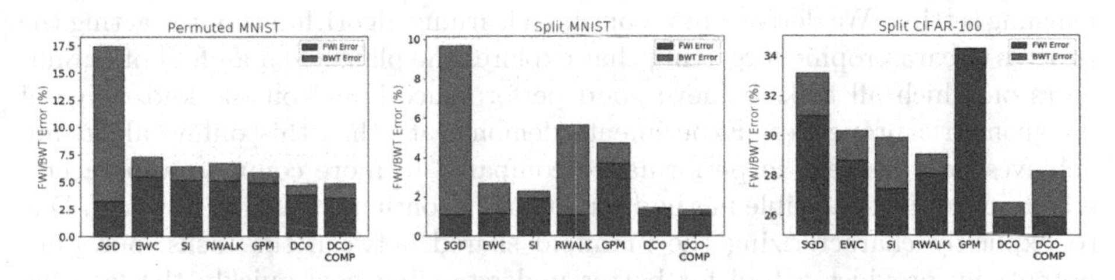

Fig. 6. FWI error and BWT error (**left:** Permuted MNIST, **middle:** Split MNIST, **right:** Split CIFAR-100).

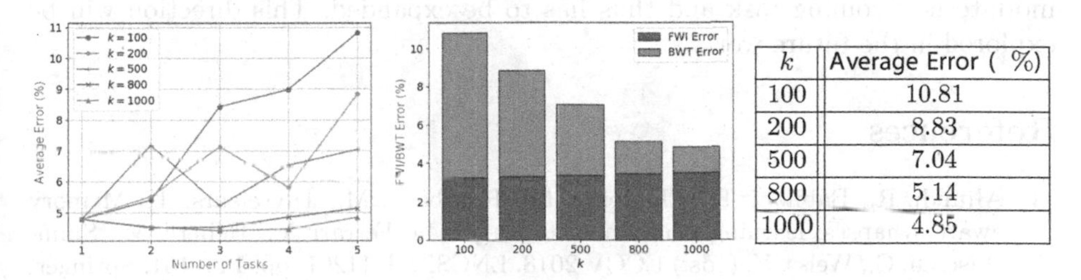

k	Average Error (%)
100	10.81
200	8.83
500	7.04
800	5.14
1000	4.85

Fig. 7. DCO-COMP on permuted MNIST (**left**: Average error versus the number of tasks; **middle**: FWI and BWT errors; **right**: Final average error versus number of prohibited directions k).

Finally, in Fig. 8 we report the results of cross-domain experiment on MNIST/Fashion MNIST. DCO(-COMP) performs favorably to GPM and outperforms all other continual learning methods.

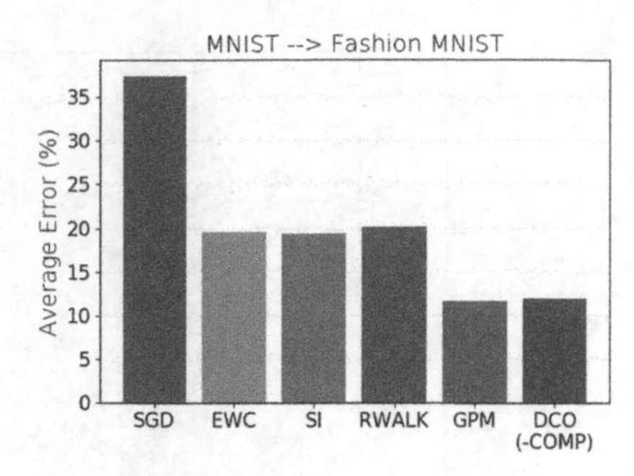

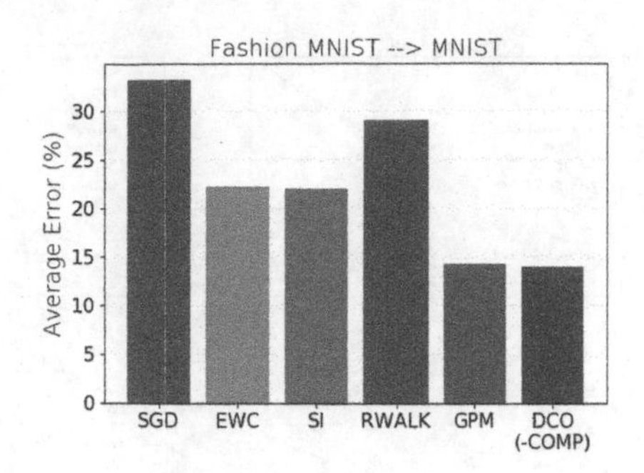

Fig. 8. Average error (**left:** MNIST → Fashion MNIST; **right:** Fashion MNIST → MNIST).

6 Conclusion

This paper elucidates the interplay between the local geometry of a deep learning optimization landscape and the quality of a network's performance in a continual learning setting. We derive a new continual learning algorithm counter-acting the process of catastrophic forgetting that explores the plausible manifold of parameters on which all tasks achieve good performance based on the knowledge of its geometric properties. Experiments demonstrate that this online algorithm achieves improvement in performance compared to more common approaches, which makes it a plausible method for solving a continual learning problem. Due to explicitly characterizing the manifold shared between the tasks, our work potentially provides a tool for better understanding how quickly the learning capacity of the network with a fixed architecture is consumed by adding new tasks and identifying the moment when the network lacks capacity to accommodate new coming task and thus has to be expanded. This direction will be explored in the future work.

References

1. Aljundi, R., Babiloni, F., Elhoseiny, M., Rohrbach, M., Tuytelaars, T.: Memory aware synapses: learning what (not) to forget. In: Ferrari, V., Hebert, M., Sminchisescu, C., Weiss, Y. (eds.) ECCV 2018. LNCS, vol. 11207, pp. 144–161. Springer, Cham (2018). https://doi.org/10.1007/978-3-030-01219-9_9
2. Aljundi, R., Chakravarty, P., Tuytelaars, T.: Expert gate: Lifelong learning with a network of experts. In: CVPR (2017)
3. Aljundi, R., Lin, M., Goujaud, B., Bengio, Y.: Gradient based sample selection for online continual learning. In: NeurIPS (2019)
4. Bao, X., Lucas, J., Sachdeva, S., Grosse, R.B.: Regularized linear autoencoders recover the principal components, eventually. In: Larochelle, H., Ranzato, M., Hadsell, R., Balcan, M., Lin, H. (eds.) Advances in Neural Information Processing Systems, vol. 33, pp. 6971–6981 (2020)
5. Bottou, L.: Online algorithms and stochastic approximations. In: Online Learning and Neural Networks. Cambridge University Press (1998)

6. Chaudhry, A., Dokania, P.K., Ajanthan, T., Torr, P.H.S.: Riemannian walk for incremental learning: understanding forgetting and intransigence. In: Ferrari, V., Hebert, M., Sminchisescu, C., Weiss, Y. (eds.) ECCV 2018. LNCS, vol. 11215, pp. 556–572. Springer, Cham (2018). https://doi.org/10.1007/978-3-030-01252-6_33
7. Chaudhry, A., Ranzato, M.A., Rohrbach, M., Elhoseiny, M.: Efficient lifelong learning with a-gem. In: ICLR (2019). https://openreview.net/forum?id=Hkf2_sC5FX
8. Chaudhry, A., Khan, N., Dokania, P.K., Torr, P.H.: Continual learning in low-rank orthogonal subspaces. arXiv preprint arXiv:2010.11635 (2020)
9. Farajtabar, M., Azizan, N., Mott, A., Li, A.: Orthogonal gradient descent for continual learning. CoRR abs/1910.07104 (2019)
10. Farquhar, S., Gal, Y.: Towards robust evaluations of continual learning. CoRR abs/1805.09733 (2018)
11. Feng, Y., Tu, Y.: How neural networks find generalizable solutions: self-tuned annealing in deep learning. CoRR abs/2001.01678 (2020)
12. Goodfellow, I.J., Mirza, M., Xiao, D., Courville, A., Bengio, Y.: An empirical investigation of catastrophic forgetting in gradient-based neural networks. In: ICLR (2014)
13. He, K., Zhang, X., Ren, S., Sun, J.: Deep residual learning for image recognition. In: CVPR (2016)
14. He, X., Sygnowski, J., Galashov, A., Rusu, A.A., Teh, Y.W., Pascanu, R.: Task agnostic continual learning via meta learning. CoRR abs/1906.05201 (2019)
15. Hou, S., Pan, X., Loy, C.C., Wang, Z., Lin, D.: Learning a unified classifier incrementally via rebalancing. In: Proceedings of the IEEE/CVF Conference on Computer Vision and Pattern Recognition (CVPR) (2019)
16. Hung, C.Y., Tu, C.H., Wu, C.E., Chen, C.H., Chan, Y.M., Chen, C.S.: Compacting, picking and growing for unforgetting continual learning. In: NeurIPS (2019)
17. Isele, D., Cosgun, A.: Selective experience replay for lifelong learning. In: AAAI (2018)
18. Kirkpatrick, J., et al.: Overcoming catastrophic forgetting in neural networks. PNAS **114**(13), 3521–3526 (2017)
19. Krizhevsky, A., Nair, V., Hinton, G.: CIFAR-10 and CIFAR-100 datasets. https://www.cs.toronto.edu/kriz/cifar.html (2009)
20. Krizhevsky, A., Sutskever, I., Hinton, G.E.: Imagenet classification with deep convolutional neural networks. In: NIPS (2012)
21. LeCun, Y., et al.: Backpropagation applied to handwritten zip code recognition. Neural Comput. **1**(4), 541–551 (1989)
22. Li, X., Zhou, Y., Wu, T., Socher, R., Xiong, C.: Learn to grow: a continual structure learning framework for overcoming catastrophic forgetting. In: ICML (2019). http://proceedings.mlr.press/v97/li19m.html
23. Li, Z., Hoiem, D.: Learning without forgetting. IEEE Trans. Pattern Anal. Mach. Intell. **40**(12), 2935–2947 (2017)
24. Liu, H., Liu, H.: Continual learning with recursive gradient optimization. In: International Conference on Learning Representations (2022). https://openreview.net/forum?id=7YDLgf9_zgm
25. Lopez-Paz, D., Ranzato, M.A.: Gradient episodic memory for continual learning. In: NeurIPS (2017)
26. Mallya, A., Davis, D., Lazebnik, S.: Piggyback: adapting a single network to multiple tasks by learning to mask weights. In: Ferrari, V., Hebert, M., Sminchisescu, C., Weiss, Y. (eds.) ECCV 2018. LNCS, vol. 11208, pp. 72–88. Springer, Cham (2018). https://doi.org/10.1007/978-3-030-01225-0_5

27. Mallya, A., Lazebnik, S.: PackNet: adding multiple tasks to a single network by iterative pruning. In: CVPR (2018)
28. McCloskey, M., Cohen, N.J.: Catastrophic interference in connectionist networks: the sequential learning problem. In: Psychology of learning and motivation, vol. 24, pp. 109–165. Elsevier (1989)
29. Parisi, G., Kemker, R., Part, J., Kanan, C., Wermter, S.: Continual lifelong learning with neural networks: a review. Neural Netw. (2018). https://doi.org/10.1016/j.neunet.2019.01.012
30. Rannen, A., Aljundi, R., Blaschko, M.B., Tuytelaars, T.: Encoder based lifelong learning. In: ICCV (2017)
31. Rao, D., Visin, F., Rusu, A., Pascanu, R., Teh, Y.W., Hadsell, R.: Continual unsupervised representation learning. In: NeurIPS (2019)
32. Riemer, M., et al.: Learning to learn without forgetting by maximizing transfer and minimizing interference. In: ICLR (2019). https://openreview.net/forum?id=B1gTShAct7
33. Ritter, H., Botev, A., Barber, D.: Online structured Laplace approximations for overcoming catastrophic forgetting. arXiv preprint arXiv:1805.07810 (2018)
34. Rostami, M., Kolouri, S., Pilly, P.K.: Complementary learning for overcoming catastrophic forgetting using experience replay. In: IJCAI (2019). https://doi.org/10.24963/ijcai.2019/463. https://doi.org/10.24963/ijcai.2019/463
35. Rusu, A.A., et al.: Progressive neural networks. CoRR abs/1606.04671 (2016)
36. Saha, G., Garg, I., Roy, K.: Gradient projection memory for continual learning. In: International Conference on Learning Representations (2021). https://openreview.net/forum?id=3AOj0RCNC2
37. Schwarz, J., et al.: Progress & compress: a scalable framework for continual learning. In: ICML (2018). http://proceedings.mlr.press/v80/schwarz18a.html
38. Shin, H., Lee, J.K., Kim, J., Kim, J.: Continual learning with deep generative replay. In: NeurIPS (2017)
39. Simonyan, K., Zisserman, A.: Very deep convolutional networks for large-scale image recognition. In: ICLR (2015)
40. Tao, X., Hong, X., Chang, X., Dong, S., Wei, X., Gong, Y.: Few-shot class-incremental learning. In: IEEE/CVF Conference on Computer Vision and Pattern Recognition (CVPR) (2020)
41. Yoon, J., Yang, E., Lee, J., Hwang, S.J.: Lifelong learning with dynamically expandable networks. In: ICLR (2018). https://openreview.net/forum?id=Sk7KsfW0-
42. Zenke, F., Poole, B., Ganguli, S.: Continual learning through synaptic intelligence. In: ICML (2017)
43. Zeno, C., Golan, I., Hoffer, E., Soudry, D.: Task agnostic continual learning using online variational Bayes. CoRR abs/1803.10123 (2018)

Newer is Not Always Better: Rethinking Transferability Metrics, Their Peculiarities, Stability and Performance

Shibal Ibrahim[1](✉), Natalia Ponomareva[2], and Rahul Mazumder[1]

[1] Massachusetts Institute of Technology, Cambridge, MA, USA
{shibal,rahulmaz}@mit.edu
[2] Google Research, New York, NY, USA

Abstract. Fine-tuning of large pre-trained image and language models on small customized datasets has become increasingly popular for improved prediction and efficient use of limited resources. Fine-tuning requires identification of best models to transfer-learn from and quantifying transferability prevents expensive re-training on *all* of the candidate models/tasks pairs. In this paper, we show that the statistical problems with covariance estimation drive the poor performance of H-score—a common baseline for newer metrics—and propose shrinkage-based estimator. This results in up to 80% absolute gain in H-score correlation performance, making it competitive with the state-of-the-art LogME measure. Our shrinkage-based H-score is 3–10 times faster to compute compared to LogME. Additionally, we look into a less common setting of target (as opposed to source) task selection. We demonstrate previously overlooked problems in such settings with different number of labels, class-imbalance ratios etc. for some recent metrics e.g., NCE, LEEP that resulted in them being misrepresented as leading measures. We propose a correction and recommend measuring correlation performance against relative accuracy in such settings. We support our findings with $\sim$ 164,000 (fine-tuning trials) experiments on both vision models and graph neural networks.

Keywords: Transferability metrics · Fine-tuning · Transfer learning · Domain adaptation · Neural networks

1 Introduction

Transfer learning is a set of techniques of using abundant somewhat related source data $p(X^{(s)}, Y^{(s)})$ to ensure that a model can generalize well to the target domain, defined as either little amount of labelled data $p(X^{(t)}, Y^{(t)})$ (supervised), and/or a lot of unlabelled data $p(X^{(t)})$ (unsupervised transfer learn-

S. Ibrahim—This work was completed as an Intern and Student Researcher at Google.

Supplementary Information The online version contains supplementary material available at https://doi.org/10.1007/978-3-031-26387-3_42.

M.-R. Amini et al. (Eds.): ECML PKDD 2022, LNAI 13713, pp. 693–709, 2023.
https://doi.org/10.1007/978-3-031-26387-3_42

ing). Transfer learning is most commonly achieved either via fine-tuning or co-training. Fine-tuning is a process of adapting a model trained on source data by using target data to do several optimization steps (for example, stochastic gradient descent) that update the model parameters. Co-training on source and target data usually involves reweighting the instances in some way or enforcing domain irrelevance on feature representation layer, such that the model trained on such combined data works well on target data. Fine-tuning is becoming increasing popular because large models like ResNet50 [11], BERT [6] etc. are released by companies and are easily modifiable. Training such large models from scratch is often prohibitively expensive for the end user.

In this paper, we are primarily interested in effectively measuring transferability before training of the final model begins. Given a source data/model, a **transferability measure** quantifies how much knowledge of source domain/model is transferable to the target model. Transferability measures are important for various reasons: they allow understanding of relationships between different learning tasks, selection of highly transferable tasks for joint training on source and target domains, selection of optimal pre-trained source models for the relevant target task, prevention of trial procedures attempting to transfer from each source domain and optimal policy learning in reinforcement learning scenarios (e.g. optimal selection of next task to learn by a robot). If a measure is capable of efficiently and accurately measuring transferability across arbitrary tasks, the problem of task transfer learning is greatly simplified by using the measure to search over candidate transfer sources and targets.

Contributions. Our contributions are three-fold:

1. We show that H-score, commonly used as a baseline for newer transferability measures, suffers from instability due to poor estimation of covariance matrices. We propose shrinkage-based estimation of H-score with regularized covariance estimation techniques from statistical literature. We show 80% absolute increase over the original H-score and show superior performance in majority cases against all newer transferability measures across various fine-tuning scenarios.
2. We present a fast implementation of our estimator that is $3-10$ times faster than state-of-the-art LogME measure.
3. We identify problems with 3 other transferability measures (NCE, LEEP and $\mathcal{N}$LEEP) in target task selection when either the number of target classes or the class imbalance varies across candidate target tasks. We propose measuring correlation against relative target accuracy (instead of vanilla accuracy) in such scenarios.

Our large set of $\sim 164,000$ fine-tuning experiments with vision models and graph convolutional networks on real-world datasets shows usefulness of our proposals.

This paper is organized as follows. Section 2 describes general fine-tuning regimes and transfer learning tasks. Section 3 discusses transferability measures. Section 4 addresses shortcomings of the pioneer transferability measure (H-Score) that arise due to unreliable estimation and proposes a new shrinkage-based estimator for the H-Score. In Sect. 5, we demonstrate problems with recent NCE,

LEEP and $\mathcal{N}$LEEP metrics and propose a way to address them. Finally, Sect. 6 presents a meta study of all metrics.

2 Transferability Setup

We consider the following fine-tuning scenarios based on existing literature.

(i) *Source Model Selection (SMS)*: For a particular target data/task, this regime aims to select the "optimal" source model (or data) to transfer-learn from, from a collection of candidate models/data.

(ii) *Target Task Selection (TTS)*: For a particular (source) model, this regime aims to find the most related target data/task.

In addition, we primarily consider two different fine-tuning strategies:

(i) *Linear fine-tuning/head only fine-tuning (LFT)*: All layers except for the penultimate layer are frozen. Only the weights of the head classifier are re-trained while fine-tuning.

(ii) *Non-linear fine-tuning (NLFT)*: Any layer can be designated as a feature extractor, up to which all the layers are frozen; the subsequent layers include nonlinear transformations and are re-trained along with the head on target data.

3 Related Work

Recent literature in transfer learning has proposed efficient transferability measures. Inspired by principles in information theory, Negative Conditional Entropy (NCE) [31] uses pre-trained source model and evaluates conditional entropy between target pseudo labels (source models' assigned labels) and real target labels. Log Expected Empirical Predictor (LEEP) [21] modifies NCE by using soft predictions from the source model. Both NCE and LEEP do not directly use feature information, hence they are not applicable for layer selection. The authors in [4] propose representing each output class by the mean of all images from that class and computing Earth Mover's distance between the centroids of the source classes and target classes.

Other works [1,5,12,18,33] proposed metrics that capture information from both the (learnt) feature representations and the real target labels. These metrics are more appealing as these can be broadly applicable for models that are pre-trained in either supervised or unsupervised fashion. They are also applicable for embedding layer selection. The authors in [18] proposed $\mathcal{N}$LEEP that fits a Gaussian mixture model on the target feature embeddings and computes the LEEP score between the probabalistic assignment of target features to different clusters and the target labels. The authors in [12] introduced TransRate—a computationally-friendly surrogate of mutual information (using coding rate) between the target feature embeddings and the target labels. H-score was proposed by [1] that takes into account inter-class feature variance and feature redundancy. [33] proposed LogME that considers an optimization problem rooted in Bayesian statistics to maximize the marginal likelihood under a linear classifier head. [5] introduced LFC to measure in-class similarity of target feature embeddings across samples.

Finally, the authors in [30] used Optimal Transport to evaluate domain distance, and combined it, via a linear combination, with NCE. To learn such a measure, a portion of target tasks were set aside, the models were transferred onto these tasks and the results were used to learn the coefficients for the combined Optimal Transport based Conditional Entropy (OTCE) metric. While the resulting metric appears to be superior over other non-composite metrics like H-score and LEEP, it is expensive to compute since it requires finding the appropriate coefficients for the combination.

4 Improved Estimation of H-score

H-score [1] is one of the pioneer measures that is often used as a baseline for newer transferability measures, which often demonstrate the improved performance. It characterizes discriminatory strength of feature embedding for classification:

$$\mathrm{H}(f) = \mathrm{tr}({\boldsymbol{\Sigma}^{f}}^{-1}\boldsymbol{\Sigma}^{z}) \tag{1}$$

where, d is the embedding dimension, $\boldsymbol{f}_i = h(\boldsymbol{x}_i^{(t)}) \in \mathbb{R}^d$ is the target feature embeddings when the feature extractor ($h : \mathbb{R}^p \to \mathbb{R}^d$) from the source model is applied to the target sample $\boldsymbol{x}_i^{(t)} \in \mathbb{R}^p$, $\boldsymbol{F} \in \mathbb{R}^{n_t \times d}$ denotes the corresponding target feature matrix, $Y = Y^{(t)} \in \mathcal{Y} = \{1, \cdots, C\}$ are the target data labels, $\boldsymbol{\Sigma}^f \in \mathbb{R}^{d\times d}$ denotes the sample feature covariance matrix of $\boldsymbol{f}$, $\boldsymbol{z} = \mathbb{E}[\boldsymbol{f}|Y] \in \mathbb{R}^d$ and $\boldsymbol{Z} \in \mathbb{R}^{n_t \times d}$ denotes the corresponding target-conditioned feature matrix, $\boldsymbol{\Sigma}^z \in \mathbb{R}^{d\times d}$ denotes the sample covariance matrix of $\boldsymbol{z}$. Intuitively, $\mathrm{H}(f)$ captures the notion that higher inter-class variance and small feature redundancy results in better transferability.

We hypothesize that the sub-optimal performance of H-Score (compared to that of more recent metrics) for measuring transferability in many of the evaluation cases, e.g., in [21], is due to lack of robust estimation of H-Score. We empirically validate this hypothesis using a synthetic classification data. We generated 1 million 1000-dimensional features with 10 classes using Sklearn multi-class dataset generation function [22]. Number of informative features is set to 500 with rest filled with random noise. We visualize the original and the population version of the H-score for different sample sizes in Fig. 1. We

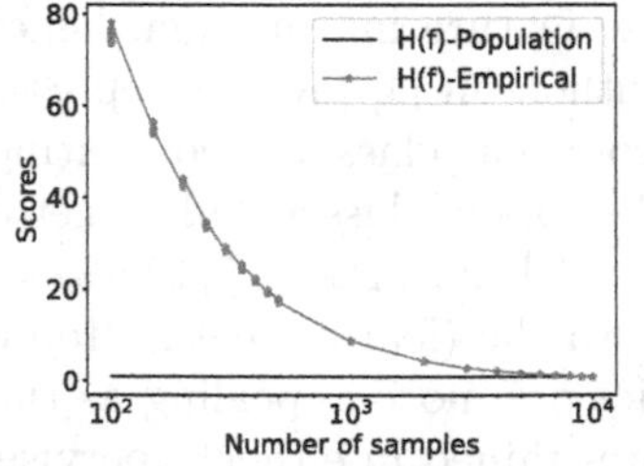

Fig. 1. Non-reliability of $\mathrm{H}(f)$. $\mathrm{H}(f)$ is $\sim 75\times$ larger than the population version of the H-Score. Population version is estimated with 10^6 samples.

observe that the original H-Score becomes highly unreliable as the number of samples decreases.

Many of the deep learning models in the context of transfer learning have high-dimensional feature embedding space—typically larger than the number of target samples. Consequently, the estimation of the two covariance matrices in H-score becomes challenging: the sample covariance matrix of the feature embedding has a large condition number[1] in small data regimes. In many cases, it cannot even be inverted. [1] used a pseudo-inverse of the covariance matrix $\boldsymbol{\Sigma}^f$. However, this method of estimating a precision matrix can be sub-optimal as inversion can amplify estimation error [16]. We propose to use well-conditioned shrinkage estimators motivated by the rich literature in statistics on the estimation of high-dimensional covariance (and precision) matrices [23]. We show that the use of such shrinkage estimators can offer significant gain in the performance of H-score in predicting transferability. In many cases, as our experiments show, the gain is so significant that H-score becomes a leading transferability measure, surpassing the performance of state-of-the-art measures.

4.1 Proposed Transferability Measure

We propose the following shrinkage based H-score:

$$\mathrm{H}_\alpha(f) = \mathrm{tr}\big(\boldsymbol{\Sigma}_\alpha^{f^{-1}} \cdot (1-\alpha)\boldsymbol{\Sigma}^z\big), \tag{2}$$

Estimating $\boldsymbol{\Sigma}_\alpha^{(f)}$. While there are several possibilities to obtain a regularized covariance matrix [23], we present an approach that considers a linear operation on the eigenvalues of the sample version of the feature embedding covariance matrix. Similar ideas of using well-conditioned plug-in covariance matrices are used in the context of discriminant analysis [10]. In particular, we improve the conditioning of the covariance matrix by considering its weighted convex combination with a scalar multiple of the identity matrix:

$$\boldsymbol{\Sigma}_\alpha^f = (1-\alpha)\boldsymbol{\Sigma}^f + \alpha\sigma \boldsymbol{I}_d \tag{3}$$

where $\alpha \in [0,1]$ is the shrinkage parameter and σ is the average variance computed as $\mathrm{tr}(\boldsymbol{\Sigma}^f)/d$. The linear operation on the eigenvalues ensures the covariance estimator is positive definite. Note that the inverse of $\boldsymbol{\Sigma}_\alpha^f$ can be computed for every α, by using the eigen-decomposition of $\boldsymbol{\Sigma}^f$. The shrinkage parameter controls the bias and variance trade-off; the optimal α needs to be selected. This distribution-free estimator is well-suited for our application as the explicit convex linear combination is easy to compute and makes the covariance estimates well-conditioned and more accurate [2,16,28].

Understanding $(\mathbf{1}-\boldsymbol{\alpha})\boldsymbol{\Sigma}^z$ The scaling factor $(1-\alpha)$ can be understood in terms of regularized covariance matrix estimation under a ridge penalty:

$$1/(1+\lambda)\cdot\boldsymbol{\Sigma}^z = \mathrm{argmin}_{\hat{\boldsymbol{\Sigma}}}\ ||\hat{\boldsymbol{\Sigma}} - \boldsymbol{\Sigma}^z||_2^2 + \lambda||\hat{\boldsymbol{\Sigma}}||_2^2 \tag{4}$$

[1] Condition number of a positive semidefinite matrix A, is the ratio of its largest and smallest eigenvalues.

where $\lambda \geq 0$ is the ridge penalty. Choosing $\lambda = \alpha/(1-\alpha)$, it becomes clear that $(1-\alpha)\boldsymbol{\Sigma}^z$ is the regularized covariance matrix.

Choice of $\boldsymbol{\alpha}$ [16] proposed a covariance matrix estimator that minimizes mean squared error loss between the shrinkage based covariance estimator and the true covariance matrix. The optimization considers the following objective:

$$\min_{\alpha,v} \ \mathbb{E}[||\boldsymbol{\Sigma}^* - \boldsymbol{\Sigma}||^2] \quad \text{s.t.} \ \ \boldsymbol{\Sigma}^* = (1-\alpha)\boldsymbol{\Sigma}^f + \alpha v \boldsymbol{I}, \quad \mathbb{E}[\boldsymbol{\Sigma}^f] = \boldsymbol{\Sigma}. \tag{5}$$

where $||\boldsymbol{A}||^2 \stackrel{\text{def}}{=} \text{tr}(\boldsymbol{A}\boldsymbol{A}^T)/d$ for a matrix $\boldsymbol{A} \in \mathbb{R}^{d\times d}$. This optimization problem permits a closed-form solution for the optimal shrinkage parameter, which is given by:

$$\alpha^* = \mathbb{E}[||\boldsymbol{\Sigma}^f - \boldsymbol{\Sigma}||^2]/\mathbb{E}[||\boldsymbol{\Sigma}^f - (\text{tr}(\boldsymbol{\Sigma})/d)\cdot \boldsymbol{I}_d||^2] \tag{6}$$

$$\simeq \min\{(1/n_t^2) \sum_{i\in[n_t]} \frac{||\boldsymbol{f}_i \boldsymbol{f}_i^T - \boldsymbol{\Sigma}^f||^2}{||\boldsymbol{\Sigma}^f - (\text{tr}(\boldsymbol{\Sigma}^f)/d)\cdot \boldsymbol{I}_d||^2}, 1\}. \tag{7}$$

where Eq. (7) defines a valid estimator (not dependent on true covariance matrix) for practical use. For proof, we refer the readers to Sects. 2.1 and 3.3 in [16].

Following the synthetic motivational example presented earlier showing the unreliability of the original H-Score, we investigate the reliability of shrinkage-based H-Score. We visualize the shrinkage-based H-Score in Fig. 2[Left]. We observe that the original H-Score becomes highly unreliable as the number of samples decreases. In contrast, the shrunken estimation of H-Score is highly stable and has a small error when compared with the population H-Score. Hence, shrinkage-based H-score seems to be a much better estimator of the "true" H-score in contrast to the empirical H-Score. We further visualize the effect of using non-optimal values of α on the shrinkage-based H-Score in Fig. 2[Right]. We can see that the shrinkage-based H-Score with optimal shrinkage α^* is much closer to the population version of the original H-Score, especially for smaller sample cases. This validates the use of α^* as computed in Eq. (7).

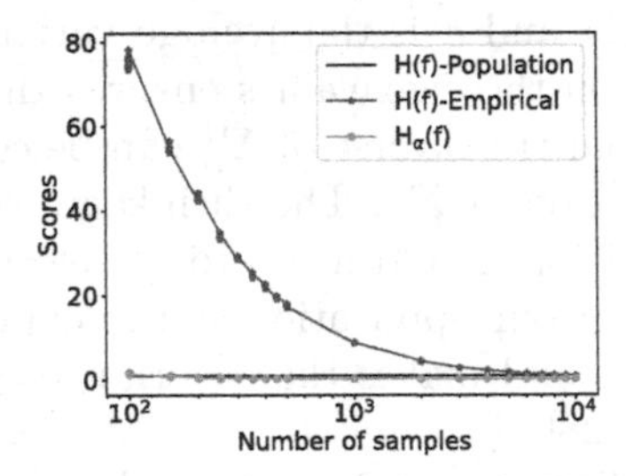

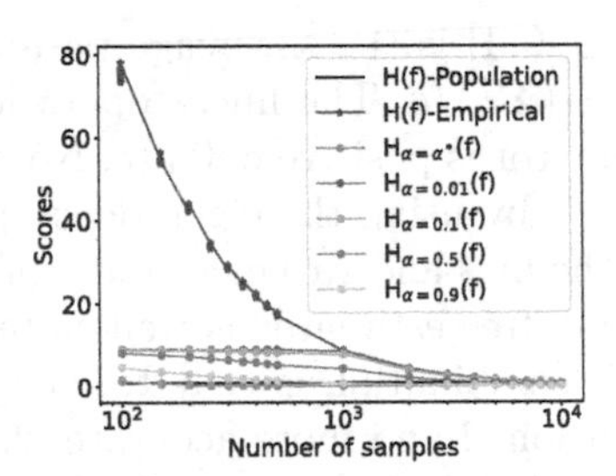

Fig. 2. [Left] Stability of $\text{H}(f)$ and our shrinkage-based $\text{H}_\alpha(f)$ with respect to number of samples. $\text{H}(f)$ is $\sim$ 75 times larger than the population version of the H-Score (estimated with a sample size of 10^6). In contrast, the shrinkage-based H-Score is significantly more reliable. [Right] Effect of α on shrunk H-Score.

Additional Discussion on same shrinkage α for the two covariances in shrinkage-based H-Score. The covariance $\boldsymbol{\Sigma}^z$ can not be shrunk independently of $\boldsymbol{\Sigma}^f$ in the estimation of $\mathrm{H}_\alpha(f)$— the two covariances are coupled by the law of total covariance:

$$\boldsymbol{\Sigma}^f = \mathbb{E}[\boldsymbol{\Sigma}^{f_Y}] + \boldsymbol{\Sigma}^z. \tag{8}$$

where $\boldsymbol{f_Y}$ denotes the feature embedding of target samples belonging to class $Y \in \mathcal{Y}$ and $\boldsymbol{\Sigma}^{f_Y} = \mathrm{Cov}(\boldsymbol{f}|Y)$ denotes the class-conditioned covariances. We write

$$(1-\alpha)\boldsymbol{\Sigma}^f = (1-\alpha)\mathbb{E}[\boldsymbol{\Sigma}^{f_Y}] + (1-\alpha)\boldsymbol{\Sigma}^z,$$

$$(1-\alpha)\boldsymbol{\Sigma}^f + \alpha\frac{\mathrm{tr}(\boldsymbol{\Sigma}^f)}{d}\boldsymbol{I}_d = (1-\alpha)\mathbb{E}[\boldsymbol{\Sigma}^{f_Y}] + \alpha\frac{\mathrm{tr}(\boldsymbol{\Sigma}^f)}{d}\boldsymbol{I}_d + (1-\alpha)\boldsymbol{\Sigma}^z, \tag{9}$$

$$\text{i.e., } \boldsymbol{\Sigma}^f_\alpha = (1-\alpha)\mathbb{E}[\boldsymbol{\Sigma}^{f_Y}] + \alpha\frac{\mathrm{tr}(\boldsymbol{\Sigma}^f)}{d}\boldsymbol{I}_d + (1-\alpha)\boldsymbol{\Sigma}^z. \tag{10}$$

Comparing Eq. 10 with Eq. 8, we see that the same shrinkage parameter α should be used when using shrinkage estimators, to preserve law of total covariance. The first two terms on the right side in Eq. (10) can be understood as shrinkage of class-conditioned covariances to the average (global) variance. The third term in Eq. (10) (e.g. $(1-\alpha)\boldsymbol{\Sigma}^z$) can then be understood as ridge shrinkage as in Eq. (4).

4.2 Challenges of Comparing $\mathrm{H}_\alpha(f)$ Across Source Models/Layers

Next, we discuss challenges of using $\mathrm{H}_\alpha(f)$ on the feature embeddings of target data derived from the source model for source model selection. The feature dimension (d) across different source models even for the penultimate layer can vary significantly e.g. from 1024 in MobileNet to 4096 in VGG19. Such differences makes source model/layer selection for fine-tuning highly problematic.

We propose dimensionality reduction of feature embeddings before computing $\mathrm{H}_\alpha(f)$. We project feature embeddings to a lower q-dimensional space, where q is taken to be the same across the K candidate models/layers and satisfies: $q \leq \min_{\boldsymbol{f}^{(1)},\boldsymbol{f}^{(2)},\ldots,\boldsymbol{f}^{(K)}} |\boldsymbol{f}^{(\cdot)}|$ where $|.|$ operator denotes the cardinality of the feature spaces. The dimensionality reduction allows for more meaningful comparison of $\mathrm{H}_\alpha(f)$ across source/target pairs; this is relevant for source/layer selection. More generally, it also allows for faster and more robust estimate for limited target samples case ($n_t < d$) for linear and nonlinear fine-tuning. In the case of nonlinear fine-tuning, the intermediate layers of visual and language models have really large $d \sim 10^5$, see Table S1 in Supplement for examples.

We consider Gaussian Random Projection, which uses a linear transformation matrix $\boldsymbol{V}$ to derive the transformed features $\hat{\boldsymbol{F}} = \boldsymbol{F}\boldsymbol{V}$; it samples components from $\mathcal{N}(0, \frac{1}{q})$ to preserve pairwise distances between any two samples of the dataset. Untrained auto-encoders (AE) are other alternatives that have been used to detect covariate shifts in input distributions by [24]. It is not known how sensitive these untrained AE are to the underlying architecture—using trained AE is less appropriate for use in transferability measurement for fine-tuning as those maybe more time-consuming than the actual fine-tuning. We demonstrate improved correlation performance of $\mathrm{H}_\alpha(f)$ with dimensionality reduction in Table 3 in Sect. 6.1 for source model selection.

4.3 Efficient Computation for Small Target Data

For small target data ($C \leq n_t < d$), the naive implementation of $\mathrm{H}_\alpha(f)$ can be very slow. We propose an optimized implementation for our shrinkage-based H-Score that exploits diagonal plus low-rank structure of $\boldsymbol{\Sigma}_\alpha^{(f)}$ for efficient matrix inversion and the low-rank structure of $\boldsymbol{\Sigma}^{(z)}$ for faster matrix-matrix multiplications. We assume $\boldsymbol{F}$ (and correspondingly $\boldsymbol{Z}$) are centered. The optimized computation of $\mathrm{H}_\alpha(f)$ is given by:

$$\mathrm{H}_\alpha(f) = (1-\alpha)/(n_t\alpha\sigma) \cdot \left(\|\boldsymbol{R}\|_F^2 - (1-\alpha) \cdot \mathrm{vec}\left(\boldsymbol{G}\right)^T \mathrm{vec}\left(\boldsymbol{W}^{-1}\boldsymbol{G}\right)\right), \quad (11)$$

where $\boldsymbol{R} = \left[\sqrt{n_1}\bar{\boldsymbol{f}}_{Y=1}, \cdots, \sqrt{n_C}\bar{\boldsymbol{f}}_{Y=C}\right] \in \mathbb{R}^{d\times C}$, $\boldsymbol{G} = \boldsymbol{F}\boldsymbol{R} \in \mathbb{R}^{n_t \times C}$, $\boldsymbol{W} = n_t\alpha\sigma\boldsymbol{I}_n + (1-\alpha)\boldsymbol{F}\boldsymbol{F}^T \in \mathbb{R}^{n_t\times n_t}$. The algorithm (and derivation) is provided in the Supplementary document. We make a timing comparison of our optimized implementation of $\mathrm{H}_\alpha(f)$ against the computational times of the state-of-the-art LogME measure in Sect. 6.3.

5 A Closer Look at NCE, LEEP and $\mathcal{N}$LEEP Measures

Next, we pursue a deeper investigation of some of the newer metrics that are reported to be superior to H-Score and bring to light what appears to be some overlooked issues with these metrics in target task selection scenario. Target task selection has received less attention than source model selection. To our knowledge, we are the first to bring to light some problematic aspects with NCE, LEEP and $\mathcal{N}$LEEP, which can potentially lead to the misuse of these metrics in measuring transferability.

These measures are sensitive to the number of target classes (C) and tend to be smaller when C is larger (see Fig. 3[Left]). Therefore, use of these measures

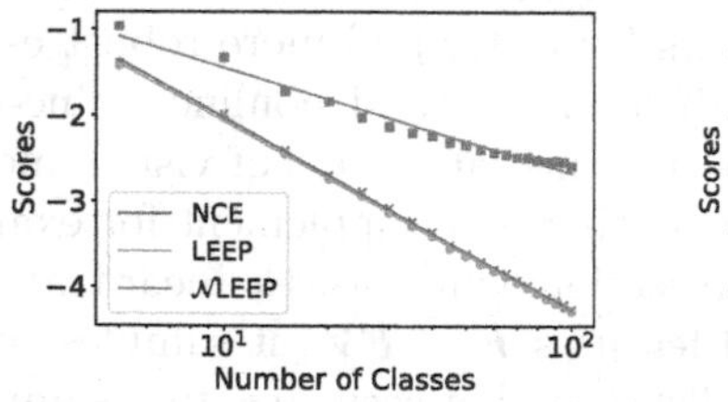

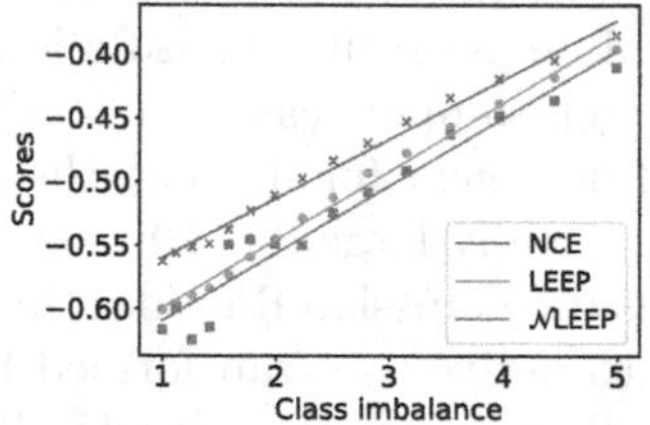

Fig. 3. Relation of NCE, LEEP & $\mathcal{N}$LEEP to [Left] number of classes (log-scale) and [Right] class imbalance, $max(n_1, n_2)/min(n_1, n_2)$, for VGG19 on CIFAR100. For [Left], we randomly select 2–100 classes. For [Right], we randomly select 2 classes and vary the class imbalances.

for target tasks with *different* C will most likely result in selecting the task with a smaller C. However, in practice, it is not always the case that transferring to a task with a smaller C is easier; for example, reframing a multiclass classification into a set of binary tasks can create more difficult to learn boundaries [8]. Furthermore, the measures are also problematic if two candidate target tasks have different degree of imbalance in their classes even if C is the same. The measures would predict higher transferability for imbalanced data regimes over balanced settings (see Fig. 3[Right]). However, imbalanced datasets are typically harder to learn. If these measures are correlated against vanilla accuracy, which tends to be higher as the imbalance increases e.g. for binary classification, the measures would falsely suggest they are good indicators of performance. Earlier work did not consider both these aspects and erroneously showed good correlation of these metrics against vanilla accuracy to show dominance of these metrics in target task selection with different C [21,30] and imbalance [30].

Here, we propose a method to ameliorate the shortcomings of NCE, LEEP and $\mathcal{N}$LEEP to prevent misuse of these measures, so that they lead to more reliable conclusions. We propose to standardize the metrics by the entropy of the target label priors, leading to the definitions in Eq. (12). This standardization considers both the class imbalance as well as number of classes through the entropy of the target label priors.

$$\begin{aligned} \text{n-NCE} &\stackrel{\text{def}}{=} 1 + \text{NCE}/\text{H}(Y), \\ \text{n-LEEP} &\stackrel{\text{def}}{=} 1 + \text{LEEP}/\text{H}(Y), \\ \text{n-}\mathcal{N}\text{LEEP} &\stackrel{\text{def}}{=} 1 + \mathcal{N}\text{LEEP}/\text{H}(Y). \end{aligned} \tag{12}$$

Our proposed normalizations in Eq. (12) ensures the normalized NCE is bounded between $[0, 1]$. For proof, see Supplementary document. n-NCE is in fact equivalent to normalized mutual information and has been extensively used to measure correlation between two different labelings/clustering of samples [32]. Given the similar behavior of LEEP and NCE to different C and class imbalance as shown in Fig. 3, we suggest the same normalization as given in Eq. (12). However, this normalization does not ensure boundedness of n-LEEP score (and by extension n-$\mathcal{N}$LEEP) in the range $[0, 1]$ as in the case of n-NCE.

For scenarios where candidate target tasks have different C, we propose an alternative evaluation criteria (*relative* accuracy) instead of vanilla accuracy—see Sect. 6 for more details. We provide empirical validation of the proposed normalization to these measures in Table 2 in Sect. 6.1. We also show that our proposed shrinkage-based H-Score is the leading metric even in these scenarios.

6 Experiments

We evaluate existing transferability measures and our proposed modifications on two class of models: vision models and graph neural networks. We study various fine-tuning regimes and data settings. We draw inspiration from [21] who consider target task selection and source model selection. The experimental setup highlights important aspects of transferability measures, e.g., dataset size for computing empirical distributions and covariance matrices, number of target classes, and feature dimension etc. Some of these aspects have been overlooked when evaluating transferability measures, leading to improper conclusions.

Evaluation Criteria. Transferability measures are often evaluated by how well they correlate with the test accuracy after fine-tuning the model on target data. Following [12,21,31], we used Pearson correlation. We include additional results with respect to rank correlations (e.g., Spearman) in Supplementary document. We argue that considering correlation with the target test accuracy is flawed in some scenarios. In particular, for target task selection, it is wrong to compare target tasks based on accuracy when C is different e.g. 5 vs 10 classes. In such a case, task with 10 classes will have a high chance of arriving at lesser test accuracy compared to that for task with 5 classes. In this case, it is more appropriate to consider the gain in accuracy achieved by the model over it's random baseline. Hence we use relative accuracy (for balanced classes): $\frac{\text{Accuracy}-1/C}{1/C}$. This measure is more effective in capturing the performance gain achieved by the same model in transferring to two domains with different C. This also highlights the limitation of NCE, LEEP and $\mathcal{N}$LEEP which are sensitive to C and tend to have smaller values with higher C; these measures do not provide useful information about how hard these different tasks when evaluated with vanilla accuracy.

Correlations marked with asterisks (*) in Tables 1, 2, 3, 4 are not statistically significant (p-value > 0.05). Larger correlations indicate better identifiability as quantified by transferability measure. Hyphen (-) indicates the computation ran out of memory or was really slow.

6.1 Case Study: Vision Models

First, we evaluate our proposals on visual classification tasks with vision models e.g., VGG19 [29], ResNet50 [11] that have been pre-trained on ImageNet. We fine-tune on subsets of CIFAR-100/CIFAR-10 data. We use Tensorflow Keras [3] for our implementation. Imagenet checkpoints come from Keras[2].

Fine-tuning with Hyperparameter Optimization. The optimal choice of hyperparameters for fine-tuning is not only target data dependent but also sensitive to the domain similarity between the source and target datasets [17]. We thus tune the hyperparameters for fine-tuning: we use Adam optimizer and tune batch size, learning rate, number of epochs and weight decay (L2 regularization on the classifier head). For validation tuning, we set aside a portion of the training data (20%) and try 100 random draws from hyperparameters' multi-dimensional grid. With this additional tuning complexity, we performed 650×100 fine-tuning experiments. See additional information and motivation in Supplement.

[2] https://keras.io/api/applications/.

Table 1. Pearson correlation of transferability measures against fine-tuned target accuracy of vision models in the context of target task selection. We compare our proposed $H_\alpha(f)$ against original $H(f)$ and state-of-the-art measures.

Strategy	Target	Model	Reg.	$H(f)$	$H_\alpha(f)$	NCE	LEEP	$\mathcal{N}$LEEP	TransRate	LFC	LogME
LFT	CIFAR-100	VGG19	S-B	–0.14*	0.81	0.67	0.65	0.81	0.56	0.76	**0.85**
			S-IB	0.03*	**0.77**	0.57	0.63	0.70	0.46	0.47	0.75
		ResNet50	S-B	0.03*	**0.87**	0.66	0.68	0.81	0.27	0.77	0.83
			S-IB	–0.10	0.79	0.56	0.57	0.70	0.44	0.52	**0.82**
	CIFAR-10	VGG19	S-B	0.00*	0.67	0.52	0.60	0.61	0.42	0.44	**0.74**
			S-IB	0.09*	0.81	0.75	0.82	0.83	0.29	0.32	**0.89**
		ResNet50	S-B	–0.29	**0.73**	0.43	0.44	0.61	–0.02*	0.57	0.71
			S-IB	0.17*	**0.89**	0.66	0.71	0.75	0.28	0.01*	0.83
NLFT	CIFAR-100	VGG19	S-B	0.17*	**0.73**	0.58	0.59	0.67	–0.03*	0.70	–
			S-IB	0.03*	0.49	0.49	0.54	**0.55**	0.48	0.17	–

Target Task Selection. We first investigate the correlation performance of our proposed estimator and existing transferability measures in the context of target task selection. We consider two small target data cases. We provide a summary of the different cases (inspired from [21]) below. Additional details are in the Supplementary document.

- *Small-Balanced Target Data (S-B):* We make a random selection of 5 classes from CIFAR-100/CIFAR-10 and sample 50 samples per class from the original train split. We repeat this exercise 50 times (with a different selection of 5 classes). We evaluate correlations of transferability measures across the 50 random experiments.
- *Small-Imbalanced Target Data (S-IB):* We make 50 random selections of 2 classes from CIFAR-100/CIFAR-10, sample between 30–60 samples from the first class and sample 5× the number of samples from the second class. This makes for a binary imbalanced classification task. We again measure performance of transferability measures against optimal target test accuracy. Note the imbalance is constant across the candidate target tasks.

Table 2. Pearson correlation of transferability measures against *relative* accuracy for large balanced CIFAR-100 dataset with different number of classes across target tasks.

Model	$H(f)$	$H_\alpha(f)$	NCE	n-NCE	LEEP	n-LEEP	$\mathcal{N}$LEEP	n-$\mathcal{N}$LEEP	TransRate	LogME
VGG19	0.88	**0.97**	–0.95	0.66	–0.95	0.66	–0.93	0.95	0.68	0.96
ResNet50	0.95	**0.98**	–0.95	–0.74	–0.95	–0.73	–0.94	–0.63	0.56	0.96

We evaluate target task selection for linear and nonlinear fine-tuning under small sample setting. The layers designated as embedding layers for nonlinear fine-tuning of VGG19 and ResNet50 is given in Table S1 in Supplementary document. We empirically compare the shrinkage-based H-score against the original measure by [1] and the state-of-the-art measures. Table 1 demonstrates 80% absolute gains in correlation performance of $H_\alpha(f)$ over $H(f)$, making it a leading metric in many cases in small target data regimes.

Next, we study a large-balanced target data setting where the number of classes varies across the target tasks. We construct the target tasks as follows: We randomly select 2–100 classes from CIFAR-100 and include all samples from the chosen classes. This constructs a collection of large balanced target tasks with different number of classes (L-B-C). We generate 50 such target datasets. In this setting, we evaluate correlation of transferability measures with *relative* target test accuracy for reasons highlighted in Sect. 5. This setting validates that the normalizations, proposed in Eq. (12) in Sect. 5, can improve the correlations of NCE, LEEP and $\mathcal{N}$LEEP when evaluated on the more appropriate relative accuracy scale. Table 2 demonstrates how various transferability measures perform on target task selection when the number of target classes **varies.** $H_\alpha(f)$ dominates the performance for both VGG19 and ResNet50 models, surpassing all transferability measures, closely followed by LogME.

Source Model Selection. Next, we study source model selection scenario for vision models. We select 9 small to large (pre-trained ImageNet) vision models: VGG19, ResNet50, ResNet101, DenseNet121, DenseNet201, Xception, InceptionV3, MobileNet, EfficientNetB0. We evaluate source model selection for linear and nonlinear fine-tuning under small sample setting. The layers designated as embedding layers for nonlinear fine-tuning of all 9 models is shown in Table S1 in Supplementary document. We sample 50 images per class from all classes available in the original train split of CIFAR-100/CIFAR-10. We designate 10 samples per class for hyperparameter tuning.

We demonstrate that $H_\alpha(f)$ is a leading metric in source model selection as well. Given that the feature dimensions vary significantly across different models in source model selection for both linear and nonlinear finetuning, we apply proposed dimensionality reduction via random projection in Sect. 4.2 for $H_\alpha(f)$ to project feature embeddings to 128-dimensional space ($q = 128$). This allows for more meaningful comparison of H-score across source models. This leads to the gains of proposed $H_\alpha(f)$ in terms of correlation in the context of source model selection as well for small samples as given in Table 3, making it again a leading metric in source model selection.

6.2 Case Study: Graph Neural Networks

We evaluate our proposals on Graph Neural Networks on Twitch Social Networks [25–27]. The datasets are social networks of gamers from the streaming service Twitch, where nodes correspond to Twitch users and links correspond to mutual friendships. There are country-specific sub-networks. We consider six sub-networks: {DE, ES, FR, RU, PTBR, ENGB}. Features describe the history

Table 3. Pearson correlation of proposed $\mathrm{H}_\alpha(f)$ without/with Random Projection (RP) for fine-tuning in source model selection of vision models in small data regimes.

Strategy	Target	H(f)	$\mathrm{H}_\alpha(f)$[No RP]	$\mathrm{H}_\alpha(f)$[RP]	NCE	LEEP	$\mathcal{N}$LEEP	TransRate	LogME
LFT	CIFAR-100	–0.190*	0.024*	**0.859**	0.825	0.839	0.852	–0.204*	0.705
	CIFAR-10	0.276*	0.277*	**0.939**	0.938	0.936	0.938	0.311*	0.923
NLFT	CIFAR-100	–0.108*	0.125*	0.879	0.967	0.976	**0.977**	–	–

of games played and the associated task is binary classification of whether a gamer streams adult content. The country specific graphs share the same node features which means that we can perform transfer learning with these datasets. Additional details about the Twitch Social Networks datasets are included in Supplementary document.

Transferability Setup. We consider a two-layered Graph Convolutional Network (GCN) [13]. The network takes the following functional form:

$$\text{logit}(\boldsymbol{x}) = \hat{\boldsymbol{G}} \cdot \text{Dropout}(\text{ReLU}(\hat{\boldsymbol{G}} \cdot \boldsymbol{x} \cdot \boldsymbol{W}^1)) \cdot \boldsymbol{W}^2, \tag{13}$$

where $\hat{\boldsymbol{G}} = \hat{\boldsymbol{D}}^{1/2}(\boldsymbol{G} + \boldsymbol{I})\hat{\boldsymbol{D}}^{1/2}$ denotes the renormalization trick from [13] when applied to the graph adjacency matrix $\boldsymbol{G} \in \mathbb{R}^{m \times m}$, $\hat{D}_{ii} = \sum_j (\boldsymbol{G} + \boldsymbol{I}_m)_{ij}$ denotes the degree of node i, $\boldsymbol{W}^1 \in \mathbb{R}^{p \times d}$ and $\boldsymbol{W}^2 \in \mathbb{R}^{d \times C}$ denote the learnable weights for the first and second layer respectively. The mapping from logits to Y can be done by applying a softmax and returning the class with the highest probability.

For studying transferability, we consider the target feature embeddings as: $\boldsymbol{F} = h(\boldsymbol{X}^{(t)}) = \hat{\boldsymbol{G}} \cdot \text{Dropout}(\text{ReLU}(\hat{\boldsymbol{G}} \cdot \boldsymbol{X}^{(t)} \cdot \boldsymbol{W}^1))$. This creates a linear transfer learning strategy, which is similar to the linear fine-tuning regime studied for vision models in Sect. 6.1. We study target task selection in Sect. 6.2 and source model selection in Supplementary document.

Pre-training Implementation. We use PyTorch Geometric [7] to setup the pre-training of GCN models. The pre-training uses a country-specific subnetwork and performs training, model selection and testing via a 64%/16%/20% split. The training considers transductive learning in graph networks. We perform 200 hyperparameter trials that tune over Adam learning rates $[10^{-5}, 10^{-1}]$, batch sizes $\{16, 32, 64\}$, L2 regularization $[10^{-4}, 1]$, dropout of 0.5 and maximum 1000 epochs with early stopping with a patience of 50.

Fine-tuning. For fine-tuning experiments, we recover the target feature embeddings by applying optimal pre-trained source model on a different subnetwork of users. Given we study linear fine-tuning in Graph Networks, we use GridSearchCV with ℓ_2-regularized Logistic Regression from sklearn [22] on (only) target train data to perform (stratified) 5-fold cross-validation. We consider 100 values for L2 regularization in the range $[10^{-5}, 10^3]$ on the log scale. The optimal L2 regularization is used to get the optimal model and the test accuracy is computed to measure correlation of transferability measures.

Table 4. Pearson correlation of transferability measures against fine-tuned target accuracy of *Graph Convolutional Networks* in Target Task selection scenario. We compare our proposed $H_\alpha(f)$ against original $H(f)$ and state-of-the-art measures.

n_t	Model	Source	H(f)	$H_\alpha(f)^a$	NCE	LEEP	$\mathcal{N}$LEEP	LFC	TransRate	LogME
500	GCN-256	DE	0.10*	0.35	0.15*	0.15*	0.40	**0.53**	-0.34	0.40
		ES	0.24	0.41	0.48	**0.50**	-0.13*	0.24*	-0.34*	0.20*
		FR	0.57	**0.61**	-0.03*	0.01*	-0.05*	0.26*	-0.30*	0.44
		RU	0.11*	**0.34**	-0.19*	-0.17*	-0.12*	0.04*	-0.21	0.11*
		PTBR	**0.37**	0.24*	0.16*	0.16*	-0.13*	-0.02*	-0.05*	0.15*
		ENGB	0.48	**0.53**	-0.12*	-0.09*	-0.12*	0.15*	-0.05*	0.32*
1000	GCN-512	DE	0.48	**0.71**	0.29*	0.44	-0.14*	0.45	0.17*	0.59
		ES	0.68	**0.78**	0.59	0.54	0.35*	0.49	-0.12*	0.59
		FR	0.59	**0.61**	0.25*	0.27*	0.04*	0.10*	0.58	0.22*
		RU	0.35	**0.44**	-0.12*	0.08*	-0.25*	0.14*	-0.07*	0.27*
		PTBR	0.67	**0.77**	0.37	0.40	0.04*	0.18*	0.10	0.50
		ENGB	**0.82**	0.81	-0.04*	-0.08*	0.17*	0.26*	0.15*	0.65

[a] We empirically observed dimensionality reduction with random projection to improve correlation performance for $H_\alpha(f)$ in this target task selection setting as well. We used $q = 128$.

Target Task Selection. We pre-train the GCN model with each of the country-specific sub-network. For each country-specific sub-network, $S \in \{$DE, ES, FR, RU, PTBR, ENGB$\}$, we construct 30 different target tasks. We exclude the specific country on which the source model is pre-trained and use the remaining countries to construct different combinations of networks as targets. For example, if the source model is pre-trained on DE, then the target tasks are given by: $\{$ES, FR, RU, PTBR, ENGB, (ES,FR), (ES,RU), $\cdots$, (ES,FR,RU,PTBR,ENGB)$\}$.

We study balanced targets in this regime. Given that the degree of imbalance varies significantly across different country-specific networks, we collect the largest balanced datasets for each target. Next, we sample $n_t = 1000$ nodes for fine-tuning and allocate the remaining nodes as test samples. We consider two different embedding sizes for GCN network in this study. We also validate our proposals when we allocate 500 samples for fine-tuning.

We present the Pearson correlation performance of transferability measures against fine-tuned target test accuracy in Table 4. The correlations demonstrate our proposed $H_\alpha(f)$ as the leading metric for target task selection for linear fine-tuning in Graph Neural Networks. We include additional results for source model selection in Supplementary document.

6.3 Timing Comparison Between LogME and $H_\alpha(f)$

We empirically investigate the computational times of $H_\alpha(f)$ when computed via our optimized implementation in Eq. (11). For this exercise, we generate synthetic multi-class classification data using Sklearn [22] multi-class dataset

Table 5. Timing comparison of LogME and our $\mathrm{H}_\alpha(f)$. All times are in ms.

n_t	d	$\lvert\mathcal{Y}\rvert = C$	LogME	$\mathrm{H}(f)$	$\mathrm{H}_\alpha(f)$
500	500	50	201	123	**22**
500	1000	50	185	376	**36**
500	5000	50	392	9680	**373**
500	1000	10	111	259	**33**
500	1000	100	268	271	**36**
100	1000	50	72	255	**16**
1000	1000	50	318	335	**75**

generation function that is adapted from [9]. We investigate different values for number of samples (n_t), feature dimension (d) and number of classes (C). For data generation, we set number of informative features to be 100 with the rest of the features filled with random noise. For LogME, we use a faster variant proposed by [34]. For a fair comparison, we do not use any dimensionality reduction in the computation of $\mathrm{H}_\alpha(f)$. Table 5 demonstrates a significant computational advantage of $\mathrm{H}_\alpha(f)$ over LogME. We observe $3-10$ times faster computational times.

7 Conclusion

We study transferability measures in the context of fine-tuning. Our contributions are three-fold. First, we show that H-score measure, commonly used as a baseline for newer transferability measures, suffers from instability due to poor estimation of covariance matrices. We propose shrinkage-based estimation of H-score with regularized covariance estimation techniques from statistical literature. We show 80% absolute increase over the original H-score and show superior performance in many cases against all newer transferability measures across various model types, fine-tuning scenarios and data settings. Second, we present a fast implementation of our estimator that provides a 3–10 times computational advantage over state-of-the-art LogME measure. Third, we identify problems with 3 other transferability measures (NCE, LEEP and $\mathcal{N}$LEEP) in target task selection (an understudied fine-tuning scenario than source model selection) when either the number of target classes or the class imbalance varies across candidate target tasks. We propose an alternative evaluation scheme that measures correlation against relative target accuracy (instead of vanilla accuracy) in such scenarios. Our large set of $\sim 164{,}000$ fine-tuning experiments with multiple vision models and graph neural networks in different regimes demonstrates usefulness of our proposals. We leave it for future work to explore how predictive various transferability measures are for co-training regimes (as opposed to fine-tuning).

References

1. Bao, Y., Li, Y., Huang, S., et al.: An information-theoretic approach to transferability in task transfer learning. In: 2019 IEEE ICIP, pp. 2309–2313 (2019)
2. Chen, Y., Wiesel, A., Eldar, Y.C., et al.: Shrinkage algorithms for MMSE covariance estimation. IEEE Trans. Signal Process. **58**(10), 5016–5029 (2010)
3. Chollet, F., et al.: Keras. https://github.com/fchollet/keras (2015)
4. Cui, Y., Song, Y., Sun, C., et al.: Large scale fine-grained categorization and domain-specific transfer learning. CoRR abs/1806.06193 (2018)
5. Deshpande, A., Achille, A., Ravichandran, A., et al.: A linearized framework and a new benchmark for model selection for fine-tuning (2021)
6. Devlin, J., Chang, M., Lee, K., et al.: BERT: pre-training of deep bidirectional transformers for language understanding. CoRR abs/1810.04805 (2018)
7. Fey, M., Lenssen, J.E.: Fast graph representation learning with PyTorch Geometric. In: ICLR Workshop on Representation Learning on Graphs and Manifolds (2019)
8. Friedman, J., Hastie, T., Tibshirani, R.: Additive logistic regression: a statistical view of boosting (With discussion and a rejoinder by the authors). Ann. Stat. **28**(2), 337–407 (2000)
9. Guyon, I.: Design of experiments for the nips 2003 variable selection benchmark (2003)
10. Hastie, T., Tibshirani, R., Friedman, J.: The Elements of Statistical Learning. SSS, Springer, New York (2009). https://doi.org/10.1007/978-0-387-84858-7
11. He, K., Zhang, X., Ren, S., et al.: Deep residual learning for image recognition. CoRR abs/1512.03385 (2015)
12. Huang, L.K., Wei, Y., Rong, Y., et al.: Frustratingly easy transferability estimation. ArXiv abs/2106.09362 (2021)
13. Kipf, T.N., Welling, M.: Semi-supervised classification with graph convolutional networks. In: ICLR 2017, Toulon, France, 24–26 April 2017. OpenReview.net (2017)
14. Kornblith, S., Shlens, J., Le, Q.V.: Do better imageNet models transfer better? In: 2019 IEEE/CVF CVPR, pp. 2656–2666 (2019)
15. Krizhevsky, A., Sutskever, I., Hinton, G.: Imagenet classification with deep convolutional neural networks. In: Neural Information Processing Systems 25 (01 2012)
16. Ledoit, O., Wolf, M.: A well-conditioned estimator for large-dimensional covariance matrices. J. Multivar. Anal. **88**(2), 365–411 (2004)
17. Li, H., Chaudhari, P., Yang, H., et al.: Rethinking the hyperparameters for fine-tuning. CoRR abs/2002.11770 (2020)
18. Li, Y., Jia, X., Sang, R., et al.: Ranking neural checkpoints. In: Proceedings of the IEEE/CVF CVPR, pp. 2663–2673 (2021)
19. Mahajan, D., Girshick, R.B., Ramanathan, V., et al.: Exploring the limits of weakly supervised pretraining. CoRR abs/1805.00932 (2018)
20. Max, A.W.: Inverting modified matrices. In: Memorandum Rept. 42, Statistical Research Group, p. 4. Princeton Univ. (1950)
21. Nguyen, C.V., Hassner, T., Seeger, M., Archambeau, C.: Leep: A new measure to evaluate transferability of learned representations (2020)
22. Pedregosa, F., Varoquaux, G., Gramfort, A., et al.: Scikit-learn: machine learning in python. J. Mach. Learn. Res. **12**, 2825–2830 (2011)
23. Pourahmadi, M.: High-dimensional covariance estimation: with high-dimensional data, vol. 882. John Wiley & Sons (2013)

24. Rabanser, S., Günnemann, S., Lipton, Z.C.: Failing loudly: an empirical study of methods for detecting dataset shift. In: NeurIPS (2019)
25. Rozemberczki, B., Allen, C., Sarkar, R.: Multi-scale attributed node embedding. J. Complex Netw. **9**(2), cnab014 (2021)
26. Rozemberczki, B., Sarkar, R.: Characteristic functions on graphs: birds of a feather, from statistical descriptors to parametric models. In: Proceedings of the 29th ACM CIKM, pp. 1325–1334. CIKM 2020, ACM, New York, NY, USA (2020)
27. Rozemberczki, B., Sarkar, R.: Twitch gamers: a dataset for evaluating proximity preserving and structural role-based node embeddings (2021)
28. Schäfer, J., Strimmer, K.: A shrinkage approach to large-scale covariance matrix estimation and implications for functional genomics. Stat. Appl. Genet. Mol. Biol. **4**, 32 (2005)
29. Simonyan, K., Zisserman, A.: Very deep convolutional networks for large-scale image recognition. CoRR abs/1409.1556 (2015)
30. Tan, Y., Li, Y., Huang, S.: OTCE: a transferability metric for cross-domain cross-task representations. CoRR abs/2103.13843 (2021)
31. Tran, A., Nguyen, C., Hassner, T.: Transferability and hardness of supervised classification tasks. In: 2019 IEEE/CVF ICCV, pp. 1395–1405 (2019)
32. Vinh, N.X., Epps, J., Bailey, J.: Information theoretic measures for clusterings comparison: variants, properties, normalization and correction for chance. J. Mach. Learn. Res. **11**(95), 2837–2854 (2010)
33. You, K., Liu, Y., Wang, J., Long, M.: LogME: practical assessment of pre-trained models for transfer learning. In: ICML (2021)
34. You, K., Liu, Y., Zhang, Z., Wang, J., Jordan, M.I., Long, M.: Ranking and tuning pre-trained models: a new paradigm of exploiting model hubs (2021)

Learning to Teach Fairness-Aware Deep Multi-task Learning

Arjun Roy[1,2](✉) and Eirini Ntoutsi[2]

[1] L3S Research Center, Leibniz University Hannover, Hanover, Germany
[2] Institute of Computer Science, Free University Berlin, Berlin, Germany
{arjun.roy,eirini.ntoutsi}@fu-berlin.de

Abstract. Fairness-aware learning mainly focuses on single task learning (STL). The fairness implications of multi-task learning (MTL) have only recently been considered and a seminal approach has been proposed that considers the fairness-accuracy trade-off for each task and the performance trade-off among different tasks. Instead of a rigid fairness-accuracy trade-off formulation, we propose a flexible approach that learns how to be fair in a MTL setting by selecting which objective (accuracy or fairness) to optimize at each step. We introduce the *L2T-FMT* algorithm that is a teacher-student network trained collaboratively; the student learns to solve the fair MTL problem while the teacher instructs the student to learn from either accuracy or fairness, depending on what is harder to learn for each task. Moreover, this dynamic selection of which objective to use at each step for each task reduces the number of trade-off weights from $2T$ to T, where T is the number of tasks. Our experiments on three real datasets show that *L2T-FMT* improves on both fairness (12–19%) and accuracy (up to 2%) over state-of-the-art approaches.

1 Introduction

Multi-Task Learning (MTL) [21] aims to leverage useful information contained in multiple tasks to help improve the generalization performance over all tasks. It is inspired by human's ability to learn multiple tasks and it has been already successfully applied in a variety of applications from natural language processing [12] to vision [3]. Many methods and deep neural network architectures [3,9,12] for MTL have been proposed, still the basic optimization problem is formulated as minimizing a weighted sum of task-specific losses, where the weights are inter-task trade-off hyperparameters used to avoid inter-task loss dominance [21].

Despite its popularity, the fairness implications of MTL have only recently come into focus [22]. The area of fairness-aware learning for Single Task Learning (STL) has received a lot of attention in the last years [15] and methods have been proposed that aim to learn correct predictions without discriminating on the

Supplementary Information The online version contains supplementary material available at https://doi.org/10.1007/978-3-031-26387-3_43.

M.-R. Amini et al. (Eds.): ECML PKDD 2022, LNAI 13713, pp. 710–726, 2023.
https://doi.org/10.1007/978-3-031-26387-3_43

basis of some protected attribute, like gender or race. Methods in this category reformulate the classification problem by explicitly incorporating the model's discrimination behavior in the objective function through e.g., regularization or constraints. The fair-MTL problem was only recently introduced and a solution, *MTA-F*, has been proposed [22] that considers *inter-task* fairness-accuracy trade-offs (as is typical in MTL) and *intra-task* performance trade-offs (as is typical in fairness-aware STL). For T tasks, such an approach requires $2T$ trade-off weights. The current practice to find the correct trade-off weights is by hyperparameter tuning [22]. The performance of MTLs are highly dependent on the correct weighting between each task's loss [21]. Tuning these weights by hand is a difficult and expensive process [13], which further gets leveraged in the fair-MTL problem.

Moreover, the trade-off weights in *MTA-F* are fixed throughout the training process. However, the inter-task trade-offs and the intra-task fairness-accuracy balance may change over the training due to e.g., factors like data batch [1].

In this paper, instead of a fixed fairness-accuracy trade-off formulation, we propose to *dynamically* select one among fairness and accuracy objectives at each training step for each task. To this end, we formulate the fair-MTL problem as a student-teacher problem and propose the Learning to Teach Fair Multi-tasking (referred to as *L2T-FMT*) algorithm. Our design inspiration comes from recent learning to teach (L2T) algorithms [6,23]. The student in our proposed algorithm is the desired MTL model, which follows the instruction of the teacher to learn from the available accuracy or fairness objectives for each task, and updates its parameters accordingly. The student sends feedback about its progress on fairness and accuracy in each task to the teacher. The teacher learns from the feedback and updates its model. This way, both student and teacher networks are trained collaboratively. Except for the dynamic intra-task loss selection, we also propose to set the inter-task parameters *dynamically* at each training step using GradNorm [1], as opposed to fixing them throughout the training process [22].

Our contributions can be summarized as follows: i) We introduce the dynamic objective selection paradigm for fair and accurate MTL. ii) We propose a new algorithm, *L2T-FMT*, based on a student-teacher framework that executes the dynamic objective selection paradigm and efficiently solves fairness-aware MTL. iii) Our dynamic objective selection results in a reduction of parameters to be learned per training step from $2T$ to T. iv) We eliminate the dependency of searching for the correct balance of inter-task trade-off weights by automatically learning the weights at each training step. v) *L2T-FMT*outperforms state-of-the-art methods by improving on both fairness and accuracy as demonstrated on real-world datasets of varying characteristics and number of tasks.

The rest of the paper is organized as follows: In Sect. 2 we review the related work. Necessary basic concepts are provided in Sect. 3. Our method is introduced in Sect. 4, followed by an experimental evaluation in Sect. 5. Conclusions and outlook are summarized in Sect. 6.

2 Related Work

Fairness-Aware Learning. A growing body of work has been proposed over the last years to address the problem of fairness and algorithmic discrimination [15] against demographic groups defined on the basis of protected attributes like gender or race. In parallel to bias mitigation methods, a plethora of fairness notions have been proposed; the interested reader is referred to [15] for a taxonomy of various fairness definitions. Statistical parity [5], equal opportunity and equalized odds [10] are among the most popular measures for measuring discrimination. In this work, we adopt equalized odds as our notion of fairness (Sect. 3.1).

Multi-task Learning. In MTL, multiple learning tasks are solved simultaneously, while exploiting commonalities and differences across the tasks [21]. There are two main categories of parameter sharing: hard vs. soft. In hard parameter sharing [1,3,9,22], model weights are shared between multiple tasks, while output layers are kept task-specific. In soft parameter sharing [12], different tasks have individual task-specific models with separate weights, but the distance between the model parameters is regularized in order to encourage the parameters to be similar. In this work, we follow the most popular hard parameter sharing approach where model weights are shared between multiple tasks.

Fairness-Aware Multi-task Learning. The fairness implications of MTL have been only recently considered: [24] studies multi-task regression to improve fairness in ranking; [18] proposes an MTL formulation to solve multi-attribute fairness on a single task. The closest to our work is the seminal work [22], which formulates the fair-MTL problem as a weighted sum of task-specific accuracy-fairness trade-offs. This formulation results in duplication of parameters, which are learned via hyperparameter tuning. Moreover, [22] introduced the concept of task-exclusive labels signifying examples that are only positive (negative) for the concerned task and negative (positive) for all the other tasks. Then, they proposed the *MTA-F* algorithm that updates the task-specific layer with the summed loss of accuracy and task-specific fairness (computed with exclusive examples) and shared layers with the summed loss of accuracy and shared fairness (computed with non-exclusive examples). In our method, we do not use the task-exclusive concept, as in the presence of a large number of tasks, the exclusive set of instances may reduce to null. Moreover, contrary to [22] that assumes a fixed intra-task accuracy-fairness trade-off, we rather *learn to choose at each step* of the training process whether the accuracy loss or the fairness loss should be used for model training. Also, instead of fixing the inter-task weights, we propose to learn the right trade-off parameters at each step using GradNorm [1].

3 Problem Setting and Basic Concepts

We assume a set of tasks $\mathcal{T} = \{1, \cdots, T\}$ sharing an input space $\mathcal{X} = U \times S$, where U is the subspace of *non-protected attributes* and S is the subspace of *protected attributes*. Each task t has its own label space Y_t. A dataset D of i.i.d. instances from the input space $X = U \times S$ and task spaces $\{Y_t\}_1^T$ is given as: $D = \{(u_i, s_i, y_i^1, \cdots y_i^T)\}_{i=1}^n$ where (u_i, s_i) is the description of instance i and $(y_i^1 \cdots y_i^T)$ are the associated class labels for tasks $1 \cdots T$. For simplicity, we assume binary tasks, i.e., $\forall t \in T : Y_t \in \{1, 0\}$, with 1 representing a positive (e.g., "granted") and 0 representing a negative (e.g.,"rejected") class. We also assume the protected subspace S to be a binary protected attribute: $S \in \{g, \overline{g}\}$, where g and $\overline{g}$ represent the *protected* and *non-protected group*, respectively.

3.1 Fairness Definition and Metric

As our fairness measure, we use `Equalized odds` (*EO*) [10], introduced for STL. For a task t, *EO* states that a classifier's prediction $\hat{Y}_t$ conditioned on ground truth Y_t must be independent from the protected attribute S. Based on [10], fairness is preserved when: $P(\hat{Y}_t = 1|S = g, Y_t = y) = P(\hat{Y}_t = 1|S = \overline{g}, Y_t = y)$, where $y \in \{1, 0\}$. A classifier satisfies the *EO* definition if the protected and non-protected groups have equal true positive rate (TPR) and false positive rate (FPR). For a task t, the FPR w.r.t. the protected group g is given by: $FPR_t(g)$: $P(\hat{Y}_t = 1|Y_t = 0, S = g)$, whereas the TPR is given by: $TPR_t(g)$: $P(\hat{Y}_t = 1|Y_t = 1, S = g)$. Similarly, we can define $FPR_t(\overline{g})$ and $TPR_t(\overline{g})$ for the non-protected group $\overline{g}$. The (absolute) differences in TPR and FPR define the violation of *EO* w.r.t. S, denoted by EO_{viol}. Given $TPR = 1 - FNR$, the violation can be also expressed in terms of FNR and FPR differences:

$$\begin{aligned} EO_{viol} &= \left|TPR_t(g) - TPR_t(\overline{g})\right| + \left|FPR_t(g) - FPR_t(\overline{g})\right| \\ &= \left|FNR_t(g) - FNR_t(\overline{g})\right| + \left|FPR_t(g) - FPR_t(\overline{g})\right| \end{aligned} \tag{1}$$

3.2 Vanilla Multi-task Learning (MTL)

Let $\mathcal{M}$ be an MTL model parameterized by the set of parameters $\theta \in \Theta$, which includes: shared parameters θ_{sh} (i.e., weights of layers shared by all tasks) and task-specific parameters θ_t (i.e., weights of the task specific layers), i.e.: $\theta = \theta_{sh} \times \theta_1 \times \cdots \times \theta_T$. An overview is given in Fig. 1.

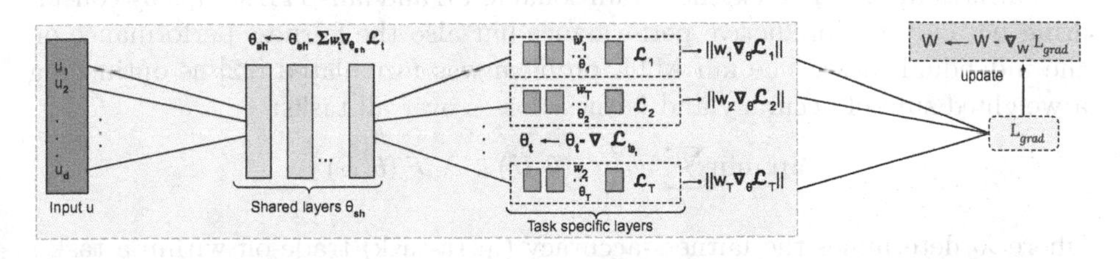

Fig. 1. An overview of (vanilla) MTL with GradNorm update

The goal of vanilla MTL training is to jointly minimise multiple loss functions, one for each task: $\underset{\theta}{\operatorname{argmin}}(\mathcal{L}_1(\theta, U), \cdots, \mathcal{L}_T(\theta, U))$. Finding a model θ that minimizes all T tasks simultaneously is hard and typically, a scalarization approach is followed [1,9,12] which turns the multi-tasking function into a single function using inter-task trade-off weights w_t, as follows:

$$\underset{\theta}{\operatorname{argmin}} \sum_t w_t \mathcal{L}_t(\theta, U) \tag{2}$$

The trade-off weights w_t signify the importance of each task; usually selected by hyperparameter tuning over a validation set. This, however, means *fixing* the weights over the whole training process. A better approach, which we follow in this work, is to vary the weights at each training step to balance the contribution of each task for optimal model training. In this direction, GradNorm [1] proposes to balance the training rates of different tasks; if a task is training relatively quickly, then its weight should decrease relative to other tasks' weights to allow other tasks to influence training. GradNorm is implemented as an $L1$ loss function between the actual and target gradient norms at each training step for each task, summed over all tasks:

$$L_{grad} = \sum_t |G_t(\theta, U) - \overline{G(\theta, U)} \times \rho_t^{\alpha}| \tag{3}$$

where $G_t(\theta, U)$ is the L2 norm of the gradient of the loss for task t w.r.t. the chosen weights θ, defined as $G_t(\theta, U) = ||\nabla_\theta w_t \mathcal{L}_t(\theta, U)||_2$; $\overline{G(\theta, U)}$ is the average gradient norm. Finally, ρ_t is the inverse training rate for task t and α is a hyperparameter of strength to pull any task back to the average training rate.

3.3 Fairness-Aware Multi-task Learning (FMTL)

Fairness has been extensively studied in the recent years for STL problems; most approaches combine accuracy and fairness losses into a single overall loss [19] as:

$$\underset{\theta}{\operatorname{argmin}} \Big(\mathcal{L}(\theta, U) + \lambda \mathcal{F}(\theta, S)\Big) \tag{4}$$

where $\mathcal{L}(\theta, U)$ is the typical *accuracy loss*, $\mathcal{F}(\theta, S)$ is the loss associated with the protected attribute S (refered to as *fairness loss*) and λ is a weight parameter that determines the fairness-accuracy trade-off.

Fairness-aware MTL extends traditional MTL and fair-STL setups by considering not only the predictive performance but also the fairness performance on the individual tasks. The fair-MTL problem was formulated [22] as optimizing a weighted sum of accuracy and fairness losses over all tasks:

$$\underset{\theta}{\operatorname{argmin}} \sum_t w_t \Big(\mathcal{L}_t(\theta, U) + \lambda_t \mathcal{F}_t(\theta, S)\Big) \tag{5}$$

where λ_t determines the fairness-accuracy (intra-task) trade-off within a task t and w_t determines the relative importance of task t (inter-task trade-off). Note that with this formulation the number of parameters required to be learned is doubled from T (for vanilla MTL, c.f., Eq. 2) to $2T$ (for FMTL c.f Eq. 5).

3.4 Deep Q-learning (DQN) and Multi-tasking DQN (MT-DQN)

Reinforcement Learning (RL) is based on learning via interaction with a single environment. At each step j, the environment observes a state $z_j \in Z$. The agent takes an action $a_j \in A$ in the environment, causing a transition to a new state $z_{j+1} \in Z$. For the transition, the agent receives a reward $R(j) \in \mathbb{R}$. The goal of the agent is to learn a policy $\pi : A \times Z \rightarrow A$ that maximizes the expected future discounted reward. State-action values (Q-values) are often used as an estimator of the expected future return. When the state-action space is large, the Q-values are typically approximated via a DNN, known as Deep Q-Network (DQN). The input to the DQN are the states, whereas the outputs are the state-action values (or, Q-values). A DQN is trained with parameters θ^Q, to minimize the loss between the *predicted Q-values* and the *target future return*:

$$L_Q(Z, A, \theta^Q, R) = E\Big[Q(z_j, \alpha_j|\theta^Q) - \Big(R(j) + \gamma \max_{\alpha'_j} Q(z'_j, \alpha'_j|\theta^Q)\Big)\Big] \quad (6)$$

where $Q(z_j, \alpha_j|\theta^Q)$ is the predicted Q-value. The term $R(j)+\gamma \max_{\alpha'_j} Q(z'_j, \alpha'_j|\theta^Q))$ is the target value defined as the sum of the direct reward $R(j)$ for transitioning from state z_j to a successor state z'_j and of the Q-value of the best successor state [16] (as predicted by the DQN).

In *multi-task reinforcement learning*, a single agent must learn to master T different environments/task, so the environments are the different tasks. The DQN for approximating the Q-values of state-action pairs is now a multi-tasking deep network (c.f., Sect. 3.2) with parameters $\theta^Q = \theta^Q_{sh} \times \theta^Q_1 \times \cdots \times \theta^Q_T$, where θ^Q_{sh} are shared across all the task learning environments, and θ^Q_t are the task exclusive parameters for learning the optimal policy in the particular environment/task. The network should learn to predict the Q-values under different learning environments/tasks, so the new objective becomes:

$$\underset{\theta^Q}{\text{argmin}} \sum_t w^Q_t L_Q(Z(t), A(t), \theta^Q, R(t)) \quad (7)$$

where w^Q_t is the learned weight for the environment/task E_t to overcome the challenge of inter-task loss dominance.

4 Learning to Teach Fairness-Aware Multi-tasking

We first present the dynamic objective selection paradigm to formulate the fair-MTL problem (Sect. 4.1). Then we introduce our *L2T-FMT*algorithm in Sect. 4.2, which is a teacher-student network framework. The student network (Sect. 4.3) learns a fair-MTL model, using advice from the teacher (Sect. 4.4) regarding the choice of the loss function.

4.1 Dynamic Loss Selection Formulation

The fair-MTL problem definition (cf. Eq. 5) introduces the λ_t parameters for combining accuracy and fairness losses for each task and therefore, the number of parameters required to be learned is doubled from T (for traditional MTL, cf. Eq. 2) to $2T$ (for fairness-aware MTL, cf. Eq. 5)). We propose to get rid of the λ dependency and rather *learn to choose at each training step* whether the accuracy loss ($\mathcal{L}_t$) or the fairness loss ($\mathcal{F}_t$) should be used to train the model on task t. Such an approach reduces the optimization problem of Eq. 5 to that of Eq. 2, which is well studied in the literature [1,9,12,21]. Moreover, it also provides the flexibility to the learner to emphasise on the objective (accuracy or fairness) that is harder to learn for each task at each training step.

The transformed learning problem is formulated as:

$$\underset{\theta,m}{\operatorname{argmin}} \sum_t w_t(m) \begin{cases} \mathcal{L}_t(\theta, U, m), \text{if task } t \text{ is trained for } \textit{accuracy} \text{ in step } m \\ \mathcal{F}_t(\theta, S, m), \text{if task } t, \text{ is trained for } \textit{fairness} \text{ in step } m \end{cases} \tag{8}$$

where m is the step[1] number that sequences until model convergence.

The decision about which loss to use is based on the teacher network, which is trained jointly with the student network (Sect. 4.2). The accuracy loss and fairness loss functions that we adopt in this work are provided hereafter.

Accuracy Loss: We adopt the negative log-likelihood loss $\mathcal{L}_t(\theta, U)$ for a task t:

$$\mathcal{L}_t(\theta, U) = -\sum_i (y_i^t \log \mathcal{M}(u_i, \theta) + (1 - y_i^t) \log(1 - \mathcal{M}(u_i, \theta))) \tag{9}$$

Fairness Loss: Several works discuss how to formulate a fairness loss function [4,7,8,17,20]. To keep the characteristic of the fairness loss function similar to the one used for the accuracy loss, we adopt the robust log-loss [19,20], which focuses on the worst-case log loss and is given by:

$$\begin{aligned} \mathcal{F}_t(\theta, U, S) = \max(&\mathcal{FNR}_t^g(\theta, U, S), \mathcal{FNR}_t^{\overline{g}}(\theta, U, S)) \\ &+ \max(\mathcal{FPR}_t^g(\theta, U, S), \mathcal{FPR}_t^{\overline{g}}(\theta, U, S)) \end{aligned} \tag{10}$$

where the negative log-likelihood loss over a group (say g) for a task t:

$$\begin{aligned} \mathcal{FNR}_t^g(\theta, U, S, Y_t) &= -\sum_i y_i^t \log \mathcal{M}(u_i, \theta | y_i^t = 1, s_i = g) \\ \mathcal{FPR}_t^g(\theta, U, S, Y_t) &= -\sum_i (1 - y_i^t) \log(1 - \mathcal{M}(u_i, \theta | y_i^t = 0, s_i = g)) \end{aligned} \tag{11}$$

[1] To note that the step information though specifically used in Eq. 8 to understand the temporal aspect of the selection, is valid for all the previous and following equations and is omitted for the rest of the paper for ease of reading.

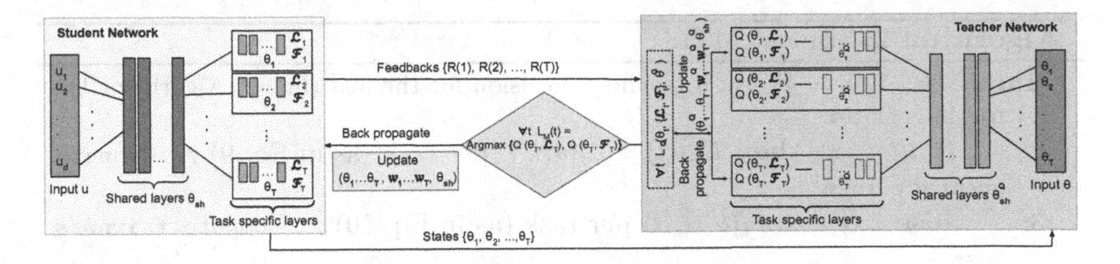

Fig. 2. The L2T-FMT architecture. Steps are in order from bottom to top

4.2 L2T-FMT Algorithm

An overview of our *Learn to Teach Fairness-aware MTL (L2T-FMT)* method is shown in Fig. 2. It consists of a student network and a teacher network which are trained collaboratively, as seen in Algorithm 1.

The *teacher network* $\mathcal{Q}$ aims at each training round to select (line 2) the best loss function for each task among the two available options: accuracy loss and fairness loss. The selected loss function is adopted by the student network $\mathcal{M}$ and is used for network update (line 3). After the update, the *student network* provides feedback (line 4) to the *teacher network* about the progress made in each of the tasks following the advice of the *teacher*. Based on the feedback, the *teacher* network updates itself (line 5). So the two networks are trained collaboratively; their functionalities are explained in the following sections.

Algorithm 1: The L2T-FMT algorithm

Input: A MTL dataset $D \in U \times S \times Y$ (see Sect. 3)
Init: Initialize student model $\mathcal{M}$ parameters: $\theta = \theta_{sh} \times \theta_1 \times \cdots \times \theta_T$ and task weights: $w_t = \frac{1}{T}$
Initialise teacher model $\mathcal{Q}$: parameters- $\theta^Q = \theta^Q_{sh} \times \theta^Q_1 \times \cdots \times \theta^Q_T$ and environment weights $w^Q = \{w^Q_t = \frac{1}{T}\}$
1: **Until** convergence **do**
2: $d = \{\underset{\alpha \in \{\mathcal{L}_t, \mathcal{F}_t\}}{\text{argmax}}\ Q(\theta_t, \alpha), \forall t\}$`// teacher decides the best loss to learn for student upon seeing` θ_t
3: Call Algorithm 2 with input (U, S, Y, d)`// Training` $\mathcal{M}$
4: Evaluate $R = \{[\mathcal{R}(\mathcal{L}_t), \mathcal{R}(\mathcal{F}_t)]\ \forall t\}$ using Eq. 12 `// generate feedback for the teacher`
5: Call Algorithm 3 with input $(\{\mathcal{M}(t)\}, \{\theta_t\}, R, A)$ `// Training` $\mathcal{Q}$
Output: $\mathcal{M}$ with learned parameters $\theta = \theta_{sh} \times \theta_1 \times \cdots \times \theta_T$

Algorithm 2: Student_MTL

Input: U, S, $Y = \{Y_t\}_1^T$, d teacher's decision for the loss (line 2, Algorithm 1)
1: **for** 1,...,T **do**
2: **if** $d(t) = \mathcal{L}_t$ **then** $L_{\mathcal{M}}(t) = \mathcal{L}_t(\theta, U)$ per task (as in Eq. 9) `// Compute accuracy loss`
3: **else** $L_{\mathcal{M}}(t) = \mathcal{F}_t(\theta, U, S)$ per task (as in Eq. 10) `// Compute fairness loss`
4: $\theta_t \leftarrow \theta_t - \eta \nabla_{\theta_t} L_{\mathcal{M}}(t)$`// Update task-specific layers params`
5: $G_t = ||\nabla_\theta w_t L_{\mathcal{M}}(t)||_2$, and $\rho_t = \frac{L_{\mathcal{M}}(t)}{E(L_{\mathcal{M}}(t))}$`// calculate gradient norm, and inverse training rate`
6: **end for**
7: Compute L_{grad} as in Eq. 3 and update W $\forall_{w_t \in W} w_t \leftarrow w_t - \eta \nabla_{w_t} L_{grad}$
8: $\theta_{sh} \leftarrow \theta_{sh} - \eta \sum_t w_t \nabla_{\theta_{sh}} L_{\mathcal{M}}(t)$ `// update the shared parameters`

4.3 Student Network

The student network $\mathcal{M}$ is a deep multi-tasking neural network with learning parameters $\theta = \theta_{sh} \times \theta_1 \times \cdots \times \theta_T$, as described in Sect. 3.2. It aims to learn to solve the fairness-aware multi-tasking problem by optimizing Eq. 8. The pseudocode is shown in Algorithm 2. For each task t, the decision of the *teacher* $\mathcal{Q}$ (given as input) about which action/loss to use is followed. Based on the decision, the selected loss (accuracy or fairness) is computed (lines 2-3) and used for the update of task-specific network layers (line 4). The task-specific weights w_t are learned using GradNorm (lines 5 and 7). Finally, the shared parameters are updated (line 8) using the updated task-specific weights and the loss function decided by the teacher network. The weight update mechanism using GradNorm (c.f., Sect. 3.2) is visualized in Fig. 1 for reference.

4.4 Teacher Network

The teacher network $\mathcal{Q}$ is a MT-DQN agent, as described in Sect. 3.4. It aims to learn to decide about which loss function, among the accuracy and fairness, the student network $\mathcal{M}$ should use. In particular, $\mathcal{Q}$ learns to predict the Q-values of accuracy-, fairness-loss/actions; the actual decision is the action with the largest Q-value (line 2 of Algorithm 1).

The pseudocode of $\mathcal{Q}$ is shown in Algorithm 3. For each task t, it estimates the Q-values of the accuracy- and fairness-loss functions/actions based on the current model parameters (line 2). The network is updated (lines 3-9) as described in Sect. 3.4. Teacher's training depends on the feedback by the student network, in the form of reward, which is computed by evaluating the student's output in terms of *Acc* (accuracy) and EO_{viol} (fairness) (line 4, Algorithm 1). The main intuition for the design of the reward function is to reward positively only if *Acc* increases and simultaneously EO_{viol} decreases for any action suggested by Q for a task t. On violation of either of the two conditions, the reward should be

Algorithm 3: Teacher_DQN

Input: $\{\mathcal{M}(t)\}$, $\{\theta_t\}$, $\{R(t) = [\mathcal{R}(\mathcal{L}_t), \mathcal{R}(\mathcal{F}_t)]\}$, $\{A(t)|A(t) = \{\mathcal{L}_t(\theta, U), \mathcal{F}_t(\theta, S)\}\}$,

1: **for** 1,...,T **do**

2: Estimate $Q(\theta_t, \mathcal{L}_t)$, $Q(\theta_t, \mathcal{F}_t)$ `// estimating Q-values for each state-action pair`

3: $\forall \alpha_t' \in A(t)$ take action α_t' in $M(t)$to produce θ_t'`// teacher takes a one look forward into the student environment`

4: Compute the loss $L_Q(\theta_t, A(t), \theta^Q, R(t))$ (as in Eq. 6)

5: $\theta_t^Q \leftarrow \theta_t^Q - \eta\nabla_{\theta_t^Q} L_Q(\theta_t, A(t), \theta^Q)$`// teacher updates the parameters of the task environment specific layers`

6: $G_t^Q = ||\nabla_\theta w_t^Q L_Q(\theta_t, A(t), \theta^Q)||_2$, and $\rho_t^Q = \frac{L_Q(\theta_t, A(t), \theta^Q)}{E(L_Q(\theta_t, A(t), \theta^Q))}$`// calculate gradients norm, and inverse training rate`

7: **end for**

8: Compute L_{grad}^Q as in Eq. 3 and update w^Q $\forall_{w_t^Q \in w^Q} w_t^Q \leftarrow w_t^Q - \eta\nabla_{w_t^Q} L_{grad}^Q$

9: $\theta_{sh}^Q \leftarrow \theta_{sh}^Q - \eta \sum_t w_t^Q \nabla_{\theta_{sh}^Q} L_Q(\theta_t, A(t), \theta^Q)$ `// update the shared parameters`

negative (the min function ensures this positive/negative property). However, the problem of intra-task dominance arises in estimation of $\mathcal{R}$ as the scales of evaluated *accuracy* and *fairness* for task t might differ. To calculate a scale-invariant reward, we take inspiration from [11], and estimate $\mathcal{R}(\alpha_t) \in \mathbb{R}$ for $\alpha_t \in \{\mathcal{L}_t, \mathcal{F}_t\}$, as the transformed evaluated output:

$$\mathcal{R}(\alpha_t) = \min(\frac{Acc(\hat{Y}_t) - Acc^{best}(t)}{Acc^{best}(t)}, \frac{EO_{viol}^{best}(t) - EO_{viol}(\hat{Y}_t)}{1 - EO_{viol}^{best}(t)}) \tag{12}$$

where $Acc(\hat{Y}_t)$, $Acc^{best}(t)$, and $EO_{viol}(\hat{Y}_t)$, $EO_{viol}^{best}(t)$ are respectively the current and best till current step *accuracy*, and *fairness* values of $\mathcal{M}$ in task t.

5 Experiments

We first evaluate the accuracy-and fairness of our *L2T-FMT*[2] in comparison to other approaches for different MTL problems (Sect. 5.2) including a more task-specific evaluation (Sect. 5.3). Next, we analyse the impact of the dynamic loss function selection by *L2T-FMT* (Section 5.4). The experimental setup including datasets, evaluation measures, and competitors is discussed in Sect. 5.1.

5.1 Experimental Setup

Datasets. We evaluate on one tabular and two visual datasets. The tabular dataset is the recently released *ACS-PUMS* [2], which comprises a superset of

[2] https://github.com/arjunroyihrpa/L2TFMT.

the popular Adult dataset from available US Census sources, and consists of 5 different well defined binary classification tasks[3]. We use *gender* as the protected attribute. For training we use the census data from the year "2018", divided into train (70%) and validation (30%) sets. For testing we use the data from the following year "2019" (both years of size $\approx 1.65M$). The visual datasets come from CelebA dataset [14] consisting of $202.5K$ celebrity facial images and 40 different binary attributes. We use the provided[4] partitioning into train (#162,770 instances), validation (#19,867 instances), and test (#19,962 instances) set. We use two different protected attributes, gender and age, resulting into two versions of the dataset, *CelebA-Gender* and *CelebA-Age*. We don't consider all 40 attributes as tasks for the MTL since some attributes are extremely skewed towards the protected or non-protected group. For example, the attribute "Mustache" is true only for 3 female instances. We set the filtering threshold to 1.5% or 2.5K instances. The filtering process reduces the number of attributes to 17 for *CelebA-Gender* and 31 for *CelebA-Age*; these are the MTL tasks. Further details on the datasets and experimental setup are provided in the Appendix.

Methods. We compared *L2T-FMT* against the following methods:

i) MTA-F: the vanilla fairness-aware MTL method [22] that minimises the weighted sum of accuracy- and fairness-losses (c.f. Eq. 5). For fairness it calculates two separate loss, one for updating θ_t and another for updating θ_{sh}. The weights w_t and λ_t are set via hyperparameter tuning.

ii) G-FMT: a variation of our *L2T-FMT* approach that always chooses *greedily* the best action/loss function among the available choices, by optimising: $argmin_\theta \sum_t w_t\{\max\{\mathcal{L}_t(\theta, U), \mathcal{F}_t(\theta, S)\}\}$.

iii) Vanilla MTL: the vanilla MTL approach that does not consider fairness but aims at minimising the weighted sum of task-specific accuracy losses (c.f. Eq. 2). The task-specific weights w_t are learned via GradNorm [1] as in Eq 3.

iv) STL: trains a separate fair-accurate model on each respective task.

Evaluation Measures. Following [22], we report on the relative performance of the `MTL` model ($Acc(t)_{mtl}$, $EO_{viol}(t)_{mtl}$) to the performance of a `STL` model trained on each respective task t ($Acc(t)_{stl}$, $EO_{viol}(t)_{stl}$). Specifically, for accuracy we report on the *average relative Acc* (*ARA*): $ARA = \frac{1}{T}\sum_t^T \frac{Acc(t)_{mtl}}{Acc(t)_{stl}}$ and for fairness, on the *average relative EO* (*AREO*): $AREO = \frac{1}{T}\sum_t^T \frac{EO_{viol}(t)_{mtl}}{EO_{viol}(t)_{stl}}$.

5.2 Overall Fairness-Accuracy Evaluation

The overall fairness (AREO) and accuracy (ARA) performance of the different methods on all the datasets is shown in Table 1. As we can see, *L2T-FMT* outperforms all the competitors in *fairness* by producing the lowest AREO scores across all the datasets; the relative reduction in discrimination w.r.t. the best

[3] https://github.com/zykls/folktables.

[4] http://mmlab.ie.cuhk.edu.hk/projects/CelebA.html.

baseline is in the range [12%–19%]. Interestingly, the second best approach in terms of AREO is our greedy variation, *G-MFT*. In terms of ARA, *L2T-FMT* is best by a small margin comparing to the best baseline with the exception of the ACS-PUMPS dataset for which Vanilla-MTL scores first. In particular, for *CelebA-Gender* and *CelebA-Age*, our *L2T-FMT* beats the best baseline by 2% and 1%, respectively, whereas for the *ACS-PUMPS* dataset *L2T-FMT*scores second with a 3% decrease comparing to the best performing *Vanilla MTL*.

To get better insights on the results, in the next section we also report on the task-specific performance using accuracy and EO_{viol} for each task.

Table 1. ARA vs AREO: Higher values better for accuracy/ARA, lower values better for discrimination/AREO . Best values in bold, second best underlined. (%) indicates our relative difference over the performance of the best baseline.

Dataset	#tasks T	Metric	Vanilla MTL	MTA-F	G-FMT	L2T-FMT
ACS-PUMS	5	ARA	**1.06**	0.97	1.01	<u>1.03 (–3%)</u>
		AREO	2.38	3.52	<u>1.50</u>	**1.21 (–19%)**
CelebA-Gender	17	ARA	0.89	<u>0.95</u>	0.86	**0.97 (2%)**
		AREO	2.72	2.29	<u>1.77</u>	**1.51 (–15%)**
CelebA-Age	31	ARA	<u>0.94</u>	0.85	0.86	**0.95 (1%)**
		AREO	2.61	1.79	<u>1.72</u>	**1.52 (–12%)**

5.3 Performance Distribution over the Tasks

We analyze the distribution of accuracy and fairness scores over the tasks for all methods using boxplots. The results are shown in Fig. 3, 4, and 5 for the *ACS-PUMS*, *CelebA-Gender* and *CelebA-Age*, respectively. For fairness a positively skewed box (median closer to Q1) with low Q1 is better, while for accuracy a negatively skewed box (median closer to Q3) with high Q3 is better.

ACS-PUMS Dataset. In Fig. 3a, we see that *L2T-FMT* has the lowest *median*, and the lowest *Q1* of EO_{viol}, with *Q3* marginally above the *STLs* but lower than all the competitors. Henceforth, it achieves the best AREO score (c.f., Table 1). In Fig. 3b, we see that *L2T-FMT* has the second highest median after G-FMT, however it has a higher *Q1* than *G-FMT* but a lower *Q1* and lower *Q3* than *vanilla MTL*. Thus, in Table 1 we find *L2T-FMT* to be second best in ARA score behind *vanilla MTL* on ACS-PUMS. *MTA-F* has the most consistent outcome of EO_{viol} with low spread over the tasks for both fairness and accuracy, but has the highest median and *Q1* of EO_{viol}, and the lowest median and lowest *Q3* of accuracy. Thus, in overall it has the worst overall performance as also seen in Table 1. *G-FMT* has the highest spread in both fairness and accuracy and is negatively skewed in accuracy with high accuracy for some tasks. *Vanilla-MTL* comes second in terms of spread and is positively skewed in EO_{viol}. However, its upper whisker is longer with a single point above Q3; this corresponds to task 1 (*Employment Status*) for which the EO_{viol} score is high.

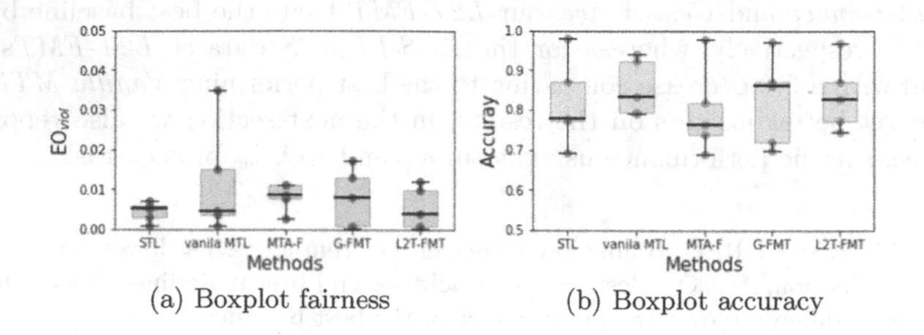

(a) Boxplot fairness (b) Boxplot accuracy

Fig. 3. ACS-PUMS dataset: Performance distribution over the tasks.

CelebA-Gender Dataset. In Fig. 4a, we see that all methods have low spread indicating their consistent performance w.r.t. fairness across the tasks. Still, *L2T-FMT* outperforms the competitors with the lowest median, *Q1*, and *Q3* values. *MTA-F* has the third best median, *Q1*, and *Q3* of EO_{viol},. As we see in Fig. 4a it has a much longer upper whisker indicating larger variance among the larger values, i.e., tasks with higher discrimination with the worst discrimination of 0.175 which corresponds to task 11 (*Narrow Eyes*). However, in accuracy all the methods vary substantially (high spread) as we see in Fig. 4b. *L2T-FMT* holds the highest median and *Q3*, but its *Q1* is marginally lower than *MTA-F* which indicates that in a few tasks *L2T-FMT* gets outperformed by *MTA-F* (second best median, *Q3*) on accuracy. This explains the ARA scores in Table 1, where *L2T-FMT* scores first followed by *MTA-F*. *G-FMT* has the second best median, *Q3*, and *Q1* of EO_{viol} score, but has the worst median, *Q3*, and *Q1* of accuracy. Thus, in Table 1 we see that *G-FMT* bags the second best AREO score but has the worst ARA score.

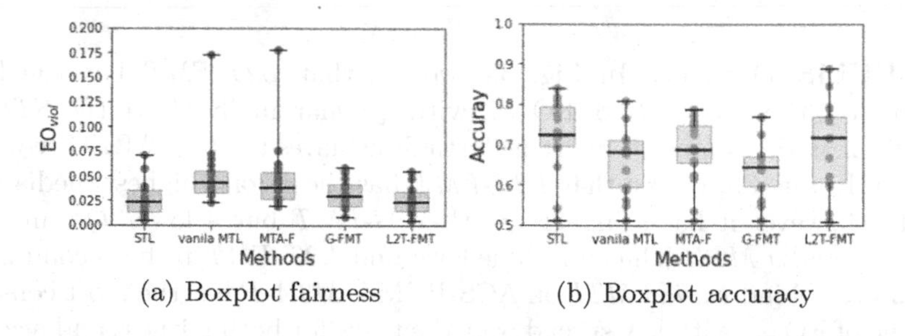

(a) Boxplot fairness (b) Boxplot accuracy

Fig. 4. CelebA-Gender: Performance distribution over the tasks.

CelebA-Age Dataset. There is large spread accuracy (Fig. 5b) across the different methods. *L2T-FMT* has the lowest median, *Q3*, and *Q1* of EO_{viol} score, and highest median, *Q3*, and *Q1* of accuracy, respectively. This reflects in Table 1 where *L2T-FMT* achieves the best AREO, and ARA scores. *MTA-F* delivers consistent fair performance over the tasks with the second best median, *Q3*, and *Q1* of EO_{viol} score almost same as *G-FMT*. This is the reason why the AREO score of *G-FMT* and *MTA-F* are nearly same, with *G-FMT* marginally ahead, making *MTA-F* the third best on AREO score. On accuracy *G-FMT* has the third best median, *Q3*, and *Q1*, having marginal improvements over *MTA-F*. *Vanilla MTL* with no fairness treatment has the worst median, *Q3*, and *Q1* of of EO_{viol} score over the tasks positioning it at the last place on AREO score, while having median, *Q3*, and *Q1* of accuracy over the tasks very similar to that of *L2T-FMT* making it a very close second on ARA score.

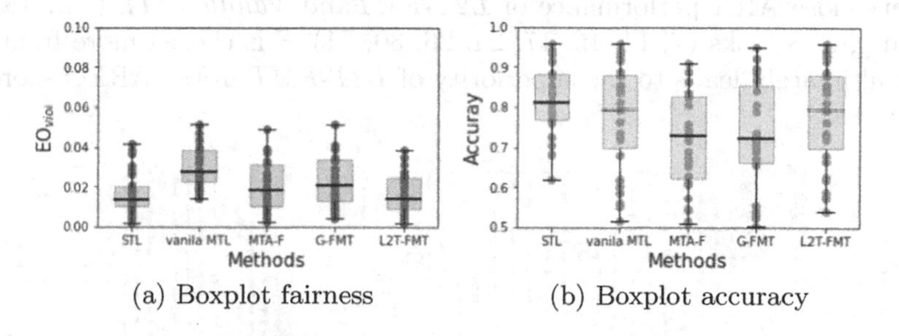

(a) Boxplot fairness (b) Boxplot accuracy

Fig. 5. CelebA-Age: Performance distribution over the tasks.

Summary. For all datasets, *L2T-FMT* stands out among the competitors with a very low median of EO_{viol}, and a very high median of accuracy. The fairness performance of *MTA-F* depends on the number of tasks; for larger MTL problems (like *CelebA-Gender and CelebA-Race*) the performance varies across the tasks including tasks with high discrimination (high upper whisker). *G-FMT* consistently delivers descent fairness performance positioning it always in the second place, however its accuracy gets affected when the number of tasks is high (CelebA-Gender and CelebA-Age). The *Vanilla MTL* without any fairness treatment on EO_{viol} score is often bad with very high upper whiskers.

5.4 Dynamic Loss Selection

We focus on the (dynamic) loss selection of the teacher network by looking at which function among the two available options: accuracy loss ($\mathcal{L}$) and fairness loss ($\mathcal{F}$) is used for each task over the training process. The results for the different datasets are shown in Fig. 6.

Regarding *ACS-PUMS* (Fig. 6a), the decision of selecting $\mathcal{L}$ or $\mathcal{F}$ is almost equally distributed over the tasks. Using *Vanilla MTL* (trained only for accuracy) as a reference method, we see in Table 1 that to achieve the best accuracy (best

ARA) in this dataset one needs to always tune with $\mathcal{L}$. However, this comes with depreciation in fairness (high AREO). Thus, the necessary balance as selected by *L2T-FMT* is required.

For *CelebA-Gender* (Fig. 6b), the accuracy loss $\mathcal{L}$ is selected more often in some tasks (2, 10, 17) and the fairness loss $\mathcal{F}$ in other tasks (6, 9, 11, 15), whereas there are also tasks with balanced selection (e.g., 1). *Vanilla MTL* that only tunes for $\mathcal{L}$ does not produce the best accuracy (c.f., Table 1); in contrast, *L2T-FMT* with dynamic loss selection achieves the best accuracy and fairness, justifying the selection (im-) balance.

For *CelebA-Age*, (Fig. 6c) we notice that for the majority of the tasks (21 out of 31) the loss selection is unevenly distributed with 13 tasks in favour of $\mathcal{L}$, and 8 in favour of $\mathcal{F}$. Interestingly, in 9 tasks (1, 5, 6, 10, 15, 18, 23, 25, 29) $\mathcal{L}$ is selected continuously over epochs at a stretch (≥ 7), signifying that in these tasks tuning for accuracy is far more important. This is also reflected in the very close ARA performance of *L2T-FMT*and *Vanilla MTL* (c.f., Table 1). Similarly, in 8 tasks (7, 14, 16, 17, 24, 26, 30, 31) $\mathcal{F}$ is chosen more frequently, which ultimately leads to the superiority of *L2T-FMT* in the AREO score.

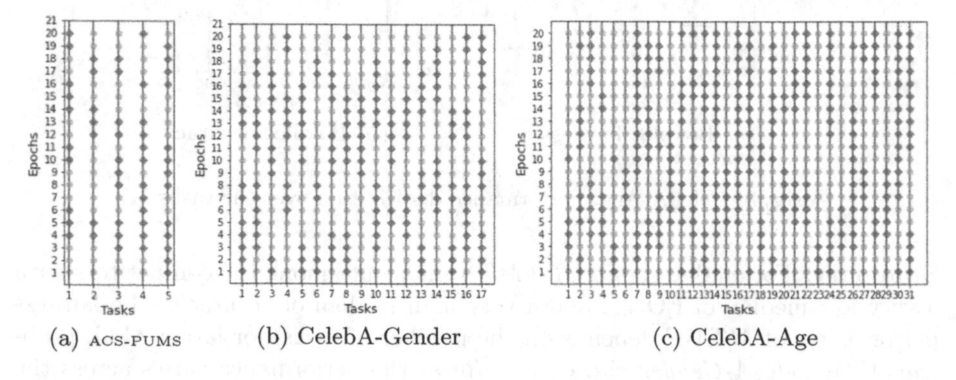

(a) ACS-PUMS (b) CelebA-Gender (c) CelebA-Age

Fig. 6. Loss function selection for each task over the training epochs. Accuracy-loss ($\mathcal{L}$) selection in orange, fairness-loss ($\mathcal{F}$) selection in blue. (Color figure online)

6 Conclusion

We proposed *L2T-FMT*, an approach for fairness-aware multi-task learning that dynamically selects for each task the best loss function to be used at each training step among the available: accuracy loss and fairness loss. Our approach is a student-teacher network framework, where the student learns to solve the fair-MTL problem while the teacher decides the action/loss function that the student should use for its update. The teacher is implemented as a DQN, whereas the student is implemented as a deep MTL network. In contrast to a rigid fairness-accuracy trade-off formulation [22], *L2T-FMT* allows for a flexible model update based on which objective (accuracy or fairness) is harder to learn for each task.

Moreover, it reduces the number of parameters to be learned to half. Our experiments on three real datasets show that *L2T-FMT* improves on both fairness and accuracy over state-of-the-art approaches. Moreover, we also show the effectiveness of learning to select the best loss in producing such favourable outcomes.

As part of our future work, instead of jointly training on all the tasks, we plan to identify the tasks that would benefit from training together. Such task groupings might differ based on whether fairness or accuracy is considered.

Acknowledgements. The work of A.Roy is supported by Volkswagen Foundation under the call "Artificial Intelligence and the Society of the Future" project "Bias and Discrimination in Big Data and Algorithmic Processing - BIAS".

References

1. Chen, Z., Badrinarayanan, V., Lee, C.Y., Rabinovich, A.: GradNorm: gradient normalization for adaptive loss balancing in deep multitask networks. In: ICM, pp. 794–803. PMLR (2018)
2. Ding, F., Hardt, M., Miller, J., Schmidt, L.: Retiring adult: new datasets for fair machine learning. In: NeurIPS 34 (2021)
3. Dong, N., Kampffmeyer, M., Voiculescu, I.: Self-supervised multi-task representation learning for sequential medical images. In: Oliver, N., Pérez-Cruz, F., Kramer, S., Read, J., Lozano, J.A. (eds.) ECML PKDD 2021. LNCS (LNAI), vol. 12977, pp. 779–794. Springer, Cham (2021). https://doi.org/10.1007/978-3-030-86523-8_47
4. Donini, M., Oneto, L., Ben-David, S., Shawe-Taylor, J., Pontil, M.: Empirical risk minimization under fairness constraints. In: NeurIPS, pp. 2796–2806 (2018)
5. Dwork, C., Hardt, M., Pitassi, T., Reingold, O., Zemel, R.: Fairness through awareness. In: ITCS, pp. 214–226 (2012)
6. Fan, Y., Tian, F., Qin, T., Li, X.Y., Liu, T.Y.: Learning to teach. In: ICLR (2018)
7. Feldman, M., Friedler, S.A., Moeller, J., Scheidegger, C., Venkatasubramanian, S.: Certifying and removing disparate impact. In: KDD, pp. 259–268 (2015)
8. Gretton, A., Borgwardt, K., Rasch, M., Schölkopf, B., Smola, A.: A kernel method for the two-sample-problem. In: NeurIPS, vol.19, pp. 513–520 (2006)
9. Guo, P., Deng, C., Xu, L., Huang, X., Zhang, Yu.: Deep multi-task augmented feature learning via hierarchical graph neural network. In: Oliver, N., Pérez-Cruz, F., Kramer, S., Read, J., Lozano, J.A. (eds.) ECML PKDD 2021. LNCS (LNAI), vol. 12975, pp. 538–553. Springer, Cham (2021). https://doi.org/10.1007/978-3-030-86486-6_33
10. Hardt, M., Price, E., Srebro, N.: Equality of opportunity in supervised learning. In: NeurIPS, vol. 29, 3315–3323 (2016)
11. Hessel, M., Soyer, H., Espeholt, L., Czarnecki, W., Schmitt, S., van Hasselt, H.: Multi-task deep reinforcement learning with popart. In: AAAI. vol. 33, pp. 3796–3803 (2019)
12. Kacupaj, E., Premnadh, S., Singh, K., Lehmann, J., Maleshkova, M.: Vogue: answer verbalization through multi-task learning. In: ECML PKDD, pp. 563–579. Springer (2021)
13. Kendall, A., Gal, Y., Cipolla, R.: Multi-task learning using uncertainty to weigh losses for scene geometry and semantics. In: Proceedings of the IEEE Conference on Computer Vision and Pattern Recognition, pp. 7482–7491 (2018)

14. Liu, Z., Luo, P., Wang, X., Tang, X.: Deep learning face attributes in the wild. In: ICCV (2015)
15. Mehrabi, N., Morstatter, F., Saxena, N., Lerman, K., Galstyan, A.: A survey on bias and fairness in machine learning. CSUR **54**(6), 1–35 (2021)
16. Mnih, V., et al.: Human-level control through deep reinforcement learning. Nature **518**(7540), 529–533 (2015)
17. Oneto, L., Donini, M., Pontil, M.: General fair empirical risk minimization. In: IJCNN, pp. 1–8. IEEE (2020)
18. Oneto, L., Doninini, M., Elders, A., Pontil, M.: Taking advantage of multitask learning for fair classification. In: AIES, pp. 227–237 (2019)
19. Padala, M., Gujar, S.: FNNC: achieving fairness through neural networks. In: IJCAI (2020)
20. Rezaei, A., Fathony, R., Memarrast, O., Ziebart, B.: Fairness for robust log loss classification. In: AAAI, vol. 34, pp. 5511–5518 (2020)
21. Vandenhende, S., Georgoulis, S., Van Gansbeke, W., Proesmans, M., Dai, D., Van Gool, L.: Multi-task learning for dense prediction tasks: a survey. TPAMI (2021)
22. Wang, Y., Wang, X., Beutel, A., Prost, F., Chen, J., Chi, E.H.: Understanding and improving fairness-accuracy trade-offs in multi-task learning. In: KDD, pp. 1748–1757 (2021)
23. Wu, L., et al.: Learning to teach with dynamic loss functions. In: NeurIPS (2018)
24. Zhao, C., Chen, F.: Rank-based multi-task learning for fair regression. In: ICDM, pp. 916–925. IEEE (2019)

Author Index

M.-R. Amini et al. (Eds.): ECML PKDD 2022, LNAI 13713, pp. 727–729, 2023.
https://doi.org/10.1007/978-3-031-26387-3

Printed in the United States
by Baker & Taylor Publisher Services

Printed in the United States
by Baker & Taylor Publisher Services